Lecture Notes in Computer Science 11544

More information about this series at http://www.springer.com/series/7407

Ilias Kotsireas · Panos Pardalos ·
Konstantinos E. Parsopoulos ·
Dimitris Souravlias · Arsenis Tsokas (Eds.)

Analysis of Experimental Algorithms

Special Event, SEA^2 2019
Kalamata, Greece, June 24–29, 2019
Revised Selected Papers

Editors
Ilias Kotsireas
Wilfrid Laurier University
Waterloo, ON, Canada

Panos Pardalos
University of Florida
Gainesville, FL, USA

Konstantinos E. Parsopoulos
University of Ioannina
Ioannina, Greece

Dimitris Souravlias
Delft University of Technology
Delft, The Netherlands

Arsenis Tsokas
University of Florida
Gainesville, FL, USA

ISSN 0302-9743 ISSN 1611-3349 (electronic)
Lecture Notes in Computer Science
ISBN 978-3-030-34028-5 ISBN 978-3-030-34029-2 (eBook)
https://doi.org/10.1007/978-3-030-34029-2

LNCS Sublibrary: SL1 – Theoretical Computer Science and General Issues

This Springer imprint is published by the registered company Springer Nature Switzerland AG
The registered company address is: Gewerbestrasse 11, 6330 Cham, Switzerland

Preface

These proceedings contain rigorously refereed versions of papers presented at SEA^2 2019, the Special Event on Analysis of Experimental Algorithms, held at the Elite Hotel City Resort, in Kalamata, Greece, during June 24–29, 2019. SEA^2 2019 is an international forum for researchers in the area of design, analysis, and experimental evaluation and engineering of algorithms, as well as in various aspects of computational optimization and its applications. SEA^2 attracted papers from both the Computer Science and the Operations Research/Mathematical Programming communities. This volume contains the best papers selected as full-length papers and orally presented at the conference. In the context of the conference, participants had the opportunity to experience firsthand the vibrant city of Kalamata, with its rich culture, exquisite cuisine, and a wide spectrum of available activities for adults and children. In addition, conference participants had the chance to visit the Ancient Messini archaeological site, located 25 kilometers north of Kalamata, which features the excavated remains of an entire ancient city, with temples, a stadium, a theater, and several other places of interest; the visit onsite took two and a half hours.

We take this opportunity to thank all the people that were involved in making this conference a success. We express our gratitude to the authors who contributed their work, to the members of the Program Committee, and to the Local Organizing Committee for helping with organizational matters. We extend our heartfelt thanks to the dozens of referees that provided constructive expert referee reports to the authors of this volume. Their reports contributed decisively to substantial improvements of the quality of the accepted papers. In addition, authors of the accepted papers were given the opportunity to integrate in their final versions of the papers, the results of discussions that took place during the conference. We specially thank Stefan Voss, Dimitrios M. Thilikos, and Sirani M. Perera for their inspiring keynote talks. The conference program featured several tracks in (a) Algorithms Engineering and Design, and (b) Computational Optimization and Operations Research.

Finally, we thank Springer for supporting the SEA^2 2019 conference with a 1,000 euros Best Paper Award. A Selection Committee has adjudicated the SEA^2 2019 Best Paper Award to the paper "Effective heuristics for matchings in hypergraphs." We are also grateful to the Kalamata Municipality for sponsoring the conference as well as additional in-kind sponsorships from the CARGO Lab (Waterloo, Canada), CAO (Florida, USA), and the High Performance Intelligent Computing and Signal Processing - HICASP Lab (Ioannina, Greece).

We hope the reader will find this volume useful both as a reference to current research and as a starting point for future work.

October 2019

Ilias Kotsireas
Panos Pardalos
Konstantinos E. Parsopoulos
Dimitris Souravlias
Arsenis Tsokas

Organization

General Chairs

Ilias S. Kotsireas	Wilfrid Laurier University, Canada
Panos M. Pardalos	University of Florida, USA

Program Committee

Konstantinos E. Parsopoulos (Chair)	University of Ioannina, Greece
Christian Blum	Artificial Intelligence Research Institute, Spain
Petros Drineas	Purdue University, USA
Andries Engelbrecht	Stellenbosch University, South Africa
Mario Guarracino	National Research Council of Italy, Italy
Valery Kalyagin	Niznhy Novgorod, Russia
Michael Khachay	Ekaterinebourg, Russia
Timoleon Kipouros	University of Cambridge, UK
Lefteris Kiroussis	University of Athens, Greece
Isaac Lagaris	University of Ioannina, Greece
Gabriel Luque	University of Málaga, Spain
Giuseppe Nicosia	University of Catania, Italy
Sotiris Nikoletseas	University of Patras, CTI, Greece
Jun Pei	Hefei University of Technology, P.R. China
Stefan Pickl	Universität der Bundeswehr München, Germany
Helena Ramalhinho Lourenço	Universitat Pompeu Fabra, Spain
Christoforos Raptopoulos	University of Patras, Greece
Steffen Rebennack	Karlsruher Institut für Technologie, Germany
Mauricio G. C. Resende	Amazon, USA
Konstantina Skouri	University of Ioannina, Greece
Georgios E. Stavroulakis	Technical University of Crete, Greece
Radzic Tomasz	King's College, London, UK
Jamal Toutouh	Massachusetts Institute of Technology, USA
Renato Umeton	Dana-Farber Cancer Institute and MIT, USA

Local Organizing Committee

Dimitris Souravlias	Delft University of Technology, The Netherlands
Bill Tatsis	University of Ioannina, Greece
Arsenios Tsokas	University of Florida, USA

Contents

Voronoi Diagram of Orthogonal Polyhedra in Two and Three Dimensions . . . 1
Ioannis Z. Emiris and Christina Katsamaki

The Complexity of Subtree Intersection Representation of Chordal Graphs and Linear Time Chordal Graph Generation . . . 21
Tınaz Ekim, Mordechai Shalom, and Oylum Şeker

Computing a Minimum Color Path in Edge-Colored Graphs . . . 35
Neeraj Kumar

Student Course Allocation with Constraints . . . 51
Akshay Utture, Vedant Somani, Prem Krishnaa, and Meghana Nasre

A Combinatorial Branch and Bound for the Min-Max Regret Spanning Tree Problem . . . 69
Noé Godinho and Luís Paquete

Navigating a Shortest Path with High Probability in Massive Complex Networks . . . 82
Jun Liu, Yicheng Pan, Qifu Hu, and Angsheng Li

Engineering a PTAS for Minimum Feedback Vertex Set in Planar Graphs . . . 98
Glencora Borradaile, Hung Le, and Baigong Zheng

On New Rebalancing Algorithm . . . 114
Koba Gelashvili, Nikoloz Grdzelidze, and Mikheil Tutberidze

Colorful Frontier-Based Search: Implicit Enumeration of Chordal and Interval Subgraphs . . . 125
Jun Kawahara, Toshiki Saitoh, Hirofumi Suzuki, and Ryo Yoshinaka

Unit Disk Cover for Massive Point Sets . . . 142
Anirban Ghosh, Brian Hicks, and Ronald Shevchenko

Improved Contraction Hierarchy Queries via Perfect Stalling . . . 158
Stefan Funke and Thomas Mendel

Constraint Generation Algorithm for the Minimum Connectivity Inference Problem . . . 167
Édouard Bonnet, Diana-Elena Fălămaş, and Rémi Watrigant

Efficient Split-Radix and Radix-4 DCT Algorithms and Applications . . . 184
Sirani M. Perera, Daniel Silverio, and Austin Ogle

Analysis of Max-Min Ant System with Local Search Applied to the Asymmetric and Dynamic Travelling Salesman Problem with Moving Vehicle 202
João P. Schmitt, Rafael S. Parpinelli, and Fabiano Baldo

Computing Treewidth via Exact and Heuristic Lists of Minimal Separators. . . . 219
Hisao Tamaki

Fast Public Transit Routing with Unrestricted Walking Through Hub Labeling 237
Duc-Minh Phan and Laurent Viennot

Effective Heuristics for Matchings in Hypergraphs 248
Fanny Dufossé, Kamer Kaya, Ioannis Panagiotas, and Bora Uçar

Approximated ZDD Construction Considering Inclusion Relations of Models 265
Kotaro Matsuda, Shuhei Denzumi, Kengo Nakamura, Masaaki Nishino, and Norihito Yasuda

Efficient Implementation of Color Coding Algorithm for Subgraph Isomorphism Problem 283
Josef Malík, Ondřej Suchý, and Tomáš Valla

Quantum-Inspired Evolutionary Algorithms for Covering Arrays of Arbitrary Strength 300
Michael Wagner, Ludwig Kampel, and Dimitris E. Simos

An Experimental Study of Algorithms for Geodesic Shortest Paths in the Constant-Workspace Model. 317
Jonas Cleve and Wolfgang Mulzer

Searching for Best Karatsuba Recurrences 332
Çağdaş Çalık, Morris Dworkin, Nathan Dykas, and Rene Peralta

Minimum and Maximum Category Constraints in the Orienteering Problem with Time Windows 343
Konstantinos Ameranis, Nikolaos Vathis, and Dimitris Fotakis

Internal Versus External Balancing in the Evaluation of Graph-Based Number Types 359
Hanna Geppert and Martin Wilhelm

Hacker's Multiple-Precision Integer-Division Program in Close Scrutiny 376
Jyrki Katajainen

Assessing Algorithm Parameter Importance Using Global Sensitivity Analysis 392
Alessio Greco, Salvatore Danilo Riccio, Jon Timmis, and Giuseppe Nicosia

A Machine Learning Framework for Volume Prediction 408
Umutcan Önal and Zafeirakis Zafeirakopoulos

Faster Biclique Mining in Near-Bipartite Graphs 424
Blair D. Sullivan, Andrew van der Poel, and Trey Woodlief

k-Maximum Subarrays for Small k: Divide-and-Conquer Made Simpler 454
Ovidiu Daescu and Hemant Malik

A Faster Convex-Hull Algorithm via Bucketing 473
Ask Neve Gamby and Jyrki Katajainen

Fixed Set Search Applied to the Minimum Weighted Vertex Cover Problem 490
Raka Jovanovic and Stefan Voß

Automated Deep Learning for Threat Detection in Luggage from X-Ray Images 505
Alessio Petrozziello and Ivan Jordanov

Algorithmic Aspects on the Construction of Separating Codes 513
Marcel Fernandez and John Livieratos

Lagrangian Relaxation in Iterated Local Search for the Workforce Scheduling and Routing Problem 527
Hanyu Gu, Yefei Zhang, and Yakov Zinder

Approximation Algorithms and an Integer Program for Multi-level Graph Spanners 541
Reyan Ahmed, Keaton Hamm, Mohammad Javad Latifi Jebelli, Stephen Kobourov, Faryad Darabi Sahneh, and Richard Spence

Author Index 563

Voronoi Diagram of Orthogonal Polyhedra in Two and Three Dimensions

Ioannis Z. Emiris[1,2] and Christina Katsamaki[1(✉)]

[1] Department of Informatics and Telecommunications, National and Kapodistrian University of Athens, Athens, Greece
{emiris,ckatsamaki}@di.uoa.gr

[2] ATHENA Research and Innovation Center, Maroussi, Greece

Abstract. Voronoi diagrams are a fundamental geometric data structure for obtaining proximity relations. We consider collections of axis-aligned orthogonal polyhedra in two and three-dimensional space under the max-norm, which is a particularly useful scenario in certain application domains. We construct the exact Voronoi diagram inside an orthogonal polyhedron with holes defined by such polyhedra. Our approach avoids creating full-dimensional elements on the Voronoi diagram and yields a skeletal representation of the input object. We introduce a complete algorithm in 2D and 3D that follows the subdivision paradigm relying on a bounding-volume hierarchy; this is an original approach to the problem. The complexity is adaptive and comparable to that of previous methods. Under a mild assumption it is $O(n/\Delta + 1/\Delta^2)$ in 2D or $O(n\alpha^2/\Delta^2 + 1/\Delta^3)$ in 3D, where n is the number of sites, namely edges or facets resp., Δ is the maximum cell size for the subdivision to stop, and α bounds vertex cardinality per facet. We also provide a numerically stable, open-source implementation in Julia, illustrating the practical nature of our algorithm.

Keywords: Max norm · Axis-aligned · Rectilinear · Straight skeleton · Subdivision method · Numeric implementation

1 Introduction

Orthogonal shapes are ubiquitous in numerous applications including raster graphics and VLSI design. We address Voronoi diagrams of 2- and 3-dimensional orthogonal shapes. We focus on the L_∞ metric which is used in the relevant applications and has been studied much less than L_2.

A *Voronoi diagram* partitions space into regions based on distances to a given set $\mathcal{S}$ of geometric objects in $\mathbb{R}^d$. Every $s \in \mathcal{S}$ is a *Voronoi site* (or simply a *site*) and its *Voronoi region* under metric μ, is $V_\mu(s) = \{x \in \mathbb{R}^d \mid \mu(s,x) < \mu(x,s'),\ s' \in \mathcal{S} \setminus s\}$. The *Voronoi diagram* is the set $\mathcal{V}_\mu(\mathcal{S}) = \mathbb{R}^d \setminus \bigcup_{s\in\mathcal{S}} V_\mu(s)$, consisting of all points that attain their minimum distance to $\mathcal{S}$ by at least two Voronoi sites. For general input, the Voronoi diagram is a collection of faces of

I. Kotsireas et al. (Eds.): SEA[2] 2019, LNCS 11544, pp. 1–20, 2019.
https://doi.org/10.1007/978-3-030-34029-2_1

dimension $0, 1, \ldots, d-1$. A face of dimension k comprises points equidistant to at least $d+1-k$ sites. Faces of dimension 0 and 1 are called *Voronoi vertices* and *Voronoi edges* respectively. The union of Voronoi edges and vertices is the *1-skeleton*. Equivalently, a Voronoi diagram is defined as the *minimization diagram* of the distance functions to the sites. The diagram is a partitioning of space into regions, each region consisting of points where some function has lower value than any other function.

In this paper, we study Voronoi diagrams in the interior of an axis-aligned *orthogonal polyhedron*; its faces meet at right angles, and the edges are aligned with the axes of a coordinate system. It may have arbitrarily high genus with holes defined by axis-aligned orthogonal polyhedra, not necessarily convex. Facets are simply connected (without holes) for simplicity. The sites are the facets on the boundary of all polyhedra.

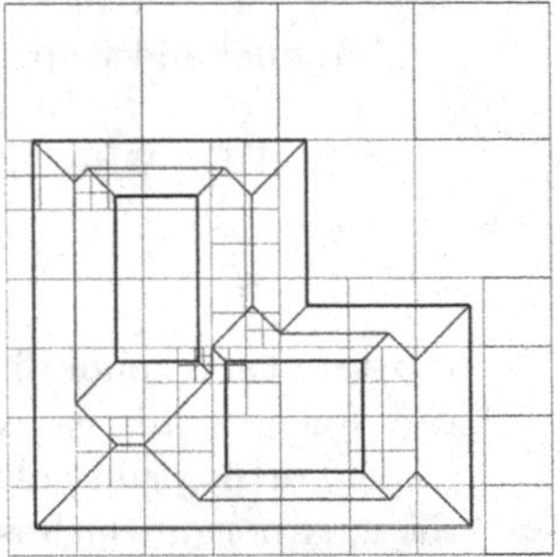

Fig. 1. Voronoi diagram of a rectilinear polygon with 2 holes. (Color figure online)

The distance of two points $x, y \in \mathbb{R}^d$ under L_∞ is $\mu_\infty(x,y) = \max_i\{|x_i - y_i|\}$ and the distance of x to a set $S \subset \mathbb{R}^d$ is $\mu_\infty(x,S) = \inf\{\mu_\infty(x,y) \mid y \in S\}$. In Fig. 1, the Voronoi diagram[1] of a rectilinear polygon with 2 holes is shown in blue. Our algorithm follows the *Subdivision Paradigm* and handles 2D and 3D sites. It reads in a region bounding all input sites and performs a recursive subdivision into cells (using quadtrees or octrees). Then, a reconstruction technique is applied to produce an isomorphic representation of the Voronoi diagram.

Previous Work. If V is the number of polyhedral vertices, the combinatorial complexity of our Voronoi diagrams equals $O(V)$ in 2D [13] and $O(V^2)$ in 3D [4]. In 3D, it is estimated experimentally to be, in general, $O(V)$ [12].

Related work in 2D concerns L_∞ Voronoi diagrams of segments. In [13], they introduce an $O(n \log n)$ sweep-line algorithm, where n is the number of segments; they offer a robust implementation for segments with $O(1)$ number of orientations. Another algorithm implemented in library CGAL [6] is incremental. The L_∞ Voronoi diagram of orthogonal polyhedra (with holes) is addressed in [12] in view of generalizing the sweep-line paradigm to 3D: in 2D it runs in $O(n \log n)$ as in [13], and in 3D the sweep-plane version runs in $O(kV)$, where $k = O(V^2)$ is the number of events.

When the diagram is restricted in the interior of a polygon or polyhedron, it serves as a skeletal representation. A skeleton reduces the dimension of the input capturing its boundary's geometric and topological properties. In particular, *straight skeletons* are very related to the L_∞ Voronoi diagram of rectilinear polygons [2]. An algorithm for the straight skeleton of a simple polygon (not necessarily rectilinear) has complexity $O(V^{\frac{17}{11}+\varepsilon})$ for fixed $\varepsilon > 0$ [9]. For x-monotone rectilinear polygons, a linear time algorithm was recently introduced [7]. In 3D,

[1] Computed by our software and visualized with `Axl` viewer.

an analogous equivalence of the straight skeleton of orthogonal polyhedra and the L_∞ Voronoi diagram exists [4] and a complete analysis of 3D straight skeletons is provided in [3]. Specifically for 3D orthogonal polyhedra, in [4] they offer two algorithms that construct the skeleton in $O(\min\{V^2 \log V, k \log^{O(1)} V\})$, where $k = O(V^2)$ is the number of skeleton features. Both algorithms are rather theoretical and follow a wavefront propagation process. Recently, the straight skeleton of a 3D polyhedral terrain was addressed [11].

A Voronoi diagram can contain full-dimensional faces, as part of a bisector. Under L_∞, when two points have same coordinate value, their bisector is full dimensional (Fig. 2a). Conventions have been adopted, to ensure bisectors between sites are not full-dimensional [6,12,13]. We address this issue in the next section. Subdivision algorithms for Voronoi diagrams are numerous, e.g. [5,8,15]; our work is related to [5,15]. These algorithms are quite efficient, since they adapt to the input, and rather simple to implement. None exists for our problem.

Our Contribution. We express the problem by means of the minimization diagram of a set of algebraic functions with restricted domain, that express the L_∞ distance of points to the boundary. The resulting Voronoi diagram, for general input, is $(d-1)$–dimensional. We focus on 2D and 3D orthogonal polyhedra with holes, where the resulting Voronoi diagram is equivalent to the straight skeleton. We introduce an efficient and complete algorithm for both dimensions, following the subdivision paradigm which is, to the best of our knowledge, the first subdivision algorithm for this problem. We compute the exact Voronoi diagram (since L_∞ bisectors are linear). The output data structure can also be used for nearest-site searching.

The overall complexity is output-sensitive, which is a major advantage. Under the 'Uniform Distribution Hypothesis' (Sect. 3.3), which captures the expected geometry of the input as opposed to worst-case behaviour, the complexity is $O(n/\Delta + 1/\Delta^2)$ in 2D, where n the number of sites (edges) and Δ the separation bound (maximum edge length of cells that guarantees termination). This bound is to be juxtaposed to the worst-case bound of $O(n \log n)$ of previous methods. In 3D, it is $O(n\,\alpha^2/\Delta^2 + 1/\Delta^3)$ where α bounds vertex cardinality per facet (typically constant). Under a further assumption (Remark 2) this bound becomes $O(V/\Delta^2 + 1/\Delta^3)$ whereas existing worst-case bounds are quasi-quadratic or cubic in V. Δ is measured under appropriate scaling for the bounding box to have edge length 1. Scaling does not affect arithmetic complexity, but may be adapted to reduce the denominators' size in rational calculations. The algorithm's relative simplicity has allowed us to develop a numerically stable software in Julia[2], a user-friendly platform for efficient numeric computation; it consists of about 5000 lines of code and is the first open-source code in 3D.

The rest of this paper is organized as follows. The next section provides structural properties of Voronoi diagrams. In Sect. 3 we introduce our 2D algorithm: the 2D and 3D versions share some basic ideas which are discussed in detail in this section. In particular, we describe a hierarchical data structure of bounding

[2] https://gitlab.inria.fr/ckatsama/L_infinity_Voronoi/.

volumes, used to accelerate the 2D algorithm for certain inputs and is necessary for the efficiency of the 3D algorithm. Then we provide the complexity analysis of the 2D algorithm. In Sect. 4 we extend our algorithm and analysis to 3D. In Sect. 5 we conclude with some remarks, examples and implementation details. Due to space limitations, omitted proofs are given in the Appendix.

2 Basic Definitions and Properties

We introduce useful concepts in general dimension. Let $\mathcal{P}$ be an orthogonal polyhedron of full dimension in d dimensions, whose boundary consists of n *simply connected* (without holes) facets; these are edges or flats in 2D and 3D, resp. Note that $\mathcal{P}$ includes the shape's interior and boundary. Now $\mathcal{S}$ consists of the *closed facets* that form the boundary of $\mathcal{P}$ including all facets of the interior polyhedra. There are as many such polyhedra as the genus. Let $V_\infty(s)$ denote the Voronoi region of site s under the L_∞ metric. Lemma 1 gives a property of standard L_∞ Voronoi diagram preserved by Definition 1.

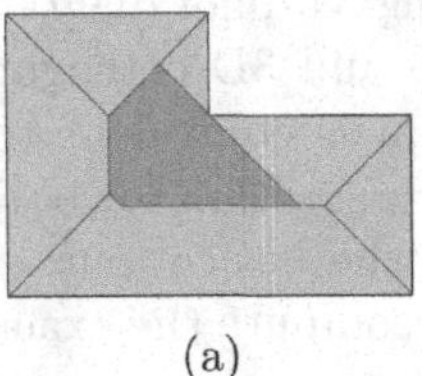

(a)

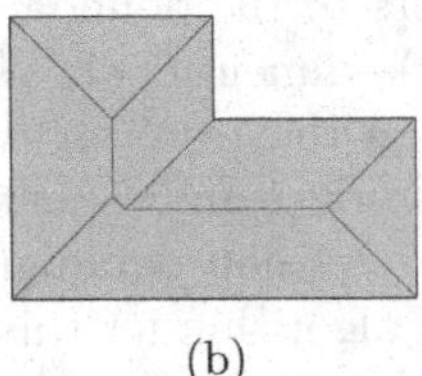

(b)

Fig. 2. Voronoi diagrams (in red): (a) standard, under L_∞, (b) under Definition 1 (Color figure online)

Lemma 1. *Let $s \in \mathcal{S}$. For every point $p \in V_\infty(s)$ it holds that $\mu_\infty(p,s) = \mu_\infty(p, \mathit{aff}(s))$, where aff$(s)$ is the affine hull of s.*

For $s \in \mathcal{S}$ let $\mathcal{H}(s)$ be the *closed* halfspace of $\mathbb{R}^d$ induced by aff(s) such that for every $p \in s$ there exists a point $q \in \mathcal{H}(s)$ s.t. $q \in int(\mathcal{P})$ and $\mu_\infty(p,q) < \epsilon$, $\forall \epsilon > 0$. We define the **(unoriented) zone** of s as $\mathcal{Z}(s) := \{p \in \mathbb{R}^d \mid \mu_\infty(p,s) = \mu_\infty(p, \text{aff}(s))\}$. The **oriented zone** of s is $\mathcal{Z}^+(s) := \mathcal{H}(s) \cap \mathcal{Z}(s)$ (Fig. 3).

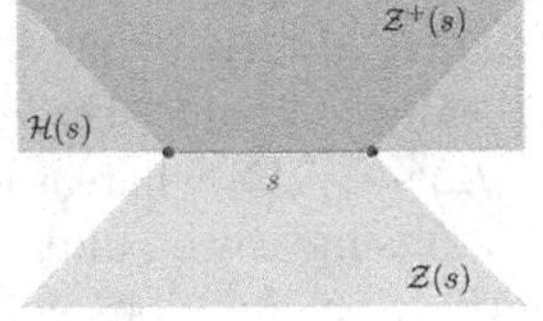

Fig. 3. $\mathcal{H}(s)$, $\mathcal{Z}(s)$, $\mathcal{Z}^+(s)$ for segment s.

We associate to s the distance function

$$D_s(\cdot) : \mathbb{R}^d \to \mathbb{R} : p \mapsto \begin{cases} \mu_\infty(p,s), & \text{if } p \in \mathcal{Z}^+(s), \\ \infty, & \text{otherwise.} \end{cases}$$

The minimization diagram of $\mathcal{D} = \{D_s \mid s \in \mathcal{S}\}$ restricted to $\mathcal{P}$ yields a Voronoi partitioning. The *Voronoi region* of s with respect to $D_s(\cdot)$ is

$$V_\mathcal{D}(s) = \{p \in \mathcal{P} \mid D_s(p) < \infty \text{ and } \forall s' \in \mathcal{S} \setminus s \; D_s(p) < D_{s'}(p)\}$$

Definition 1. *The Voronoi diagram of $\mathcal{P}$ w.r.t. $\mathcal{D}$ is $\mathcal{V}_{\mathcal{D}}(\mathcal{P}) = \mathcal{P} \setminus \bigcup_{s\in\mathcal{S}} V_{\mathcal{D}}(s)$.*

This means one gets the Voronoi diagram of Fig. 2b. Clearly $\mathcal{P} \subset \bigcup_{s\in\mathcal{S}} \mathcal{Z}^+(s)$ (Fig. 2a). Denoting by $\overline{X}$ the closure of a set X, then $V_\infty(s) \subseteq V_{\mathcal{D}}(s) \subseteq \overline{V_\infty(s)} \subseteq \mathcal{Z}^+(s)$. The bisector of $s, s' \in \mathcal{S}$ w.r.t. $\mathcal{D}$ is $\text{bis}_{\mathcal{D}}(s, s') = \{x \in \mathbb{R}^d \mid D_s(x) = D_{s'}(x) < \infty\}$. Then $\text{bis}_{\mathcal{D}}(s, s') \subset \text{affbis}(s, s')$, where $\text{affbis}(s, s')$ denotes the L_∞ (affine) bisector of $\text{aff}(s), \text{aff}(s')$. In 2D (resp. 3D) if sites have not the same affine hull, bisectors under $\mathcal{D}$ lie on lines (resp. planes) parallel to one coordinate axis (resp. plane) or to the bisector of two perpendicular coordinate axes (resp. planes). Although the latter consists of two lines (resp. planes), $\text{bis}_{\mathcal{D}}$ lies only on one, and it can be uniquely determined by the orientation of the zones. Degeneracy of full-dimensional bisectors, between sites with the same affine hull, is avoided by infinitesimal perturbation of corresponding sites. This is equivalent to assigning priorities to the sites; the full dimensional region of the former diagram is 'to the limit' assigned to the site with the highest priority (Fig. 4b). Such a perturbation always exists, both for 2D [12, Lem. 13] and 3D [12, Lem. 31].

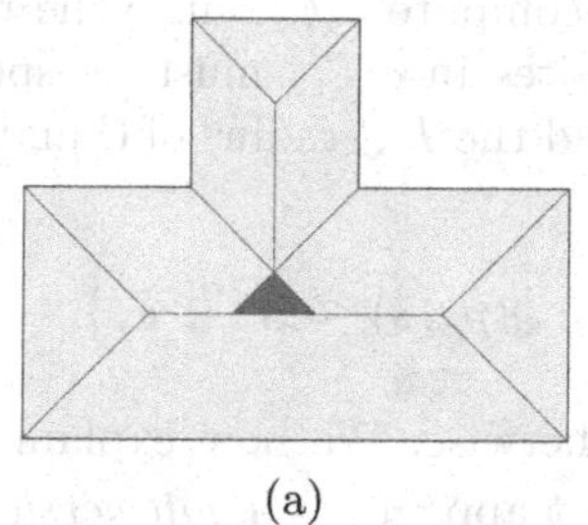
(a)

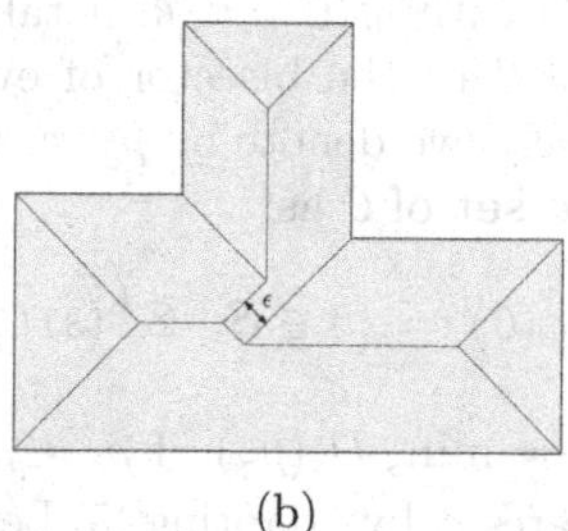

(b)

Fig. 4. (a) 2D Voronoi diagram for polygon with colinear edges. (b) 1D Voronoi diagram after infinitesimal perturbation of edges, where $\epsilon \to 0^+$.

Set X is *weakly star shaped* with respect to $Y \subseteq X$ if $\forall x \in X$, $\exists y \in Y$ such that the segment (x, y) belongs to X.

Lemma 2. *For every $s \in S$, $\overline{V_{\mathcal{D}}(s)}$ is weakly star shaped with respect to s.*

Therefore, since every s is simply connected, from Lemma 2 $\overline{V_{\mathcal{D}}(s)}$ is *simply connected* and $V_{\mathcal{D}}(s)$ is also simply connected. Let the *degree* of a Voronoi vertex be the number of sites to which it is equidistant. If the degree is $> d + 1$, the vertex is *degenerate.* Lemma 3 is nontrivial: in metrics like L_2 degree is arbitrarily large. For $d = 2, 3$ this bound is tight [12].

Lemma 3. *(a) The maximum degree of a Voronoi vertex is less than or equal to $2^d d$. (b) When $d = 2$, a Voronoi vertex cannot have degree 7.*

3 Subdivision Algorithm in Two Dimensions

Given manifold rectilinear polygon $\mathcal{P}$, i.e. every vertex being shared by exactly two edges, the input consists of $\mathcal{S}$ and a box $\mathcal{C}_0$ bounding $\mathcal{P}$. Non-manifold vertices can be trivially converted to manifold with an infinitesimal perturbation. Subdivision algorithms include two phases. First, recursively subdivide $\mathcal{C}_0$ to 4 identical cells until certain criteria are satisfied, and the diagram's topology can be determined in $O(1)$ inside each cell. The diagram is reconstructed in the second phase.

3.1 Subdivision Phase

We consider subdivision cells as closed. Given cell $\mathcal{C}$, let $\phi(\mathcal{C})$ be the set of sites whose closed Voronoi region intersects $\mathcal{C}$: $\phi(\mathcal{C}) = \{s \in \mathcal{S} \mid \overline{V_{\mathcal{D}}(s)} \cap \mathcal{C} \neq \emptyset\}$. For point $p \in \mathcal{P}$ we define its **label set** $\lambda(p) = \{s \in \mathcal{S} \mid p \in \overline{V_{\mathcal{D}}(s)}\}$. When $p \in \mathcal{P}^c$, where $\mathcal{P}^c$ is the complement of $\mathcal{P}$, then $\lambda(p) = \emptyset$. The computation of $\phi(\mathcal{C})$ is hereditary, since $\phi(\mathcal{C}) \subseteq \phi(\mathcal{C}')$, if $\mathcal{C}'$ is the parent of $\mathcal{C}$. But it is rather costly; given $\phi(\mathcal{C}')$ with $|\phi(\mathcal{C}')| = \kappa$, it takes $O(\kappa^2)$ to compute $\phi(\mathcal{C})$, since the relative position of $\mathcal{C}$ to the bisector of every pair of sites in $\phi(\mathcal{C}')$ must be specified. Alternatively, we denote by $p_{\mathcal{C}}$, $r_{\mathcal{C}}$ the center and the L_∞-radius of $\mathcal{C}$ and define the **active set** of $\mathcal{C}$ as:

$$\widetilde{\phi}(\mathcal{C}) := \{s \in \mathcal{S} \mid \mathcal{Z}^+(s) \cap \mathcal{C} \neq \emptyset, \text{ and } \mu_\infty(p_{\mathcal{C}}, s) \leq 2r_{\mathcal{C}} + \delta_{\mathcal{C}}\},$$

where $\delta_{\mathcal{C}} = \min_s D_s(p_{\mathcal{C}})$, if $p_{\mathcal{C}} \in \mathcal{P}$, and 0 otherwise. We now explain how $\widetilde{\phi}$ approximates ϕ by adapting [5, Lem. 2], where $\widetilde{\phi}$ appears as a *soft version* of ϕ.

Lemma 4. *(a) For every cell $\mathcal{C}$, $\phi(\mathcal{C}) \subseteq \widetilde{\phi}(\mathcal{C})$. (b) For a sequence of cells $(\mathcal{C})_i$ monotonically convergent to point $p \in \mathcal{P}$, $\widetilde{\phi}(\mathcal{C}_i) = \phi(p)$ for i large enough.*

One can easily verify $\widetilde{\phi}(\mathcal{C}) \subseteq \widetilde{\phi}(\mathcal{C}')$, therefore the complexity of computing $\widetilde{\phi}(\mathcal{C})$ is linear in the size of $\widetilde{\phi}(\mathcal{C}')$. The algorithm proceeds as follows: For each subdivision cell we maintain the label sets of its corner points and of its central point, and $\widetilde{\phi}$. The subdivision of a cell stops whenever at least one of the *termination criteria* below holds (checked in turn). Upon subdivision, we propagate $\widetilde{\phi}$ and the label sets of the parent cell to its children. For every child we compute the remaining label sets and refine its active set. Let M be the maximum degree of a Voronoi vertex ($M \leq 8$).

Termination Criteria: (T1) $\mathcal{C} \subseteq V_{\mathcal{D}}(s)$ for some $s \in \mathcal{S}$; (T2) $int(\mathcal{C}) \cap \mathcal{P} = \emptyset$; (T3) $|\widetilde{\phi}(\mathcal{C})| \leq 3$; (T4) $|\widetilde{\phi}(\mathcal{C})| \leq M$ and the sites in $\widetilde{\phi}(\mathcal{C})$ define a unique Voronoi vertex $v \in \mathcal{C}$.

When (T1) holds, $\mathcal{C}$ is contained in a Voronoi region so no part of the diagram is in it. (T2) stops the subdivision when the open cell is completely outside the polygon. If (T3) holds, we determine in $O(1)$ time the diagram's topology in

$\mathcal{C}$ since there are ≤ 3 Voronoi regions intersected. (T4) stops cell subdivision if it contains a single degenerate Voronoi vertex. The process is summarized in Algorithm 1.

Algorithm 1. Subdivision2D($\mathcal{P}$)

root $\leftarrow$ bounding box of $\mathcal{P}$
$Q \leftarrow$ root
while $Q \neq \emptyset$ **do**
    $\mathcal{C} \leftarrow$ pop(Q)
    Compute $\widetilde{\phi}(\mathcal{C})$ and the label sets of the vertices and the central point.
    if (T1) $\vee$ (T2) $\vee$ (T3) $\vee$ (T4) **then**
        return
    else
        Subdivide $\mathcal{C}$ into $\mathcal{C}_1, \mathcal{C}_2, \mathcal{C}_3, \mathcal{C}_4$
        $Q \leftarrow Q \cup \{\mathcal{C}_1, \mathcal{C}_2, \mathcal{C}_3, \mathcal{C}_4\}$
    end if
end while

Theorem 1. *Algorithm 1 halts.*

Proof. Consider an infinite sequence of boxes $\mathcal{C}_1 \supsetneq \mathcal{C}_2 \supsetneq \ldots$ such that none of the termination criteria holds. Since (T1) and (T2) do not hold for any $\mathcal{C}_i$ with $i \geq 1$, the sequence converges to a point $p \in \mathcal{V}_\mathcal{D}(\mathcal{P})$. From Lemma 4(b), there exists $i_0 \in \mathbb{N}$ such that $\widetilde{\phi}(\mathcal{C}_{i_0}) = \phi(p) = \lambda(p)$. Since $|\lambda(p)| \leq 8$, (T4) will hold. □

Lemma 5. *For a subdivision cell $\mathcal{C}$, let $v_1, \ldots, v_4$ its corner vertices. For $s \in \mathcal{S}$, $\mathcal{C} \subseteq V_\mathcal{D}(s)$ if and only if $v_1, \ldots, v_4 \in V_\mathcal{D}(s)$.*

Hence one decides (T1) by checking the vertices' labels. (T2) is valid for $\mathcal{C}$ iff $\lambda(p_c) = \emptyset$ and $\forall s \in \widetilde{\phi}(\mathcal{C})$, $s \cap int(\mathcal{C}) = \emptyset$. For (T4), the presence of a Voronoi vertex in $\mathcal{C}$ is verified through constructor `VoronoiVertexTest`: given $\mathcal{C}$ with $|\widetilde{\phi}(\mathcal{C})| \geq 3$, the affine bisectors of sites in $\widetilde{\phi}(\mathcal{C})$ are intersected. If the intersection point is in $\mathcal{C}$ and in $\mathcal{Z}^+(s)$ for every $s \in \widetilde{\phi}(\mathcal{C})$ then it is a Voronoi vertex. We do not need to check whether it is in $\mathcal{P}$ or not; since (T1) fails for $\mathcal{C}$, if $v \notin \mathcal{P}$, there must be s intersecting $\mathcal{C}$ such that $v \notin \mathcal{Z}^+(s)$: contradiction.

3.2 Reconstruction Phase

We take the quadtree of the subdivision phase and output a planar straight-line graph (PSLG) $G = (V, E)$ representing the Voronoi diagram of $\mathcal{P}$. G is a (vertex) labeled graph and its nodes are of two types: *bisector nodes* and *Voronoi vertex nodes*. Bisector nodes span Voronoi edges and are labeled by the two sites to which they are equidistant. Voronoi vertex nodes correspond to Voronoi vertices and so are labeled by at least 3 sites. We visit the leaves of the quadtree and, whenever the Voronoi diagram intersects the cell, bisector or vertex nodes are introduced. By connecting them accordingly with straight-line edges, we obtain the exact Voronoi diagram and not an approximation. We process leaves with $|\widetilde{\phi}(\cdot)| \geq 2$ that do not satisfy (T1) nor (T2).

Cell with Two Active Sites. When $\widetilde{\phi}(\mathcal{C}) = \{s_1, s_2\}$, $\mathcal{C}$ intersects $V_{\mathcal{D}}(s_1)$ or $V_{\mathcal{D}}(s_2)$ or both. The intersection of $\text{bis}_{\mathcal{D}}(s_1, s_2)$ with the cell, when non empty, is part of the Voronoi diagram: for each $p \in \text{bis}_{\mathcal{D}}(s_1, s_2) \cap \mathcal{C}$ it holds that $D_{s_1}(p) = D_{s_2}(p)$ and $\lambda(p) \subseteq \widetilde{\phi}(\mathcal{C}) = \{s_1, s_2\}$. Therefore $p \in \overline{V_{\mathcal{D}}(s_1)} \cap \overline{V_{\mathcal{D}}(s_2)}$.

Remark 1. If there is no Voronoi vertex in $\mathcal{C}$ and $p_1, p_2 \in \text{bis}_{\mathcal{D}}(s_1, s_2) \cap \mathcal{C}$ for $s_1, s_2 \in \widetilde{\phi}(\mathcal{C})$, then $p_1p_2 \subset \text{bis}_{\mathcal{D}}(s_1, s_2)$.

Since $\text{bis}_{\mathcal{D}}(s_1, s_2) \subset \text{affbis}(s_1, s_2)$ we intersect the affine bisector with the boundary of the cell. An intersection point $p \in bis_{\infty}(\text{aff}(s_1), \text{aff}(s_2))$ is in $bis_{\mathcal{D}}(s_1, s_2)$ iff $p \in \mathcal{Z}^+(s_1) \cap \mathcal{Z}^+(s_2)$. If intersection points are both in $\mathcal{Z}^+(s_1) \cap \mathcal{Z}^+(s_2)$, we introduce a bisector node in the middle of the line segment joining them, labeled by $\{s_1, s_2\}$. When only one intersection point is in $\mathcal{Z}^+(s_1) \cap \mathcal{Z}^+(s_2)$, then s_1, s_2 must intersect in $\mathcal{C}$. Introduce a bisector node at the intersection point labeled by $\{s_1, s_2\}$.

Cell with 3 Active Sites or More. When $|\widetilde{\phi}(\mathcal{C})| = 3$ and the `VoronoiVertexTest` finds a vertex in $\mathcal{C}$ or when $|\widetilde{\phi}(\mathcal{C})| \geq 4$ (a vertex has already been found), we introduce a Voronoi vertex node at the vertex, labeled by corresponding sites. In the presence of corners of $\mathcal{P}$ in $\mathcal{C}$, bisector nodes are introduced and connected to the vertex node (Fig. 5).

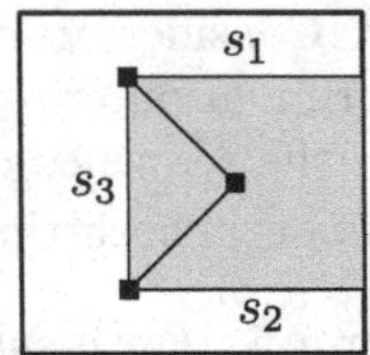

Fig. 5. A Voronoi vertex node connected with two bisector nodes.

If no Voronoi vertex is in $\mathcal{C}$, we repeat the procedure described in previous paragraph for each pair of sites. Even if a bisector node is found, it is not inserted at the graph if it is closer to the third site.

Connecting the Graph Nodes. The remaining graph edges must cross two subdivision cells. We apply "dual marching cubes" [14] to enumerate pairs of neighboring cells in time linear in the size of the quadtree: cells are neighboring if they share a facet. Let v_1, v_2 be graph nodes in neighboring cells. We connect them iff:

- v_1, v_2 are bisector nodes and $\lambda(v_1) = \lambda(v_2)$.
- v_1 is a bisector node, v_2 is a Voronoi vertex node and $\lambda(v_1) \subset \lambda(v_2)$.
- v_1, v_2 are Voronoi vertex nodes, $|\lambda(v_1) \cap \lambda(v_2)| = 2$ and $v_1v_2 \subset \mathcal{P}$.

See Fig. 6 for an example where v_1, v_2 are Voronoi vertex nodes with $|\lambda(v_1) \cap \lambda(v_2)| = 2$ and $v_1v_2 \not\subset \mathcal{P}$.

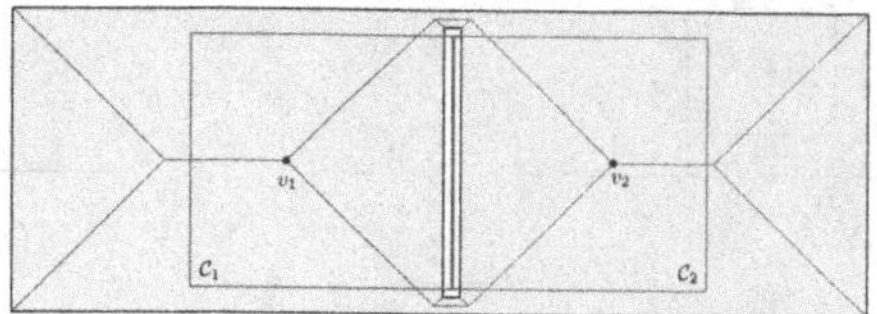

Fig. 6. $\mathcal{C}_1, \mathcal{C}_2$ are two neighboring subdivision cells and the Voronoi vertices v_1, v_2 have two common sites as labels but are not connected with a Voronoi edge.

Theorem 2 (Correctness). *The output graph is isomorphic to* $\mathcal{V}_\mathcal{D}(\mathcal{P})$.

Proof. We need to prove that the nodes in the graph are connected correctly. Let neighboring cells $\mathcal{C}_1, \mathcal{C}_2$ and v_1, v_2 graph nodes in each of them respectively. If v_1, v_2 are bisector nodes and $\lambda(v_1) = \lambda(v_2)$, then the line segment v_1v_2 is in $bis_\mathcal{D}(s_1, s_2)$, for $s_1, s_2 \in \lambda(v_1)$, and on the Voronoi diagram (Remark 1). If v_1 is a bisector node and v_2 is a Voronoi vertex node s.t. $\lambda(v_1) \subsetneq \lambda(v_2)$, then $v_1v_2 \subset \text{bis}_\infty(s_1, s_2)$. If the segment v_1v_2 is not on the Voronoi diagram then, there is a Voronoi vertex node different than v_2 in $\mathcal{C}_1$ or $\mathcal{C}_2$; contradiction. At last, let v_1 and v_2 be Voronoi vertex nodes such that their labels have two sites in common, say s, s', and the edge $v_1v_2 \subset \mathcal{P}$. Vertices v_1, v_2 are both on the boundary of $\overline{\mathcal{V}_\mathcal{D}(s)} \cap \overline{\mathcal{V}_\mathcal{D}(s')}$. Since $v_1v_2 \subset \mathcal{P}$, if it does not coincide with the Voronoi edge equidistant to s, s', then both v_1, v_2 must also be on the boundary of a Voronoi region other than $\overline{\mathcal{V}_\mathcal{D}(s)}$ and $\overline{\mathcal{V}_\mathcal{D}(s')}$. This leads to a contradiction. □

3.3 Primitives, Data-Structures, Complexity

Assuming the input vertices are rational, Voronoi vertices are rational [13]. Computing Voronoi vertices, and intersections between affine bisectors and cell facets require linear operations, distance evaluations and comparisons. Therefore, they are exact. The above operations, computing $\widetilde{\phi}$ and deciding site-cell intersection are formulated to allow for a direct extension to 3D. In the sequel we discuss design of predicates, computation of label sets and construction of a Bounding Volume Hierarchy. We also provide a complexity analysis of the algorithm.

Membership in $\mathcal{H}(s)$ is trivial to decide, thus we focus on predicates that decide membership in $\mathcal{Z}(s)$. Given $p \in \mathbb{R}^2$ and $s \in \mathcal{S}$, let $pr_{\text{aff}(s)}(p)$ the projection of p to aff(s) and $I_{p,s}$ the $1d$−interval on aff(s) centered at $pr_{\text{aff}(s)}(p)$ with radius $\mu_\infty(p, \text{aff}(s))$. `inZone(p,s)` decides if $p \in \mathcal{Z}(s)$; this holds iff $I_{p,s} \cap s \neq \emptyset$ (Fig. 7). Given $\mathcal{C}$, $s \in \mathcal{S}$, `ZoneInCell(s,C)` decides if $\mathcal{Z}(s) \cap \mathcal{C} \neq \emptyset$. For this evaluation see Lemma 6 and Fig. 8.

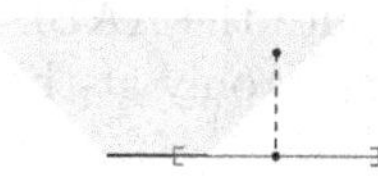

Fig. 7. Test performed by `inZone(p,s)`.

Lemma 6. *Let* $s \in \mathcal{S}$, f_1, f_2 *the two facets of* $\mathcal{C}$ *parallel to aff(s)*, $\rho_i = \mu_\infty(f_i, \mathit{aff}(s))$ *for* $i = 1, 2$ *and* $p'_c = pr_{\mathit{aff}(s)}(p_c)$. *Then,* $\mathcal{Z}(s) \cap \mathcal{C} \neq \emptyset$ *iff* $\exists i \in \{1, 2\}$ *s.t.* $B_\infty(p'_c, r_c + \rho_i) \cap s \neq \emptyset$.

Fig. 8. Illustration of test performed by `ZoneInCell`

To decide if $s \cap \mathcal{C} \neq \emptyset$ and if $s \cap int(\mathcal{C}) \neq \emptyset$, we use `isIntersecting`$(s, \mathcal{C})$ and `isStrictlyIntersecting`$(s, \mathcal{C})$ respectively. Design is trivial. All these predicates are computed in $O(1)$.

Computing Label Sets. If $p \in \mathcal{P} \cap \mathcal{C}$ then its closest sites are in $\widetilde{\phi}(\mathcal{C})$. Deciding if $p \in \mathcal{P}$ is done by `LocationTest`, which identifies position based on the sites that intersect $\mathcal{C}$: among these we select those with minimum L_∞ distance to p and for whom `inZone`(p, s) is true. If a convex (resp. concave) corner w.r.t. the interior of $\mathcal{P}$ is formed by these sites then $p \in \mathcal{P}$ iff it belongs to the intersection (resp. union) of the oriented zones. If no corner is formed or even if $\mathcal{C}$ is not intersected by any site, decision is trivial. This takes $O(|\widetilde{\phi}(\mathcal{C})|)$.

Bounding Volume Hierarchy. We decompose $\mathcal{P}$ into a collection of rectangles such that any two of them have disjoint interior. We construct a kd-tree on the reflex vertices of the polygon, splitting always at a vertex. An orthogonal polygon with h holes, has $r = n/2 + 2(h-1)$ reflex vertices. The kd-tree subdivides the plane into at most $r+1$ regions. Every terminal region contains a nonempty collection of disjoint rectangles. Let t be the maximum number of such rectangles. Using this decomposition, we construct a *Bounding Volume Hierarchy* (BVH) [1,10]. It is a tree structure on a set of objects stored at the leaves along with their bounding volume while internal nodes store information of their descendants' bounding volume. Two important properties are *minimal volume* and *small overlap*. They are achieved by using the Axis Aligned Bounding Box (AABB) as bounding shape and building the BVH by bottom-up traversing the constructed kd-tree: at every leaf of the kd-tree we compute the AABB of its rectangles (namely a *terminal bounding box*) and for every internal node we compute the AABB of its children. The bounding volumes of a node's children intersect only at their boundary. Space complexity is linear in tree size.

Rectangle-Intersection Queries: Given query rectangle Q the data structure reports all rectangles in the decomposition overlapping with Q. Starting from the root, for every internal node, we check whether Q intersects its bounding rectangle or not. In the latter case the data structure reports no rectangles. In the former, we check the position of Q relative to the bounding boxes of the node's children so as to decide for each one if it should be traversed or not. We continue similarly: when we reach a terminal bounding box, we check the position of Q relative to every rectangle in it. Let k be the number of terminal bounding boxes intersected by Q. Following [1] we show:

Theorem 3. *Rectangle intersection queries are answered in $O(k \lg r + kt)$.*

Proof. Let Q a rectangle-intersection query and v an internal node of the BVH tree visited during the query. We distinguish two cases; first, the subtree rooted at v contains a terminal bounding box that intersects Q. There are $O(k)$ such nodes at each level. Otherwise, Q intersects with the bounding rectangle V stored at v but does not intersect any terminal bounding box of the subtree rooted at v. There are at least two such terminal bounding boxes, say b and b'. Since Q does not intersect b there is a line ℓ passing through a facet of Q separating Q from b. Similarly, there exists a line ℓ' passing through a facet of Q that separates it from b'. W.l.o.g. there is a choice of b, b' such that ℓ and ℓ' are distinct: if all the terminal bounding boxes of the subtree can be separated by the same line, then V cannot intersect Q. If ℓ, ℓ' are perpendicular, then their intersection also intersects V. Since the bounding boxes of each level are strictly non-overlapping, every vertex of Q intersects a constant number of them (up to 4). So, there is a constant number of such nodes at a given level. When ℓ, ℓ' are parallel and no vertex of Q intersects V, then the terminal bounding rectangles of the subtree can be partitioned to those separated by ℓ from Q and to those separated by ℓ' from Q. For these distinct sets of terminal bounding boxes to be formed, there must occur a split of V by a line parallel and in between ℓ, ℓ'. So there is a reflex vertex of the polygon in $V \cap Q$, causing this split. But $V \cap Q \cap \mathcal{P} = \emptyset$; a contradiction. So there are $O(k)$ internal nodes visited at each level of tree. The visited leaf nodes correspond to the $O(k)$ terminal bounding boxes that intersect Q and since each of them encloses at most t rectangles, the additional amount of operations performed equals $O(kt)$. Summing over all levels of the tree yields a total query complexity of $\sum_{i=0}^{\lceil \lg r \rceil} O(k) + O(kt) = O(k \lg r + kt)$. □

Point Queries: Given $p \in \mathbb{R}^2$, we report on the rectangles of the decomposition in which p lies inside (at most 4 rectangles). When zero, the point lies outside the polygon. Since it is a special case of a rectangle-intersection query, the query time complexity is $O(\lg r + t)$.

Complexity. Analysis requires a bound on the height of the quadtree. The edge length of the initial bounding box is supposed to be 1 under appropriate scaling. Let *separation bound* Δ be the maximum value s.t. for every cell of edge length $\geq \Delta$ at least one termination criterion holds. Then, the maximum tree height is $L = O(\lg(1/\Delta))$. Let β be the minimum distance of two Voronoi vertices, and γ the relative thickness of $\mathcal{P}^c$, i.e. the minimum diameter of a maximally inscribed L_∞-ball in $\mathcal{P}^c$, where $\mathcal{P}^c$ is the complement of $\mathcal{P}$.

Theorem 4. *Separation bound Δ is $\Omega(\min\{\gamma, \beta\})$, where the asymptotic notation is used to hide some constants.*

Proof. The algorithm mainly subdivides cells that intersect $\mathcal{V}_\mathcal{D}(\mathcal{P})$, since a cell inside a Voronoi region or outside $\mathcal{P}$ is not subdivided (Termination criteria (T1), (T2)). Most subdivisions occur as long as non neighboring Voronoi regions are "too close". Consider $\mathcal{C}$ centered at $p_\mathcal{C} \in V_\mathcal{D}(s)$ and site $s' \neq s$, with $V_\mathcal{D}(s), V_\mathcal{D}(s')$ non neighboring. For $r_\mathcal{C} < \frac{\mu_\infty(p_\mathcal{C}, s') - \mu_\infty(p_\mathcal{C}, s)}{2}$ site s' is not in $\widetilde{\phi}(\mathcal{C})$. It holds that

$\mu_\infty(p_c, s') - \mu_\infty(p_c, s) \geq \zeta(s, s')$, where $\zeta(s, s') = \min\{\mu_\infty(p, q) \mid p \in \overline{V_\mathcal{D}(s)}, q \in \overline{V_\mathcal{D}(s')}\}$, i.e. the minimum distance of the closure of the two Voronoi regions. When $\overline{V_\mathcal{D}(s)}, \overline{V_\mathcal{D}(s')}$ are connected with a Voronoi edge, $\zeta(s, s') = \Omega(\beta)$. When a minimum cell size of $\Omega(\beta)$ is not sufficient for s' to not belong in $\widetilde{\phi}(\mathcal{C})$, then there is a hole between $\overline{V_\mathcal{D}(s)}, \overline{V_\mathcal{D}(s')}$ and $\zeta(s, s')$ is $\Omega(\gamma)$ in this case. □

This lower bound is tight: in Fig. 9a for $\Delta = 0.8125\beta$, and in Fig. 9b for $\Delta = \gamma$. Next we target a realistic complexity analysis rather than worst-case. For this, assume the site distribution in $\mathcal{C}_0$ is "sufficiently uniform". Formally:

Uniform Distribution Hypothesis (UDH): For L_∞ balls $A_1 \subseteq A_0 \subset \mathcal{C}_0$, let N_0 (resp. N_1) be the number of sites intersecting A_0 (resp. A_1). We suppose $N_1/N_0 = O(vol(A_1)/vol(A_0))$, where $vol(\cdot)$ denotes the volume of a set in $\mathbb{R}^d$, d being the dimension of $\mathcal{C}_0$.

Theorem 5. *Under UDH the algorithm complexity is $O(n/\Delta + 1/\Delta^2)$, where n is the total number of boundary edges (including any holes).*

Proof. At each node, refinement and checking the termination criteria run in time linear in the size of its parent's active set. At the root $|\widetilde{\phi}(\mathcal{C}_0)| = n$. The cardinality of active sets decreases as we move to the lower levels of the quadtree: Let $A(p, d, R) = \{q \in \mathbb{R}^2 \mid d \leq \mu_\infty(p, q) \leq 2R + d\}$. For cell $\mathcal{C}$ and $s \in \widetilde{\phi}(\mathcal{C})$, $s \cap A(p_c, \delta_c, r_c) \neq \emptyset$. Let $E = vol(A(p_c, \delta_c, r_c))$. For $\mathcal{C}_1$ a child of $\mathcal{C}$ and $s_1 \in \widetilde{\phi}(\mathcal{C}_1)$, $s_1 \cap A(p_{c_1}, \delta_{c_1}, r_{c_1}) \neq \emptyset$. Since $B_\infty(p_c, \delta_c)$ is empty of sites and may intersect with $A(p_{c_1}, \delta_{c_1}, r_{c_1})$, we let $E_1 = vol(A(p_{c_1}, \delta_{c_1}, r_{c_1}) \setminus (A(p_{c_1}, \delta_{c_1}, r_{c_1}) \cap B_\infty(p_c, \delta_c)))$. We prove that in any combination of $\delta_c, \delta_{c_1}, r_c$ it is $E_1 \leq E/2$. Under *UDH*, a cell at tree level i has $|\widetilde{\phi}(\mathcal{C}_i)| = O(n/2^i)$. Computation per tree level, is linear in sum of active sets' cardinality, therefore summing over all levels of the tree, we find the that complexity of the subdivision phase is $O(n/\Delta)$. The complexity of the reconstruction phase is $O(\tilde{n})$, where $\tilde{n}$ is the number of leaf nodes in the quadtree, which is in turn $O(1/\Delta^2)$. This allows to conclude. □

Queries in the BVH can be used to compute label sets and the active set of a cell. Assume the number of segments touching a rectangle's boundary is $O(1)$, which is the typical case. Then, we prove the following.

Lemma 7. *We denote by $\mathcal{C}'$ the parent of $\mathcal{C}$ in the subdivision. Using BVH accelerates the refinement of $\mathcal{C}$ if $|\widetilde{\phi}(\mathcal{C}')|/|\widetilde{\phi}(\mathcal{C})| = \Omega(\lg n + t)$.*

4 Subdivision Algorithm in Three Dimensions

Let $\mathcal{P}$ be a manifold orthogonal polyhedron: every edge of $\mathcal{P}$ is shared by exactly two and every vertex by exactly 3 facets. For non-manifold input we first employ trihedralization of vertices, discussed in [4,12]. Input consists of $\mathcal{S}$ and bounding box $\mathcal{C}_0$ of $\mathcal{P}$. An octree is used to represent the subdivision.

The main difference with the 2D case is that Voronoi sites can be nonconvex. As a consequence, for site s, $\mathcal{Z}^+(s)$ is not necessarily convex and therefore

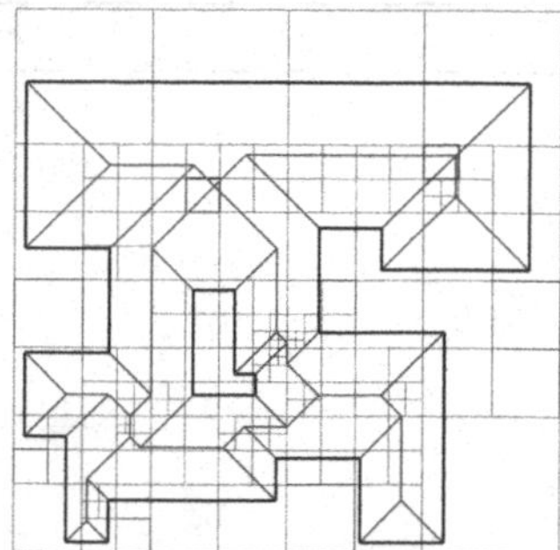

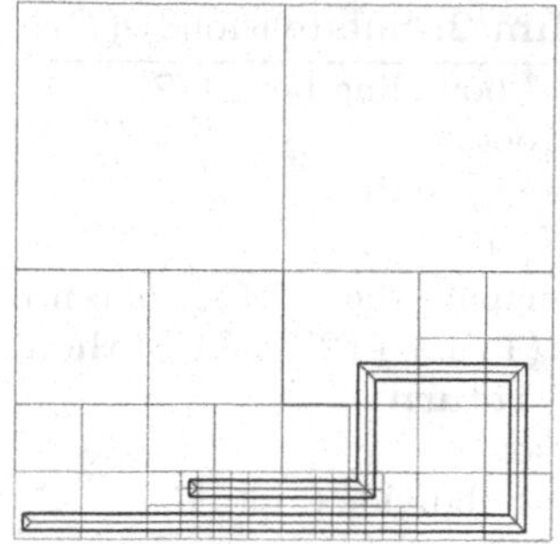

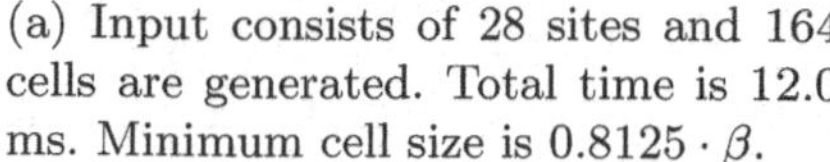

(a) Input consists of 28 sites and 164 cells are generated. Total time is 12.0 ms. Minimum cell size is $0.8125 \cdot \beta$.

(b) Input consists of 12 sites and 64 cells are generated. Total time is 5.6 ms. Minimum cell size is γ.

Fig. 9. The 1-skeleton of the Voronoi diagram is shown in blue. (Color figure online)

the distance function $D_s(\cdot)$ cannot be computed in $O(1)$: it is not trivial to check membership in $\mathcal{Z}^+(s)$. It is direct to extend the 2D algorithm in three dimensions. However, we examine efficiency issues.

For an efficient computation of the basic predicates (of Sect. 3.3), we **preprocess** every facet of the polyhedron and decompose it to a collection of rectangles. Then a BVH on the rectangles is constructed. The basic operation of all these predicates in 2D is an overlap test between an interval and a segment in 1D. In 3D, the analog is an overlap test between a 2D rectangle and a site (rectilinear polygon). Once the BVH is constructed for each facet, the rectangle-intersection query takes time logarithmic in the number of facet vertices (Theorem 3).

Subdivision. The active set $\widetilde{\phi}$, ϕ and the label set of a point are defined as in to 2D. Most importantly, Lemma 4 is valid in 3D as well. The algorithm proceeds as follows: We recursively subdivide $\mathcal{C}_0$ into 8 identical cells. The subdivision of a cell stops whenever at least one of the termination criteria below holds. For each cell of the subdivision we maintain the label set of its central point and $\widetilde{\phi}$. Upon subdivision, we propagate $\widetilde{\phi}$ from a parent cell to its children for further refinement. We denote by M the maximum degree of a Voronoi vertex ($M \leq 24$).

3D Termination Criteria: (T1') $int(\mathcal{C}) \cap \mathcal{P} = \emptyset$, (T2') $|\widetilde{\phi}(\mathcal{C})| \leq 4$, (T3') $|\widetilde{\phi}(\mathcal{C})| \leq M$ and the sites in $\widetilde{\phi}(\mathcal{C})$ define a unique Voronoi vertex $v \in \mathcal{C}$.

Subdivision is summarized in Algorithm 2. (T1') is valid for $\mathcal{C}$ iff $\lambda(p_{\mathcal{C}}) = \emptyset$ and $\forall s \in \widetilde{\phi}(C)$ it holds that $s \cap int(\mathcal{C}) = \emptyset$. Detecting a Voronoi vertex in $\mathcal{C}$ proceeds like in 2D. A Voronoi vertex is equidistant to at least 4 sites and there is a site parallel to each coordinate hyperplane among them. (T1) used in 2D is omitted, for it is not efficiently decided: labels of the cell vertices cannot guarantee that $\mathcal{C} \subseteq V_{\mathcal{D}}(s)$. However, as the following lemma indicates, termination of the subdivision is not affected: cells contained in a Voronoi region whose radius is $< r^*$, where r^* is a positive constant, are not subdivided.

Algorithm 2. Subdivision3D($\mathcal{P}$)

root ← bounding box of $\mathcal{P}$
Q ← root
while $Q \neq \emptyset$ **do**
    $\mathcal{C} \leftarrow$ pop(Q)
    Compute the label set of central point and $\widetilde{\phi}(\mathcal{C})$.
    if (T1') ∨ (T2') ∨ (T3') **then**
        return
    else
        Subdivide $\mathcal{C}$ into $\mathcal{C}_1, \ldots, \mathcal{C}_8$
        $Q \leftarrow Q \cup \{\mathcal{C}_1, \ldots, \mathcal{C}_8\}$
    end if
end while

Lemma 8. *Let $\mathcal{C} \subseteq V_{\mathcal{D}}(s)$. There is $r^* > 0$ s.t. $r_{\mathcal{C}} < r^*$ implies $\widetilde{\phi}(\mathcal{C}) = \{s\}$.*

Theorem 6. *Algorithm 2 halts.*

Reconstruction. We construct a graph $G = (V, E)$, representing the 1-skeleton of the Voronoi diagram. The nodes of G are of two types, *skeleton nodes* and *Voronoi vertex nodes*, and are labeled by their closest sites. Skeleton nodes span Voronoi edges and are labeled by 3 or 4 sites. We visit the leaves of the octree and process cells with $|\widetilde{\phi}(\mathcal{C})| \geq 3$ and that do not satisfy (T1'). We introduce the nodes to the graph as in 2D. Graph edges are added between corners and Voronoi vertex nodes inside a cell. We run dual marching cubes (linear in the octree size) and connect graph nodes v_1, v_2 located in neighboring cells, iff:

- v_1, v_2 are skeleton nodes and $\lambda(v_1) = \lambda(v_2)$, or
- v_1 is a skeleton node, v_2 is a Voronoi vertex node and $\lambda(v_1) \subset \lambda(v_2)$, or
- v_1, v_2 are Voronoi vertex nodes, $|\lambda(v_1) \cap \lambda(v_2)| = 3$ and $v_1 v_2 \subset \mathcal{P}$.

Theorem 7 (Correctness). *The output graph is isomorphic to the 1-skeleton of the Voronoi diagram.*

Primitives. Deciding membership in $\mathcal{H}(\cdot)$ is trivial. The predicates of Sect. 3.3 extend to 3D and the runtime of each is that of a rectangle-intersection query on the BVH constructed for the corresponding site at preprocessing: Let $pr_{\mathrm{aff}(s)}(p)$ be the projection of p to aff(s) and $B_{p,s}$ the $2d$−box on aff(s) centered at $pr_{\mathrm{aff}(s)}(p)$ with radius $\mu_\infty(p, \mathrm{aff}(s))$. Then, $p \in \mathcal{Z}(s)$ iff $B_{p,s} \cap s \neq \emptyset$. A query with $B_{p,s}$ is done by `inZone`(p, s). For `ZoneInCell`$(p, \mathcal{C})$ we do a query with $\overline{B_\infty(p'_c, r_c + \rho_i)}$ where $p'_c = pr_{\mathrm{aff}(s)}(p_c)$, $\rho_i = \mu_\infty(f_i, \mathrm{aff}(s))$ and f_1, f_2 the two facets of $\mathcal{C}$ parallel to aff(s). Queries with $\overline{B_\infty(p'_c, r_c)}$ are also performed by `isIntersecting`$(s, \mathcal{C})$ and `isStrictlyIntersecting`$(s, \mathcal{C})$. When computing *label sets*, `LocationTest` is slightly modified, since the corners used to identify the position of a point can also be formed by 3 sites.

Complexity. Under appropriate scaling so that the edge length of $\mathcal{C}_0$ be 1, if Δ is the separation bound, then the maximum height of the octree is $L = O(\lg(1/\Delta))$.

The algorithm mainly subdivides cells intersecting $\mathcal{P}$, unlike the 2D algorithm that mainly subdivides cells intersecting $\mathcal{V}_{\mathcal{D}}(\mathcal{P})$, because a criterion like (T1) is missing. This absence does not affect tree height, since, by the proof of Lemma 8, the minimum cell size is same as when we separate sites whose regions are non-neighboring (handled by Theorem 4). If β is the minimum distance of two Voronoi vertices and γ the relative thickness of $\mathcal{P}^c$, taking $\Delta = \Omega(\min\{\beta, \gamma\})$ suffices, as in 2D (Theorem 4).

Theorem 8. *Under UDH and if n is the number of polyhedral facets, α the maximum number of vertices per facet and t_α the maximum number of rectangles in a BVH leaf, the algorithm's complexity is $O(n\,\alpha(\lg\alpha + t_\alpha)/\Delta^2 + 1/\Delta^3)$.*

Proof. We sum active sets' cardinalities of octree nodes, since refining a cell requires a number of rectangle-intersection queries linear in the size of its parent's active set. Let cell $\mathcal{C}$ and its child $\mathcal{C}_1$. Any $s \in \widetilde{\phi}(\mathcal{C})$ satisfies $\delta_c \leq \mu_\infty(p_c, s) \leq 2r_c + \delta_c$. We denote by E the volume of the annulus $\{q \in \mathbb{R}^3 \mid \delta_c \leq \mu_\infty(p_c, q) \leq 2r_c + \delta_c\}$ and by E_1 the volume of the respective annulus for $\mathcal{C}_1$, minus the volume of the annulus' intersection with $B_\infty(p_c, \delta_c)$. It is easy to show $E_1 \leq E/2$. Under *UDH*, we sum all levels and bound by $O(4^L n)$ the number of rectangle-intersection queries. Using Theorem 3, we find that the complexity of the subdivision phase is $O(n\,\alpha(\lg\alpha + t_\alpha)/\Delta^2)$. The complexity of the reconstruction phase is $O(\tilde{n})$, where $\tilde{n}$ is the number of leaf nodes in the octree, which is in turn $O(1/\Delta^3)$. This allows to conclude. □

Remark 2. $t_\alpha = O(\alpha)$ so the bound of Theorem 8 is $O(n\alpha^2/\Delta^2 + 1/\Delta^3)$. Let V be the number of input vertices. It is expected that $n\alpha = O(V)$; also α is usually constant. In this case, the complexity simplifies to $O(V/\Delta^2 + 1/\Delta^3)$.

5 Implementation and Concluding Remarks

Our algorithms were implemented in Julia and use the algebraic geometric modeler `Axl` for visualization. They are available in https://gitlab.inria.fr/ckatsama/L_infinity_Voronoi. They are efficient in practice and their performance scales well in thousands of input sites. Some examples with runtimes are given in Figs. 9 and 10. All experiments were run on a 64-bit machine with an Intel(R) Core(TM) i7-8550U CPU @1.80 GHz and 8.00 GB of RAM.

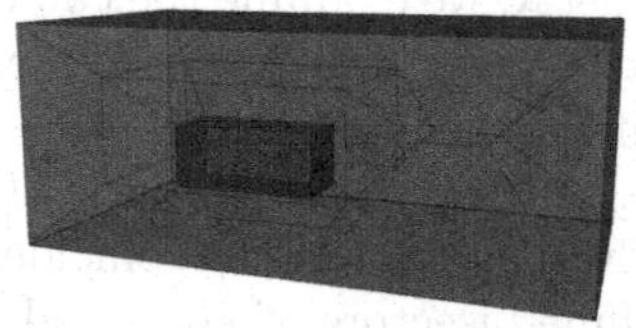

Fig. 10. The 1-skeleton of the Voronoi diagram is shown in blue. Input consists of 12 sites and 386 cells are generated (not shown). Total time is 94.8 ms. (Color figure online)

Acknowledgements. We thank Evanthia Papadopoulou for commenting on a preliminary version of the paper and Bernard Mourrain for collaborating on software. Both authors are members of AROMATH, a joint team between INRIA Sophia-Antipolis (France) and NKUA.

Appendix A Omitted Proofs

Lemma 1. *Let $s \in \mathcal{S}$. For every point $p \in V_\infty(s)$ it holds that $\mu_\infty(p, s) = \mu_\infty(p, aff(s))$, where $aff(s)$ is the affine hull of s.*

Proof. Assume w.l.o.g. that $s \subset \{x \in \mathbb{R}^d \mid x_k = c\}$, $k \in [d]$ and $c \in \mathbb{R}$. If $\mu_\infty(p, s) \neq \mu_\infty(p, \text{aff}(s))$, then $\mu_\infty(p, s) = \inf\{\max_{i\in[d]\setminus k}\{|p_i - q_i|\} \mid \forall q \in s\}$ and there is $q \in \partial s$ such that $\mu_\infty(p, s) = \mu_\infty(p, q) = |p_j - q_j|$, $j \in [d] \setminus k$. To see this, suppose on the contrary that q is in the interior of s. Then, we can find $q' \in s$ ε-close to q such that $|p_j - q'_j| = |p_j - q_j| - \varepsilon \Rightarrow \mu_\infty(p, q') = \mu_\infty(p, q) - \varepsilon$, for any $\varepsilon > 0$. This leads to a contradiction. Therefore, there is a site $s' \neq s$ with $q \in s'$. Since $p \in V_\infty(s)$, then $\mu_\infty(p, s) < \mu_\infty(p, s') \leq \mu_\infty(p, q)$; contradiction. □

Lemma 2. *For every $s \in S$, $\overline{V_\mathcal{D}(s)}$ is weakly star shaped with respect to s.*

Proof. Let $p \in \overline{V_\mathcal{D}(s)}$ and $\rho = D_s(p) = \mu_\infty(p, q)$, $q \in s$. The open ball $B_\infty(p, \rho)$ centered at p with radius ρ is empty of sites. For $t \in (0, 1)$, let $w = tp + (1 - t)q$ on the line segment (p, q). Then, since for every $i \in [d]$ it is $|w_i - q_i| = t|p_i - q_i|$, it holds that $w \in \mathcal{Z}^+(s)$ and $D_s(w) = t\rho$. If $w \notin \overline{V_\mathcal{D}(s)}$ there is a site s' such that $D_{s'}(w) < t\rho$. But $B_\infty(w, t\rho) \subset B_\infty(p, \rho)$ and s' intersects $B_\infty(p, \rho)$, leading to a contradiction. □

Lemma 3. *(a) The maximum degree of a Voronoi vertex is less than or equal to $2^d d$. (b) When $d = 2$, a Voronoi vertex cannot have degree 7.*

Proof. (a) Consider the vertex placed at the origin; 2^d orthants are formed around the vertex. To obtain the maximum number of Voronoi regions in each orthant, we count the maximum number of Voronoi edges in the interior of an orthant that have this Voronoi vertex as endpoint; at most one such edge can exist in each orthant. Since these Voronoi edges are equidistant to d sites, result follows.

(b) Let $v^* = (x^*, y^*)$ be a Voronoi vertex of degree 7. Since 7 Voronoi edges meet at v^*, due to symmetry, we examine the two cases of Fig. 11. When the configuration of Voronoi regions around the vertex is like in Fig. 11a, then s_1 is a horizontal segment and s_2, s_7 are vertical. Then, $\text{aff}(s_2), \text{aff}(s_7) \subset \{(x, y) \in \mathbb{R}^2 \mid x > x^*\}$. Since $v^* \in \mathcal{Z}^+(s_2) \cap \mathcal{Z}^+(s_7)$ and is equidistant to both s_2 and s_7, the affine hulls of s_2, s_7 coincide. Then, whichever is the orientation of s_1, the affine bisectors of s_1, s_2 and s_1, s_7 cannot meet like in Fig. 11a. When like in Fig. 11b, since b_3 is vertical, s_1 is vertical. But since b_1 is horizontal, s_1 must be horizontal; a contradiction. □

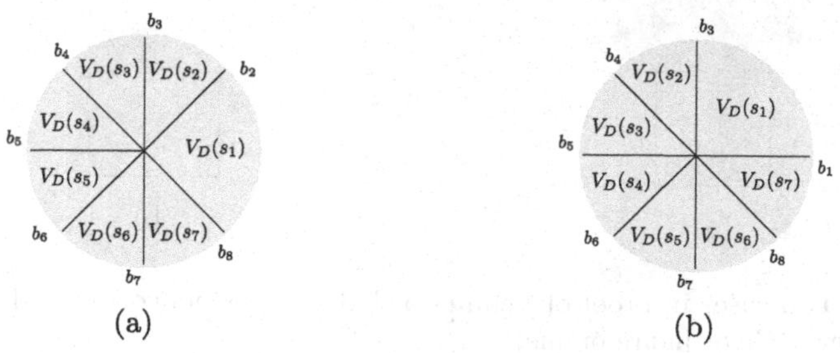

Fig. 11. The two cases in proof of Lemma 3.

Lemma 4. *(a) For every cell* $\mathcal{C}$, $\phi(\mathcal{C}) \subseteq \widetilde{\phi}(\mathcal{C})$.*(b) For a sequence of cells* $(\mathcal{C})_i$ *monotonically convergent to point* $p \in \mathcal{P}$, $\widetilde{\phi}(\mathcal{C}_i) = \phi(p)$ *for* i *large enough.*

Proof. (a) If $\phi(\mathcal{C}) = \emptyset$, assertion follows trivially. Let $s \in \phi(\mathcal{C})$ and $p \in \mathcal{C} \cap \overline{V_\mathcal{D}(s)}$. It holds that $D_s(p) \leq D_{s'}(p) \Rightarrow \mu_\infty(p, s) \leq \mu_\infty(p, s')$ for every $s' \in \mathcal{S}$. We distinguish two cases according to the position of $p_\mathcal{C}$ relatively to $\mathcal{P}$. If $p_\mathcal{C} \in \mathcal{P}$ and $p_\mathcal{C} \in \overline{V_\mathcal{D}(s^*)}$, then:

$$\begin{aligned}\mu_\infty(p_\mathcal{C}, s) &\leq \mu_\infty(p_\mathcal{C}, p) + \mu_\infty(p, s) \leq \mu_\infty(p_\mathcal{C}, p) + \mu_\infty(p, s^*)\\ &\leq 2\mu_\infty(p_\mathcal{C}, p) + \mu_\infty(p_\mathcal{C}, s^*) \leq 2r_\mathcal{C} + \mu_\infty(p_\mathcal{C}, s^*).\end{aligned}$$

Otherwise, if $p_\mathcal{C} \notin \mathcal{P}$, since $\mathcal{C} \cap \mathcal{P} \neq \emptyset$, there is a site s' intersecting $\mathcal{C}$ such that $\mu_\infty(p, s') \leq r_\mathcal{C}$. Therefore, $\mu_\infty(p_\mathcal{C}, s) \leq \mu_\infty(p_\mathcal{C}, p) + \mu_\infty(p, s) \leq \mu_\infty(p_\mathcal{C}, p) + \mu_\infty(p, s') \leq 2r_\mathcal{C}$.

(b) There exists $i_0 \in \mathbb{N}$ such that for $i \geq i_0$ $\mathcal{C}_i \cap \mathcal{P} \neq \emptyset$. Therefore, for $i \gg i_0$, since $p_{\mathcal{C}_i} \to p$ and $r_{\mathcal{C}_i} \to 0$, for every $s \in \mathcal{C}_i$, (a) implies that $s \in \lambda(p) = \phi(p)$. Since $\phi(p) \subseteq \widetilde{\phi}(\mathcal{C}_i)$, result follows. □

Lemma 5. *For a subdivision cell* $\mathcal{C}$*, let* $v_1, \ldots, v_4$ *its corner vertices. For* $s \in \mathcal{S}$, $\mathcal{C} \subseteq V_\mathcal{D}(s)$ *if and only if* $v_1, \ldots, v_4 \in V_\mathcal{D}(s)$.

Proof. Let $v_1, \ldots, v_4 \in V_\mathcal{D}(s)$ and $p \in \mathcal{C}$. Then, $p \in \mathcal{Z}^+(s)$, since $\mathcal{Z}^+(s)$ is convex in 2D and $v_1, \ldots, v_4 \in \mathcal{Z}^+(s)$. For $i = 1, \ldots, 4$ the open ball $B_i := B_\infty(v_i, \mu_\infty(v_i, s))$ is empty of sites. Since $B_\infty(p, \mu_\infty(p, \text{aff}(s))) \subset \cup_{i\in[4]} B_i$ it holds that $\mu_\infty(p, \mathcal{P}) \geq \mu_\infty(p, \text{aff}(s)) = \mu_\infty(p, s)$. On the other hand, $\mu_\infty(p, \mathcal{P}) \leq \mu_\infty(p, s)$. So, if $p \notin V_\mathcal{D}(s)$ there is a site s' s.t. $D_{s'}(p) = D_s(p)$ and $p \in \mathcal{V}_\mathcal{D}(\mathcal{P})$. Therefore, since Voronoi regions are simply connected and Voronoi edges are straight lines, p must be on the boundary of $\mathcal{C}$. The two possible configurations are shown in Fig. 12 and are contradictory; for the first, use an argument similar to that of Lemma 3(b) to show that the Voronoi edges separating the yellow-blue, and yellow-green Voronoi regions, cannot meet like in the Fig. 12a. For the second, notice that this cannot hold since the cell is square. We conclude that $\mathcal{C} \subseteq V_\mathcal{D}(s)$. The other direction is trivial. □

Fig. 12. The two cases in proof of Lemma 5. Different colors correspond to different Voronoi regions. (Color figure online)

Lemma 6. *Let $s \in \mathcal{S}$, f_1, f_2 the two facets of $\mathcal{C}$ parallel to aff(s), $\rho_i = \mu_\infty(f_i, aff(s))$ for $i = 1, 2$ and $p'_c = pr_{aff(s)}(p_c)$. Then, $\mathcal{Z}(s) \cap \mathcal{C} \neq \emptyset$ iff $\exists i \in \{1,2\}$s.t.$\}$ $B_\infty(p'_c, r_c + \rho_i) \cap s \neq \emptyset$.*

Proof. $\mathcal{Z}(s) \cap \mathcal{C} \neq \emptyset$ iff $\mathcal{Z}(s) \cap f_i \neq \emptyset$ for at least one $i \in \{1,2\}$: Let $p \in \mathcal{Z}(s) \cap \mathcal{C}$ s.t. $p \notin f_1 \cup f_2$ and $\mathrm{pr}_{f_i}(p)$ be the projection of p on f_i. There exists $i \in \{1,2\}$ s.t. $\mu_\infty(\mathrm{pr}_{f_i}(p), \mathrm{aff}(s)) > \mu_\infty(p, \mathrm{aff}(s))$. Then, $\mathrm{pr}_{f_i}(p) \in \mathcal{Z}(s)$.

It holds that $\mathcal{Z}(s) \cap f_i \neq \emptyset$ iff $B_\infty(p'_c, r_c + \rho_i) \cap s \neq \emptyset$: Let $q \in \mathcal{Z}(s) \cap f_i$ and q' its projection on aff(s). Then $\mu_\infty(q', s) \leq \mu_\infty(q, \mathrm{aff}(s)) = \rho_i$ and $\mu_\infty(p'_c, q') \leq r_c$. We deduce that $B_\infty(p'_c, r_c + \rho_i) \cap s \neq \emptyset$, since $\mu_\infty(p'_c, s) \leq \mu_\infty(p'_c, q') + \mu_\infty(q', s) \leq r_c + \rho_i$. For the inverse direction, let $B_\infty(p'_c, r_c + \rho_i) \cap s \neq \emptyset$ and q' in s s.t. $\mu_\infty(p'_c, q') \leq r_c + \rho_i$. Let q be its projection on aff(f_i). If $q \in f_i$ we are done. Otherwise, q is at L_∞ distance from f_i equal to $\mu_\infty(p'_c, q') - r_c$, attained at a boundary point $q'' \in f_i$. Then, $\rho_i \leq \mu_\infty(q'', s) \leq \mu_\infty(q'', q') = \max\{\rho_i, \mu_\infty(p'_c, q') - r_c\} = \rho_i$. It follows that $q'' \in \mathcal{Z}(s)$. □

Lemma 7. *We denote by $\mathcal{C}'$ the parent of $\mathcal{C}$ in the subdivision. Using BVH accelerates the refinement of $\mathcal{C}$ if $|\widetilde{\phi}(\mathcal{C}')|/|\widetilde{\phi}(\mathcal{C})| = \Omega(\lg n + t)$.*

Proof. A label set $\lambda(p)$ is determined by performing a point and a rectangle-intersection query; once the point is detected to lie inside a rectangle R_0 of a leaf T_0 we find an initial estimation d_0 of $\mu_\infty(p, \mathcal{P})$. Since the closest site to p may be on another leaf, we do a rectangle-intersection query centered at p with radius d_0. The closest site(s) to p are in the intersected leaves. Thus, finding $\lambda(p)$ takes $O(k(p, d_0) \lg r + k(p, d_0)t)$ (Theorem 3), where $k(p, d_0) = O(1)$ is the number of BVH leaves intersected by $\overline{B_\infty(p, d_0)}$. Computing the sites in $\widetilde{\phi}(\mathcal{C})$ is accelerated if combined with a rectangle intersection query to find segments at L_∞-distance $\leq 2r_c + \delta_c$ from p_c. Let k_c be the number of BVH leaves intersected by this rectangle-intersection query. We obtain a total refinement time for the cell equal to $O(k_c \lg r + k_c t)$. Since $k_c = O(|\widetilde{\phi}(\mathcal{C})|)$ and $r = O(n)$ the lemma follows. □

Lemma 8. *Let $\mathcal{C} \subseteq V_{\mathcal{D}}(s)$. There is $r^* > 0$ s.t. $r_c < r^*$ implies $\widetilde{\phi}(\mathcal{C}) = \{s\}$.*

Proof. Let $s' \in \mathcal{S} \setminus s$. We will prove that if $\mathcal{Z}^+(s') \cap \mathcal{C} \neq \emptyset$, it holds that $\delta_c < \mu_\infty(p_c, s')$. Therefore there is $r(s') > 0$ such that $2r(s') + \delta_c < \mu_\infty(p_c, s')$.

Let r^* be the minimum of these radii for every site different than s. When $r_c < r^*$, it holds that $\widetilde{\phi}(\mathcal{C}) = \{s\}$.

Suppose that $\delta_c = \mu_\infty(p_c, s') = \mu_\infty(p_c, q)$ for $q \in s'$. We denote by $v_1, \ldots, v_4$ the vertices of $\mathcal{C}$. Since $\cup_{i\in[4]} B_\infty(v_i, \mu_\infty(v_i, s))$ is empty of sites, $q \in \text{aff}(s)$. Also, s' cannot be a subset of aff(s), for p_c will be in $\mathcal{Z}^+(s')$, a contradiction. So, s' is perpendicular and adjacent to s. By hypothesis $\mathcal{Z}^+(s') \cap \mathcal{C} \neq \emptyset$ and since for adjacent sites it holds that $\mathcal{Z}^+(s) \cap \mathcal{Z}^+(s') = \text{bis}_\mathcal{D}(s, s')$, the Voronoi face separating sites s, s' intersects $\mathcal{C}$ which is a contradiction. Thus, there is no site s' with $\mathcal{Z}^+(s') \cap \mathcal{C} \neq \emptyset$ and $\mu_\infty(p_c, s') = \delta_c$. □

References

1. Agarwal, P.K., de Berg, M., Gudmundsson, J., Hammar, M., Haverkort, H.J.: Box-trees and R-trees with near-optimal query time. Discrete Comput. Geom. **28**(3), 291–312 (2002). https://doi.org/10.1007/s00454-002-2817-1
2. Aichholzer, O., Aurenhammer, F., Alberts, D., Gärtner, B.: A novel type of skeleton for polygons. J. Univ. Comput. Sci. **1**, 752–761 (1995)
3. Aurenhammer, F., Walzl, G.: Straight skeletons and mitered offsets of nonconvex polytopes. Discrete Comput. Geom. **56**(3), 743–801 (2016)
4. Barequet, G., Eppstein, D., Goodrich, M., Vaxman, A.: Straight skeletons of three-dimensional polyhedra. In: Proceedings of the Twenty-fifth ACM Annual Symposium on Computational Geometry, pp. 100–101. ACM Press, Aarhus, Denmark, (2009). https://doi.org/10.1145/1542362.1542384.
5. Bennett, H., Papadopoulou, E., Yap, C.: Planar minimization diagrams via subdivision with applications to anisotropic Voronoi diagrams. Comput. Graph. Forum **35**(5), 229–247 (2016)
6. Cheilaris, P., Dey, S.K., Gabrani, M., Papadopoulou, E.: Implementing the L_∞ segment voronoi diagram in CGAL and applying in VLSI pattern analysis. In: Hong, H., Yap, C. (eds.) ICMS 2014. LNCS, vol. 8592, pp. 198–205. Springer, Heidelberg (2014). https://doi.org/10.1007/978-3-662-44199-2_32
7. Eder, G., Held, M., Palfrader, P.: Computing the straight skeleton of an orthogonal monotone polygon in linear time. In: European Workshop on Computational Geometry, Utrecht, March 2019. www.eurocg2019.uu.nl/papers/16.pdf
8. Emiris, I.Z., Mantzaflaris, A., Mourrain, B.: Voronoi diagrams of algebraic distance fields. J. Comput. Aided Des. **45**(2), 511–516 (2013). Symposium on Solid Physical Modeling 2012
9. Eppstein, D., Erickson, J.: Raising roofs, crashing cycles, and playing pool: applications of a data structure for finding pairwise interactions. Discrete Comput. Geom. **22**, 58–67 (1998)
10. Haverkort, H.J.: Results on geometric networks and data structures. Ph.D. thesis, Utrecht University (2004). http://igitur-archive.library.uu.nl/dissertations/2004-0506-101707/UUindex.html
11. Held, M., Palfrader, P.: Straight skeletons and mitered offsets of polyhedral terrains in 3D. J. Comput. Aided Des. Appl. **16**, 611–619 (2018)
12. Martínez, J., Garcia, N.P., Anglada, M.V.: Skeletal representations of orthogonal shapes. Graph. Models **75**(4), 189–207 (2013)

13. Papadopoulou, E., Lee, D.: The L_∞ Voronoi diagram of segments and VLSI applications. Int. J. Comput. Geom. Appl. **11**(05), 503–528 (2001)
14. Schaefer, S., Warren, J.: Dual marching cubes: primal contouring of dual grids. Comput. Graph. Forum **24**(2), 195–201 (2005)
15. Yap, C., Sharma, V., Jyh-Ming, L.: Towards exact numerical Voronoi diagrams. In: IEEE International Symposium on Voronoi Diagrams in Science and Engineering (ISVD), New Brunswick, NJ, June 2012

The Complexity of Subtree Intersection Representation of Chordal Graphs and Linear Time Chordal Graph Generation

Tınaz Ekim[1], Mordechai Shalom[2](✉), and Oylum Şeker[1]

[1] Department of Industrial Engineering, Bogazici University, Istanbul, Turkey
{tinaz.ekim,oylum.seker}@boun.edu.tr

[2] TelHai Academic College, Upper Galilee, 12210 Qiryat Shemona, Israel
cmshalom@telhai.ac.il

Abstract. It is known that any chordal graph on n vertices can be represented as the intersection of n subtrees in a tree on n nodes [5]. This fact is recently used in [2] to generate random chordal graphs on n vertices by generating n subtrees of a tree on n nodes. It follows that the space (and thus time) complexity of such an algorithm is at least the sum of the sizes of the generated subtrees assuming that a tree is given by a set of nodes. In [2], this complexity was mistakenly claimed to be linear in the number m of edges of the generated chordal graph. This error is corrected in [3] where the space complexity is shown to be $\Omega(mn^{1/4})$. The exact complexity of the algorithm is left as an open question.

In this paper, we show that the sum of the sizes of n subtrees in a tree on n nodes is $\Theta(m\sqrt{n})$. We also show that we can confine ourselves to contraction-minimal subtree intersection representations since they are sufficient to generate every chordal graph. Furthermore, the sum of the sizes of the subtrees in such a representation is at most $2m + n$. We use this result to derive the first linear time random chordal graph generator. In addition to these theoretical results, we conduct experiments to study the quality of the chordal graphs generated by our algorithm and compare them to those in the literature. Our experimental study indicates that the generated graphs do not have a restricted structure and the sizes of maximal cliques are distributed fairly over the range. Furthermore, our algorithm is simple to implement and produces graphs with 10000 vertices and 4.10^7 edges in less than one second on a laptop computer.

Keywords: Chordal graph · Representation complexity · Graph generation

1 Introduction

Chordal graphs are extensively studied in the literature from various aspects which are motivated by both theoretical and practical reasons. Chordal graphs

The first author acknowledges the support of the Turkish Academy of Science TUBA GEBIP award.

I. Kotsireas et al. (Eds.): SEA^2 2019, LNCS 11544, pp. 21–34, 2019.
https://doi.org/10.1007/978-3-030-34029-2_2

have many application areas such as sparse matrix computations, database management, perfect phylogeny, VLSI, computer vision, knowledge based systems, and Bayesian networks (see e.g. [6,8,10]). Consequently, numerous exact/heuristic/parameterized algorithms have been developed for various optimization and enumeration problems on chordal graphs. The need for testing and comparing these algorithms motivated researchers to generate random chordal graphs [1,7,9]. A more systematic study of random chordal graph generators has been initiated more recently in [2,3]. The generic method developed in these papers is based on the characterization of chordal graphs as the intersection graph of subtrees of a tree [5], to which we will refer as a *subtree intersection representation.* In this method a chordal graph on n vertices and m edges is generated in three steps:

1. Generate a tree T on n nodes uniformly at random.
2. Generate n non-empty subtrees $\{T_1, \ldots, T_n\}$ of T.
3. Return the intersection graph G of $\{V(T_1), \ldots, V(T_n)\}$.

Three methods for generating subtrees in Step 2. have been suggested. In all these methods, every node of every subtree is generated. Steps 1. and 3. being linear in the size of G, the time and space complexities of the algorithm are dominated by the sum of the sizes of the subtrees generated at Step 2. which is $\sum_{i=1}^{n} |V(T_i)|$. In [2], this complexity was mistakenly claimed to be linear in the size of G, that is $O(n+m)$. In [3], this mistake is corrected by showing that $\sum_{i=1}^{n} |V(T_i)|$ is $\Omega(mn^{1/4})$, leaving the upper bound as an open question. This question is crucial for the complexity of any chordal graph generator that produces every subtree intersection representation on a tree of n nodes. In this paper, we investigate the complexity of subtree intersection representations of chordal graphs. We show that $\sum_{i=1}^{n} |V(T_i)|$ is $\Theta(m\sqrt{n})$. In other words, we both improve the lower bound of $\Omega(mn^{1/4})$ given in [3] and provide a matching upper bound. On the other hand, we show that the size of a "contraction-minimal" representation is linear, more precisely, at most $2m+n$. This result plays the key role in developing a linear time chordal graph generator, the first algorithm in the literature having this time complexity, to the best of our knowledge. Our algorithm is also simple to implement. Our experiments indicate that it produces graphs for which the maximal clique sizes are distributed fairly over the range. Furthermore, the running time of the algorithm clearly outperforms the existing ones: graphs with 10000 vertices and 4.10^7 edges are generated in less than one second on a personal computer.

We proceed with definitions and preliminaries in Sect. 2. Then, for technical reasons, we first consider contraction-minimal representations in Sect. 3. We develop our linear time random chordal graph generator in Sect. 3.1. We study the variety of chordal graphs generated by this algorithm in Sect. 3.2; to this end, based on similar studies in the literature, we analyze maximal cliques of the generated graphs. We proceed with the complexity of arbitrary subtree intersection representations in Sect. 4. We conclude in Sect. 5 by suggesting further research.

2 Preliminaries

Graphs: We use standard terminology and notation for graphs, see for instance [4]. We denote by $[n]$ the set of positive integers not larger than n. Given a simple undirected graph G, we denote by $V(G)$ the set of vertices of G and by $E(G)$ the set of the edges of G. We use $|G|$ as a shortcut for $|V(G)|$. We denote an edge between two vertices u and v as uv. We say that (a) the edge $uv \in E(G)$ is *incident* to u and v, (b) u and v are the endpoints of uv, and (c) u and v are adjacent to each other. We denote by $G_{/e}$ the graph obtained from G by contracting the edge e. A *chord* of a cycle C of a graph G is an edge of G that connects two vertices that are non-adjacent in C. A graph is *chordal* if it contains no induced cycles of length 4 or more. In other words, a graph is chordal if every cycle of length at least 4 contains a chord. A vertex v of a graph G is termed *simplicial* if the subgraph of G induced by v and its neighbors is a complete graph. A graph G on n vertices is said to have a *perfect elimination order* if there is an ordering $v_1, \ldots, v_n$ of its vertices, such that v_i is simplicial in the subgraph induced by the vertices $\{v_1, \ldots, v_i\}$ for every $i \in [n]$. It is known that a graph is chordal if and only if it has a perfect elimination order [6].

Trees, Subtrees and Their Intersection Graphs: Let $\mathcal{T} = \{T_1, \ldots, T_n\}$ be a set of subtrees of a tree T. Let $G = (V, E)$ be a graph over the vertex set $\{v_1, \ldots, v_n\}$ where v_i represents T_i and such that v_i and v_j are adjacent if and only if T_i and T_j have a common node. Then, G is termed as the *vertex-intersection graph* of $\langle T, \mathcal{T} \rangle$ and conversely $\langle T, \mathcal{T} \rangle$ is termed a subtree intersection representation, or simply a *representation* of G. Whenever there is no confusion, we will denote the representation $\langle T, \mathcal{T} \rangle$ simply as $\mathcal{T}$. We also denote the intersection graph of $\mathcal{T}$ as $G(\mathcal{T})$. Gavril [5] showed that a graph is chordal if and only if it is the vertex-intersection graph of subtrees of a tree. Throughout this work, we refer to the vertices of T as *nodes* to avoid possible confusion with the vertices of $G(\mathcal{T})$.

Let G be a chordal graph with a representation $\langle T, \mathcal{T} \rangle$, and v be a node of T. We denote as $\mathcal{T}_v$ the set of subtrees in $\mathcal{T}$ that contain the node v, i.e. $\mathcal{T}_v \stackrel{def}{=} \{T_i \in \mathcal{T} | v \in V(T_i)\}$. Clearly, the set $\mathcal{T}_v$ corresponds to a clique of G. It is also known that, conversely, every maximal clique of G corresponds to $\mathcal{T}_v$ for some node v of T.

Two sets of subtrees $\mathcal{T}$ and $\mathcal{T}'$ are *equivalent* if $G(\mathcal{T}) = G(\mathcal{T}')$. Let $\mathcal{T}$ be a set of subtrees of a tree T and e an edge of T. We denote by $\mathcal{T}_{/e}$ the set of subtrees of $T_{/e}$ that is obtained by contracting the edge e of every subtree in $\mathcal{T}$ that contains e. A set of subtrees is *contraction-minimal* (or simply *minimal*) if for every edge e of T we have $G(\mathcal{T}_{/e}) \neq G(\mathcal{T})$.

Through the rest of this work, G is a chordal graph with vertex set $[n]$ and m edges, T is a tree on $t \leq n$ nodes and $\mathcal{T} = \{T_1, \ldots, T_n\}$ is a set of subtrees of T such that $G(\mathcal{T}) = G$ and $\langle T, \mathcal{T} \rangle$ is contraction-minimal. We also denote by $t_j \stackrel{def}{=} |\mathcal{T}_j|$ the number of subtrees in $\mathcal{T}$ that contain the node j of T. Recall that

$\mathcal{T}_j$ corresponds to a clique of G. The nodes of T are numbered such that

$$t_t = \max\{t_j | j \in [t]\}$$

and all the other vertices are numbered according to a bottom-up order of T where t is the root.

In what follows, we first analyze contraction-minimal representations, then proceed into the analysis of the general case.

3 Contraction-Minimal Representations

We first show that the size of a contraction-minimal representation is at most $2m+n$. Based on this result, we derive a linear time algorithm to generate random chordal graphs. In the second part of this section, we conduct experiments to compare chordal graphs obtained by our algorithm to those in the literature. Our experimental study indicates that our method is faster than existing methods in the literature. Our algorithm produces graphs with 10000 vertices and 4.10^7 edges in less than one second on a personal computer. In addition, the generated graphs do not have a restricted structure as far as the size of their maximal cliques are concerned.

3.1 Chordal Graph Generation in Linear Time

The following observation plays an important role in our proofs as well as in our chordal graph generation algorithm.

Observation 1. *$\langle T, \mathcal{T} \rangle$ is minimal if and only if for every edge jj' of T, none of $\mathcal{T}_j$ and $\mathcal{T}_{j'}$ contains the other (i.e., both $\mathcal{T}_j \setminus \mathcal{T}_{j'}$ and $\mathcal{T}_{j'} \setminus \mathcal{T}_j$ are non-empty).*

Proof. Suppose that $\mathcal{T}_j \subseteq \mathcal{T}_{j'}$ for some edge jj' of T and let v be the node obtained by the contraction of the edge jj'. Then every pair of subtrees that intersect on j also intersect on j'. Thus, they intersect also on v (after the contraction of jj'). Conversely, every pair of subtrees that intersect on v contains at least one of j, j'. By our assumption, they contain j', thus they intersect on j'. Therefore, a pair of subtrees intersect in $\mathcal{T}$ if and only if they intersect in $\mathcal{T}_{/jj'}$. Therefore, $G(\mathcal{T}_{/jj'}) = G(\mathcal{T})$, thus $\langle T, \mathcal{T} \rangle$ is not contraction-minimal.

Now suppose that $\mathcal{T}'_j \setminus \mathcal{T}_j \neq \emptyset$ and $\mathcal{T}_j \setminus \mathcal{T}'_j \neq \emptyset$ for every edge jj' of T. Then, for every edge jj' of T there exists a subtree that contains j but not j' and another subtree that contains j' but not j. These two subtrees do not intersect, but they intersect on v after the contraction of jj'. Therefore, $G(\mathcal{T}_{/jj'}) \neq G(\mathcal{T})$ for every edge jj' of T. We conclude that $\langle T, \mathcal{T} \rangle$ is contraction-minimal.

Lemma 1. *Let $\langle T, \mathcal{T}\rangle$ be a minimal representation of some chordal graph G on n vertices and m edges. There exist numbers $s_1, \ldots, s_t$ such that*

$$\forall j \in [t] \quad s_j \geq 1, \tag{1}$$

$$\sum_{j=1}^{t} s_j = n, \tag{2}$$

$$s_j \leq t_j \leq \sum_{i=j}^{t} s_i \tag{3}$$

$$2\sum_{j=1}^{t} s_j t_j - \sum_{j=1}^{t} s_j^2 = 2m + n \tag{4}$$

Proof. Consider the following pruning procedure of T that implies a perfect elimination order for G. We first remove the leaf $j = 1$ from T and all the simplicial vertices of G which are represented by the subtrees $T_i \in \mathcal{T}$ that consist of the leaf $j = 1$. There is at least one such subtree by Observation 1. We continue in this way for every $j \in [t]$ until both T and G vanish. Let G^j be the remaining graph at step j before node j is removed, and s_j be the number of simplicial vertices of G^j eliminated with the removal of node j. Clearly, the numbers s_j satisfy relations (1) and (2). Recall that $t_j = |\mathcal{T}_j|$. To see that (3) holds, observe that the number of subtrees removed at step j is at most the number of subtrees containing node j. Observe also that $t_j \leq n - \sum_{i=1}^{j-1} s_i$ as the subtrees eliminated at prior steps do not contain the node j by the choice of the nodes to be removed at every step.

To show (4), let e_j be the number of edges of G that have been eliminated at phase j of the pruning procedure (during which node j of T is removed). We recall that a clique of s_j vertices is removed, each vertex of which is adjacent to $t_j - s_j$ other vertices of G. Therefore, $e_j = s_j(s_j - 1)/2 + s_j(t_j - s_j)$, i.e., $2e_j + s_j = 2s_j t_j - s_j^2$. Summing up over all $j \in [t]$ we get

$$2\sum_{j=1}^{t} s_j t_j - \sum_{j=1}^{t} s_j^2 = \sum_{j=1}^{t}(2e_j + s_j) = 2m + n.$$

We are now ready to prove the main result of this section.

Theorem 1. *If $\langle T, \mathcal{T}\rangle$ is a minimal representation of some chordal graph G on n vertices and m edges then*

$$\sum_{i=1}^{n} |V(T_i)| \leq 2m + n.$$

Proof. We first note that $\sum_{i=1}^{n} |V(T_i)| = \sum_{j=1}^{t} t_j$ since every node of every subtree T_i contributes one to both sides of the equation. We conclude as follows using Lemma 1:

$$\sum_{j=1}^{t} t_j \leq \sum_{j=1}^{t} s_j t_j \leq 2\sum_{j=1}^{t} s_j t_j - \sum_{j=1}^{t} s_j^2 = 2m + n$$

where the first inequality and the last equality hold by relations (1) and (4) of Lemma 1 and the second inequality is obtained by replacing t_j with $t_j + (t_j - s_j)$ and noting that $t_j - s_j \geq 0$.

We now present algorithm GenerateContractionMinimal that generates a random contraction-minimal representation together with the corresponding chordal graph where every contraction-minimal representation has a positive probability to be returned. It creates a tree T at random by starting from a single node and every time adding a leaf v adjacent to some existing node u that is chosen uniformly at random. Every time a node v is added, the algorithm performs the following: (a) a non-empty set of subtrees consisting of only v is added to $\mathcal{T}$, (b) a proper subset of the subtrees in $\mathcal{T}_u$ (i.e., those containing u) is chosen at random and every subtree of it is extended by adding the node v and the edge uv, (c) the graph G is extended to reflect the changes in $\mathcal{T}$. A pseudo code of the algorithm is given in Algorithm 1.

Algorithm 1. GenerateContractionMinimal

Require: $n \geq 1$
Ensure: A contraction-minimal representation $\langle T, \mathcal{T} \rangle$ with $|\mathcal{T}| = n$, and $G = G(\mathcal{T})$.
$\mathcal{T} \leftarrow \emptyset$.
$G \leftarrow (\emptyset, \emptyset)$.
$v \leftarrow$ NewNode.
$T \leftarrow (\{v\}, \emptyset)$.
while $|\mathcal{T}| < n$ **do**
  Pick a node u of T uniformly at random.
  $v \leftarrow$ NewNode.
  $T \leftarrow (V(T) + v,\ E(T) + uv)$.
  Pick a proper subset $\mathcal{T}'$ of $\mathcal{T}_u$ at random.
  for all $T_i \in \mathcal{T}'$ **do**
    $T_i \leftarrow (V(T_i) + v,\ E(T_i) + uv)$.
    $E(G) \leftarrow E(G) \cup \{i\} \times K_v$.
return $(\langle T, \mathcal{T} \rangle, G)$.

function NewNode
  Pick a number $k \in [n - |\mathcal{T}|]$ at random.
  $v \leftarrow$ a new node.
  $U \leftarrow (\{v\}, \emptyset)$. ▷ A tree with a single vertex
  $\mathcal{T} \leftarrow \mathcal{T} \cup k$ copies of U.
  $K_v \leftarrow$ a clique on k vertices.
  $G \leftarrow G \cup K_v$.
  return v.

Theorem 2. *Algorithm* GENERATECONTRACTIONMINIMAL *generates a chordal graph in linear time. Moreover, it generates any chordal graph on n vertices with strictly positive probability.*

Proof. The algorithm creates the tree T incrementally, and the subtrees of $\mathcal{T}$ are created and extended together with T. More precisely, the set of subtrees containing a node of T is not altered after a newer node is created. Note that, however, the subtrees themselves might be altered to contain newer nodes. Consider an edge uv of T where v is newer than u. The sets $\mathcal{T}_u$ and $\mathcal{T}_v$ of the subtrees containing u and v respectively, are established by the end of the iteration that creates v. Since only a proper subset of $\mathcal{T}_u$ is chosen to be extended to v, at least one subtree in $\mathcal{T}_u$ does not contain v. Furthermore, since v is created with a non-empty set of subtrees containing it, and none of these subtrees may contain u, there is at least one subtree in $\mathcal{T}_v$ that does not contain u. Therefore, by Observation 1, we conclude that $\langle T, \mathcal{T}\rangle$ is contraction-minimal.

We proceed with the running time of the algorithm. It is well known that the addition of a single node and the addition of a single edge to a graph can be done in constant time. We observe that the number of operations performed by NEWNODE is $k + \binom{k}{2} = |V(K_v)| + |E(K_v)|$. Therefore, the number of operations performed in all invocations of NEWNODE is $n + \sum_{v \in V(T)} |E(K_v)|$. Let uv be the edge added to T at some iteration of GENERATECONTRACTIONMINIMAL. We now observe that the number of other operations (i.e., except the invocation of NEWNODE) performed during this iteration is exactly $|E(G) \cap (K_u \times K_v)|$. We conclude that the number of operations of the algorithm is proportional to $n + |E(G)|$.

Consider a contraction-minimal representation $\langle \bar{T}, \bar{\mathcal{T}}\rangle$ with $\left|\bar{\mathcal{T}}\right| = n$ and let $\bar{t} = \left|\bar{T}\right|$. It remains to show that the algorithm returns $\langle \bar{T}, \bar{\mathcal{T}}\rangle$ with a positive probability.

We show by induction on j, that at the beginning of iteration j (the j'th time the algorithm executes Line 6) $\langle T, \mathcal{T}\rangle$ is a sub-representation of $\langle \bar{T}, \bar{\mathcal{T}}\rangle$ with positive probability. That is, T is a subtree of $\bar{T}$ and $\mathcal{T}$ consists of the non-empty intersections of the subtrees of $\bar{\mathcal{T}}$ with T (formally, $\mathcal{T} = \{\bar{T}_i[V(T)] \mid \bar{T}_i \in \bar{\mathcal{T}}\} \setminus \{(\emptyset, \emptyset)\}$) with positive probability. Let $\bar{v}$ be a node of $\bar{T}$, and let $\bar{k} = \left|\bar{\mathcal{T}}_{\bar{v}}\right|$. Clearly, $\bar{k} \leq n$. With probability $1/n$ the algorithm will start by creating a node with $\bar{k}$ trivial subtrees in which case $\langle T, \mathcal{T}\rangle$ is a sub-representation of $\langle \bar{T}, \bar{\mathcal{T}}\rangle$. Therefore, the claim holds for $j = 1$. Now suppose that $\langle T, \mathcal{T}\rangle$ is a sub-representation of $\langle \bar{T}, \bar{\mathcal{T}}\rangle$ at the beginning of iteration j with probability $p > 0$. If $T = \bar{T}$ then $\langle T, \mathcal{T}\rangle = \langle \bar{T}, \bar{\mathcal{T}}\rangle$, thus $|\mathcal{T}| = \left|\bar{\mathcal{T}}\right| = n$ and the algorithm does not proceed to iteration j. Otherwise, T is a proper subtree of $\bar{T}$, i.e. $\bar{T}$ contains an edge uv with $u \in V(T)$ and $v \notin V(T)$. At iteration j, u will be chosen with probability $1/|V(T)| \geq 1/n$ by the algorithm and the edge uv will be added to T, ensuring that T is a subtree of $\bar{T}$ with probability at least p/n at the end of iteration j. The number $\bar{k}$ of subtrees in $\bar{\mathcal{T}}$ that contain v but not u is at most $n - |\mathcal{T}|$. Since $\langle \bar{T}, \bar{\mathcal{T}}\rangle$ is contraction-minimal, we have $\bar{k} \geq 1$. Therefore, $\bar{k} \in [n - |\mathcal{T}|]$ and the algorithm creates $\bar{k}$ trivial subtrees in v with probability $1/(n - |\mathcal{T}|) > 1/n$. Since $\langle \bar{T}, \bar{\mathcal{T}}\rangle$ is contraction-minimal, the set of subtrees in $\bar{\mathcal{T}}$ that contain both

u and v is a proper subset of $\bar{\mathcal{T}}_u = \mathcal{T}_u$. The algorithm chooses this proper subset with probability $1/(2^{|\mathcal{T}_u|} - 1)$ and adds the edge uv to each of them. We conclude that at the end of iteration j (thus at the beginning of iteration $j+1$), $\langle T, \mathcal{T} \rangle$ is a sub-representation of $\langle \bar{T}, \bar{\mathcal{T}} \rangle$ with probability at least $p/(n^2 2^n) > 0$.

3.2 Experimental Studies

In this section, we present our experimental results to demonstrate the computational efficiency of GenerateContractionMinimal and to provide some insight into the distribution of chordal graphs it generates. We implemented the presented algorithm in C++, and executed it on a laptop computer with 2.00-GHz Intel Core i7 CPU. The implementation of the algorithm spans only 70 lines of C++ code. Our source code is available in http://github.com/cmshalom/ChordalGraphGeneration.

Following the approach of works [2,3,9], we consider the characteristics of the maximal cliques of the returned graph. Table 1 provides a summary of the computational results of our algorithm. The first column reports the number of vertices n. For every value of n, we use four different average edge density values of 0.01, 0.1, 0.5, and 0.8, where edge density is defined as $\rho = \frac{m}{n(n-1)/2}$ with m being the number of edges in the graph. For each pair of values n, ρ, we performed ten independent runs and reported the average values across those ten runs. The table exhibits the number of connected components, the number of maximal cliques, and the minimum, maximum, and average size of the maximal cliques along with their standard deviation. The rightmost column shows the time in seconds that it takes the algorithm takes to construct one graph. In order to achieve the desired edge density values, we discarded the graphs that turned out to be outside the range $[(1-\epsilon)\rho, (1+\epsilon)\rho]$, for $\epsilon = 0.05$. For $\rho \leq 0.1$, we adjusted the upper bound at Line 14 in function NewNode so that graphs with small edge densities are obtained more probably.

Algorithm GenerateContractionMinimal produces connected chordal graphs for $\rho \geq 0.1$. When the average edge density is 0.01, the average number of connected components decreases as n increases. A minimum clique size of 1 for $\rho = 0.01$ and $n = 1000$ implies that the disconnectedness of the graphs is due to the existence of isolated vertices. As for the running time of the algorithm, the linear time complexity shown in Theorem 2 clearly manifests itself in the amount of time it takes to construct a chordal graph. The rightmost column of Table 1 shows that our algorithm constructs a chordal graph in less than one second on average, even when $n = 10000$ and $\rho = 0.797$, i.e., $m > 4 \cdot 10^7$.

We compare our results to those of the two other methods from the literature. The first one is algorithm ChordalGen proposed in Şeker *et al.*'s work [3], which is based on the subtree intersection representation of chordal graphs. This algorithm is presented along with three alternative subtree generation methods. Here, we only consider algorithm ChordalGen together with the subtree generation method called GrowingSubtree, because this one is claimed to stand out as compared to the other presented methods, as far as the distribution of maximal clique sizes are concerned. The second algorithm we compare to is

Table 1. Experimental results of algorithm GENERATECONTRACTIONMINIMAL

n	Density	# conn. comp.s	# maximal cliques	Min clique size	Max clique size	Avg clique size	Sd of clique sizes	Time to build
1000	0.010	24.6	422.2	1.0	17.2	5.8	2.9	0.002
	0.100	1.0	62.8	7.3	125.0	43.3	26.8	0.003
	0.500	1.0	9.5	92.5	520.0	255.7	127.1	0.008
	0.780	1.0	6.5	167.2	767.4	393.1	201.0	0.010
2500	0.010	6.3	582.8	1.1	38.8	12.1	6.7	0.005
	0.100	1.0	70.7	12.4	311.9	101.4	65.8	0.010
	0.507	1.0	10.0	227.5	1357.7	617.6	330.4	0.030
	0.808	1.0	7.2	379.9	1997.9	931.1	535.3	0.049
5000	0.010	1.7	703.6	1.8	75.7	21.8	12.9	0.014
	0.102	1.0	78.5	22.7	635.1	196.1	131.0	0.035
	0.503	1.0	9.7	457.6	2479.1	1236.1	636.2	0.166
	0.796	1.0	7.2	775.5	3986.7	1914.0	1058.4	0.204
10000	0.010	1.4	825.0	2.8	146.4	40.4	25.3	0.036
	0.100	1.0	90.0	32.0	1385.9	356.9	266.5	0.132
	0.499	1.0	9.9	829.4	5343.8	2407.2	1366.3	0.659
	0.797	1.0	9.2	902.7	8125.6	3065.4	2243.1	0.952

Andreou *et al.*'s algorithm [1]. This algorithm is also used in [3] for comparison purposes, and we refer to the implementation therein. In order to obtain results comparable to those given in [3], we use the same n, ρ value pairs in our experiments.

We now compare the results in Table 1 to those reported in [3] for algorithm CHORDALGEN with GROWINGSUBTREE and Andreou *et al.*'s algorithm. We observe that the number of maximal cliques of the graphs produced by GENERATECONTRACTIONMINIMAL is usually lower than the others, and inevitably, their average clique sizes are higher than the others. The most notable difference of our algorithm from the others is its running time. Whereas a running time analysis of Andreou *et al.*'s algorithm has not been given in [1], the average running time of our implementation of their algorithm is of 477.1 s per generated graph, excluding graphs on 10000 vertices for which the algorithm was extremely slow to experiment. The average running times of our implementation of algorithm CHORDALGEN is 93.2, 4.7, 182.6 s with the subtree generation methods GROWINGSUBTREE, CONNECTINGNODES, and PRUNEDTREE, respectively. Algorithm GENERATECONTRACTIONMINIMAL, however, achieves an average running time of 0.14 s.

In our next set of experimental results, we investigate the distribution of the sizes of maximal cliques to get some visual insight into the structure of the chordal graphs produced. Figure 1 shows the average number of maximal cliques across ten independent runs for $n = 1000$ vertices and four edge density values. The figure is comprised of three rows, each row describing the result of the experiments on one algorithm; algorithms GENERATECONTRACTIONMINIMAL, CHORDALGEN combined with GROWINGSUBTREE method, and the implementation of Andreou *et al.*'s algorithm [1] as given in [3]. Every row consists of four histograms corresponding to four different average edge density values $\rho = 0.01, 0.1, 0.5$, and 0.8. The bin width of the histograms is taken as five; that is, frequencies of maximal clique sizes are summed over intervals of width five (from one to five, six to ten, etc.) and divided by the number of runs (i.e., ten) to obtain the average values. For a given n and average edge density value, we keep the ranges of x-axes the same in order to make the histograms comparable. The y-axes, however, have different ranges because maximum frequencies in histograms vary considerably.

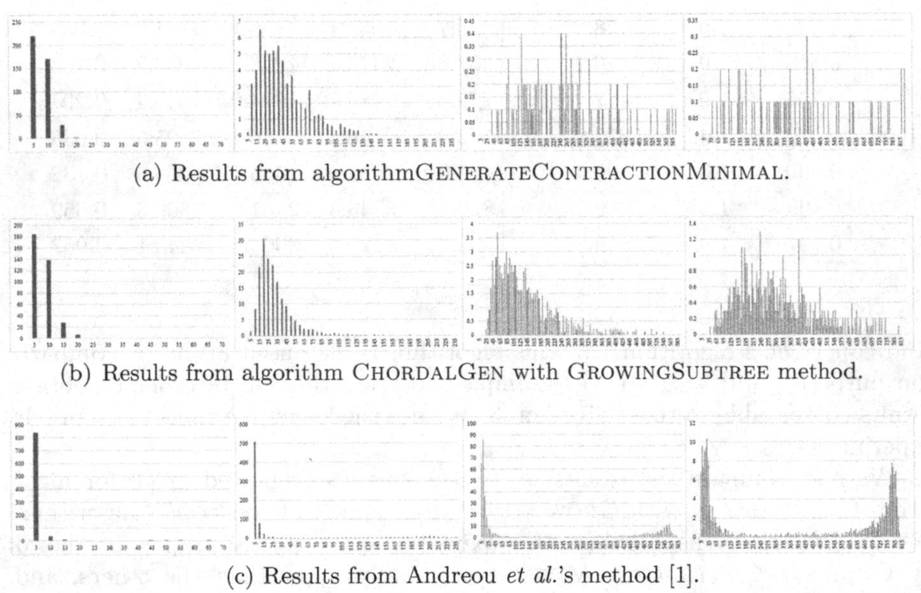

(a) Results from algorithmGENERATECONTRACTIONMINIMAL.

(b) Results from algorithm CHORDALGEN with GROWINGSUBTREE method.

(c) Results from Andreou *et al.*'s method [1].

Fig. 1. Histograms of maximal clique sizes for $n = 1000$ and average edge densities 0.01, 0.1, 0.5, and 0.8 (from left to right).

The histograms in Fig. 1(a) reveal that the sizes of maximal cliques of graphs produced by our algorithm are not clustered around specific values; they are distributed fairly over the range. The shapes of the histograms for average edge densities 0.01 and 0.1 are similar for our algorithm and algorithm CHORDALGEN, as we observe from the first half of Fig. 1(a) and (b). For higher densities (as we proceed to the right), the sizes of maximal cliques are distributed more uniformly in the graphs generated by our algorithm; there is no obvious mode of the distribution. In the graphs produced by Andreou *et al.*'s method, the vast majority of maximal cliques have up to 15 vertices when the average edge densities are 0.01 and 0.1. As we increase the edge density, frequencies of large-size maximal cliques become noticeable relative to the dominant frequencies of small-size maximal cliques. In any case, the range outside its extremes is barely used.

For brevity, we do not present the histograms for every n-value we consider in this study. Having presented the histograms for the smallest value of n we consider, next we provide the set of results for a larger value of n. The implementation of Andreou *et al.*'s algorithm [1] turned out to be too slow to allow testing graphs on 10000 vertices in a reasonable amount of time. In order to present a complete comparison with the methods we look at from the literature, we provide the results for the next largest value of n in Fig. 2. From the histograms in Fig. 2, we observe that the general distribution of maximal clique sizes do not change much with the increase in the number of vertices. Maximal clique sizes of chordal graphs produced by our algorithm are not confined to a limited area; they are distributed fairly over the range.

To summarize, our experiments show that GENERATECONTRACTIONMINIMAL is by far faster than the existing methods in practice, in accordance with our theoretical bounds. Moreover, our inspection of the generated graphs in terms of their maximal cliques shows that the algorithm produces chordal graphs with no restricted structure.

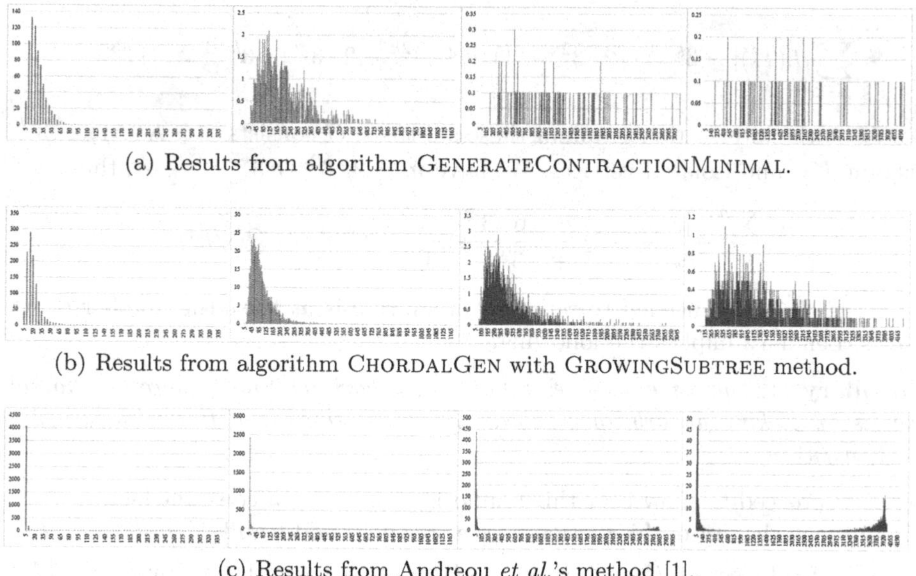

(a) Results from algorithm GENERATECONTRACTIONMINIMAL.

(b) Results from algorithm CHORDALGEN with GROWINGSUBTREE method.

(c) Results from Andreou *et al.*'s method [1].

Fig. 2. Histograms of maximal clique sizes for $n = 5000$ and average edge densities 0.01, 0.1, 0.5, and 0.8 (from left to right).

4 Arbitrary Representations

We start this section by showing that the upper bound of Theorem 1 does not hold for arbitrary representations on trees with n nodes. We denote by $L(T)$ the set of leaves of a tree T.

Lemma 2. *Let $\mathcal{T}' = \{T'_1, \ldots, T'_n\}$ be a set of subtrees on a tree with n nodes and m be the number of edges of $G(\mathcal{T}')$. Then*

$$\sum_{i=1}^{n} |L(T'_i)| \text{ is } \Omega(m\sqrt{n}).$$

Proof. Let k be a non-negative integer, and $n = 6 \cdot 3^{2k}$. Let T' be a tree on n nodes $\{v_1, \ldots, v_{4 \cdot 3^{2k}}, u_1, \ldots, u_{2 \cdot 3^{2k}}\}$, where the nodes $\{v_1, \ldots, v_{4 \cdot 3^{2k}}\}$ induce a path and the nodes $\{u_1, \ldots, u_{2 \cdot 3^{2k}}\}$ induce a star with center u_1. The representation contains the following subtrees:

- S_1: 2 trivial paths on every node in $\{v_1, \ldots, v_{2 \cdot 3^{2k}}\}$, for a total of $4 \cdot 3^{2k}$ paths,
- S_2: 3^{k+1} copies of the star on nodes $\{u_1, \ldots, u_{2 \cdot 3^{2k}}\}$, and
- S_3: $2 \cdot 3^{2k} - 3^{k+1}$ disjoint trivial paths on part of the nodes in $\{v_{2 \cdot 3^{2k}+1}, \ldots, v_{4 \cdot 3^{2k}}\}$. Note that the number of these paths is less than the number $2 \cdot 3^{2k}$ of nodes in the path, thus disjointness can be achieved.

The number of subtrees is $4 \cdot 3^{2k} + 3^{k+1} + 2 \cdot 3^{2k} - 3^{k+1} = 6 \cdot 3^{2k} = n$, as required. As for the total number of leaves, we have:

$$\sum_{i=1}^{n} |L(T'_i)| \geq 3^{k+1} \cdot (2 \cdot 3^{2k} - 1) + 4 \cdot 3^{2k} + 2 \cdot 3^{2k} - 3^{k+1} \geq 6 \cdot 3^{3k}.$$

Let G be the intersection graph of these subtrees. G consists of a $K_{3^{k+1}}$, $2 \cdot 3^{2k}$ disjoint K_2s and isolated vertices. We have $m = 2 \cdot 3^{2k} + 3^{k+1}\frac{3^{k+1}-1}{2}$, thus

$$\frac{\sum_{i=1}^{n} |L(T'_i)|}{m} \geq \frac{6 \cdot 3^{3k}}{\frac{13}{2} \cdot 3^{2k} - \frac{3}{2}3^{k}} = \Omega(3^k) = \Omega(\sqrt{n}).$$

Since the space needed to represent a subtree is at least the number of its leaves, Lemma 2 implies the following:

Corollary 1. *The time complexity of any algorithm that generates chordal graphs by picking an arbitrary subtree representation on a tree with n nodes is $\Omega(m\sqrt{n})$.*

We now proceed to show that this bound is tight up to a constant factor.

Through the rest of this section T' is a tree on n nodes and $\mathcal{T}' = \{T'_1, \ldots, T'_n\}$ is a set of subtrees of T' such that $G(\mathcal{T}') = G$. We also denote by $t'_j \stackrel{def}{=} |\mathcal{T}'_j|$ the number of subtrees in $\mathcal{T}'$ that contain the node j of T'. We assume that $\langle T, \mathcal{T} \rangle$ is a contraction-minimal representation of G obtained from $\langle T', \mathcal{T}' \rangle$ by zero or more successive contraction operations. Then, T has $t \leq n$ nodes and the *multiplicity* of j (with respect to T'), denoted by k_j, is the number of contractions effectuated in T' in order to obtain node j in T, plus one.

Lemma 3. *With the above notations, we have*

$$\forall j \in [t] \quad k_j \geq 1 \tag{5}$$

$$\sum_{j=1}^{t} k_j = n. \tag{6}$$

$$\sum_{i=1}^{n} |V(T_i')| \leq \sum_{j=1}^{t} k_j t_j. \tag{7}$$

Proof. Relations (5) and (6) hold by definition of t and multiplicity.

When an edge jj' of T' is contracted to a node v we have $k_v = k_j + k_{j'}$ and $t_v = \max\{t_j, t_{j'}\}$. Therefore,

$$k_v t_v = (k_j + k_{j'})t_v \geq k_j t_j + k_{j'} t_{j'}.$$

Using the above fact and noting that we have $\sum_{i=1}^{n} |V(T_i')| = \sum_{j=1}^{n} t_j'$, inequality (7) follows by induction on the number of contractions.

Recall that our task is to find an upper bound on the sum of the sizes of subtrees in a representation on a tree with n nodes of a chordal graph on n vertices. Relation (7) allows us to focus on the sum of $k_j t_j$ values (in a minimal representation) in order to achieve this goal. In what follows, we treat this task as an optimization problem under a given set of constraints. Thus, the following lemma whose proof is in the full version of the paper should be read independently from graph theoretic interpretations of each parameter.

Lemma 4. *Let t, $s_1, \ldots, s_t$, $t_1, \ldots t_t$ and $k_1, \ldots, k_t$ be numbers that satisfy (1), (2), (3), (4), (5), and (6). Then*

$$\rho \overset{def}{=} \frac{\sum_{j=1}^{t} k_j t_j}{2m+n} \text{ is } \mathcal{O}(\sqrt{n}).$$

We can now infer the main theorem of this section.

Theorem 3. *Let $\langle T', \mathcal{T}' \rangle$ be a representation of a chordal graph G where T' has n nodes and G has n vertices and m edges. Then we have*

$$\sum_{i=1}^{n} |V(T_i')| \text{ is } \Theta(m\sqrt{n}).$$

Proof. Let $\langle T, \mathcal{T} \rangle$ be a minimal representation of G obtained from $\langle T', \mathcal{T}' \rangle$. Then $\langle T, \mathcal{T} \rangle$ satisfies (1), (2), (3), (4) by Lemma 1, and (5), (6), (7) hold by Lemma 3. The lower and upper bounds provided in Lemmas 2 and 4 respectively allows us to conclude the proof.

5 Conclusion

In this work, we present a linear time algorithm to generate random chordal graphs. To the best of our knowledge, this is the first algorithm with this time complexity. Our algorithm is fast in practice and simple to implement. We also show that the complexity of any random chordal graph generator which produces any subtree intersection representation on a tree of n nodes with positive probability is $\Omega(m\sqrt{n})$.

We conducted experiments to analyze the distribution of the sizes of the maximal cliques of the generated chordal graphs and concluded that our method generates fairly varied chordal graphs with respect to this measure. It should be noted that, however, we do not know the distribution of the maximal clique sizes over the space of all chordal graphs of a given size.

We have shown that every chordal graph on n vertices is returned by our algorithm with positive probability. The development of an algorithm that generates chordal graphs uniformly at random is subject of further research.

References

1. Andreou, M.I., Papadopoulou, V.G., Spirakis, P.G., Theodorides, B., Xeros, A.: Generating and radiocoloring families of perfect graphs. In: Nikoletseas, S.E. (ed.) WEA 2005. LNCS, vol. 3503, pp. 302–314. Springer, Heidelberg (2005). https://doi.org/10.1007/11427186_27
2. Şeker, O., Heggernes, P., Ekim, T., Taşkın, Z.C.: Linear-time generation of random chordal graphs. In: Fotakis, D., Pagourtzis, A., Paschos, V.T. (eds.) CIAC 2017. LNCS, vol. 10236, pp. 442–453. Springer, Cham (2017). https://doi.org/10.1007/978-3-319-57586-5_37
3. Şeker, O., Heggernes, P., Ekim, T., Taşkın, Z.C.: Generation of random chordal graphs using subtrees of a tree. arXiv preprint arXiv:1810.13326 (2018)
4. Diestel, R.: Graph Theory. GTM, vol. 173, 4th edn. Springer, Heidelberg (2017)
5. Gavril, F.: The intersection graphs of subtrees in trees are exactly the chordal graphs. J. Comb. Theory **16**, 47–56 (1974)
6. Golumbic, M.C.: Algorithmic Graph Theory and Perfect Graphs. Annals of Discrete Mathematics, vol. 57. North-Holland Publishing Co., Amsterdam (2004)
7. Markenzon, L., Vernet, O., Araujo, L.H.: Two methods for the generation of chordal graph. Ann. Oper. Res. **157**(1), 47–60 (2008)
8. Pearl, J.: Probabilistic Reasoning in Intelligent Systems: Networks of Plausible Inference. Morgan Kaufmann, Burlington (2014)
9. Pemmaraju, S.V., Penumatcha, S., Raman, R.: Approximating interval coloring and max-coloring in chordal graphs. J. Exp. Algorithms **10**, 2–8 (2005)
10. Rose, D.J.: A graph-theoretic study of the numerical solution of sparse positive definite systems of linear equation. In: Graph Theory and Computing, pp. 183–217 (1972)

Computing a Minimum Color Path in Edge-Colored Graphs

Neeraj Kumar(✉)

Department of Computer Science,
University of California, Santa Barbara, USA
neeraj@cs.ucsb.edu

Abstract. In this paper, we study the problem of computing a *min-color* path in an *edge-colored* graph. More precisely, we are given a graph $G = (V, E)$, source s, target t, an assignment $\chi : E \rightarrow 2^{\mathcal{C}}$ of edges to a set of colors in $\mathcal{C}$, and we want to find a path from s to t such that the number of unique colors on this path is minimum over all possible $s-t$ paths. We show that this problem is hard (conditionally) to approximate within a factor $O(n^{1/8})$ of optimum, and give a polynomial time $O(n^{2/3})$-approximation algorithm. We translate the ideas used in this approximation algorithm into two simple greedy heuristics, and analyze their performance on an extensive set of synthetic and real world datasets. From our experiments, we found that our heuristics perform significantly better than the best previous heuristic algorithm for the problem on all datasets, both in terms of path quality and the running time.

1 Introduction

An *edge colored* graph $G = (V, E, \mathcal{C}, \chi)$ comprises of an underlying graph $G = (V, E)$, and a set of colors $\mathcal{C}$ such that each edge $e \in E$ is assigned a subset $\chi(e) \subseteq \mathcal{C}$ of colors. For any path π in this graph, suppose we define its cost to be $|\chi(\pi)|$ where $\chi(\pi) = \bigcup_{e \in \pi} \chi(e)$ is the set of colors used by this path. In this paper, we study the natural problem of computing a path π from a source s to some target t such that its cost, that is the number of colors used by π, is minimized. The problem is known to be NP-hard, and by a reduction from SET-COVER is also hard to approximate within a factor $o(\log n)$.

The problem was first studied by Yuan et al. [19] and was motivated by applications in maximizing the reliability of connections in mesh networks. More precisely, each network link is assigned one or more colors where each color corresponds to a given failure event that makes the link unusable. Now if the probability of all the failure events is the same, a path that minimizes the number of colors used has also the least probability of failure. Therefore, the number of colors used by a minimum color path can be used as a measure for 'resilience' of the network. This has also been applied in context of sensor networks [2] and attack graphs in computer security [14]. Apart from resilience, the minimum color

This work was supported by NSF under Grant CCF-1814172.

I. Kotsireas et al. (Eds.): SEA^2 2019, LNCS 11544, pp. 35–50, 2019.
https://doi.org/10.1007/978-3-030-34029-2_3

path problem can also be used to model licensing costs in networks. Roughly speaking, each link can be assigned a set of colors based on the providers that operate the link, and a minimum color path then corresponds to a minimum number of licenses that are required to ensure connectivity between two given nodes. More generally, the problem applies to any setting where colors can be thought of as "services" and we only need to pay for the first usage of that service. The problem was also studied by Hauser [12] motivated by robotics applications. In such settings, colors are induced by geometric objects (obstacles) that block one or more edges in a path of the robot. Naturally, one would like to remove the minimum number of obstacles to find a clear path, which corresponds to the colors used by a minimum color path.

The problem has also gathered significant theoretical interest. If each edge of the graph is assigned exactly one color (called its *label*), the problem is called *min-label* path and was studied in [11]. They gave an algorithm to compute an $O(\sqrt{n})$-approximation and also show that it is hard to approximate within $O(log^c n)$ for any fixed constant c, and n being the number of vertices. Several other authors have also studied related problems such as minimum label spanning tree and minimum label cut [8,16]. The min-color path problem on *vertex-colored* graph was recently studied in [1] where they gave an $O(\sqrt{n})$-approximation algorithm. Indeed, one can transform an edge-colored graph into a vertex-colored graph by adding a vertex of degree two on each edge e and assigning it the set of colors $\chi(e)$. However, this does not gives a sublinear approximation in n as the number of vertices in the transformed graph can be $\Omega(n^2)$.

1.1 Our Contribution

In this work, we make progress on the problem by improving the known approximation bounds and by designing fast heuristic algorithms.

- By a reduction from the *minimum k-union* problem [6] which was recently shown to be hard to approximate within a factor of $O(n^{1/4})$, we show that min-color path cannot be approximated within a factor $O(n^{1/8})$ of optimum on edge-colored graphs. This also implies improved lower bounds for min-label path, as well as min-color path on *vertex-colored* graphs.
- We give an $O(n^{2/3})$-approximation algorithm for min-color path problem on edge-colored graphs. If the number of colors on each edge is bounded by a constant, the algorithm achieves an approximation factor of $O((\frac{n}{OPT})^{2/3})$, where OPT is the number of colors used by the optimal path.
- We translate the ideas from the above approximation algorithm into two greedy heuristics and analyze its performance on a set of synthetic and real-world instances [15]. Although similar greedy heuristics were proposed by the previous work [19] and have been shown to perform well on *randomly generated* colored graphs; a holistic analysis of such algorithms on more challenging and realistic instances seems to be lacking. We aim to bridge this gap by identifying the characteristics of challenging yet realistic instances that helps us design a set of synthetic benchmarks to evaluate our algorithms. From our

experiments, we found that our heuristics achieve significantly better performance than the heuristic from [19] while being significantly (up to 10 times) faster. We also provide an ILP formulation for the problem that performs reasonably well in practice. All source code and datasets have been made available online on github [17].

The rest of the paper is organized as follows. In Sect. 2, we discuss our hardness reduction. The details of $O(n^{2/3})$–approximation is given in Sect. 3. We discuss our heuristic algorithms and experiments in Sect. 4. For the rest of our discussion, unless stated otherwise, we will use the term colored graph to mean an *edge-colored* graph and our goal is to compute a min-color path on such graphs. On some occasions, we will also need to refer to the *min-label* path problem, which is a special case of min-color path when all edges have exactly one color.

2 Hardness of Approximation

By a simple reduction from SET-COVER, it is known that the min-color path problem is hard to approximate to a factor better than $o(\log n)$, where n is the number of vertices. This was later improved by Hassin et al. [11], where they show that the min-label path (and therefore the min-color path) problem is hard to approximate within a factor $O(\log^c n)$ for any fixed constant c. In this section, we work towards strengthening this lower bound. To this end, we consider the *minimum k-union* problem that was recently studied in [6].

In the *minimum k-union* problem, we are given a collection $\mathcal{S}$ of m sets over a ground set U and the objective is to pick a sub-collection $\mathcal{S}' \subseteq \mathcal{S}$ of size k such that the union of all sets in $\mathcal{S}'$ is minimized. The problem is known to be hard to approximate within a factor $O(m^{1/4})$. However, it is important to note that this lower bound is conditional, and is based on the so-called DENSE VS RANDOM conjecture being true [6]. The conjecture has also been used to give lower bound guarantees for several other problems such as Densest k-subgraph [3], Lowest Degree 2-Spanner, Smallest m-edge subgraph [5], and Label cover [7].

In the following, we will show how to transform an instance of minimum k-union problem to an instance of min-color path. More precisely, given a collection $\mathcal{S}$ of m sets over a ground set U and a parameter k, we will construct a colored graph $G = (V, E, \mathcal{C}, \chi)$ with two designated vertices s, t, such that a solution for min-color path on G corresponds to a solution of minimum k-union on $\mathcal{S}$ and vice versa. We construct G in three steps (See also Fig. 1).

- We start with a path graph G' that has $m+1$ vertices and m edges. Next, we create $m-k+1$ copies of G' and arrange them as rows in a $(m-k+1)\times(m+1)$-grid, as shown in Fig. 1.
- So far we only have horizontal edges in this grid of the form $(v_{ij}, v_{i(j+1)})$. Next, we will add *diagonal edges* of the form $(v_{ij}, v_{(i+1)(j+1)})$ that basically connect a vertex in row i to its right neighbor in row $i+1$.

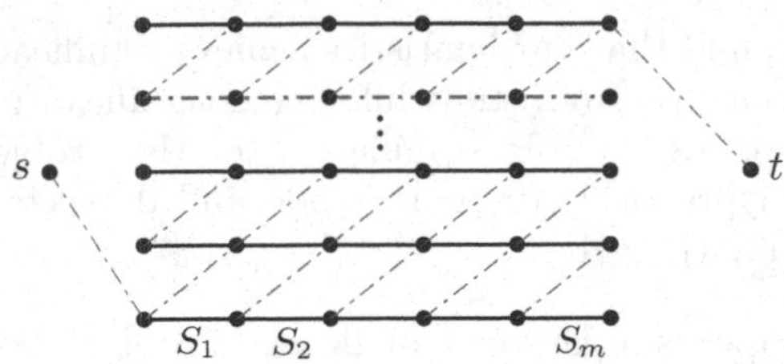

Fig. 1. Reducing minimum k-union to min-color path. The dashed edges are uncolored. The horizontal edge $(v_{ij}, v_{i(j+1)})$ is assigned color corresponding to the set S_j in all rows i.

- Finally, we add the vertices s, t and connect them to bottom-left vertex v_{11} and the top-right vertex $v_{(m-k+1)(m+1)}$, respectively.

We will now assign colors to our graph G. For the set of possible colors $\mathcal{C}$, we will use the ground set U and assign every subset $S_j \in \mathcal{S}$ to horizontal edges of G from left to right. More precisely,

- The diagonal edges of G are not assigned any color.
- Every horizontal edge that connects a node in column j to $j+1$ gets assigned the set of color S_j, for all $j \in 1, 2, \ldots, m$. That is $\chi(v_{ij}, v_{i(j+1)}) = S_j$, for all $i \in \{1, 2, \ldots, m-k+1\}$.

We make the following claim.

Lemma 1. *Assuming that minimum k-union problem is hard to approximate within a factor $O(m^{1/4})$ of optimal, the min-color path problem cannot be approximated within a factor of $O(n^{1/8})$, where m is the number of sets in the collection and n is the number of vertices in G.*

Proof. Consider any $s-t$ path π in G. Without loss of generality, we can assume that π is simple and moves monotonically in the grid. This holds because if π moves non-monotonically in the grid, then we can replace the non-monotone subpath by a path that uses the same (or fewer) number of colors. Now observe that in order to get from s to t, the path π must make a horizontal displacement of m columns and a vertical displacement of $m-k$ rows. Since the vertical movement is only provided by diagonal edges and π can only take at most $m-k$ of them, it must take k horizontal edges. If π is the path that uses minimum number of colors, then the sets S_j corresponding to the horizontal edges taken by π must have the minimum size union and vice versa.

Now suppose it was possible to approximate the min-color path problem within a factor $O(n^{1/8})$ of optimal. Then given an instance of minimum k-union, we can use the above reduction to construct a graph G that has $n = O(m^2)$ vertices and run this $O(n^{1/8})$ approximation algorithm. This will give us a path π that uses at most $O(m^{1/4})r$ colors, where r is the minimum number of colors used. As shown above, r must also be the number of elements in an optimal solution of minimum k-union. Therefore, we can compute a selection of k sets that have union at most $O(m^{1/4})$ times optimal, which is a contradiction. □

Note that we can also make G vertex-colored: subdivide each horizontal edge e by adding a vertex v_e of degree two, and assign the set of colors $\chi(e)$ to v_e. Observe that since the graph G we constructed above had $O(n)$ edges, the same lower bound also translates to min-color path on vertex-colored graph.

Corollary 1. *Min-color path on vertex-colored graphs is hard to approximate within a factor $O(n^{1/8})$ of optimal.*

One can also obtain a similar bound for min-label path.

Lemma 2. *Min-label path is hard to approximate within a factor $O((\frac{n}{OPT})^{1/8})$, where OPT is the number of colors used by the optimal path.*

3 An $O(n^{2/3})$– Approximation Algorithm

In this section, we describe an approximation algorithm for our problem that is sublinear in the number of vertices n. Note that if the number of colors on each edge is at most one, there exists an $O(\sqrt{n})$ approximation algorithm [11]. However, their technique critically depends on the number of colors on each edge to be at most one, and therefore cannot be easily extended to obtain a sublinear approximation[1].

An alternative approach is to transform our problem into an instance of min-color path on vertex-colored graphs by adding a vertex of degree two on each edge e and assigning the colors $|\chi(e)|$ to this vertex. Applying the $O(\sqrt{|V|})$-approximation from [1], easily gives an $O(\sqrt{|E|})$-approximation for our problem, which is sub-linear in n if the graph is *sparse*, but can still be $\Omega(n)$ in the worst case. To address this problem, we apply the technique of Goel et al. [9] where the idea is to partition the graph G into *dense* and *sparse* components based on the degree of vertices (Step 1 of our algorithm). We consider edges in both these components separately. For edges in dense component, we simply discard their colors, whereas for edges in the sparse components, we use a pruning strategy similar to [1] to discard a set of colors based on their occurrence. Finally, we show that both these pieces combined indeed compute a path with small number of colors. We start by making a couple of simple observations that will be useful.

First, we assume that the number of colors used by the optimal path (denoted by k) is given to our algorithm as input. Since k is an integer between 1 and $|\mathcal{C}|$, it is easy to see that an α-approximation for this version gives an α-approximation

[1] This is because if each edge has at most one color, the pruning stage in algorithm from [11] can be phrased as a maximum coverage problem, for which constant approximations are known. However even if number of colors is exactly two, the pruning stage becomes a variant of the *densest k-subgraph* problem which is hard to approximate within a factor of $\Omega(n^{1/4})$ of optimum.

for min-color path. This holds as we can simply run the α-approximation algorithm $|\mathcal{C}|$ times, once for each value of k and return the best path found.

Now, since we have fixed k, we can remove all edges from the graph that contain more than k colors, as a min-color path will never use these edges. Since each edge in G now contains at most k colors, we have the following lemma.

Lemma 3. *Any $s-t$ path of length ℓ uses at most $k\ell$ colors and is therefore an ℓ-approximation.*

This suggests that if there exists a path in G of small length, we readily get a good approximation. Note that the diameter of a graph $G = (V, E)$ is bounded by $\frac{|V|}{\delta(G)}$, where $\delta(G)$ is the minimum degree over vertices in G. So if the graph is dense, that is, degree of each vertex is high enough, the diameter will be small, and any path will be a good approximation (Lemma 3).

We are now ready to describe the details of our algorithm. We outline the details for the most general case when the number of colors on each edge is bounded by a parameter $z \leq k$. If z is a constant, the algorithm achieves slightly better bounds.

Algorithm: Approximate k-Color Path. The input to our algorithm is a colored graph $G = (V, E, \mathcal{C}, \chi)$, two fixed vertices s and t, the number of colors k and a threshold β (which we will fix later) for deciding if a vertex belongs to a dense component or a sparse component. Note that all edges of G have at most $z \leq k$ colors on them.

1. First, we will classify the vertices of G as lying in sparse or dense component. To do this, we include vertices of degree at most β to the *sparse* component and remove all edges adjacent to it. Now we repeat the process on the modified graph until no such vertex exists. Finally, we assign the remaining vertices to the *dense* component, and restore G to be the original graph.
2. For all edges $e = (u, v)$ such that both u, v lie in the dense component, discard its colors. That is set color $\chi(e) = \emptyset$.
3. Now, consider the set of edges that have at least one endpoint in the sparse component, call them *critical edges.* Note that the number of such edges is at most $n\beta$.
4. Remove every color c_i that occurs on at least $\sqrt{\frac{zn\beta}{k}}$ critical edges. That is, set $\chi(e) = \chi(e) \setminus \{c_i\}$, for all edges $e \in E$.
5. Let G' be the colored graph obtained after above modifications. Using $|\chi(e)|$ as weight of the edge e, run Dijkstra's algorithm to compute a minimum weight $s-t$ path π in G'. Return π.

It remains to show that the algorithm above indeed computes an approximately good path. We will prove this in two steps. First, we make the following claim.

Lemma 4. *The number of colors that lie on the path π in the modified colored graph G' is at most $\sqrt{zkn\beta}$.*

Proof. Observe that each color appears on no more than $\sqrt{\frac{zn\beta}{k}}$ edges of G'. Now consider the optimal path π^* in G that uses k colors. Since each of these k colors contribute to the weight of at most $\sqrt{\frac{zn\beta}{k}}$ edges of π^*, the weight of the path π^* in G' is at most $(k \cdot \sqrt{\frac{zn\beta}{k}}) = \sqrt{zkn\beta}$. Therefore, the minimum weight $s-t$ path π will use no more than $\sqrt{zkn\beta}$ colors. □

Lemma 5. *The number of colors that lie on the path π in the original colored graph G is $O(\frac{zn}{\beta} + \sqrt{zkn\beta})$.*

Proof. To show this, we will first bound the number of colors of π that we may have discarded in Steps 2 and 4 of our algorithm.

Consider a connected dense component C_i. Now let G_i be the subgraph induced by vertices in C_i. Since the degree of each vertex in G_i is at least β, the diameter of G_i is at most $\frac{n_i}{\beta}$, where n_i is the number of vertices in the component C_i. Observe that since the weight of all edges of C_i is zero in G', we can safely assume that π only enters C_i at most once. This holds because if π enters and exits C_i multiple times, we can simply find a shortcut from the first entry to last exit of weight zero, such a shortcut always exists because C_i is connected. Therefore π contains at most $\frac{n_i}{\beta}$ edges and uses at most $\frac{zn_i}{\beta}$ colors in the component C_i. Summed over all components, the total number of colors discarded in Step 2 that can lie on π is at most $z\sum_i \frac{n_i}{\beta} \leq \frac{zn}{\beta}$. Next, we bound the number of colors discarded in Step 4. Observe that since each critical edge contains at most z colors, the total number of occurrences of all colors on all critical edges is $zn\beta$. Since we only discard colors that occur on more than $\sqrt{\frac{zn\beta}{k}}$ edges, the total number of discarded colors is bounded by $\left(zn\beta/\sqrt{\frac{zn\beta}{k}}\right) = \sqrt{zkn\beta}$.

Summing these two bounds with the one from Lemma 4, we achieve the claimed bound. □

The bound from Lemma 5 is minimized when $\beta = (\frac{zn}{k})^{1/3}$. This gives the total number of colors used to be $O((\frac{zn}{k})^{2/3}) \cdot k)$ and therefore, an approximation factor of $O((\frac{zn}{k})^{2/3})$. If the number of colors z on each edge is bounded by a constant, we get an approximation factor of $O((\frac{n}{k})^{2/3})$. Otherwise, we have that $z \leq k$, which gives an $O(n^{2/3})$-approximation.

Theorem 1. *There exists a polynomial time $O(n^{2/3})$-approximation algorithm for min-color path in an edge-colored graphs $G = (V, E, \mathcal{C}, \chi)$. If the number of colors on each edge is bounded by a constant, the approximation factor can be improved to $O((\frac{n}{OPT})^{2/3})$.*

4 Fast Heuristic Algorithms and Datasets

In this section we will focus on designing fast heuristic algorithms for the minimum color path problem. Given a colored graph $G = (V, E, \mathcal{C}, \chi)$, one natural heuristic is to use Dijkstra's algorithm as follows: simply replace the set of colors $\chi(e)$ on each edge e by their cardinalities $|\chi(e)|$ as weights and then compute a minimum weight path in this graph.

Building upon this idea, Yuan et al. [19] proposed a greedy strategy where they start with a path computed by Dijkstra's algorithm as above, and iteratively select the color that improves the path found so far by maximum amount. More precisely, for each color $c \in \mathcal{C}$, decrement the weight of each edge on which c occurs by one, and compute a path using Dijkstra's algorithm. Now, select the color that improves the path found so far by maximum amount in terms of number of colors used. Keep the weight of edges with selected color to their decremented value and repeat the process until the path can no longer be improved.

This heuristic was called Single-Path All Color Optimization Algorithm (SPACOA) in their paper and was shown to achieve close to optimal number of colors on *uniformly colored* random graphs[2]. We argue that although their heuristic performs well on such instances, there still is a need to design and analyze algorithms on a wider range of more realistic instances. This holds because of two reasons. First, in most practical applications where the min-color path is used, the distribution of colors is typically not uniform. For instance, in network reliability applications where colors correspond to a failure event, it is likely that a specific failure event is more common (occurs on more edges) than the other. Similarly, in a network topology setting [15], where colors correspond to ISPs, some providers have larger connectivity than the others. Second, in most of these applications, existence of an edge between two nodes typically depends on proximity of nodes (imagine wireless routers or sensor networks) which is also not accurately captured by random graphs.

Moreover, we note that due to their structural properties (such as small diameters) uniformly colored random graphs are not good instances to measure the efficacy of heuristic algorithms because on these instances the number of colors used by a color oblivious Dijkstra's algorithm is also quite close to optimal, and as such there is little room for improvement. (See also Table 1). In the next two sections, we aim to construct synthetic instances for the min-color path problem that are more challenging and at the same time realistic. Thereafter, we present a couple of greedy heuristic algorithms and analyze their performance on these synthetic and some real world instances. We will use the SPACOA heuristic from [19] as a benchmark for our comparisons. We begin by analyzing min-color paths in uniformly colored random graphs and explain why a color oblivious algorithm such as Dijkstra performs so well. This gives some useful insights into characteristics of hard instances.

4.1 Min-color Path in Uniformly Colored Random Graphs

We begin by analyzing uniformly colored random graphs where given a random graph, colors are assigned uniformly to its edges [19]. That is, for each edge in

[2] We assume that the random graph is constructed under $G(n, p)$ model, that is an edge exists between a pair of vertices with probability p.

the graph, a color is picked uniformly at random from the set $\mathcal{C}$ of all colors, and assigned to that edge. We note that a colored graph G is likely to be a 'hard instance' for min-color path if there exists an $s-t$ path in G with small number of colors, and the expected number of colors on any $s-t$ path is much larger, so that a color oblivious algorithm is 'fooled' into taking one of these paths. We observe that in randomly generated colored graphs as above, this is quite less likely to happen, which is why the paths computed by a color oblivious Dijkstra's algorithm are still quite good.

To see this, consider an $s-t$ path π of length ℓ in G. Observe that since the colors are independently assigned on each edge, it is equivalent to first fix a path and then assign colors to its edges. Let $p_{i\pi}$ be the probability that color i appears on some edge of π. The probability that color i does not appear on any edge of π is $(1-\frac{1}{|\mathcal{C}|})^{\ell}$ and therefore $p_{i\pi} = 1-(1-\frac{1}{|\mathcal{C}|})^{\ell}$, for each color i. In other words, we can represent the occurrence of a color i on the path π by a Bernoulli random variable with a success probability $p_{i\pi}$. The number of colors on this path π will then correspond to the number of successes in $|\mathcal{C}|$ such trials, which follows the binomial distribution $B(|\mathcal{C}|, p_{i\pi})$. The expected number of colors on the path is given by $|\mathcal{C}|p_{i\pi}$ which clearly increases as the length of the path increases. The probability that the number of colors on π is k is given by $\binom{|\mathcal{C}|}{k} \cdot p_{i\pi}^{k} \cdot (1-p_{i\pi})^{|\mathcal{C}|-k}$. For example, if the number of colors $|\mathcal{C}| = 50$, then the probability that a path of length 20 uses a small number, say 5, colors is about 10^{-4}.

Therefore, in order to construct colored graph instances where there is significant difference between the paths computed by a color oblivious algorithm such as Dijkstra and the optimal path, we need to ensure that (a) there are a large number of paths between the source vertex s and destination vertex t (b) the vertices s and t are reasonably far apart. The first condition maximizes the probability that there will be an $s-t$ path with a small number of colors. The second condition ensures that the expected number of colors on any $s-t$ path is large, and it is quite likely that a color oblivious algorithm is fooled into taking one of these expensive paths.

4.2 Constructing Hard Instances

We construct our instances in two steps. First, we show how to construct the underlying graph $G=(V,E)$ and next describe how to assign colors on edges of G. We begin by assuming that unless otherwise stated, the vertices s and t are always assigned to be the pair of vertices that are farthest apart in G, that is, they realize the diameter of G. The idea now is to construct graphs that have large diameters (so that s, t are reasonably separated), are 'locally' dense (so that there is a large number of $s-t$ paths) and capture application scenarios for min-color path problems.

- **Layered Graphs.** These graphs comprise of n nodes arranged in a $k \times (n/k)$ grid. Each column consists of k nodes that form a layer and consecutive layers are fully connected. More precisely, a node v_{ij} in column j is connected to

all nodes v_{lj+1} in the next column and the for all $l = \{1, \ldots, k\}$. All vertices in the first column are connected to the source s, and the last column are connected to t. Such graphs are known to appear in design of centralized telecommunication networks [10], task scheduling, or software architectures.
- **Unit Disk Intersection Graphs.** These graphs comprise of a collection of n unit disks randomly arranged in a rectangular region. The graph is defined as usual, each disk corresponds to a vertex and is connected to all the other disks it intersects. Since the edges only exist between vertices that are close to each other, disk intersection graphs tend to have large diameters proportional to the dimensions of the region they lie in. These graphs appear quite frequently in ad-hoc wireless communication networks [13]
- **Road Networks.** These graphs intend to capture applications of min-color path to transportation networks such as logistics, where colors may correspond to trucking companies that operate between certain cities, and one would like to compute a path with fewest number of contracts needed to send cargo between two cities. For these graphs, we simply use the well-known road network datasets such as the California road network from [18]. As one may expect, road networks also tend to have large diameters.

Next, we assign colors to edges of the graph G. To keep things simple, we will assign colors to edges of G independently. We consider edges of G one by one and assign them up to z colors, by sampling the set of colors z times. However, in order to also capture that some colors are more likely to occur than others, we sample the colors from a truncated normal distribution as follows. We start with a normal distribution with mean $\mu = 0.5$, standard deviation $\sigma = 0.16$ (so that $0 < \mu \pm 3\sigma < 1$) and scale it by the number of colors. Now we sample numbers from this distribution rounding down to the nearest integer. With high probability, the sampled color indices will lie in the valid range $[0, |\mathcal{C}|)$, otherwise the sample returns an empty color set.

Table 1. Number of colors used by Dijkstra vs best known solutions on various colored graph instances. Note the higher difference between Dijkstra and optimal values for our instances.

Instance	Dijkstra	Best known	Remarks
Layered	43.38	17.6	$k = 4$ nodes per layer
Unit-disk	34.66	13.88	$n = 1000$ random disks in a 10×100 rectangle
Road-network	366	246	1.5 M nodes, 2.7 M edges, 500 colors
Uniform-col [19]	11.45	9.5	edges added with $p = \log n/2n$

Finally, for the sake of comparison, we also include the randomly generated colored graphs (Uniform-col) from [19]. The number of colors used by Dijkstra's algorithm and the optimal number of colors are shown in Table 1. For all the datasets except Road-network, the number of nodes is 1000, the number of colors is 50 and the number of samples per edge was 3. The reported values are averaged over 20 runs. For the Uniform-col instances, the probability of adding edges p was chosen so that the difference between colors used by Dijkstra's algorithm and the optimal is maximized.

4.3 ILP Formulation

We will now discuss an ILP formulation to solve the min-color path problem exactly. Given a colored graph $G = (V, E, \mathcal{C})$, the formulation is straightforward. We have a variable c_i for each color $i \in \mathcal{C}$, and another variable e_j for each edge $j \in E$. The objective function can be written as:

$$\begin{aligned}
&\mathit{minimize} \sum_i c_i && \text{subject to} \\
&c_i \geq e_j && i \in \chi(j) \text{ (color } i \text{ lies on edge } j) \quad (1) \\
&\sum_{j \in out(v)} e_j - \sum_{j \in in(v)} e_j = \begin{Bmatrix} 1, & v = s \\ -1, & v = t \\ 0, & v \neq s, t \end{Bmatrix} && \forall v \in V \quad (2) \\
&c_i, e_j \in \{0, 1\}
\end{aligned}$$

The first set of constraints (1) ensure that whenever an edge is picked, its colors will be picked as well. The second set of constraints (2) ensure that the set of selected edges form a path. We implemented the above formulation using Gurobi MIP solver (version 8.0.1) and found that they run surprisingly fast (within a second) on Uniform-col instances from Table 1. However, the solver tends to struggle even on small instances (about a hundred nodes) of all other datasets, suggesting that these instances are indeed challenging. In the next section, we will discuss a couple of heuristic algorithms that can compute good paths reasonably fast, and later in Sect. 4.5 compare their results with the optimal values computed by the ILP solver for some small instances.

4.4 Greedy Strategies

We begin by noting that a reasonably long path that uses small number of colors must repeat a lot of its colors. Therefore, the primary challenge is to identify the set of colors that are likely to be repeated on a path, and "select" them so that they are not counted multiple times by a shortest path algorithm. This selection is simulated by removing the color from the colored graph, so that subsequent runs of shortest path algorithm can compute potentially better paths. Inspired from the approximation algorithm of Sect. 3, our first heuristic Greedy-Select simply selects the colors greedily based on the number of times they occur on edges of G and returns the best path found. We outline the details below.

Algorithm:* Greedy-Select *Colors. The input to the algorithm is a colored graph $G = (V, E, \mathcal{C}, \chi)$, two fixed vertices s, t and it returns an $s - t$ path π.

1. Find an initial path π_0 by running Dijkstra's algorithm on G with weight of each edge $e = |\chi(e)|$. Let the number of the colors used by π_0 is K, an upperbound on number of colors our paths can use. Set the path $\pi = \pi_0$.
2. Initialize $i = 1$, and set $G_0 = G$ the original colored graph.
3. Remove the color $c_{\max}$ that appears on maximum number of critical edges of G_{i-1}. That is, set $\chi(e) = \chi(e) - \{c_{\max}\}\ \forall e \in E$. Let G_i be the colored graph obtained.
4. Compute the minimum weight path π_i in G_i with weight of each edge $e = |\chi(e)|$ using Dijkstra's algorithm.
5. Let K' be the number of colors on π_i in the *original* colored graph G. If $K' < K$, update the upperbound $K = K'$ and set $\pi = \pi_i$.
6. if $i < K$, set $i = i + 1$ and return to Step 3. Otherwise return path π.

Although the above algorithm runs quite fast and computes good paths, we can improve the path quality further by the following observation. Consider a color c_i that occurs on a small number of edges, then using an edge that contains c_i (unless absolutely necessary) can be detrimental to the path quality, as we may be better off picking edges with colors that occur more frequently. This suggests an alternative greedy strategy: we try to guess a color that the path is *not likely* to use, discard the edges that contain that color, and repeat the process until s and t are disconnected. This way we arrive at a small set of colors from the opposite direction, by iteratively discarding a set of 'expensive' candidates. To decide which color to discard first, we can again use their number of occurrences on edges of G – a small number of occurrences means a small number of edges are discarded and $s - t$ are more likely to remain connected. However, we found that this strategy by itself is not as effective as GREEDY-SELECT, but one can indeed combine both these strategies together into the GREEDY-PRUNE-SELECT heuristic, that is a little slower, but computes even better paths.

Algorithm:* Greedy-Prune-Select *Colors. The input is a colored graph $G = (V, E, \mathcal{C}, \chi)$, two fixed vertices s, t, and a parameter *threshold* that controls the number of times we invoke GREEDY-SELECT heuristic. The output is an $s - t$ path π.

1. For each color $c \in \mathcal{C}$, initialize *preference*(c) to be number of edges it occurs on. Initialize $i = 0$, $G_0 = G$ to be the initial graph, and $\mathcal{C}_0 = \mathcal{C}$ to be the initial set of candidate colors that can be discarded.
2. Run GREEDY-SELECT on G_0 to find an initial path π_0 to improve upon. Record the number of edges $M = |E|$ in the graph at this point.
3. Repeat the following steps until $\mathcal{C}_i$ is empty:
   (a) Pick a color $c_i \in \mathcal{C}_{i-1}$ such that *preference*(c_i) is minimum, and remove all edges e such that $c_i \in \chi(e)$. Let G_i be the graph obtained, and $\mathcal{C}_i = \mathcal{C}_{i-1} \setminus \{c_i\}$.
   (b) If s, t are disconnected in G_i, restore the discarded edges. That is set $G_i = G_{i-1}$. Set $i = i + 1$ and return to Step 3.
   (c) Otherwise, remove all edges from G_i that do not lie in the same connected component as s, t. Update *preference* of all colors that lie on these discarded edges.

(d) If the graph G_i has changed significantly, that is $M - |E_i| \geq$ *threshold* or if this is the last iteration, run GREEDY-SELECT again to compute the path π_i. Update $M = |E_i|$.

(e) Set $i = i + 1$ and return to Step 3.

4. Return the path π_i that is best in terms of number of colors.

The running time is typically dominated by the number of calls to GREEDY-SELECT. In our experiments, we set *threshold* $= 0.25|E|$ which guarantees that we only make a small number of calls to GREEDY-SELECT. Theoretically, GREEDY-SELECT runs in $O(|\mathcal{C}| \cdot D)$ time, where D is the running time of Dijkstra's algorithm. An implementation of GREEDY-PRUNE-SELECT using BFS to test connectivity runs in $O(|\mathcal{C}| \cdot (|V| + |E|)) + O(|\mathcal{C}| \cdot D)$ time, which is also $O(|\mathcal{C}| \cdot D)$. This is an order of magnitude better than SPACOA heuristic that has a worst-case running time of $O(|\mathcal{C}|^2 \cdot D)$.

4.5 Experiments and Results

We will now discuss the performance of above heuristic algorithms on our datasets. We compare our results with the values computed by the ILP solution (on small instances) and the SPACOA heuristic from [19]. In summary, we found that both our heuristics compute paths that are much better than SPACOA heuristic from [19], while also being significantly faster. The paths computed by GREEDY-PRUNE-SELECT are almost always significantly better than GREEDY-SELECT and the difference especially shows on larger datasets. The results are shown in Tables 2, 3, 4 and 5 averaged over five runs with the exception of real-world instances. Runtimes longer than one hour are marked with ∞. All code was written in C++ using the OGDF graph library [4] and executed on a standard linux machine (Ubuntu 16.04) running on Intel(R) Core(TM) i5-4460S CPU @ 2.90 GHz with 16 GB RAM.

Layered Graph Instances. We run our algorithms on a 4×125 layered graph instance with 50 colors on a 4×2500 instance with 500 colors. As the number of layers grows from 125 to 2500, these instances get progressively challenging for the ILP solver due to a large number of candidate paths. As expected, the SPACOA runs really slow on large instances as it needs to try a lot of colors per iteration. There is reasonable difference between the quality of paths computed by GREEDY-PRUNE-SELECT and GREEDY-SELECT especially on larger instances. The results are shown in Table 2.

Unit-Disk Instances. We run our algorithms on two sets of instances, with 500 nodes (disks) in a 10×50 rectangle, and a 10^4 nodes in a 10×1000 rectangle. The rectangles are chosen narrow so that the graph has a large diameter. The behavior is quite similar to layered graphs. The results are shown in Table 3.

Real-World Instances. Next, we focus on a couple of real-world examples. Our first instance is the California road network [18] that has 1.5 M nodes and 2.7 M edges. The graph however was not colored to begin with, so we color

Table 2. Path quality and running time on layered graph instances.

Algorithm	Colors used	Time taken (ms)	Colors used	Time taken (ms)
	$4 \times 125 = 0.5\,k$ nodes		$4 \times 2500 = 10\,k$ nodes	
Dijkstra	36.8	0.6	441.8	23.6
SPACOA	33.6	65	396	127×10^3
Greedy-Select	18.2	12.6	185.6	3.5×10^3
Greedy-Prune-Select	17.2	49	173	12.5×10^3
ILP	16.4	707×10^3	∞	∞

Table 3. Path quality and running time on Unit disk graph instances.

Algorithm	Colors used	Time taken (ms)	Colors used	Time taken (ms)
	$4 \times 125 = 0.5\,k$ nodes		$4 \times 2500 = 10\,k$ nodes	
Dijkstra	28.8	1	357.8	38
SPACOA	23	124.8	333.6	41.4×10^3
Greedy-Select	14.2	13	145.6	4.7×10^3
Greedy-Prune-Select	13.4	55	134	17.6×10^3
ILP	12.6	1176×10^3	∞	∞

it artificially by assigning 500 colors from the truncated normal distribution as explained before. Our second instance is from the Internet Topology Zoo [15], a manually compiled dataset of connectivity of internet service providers over major cities of the world. We translate this to our colored graph model, the cities naturally correspond to nodes, providers correspond to colors, and a color is assigned to an edge if the corresponding provider provides connectivity between these two cities. This graph has $5.6\,k$ nodes, $8.6\,k$ edges and 261 colors, with an average of 1.44 colors per edge.

Table 4. Path quality and running time on some real world instances.

Algorithm	Colors used	Time taken (ms)	Colors used	Time taken (ms)
	CA Road Network		Internet topology	
Dijkstra	366	3.068×10^3	7	26
SPACOA	355	3.12×10^6	4	3111
Greedy-Select	251	0.73×10^6	5	29
Greedy-Prune-Select	246	2.71×10^6	4	286
ILP	∞	∞	4	1817

The road-network instance due to its size is challenging to all algorithms. On the other hand, the internet-topology dataset seems quite easy for all the instances. One possible explanation for this is that although the number of nodes is large, the graph has a lot of connected components and a small diameter. This limits the space of candidate paths making all algorithms (particularly the ILP solver) quite fast.

Uniform-Col Instances. These instances are the same as one from [19] and have been mostly included for the sake of comparison. We run our algorithms on a Uniform-Col instance with 10^3 nodes and 50 colors, and another instance with 10^4 nodes and 500 colors.

Table 5. Path quality and running time on Uniform-Col instances.

Algorithm	Colors used	Time taken (in ms)	Colors used	Time taken (in ms)
	10^3 nodes		10^4 nodes	
Dijkstra	11.45	0.95	11.7	15
SPACOA	9.95	75.45	11.2	8664
Greedy-Select	10.4	9.2	11.5	150
Greedy-Prune-Select	10.3	46.3	11.4	1919
ILP	9.5	3913.8	–	–

The SPACOA heuristic performs marginally better than our heuristics on these examples. The primary reason for this is that the difference between optimal solution and that computed by Dijkstra's algorithm is really small (about 2), and the SPACOA heuristic typically overcomes this difference in just one iteration by trying all colors and picking the one that gives the best path. The cases in which SPACOA heuristic struggles to find good paths is when it has to try multiple iterations to bridge the gap between Dijkstra's algorithm and optimal, and gets stuck in a local minima. That is less likely to happen when the difference between optimal and Dijkstra value is small.

5 Conclusion

In this paper, we made progress on the min-color path problem by showing that under plausible complexity conjectures, the problem is hard to approximate within a factor $O(n^{1/8})$ of optimum. We also provide a simple $O(n^{2/3})$-approximation algorithm and designed heuristic algorithms that seem to perform quite well in practice. A natural open question is to see if these bounds can be improved further. The log-density framework has been useful in designing tight approximation bounds for related problems such as minimum k-union [6] and densest k-subgraph [3]. It would be interesting to see if those techniques can be applied to min-color path.

References

1. Bandyapadhyay, S., Kumar, N., Suri, S., Varadarajan, K.: Improved approximation bounds for the minimum constraint removal problem. In: APPROX 2018. LIPIcs, vol. 116, pp. 2:1–2:19 (2018)
2. Bereg, S., Kirkpatrick, D.: Approximating barrier resilience in wireless sensor networks. In: Dolev, S. (ed.) ALGOSENSORS 2009. LNCS, vol. 5804, pp. 29–40. Springer, Heidelberg (2009). https://doi.org/10.1007/978-3-642-05434-1_5
3. Bhaskara, A., Charikar, M., Chlamtac, E., Feige, U., Vijayaraghavan, A.: Detecting high log-densities: an $O(n^{1/4})$ approximation for densest k-subgraph. In: Proceedings of the 42nd STOC, pp. 201–210. ACM (2010)
4. Chimani, M., Gutwenger, C.: The Open Graph Drawing Framework (OGDF)
5. Chlamtac, E., Dinitz, M., Krauthgamer, R.: Everywhere-sparse spanners via dense subgraphs. In: Proceedings of the 53rd FOCS, pp. 758–767 (2012)
6. Chlamtáč, E., Dinitz, M., Makarychev, Y.: Minimizing the union: tight approximations for small set bipartite vertex expansion. In: Proceedings of the 28th SODA, pp. 881–899 (2017)
7. Chlamtáč, E., Manurangsi, P., Moshkovitz, D., Vijayaraghavan, A.: Approximation algorithms for label cover and the log-density threshold. In: Proceedings of the 28th SODA, pp. 900–919 (2017)
8. Fellows, M.R., Guo, J., Kanj, I.: The parameterized complexity of some minimum label problems. J. Comput. Syst. Sci. **76**(8), 727–740 (2010)
9. Goel, G., Karande, C., Tripathi, P., Wang, L.: Approximability of combinatorial problems with multi-agent submodular cost functions. In: Proceedings of the 50th FOCS, pp. 755–764 (2009)
10. Gouveia, L., Simonetti, L., Uchoa, E.: Modeling hop-constrained and diameter-constrained minimum spanning tree problems as steiner tree problems over layered graphs. Math. Program. **128**(1–2), 123–148 (2011)
11. Hassin, R., Monnot, J., Segev, D.: Approximation algorithms and hardness results for labeled connectivity problems. J. Comb. Optim. **14**(4), 437–453 (2007)
12. Hauser, K.: The minimum constraint removal problem with three robotics applications. Int. J. Robot. Res. **33**(1), 5–17 (2014)
13. Huson, M.L., Sen, A.: Broadcast scheduling algorithms for radio networks. In: Proceedings of MILCOM 1995, vol. 2, pp. 647–651. IEEE (1995)
14. Jha, S., Sheyner, O., Wing, J.: Two formal analyses of attack graphs. In: Computer Security Foundations Workshop, pp. 49–63. IEEE (2002)
15. Knight, S., Nguyen, H.X., Falkner, N., Bowden, R., Roughan, M.: The internet topology zoo. IEEE J. Sel. Areas Commun. **29**(9), 1765–1775 (2011)
16. Krumke, S.O., Wirth, H.C.: On the minimum label spanning tree problem. Inf. Process. Lett. **66**(2), 81–85 (1998)
17. Kumar, N.: Minimum color path experiments: source code and datasets (2019). https://doi.org/10.5281/zenodo.3382340
18. Leskovec, J., Lang, K.J., Dasgupta, A., Mahoney, M.W.: Community structure in large networks: natural cluster sizes and the absence of large well-defined clusters. Internet Math. **6**(1), 29–123 (2009)
19. Yuan, S., Varma, S., Jue, J.P.: Minimum-color path problems for reliability in mesh networks. In: INFOCOM 2005, vol. 4, pp. 2658–2669 (2005)

Student Course Allocation with Constraints

Akshay Utture[2(✉)], Vedant Somani[1], Prem Krishnaa[1], and Meghana Nasre[1]

[1] Indian Institute of Technology Madras, Chennai, India
{vedant,jpk,meghana}@cse.iitm.ac.in
[2] University of California, Los Angeles, USA
akshayutture@ucla.edu

Abstract. Real-world matching scenarios, like the matching of students to courses in a university setting, involve complex downward-feasible constraints like credit limits, time-slot constraints for courses, basket constraints (say, at most one humanities elective for a student), in addition to the preferences of students over courses and vice versa, and class capacities. We model this problem as a many-to-many bipartite matching problem where both students and courses specify preferences over each other and students have a set of downward-feasible constraints. We propose an Iterative Algorithm Framework that uses a many-to-one matching algorithm and outputs a many-to-many matching that satisfies all the constraints. We prove that the output of such an algorithm is Pareto-optimal from the student-side if the many-to-one algorithm used is Pareto-optimal from the student side. For a given matching, we propose a new metric called the Mean Effective Average Rank (MEAR), which quantifies the goodness of allotment from the side of the students or the courses. We empirically evaluate two many-to-one matching algorithms with synthetic data modeled on real-world instances and present the evaluation of these two algorithms on different metrics including MEAR scores, matching size and number of unstable pairs.

1 Introduction

Consider an academic institution where each semester students choose elective courses to credit in order to meet the credit requirements. Each student has a fixed number of credits that need to be satisfied by crediting electives. Each course has a credit associated with it and a capacity denoting the maximum number of students it can accommodate. Both students and courses have a strict preference ordering over a subset of elements from the other set. In addition, it is common to have constraints like a student wanting to be allotted at most one course from a basket of courses, or a student not wanting to be allotted to time-conflicting courses. The goal is to compute an *optimal* assignment of elective courses to students satisfying the curricular restrictions while respecting the course capacities and satisfying the credit requirements.

A. Utture—Part of this work was done when the author was a Dual Degree student at the Indian Institute of Technology Madras.

I. Kotsireas et al. (Eds.): SEA[2] 2019, LNCS 11544, pp. 51–68, 2019.
https://doi.org/10.1007/978-3-030-34029-2_4

We model the course allocation problem as a many-to-many bipartite matching problem with two sided preferences where one side of the bipartition allows *downward feasible constraints* (as described by [7]) and the other side of the bipartition has capacity constraints. Formally, we have a bipartite graph $G = (S \cup C, E)$ where S represents the set of students, C represents the set of courses and an edge $(s, c) \in E$ denotes that the course c can be assigned to the student s. Each course $c \in C$ has an integer valued capacity represented by $q(c)$. Each vertex $x \in S \cup C$ ranks its neighbours in G in a strict order and this ordering is called the preference list of the vertex x. Each student s defines a set of *downward feasible constraints* $X(s)$ over the neighbours in G. A set of downward feasible constraints is one in which if $C \subset C'$, and C' is feasible, then C must be feasible. In other words, if a set of courses is feasible for a student, any subset of those courses should also be feasible. There are many types of constraints expressible as downward feasible constraints.

1. **Student Capacity:** A student may have an upper limit on the number of courses to take in a semester.
2. **Student Credits:** A student may specify the maximum number of credits to be allotted. This assumes that each course has a variable number of credits.
3. **Time slots:** Each course runs in a particular time slot, and courses with overlapping time slots cannot be assigned to a student.
4. **Curricular constraints:** A student may specify an upper limit on a subset of courses that are of interest to him. This upper limit denotes the maximum number of courses that can be assigned to him from the subset. For instance, in a semester, a student may want to be assigned at most two Humanities electives amongst the multiple Humanities electives in his preference list.

There exist natural examples of constraints like minimum number of students required in a course (for the course to be operational) or pre-requisites on courses which *cannot* be expressed as downward feasible constraints.

Figure 1 shows an example instance of the student course allocation problem with downward feasible constraints. Here there are three students and three courses and the preference lists of the participants can be read from the figure. We assume all courses have uniform credits and therefore students specify a maximum capacity on the number of desired courses. Each student has a capacity of two. Each course can accommodate at most two students. Finally, for each student, there is a constraint that the student can be allotted at most one of the two courses $\{c_2, c_3\}$. Such a constraint could be imposed if the courses c_2 and c_3 run in overlapping time-slots. The student s_3 has a constraint that he can be allotted at most one of $\{c_1, c_3\}$. All these constraints (tabulated in Fig. 1), including the capacity constraints, are downward feasible constraints.

A matching M in a student course allocation setting is a subset of the edges of the underlying bipartite graph; matching M is said to be feasible if M respects all downward feasible constraints specified. As seen in Fig. 1 an instance may admit multiple feasible matchings. This motivates the need for a notion of optimality in order to select a matching from the set of all feasible matchings.

Student	Capacity	Student Preference List	Course	Capacity	Course Preference List
s_1	2	c_1, c_2, c_3	c_1	2	s_1, s_2, s_3
s_2	2	c_1, c_2, c_3	c_2	2	s_1, s_2, s_3
s_3	2	c_1, c_2, c_3	c_3	2	s_1, s_2, s_3
Constraints: Each student can take only 1 of $\{c_2, c_3\}$. s_3 can take only 1 of $\{c_1, c_3\}$					

Fig. 1. Each student s_i, for $i = 1, 2, 3$ prefers c_1 followed by c_2 followed by c_3. Each course has the same preference order. Matchings $M_1 = \{(s_1, c_1), (s_2, c_1), (s_1, c_2), (s_2, c_2), (s_3, c_3)\}$ and $M_2 = \{(s_1, c_1), (s_2, c_1), (s_3, c_2), (s_1, c_2), (s_2, c_3)\}$ are both feasible in this instance.

To the best of our knowledge, this problem setting has not be considered in the literature; settings with a subset of these constraints have been studied before. Cechlarova et al. [7] study a similar problem for one-sided preferences where they consider computing pareto-optimal matchings (defined below). In the two sided preference list model, *laminar classifications* [15,19] have been studied which can be defined as follows. Each vertex $x \in S \cup C$ specifies a family of sets called classes over the neighbours of x in G. A class J_x^i is allowed to have an upper quota $q^+(J_x^i)$ and a lower quota $q^-(J_x^i)$ which specifies the maximum and minimum number of vertices from J_x^i that need to be matched to x in any feasible matching. The family of classes for a vertex x is laminar if for any two classes of a vertex either one is contained inside the other or they are disjoint. Note that if there are no lower quotas associated with classes, then a laminar classification is a special case of downward feasibility. In fact, downward feasibility allows for non-laminar classes and course credits but as noted earlier, cannot capture lower quotas. Huang [19] and later Fleiner and Kamiyama [15] extended the notion of *stability* for laminar classifications. Their results show that in a many-to-many bipartite matching problem with two sided preferences and two sided laminar classifications with upper and lower quotas, it is possible to decide in polynomial time whether there exists a stable matching. In contrast, if classifications are non-laminar, even without lower quotas, it is NP-Complete to decide whether a stable matching exists [19]. Below we discuss two well-studied notions of optimality.

Stability: In the presence of two sided preferences, stability is the de-facto notion of optimality. We recall the definition of stability as in [15] which is applicable to our setting as well. Let M be a feasible matching in an instance G of the student course allocation problem with downward feasible constraints. For a vertex $u \in S \cup C$ let $M(u)$ denote the set of partners assigned to u in M. An edge $(s, c) \in E \setminus M$ is blocking w.r.t. M if both the conditions below hold:

- Either $M(s) \cup \{c\}$ is feasible for s or there exists a $c' \in M(s)$ such that s prefers c over c' and $(M(s) \setminus \{c'\}) \cup \{c\}$ is feasible for s, and
- Either $M(c) \cup \{s\}$ is feasible for c or there exists an $s' \in M(c)$ such that c prefers s over s' and $(M(c) \setminus \{s'\}) \cup \{s\}$ is feasible for c.

A matching M is stable if no pair $(s, c) \in E \setminus M$ blocks M. It is easy to verify that the matching M_1 in Fig. 1 is stable whereas the matching M_2 is unstable since the pair (s_2, c_2) blocks M_2.

Student Pareto-optimality: We now recall the definition of pareto-optimality from [7] which we consider for the student side. A student s prefers a set $M(s)$ over another set $M'(s)$ if $M(s)$ is lexicographically better than $M'(s)$; we denote this as $M(s) >_s M'(s)$. A matching M dominates another matching M' if there exists at least one student s such that $M(s) >_s M'(s)$ and for all students $s' \in S \setminus s$, we have $M(s') \geq'_s M'(s')$. A matching M is pareto-optimal if there is no matching M' that dominates M. Since we consider the domination only from the student side, we call this student pareto-optimality. However, since there is no ambiguity, we will simply denote it as pareto-optimal. For the example in Fig. 1, both M_1 and M_2 are pareto-optimal.

We now contrast these two notions via a simple example as shown in Fig. 2. There is only one student and three course and the credits of the student and courses as well as preferences of the student can be read from the figure. There are two feasible matchings, M_s and M_p, both of which are stable, yet only M_p is pareto-optimal. Note that M_p matches s to its top choice and satisfies the credit requirements. The example illustrates that in the presence of credits, stability may not be the most appealing notion of optimality. However, the only Pareto-optimal matching is M_p is more suitable in this scenario.

Student Preference List (s)	c_1, c_2, c_3
Max Credit Limit (s)	2
Credits(c_1) = 2; Credits(c_2) = Credits(c_3) = 1	

Fig. 2. Stability versus pareto-optimality. $M_s = \{(s, c_2), (s, c_3)\}$, $M_p = \{(s, c_1)\}$.

Our Contributions

- We provide an efficient algorithm for the many-to-many matching problem with two sided preferences, downward feasible constraints on one side, and capacity constraints on the other. Our framework can be easily extended to allow downward feasible constraints on both sides (Sect. 2).
- We prove that if the many-to-one matching algorithm used in the framework is student-pareto-optimal then our output for the many-to-many matching problem is also student pareto-optimal (Sect. 3).
- We introduce a new evaluation metric Mean Effective Average Rank (MEAR) score, a variation of the average rank metric, to quantify the quality of the matchings produced in a such a problem setting (Sect. 4).
- We empirically evaluate the well-known Gale-Shapley [16] stable matching algorithm and First Preference Allotment algorithm in the Iterative Algorithm Framework, using synthetic data-sets modeled on real-world instances, on different metrics including MEAR scores, matching size and number of unstable pairs (Sect. 5).

Related Work: The closest to our work is the work by Cechlarova et al. [7] where they consider the many-to-many matchings under downward feasible constraints. They give a characterization and an algorithm for Pareto-optimal many-to-many matchings under one sided preferences Variants of this problem deal with constraints like course prerequisites [6], lower-quotas for courses [8], or ties in the preference list [5]. Our problem setting deals with a similar setup but allows two-sided preferences.

The Gale and Shapley algorithm used by us is a classical algorithm to compute stable matchings in the well-studied Hospital Residents (HR) problem [16]. The HR problem is a special case of our problem with no downward feasible constraints, and a unit capacity for every student. Variants of the HR problem with constraints include allowing lower quotas [18], class constraints [15,19] and exchange-stability [9,21]. Other variations allow multiple partners [3,27], couples [25], colleague preferences [11] and ties [20,22–24]. The algorithms for the models considered above have strong mathematical guarantees, unlike our solution. However, the constraints studied do not capture all the complexities of the course allocation problem considered in our setting.

An empirical analysis of matching algorithms has been done in the context of the National Residency Matching Program in the U.S. [1,12–14,29,31], and its counter-part in the U.K. [2,4,30]. Manlove et al. [28] introduce a constraint programming model to solve the Hospital Residents problem with couples and justify its quality by its empirically obtained low execution time and number of blocking pairs. Krishnapriya et al. [26] empirically study the quality of the matchings produced by a popular matching on metrics of practical importance, like size, number of blocking pairs and number of blocking residents. Giannakopoulos et al. [17] give a heuristic algorithm for the NP-Hard equitable stable marriage problem, and empirically show that it outputs high-quality matchings. Diebold et al. [10] conduct a field experiment to understand the benefits of the efficiency-adjusted deferred acceptance mechanism as compared to the Gale-Shapley student optimal stable marriage mechanism. In this paper, we use a similar experimental approach to justify the practical importance of our proposed algorithm.

2 Algorithm Description

In this section, we present a framework called the Iterative Algorithm Framework (IAF for short) into which one can insert any many-to-one matching algorithm with two-sided preferences, and get a concrete algorithm which solves the many-to-many student-course matching problem with two-sided preferences and downward feasible constraints. As an example, we show two such many-to-one matching algorithms, namely the Gale-Shapley algorithm [16] and the First Preference Allotment. Finally, we suggest a simple extension to deal with downward feasible constraints on both sides of the bipartition.

2.1 Iterative Algorithm Framework

Algorithm. 1 gives the pseudo-code for the Iterative Algorithm Framework (IAF). As input, we require the set of students (S), the set of courses (C), the set of constraints $X(s)$ which determines what subset of courses from C is feasible for student s, the preference list of each student s ($P(s)$), the preference list of each course c ($P(c)$), and the capacity of each course c (q(c)). The framework outputs for each student s the set of allotted courses (M(s)).

The residual capacity of a course (denoted by $r(c)$) is the remaining capacity of a course at some point in the algorithm. It is initialized to its total capacity $q(c)$. Initialize the set of allotted courses $M(s)$ of each student s to be the empty set. We also maintain for every student s a reduced preference list ($R(s)$) which is intialized to the original preference list ($P(s)$). During the course of the algorithm the reduced preference list contains the set of courses in the preference list which have not yet been removed by the algorithm.

The framework is iterative and as long as some student has a non-empty reduced preference list, it invokes the *manyToOneMatch* function. The *manyToOneMatch* function is invoked with the capacities of the courses being set to the residual capacities. The *manyToOneMatch* function can be substituted with any many-to-one matching algorithm with two-sided preferences like the Gale-Shapley algorithm. An important characteristic of the IAF is that the allotments made at the end of the *manyToOneMatch* function are frozen, and matched student-course pairs cannot get unmatched in a future iteration. This characteristic is essential to ensure the termination of the algorithm. After the execution of the *manyToOneMatch* function, the residual capacities of the courses are recalculated based on the allotment in *manyToOneMatch*.

In the remaining part of the loop the algorithm removes preferences which can no longer be matched given the current set of allotted courses. If the course c was allotted to student s in this iteration, we remove c from the preference list of s. Additionally, we remove courses with no residual capacity, because all allotments up to this point are frozen and none of the residual capacity is going to free up in a future iteration. In order to maintain feasibility in the future iterations, we also need to remove all courses on a student's preference list which are infeasible with the current partial allotment of courses. In this part, it is implicit that if a course c is removed from the preference list of student s, even s is removed from the preference list of c.

2.2 Gale-Shapley Algorithm in the Iterative Algorithm Framework

The many-to-one Gale-Shapley algorithm [16] is one option for the *manyToOneMatch* function in Algorithm. 1. It works in a series of rounds until each student has exhausted his preference list. In each round, every unallotted student applies to his most preferred course that he has not applied to before, and if the course is either not full or prefers this student to its least preferred allotted student, the course will provisionally accept this student (and reject its least preferred allotted student in case it was full). Consider applying the Gale-Shapley algorithm with the IAF to the example shown in Fig. 1. Table 1 gives the partial allotments made after each iteration. We note that the allotments made in the first iteration are frozen and cannot be modified in the next iteration.

In this example, the student-optimal stable matching returned by the Gale-Shapley algorithm is pareto-optimal among the students it allots. However, this is not always true. Figure 3 shows an example where the student-optimal stable matching is $\{(s_1, c_2), (s_3, c_1), (s_4, c_1)\}$. This is not student pareto-optimal

Data: S = set of students, C = set of courses, $X(s)$ = constraints for student s, $P(s)$ = preference list for student s, $P(c)$ = preference list for each course c, $q(c)$ = capacity for each course c
Result: For each student s, $M(s)$ = set of allotted courses

```
Let r(c) = q(c), ∀c ∈ C ;
Let M(s) = ∅, ∀s ∈ S ;
Let R(s) = P(s), ∀s ∈ S ;
Let R(c) = P(c), ∀c ∈ C ;
while ∃s ∈ S, R(s) ≠ ∅ do
    Invoke manyToOneMatch() using the residual capacities ;
    Freeze every allotment (s, c) made in manyToOneMatch() ;
    /* (s,c) cannot be unmatched in a future iteration */
    Calculate the new r(c), ∀c ∈ C ;
    foreach student s ∈ S, R(s) ≠ ∅ do
        c = course allotted to s in manyToOneMatch() of current iteration ;
        Remove c from R(s) and s from R(c);
        From R(s) remove every c' such that r(c') = 0 ;
        /* Also remove corresponding s from R(c') */
        From R(s) remove every c' such that M(s) ∪ {c'} is not feasible
        according to X(s) ;
        /* Also remove corresponding s from R(c') */
    end
end
```

Algorithm 1: Iterative algorithm framework

Table 1. Gale-Shapley+IAF algorithm for the example in Fig. 1

Iteration	Allotment in current Iteration	Partial allotment so far
1	$\{(s_1, c_1), (s_2, c_1), (s_3, c_2)\}$	$\{(s_1, c_1), (s_2, c_1), (s_3, c_2)\}$
2	$\{(s_1, c_2), (s_2, c_3)\}$	$\{(s_1, c_1), (s_2, c_1), (s_3, c_2), (s_1, c_2), (s_2, c_3)\}$

because (s_1, s_3) can exchange their partners and both be better off. We discuss why we might still use the Gale-Shapley algorithm with the IAF in Sect. 5.

2.3 First Preference Allotment in the Iterative Algorithm Framework

Another option for the *manyToOneMatch* function in Algorithm. 1 is the First Preference Allotment. The First Preference Allotment is a simple many-to-one allotment, where each student is temporarily allotted to the first course on his or her preference list. If a course c, with capacity $q(c)$ is oversubscribed by k (i.e. $q(c) + k$ students are temporarily allotted to it), c rejects its k least preferred students. The algorithm terminates here, and the students who get rejected from courses which are oversubscribed do not apply again to other courses on their preference list. Allowing unallotted students to apply to the remaining courses on their preference list will result in a student-optimal stable marriage which is not pareto-optimal among the students it allots. The First Preference Allotment algorithm, on the other hand, is pareto-optimal among the subset of students

Student	Student Preference List	Course	Course Capacity	Course Preference List
s_1	c_1, c_2	c_1	2	s_4, s_3, s_1
s_2	c_2	c_2	1	s_1, s_2, s_3
s_3	c_2, c_1			
s_4	c_1			

Fig. 3. Student-optimal stable matching is not pareto-optimal for students

it matches (even though not pareto-optimal among the entire student set), and hence results in a pareto-optimal matching when inserted into the IAF.

Consider applying this algorithm to the example in Fig. 1. Table. 2 gives the partial allotments after each iteration of the First Preference Allotment algorithm when used with the IAF. In the first iteration, all students apply to their top choice course (c_1 for all), which accepts $\{s_1, s_2\}$ and rejects its worst preferred student (s_3). The allotment so far is frozen, and in the next iteration, all students again apply to their next top choice course (c_2 for all). Again, since c_2 only has a capacity of 2, it accepts $\{s_1, s_2\}$ and rejects its worst preferred student (s_3). In the final iteration, s_3 is the only student left with any capacity, and it applies to its next top choice course (c_3) and gets accepted.

Table 2. First Preference Allotment+IAF algorithm for the example in Fig. 1

Iteration	Allotment in current Iteration	Partial allotment so far
1	$\{(s_1, c_1), (s_2, c_1)\}$	$\{(s_1, c_1), (s_2, c_1)\}$
2	$\{(s_1, c_2), (s_2, c_2)\}$	$\{(s_1, c_1), (s_2, c_1), (s_1, c_2), (s_2, c_2)\}$
3	$\{(s_3, c_3)\}$	$\{(s_1, c_1), (s_2, c_1), (s_1, c_2), (s_2, c_2), (s_3, c_3)\}$

2.4 Extending the Iterative Algorithm Framework to additionally allow downward feasible constraints for courses

We can extend the IAF for the case where both students and courses express downward feasible constraints over each other. The *manyToOneMatch* method now needs a one-to-one matching algorithm, and line 13 in Algorithm 1 needs to additionally remove preferences which are infeasible for the course constraints. The proof of student pareto-optimality is almost identical to the one shown in Sect. 3.2 for Algorithm 1, and is hence skipped for brevity. Also, since the problem is now symmetric from the student and course sides, we can similarly show course pareto-optimality if the *manyToOneMatch* method is pareto-optimal among the courses it allots.

3 Theoretical Guarantees

In this section, we first characterize a pareto-optimal matching, and then prove that if the many-to-one matching used in the *manyToOneMatch* method of the

Iterative Algorithm Framework is pareto-optimal among the subset of students it allots, then the final many-to-many matching given by the Iterative Algorithm Framework is also pareto-optimal. Note that the condition of requiring the *manyToOneMatch* method to be pareto-optimal among the subset of students allotted is weaker than requiring it to be pareto-optimal among the entire set of students, and hence gives us more flexibility in picking a many-to-one matching.

3.1 Characterization of Pareto Optimality in the Many-to-Many Setting

Cechlarova et al. [7] prove that a matching is pareto-optimal if and only if it is Maximal, Trade-In Free, and Coalition Free. These terms are defined as follows.

1. *Maximal:* M is maximal if no student-course pair (s, c) exists such that $r(c) > 0$, and $M(s) \cup \{c\}$ is feasible.
2. *Trade-In Free:* M is trade-in free if no student-course pair (s, c) exists such that $r(c) > 0$, and $M(s) \setminus C' \cup \{c\}$ is feasible, where s prefers c over every $c_1 \in C'$. (i.e. s can feasibly trade the lower preferred C' for c)
3. *Coalition:* M is coalition free if there exists no set of students $S' \subseteq S$ who can exchange courses with one another, (and drop some lower preferred courses if needed to maintain feasibility) to get a new matching M', in which lexicographically $M'(s) > M(s), \forall s \in S'$

3.2 Proof of Pareto Optimality from the Student Side

To prove pareto-optimality, we show that the allotment is Maximal, Trade-In Free and Coalition Free. Theorems in this section use the assumption that the *manyToOneMatch* method is pareto-optimal among the subset of students it allotted a course.

Lemma 1. *The matching given by Algorithm 1 is Maximal.*

Proof. Let (s, c) be a student-course pair violating maximality. Since Algorithm 1 terminates with $R(s) = \emptyset$, $(s, c) \notin R(s)$. Since a student-course pair only gets removed from $R(s)$ on Lines 12 and 13, (s, c) must satisfy at least one of these: i) $r(c) = 0$ or ii) $M(s) \cup \{c\}$ is not feasible. Hence (s, c) cannot violate maximality.

Lemma 2. *The matching given by Algorithm 1 is Trade-in Free.*

Proof. Assume that the allotment is not Trade-in free. Let s be a student who wants to trade-in the set of courses C' for the course c, which s prefers over every course in C'. By the definition of trading-in stated above, $r(c) > 0$ at the end of the algorithm. $r(c) > 0$ holds throughout the algorithm because $r(c)$ never increases after an iteration. Consider the start of the first iteration where some $c_1 \in C'$ was allotted to s. By the definition of trading-in, c is feasible with $M(s) \setminus C'$, and hence should have been feasible at this point. The allotment of (s, c_1) instead of (s, c) is a contradiction to the fact that the *manyToOneMatch* method is pareto-optimal among the students it allots. Hence the assumption that the allotment is not Trade-in free must be false.

Lemma 3. *The matching given by Algorithm 1 is Coalition Free.*

Proof. Let all arithmetic here be modulo n. Assume that K $= ((s_1, c_1), \ldots, (s_n, c_n))$ is a coalition in M. Consider the first iteration where one of the coalition pairs was allotted. The entire coalition could not have been alotted in this iteration, because *manyToOneMatch* will not violate pareto-optimality. Hence $\exists(s_k, c_k) \in K$ such that it was allotted in this iteration, but (s_{k+1}, c_{k+1}) was not.

$r(c_{k+1}) > 0$ throughout this iteration because $r(c)$ does not increase for all c, and (s_{k+1}, c_{k+1}) got matched in a later iteration. During the current iteration, s_k selected c_k instead of the more preferred c_{k+1}, even though $r(c_{k+1}) > 0$ throughout this iteration. This contradicts the condition that *manyToOneMatch* is pareto-optimal among the students it allots in an iteration. Hence the assumption that a coalition exists must be false.

3.3 Time Complexity

Let n be the total number of students and courses, and m be the sum of the sizes of the preference lists. The outer while loop in Algorithm. 1 runs for $\mathcal{O}(m)$ iterations in the worst case because at least 1 allotment happens in each iteration. The inner foreach loop runs through each student's preference list a constant number of times, and hence has complexity $\mathcal{O}(m)$ (assuming that the feasibility checking of the constraints for a preference list of length l is $\mathcal{O}(l)$). The complexity of the *manyToOneMatch* method depends on the algorithm inserted. Hence, the overall complexity of the Iterative Algorithm Framework is $\mathcal{O}(m) * (\mathcal{O}(m) + Complexity(manyToOneMatch))$.

However, for some many-to-one algorithms like the Gale-Shapley, we can obtain a tighter bound on the number of iterations of the outer while loop. Each iteration of the Gale-Shapley algorithm runs in $\mathcal{O}(m)$ time and allots each student at least one course on his or her preference list (unless all the courses on the preference list are full), and hence the number of while loop iterations is $\mathcal{O}(n)$, bringing the total complexity to $\mathcal{O}(mn)$. The complexity of the *manyToOneMatch* method with the First Preference Allotment algorithm is $\mathcal{O}(m)$, since each student only applies to his top choice course and each course checks its preference list once, resulting in an overall complexity of $\mathcal{O}(m^2)$

4 Evaluation Metrics

In this section we look at some evaluation metrics used to quantify the quality of a matching produced in the presence of downward feasible constraints.

4.1 Mean Effective Average Rank

We define a new metric called Mean Effective Average Rank (MEAR for short) for quantifying the goodness of a matching in this problem setting. We discuss the MEAR in the context of students (shortened to MEAR-S), but it is applicable for

Student Preference List	c_1, c_2, c_3, c_4, c_5
Constraint	At most 1 out of $\{c_1, c_2\}$
Allotment for student	$\{c_1, c_4\}$
Reduced Ranks (in brackets)	$c_1(1), c_2(\textit{removed}), c_3(2), c_4(3), c_5(4)$
Effective Average Rank (EAR)	$(1+3)/(2^2) = 1$

Fig. 4. Example for effective average rank calculation

courses as well (shortened to MEAR-C). MEAR is a mean of the Effective Average Rank (EAR) over all the students. EAR is a variation of the per student average rank and is defined in Eq. 2. The definition uses the Reduced Rank (see Eq. 1) which is the actual rank in the preference list, minus the number of courses above this preference which were removed due to some infeasibility with a higher ranked course. The intuition behind using the Reduced Ranks in the definition of EAR is that if a student has expressed the constraint that he be allotted only one of the first ten courses on his preference list, and he gets allotted his first and eleventh choice, then the sum of ranks is $(1+10=11)$ 11, but the sum of Reduced Ranks $(1+2=3)$ is 3, which is a more accurate representation of the allotment.

$$ReducedRank_{M(s)}(c) = Rank(c) - |\{c' : c' \in T(s), c' >_s c\}| \tag{1}$$

$$EAR(s) = \frac{\sum_{c \in M(s)} ReducedRank_s(c)}{|M(s)|^2} \tag{2}$$

Here $M(s)$ is the set of courses allotted to s, $T(s)$ is the set of courses removed due to infeasibility with higher ranked courses in $M(s)$, and the notation $c' >_s c$ means that student s prefers course c' over c.

Average Rank uses $|M(s)|$ in the denominator, but Eq. 2 uses its square instead, for normalization. For a student allotted her first 5 choices, the sum of allotted Reduced Ranks is $(1+2+3+4+5 = 15)$. Dividing by the number of courses, gives us $(15/5 = 3)$, whereas a student allotted only his second preference gets an average of 2. An Average Rank of 3 sounds like a poor outcome for the student, but the student got all his top 5 choices. Hence dividing by $|M(s)|^2$ gives us $(15/25 = 0.6)$ which represents the quality of the allotment more accurately.

Consider the example in Fig. 4, to understand the EAR calculation for student s. The table lists the preferences, and the constraint is that only one of $\{c_1, c_2\}$ can be allotted to s. $M(s) = \{c_1, c_4\}$. The Reduced Rank of c_1 is 1, and since c_2 got removed because of the allotment to c_1, the reduced ranks of c_3, c_4 and c_5 are 2, 3 and 4 respectively. Finally, the EAR is the sum of Reduced Ranks of $M(s)$, which is $(1+3)$ divided by $|M(s)|^2$ or (2^2).

Some properties of the MEAR-score are as follows. A lower EAR score implies a better allotment. The lowest EAR value (and MEAR value) possible is 0.5. If the top k courses are allotted, $EAR = (k*(k+1))/2k^2$, which approaches 0.5 as k approaches infinity. EAR favours larger matches because of the square term in the denominator. MEAR favours fairer allotments because an average is taken over all students irrespective of the number of courses allotted to each student. This is exemplified in Fig. 1, where M_1 and M_2 have a similar total sum of ranks across all students, but M_2 gets a lower MEAR score (1.25 instead of 1.5).

4.2 Other Metrics

Even though MEAR gives a single number to assess the quality of a matching, there are other metrics of interest like the allotment size and number of unstable pairs. Another possible metric is the exchange blocking pairs (studied in a similar context by [21] and [9]), defined as the number of student pairs who can exchange one of their allotted courses and both be better off. By definition, a pareto-optimal matching will not have exchange blocking pairs, but if a non pareto-optimal many-to-one matching is used with the Iterative Algorithm Framework, this becomes an interesting metric. Exchange-blocking pairs and unstable-pairs are important metrics because they quantify the dissatisfaction among students. For example, two students forming an exchange-blocking pair will want to swap their courses because they can both be better off.

5 Experimental Results

In this section we empirically evaluate the quality of the matchings produced by the Gale-Shapley algorithm and First Preference Allotment algorithms in the *manyToOneMatch* method of the Iterative Algorithm Framework (shortened as GS+IAF and FP+IAF respectively) and compare these results against a Maximum Cardinality Matching (shortened as MCM), which serves as a baseline. Other baselines are not available since we are not aware of alternative algorithms in this setting. Using an input generator (a modified version of the one used by [26]), we first generate synthetic-data which models common downward feasible constraints in the student-course matching scenario, and then study the effect of varying parameters like the instance size, competition for courses and preference list lengths. The matching sizes are reported as a fraction of the MCM size and the unstable pairs are reported as a fraction of the number of unallotted pairs. All numbers reported in this section are averaged over 10 instances. All source code and instructions on reproducing the experiments are available at https://github.com/ved5288/student-course-allocation-with-constraints.

5.1 Input Data Generator

We generate two kinds of data sets: the *Shuffle Dataset* and *Master Dataset*. The generation of these datasets is identical, except that the Shuffle dataset represents one extreme of the possible skew in preference ordering where preference orders are fully random, and the Master dataset represents the other extreme where preference orders are identical. In the Master dataset, there exists a universal ordering among courses, and all student preference lists respect this relative ordering among the courses in their preference list. All course preference lists use a similar universal ordering of students.

The Shuffle Dataset is generated as follows. The number of students, courses and the range of preference list sizes (default is [3,12]) are taken as input. For each student, the set of acceptable courses is chosen according to a geometric distribution (with parameter 0.1) among the courses. The preference ordering is random. The credits of a course and the student credit limit follow geometric distributions (with parameter 0.1) among the values (5, 10, 15, 20) and (40, 50, 60, 70) respectively. The average capacity of a course is adjusted so that the number of seats offered by courses is 1.5× the total demand from students. A set of universal class constraints across all students mimics the type of constraint where no student can take time-overlapping courses. There are 20 such constraints, and in each of them, a student can pick at most k_1 (random integer in [1,3]) courses from a set of $(n_{courses}/20)$ courses. Individual class constraints (unique to each student) mimic the constraint of a student wanting at most p_1 courses from a set of p_2 courses (eg. a student wants at most 2 out of the 7 humanities courses on his/her preference list), where p_2 is a random integer from [2, preference list size] and p_1 is a random integer from $[1,p_2 - 1]$. The number of such constraints per student is a random integer from [0, preference list size]. The values used for the generator are chosen to mimic the typical values in a university setting.

5.2 Effect of Varying Instance Size

Figure 5 shows the effect of varying the number of students and courses on MEAR-S for GS+IAF, FP+IAF and MCM. On the Shuffle Dataset, GS+IAF (mean = 0.79) and FP+IAF (mean = 0.79) perform similarly, and both clearly outperform the MCM (mean = 1.42). On the Master Dataset, the GS+IAF (mean = 1.38) performs slightly better FP+IAF (mean = 1.52), which in turn performs better than the MCM (mean = 1.83). Increasing the size of the matching does not affect this trend on either data set. Tables 3 and 4 show the effect of varying instance size on other parameters for the two datasets. GS+IAF and FP+IAF give similar scores on all metrics. The GS+IAF, however, gives a few exchange-blocking pairs because the Gale-Shapley algorithm is not pareto-optimal among the students it allots. Both IAF algorithms outperform MCM in the Unstable Pairs and Exchange Blocking Pairs (almost 0 for both IAF algorithms) by a huge margin, but do slightly worse on MEAR-C and, as expected, the matching size. The MEAR-C is higher because the GS+IAF and FP+IAF algorithms are inherently biased towards the student side. It is impossible to get an Exchange-blocking pair for any algorithm used on the Master Dataset, and hence it is not shown in Table 4. Any 2 courses common to 2 student preference lists will be listed in the same relative order, and exchanging them will leave exactly 1 student worse off.

5.3 Effect of Increasing Competition

Figure 6 shows the effect of varying competition on the MEAR-S metric. The number of students is varied, while the number of courses is kept constant at 250. On the Shuffle Dataset, FP+IAF (mean = 0.98) outperforms the GS+IAF

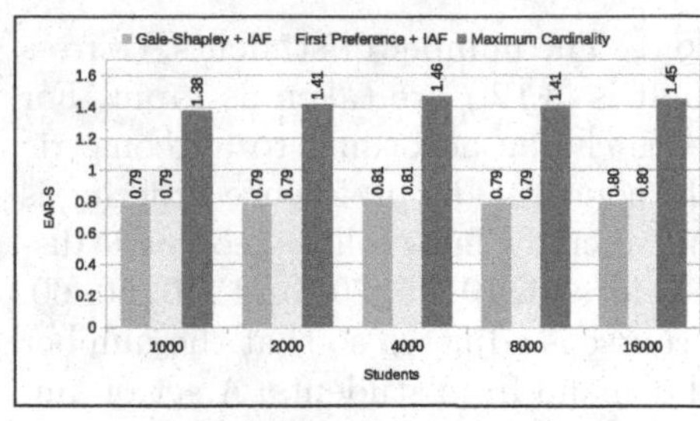

(a) Shuffle Dataset

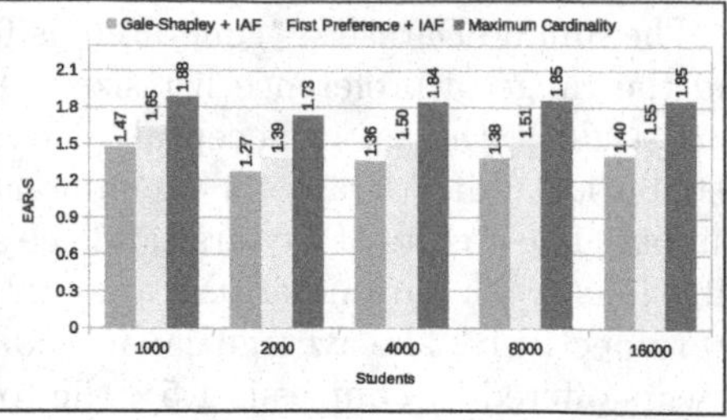

(b) Master Dataset

Fig. 5. MEAR-S values for varying instance sizes (student-course ratio is 1:6)

Table 3. Effect of varying instance size for the Shuffle Dataset

Students	MEAR-C			Matching size			Unstable-Pair Ratio			Ex. Blocking Pairs		
	GS+I	FP+I	MCM	GS+I	FP+I	MCM	GS+I	FP+I	MCM	GS+I	FP+I	MCM
1000	1.42	1.43	1.23	0.87	0.87	1.0	0.11	0.11	0.38	0.3	0	69.9
2000	1.39	1.40	1.23	0.87	0.87	1.0	0.10	0.10	0.39	0.6	0	70.9
4000	1.49	1.49	1.31	0.86	0.86	1.0	0.10	0.10	0.40	0.4	0	68.9
8000	1.39	1.40	1.21	0.87	0.87	1.0	0.10	0.10	0.38	0.1	0	50.8
16000	1.44	1.45	1.26	0.87	0.87	1.0	0.10	0.10	0.38	0	0	66.3

(mean = 1.17) on the larger inputs, and both clearly outperform the Maximum Cardinality Matching (mean = 1.92). On the Master Dataset, the FP+IAF (mean = 2.17) performs slightly better than GS+IAF (mean = 2.19), which in turn performs better than the MCM (mean = 2.69). Tables 5 and 6 show the effect of varying student competition on other parameters for the two datasets. With increasing competition, MCM deteriorates significantly on EAR-C and Exchange-blocking pairs, and loses most of its advantage in matching size. FP+IAF on the other hand scales well on almost all measures (including EAR-S) and hence is the appropriate choice for a high-competition scenario.

5.4 Effect of Varying Preference List Sizes

Figure 7 shows the effect of varying preference list sizes on the MEAR-S metric, on an input of 8000 students and 1333 courses. On the Shuffle Dataset, GS+IAF (mean = 0.81) and FP+IAF (mean = 0.81) clearly outperform the MCM (mean = 1.34). On the Master Dataset, the GS+IAF (mean = 1.33) performs slightly better than FP+IAF (mean = 1.45), which in turn outperforms the MCM (mean = 1.68). The observations here are qualitatively identical to those obtained by varying the instance size (Sect. 5.2) (Tables 7 and 8).

Table 4. Effect of varying instance size for the Master Dataset

Students	MEAR-C			Matching size			Unstable-Pair Ratio		
	GS+I	FP+I	MCM	GS+I	FP+I	MCM	GS+I	FP+I	MCM
1000	2.10	2.05	1.27	0.82	0.82	1.0	0.17	0.24	0.57
2000	1.99	1.97	1.17	0.84	0.84	1.0	0.14	0.20	0.56
4000	2.07	2.05	1.23	0.84	0.84	1.0	0.14	0.20	0.57
8000	2.04	2.04	1.24	0.84	0.84	1.0	0.15	0.21	0.57
16000	2.17	2.17	1.23	0.79	0.79	1.0	0.15	0.23	0.57

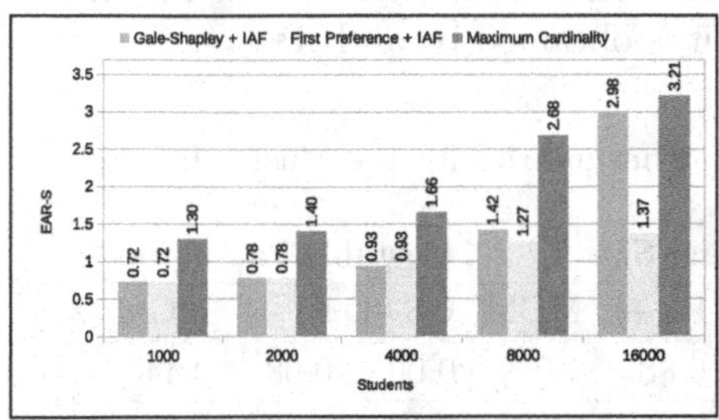

(a) Shuffle Dataset

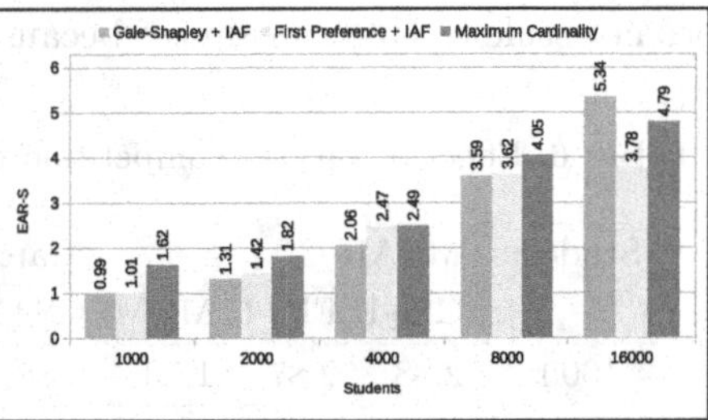

(b) Master Dataset

Fig. 6. MEAR-S values for varying competition for courses (by varying the number of students while keeping the number of courses constant at 250)

Table 5. Effect of varying competition for the Shuffle Dataset

Students	MEAR-C			Matching size			Unstable-Pair Ratio			Ex. Blocking Pairs		
	GS+I	FP+I	MCM	GS+I	FP+I	MCM	GS+I	FP+I	MCM	GS+I	FP+I	MCM
1000	1.40	1.40	1.27	0.90	0.90	1.0	0.02	0.02	0.34	0	0	66.5
2000	1.37	1.37	1.20	0.87	0.87	1.0	0.09	0.09	0.39	0	0	131.6
4000	1.43	1.46	1.41	0.85	0.84	1.0	0.24	0.24	0.53	6.7	0	321
8000	1.58	1.87	2.57	0.94	0.93	1.0	0.32	0.38	0.75	181.1	0	3598.2
16000	1.05	2.59	5.11	0.99	0.98	1.0	0.10	0.40	0.89	6287.4	0	5554.9

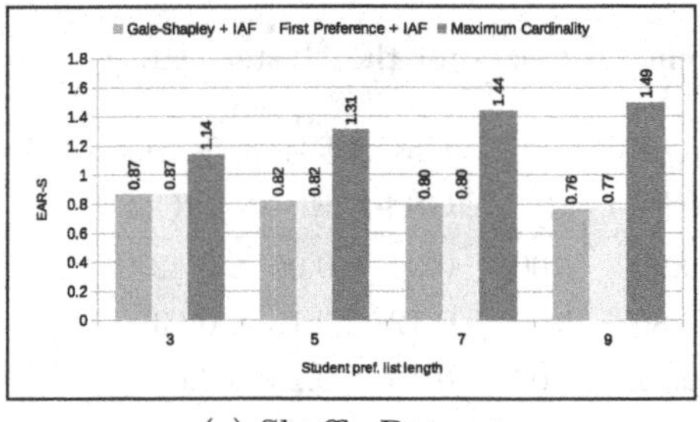

(a) Shuffle Dataset

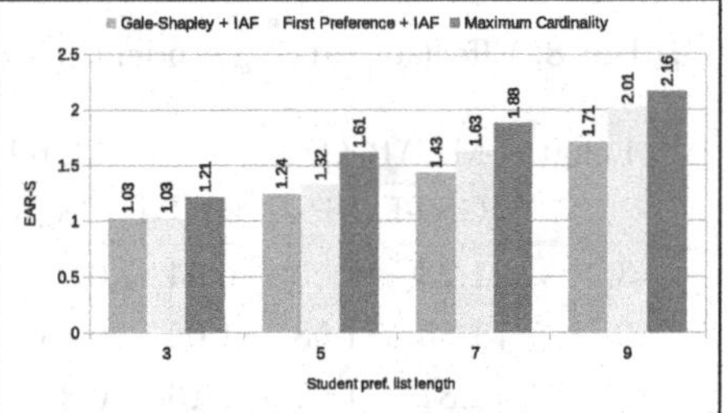

(b) Master Dataset

Fig. 7. MEAR-S values for varying student preference list sizes

5.5 Discussion

From the preceding observations, we can conclude that GS+IAF and FP+IAF give similar results on the Shuffle Dataset, but GS+IAF does better on the Master Dataset. Even though it does not have a theoretical pareto-optimal guarantee, GS+IAF gives a negligible number of exchange-blocking pairs. It also consistently fares better on the Unstable Pairs metric because it avoids Unstable Pairs within an iteration. Hence, these observations empirically justify the use of GS+IAF, even though it may not have the same guarantees as FP+IAF. MCM on the other hand clearly outputs inferior matchings for the Shuffle Dataset on all parameters except size. Since all preference lists are similar in the Master Dataset, ignoring preferences does not hurt much, and MCM fares decently. Note that the MCM does not scale to larger instances because it is obtained by an ILP.

Table 6. Effect of varying competition on other metrics for the Master Dataset

Students	MEAR-C			Matching Size			Unstable-Pair Ratio		
	GS+I	FP+I	MCM	GS+I	FP+I	MCM	GS+I	FP+I	MCM
1000	2.88	2.87	1.50	0.88	0.88	1.0	0.06	0.08	0.44
2000	2.39	2.44	1.28	0.84	0.85	1.0	0.13	0.19	0.56
4000	1.67	1.64	1.38	0.80	0.80	1.0	0.24	0.39	0.79
8000	1.90	2.00	2.56	0.91	0.91	1.0	0.25	0.54	0.90
16000	2.39	3.31	5.34	0.98	0.98	1.0	0.16	0.59	0.95

Table 7. Effect of varying student preference list sizes for the Shuffle Dataset

Length	MEAR-C			Matching Size			Unstable-Pair Ratio			Ex. Blocking Pairs		
	GS+I	FP+I	MCM	GS+I	FP+I	MCM	GS+I	FP+I	MCM	GS+I	FP+I	MCM
3	0.97	0.98	0.89	0.93	0.93	1.00	0.08	0.08	0.34	0	0	2.5
5	1.14	1.15	1.01	0.90	0.90	1.00	0.11	0.11	0.43	0.2	0	20
7	1.34	1.34	1.17	0.87	0.87	1.00	0.12	0.12	0.43	0.5	0	53.6
9	1.44	1.46	1.23	0.86	0.84	1.00	0.11	0.12	0.41	0.1	0	81.3

Table 8. Effect of varying student preference list sizes for the Master Dataset

Length	MEAR-C			Matching Size			Unstable-Pair Ratio		
	GS+I	FP+I	MCM	GS+I	FP+I	MCM	GS+I	FP+I	MCM
3	1.23	1.23	0.94	0.92	0.92	1.00	0.04	0.06	0.42
5	1.55	1.58	1.02	0.86	0.86	1.00	0.12	0.16	0.63
7	1.87	1.92	1.16	0.83	0.82	1.00	0.16	0.22	0.63
9	2.40	2.42	1.39	0.81	0.80	1.00	0.18	0.26	0.60

6 Conclusion

In this paper, we presented the Iterative Algorithm Framework to solve the many-to-many matching problem with two-sided preferences and downward feasible constraints. We proved that if the many-to-one *manyToOneMatch* subroutine is pareto-optimal among the students it allots, the framework outputs a pareto-optimal matching satisfying all the constraints. To quantify the quality of a matching, we introduced the Mean Effective Average Rank (MEAR) measure. We showed that the Gale-Shapley and First Preference Allotment algorithms used with the Iterative Algorithm Framework, get significantly higher student MEAR-scores on two different synthetic datasets, and these results hold even when the instance size, competition or preference list lengths are changed. The Iterative Algorithm Framework algorithms also perform significantly better on other metrics like the number of unstable pairs and exchange-blocking pairs. In future, it would be interesting to study constraints which are not downward feasible – for instance lower quotas, course-pre-requisites.

References

1. National Residency Matching Program. https://www.nrmp.org
2. Scottish Foundation Association Scheme. https://www.matching-in-practice.eu/the-scottish-foundation-allocation-scheme-sfas
3. Bansal, V., Agrawal, A., Malhotra, V.S.: Polynomial time algorithm for an optimal stable assignment with multiple partners. Theor. Comput. Sci. **379**(3), 317–328 (2007)
4. Biró, P., Irving, R.W., Schlotter, I.: Stable matching with couples: An empirical study. J. Exp. Algorithmics **16**, 1.2:1.1–1.2:1.27 (2011)
5. Cechlárová, K., et al.: Pareto optimal matchings in many-to-many markets with ties. Theor. Comput. Syst. **59**(4), 700–721 (2016)
6. Cechlárová, K., Klaus, B., Manlove, D.F.: Pareto optimal matchings of students to courses in the presence of prerequisites. Discrete Optim. **29**, 174–195 (2018)
7. Cechlárová, K., Eirinakis, P., Fleiner, T., Magos, D., Mourtos, I., Potpinková, E.: Pareto optimality in many-to-many matching problems. Discrete Optim. **14**, 160–169 (2014)
8. Cechlárová, K., Fleiner, T.: Pareto optimal matchings with lower quotas. Math. Soc. Sci. **88**, 3–10 (2017)
9. Cechlárová, K., Manlove, D.F.: The exchange-stable marriage problem. Discrete Appl. Math. **152**(1), 109–122 (2005)
10. Diebold, F., Aziz, H., Bichler, M., Matthes, F., Schneider, A.: Course allocation via stable matching. Bus. Inf. Syst. Eng. **6**(2), 97–110 (2014)
11. Dutta, B., Massó, J.: Stability of matchings when individuals have preferences over colleagues. J. Econ. Theory **75**(2), 464–475 (1997)
12. Roth, E.A.: The effects of the change in the NRMP matching algorithm. national resident matching program. JAMA J. Am. Med. Assoc. **278**, 729–732 (1997)
13. Roth, E.A., Peranson, E.: The redesign of the matching market for american physicians: some engineering aspects of economic design. Am. Econ. Rev. **89**, 748–780 (1999)

14. Echenique, F., Wilson, J.A., Yariv, L.: Clearinghouses for two-sided matching: an experimental study. Quant. Econ. **7**(2), 449–482 (2016)
15. Fleiner, T., Kamiyama, N.: A matroid approach to stable matchings with lower quotas. In: Proceedings of the Twenty-third Annual ACM-SIAM Symposium on Discrete Algorithms, SODA 2012 pp. 135–142 (2012)
16. Gale, D., Shapley, L.S.: College admissions and the stability of marriage. Am. Math. Mon. **69**(1), 9–15 (1962)
17. Giannakopoulos, I., Karras, P., Tsoumakos, D., Doka, K., Koziris, N.: An equitable solution to the stable marriage problem. In: Proceedings of the 2015 IEEE 27th International Conference on Tools with Artificial Intelligence ICTAI, pp. 989–996. ICTAI 2015 (2015)
18. Hamada, K., Iwama, K., Miyazaki, S.: The hospitals/residents problem with lower quotas. Algorithmica **74**(1), 440–465 (2016)
19. Huang, C.C.: Classified stable matching. In: Proceedings of the Twenty-first Annual ACM-SIAM Symposium on Discrete Algorithms, SODA 2010, pp. 1235–1253 (2010)
20. Irving, R.W.: Stable marriage and indifference. Discrete Appl. Math. **48**(3), 261–272 (1994)
21. Irving, R.W.: Stable matching problems with exchange restrictions. J. Comb. Optim. **16**(4), 344–360 (2008)
22. Irving, R.W., Manlove, D.F., O'Malley, G.: Stable marriage with ties and bounded length preference lists. J. Discrete Algorithms **7**(2), 213–219 (2009)
23. Irving, R.W., Manlove, D.F., Scott, S.: The hospitals/residents problem withties. In: Algorithm Theory - SWAT 2000, pp. 259–271 (2000)
24. Iwama, K., Miyazaki, S., Morita, Y., Manlove, D.: Stable marriage with incomplete lists and ties. In: Wiedermann, J., van Emde Boas, P., Nielsen, M. (eds.) ICALP 1999. LNCS, vol. 1644, pp. 443–452. Springer, Heidelberg (1999). https://doi.org/10.1007/3-540-48523-6_41
25. Klaus, B., Klijn, F.: Stable matchings and preferences of couples. J. Econ. Theory **121**(1), 75–106 (2005)
26. Krishnapriya, A. M., Nasre, M., Nimbhorkar, P., Rawat, A.: How good are popular matchings? In: Proceedings of the 17th International Symposium on Experimental Algorithms, SEA 2018, L'Aquila, Italy, 27–29 June 2018, pp. 9:1–9:14 (2018)
27. Malhotra, V.S.: On the stability of multiple partner stable marriages with ties. In: Albers, S., Radzik, T. (eds.) ESA 2004. LNCS, vol. 3221, pp. 508–519. Springer, Heidelberg (2004). https://doi.org/10.1007/978-3-540-30140-0_46
28. Manlove, D.F., McBride, I., Trimble, J.: "almost-stable" matchings in the hospitals/residents problem with couples. Constraints **22**(1), 50–72 (2017)
29. Peranson, E., Randlett, R.R.: The NRMP matching algorithm revisited. Acad. Med. **70**, 477–484 (1995)
30. Roth, A.E.: A natural experiment in the organization of entry-level labor markets: regional markets for new physicians and surgeons in the united kingdom. Am. Econ. Rev. **81**(3), 415–440 (1991)
31. Williams, K.J.: A reexamination of the NRMP matching algorithm national resident matching program. Academic medicine: journal of the Association of American Medical Colleges **70**, 470–476 (1995). discussion 490

A Combinatorial Branch and Bound for the Min-Max Regret Spanning Tree Problem

Noé Godinho(✉) and Luís Paquete

CISUC, Department of Informatics Engineering, University of Coimbra, Pólo II, 3030-290 Coimbra, Portugal
{noe,paquete}@dei.uc.pt

Abstract. Uncertainty in optimization can be modeled with the concept of scenarios, each of which corresponds to possible values for each parameter of the problem. The min-max regret criterion aims at obtaining a solution minimizing the maximum deviation, over all possible scenarios, from the optimal value of each scenario. Well-known problems, such as the shortest path problem and the minimum spanning tree, become NP-hard under a min-max regret criterion. This work reports the development of a branch and bound approach to solve the Minimum Spanning Tree problem under a min-max regret criterion in the discrete scenario case. The approach is tested in a wide range of test instances and compared with a generic pseudo-polynomial algorithm.

Keywords: Min-max regret criterion · Multi-objective optimization · Minimum Spanning Tree · Branch and bound

1 Introduction

The *min-max regret* formulation of an optimization problem deals with the existence of uncertainty on the objective function coefficients [10]. This uncertainty composes scenarios and the goal is to find a solution that minimizes the deviation between the value of the solution and the value of the optimal solution for each scenario. This way, the anticipated regret of a wrong decision is taken into account into the problem.

In this article, we consider a branch and bound algorithm to solve the *min-max regret minimum spanning tree* (MMR-MST) problem in the discrete scenario case. Let $G = (V, E)$ be a undirected edge-weighted graph with vertex set V and edge set E, and let n and m denote the number of vertices and edges, respectively. Let $\mathcal{T}$ denote the set of all spanning trees in G and let $S = \{s_1, \ldots, s_k\}$ be a set of k scenarios. For each scenario $s \in S$, each edge $e \in E$ has an associated positive integer cost $w(e, s)$. We assume that G is connected and simple. For a tree $T \in \mathcal{T}$, its cost $w(T, s)$ is the sum of the costs of its edges for scenario $s \in S$ as follows:

$$w(T, s) := \sum_{e \in T} w(e, s) \tag{1}$$

I. Kotsireas et al. (Eds.): SEA² 2019, LNCS 11544, pp. 69–81, 2019.
https://doi.org/10.1007/978-3-030-34029-2_5

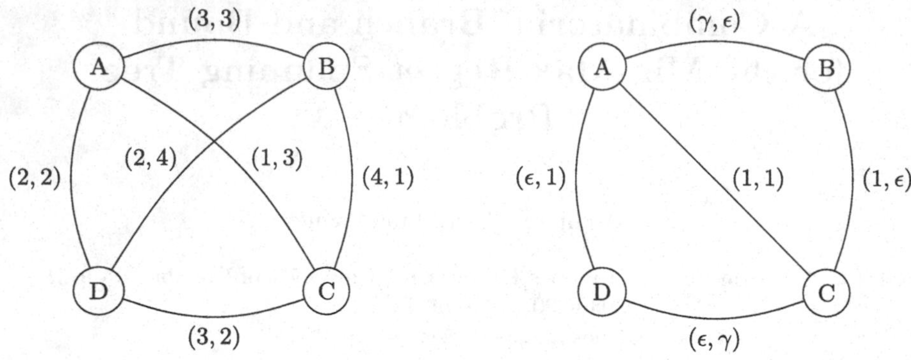

Fig. 1. Two instances for Example 1 (left) and Example 2 (right)

Let T_s^* denote the minimum spanning tree for scenario s. Note that T_s^* can be found in polynomial time with classical algorithms such as Prim's or Kruskal's Algorithm. The goal in the MMR-MST problem is to find a spanning tree that has the least deviation from the minimum spanning trees for every scenario in terms of total cost. Let $T \in \mathcal{T}$ be a spanning tree of graph G and $R(T)$ be the regret function, i.e.

$$R(T) := \max_{s \in S} \ (w(T, s) - w(T_s^*, s)) \tag{2}$$

Then, the MMR-MST problem consists of minimizing the regret function among all spanning trees, i.e.,

$$\min_{T \in \mathcal{T}} R(T) \tag{3}$$

Example 1. Figure 1, left, shows a graph with 4 vertices and two scenarios, $S = \{s_1, s_2\}$, where the tuple at each edge indicates its cost in the first and in the second scenario, respectively. Then, we have that $T_{s_1}^* = \{\{A, C\}, \{A, D\}, \{B, D\}\}$ with costs $w(T_{s_1}^*, s_1) = 5$ and $w(T_{s_1}^*, s_2) = 9$ and $T_{s_2}^* = \{\{A, D\}, \{B, C\}, \{C, D\}\}$ with costs $w(T_{s_2}^*, s_1) = 9$ and $w(T_{s_2}^*, s_2) = 5$. The optimal spanning tree T^* for the MMR-MST problem is $T^* = \{\{A, D\}, \{A, C\}, \{B, C\}\}$ with costs $w(T^*, s_1) = 7$ and $w(T^*, s_2) = 6$ and $R(T^*) = \max(7 - 5, 6 - 5) = 2$.

Note that the regret value of the minimum spanning tree for each scenario may be arbitrarily far from the optimal regret value, as shown in the following example.

Example 2. Figure 1, right, shows a graph with 4 vertices and two scenarios, $S = \{s_1, s_2\}$. Assume that $0 < \epsilon < 1$ and $\gamma > 1$ is a large integer value. Then, we have that $T_{s_1}^* = \{\{A, D\}, \{B, C\}, \{C, D\}\}$ with costs $w(T_{s_1}^*, s_1) = 1 + 2\epsilon$ and $w(T_{s_1}^*, s_2) = \gamma + 1 + \epsilon$, and $T_{s_2}^* = \{\{A, B\}, \{A, D\}, \{B, C\}\}$ with costs $w(T_{s_2}^*, s_1) = \gamma + 1 + \epsilon$ and $w(T_{s_2}^*, s_2) = 1 + 2\epsilon$. Then, $R(T_{s_1}^*) = R(T_{s_2}^*) = \gamma - \epsilon$. The spanning tree T^* with optimal regret value is $T^* = \{\{A, C\}, \{A, D\}, \{B, C\}\}$ with costs $w(T^*, s_1) = w(T^*, s_2) = 2 + \epsilon$ and $R(T^*) = 1 - \epsilon$.

The discrete case differs from the interval scenario case, in which the cost of each edge is defined by an interval; see formulation and algorithms for the MMR-MST problem in the interval case in [11,12,17] as well as its application to other optimization problems [5,13]. Unfortunately, it is not possible to use these approaches to solve the discrete case.

Solution approaches for the discrete case are rather scarce. It is known that if the number of scenarios is bounded by a constant, then the MMR-MST problem is weakly NP-hard [2]. However, if the number of scenarios is unbounded, the problem becomes strongly NP-hard [2]. A general pseudo-polynomial time algorithm is discussed in [2], which consists of solving a sequence of feasibility problems on a transformed graph that is obtained by scalarizing the costs of each edge. Each feasibility problem is defined by a given parameter v and consists of determining, for increasing value of this parameter, if there exists a solution with value v; once a spanning tree is found with value v, then it is also optimal for the original problem with respect to the min-max regret criterion. Since the time complexity depends of the range of parameter v, this approach becomes pseudo-polynomial if the feasibility problem can be solved in (pseudo-)polynomial time.

The MMR-MST problem is closely related to the multiobjective minimum spanning tree problem, that is, at least one optimal spanning tree for the MMR-MST problem is also optimal for the multiobjective version [2]. This result suggests that solving the latter would allow to solve the MMR-MST problem. However, the multiobjective minimum spanning tree is also known to be NP-hard [15]. Still, this relation may be useful for deriving search strategies for the MMR-MST. In our case, we derive a pruning technique for this problem using techniques that are known to work very well in implicit enumeration approaches for multiobjective optimization [7,16]. In particular, we derive a lower bound that combines the bounds for each scenario and use it within a branch and bound approach to discard partial spanning trees that provably do not lead to an optimal spanning tree. Moreover, we develop an upper bound that consists of a solution to a scalarized problem of the multiobjective minimum spanning tree problem.

2 A Framework for a Branch and Bound

To solve the MMR-MST problem within the framework of a branch and bound algorithm for a given graph $G = (V, E)$, we define a subproblem P in terms of a pair (C_P, D_P), where C_P is the set of edges of a subtree rooted at a vertex $v \in V$ and D_P is another set of edges disjoint with C_P. If, in a subproblem P, the set C_P is a spanning tree of G, then P is a terminal subproblem and C_P is a solution to the original problem. Otherwise, if $E \setminus (C_P \cup D_P) \neq \emptyset$, the branch and bound selects an edge e from this set such that e is incident to another edge in C_P and defines two child subproblems: (i) augmenting C_P with e; (ii) augmenting D_P with e. Note that subproblem P is solved if both children subproblems are also solved.

In the next section, we discuss extensions of this framework by deriving lower and upper bounding procedures, which should reduce the number of subproblems to visit and, consequently, to improve the running time.

2.1 A Lower Bound

For a subproblem P defined by (C_P, D_P) and a scenario $s \in S$, we define the cost $w(P, s)$ as follows

$$w(P, s) := \sum_{e \in C_P} w(e, s) \tag{4}$$

Let C_P^* be a spanning tree that minimizes the min-max regret function for which $C_P \subseteq C_P^*$ and let $R(C_P^*)$ be its regret function value. A trivial lower bound can be derived as follows:

$$LB1(P) := \max_{s \in S} (w(P, s) - w(T_s^*, s)) \tag{5}$$

where T_s^* denotes a minimal spanning tree of the edge set E for scenario $s \in S$.

A tighter lower bound can be obtained. Let $\hat{T}_s$ be a minimum spanning tree of the edge set $E \setminus (C_P \cup D_P)$ for the scenario $s \in S$ with value $w(\hat{T}_s, s)$. Then, we have the following inequality:

$$LB1(P) \leq LB2(P) := \max_{s \in S} \left(w(P, s) + w(\hat{T}_s, s) - w(T_s^*, s) \right) \leq R(C_P^*) \tag{6}$$

Note that finding $\hat{T}_s$ involves solving a minimum spanning tree problem. However, this also implies that $LB2$ is more computational demanding than $LB1$.

2.2 An Upper Bound

We build a new graph $G' = (V, E)$, where the cost of each edge e is a weighted sum of the edge costs for the several scenarios in graph G. Then, the minimum spanning tree of G' is given as follows.

$$T_\delta^* := \arg\min_{T \in \mathcal{T}} \sum_{e \in T} \sum_{s \in S} \delta_s \cdot w(e, s) \tag{7}$$

where $\sum_{s \in S} \delta_s = 1$. This is, in fact, a scalarization of the multiobjective minimum spanning tree problem [6]. Since T_δ^* is a spanning tree, we have the following inequality

$$R(T^*) \leq R(T_\delta^*) \tag{8}$$

where T^* denotes an optimal spanning tree for the MMR-MST problem.

Unfortunately, an optimal solution for this problem may not be optimal for any scalarized problem in the form of Eq. (7), as shown in the following example.

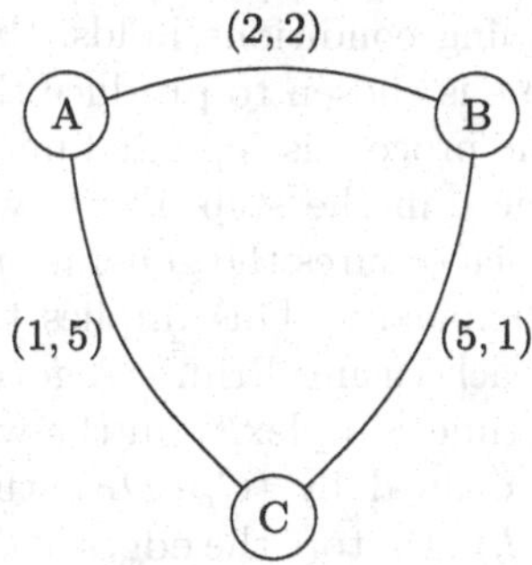

Fig. 2. An instance for Example 3

Example 3. Figure 2 shows a graph with 3 vertices and two scenarios, $S = \{s_1, s_2\}$. There exists three solutions: $T_1 = \{(A,B),(A,C)\}$ $T_2 = \{(A,B),(B,C)\}$ and $T_3 = \{(A,C),(B,C)\}$, with $w(T_1,s_1) = w(T_2,s_2) = 3$, $w(T_1,s_2) = w(T_2,s_1) = 7$ and $w(T_3,s_1) = w(T_3,s_2) = 6$. Then, the spanning tree with optimal regret value is T_3 with $R(T_3) = 3$. However, there exists no non-negative vector $\delta = (\delta_1, \delta_2)$ for which T_3 is optimal for the scalarized problem in Eq. (7).

This example shows that general approaches that enumerate all optimal solutions to scalarized problems, such as in [14], may fail to find the optimal solution for the min-max regret version.

3 A Combinatorial Branch and Bound Algorithm

Let $\mathcal{P}$ denote the set of active subproblems, which is initially $\{(\emptyset,\emptyset)\}$. At the beginning, the incumbent solution is defined as $(R^b, T^b) := (\infty, \emptyset)$, where T^b stands for the best spanning tree obtained so far and $R^b = R(T^b)$ is the regret function value of that spanning tree. Alternatively, R^b can be an upper bound computed in a pre-processing step as described in Sect. 2.2.

For a given subproblem P, the algorithm proceeds as follows. First, the feasibility of subproblem P is verified. Note that due to the edges in D_P, P may not be able to form a spanning tree, and, in that case, the subproblem is terminated. Next, the algorithm determinates if P is a terminal subproblem, in which case, the incumbent is updated if necessary and the subproblem terminates. If none of the two cases above applies, then, the following condition is verified

$$LB1(P) \geq R^b \tag{9}$$

If it fails, the following condition is also verified

$$LB2(P) \geq R^b \tag{10}$$

If any of the two above pruning conditions holds, the subproblem is also terminated. Otherwise, an edge e is chosen to produce the two child subproblems, which are added to $\mathcal{P}$ and the process is repeated until $\mathcal{P}$ is empty.

For the choice of the edge e in the step above, we follow the enumeration approach described in [8], which ensures the generation of unique spanning trees in $O(n + m + n|\mathcal{T}|)$ time complexity. This implies that, for complete graphs, our branch and bound approach (using Prim's Algorithm with Binary heap to compute $LB2$) has $O(n^{n+1})$ time complexity in the worst case.

For a given subproblem defined by (C_P, D_P), the enumeration algorithm keeps three lists, L_C, L_D and L_E, to store the edges in C_P, D_P, and $E\backslash(C_P \cup D_P)$, respectively. Then, the algorithm is called recursively using a depth-first search traversal to find all spanning trees that contain the subtree represented by the edges in C_P. It starts by selecting an edge e in L_E that is incident to an edge in L_C, adding it to the latter set. Then, all spanning trees that contain edge e are found recursively. Next, this edge is removed from L_C and inserted into L_D, another edge as above is selected and the same steps are repeated until a *bridge* is found in the graph $\bar{G}$ with edge set E excluding the edges in L_D, that is, $\bar{G}$ becomes disconnected if the selected edge is removed. This implies that all spanning trees containing the subtree defined by C_P have been found. The authors in [8] discuss an efficient method to detect a bridge in $\bar{G}$ based on the depth-first search tree that is constructed recursively, as well as appropriate data structures for L_C, L_D and L_E. In our branch and bound, the bounding conditions (9) and (10) are tested whenever a new edge is inserted into L_P.

4 Numerical Experiments

In the following, we report an experimental analysis of our approach on a wide range of instances of the MMR-MST problem. Three types of experiences were conducted: First, we analyse the performance of our approach with different pruning strategies on complete graphs, which corresponds to a worst case in practical performance, and two scenarios. Then, we study the approach with the best pruning strategy on different type of graphs with different edge densities and up to four scenarios. Finally, we compare it with the state-of-the-art algorithm proposed in [3].

The instances size ranged from 5 to 20 vertices. For the second experiment we consider three different type of instances:

- **Random:** Given n vertices, each pair of vertices has a given probability of being connected by an edge.
- **Bipartite:** Given two sets of vertices, of sizes $\lfloor n/2 \rfloor$ and $\lceil n/2 \rceil$, respectively, each pair of vertices, each of which in a distinct set, has a given probability of being connected by an edge.
- **Planar:** Given n points in the two-dimensional plane, the edges correspond to a Delaunay triangulation.

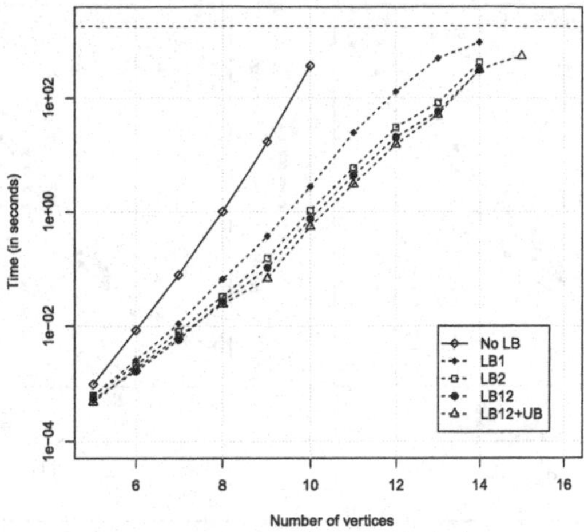

Fig. 3. Average CPU-time (in seconds) for different pruning strategies

The edge weights are generated randomly according to an uniform distribution within an predefined interval, bounded from above by 1000. We generated 10 instances for each size. The branch and bound was implemented in C++, and compiled with g++, version 5.4.0. The optimal solution for each scenario and the computation of $LB2$ were obtained with our implementation of Prim's algorithm. (it took less than 0.001 s on instances of 300 vertices) The implementations were tested in a computer with Ubuntu 16.04 operating system, 4 GB RAM and a single core CPU with 2 virtual threads and a clock rate of 2 GHz. For each run, we collected the CPU-time taken by each approach. We defined a time limit of 1800 s.

4.1 Effect of Pruning

We investigated the effect of the pruning conditions described in Sect. 2.1. We considered four versions: using only LB1, using only LB2, using both lower bounds sequentially, as explained in Sect. 3, with and without the upper bound explained in Sect. 2.2 (LB12 and LB12+UB, respectively). For the latter, some preliminaries experiments suggested that a good performance would be obtained by considering the minimum regret value from a sequence of $k+1$ scalarized problems with the following scheme for $\delta = (\delta_1, \ldots, \delta_k)$, given a number k of scenarios:

$$\delta^1 := \left(\frac{1}{k}, \frac{1}{k}, \ldots, \frac{1}{k}\right) \qquad \delta^2 := \left(\frac{k+1}{2k}, \frac{1}{2k}, \ldots, \frac{1}{2k}\right)$$

$$\delta^3 := \left(\frac{1}{2k}, \frac{k+1}{2k}, \ldots, \frac{1}{2k}\right) \quad \ldots \quad \delta^{k+1} := \left(\frac{1}{2k}, \frac{1}{2k}, \ldots, \frac{k+1}{2k}\right)$$

We also tested a further approach that does not test any pruning condition in order to be used as a reference (No LB). Figure 3 shows the mean CPU-time for

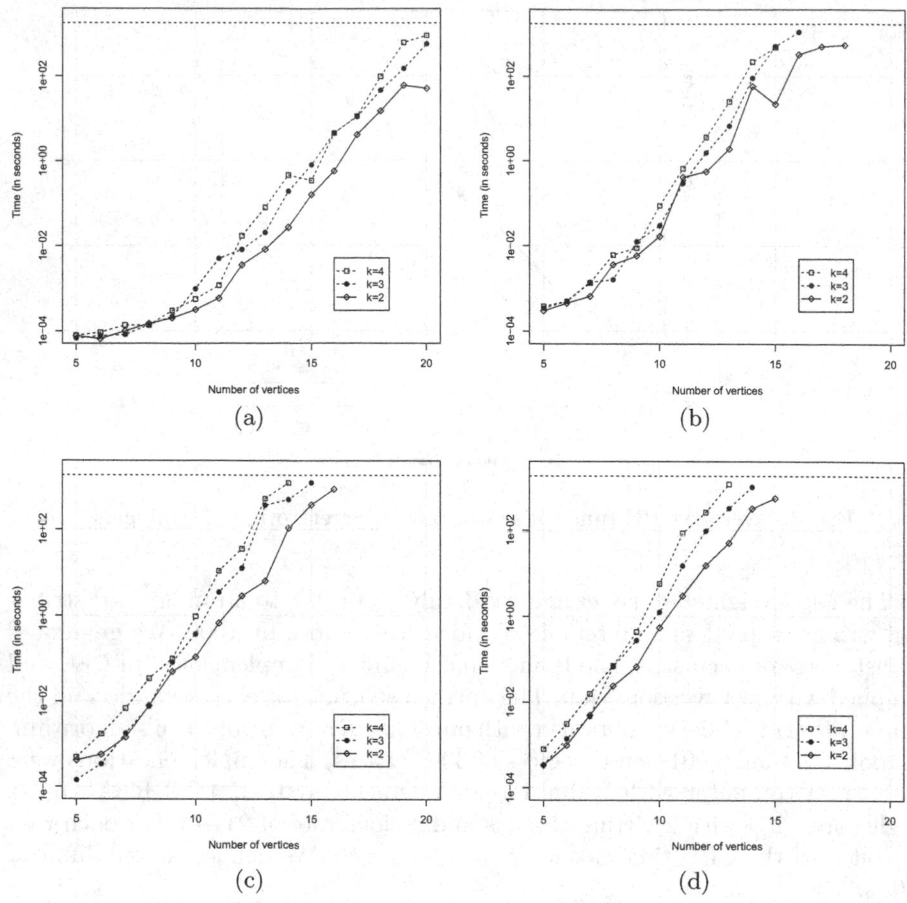

Fig. 4. Average CPU-time (in seconds) of variant LB12+UB for different number of scenarios and edge densities 0.25 (a), 0.5 (b), 0.75 (c) and complete graphs (d) on random instances

the branch and bound implementations. The results indicate that the pruning test is very effective; for instances with 10 vertices, the branch and bound implementation is approximately 100 faster than performing a complete enumeration of all spanning trees. In addition, as the instance size grows, the use of both lower bounds gives better performance than each isolatedly. Note that, for the instance sizes considered, computing LB2 amounts to a small overhead. Moreover, our approach performs approximately 20% faster with the upper bound (LB12+UB). Note that the slope change on the right tail of the lines is due to a truncated mean, since some of the runs exceeded the time limit of 1800 s (horizontal dashed line) on larger instances (1/10 and 6/10 for $n = 13$ and $n = 14$, respectively, with LB1, 1/10 for $n = 14$ with LB2 and LB12, and 6/10 for $n = 15$ with L12+UB).

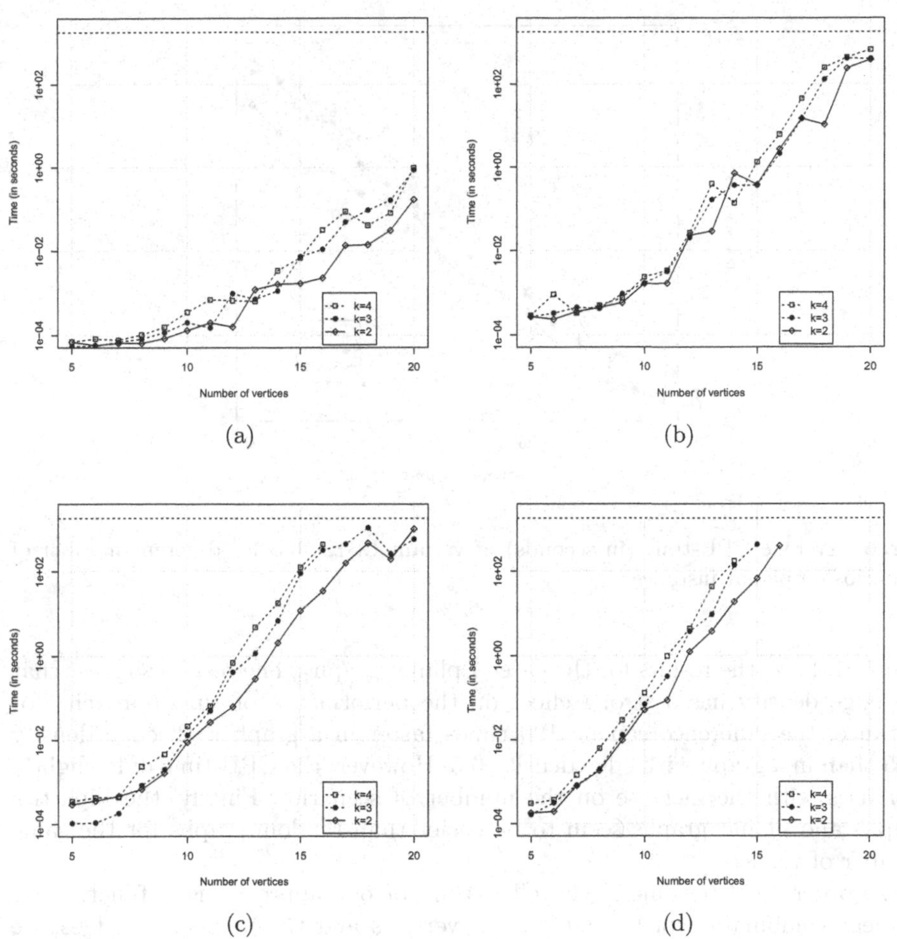

Fig. 5. Average CPU-time (in seconds) of variant LB12+UB for different number of scenarios and edge densities 0.25 (a), 0.5 (b), 0.75 (c) and complete graphs (d) on bipartite instances

4.2 Effect of Edge Density and Number of Scenarios

In the second experiment, we generated random, planar and bipartite instances with 2, 3 and 4 scenarios and with edge densities $d = \{0.25, 0.5, 0.75\}$ for random and bipartite instances; enough edges where added to the final graph in order to ensure connectedness between any pair of vertices. In addition, we also considered complete random and complete bipartite graphs. Figures 4 and 5 show the results of the branch and bound implementation with pruning strategy LB12+UB on random and bipartite graphs, respectively, with different edge densities and complete graphs (a–d, respectively) and with different scenarios.

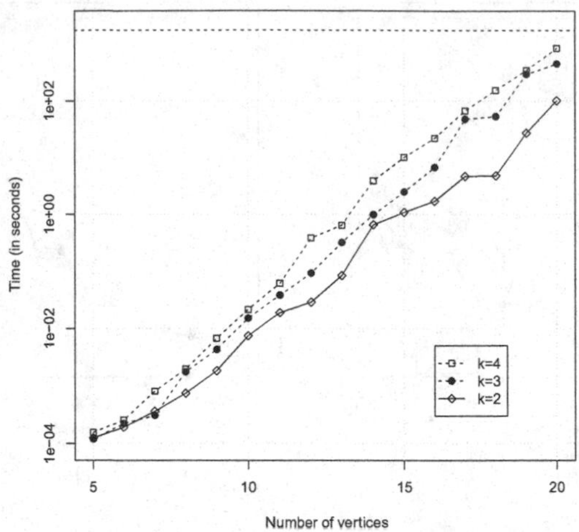

Fig. 6. Average CPU-time (in seconds) of variant LB12+UB for different number of scenarios on planar instances

Figure 6 shows the results for the case of planar graphs. The results suggest that the edge density has a strong effect on the performance of our approach. For instance, the difference can be 100 times faster in a graph with edge density 0.25 than in a graph with edge density 0.5. However, the CPU-time only slightly increases with the increase on the number of scenarios. Finally, the bipartite graphs and planar graphs seem to be easier than random graphs for the same number of vertices.

In order to determine if the CPU-time of our approach is a function on a linear combination of the number of vertices and the number of edges, we performed a multiple linear regression with n and m as independent variables and CPU-time as dependent variable. Given that the results did not suggest a strong dependency of CPU-time with respect to the number of scenarios, we did not consider the later as third independent variable. Only CPU-times of runs that ended before the cut-off time of 1800 s were considered in the regression. A Boxcox procedure suggested a logarithmic transformation on the CPU-time, which corresponds to an exponential relationship between CPU-time required by our approach with respect to a linear combination of n and m. We arrived to the following model:

$$\log t = -11.64 + 0.26n + 0.046m$$

with $R^2 = 0.91$, which suggested a good fit. Figure 7 shows the CPU-time collected in all runs (in logarithmic scale) vs. linear combination on the number of vertices and edges, as suggested by the model above.

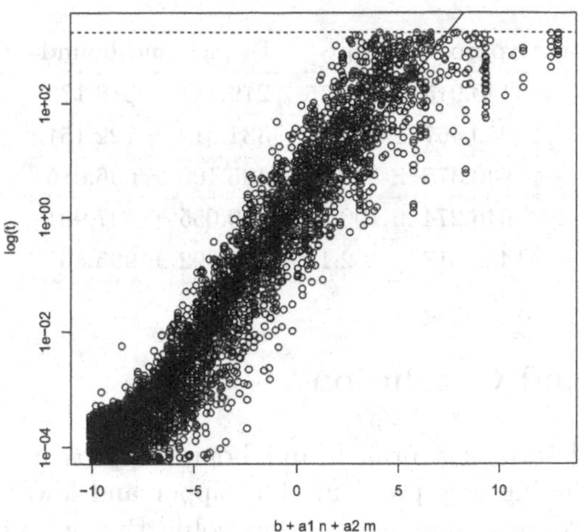

Fig. 7. CPU-time in logarithm scale vs. linear combination on the number of vertices and edges and regression line

4.3 Comparison with the Approach in [3]

We compared our approach LB1+2+UB to the generic pseudo-polynomial algorithm in [3], as described in Sect. 1. Instead of solving each feasibility problem, we implemented the approach described in [4] as suggested in [3], which allows to retrieve the number of spanning trees for each possible value from the coefficients of a polynomial expression. This expression is retrieved from the determinant of a polynomial matrix of size $(n-1) \times (n-1)$ built from the adjacency matrix of the transformed graph instance. Therefore, once the polynomial expression is obtained, each feasibility problem for a parameter v is solved by verifying whether the coefficient of the v-th term of the polynomial expression is positive. We used the `GiNaC C++` libraries [1] to generate the polynomial matrix as well as to compute the polynomial determinant. The overall implementation was implemented in `C++` and tested in the same computational environment. The main bottleneck of this approach is on computation of the determinant, which did not allow us to test for $v > 8$ in less than one hour of CPU-time. Table 1 shows the average and standard deviation for both approaches for 100 instances of size 15 and maximum edge weight value ranging from 4 to 8 (column max). It is possible to observe a significant difference between both algorithms when the maximum edge weight is 7 and 8 and that our approach is less sensible to the change of this instance parameter.

Table 1. Average and standard deviation of CPU-time for the pseudo-polynomial approach in [3] and our branch and bound

max	Approach in [3]	Branch and bound
4	140.915 ± 11.375	219.141 ± 518.135
5	289.467 ± 26.243	331.819 ± 722.151
6	540.375 ± 51.108	405.109 ± 696.086
7	946.274 ± 100.143	599.055 ± 947.901
8	1452.317 ± 162.189	657.392 ± 995.465

5 Discussion and Conclusion

In this article, we described a branch and bound approach for the min-max regret minimum spanning tree problem. The upper and lower bounding techniques are based on known approaches to multiobjective optimization and they can be applied to other combinatorial optimization problems, such as shortest path or knapsack problem. The experimental analysis shows that the sequential use of the two lower bounds proposed in this article, together with an upper bound, brings considerable improvement in terms of running time. In addition, we experimentally compared our approach with a pseudo-polynomial algorithm. The results suggest that our approach may be preferable in terms of running time for increasing maximum edge cost.

Our branch and bound algorithm should perform even faster if the lower bound $LB2$ is computed in an incremental manner using the lower bound computed in the previous recursive call. If, for a given edge $e \in E \setminus (C_P \cup D_P)$, the decision is either to augment C_P or D_P with e (see Sect. 2) and if e is incident to a leaf of the spanning tree T' corresponding to the previous lower bound, the new lower bound can easily be computed by removing the cost of the edge that connects that leaf to T'. This option was not considered in our implementation since the computation of $LB2$ was not a bottleneck for the instance sizes considered in our experimental analysis. However, one main drawback of our approach is the large variance on the running time. We observed that a few runs took much more time than the remaining, which may be related to a less tight upper bound. One possibility of reducing this variance is to perform random restarts, as suggested in [9], and/or to consider several upper bounds simultaneously, obtained by different scalarizations in Eq. (7).

Acknowledgments. This work was carried out in the scope of the MobiWise project: From mobile sensing to mobility advising (P2020 SAICTPAC/0011/2015), co-financed by COMPETE 2020, Portugal 2020 - POCI, European Union's ERDF.

References

1. Ginac C++ libraries. https://www.ginac.de/
2. Aissi, H., Bazgan, C., Vanderpooten, D.: Approximation complexity of min-max (regret) versions of shortest path, spanning tree, and knapsack. In: Brodal, G.S., Leonardi, S. (eds.) ESA 2005. LNCS, vol. 3669, pp. 862–873. Springer, Heidelberg (2005). https://doi.org/10.1007/11561071_76
3. Aissi, H., Bazgan, C., Vanderpooten, D.: Pseudo-polynomial time algorithms for min-max and min-max regret problems. In: Proceedings of the 5th International Symposium on Operations Research and Its Applications, pp. 171–178 (2005)
4. Barahona, J.F., Pulleyblank, R.: Exact arborescences, matching and cycles. Discrete Appl. Math. **16**, 91–99 (1987)
5. Chassein, A., Goerigk, M.: On the recoverable robust traveling salesman problem. Optim. Lett. **10**(7), 1479–1492 (2016)
6. Ehrgott, M.: Muticriteria Optimization, 2nd edn. Springer, Heidelberg (2005). https://doi.org/10.1007/3-540-27659-9
7. Figueira, J.R., Paquete, L., Simões, M., Vanderpooten, D.: Algorithmic improvements on dynamic programming for the bi-objective {0, 1} knapsack problem. Comput. Optim. Appl. **56**(1), 97–111 (2013)
8. Harold, N.G., Myers, E.W.: Finding all spanning trees of directed and undirected graphs. SIAM J. Comput. **7**(3), 280–287 (1978)
9. Gomes, C.P., Selman, B., Crato, N., Kautz, H.: Heavy-tailed phenomena in satisfiability and constraint satisfaction problems. J. Autom. Reason. **24**(1–2), 67–100 (2000)
10. Kouvelis, P., Yu, G.: Robust Discrete Optimization and Its Applications. Kluwer Academic Publishers, Dordrecht (1997)
11. Makuchowski, M.: Perturbation algorithm for a minimax regret minimum spanning tree problem. Oper. Res. Decis. **24**(1), 37–49 (2014)
12. Montemanni, R., Gambardella, L.M.: A branch and bound algorithm for the robust spanning tree problem with interval data. Eur. J. Oper. Res. **161**(3), 771–779 (2005)
13. Montemanni, R., Gambardella, L.M., Donati, A.V.: A branch and bound algorithm for the robust shortest path problem with interval data. Oper. Res. Lett. **32**(3), 225–232 (2004)
14. Przybylski, A., Gandibleux, X., Ehrgott, M.: A recursive algorithm for finding all nondominated extreme points in the outcome set of a multiobjective integer programme. INFORMS J. Comput. **22**(3), 371–386 (2010)
15. Serafini, P.: Some considerations about computational complexity for multi objective combinatorial problems. In: Jahn, J., Krabs, W. (eds.) Recent Advances and Historical Development of Vector Optimization. LNEMS, vol. 294, pp. 222–232. Springer, Heidelberg (1987). https://doi.org/10.1007/978-3-642-46618-2_15
16. Sourd, F., Spanjaard, O.: A multiobjective branch-and-bound framework: application to the biobjective spanning tree problem. INFORMS J. Comput. **20**(3), 472–484 (2008)
17. Yaman, H., Karasan, O.E., Pinar, M.Ç.: The robust spanning tree problem with interval data. Oper. Res. Lett. **29**, 31–40 (2001)

Navigating a Shortest Path with High Probability in Massive Complex Networks

Jun Liu[1,2(✉)], Yicheng Pan[3], Qifu Hu[1,2], and Angsheng Li[3]

[1] State Key Laboratory of Computer Science, Institute of Software, Chinese Academy of Sciences, Beijing, China
{ljun,huqf}@ios.ac.cn

[2] University of Chinese Academy of Sciences, Beijing, China

[3] State Key Laboratory of Software Development Environment, Beihang University, Beijing, China
{yichengp,angsheng}@buaa.edu.cn

Abstract. In this paper, we study the problem of point-to-point shortest path query in massive complex networks. Nowadays a breadth first search in a network containing millions of vertices may cost a few seconds and it can not meet the demands of real-time applications. Some existing landmark-based methods have been proposed to solve this problem in sacrifice of precision. However, their query precision and efficiency is not high enough. We first present a notion of *navigator*, which is a data structure constructed from the input network. Then *navigation algorithm* based on the navigator is proposed to solve this problem. It effectively navigates a path only using local information of each vertex by interacting with navigator. We conduct extensive experiments in massive real-world networks containing hundreds of millions of vertices. The results demonstrate the efficiency of our methods. Compared with previous methods, ours can navigate a shortest path with higher probability in less time.

Keywords: Navigation · Shortest path · Massive networks

1 Introduction

With the rapid development of Internet technology, a large amount of data will be produced every day. The data contains much valuable information but analysing it is a great challenge due to their tremendous amounts. Graph is one of the effective ways to organize the data because it contains the relationship between entities. There are various kinds of graphs in real world. For example, in a social graph, users are considered as vertices and edges represent friend relationships between each other; in a web graph, web-pages are regarded as vertices and an edge between two vertices represents that one web-page can be accessed from the other via the hyper-links.

This work is supported by the National Basic Research Program of China No.2014CB340302 and the National Nature Science Foundation of China No.61772503.

I. Kotsireas et al. (Eds.): SEA[2] 2019, LNCS 11544, pp. 82–97, 2019.
https://doi.org/10.1007/978-3-030-34029-2_6

Shortest path or distance computation is a greatly important and fundamental problem in graphs. It has various applications. The distance in social graphs indicates how close the relationship is. The shortest path and its length in metabolic networks are used to identify an optimal pathway and valid connectivity [17]. There are also indirect applications of shortest path and distance such as locating influential users in networks [2] and detecting information source based on a spreading model [24,25] etc.

1.1 Related Works

In the last decades, many methods have been proposed both from theoretical and experimental communities to answer the distance query efficiently. In general, an auxiliary data structure called *distance oracle* is constructed during the preprocessing phase and, then it is used to answer the distance query. Thorup and Zwick [19] showed that for an undirected weighted graph $G = (V, E)$ with n vertices and m edges and for every natural number $k \geq 1$, there is a data structure S that is constructed from G in time $O(kmn^{1/k})$ and space $O(kn^{1+\frac{1}{k}})$ such that every distance query can be answered in time $O(k)$, and the approximate distance returned from the query is within $(2k - 1)$ times of the length of the shortest paths. Wulff-Nilson [21] improved the query time in Thorup and Zwick [19] from $O(k)$ to $O(\log k)$. Chechik [5] further improved the query time in [21] from $O(\log k)$ to $O(1)$. Derungs, Jacob and Widmayer [8] proposed the notion of index graph and a path finding algorithm such that both the index set and the algorithm run in sublinear time for any pair of vertices. The algorithm finds a path between the vertices in the graph by interacting with the index set such that the length of the result path is at most $O(\log n)$ times of the length of the shortest paths.

Among the experimental community, some algorithms have been proposed to answer the shortest distance query approximately [11,15,16,20]. In general, they also construct an auxiliary data structure from G. Some vertices are selected heuristically as landmarks. The distances from every vertex to the landmarks are computed and recorded. Then for every vertex pair, their distance can be estimated by using the stored information combined with the triangle inequality. If a shortest path is wanted [11,20], then the paths rather than only the distances have to be stored in the auxiliary data structure.

To measure the shortest path or distance in computer networks, *graph coordinate systems* [6,22,23] have been proposed in which a metric space is embedded in the graph and each vertex is assigned a coordinate. In [6,22,23], they embed each vertex into a metric space such as Hyperbolic space and Euclidean space and then their metric functions are used to estimate the distance between any pair of vertices.

Search-based A* method *ALT* [10] and other goal-oriented pruning strategies [13,14] were proposed for road networks to accelerate the shortest path query. They use the special properties of road networks such as near planarity, low vertex-degree and the presence of a hierarchy based on the importance of roads. However, it is different from the complex networks used in our experiments.

1.2 Our Methods

We present a notion named *navigator*. It serves in a *navigation algorithm*, in which, we are able to find the approximately shortest path efficiently from any vertex to another by interacting with the navigator using only *local information*. Besides the target, the local information involves only the neighbours of current vertex during the path exploration. The distance of any neighbour to the target can be queried by the navigator and finally a path can be explored by this interaction. We firstly present the formal definition of navigator as follows.

Definition 1 *(Navigator). Given a connected graph $G = (V, E)$,* ***Navigator*** *is a kind of data structure constructed from G which satisfies the conditions that, for an arbitrary vertex pair $u, v \in V(u \neq v)$,*

(i) it gives an approximate answer $\delta(u, v)$ for the distance query efficiently.
(ii) there always exists at least a neighbour w of u such that $\delta(w, v) < \delta(u, v)$.

Actually, the first property of navigator is the goal of *distance oracle*. However, the greedy strategy, i.e., picking in each step the neighbour that has the shortest distance to the target answered by distance oracle, may *not* lead to a path to the target. For example, in a graph with a cut vertex u whose removal makes the graph into two parts S and T, for two vertices $s \in S$ and $t \in T$, note that every path from s to t passes through u. Since the distances answered by distance oracle are approximate, it is possible that a neighbour of u in S has smaller distance (to t) answers than all those in T, which makes the greedy strategy choose that vertex when it reaches u and thus fails in finding a path from s to t.

To overcome this defect, we add the second condition to navigator's definition. Then under this condition, the greedy strategy will get a smaller distance answer than that in the previous step, which guarantees the halt of the greedy strategy. If a distance oracle can give an exact answer to the distance query, then it will be a natural navigator. However, as far as we know, exact distance oracles are only studied in some special class of graphs such as planar graphs [7,9]. Most approximate distance oracles may not meet the second demand. By using the navigator, a navigation algorithm is formulated naturally as follows.

Definition 2 *(Navigation algorithm). Given a connected graph $G = (V, E)$, and a navigator constructed from G.* ***Navigation algorithm*** *is conducted as follows: during the path exploration, each vertex in each step only has the knowledge of the set of its local neighbours and the target. Then it selects the nearest neighbour to the target as the vertex in next step through interacting with the navigator.*

In this paper, we propose some novel navigation algorithms based on navigator to solve the point-to-point shortest path query problem, which will be introduced in Sect. 3.2. Extensive experiments have been conducted in massive networks. The results demonstrate that our methods have higher precision and better query efficiency. In some instances, the query time of our method is about 10 times faster than previous methods and the precision of the query results is about 20% higher on average.

1.3 Outlines

In Sect. 2, we give an overview of some existing landmark-based methods. Then we present our algorithms in Sect. 3 in detail. In Sect. 3.1, we will describe the method to construct the navigator. Then the navigation algorithm will be introduced in Sect. 3.2. In Sect. 4, we show the experimental results. Finally, we give our conclusion in Sect. 5.

2 Preliminary

Let $G = (V, E)$ denote a graph with $n = |V|$ vertices and $m = |E|$ edges. For simplicity of description, we shall consider an undirected and unweighted graph, although our method can be applied to weighted directed graphs as well.

A *path* $\pi(s,t)$ of length l between two vertices $s, t \in V$ is defined as a sequence $\pi(s,t) = (s, u_1, u_2, ..., u_{l-1}, t)$. We denote the length l of a path $\pi(s,t)$ as $|\pi(s,t)|$. The concatenation of two paths $\pi(s,r) = (s, ..., r)$ and $\pi(r,t) = (r, ..., t)$ is the combined path $\pi(s,t) = \pi(s,r) + \pi(r,t) = (s, ..., r, ..., t)$. The shortest distance $d_G(s,t)$ between vertices s and t in G is defined as the length of the shortest path between s and t. Then it satisfies the triangle inequality: for any $r \in V$,

$$|d_G(s,r) - d_G(r,t)| \leq d_G(s,t) \leq d_G(s,r) + d_G(r,t). \tag{1}$$

These two bounds can both be used to estimate the shortest distance between any pair of vertices. But previous work [15] indicates that lower-bound estimates are not as accurate as the upper-bound ones. Furthermore, If a vertex set $R = \{r_1, r_2, ..., r_k\}$ is selected as the landmarks, then more accurate estimation can be obtained through:

$$\delta(s,t) = \min_{r_i \in R}\{d_G(s,r_i) + d_G(r_i,t)\}. \tag{2}$$

Note that this approach only allows us to compute an approximate distance, but does not provide a way to retrieve the path itself. In order to retrieve a path, a direct way is the concatenation of the path to the landmark in the shortest path trees $T = \{T_1, T_2, ..., T_k\}$ rooted by $R = \{r_1, r_2, ..., r_k\}$ respectively. So during the construction of each T_i, in each vertex v, we not only store the distance to the landmark r_i, but also the shortest path $P_{T_i}(v, r_i)$ in T_i. Then the path can be estimated by:

$$|\pi^T(s,t)| = \min_{r_i}\{|\pi^{T_i}(s,r_i) + \pi^{T_i}(r_i,t)|\}, \tag{3}$$

However, it is not accurate enough. Some improvements have been proposed [11,20]. They utilize the constructed shortest path trees further, such as finding the lowest common ancester and possible shortcut, eliminating a possible cycle. Their best techniques are called *TreeSketch* [11] and *Landmarks-BFS* [20], both of which use online search to improve the accuracy. *TreeSketch* is a sketch-based method. The sketch of each vertex v stores the shortest path to the landmarks and it can be organized as a tree rooted by v denoted by T_v, in which the landmarks are the leaf vertices. For a query $q = (s, t)$, *TreeSketch* performs a bidirectional expansion on T_s and T_t following a breadth-first search order. Let V_s and V_t denote the sets of visited vertices from two sides, respectively. Consider $u \in T_s$ and $v \in T_t$ that are two vertices under expansion in the current iteration. *TreeSketch* checks if there is an edge from u to a vertex in V_t or from a vertex in V_s to v. If it is true, then a path between s and t is found and added to a queue Q. Denote the length of the shortest path in Q by $l_{shortest}$, the algorithm terminates if $d(s,u)+d(v,t) \geq l_{shortest}$. For a query between s and t, *Landmarks-BFS* uses the vertices in the set of paths $(\pi^{T_1}(s,r_1) + \pi^{T_1}(r_1,t), \pi^{T_2}(s,r_2) + \pi^{T_2}(r_2,t), ..., \pi^{T_k}(s,r_k) + \pi^{T_k}(r_k,t))$ to construct an induced subgraph G' from G and then a BFS routine is conducted in G' to find a path.

3 Algorithm Description

In this section, we give a detailed description of our methods. Firstly we introduce a method to construct the navigator and then describe several navigation algorithms.

3.1 Navigator

As described in Definition 1, a navigator has to be able to answer the distance query efficiently, the same as distance oracle does. For the second condition, let us consider the shortest path tree T of G. We say that the tree path (or T-path π^T when the tree is denoted by T) between two vertices is the unique path between them on the tree. The length of this tree path is called the tree distance. Note that the T-paths in any branch of T are exactly shortest paths in G and the distance in T can be easily and efficiently obtained.

Suppose that we are given k shortest path trees (SPTs) $T_1, T_2, \cdots, T_k$ of G. Denote by k-SPT the set of these k SPTs. The distance $\delta(s,t)$ in k-SPT can be defined as:

$$\delta(s,t) = \min_{T_i \in T} |\pi^{T_i}(s,t)| = \min_{c_i}\{|\pi^{T_i}(s,c_i) + \pi^{T_i}(c_i,t)|\}, \tag{4}$$

where c_i is the lowest common ancestor (LCA) between s and t in T_i.

Define the answer to the distance query between two vertices u and v as the distance $\delta(u,v)$ in k-SPT. Then we have the following lemma.

Lemma 1. *For any positive integer k, k-SPT is a navigator.*

Proof. Since the k-SPT answers the distance query by returning the shortest tree distance among the k trees, the first condition of the navigator's definition is satisfied. For the second condition, at any step, suppose that vertex u is being visited and T is the SPT that contains the shortest tree distance. So the neighbour of u that is also on the (unique) path from u to the target on T has shorter tree distance. This meets the second condition of navigator's definition.

Obviously, k-SPT can be constructed in time $O(km)$. Owing to the frequent interactions in the navigation algorithm, the navigator has to be able to answer the distance query fast. We store two parts in each vertex for each SPT, in which the first one is the parent vertex in SPT and the second is the distance to the root as illustrated in Fig. 1. So the space complexity is $O(kn)$. In each T_i rooted by r_i, we store a label $L_i(v) = \{(p_i[v], d_G(v, r_i))\}$ in each vertex v, where $p_i[v]$ indicates the parent vertex in T_i. Then $L(v) = \bigcup L_i(v)$ and $L = \bigcup L(v)$. L is our navigator.

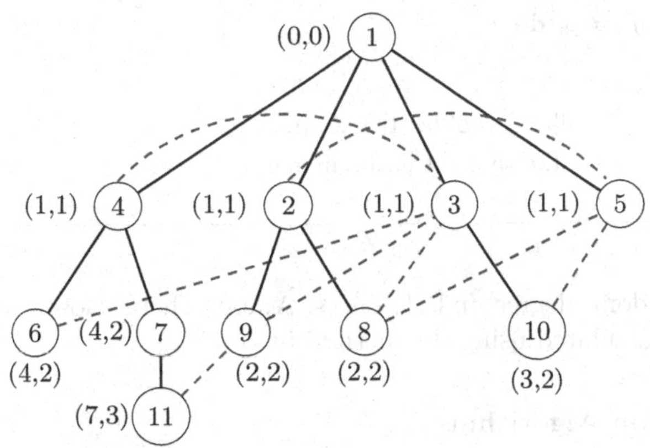

Fig. 1. An example of a navigator containing one SPT T of G. In this figure, the solid lines are the edges in T and the dotted lines are the edges of the graph that are outside of T. The first number in the label of each vertex indicates the parent vertex in T and the second one indicates the shortest distance to the root of T.

Algorithm 1 describes the details of distance query process in the navigator. Actually, it computes the k-SPT distance using the labels stored in each vertex. $L[v][i][0]$ indicates the parent vertex $p_i[v]$ in T_i and $L[v][i][1]$ indicates the distance to r_i in vertex v. For each T_i, we search up to the root using the parent vertex stored in the label both from s and t to find the lowest common ancestor between them. Then we will get the shortest path distance in T_i. Finally we choose a smallest distance in all k-SPT as the output of T-paths distance δ. Suppose that the diameter of the network is D, then the time complexity of navigator is $O(k \cdot D)$. In many real-world complex networks, D is very small. So this operation is quite efficient. Note that if s and t are in the same branch of some T_i, the distance in T_i is already the shortest and we will directly return this distance. In our experiments, this happens quite often for real-world complex networks, which speeds up the query.

If navigator can give a more accurate answer, then navigation algorithm will find a shorter path. Heuristically for k-SPT, the roots play a critical role for its accuracy. Some landmark-based methods [11,15,20] aim to select some vertices with high centrality. We also follow these strategies. Here we choose three

Algorithm 1. compute the approximate distance $\delta(s,t)$ through navigator.

Input: navigator L, vertex s and t.

procedure NAVIGATOR(s,t)
    for $i \in \{1, 2, \cdots, k\}$ **do**
        $p_s \leftarrow L[s][i][0]$, $p_t \leftarrow L[t][i][0]$.
        $d_s \leftarrow L[s][i][1]$, $d_t \leftarrow L[t][i][1]$.
        while $d_s < d_t$ **do**
            $d_t \leftarrow d_t - 1$, $p_t \leftarrow L[p_t][i][0]$.
        while $d_t < d_s$ **do**
            $d_s \leftarrow d_s - 1$, $p_s \leftarrow L[p_s][i][0]$.
        while $p_s \neq p_t$ **do**
            $p_s \leftarrow L[p_s][i][0]$, $p_t \leftarrow L[p_t][i][0]$.
            $d_s \leftarrow d_s - 1$.
        $\delta_i(s,t) = L[s][i][1] + L[t][i][1] - 2 * d_s$.
    Denote $\delta(s,t)$ is the shortest path among all $\delta_i(s,t)$.
    return $\delta(s,t)$

centralities: random, degree and closeness. Among them, closeness centrality is an estimation calculated using the method in [15].

3.2 Navigation Algorithms

In this section, we will introduce some *navigation algorithms* based on navigators. As described in Definition 2, in each step of the path exploration, the vertex selects one or several vertices which are closest to the target among all its neighbours, then it decides which of them to be the vertices in next step. We consider two extreme strategies. The first is that the vertex randomly selects one of them to be the next hop and it will eventually find a unique path. We call this method *navigating single path* (NSP). The second is that the vertex selects all of them and this may lead to multiple paths with the same length. We call this *navigating multiple paths* (NMP). There is only one vertex during each step in NSP but two or more vertices may exist during each step in NMP. So it may cost more time but get better results in NMP. Furthermore, two techniques for improvement will be introduced. One is navigation from both two directions successively. The second is that we collect the vertices explored during the navigation process and construct an induced subgraph on them, in which a BFS routine is conducted to find a path. Both two techniques obviously improve the precision of navigation algorithm.

Navigating Single Path. Algorithm 2 shows an overview on how to navigate a single path. The variable *dist* maintains the distance from current vertex to the target. It will continue to decrease to zero when reaching the target. During each step, the vertex finds a next hop closest to the target among its neighbours.

It seems that someone is walking along the tree path and meanwhile searching among his neighbours in order to find a shortcut to the target. Figure 2 illustrates

Algorithm 2. Navigation single path from source s to target t

Input: Graph $G = (V, E)$, navigator L, vertex s and t.
Output: A path $\pi(s,t)$ between s and t.

```
current_hop ← s.
π(s,t) ← (s).
Query the distance δ(s,t) between s and t through navigator.
dist ← δ(s,t).
while current_hop ≠ t do
    for each neighbour w of vertex current_hop do
        Query the distance δ(w,t) between w and t through navigator.
        if δ(w,t) < dist then
            dist ← δ(w,t).
            next_hop ← w.
    current_hop ← next_hop.
    Append vertex current_hop to π(s,t).
```

an example of Algorithm 2. When we conduct navigation in this graph from source vertex 6 to target vertex 10, we will get a path $(6, 3, 10)$ whose length is 2 because vertex 3 is nearer to the target 10 than vertex 4 in this SPT among vertex 6's neighbours $\{3, 4\}$. Denote by $N(p)$ the neighbours set of the vertices in the path p found by Algorithm 2 and by D the diameter of network G. k-SPT gives an answer in time $O(k \cdot D)$. Then the overall time complexity of Algorithm 2 is $O(k \cdot D \cdot |N(p)|)$.

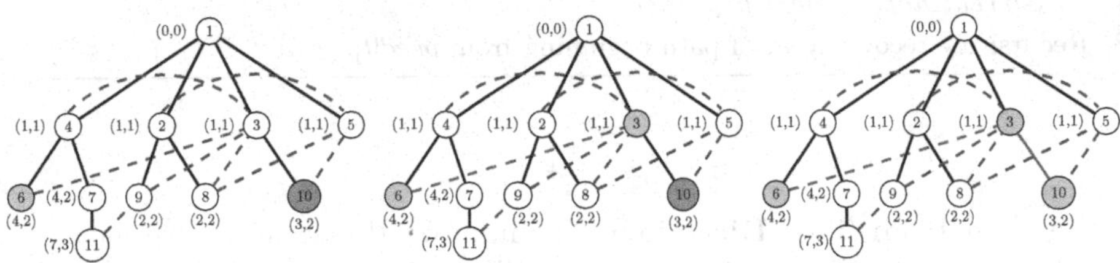

Fig. 2. An example of a navigation process from one direction using a navigator containing one SPT. The red vertices indicate the target and the green vertices indicate the traversed vertices. The navigation process is conducted from left to right. At first, vertex 6 interacts with the navigator to obtain the distance between its neighbours and target 10. Then it selects vertex 3 as the next hop and finally a path to the target 10 is found. The red edges indicate a path from source 6 to target 10. (Color figure online)

Navigating Multiple Paths. Algorithm 3 shows the detail of the algorithm of navigating multiple paths. The difference from Algorithm 2 is that during each step, the current vertex v selects all the vertices who are closest to the target among v's neighbours as the vertices in next step. Then each vertex in the next step repeats the navigation process until reaching the target. So NMP may find multiple paths and explore more vertices to find more shortcuts in order to obtain the shorter paths. In Algorithm 3, *pred*[v] is used to indicate vertex v's predecessors in the multiple paths that we find. So we are able to recover all the paths from s to t by getting t's predecessors recursively.

Algorithm 3. Navigating multiple paths from source s to target t

Input: Graph $G = (V, E)$, navigator L, vertex s and t.
Output: A set of path $\pi(s, t)$.

```
current_hops ← {s}.
Query the distance δ(s, t) between s and t through navigator.
dist ← δ(s, t).
while t ∉ current_hops do
    next_hops ← ∅.
    for each v ∈ current_hops do
        for each neighbour w of vertex v do
            Query the distance δ(w, t) between w and t through navigator.
            if δ(w, t) < dist then
                dist ← δ(w, t).
                next_hops ← {w}.
                pred[w] ← {v}.
            else
                if δ(w, t) = dist then
                    next_hops ← next_hops ∪ {w}.
                    pred[w] ← pred[w] ∪ {v}.
    current_hops ← next_hops.
Recursively recover a set of path beginning from pred[t].
```

Navigation from Two Directions. To improve the accuracy, we navigate from both two directions and may get two different paths and then choose the shorter one as the answer. The two methods described above can both adopt this optimization technique. We call these two methods with improvement *Two-way Navigating Single Path* (TNSP) and *Two-way Navigating Multiple Paths* (TNMP), respectively.

We compare our methods with previous ones by illustrating an example in Fig. 3. In *TreeSketch*, the search proceeds along the path in the tree and it gets a path between vertex 11 and 2 $(11, 7, 4, 1, 2)$ in Fig. 3. In *Landmarks-BFS*, it also only utilizes the vertices in the tree and get a path $(11, 7, 4, 1, 2)$. Different from them, our methods can utilize the vertices outside the path in the tree and find more short cuts to get a shorter path.

BFS in Induced Subgraph After Navigation. When we conduct a two-way navigating single path (TNSP) process, we collect the explored neighbours

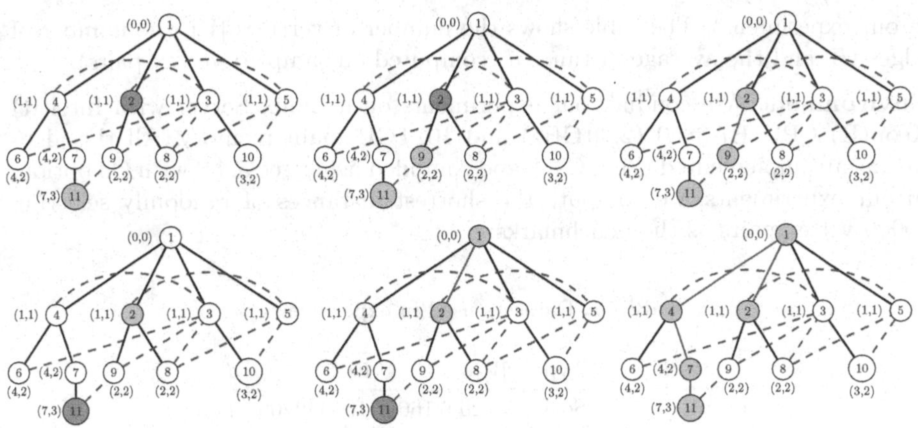

Fig. 3. An example of navigation process (TNSP) from two directions using navigator containing one SPT. The first three subfigures in the first row indicate the navigation process from 11 to 2. The last three subfigures in the second row indicate the navigation process from 2 to 11. When we conduct navigation in G from vertex 11 to vertex 2, we finally get a path $(11, 9, 2)$ whose length is 2, while when navigating from vertex 2 to 11, we get a path $(2, 1, 4, 7, 11)$ whose length is 4. They are two totally different paths.

of the two found paths p_1 and p_2. Then we construct an induced subgraph $G_{V'}$ on vertex set $V' = p_1 \cup p_2 \cup N(p_1) \cup N(p_2)$. Finally we conduct a BFS routine in $G_{V'}$ to find a shorter path. Although it is a bit more complex and costly, it will significantly improve the precision of the results because more vertices and edges are explored. We call this technique N+. The time complexity of N+ is linear with the size of $G_{V'}$.

4 Experiments

In this section, we present the results of experiments. In our experiments, we mainly consider three performance indicators: *precision*, *stretch ratio* and *querying time*. We adopt the same way of constructing the shortest path tree, so we omit the comparison of the indexing time and size of different methods. In the label of each vertex, we use 4 bytes to store the vertices ID and 1 byte to store the shortest distance to the root. We compare with two previous methods: *TreeSketch* (TS) [11], *Landmarks-BFS* (LB) [20].

4.1 Datasets and Environment Description

Datasets. The datasets used in our experiments can be found at Koblenz Network Collection [1], Laboratory for Web Algorithmics [3, 4], Network Repository [18] and Stanford Large Network Dataset Collection [12]. We treat all graphs as undirected graphs. Table 1 describes the detailed information of datasets used in our experiments. The table shows the number of vertices $|V|$, the number of edges $|E|$ and the average distance $\bar{d}$ (computed on sampled vertex pairs).

Environment. We conduct our experiments on a Linux Server with Intel(R) Xeon(R) CPU E7-8850 (2.30 GHz) and 1024 GB main memory. All the algorithms are implemented using C++ and compiled using gcc 5.4.0 with O3 option. In our experiments, we compute the shortest distances of randomly selecting 10000 vertex pairs as the benchmarks.

Table 1. Datasets used in experiments

Network	Type	$\|V\|$	$\|E\|$	$\bar{d}$
orkut	Social	2997166	106349209	4.25
uk-2002	Web	18459128	261556721	7.61
it-2004	web	29905049	698354804	5.9
socfb-konect	Social	58790782	92208195	7.86
soc-friendster	Social	65608366	1806067135	4.99
uk-2007	Web	105060524	3315357859	6.9

4.2 Evaluation Metrics

Precision. Denote the query set by Q. The precision of a method $\mathcal{M}$ for a network G is the frequency of the vertex pair $(s,t) \in Q$ satisfying $\delta^{\mathcal{M}}(s,t) = d_G(s,t)$. It is strict because it requires that the method actually finds the shortest path.

Stretch Ratio. For a vertex pair (s,t), the *stretch ratio* of a network G for a method $\mathcal{M}$ can be defined as $R(s,t) = \delta^{\mathcal{M}}(s,t)/d_G(s,t)$, Then for the query set Q, the *average stretch (AS) ratio* can be defined as: $R_{AS} = \sum_{(s,t)\in Q} R(s,t)/|Q|$. Also, the *max stretch (MS) ratio* can be defined as, $R_{MS} = \max_{(s,t)\in Q} R(s,t)$.

4.3 Analysis

In this section, we give a detailed analysis about the experimental results. For the problem studied in this paper, we mainly focus on the accuracy and efficiency. The performance of all methods is related to two factors: the number of SPT k, the strategy of roots of SPT. For convenient comparison with other methods, we fix $k = 20$ and select the same roots following the strategy of descending degree. We list these comparison results in Tables 2, 3 and 4 in which three performance indicators will be compared: *precision, stretch ratio, query time*. Then the performance of the navigator will be listed in Table 5. Finally we will present the precision comparison on different k and different strategies of selecting roots of SPT in Fig. 4.

Table 2. Precision of all methods when $k = 20$ and selecting the roots as the descending degree of vertices.

Networks	Precision						
	TS	LB	NSP	NMP	TNSP	TNMP	N+
soc-orkut	0.58	0.60	0.60	**0.69**	**0.69**	**0.79**	**0.90**
uk-2002-lcc	0.67	0.70	0.69	**0.77**	**0.81**	**0.88**	**0.92**
it-2004-lcc	0.52	0.54	0.56	**0.63**	**0.70**	**0.77**	**0.92**
socfb-konect	0.23	0.25	0.41	**0.51**	**0.55**	**0.66**	**0.60**
soc-friendster	0.20	0.21	0.26	**0.43**	**0.38**	**0.58**	**0.84**
uk-2007	0.39	0.41	0.51	**0.60**	**0.65**	**0.75**	**0.85**

Precision. From Table 2, we know that our methods NMP, TNSP, TNMP and N+ have higher precision than previous methods in massive networks and NSP has comparable precision. Among them, N+ has the best precision which is about 30% higher than *Landmarks-BFS* and reaches up to 80% in most networks when we construct $k = 20$ SPTs. Furthermore, we can nearly get a relationship of the precision among all the methods: TS < LB < NSP < NMP < TNSP < TNMP < N+ from Table 2. The reason is that our methods can use more edges not only in the k-SPT than previous methods.

Stretch Ratio. Table 3 presents the average stretch ratio and max stretch ratio of all methods. Overall, our methods not only have lower average stretch ratio

Table 3. Stretch ratio of all methods when $k = 20$ and selecting the roots as the descending degree of vertices.

Networks		TS	LB	NSP	TNSP	NMP	TNMP	N+
soc-orkut	R_{AS}	1.13	1.12	**1.11**	**1.08**	**1.08**	**1.06**	**1.02**
	R_{MS}	2.00	2.00	2.00	2.00	2.00	2.00	**1.50**
uk-2002-lcc	R_{AS}	1.05	1.05	1.05	**1.03**	**1.03**	**1.02**	**1.01**
	R_{MS}	2.00	2.00	2.00	**1.50**	2.00	**1.43**	**1.50**
it-2004-lcc	R_{AS}	1.13	1.12	**1.10**	**1.08**	**1.06**	**1.04**	**1.01**
	R_{MS}	2.33	2.33	**2.00**	**2.00**	**2.00**	**1.75**	**1.75**
socfb-konect	R_{AS}	1.19	1.19	**1.12**	**1.09**	**1.08**	**1.06**	**1.07**
	R_{MS}	2.50	2.50	**1.83**	**1.83**	**1.67**	**1.67**	**1.67**
soc-friendster	R_{AS}	1.22	1.22	**1.18**	**1.13**	**1.14**	**1.09**	**1.03**
	R_{MS}	2.00	2.00	2.00	**1.67**	2.00	**1.67**	**1.60**
uk-2007	R_{AS}	1.13	1.13	**1.10**	**1.07**	**1.06**	**1.04**	**1.02**
	R_{MS}	2.00	2.00	2.00	**1.60**	**1.80**	**1.60**	**1.60**

but also decrease the max stretch ratio R_{MS} to below 2 in all the networks of our experiments.

Query Time. From Table 4, we know that our methods except N+ is faster than existing methods in most networks. They run in milliseconds in networks containing tens of millions vertices. A BFS routine costs tens of hundreds of milliseconds. Its speedup reaches dozens of times. The method N+ has the lowest query efficiency but is still faster than BFS.

Table 4. Query time of all methods when $k = 20$ and selecting the roots as the descending degree of vertices.

Networks	Query time [ms]							BFS [ms]
	TS	LB	NSP	TNSP	NMP	TNMP	N+	
soc-orkut	4.49	2.00	**1.49**	**5.14**	**2.69**	**10.25**	38.80	238.49
uk-2002-lcc	28.21	15.86	**2.14**	**3.76**	**4.23**	**7.51**	44.82	759.49
it-2004-lcc	70.85	51.80	**9.27**	**15.18**	**18.22**	**30.32**	199.44	1176.42
socfb-konect	88.10	44.09	**2.58**	**5.74**	**5.06**	**11.98**	57.35	1540.80
soc-friendster	224.99	111.78	**5.64**	**33.97**	**11.20**	**70.39**	173.27	12321.30
uk-2007	170.85	98.09	**18.24**	**27.84**	**36.73**	**56.17**	479.11	5967.10

Performance on Navigator. We present the construction cost and query efficiency of navigator in Table 5 under different strategies. From Table 5 we know that *closeness* obviously costs nearly 5 times more than *degree* and *random* and the precision is also higher than other two strategies. Note that in some networks, navigator can give a correct answer with probability higher than 60% if we choose the *closeness* strategy.

Table 5. The cost and query time of navigator under different strategies when $k = 20$. NS indicates the size of navigator. CT indicates the construction time of navigator. QT indicates the query time of navigator.

Networks	NS [MB]	CT [s]			QT [ms]			Precision		
		deg	ran	close	deg	ran	close	deg	ran	close
soc-orkut	457.33	37	38	174.19	0.01	0.01	0.01	0.45	0.34	0.53
uk-2002-lcc	2816.64	100	93	274.84	0.01	0.01	0.01	0.41	0.43	0.61
it-2004-lcc	4563.15	165	180	526.48	0.01	0.01	0.01	0.29	0.46	0.59
socfb-konect	8970.76	304	291	790.33	0.01	0.02	0.01	0.16	0.10	0.23
soc-friendster	10011.04	1669	1678	5933.66	0.03	0.03	0.03	0.06	0.03	0.09
uk-2007	16030.96	861	855	2846.01	0.01	0.01	0.01	0.25	0.45	0.60

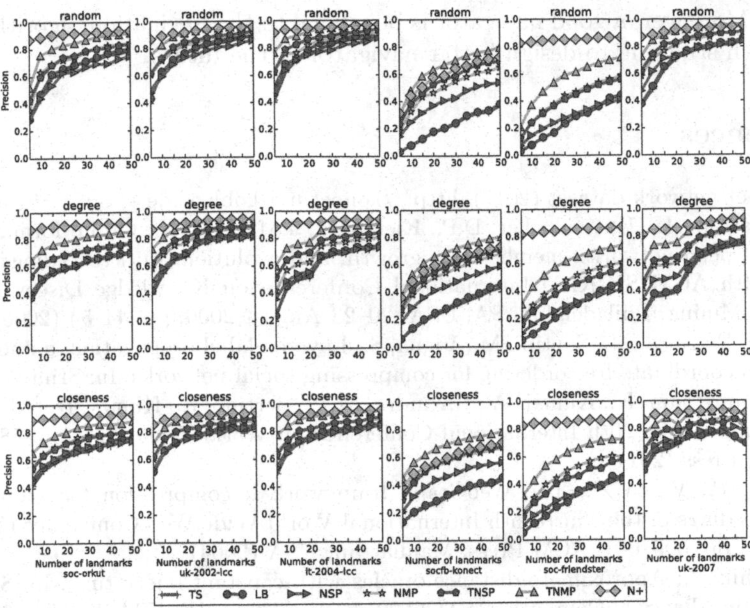

Fig. 4. Precision of all methods when using different number of roots and selecting roots with different strategies.

Performance on the Number and Strategy of Selecting Roots. We construct up to $k = 50$ shortest path trees in navigator. From Fig. 4, we know that the precision of each method is higher when k is larger and if we select the landmarks following the closeness centrality strategy, the precision of the method will be higher. In general, the curve of our methods in Fig. 4 is all above the curve of previous methods in most networks. When $k = 50$, the precision of our methods reaches 80% in most networks. Among them, the precision of N+ even reaches 90% in most networks. Overall, *closeness* strategy for selecting roots of SPT can get higher query precision.

5 Conclusion

In this paper, we propose navigation algorithms based on navigator for shortest path queries in massive complex networks. Compared with the previous methods, extensive experimental results show that our methods TMP, TNSP and TNMP have better performance not only in precision which means to find a shortest path with higher probability, but also in query efficiency which means to answer the query faster. To achieve higher precision, we propose the method N+ and it navigates a shortest path with probability higher than 80% in most real networks. Based on the feature of navigation algorithms, we believe that it is not hard to design a corresponding distributed algorithm and we will work on it in the next project. Moreover, a good navigator is beneficial to the navigation algorithm. So it is worth studying to design better navigator in the future.

References

1. Konect network dataset (2017). http://konect.uni-koblenz.de
2. Backstrom, L., Huttenlocher, D.P., Kleinberg, J.M., Lan, X.: Group formation in large social networks: membership, growth, and evolution. In: Proceedings of the Twelfth ACM SIGKDD International Conference on Knowledge Discovery and Data Mining, Philadelphia, PA, USA, 20–23 August 2006, pp. 44–54 (2006)
3. Boldi, P., Rosa, M., Santini, M., Vigna, S.: Layered label propagation: a multiresolution coordinate-free ordering for compressing social networks. In: Srinivasan, S., Ramamritham, K., Kumar, A., Ravindra, M.P., Bertino, E., Kumar, R. (eds.) Proceedings of the 20th International Conference on World Wide Web, pp. 587–596. ACM Press (2011)
4. Boldi, P., Vigna, S.: The WebGraph framework I: compression techniques. In: Proceedings of the Thirteenth International World Wide Web Conference (WWW 2004), pp. 595–601. ACM Press, Manhattan, USA (2004)
5. Chechik, S.: Approximate distance oracles with constant query time. In: Symposium on Theory of Computing, STOC 2014, New York, NY, USA, 31 May–03 June 2014, pp. 654–663 (2014)
6. Cheng, J., Zhang, Y., Ye, Q., Du, H.: High-precision shortest distance estimation for large-scale social networks. In: 35th Annual IEEE International Conference on Computer Communications, INFOCOM 2016, San Francisco, CA, USA, 10–14 April 2016, pp. 1–9 (2016)
7. Cohen-Addad, V., Dahlgaard, S., Wulff-Nilsen, C.: Fast and compact exact distance oracle for planar graphs. In: 58th IEEE Annual Symposium on Foundations of Computer Science, FOCS 2017, Berkeley, CA, USA, 15–17 October 2017, pp. 962–973 (2017)
8. Derungs, J., Jacob, R., Widmayer, P.: Approximate shortest paths guided by a small index. Algorithmica **57**(4), 668–688 (2010)
9. Gawrychowski, P., Mozes, S., Weimann, O., Wulff-Nilsen, C.: Better tradeoffs for exact distance oracles in planar graphs. In: Proceedings of the Twenty-Ninth Annual ACM-SIAM Symposium on Discrete Algorithms, SODA 2018, New Orleans, LA, USA, 7–10 January 2018, pp. 515–529 (2018)
10. Goldberg, A.V., Harrelson, C.: Computing the shortest path: a search meets graph theory. In: Proceedings of the Sixteenth Annual ACM-SIAM Symposium on Discrete Algorithms, SODA 2005, Vancouver, British Columbia, Canada, 23–25 January 2005, pp. 156–165 (2005)
11. Gubichev, A., Bedathur, S.J., Seufert, S., Weikum, G.: Fast and accurate estimation of shortest paths in large graphs. In: Proceedings of the 19th ACM Conference on Information and Knowledge Management, CIKM 2010, Toronto, Ontario, Canada, 26–30 October 2010, pp. 499–508 (2010)
12. Leskovec, J., Krevl, A.: SNAP Datasets: Stanford large network dataset collection, June 2014. http://snap.stanford.edu/data
13. Maue, J., Sanders, P., Matijevic, D.: Goal-directed shortest-path queries using precomputed cluster distances. ACM J. Exp. Algor. **14**, 27 pages (2009). Article 3.2

14. Möhring, R.H., Schilling, H., Schütz, B., Wagner, D., Willhalm, T.: Partitioning graphs to speed up Dijkstra's Algorithm. In: Nikoletseas, S.E. (ed.) WEA 2005. LNCS, vol. 3503, pp. 189–202. Springer, Heidelberg (2005). https://doi.org/10.1007/11427186_18
15. Potamias, M., Bonchi, F., Castillo, C., Gionis, A.: Fast shortest path distance estimation in large networks. In: Proceedings of the 18th ACM Conference on Information and Knowledge Management, CIKM 2009, Hong Kong, China, 2–6 November 2009, pp. 867–876 (2009)
16. Qiao, M., Cheng, H., Chang, L., Yu, J.X.: Approximate shortest distance computing: a query-dependent local landmark scheme. In: IEEE 28th International Conference on Data Engineering (ICDE 2012), Washington, DC, USA (Arlington, Virginia), 1–5 April, 2012, pp. 462–473 (2012)
17. Rahman, S.A., Schomburg, D.: Observing local and global properties of metabolic pathways: "load points" and "choke points" in the metabolic networks. Bioinformatics **22**(14), 1767–1774 (2006)
18. Rossi, R.A., Ahmed, N.K.: The network data repository with interactive graph analytics and visualization (2015). http://networkrepository.com
19. Thorup, M., Zwick, U.: Approximate distance oracles. J. ACM **52**(1), 1–24 (2005)
20. Tretyakov, K., Armas-Cervantes, A., García-Bañuelos, L., Vilo, J., Dumas, M.: Fast fully dynamic landmark-based estimation of shortest path distances in very large graphs. In: Proceedings of the 20th ACM Conference on Information and Knowledge Management, CIKM 2011, Glasgow, United Kingdom, 24–28 October 2011, pp. 1785–1794 (2011)
21. Wulff-Nilsen, C.: Approximate distance oracles with improved query time. In: Proceedings of the Twenty-Fourth Annual ACM-SIAM Symposium on Discrete Algorithms, SODA 2013, New Orleans, Louisiana, USA, 6–8 January 2013, pp. 539–549 (2013)
22. Zhao, X., Sala, A., Zheng, H., Zhao, B.Y.: Efficient shortest paths on massive social graphs. In: 7th International Conference on Collaborative Computing: Networking, Applications and Worksharing, CollaborateCom 2011, Orlando, FL, USA, 15–18 October 2011, pp. 77–86 (2011)
23. Zhao, X., Zheng, H.: Orion: shortest path estimation for large social graphs. In: 3rd Workshop on Online Social Networks, WOSN 2010, Boston, MA, USA, 22 June 2010 (2010)
24. Zhou, C., Lu, W., Zhang, P., Wu, J., Hu, Y., Guo, L.: On the minimum differentially resolving set problem for diffusion source inference in networks. In: Proceedings of the Thirtieth AAAI Conference on Artificial Intelligence, Phoenix, Arizona, USA, 12–17 February 2016, pp. 79–86 (2016)
25. Zhu, K., Ying, L.: Information source detection in the SIR model: a sample-path-based approach. IEEE/ACM Trans. Netw. **24**(1), 408–421 (2016)

Engineering a PTAS for Minimum Feedback Vertex Set in Planar Graphs

Glencora Borradaile, Hung Le(✉), and Baigong Zheng

Oregon State University, Corvallis, OR, USA
{glencora,lehu,zhengb}@oregonstate.edu

Abstract. We investigate the practicality of approximation schemes for optimization problems in planar graphs based on balanced separators. The first polynomial-time approximation schemes (PTASes) for problems in planar graphs were based on balanced separators, wherein graphs are recursively decomposed into small enough pieces in which optimal solutions can be found by brute force or other methods. However, this technique was supplanted by the more modern and (theoretically) more efficient approach of decomposing a planar graph into graphs of bounded treewidth, in which optimal solutions are found by dynamic programming. While the latter approach has been tested experimentally, the former approach has not.

To test the separator-based method, we examine the minimum feedback vertex set (FVS) problem in planar graphs. We propose a new, simple $O(n \log n)$-time approximation scheme for FVS using balanced separators and a *linear kernel*. The linear kernel reduces the size of the graph to be linear in the size of the optimal solution. In doing so, we correct a reduction rule in Bonamy and Kowalik's linear kernel [11] for FVS. We implemented this PTAS and evaluated its performance on large synthetic and real-world planar graphs. Unlike earlier planar PTAS engineering results [8,36], our implementation guarantees the theoretical error bounds on all tested graphs.

Keywords: Feedback vertex set · Planar graph algorithms · Approximation schemes · Algorithm engineering

1 Introduction

A *polynomial-time approximation scheme* (PTAS) is a $(1 + \epsilon)$-approximation algorithm that runs in polynomial time for any fixed parameter $\epsilon > 0$. The development of PTASes for optimization problems in planar graphs has been an active area of study for decades, with the development of PTASes for increasing complicated problems, including maximum independent set [33], minimum vertex cover [5], TSP [4,29], Steiner tree [12] and Steiner forest [7,20].

This material is based upon work supported by the National Science Foundation under Grant No. CCF-1252833.

I. Kotsireas et al. (Eds.): SEA² 2019, LNCS 11544, pp. 98–113, 2019.
https://doi.org/10.1007/978-3-030-34029-2_7

The first PTAS for a planar graph optimization problem was for maximum independent set [33], and was an early application of Lipton and Tarjan's balanced planar separators [32]: recursively apply the balanced separator, deleting the vertices of the separator as you go, until each piece is small enough to find a maximum independent set by brute force. This balanced-separator technique was shown applicable to a host of other planar graph optimization problems, including minimum vertex cover [14], TSP [4], minimum-weight connected vertex cover [16] and minimum-weight connected dominating set [16].

Lipton and Tarjan mentioned in their paper [33] that they could obtain "an $O(n \log n)$ algorithm with $O(1/\sqrt{\log \log n})$ relative error" for maximum independent set in planar graphs. By setting $\epsilon = c \log \log n$ where c is some constant, one would obtain a PTAS for maximum independent set only when n is very large, i.e. $2^{2^{O(1/\epsilon^2)}}$. This is critiqued in literature. For example, Chiba, Nishizeki and Saito [15] said the following:

> "It should be noted that, although the algorithm of Lipton and Tarjan can also guarantee the worst case ratio $\frac{1}{2}$, the number n of vertices must be quite huge, say $2^{2^{400}}$, so the algorithm is not practical."

And this was cited by Baker in her seminal paper [5] and by Demaine and Hajiaghayi [18] as the main disadvantage of separator-based technique.

Modern PTAS design in planar graphs instead decompose the graph into pieces of bounded treewidth and then use dynamic programming over the resulting tree decomposition to solve the problem. This technique has been used for most of the problems listed above and produces running times of the form

$$O\left(2^{1/\epsilon^{O(1)}} n^{O(1)}\right) \tag{1}$$

which are described as (theoretically) efficient. Indeed, for strongly NP-hard problems, these running times are optimal, up to constants.

Most likely for the above two reasons, experimental study of planar PTASes has been entirely restricted to treewidth-based techniques [8,36] rather than separator-based techniques. However, we argue that the separator-based algorithm can be practical:

1. Although the decomposition step in the separator-based method is comparatively more complicated, balanced separators have been experimentally shown to be highly practical [2,23,25].
2. The hidden constants in treewidth-based PTASes are usually very large [8,36], while the hidden constants in separator-based method are very reasonable.
3. The practical bottleneck for treewidth-based PTASes is decidedly the dynamic programming step [8,34,36]; further, the dynamic programs are complicated to implement and require another level of algorithmic engineering to make practical, which may be avoided in some separator-based PTASes.
4. Modern planar PTAS designers have critiqued separator-based methods as only able to obtain a good approximation ratio in impractically large graphs [5,15,18] as mentioned above. However, as we will illustrate for FVS, one can relate the error to largest component in the resulting decomposition instead of the original graph, overcoming this issue.

5. Separator-based PTASes require the input graph to be linear in the size of an optimal solution in order to bound the error [24]. However, one can use linear-based kernels to achieve this, as mentioned by Demaine, Hajiaghayi and Kawarabayashi [17] and we illustrate herein for FVS. Given the recent explosion of kernelization algorithms [10,21,22], we believe that separator-based PTASes may experience a renewal of research interest.

1.1 Case Study: Feedback Vertex Set

We illustrate our ideas on the *minimum feedback vertex set* (FVS) problem which asks for a minimum set of vertices in an undirected graph such that after removing this set leaves an acyclic graph (a forest). We use this problem for three reasons: First, FVS is a classical optimization problem: it is one of Karp's 21 original NP-Complete problems [28] and has applications in disparate areas, from deadlock recovery in operating systems to reducing computation in Bayesian inference [6]. Second, there are no existing PTAS implementations for this problem. Third, the existing (theoretical) PTASes for this problem are either too complicated to implement (such as using the bidimensionality framework [18], relying on dynamic programming over tree decompositions) or not sufficiently efficient (such as the local search PTAS [31] with running time of the form $n^{O(1/\epsilon)}$).

To this end, we propose a simple-to-implement $O(n \log n)$ PTAS for FVS in planar graphs that uses a linear kernel to address issues (4) and (5) above. Our experimental results show that our PTAS can find good solutions and achieve better approximation ratios than those theoretical guaranteed on all the graphs in our test set.

Theoretical Contributions. Our theoretical contributions are two-fold. First, we show that a reduction rule of Bonamy and Kowalik's $13k$-kernel [11] for planar FVS is incorrect. We offer a correction based on a reduction rule from Abu-Khzam and Khuzam's linear kernel [1] for the same problem (Sect. 3). Second, we show a simple separator-based PTAS for FVS, which runs in $O(n \log n)$ time (Sect. 4). Our PTAS starts with a linear kernel for planar FVS and then applies balanced separators recursively to decompose the kernel into a set of small subgraphs, in which we solve FVS optimally. Compared to existing PTASes for planar FVS, our PTAS has several advantages:

1. It only relies on two simple algorithmic black boxes: kernelization, which consists of a sequence of simple reduction rules, and balanced separators, which are known to be practical [2,23,25].
2. It has very few parameters to optimize.
3. The constants hidden by the asymptotic running time and approximation ratio are practically small.
4. Its running time is theoretically efficient.

We believe that our approach may be applied more generally: minor-free graphs admit balanced separators of sublinear size [3] and many problems admit linear kernels in minor-free graphs [10,21,22].

Engineering Contributions. We implemented our PTAS, tested this on planar graphs (of up to 6 million vertices), and discuss limitations. It is not our aim to beat the current best heuristics and exact solvers for FVS, although we do compare our PTAS with Becker and Geiger's 2-approximation algorithm [9]. We further adapted our PTAS and introduced some heuristics (Sect. 5) to improve the efficiency of our implementation and the solution quality of our algorithm. With a computational lower bound, we evaluate the approximation ratio empirically on each testing graph, illustrating that our implementation achieves better approximation ratios than the theoretically guaranteed approximation ratio on all testing graphs (Sect. 6).

Our implementation differs from previous implementations of PTASes for planar graph optimization problems (TSP [8] and Steiner tree [36]) in the sense that our implementation guarantees the theoretical approximation ratio. Previous PTAS implementations sacrificed approximation guarantees to improve the running time, because the constants hidden in the asymptotic running time (Eq. (1)) are too large: the dynamic programming table is very expensive to compute. Our simple PTAS (albeit for a problem in a different class) avoids this issue. Our work suggests a different direction from these earlier PTAS engineering papers: in order to implement a practical PTAS, return to the (theoretical) drawing board and design a simpler algorithm.

2 Preliminaries

All graphs considered in this paper are undirected and planar, and can have parallel edges and self-loops. We denote by $V(G)$ and $E(G)$ the vertex set and edge set of graph G. If there is only one edge between two vertices, we say that edge is an *single-edge*. If there are two parallel edges between two vertices, we say the two edges form a *double-edge*. The *degree* of a vertex is the number of edges incident to the vertex. We denote by $G[X]$ the subgraph induced by the subset $X \subseteq V(G)$. We use $OPT(G)$ to represent an arbitrary minimum feedback vertex set in graph G.

2.1 Balanced Separator

A *separator* is a set of vertices whose removal will partition the graph into two parts. A separator is α*-balanced* if those two parts each contain at most an α-fraction of the original vertex set. Lipton and Tarjan [32] first introduced a separator theorem for planar graphs, which says a planar graph with n vertices admits a $\frac{2}{3}$-balanced separator of size at most $2\sqrt{2n}$; they gave a linear-time algorithm to compute such a balanced separator. This algorithm computes a breadth-first search (BFS) tree for the graph and labels the vertices according to their level in the BFS tree. They prove that a balanced-separator is given by one of three cases:

(P1) The vertices of a single BFS level.
(P2) The vertices of two BFS levels.
(P3) The vertices of two BFS levels plus the vertices of a fundamental cycle[1] with respect to the BFS tree that does not intersect the two BFS levels.

2.2 Kernelization Algorithm

A parameterized decision problem with a parameter k admits a *kernel* if there is a polynomial time algorithm (where the degree of the polynomial is independent of k), called a *kernelization algorithm*, that outputs a decision-equivalent instance whose size is bounded by some function $h(k)$. If the function $h(k)$ is linear in k, then we say the problem admits a *linear* kernel.

Bonamy and Kowalik give a $13k$-kernel for planar FVS [11] which consists of a sequence of 17 reduction rules. Each rule replaces a particular subgraph with another (possibly empty) subgraph, and possibly marks some vertices that must be in an optimal solution. The algorithm starts by repeatedly applying the first five rules to the graph and initializes two queues: queue Q_1 contains some vertex pairs that are candidates to check for Rule 6 and queue Q_2 contains vertices that are candidates to check for the last five rules. While Q_1 is not empty, the algorithm repeatedly applies Rule 6, reducing $|Q_1|$ in each step. Then the algorithm repeatedly applies the remaining rules in order, reducing $|Q_2|$ until Q_2 is empty. After applying any rule, the algorithm updates both queues as necessary, and will apply the first five rules if applicable. We provide the first 11 rules in the appendix. See the original paper [11] for full details of all the reduction rules.

3 Corrected Reduction Rule

In this section, we provided a corrected Reduction Rule 6 in Bonamy and Kowalik's kernelization algorithm, denoted by `BK`. This Rule and a counterexample is illustrated in Fig. 1.

To provide a corrected rule, we need the following assistant reduction rules; the first two are from `BK`, the third is from Abu-Khzam and Khuzam's kernelization algorithm [1], denoted by `AK`, and the last is generalized from an `AK` reduction rule.

`BK` **Reduction Rule 3.** If a vertex u is of degree two, with incident edges uv and uw, then delete u and add the edge vw. (Note that if $v = w$ then a loop is added.)

[1] Given a spanning tree for a graph, a fundamental cycle consists of a non-tree edge and a path in the tree connecting the two endpoints of that edge.

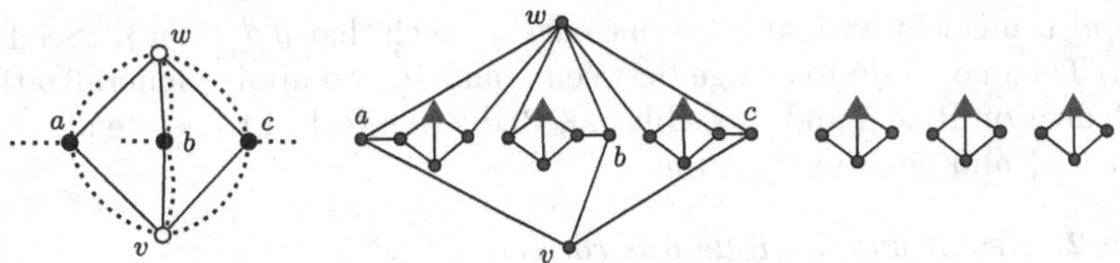

Fig. 1. BK Reduction Rule 6 and a counterexample. Left: Rule 6 removes the solid edges and labeled vertices and assumes the empty vertices are in the optimal solution. The dashed edges are optional connection in the graph (either edges or more complex components connecting the endpoints as in Center). Center: A counterexample for BK Reduction Rule 6, whose optimal solution consists of 4 vertices: w and the three triangle vertices. Right: The result of applying Reduction Rule 6, whose optimal solution consists of the three triangle vertices. Together with w and v is a FVS of size 5.

BK **Reduction Rule 4.** If a vertex u has exactly two neighbors v and w, edge uv is double, and edge uw is simple, then delete v and u, and add v into the optimal solution.

AK **Reduction Rule 4.** If a vertex u has exactly two neighbors v and w such that uv, uw and vw are double-edges, then add both v and w to the optimal solution, and delete u, v and w.

Generalization of AK **Reduction Rule 5.** Refer to Fig. 2. Let vw be a double-edge in G and let a be a vertex whose neighbors are v, w and x such that ax is a single-edge.

(1) If both av and aw are single-edges, then delete a and add edges vx and wx.
(2) If only one of av and aw, say aw, is double-edge, then delete a and w, add edge vx, and add w to the optimal solution.
(3) If both av and aw are double-edges, then delete a, w and v, and add v and w to the optimal solution.

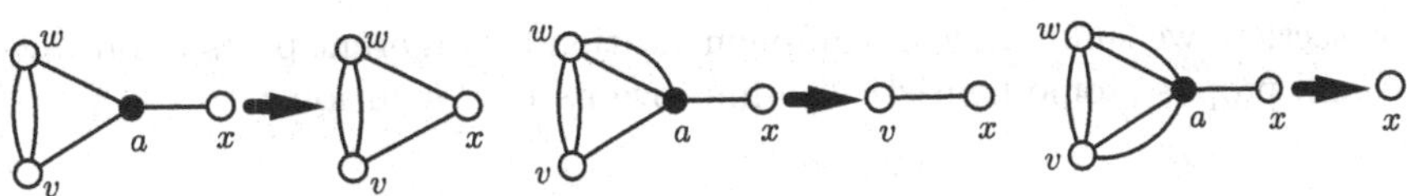

Fig. 2. Generalization of AK Reduction Rule 5.

Lemma 1. *The Generalization of AK Reduction Rule 5 is correct.*

Now we are ready to give the new Reduction Rule 6.

New Reduction Rule 6. Assume that there are five vertices a, b, c, v, w such that (1) both v and w are neighbors of each of a, b, c and (2) each vertex $x \in$

$\{a, b, c\}$ is incident with at most one edge xy such that $y \notin \{v, w\}$. (See Fig. 1: Left.) Then add a double-edge between v and w, and apply Generalization of AK Reduction Rule 5 and (possibly) AK Reduction Rule 4 to delete all vertices in $\{a, b, c\}$ and possibly v and w.

Lemma 2. *New Reduction Rule 6 is correct.*

Herein, we denote the corrected kernelization algorithm by BK.

4 Polynomial-Time Approximation Scheme

In this section, we present an $O(n \log n)$ PTAS for FVS in planar graphs using a linear kernel and balanced separators, and show that we can obtain a lower bound of an optimal solution from its solution. Our PTAS consists of the following four steps.

1. Compute a linear kernel H for the original graph G by algorithm BK, that is, $|V(H)|$ is at most $c_1|OPT(H)|$ for some constant c_1.
2. Decompose the kernel H by recursively applying the separator algorithm and remove the separators until each resulting graph has at most r vertices for some constant r. The union of all the separators has at most $c_2|V(H)|/\sqrt{r} = \epsilon|OPT(H)|$ vertices for r chosen appropriately.
3. Solve the problem optimally for all the resulting graphs.
4. Let U_H be the union of all separators and all solutions of the resulting graphs. Lift U_H to a solution U_G for the original graph. The lifting step involves unrolling the kernelization in step 1.

We can prove the above algorithm is a PTAS and obtain the following theorem whose proof is deferred to the full version.

5 Engineering Considerations

In this section, we summarize our implementation for different parts of our algorithm and propose some heuristics to improve its final solution.

5.1 Kernelization Algorithm

The original kernelization algorithm BK consists of a sequence of 17 reduction rules. The first 12 rules are simple and sufficient to obtain a $15k$-kernel [11]. Since the remaining rules do not improve the kernel by much, and since Rule 12 is a rejecting rule[2], we only implement the first 11 rules (provided in the appendix) with one rule corrected, all of which are local and independent of the parameter k.

[2] This is to return a trivial no-instance for the decision problem when the resulting graph has more than $15k$ vertices.

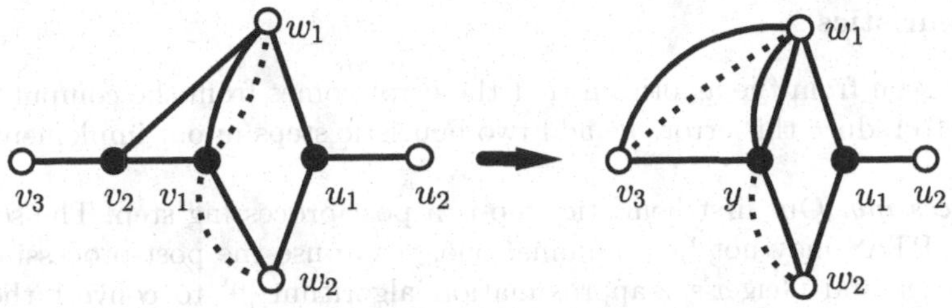

Fig. 3. Reduction rule 8 replaces the left subgraph with the right subgraph.

The original algorithm BK is designed for the decision problem, so when applying it to the optimization problem, we need some modifications to be able to convert a kernel solution to a feasible solution for the original graph, and this converting is called *lifting*. If a reduction rule does not introduce new vertices, then the lifting step for it will be trivial. Otherwise, we need to handle the vertices introduced by reduction steps, if they appear in the kernel solution. Among all the implemented reduction rules, there are two rules, namely Rule 8 and Rule 9, that introduce new vertices into the graph. Since avoiding new vertices can improve the efficiency of our implementation, we tried to modify these two rules and found that we can avoid adding new vertices in Rule 8 (Fig. 3) by the following lemma.

Lemma 3. *The new vertex introduced by Rule 8 can be replaced by a vertex from the original graph.*

For Rule 9, it seems that adding a new vertex is unavoidable, so we record the related vertices in each application of Rule 9 in the same order as we apply it. During the lifting step, we check those vertices in the reverse order to see if the recorded vertices are also in the solution. If there are involved vertices in the solution, we modify the solution according to the reverse Rule 9.

We remark that the original algorithm runs in expected $O(n)$ time, and each rule can be detected in $O(1)$ time if we use a hash table. However, we found that using a balanced binary search tree instead of a hash table gives better practical performance.

5.2 Balanced Separators

We followed a textbook version [30] of the separator algorithm, and our implementation guarantees the $2\sqrt{2n}$ bound for the size of the separator. We remark that we did not apply heuristics in our implementation for the separator algorithm. This is because we did not observe separator size improvement by some simple heuristics in the early stage of this work, and these heuristics may slow down the separator algorithm. Since our test graphs are large (up to 6 million vertices) and we will apply the algorithm recursively in our PTAS, these heuristics may slow down our PTAS even more.

5.3 Heuristics

It can be seen from the algorithm that the error comes from the computed separators. To reduce this error, we add two heuristic steps in our implementation.

Post-processing. Our first heuristic step is a post-processing step. The solution from our PTAS may not be a minimal one, so we use the post-processing step from Becker and Geiger's 2-approximation algorithm [9] to convert the final solution of the PTAS to a minimal one. This involves iterating through the vertices in the solution and trying to remove redundant vertices from the solution while maintaining feasibility. In fact, we only need to iterate through the vertices in separators, since vertices in the optimal solutions of small graphs are needed for feasibility.

Kernelization During Decomposition Step. Our second heuristic step is to apply the kernelization algorithm `BK` right after we compute a separator in the decomposition step of our PTAS. Note that there is a decomposition tree corresponding to the decomposition step, where each node corresponds to a subgraph that is applied the separator theorem. To apply this heuristic, we need to record the whole decomposition tree with all the corresponding separators such that we can lift the solutions in the right order. For example, if we want to lift a solution for a subgraph G_w corresponding to some node w in the decomposition tree, we first need to lift all solutions for the subgraphs corresponding to the children of node w in the decomposition tree.

6 Experimental Results

In this section, we evaluate the performance of our algorithm. We implemented it in `C++` and the code is compiled with `g++` (version 4.8.5) on `CentOS` (version 7.3.1611) operating system. Our PTAS implementation is built on Boyer's implementation[3] of Boyer and Myrvold's planar embedding algorithm [13]. We find trivial exact algorithm cannot solve graphs of size larger than 30 in short time, so we combine our implementation for the kernelization algorithm `BK` and Iwata and Imanish's implementation for a *fixed-parameter tractable* (FPT) algorithm as our exact algorithm in our experiment. Their algorithm is implemented[4] in `Java` and includes a linear-time kernel [26] and a branch-and-bound based FPT algorithm [27] for FVS in general graphs. The `Java` version in our machine is 1.8.0 and our experiments were performed on a machine with Intel(R) Xeon(R) CPU (2.30 GHz) running `CentOS` (version 7.3.1611) operating system.

We implemented three variants of our PTAS:

- the **vanilla** variant is a naive implementation of our PTAS, for which no heuristic is applied;

[3] http://jgaa.info/accepted/2004/BoyerMyrvold2004.8.3/planarity.zip.
[4] https://github.com/wata-orz/fvs.

- the **minimal** variant applies the post-processing heuristic to our PTAS, which will remove redundant vertices in separators;
- the **optimized** variant applies both heuristics to our PTAS, which will apply kernelization algorithm whenever each separator is computed and removed during the decomposition step, and will return a minimal final solution.

To evaluate their performance, we collect three different classes of planar graphs:

- The **random** graphs are random planar graphs generated by `LEDA` (version 6.4) [35];
- The **triangu** graphs are triangulated random graphs generated by `LEDA`, whose outer faces are not triangulated;
- The **networks** graphs (including **NY**, **BAY**, **COL**, **NW**, **CAL**, **FLA**, **LKS**, **E** and **W**) are road networks used in the 9th DIMACS Implementation Challenge-Shortest Paths [19]. We interpret each graph as a straight-line embedding and we add vertices whenever two edges intersect geometrically.

Since we are interested in the performance of the algorithms in large planar graphs, the synthetic graphs we generated have at least 600,000 vertices. And the real network graphs have at least 260,000 vertices.

6.1 Runtime

To evaluate the runtime of our implementation, we first need to fix the parameter in our algorithm, which is the maximum size of subgraphs in the decomposition step. During our experiment, we find that our exact algorithm cannot optimally solve some graphs with 120 vertices in one hour. So we set the maximum size as 120 in the decomposition step and run the three variants on all the testing graphs. The total runtime is summarized in Table 1. We notice the runtime of the post-processing step is less than 1% of the total runtime on all graphs. So **vanilla** and **minimal** almost have the same runtime, and we omit the runtime of **vanilla** in the table. We can observe in the table that the total runtime of **optimized** is about twice of that of **minimal** on all graphs.

Although the theoretical runtime of our PTAS is $O(n \log n)$, it seems that our results in Table 1 does not exhibit this fact. The main reason for this is that the third step of our PTAS (solving small subgraphs optimally with the exact algorithm) dominates the total runtime, which takes over 85% of total runtime in **minimal** and over 70% of total runtime in **optimized** on all graphs. The runtime of the third step depends on the kernel size and the structures of different graph classes. The kernel size will affect the number of subgraphs we need to solve in the third step, and the structures of different graph classes will affect the runtime of the exact algorithm. We find that even if two graphs of different classes have similar size, their kernel size may be very different, such as **FLA** and **triangu3**. And even if the two graphs have similar kernel size, the total runtime of the exact algorithm can be very different, such as **NY** and **random1**. This explains why our results seem different from the theoretical runtime, and implies that a better way to analyze the runtime is to consider the running time of each step on each graph class.

Table 1. Performance of different algorithms.

graph	vertices	kernel size	lower bound	minimal		vanilla	optimized		2-approx	
				time (s)	approx ratio	approx ratio	time (s)	approx ratio	time (s)	approx ratio
NY	264953	132631	25290	341	1.713	1.936	769	1.692	9	1.649
BAY	322694	90696	23111	244	1.513	1.688	429	1.498	9	1.480
COL	437294	95436	27711	272	1.429	1.573	382	1.416	13	1.415
NW	1214463	216994	69926	765	1.377	1.514	1117	1.364	37	1.369
FLA	1074167	317630	89446	1068	1.421	1.567	2175	1.409	33	1.407
CAL	1898842	495180	132355	1843	1.482	1.649	3622	1.466	58	1.454
LKS	2763392	784644	177873	3035	1.606	1.815	8039	1.587	91	1.558
E	3608115	899206	235151	4841	1.489	1.665	10912	1.469	112	1.456
W	6286759	1488420	418660	11229	1.441	1.599	21084	1.427	195	1.421
triangu1	600000	519733	185025	2392	1.258	1.321	3997	1.257	55	1.264
triangu2	800000	698248	245458	4314	1.263	1.329	8046	1.262	67	1.268
triangu3	1000000	878400	305411	7786	1.266	1.334	13873	1.265	82	1.270
triangu4	1200000	1060063	364631	12529	1.268	1.337	21753	1.265	99	1.270
triangu5	1400000	1244517	422415	18603	1.272	1.343	33450	1.269	118	1.272
random1	699970	121121	192863	2893	1.000	1.000	2591	1.000	80	1.004
random2	1197582	108619	284192	4006	1.000	1.000	3849	1.000	118	1.005
random3	1399947	244563	385738	7548	1.000	1.000	7288	1.000	154	1.004
random4	1999760	305610	538498	12270	1.000	1.000	12255	1.000	233	1.004
random5	2199977	426043	617506	15694	1.000	1.000	16449	1.000	264	1.004
random6	873280	8124	86796	781	1.000	1.000	812	1.000	58	1.006
random7	1061980	11502	111324	1095	1.000	1.000	1140	1.000	65	1.006
random8	1227072	12538	124635	1192	1.000	1.000	1252	1.000	74	1.006
random9	1520478	18921	166737	1729	1.000	1.000	1792	1.000	93	1.006
random10	2050946	40828	269315	3448	1.000	1.000	3615	1.000	140	1.006

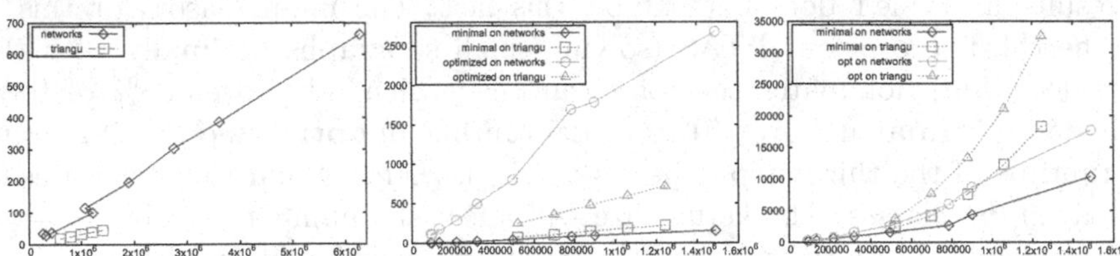

Fig. 4. Runtime of different steps in our PTAS variants. The Y axis represents the runtime in seconds. Left: runtime of the first step (kernelization), where X axis represents the input graph size. Middle: runtime of the second step (decomposition), where X axis represents the kernel size. Right: runtime of the third step (exact algorithm), where X axis represents the kernel size.

To understand the detailed runtime of our PTAS, we record the runtime of each step on two graph classes: **networks** and **triangu**. These results are shown in Fig. 4. In this figure, we can see the runtime of the first two steps on each graph class accords with the theoretical $O(n \log n)$ runtime, while the runtime of the third step seems super-linear. We think this is because of the variance of the runtime of our exact algorithm. During our experiment, we notice that our exact algorithm can solve most subgraphs in 30 s, but it needs more than 60 s to solve some (relatively large) subgraphs. When there are more subgraphs, the hard instance may appear and they cost more time.

6.2 Solution Quality

We evaluate the solution quality of our PTAS variants in two aspects: compare the solution with a lower bound of the optimal solution and with solution of another approximation algorithm. For this purpose, we implemented Becker and Geiger's 2-approximation algorithm [9], denoted by **2-approx**, and compare our PTAS with this algorithm on solution quality. This 2-approximation algorithm works for vertex-weighted FVS in general graphs and consists of two steps: computes a greedy solution and removes redundant vertices to obtain a minimal solution. The lower bound of an optimal solution can be obtained from a partial solution of our PTAS for each graph. We compare the solutions of our implementation with this lower bound, and obtain an estimated approximation ratio on each graph. The lower bound and the estimated approximation ratio of different algorithms on all graphs are given in Table 1.

We can see in the Table 1 that even **vanilla** can find good solution for all graphs, and achieve approximation ratio better than 2. This approximation ratio is better than the theoretical approximation ratio of our PTAS obtained from the kernel size-bound and separator size-bound. Further, our implementation can find an almost optimal solution on all random graphs. This is because the kernelization algorithm significantly reduce the size of input graph, and the resulting kernel consists of many small disconnected components that can be solved optimally by our exact algorithm. We observe that for **networks** and **triangu** graphs, the **minimal** variant can find better solution than **vanilla**,

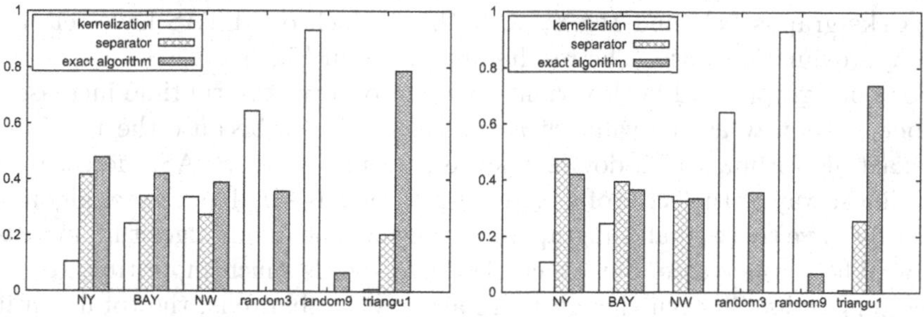

Fig. 5. Decompose PTAS solutions on different graphs. The X axis represents different graphs, and Y axis represents the percentage in the whole solution. Left: solutions from **minimal**. Right: solutions from **optimized**.

and the **optimized** variant provides the best solutions among the three variants, which implies the two heuristics both help improve the solutions. Since **vanilla** works very well on **random** graphs, the improvement from the two heuristics is negligible.

Comparing with **2-approx**, we can see that our PTAS variants work better on **random** and **triangu** graphs; for **networks**, our PTAS can find very competitive solutions. The main reason that our algorithm cannot find smaller solutions than **2-approx** on **networks** is that the subgraphs obtained from the decomposition step are still not large enough, which implies a large fraction of the solution contains the vertices of the separators.

To better understand the solutions from our PTAS, we analyze the fraction of each component in the solutions. In Fig. 5, we show this information for some typical graphs of each class. Comparing different graph classes, we can see the contributions of three components in the final solution of **networks** graphs are more averaged, while the solutions for other two classes are contributed by just one component: for **random** graphs, most solution vertices are collected by the kernelization algorithm, and for the **triangu** graphs, most solution vertices are computed by the exact algorithm. This also agrees with the kernel size information in Table 1: kernelization algorithm produce much smaller kernel for **random** graphs than the **triangu** graphs. There is not much difference between the two variants **minimal** and **optimized**, although we can observe the solution computed by the exact algorithm in **optimized** takes a litter smaller fraction than that in **minimal**. This is because the kernelization algorithm during decomposition in **optimized** reduces the size of kernel, and then the number of resulting subgraphs.

6.3 Effects of Parameters on Performance

Recall that the largest size r of the subgraphs in the decomposition step is the only parameter in our PTAS. Now we analyze the effect of this parameter on the performance of our PTAS. For this purpose, we set a time limit for our exact algorithm as one hour, and starting with $r = 20$, we increase the value of r by 5 each time until our implementation cannot find a feasible solution, which is caused by the fact that the exact algorithm cannot solve some subgraphs of size r in the time limit. We test two variants **minimal** and **optimized** on two **networks** graphs **NY** and **BAY**, on both of which our PTAS have relatively large approximation ratio. The results are shown in Fig. 6.

For both graphs and both variants, we can observe the runtime increases in a super-linear way as the value of r increases. This is because the runtime of our exact algorithm, which dominates the runtime of our PTAS, increases in a super-linear way as the value of r increases. When r is small, our exact algorithm can easily solve those small subgraphs. So the runtime of our algorithm increases slowly. When r increases, our exact algorithm needs much more time to solve larger subgraphs. So even though there are fewer subgraphs, the total runtime of the third step increases quickly as r increases. We also notice that the largest r solvable by **optimized** is smaller than **minimal**. We think this is because

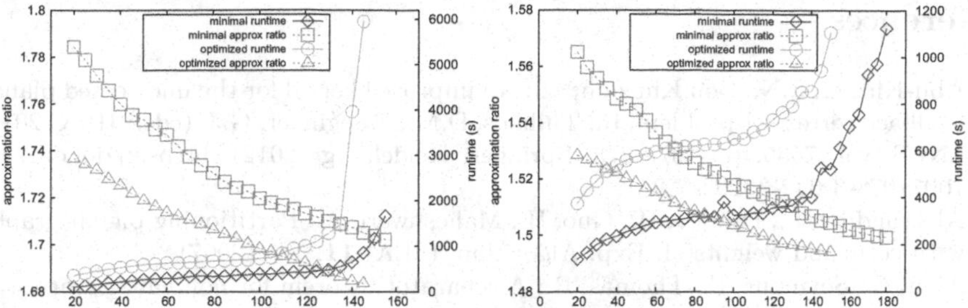

Fig. 6. Effects of region size r on the performance of **minimal** and **optimized**. The X axis represents the value of r, the left Y axis represents the estimated approximation ratio (the lower bound is obtained by setting $r = 120$) and the right Y axis represents the runtime in seconds. Left: results on graph **NY**. Right: results on graph **BAY**.

the subgraphs in **optimized** are usually harder to solve exactly than those in **minimal**. And this is one reason that explains the fact that **optimized** always needs more time than **minimal** for the same value of r.

Figure 6 also shows that the approximation ratio of both variants decreases in a sub-linear way as r increases on both graphs. This accords with the theoretical relationship between r and the error parameter ϵ: $\epsilon = O(1/\sqrt{r})$. With the same value of r, we can see **optimized** always achieves smaller approximation ratio, implying that our heuristic works well for different value of r.

7 Conclusions

We proposed an $O(n \log n)$ time PTAS for the minimum feedback vertex set problem in planar graphs. We also implemented this algorithm based on a corrected linear kernel for this problem and evaluated its performance on some large planar graphs. Our results show that our PTAS can achieve better approximation ratio than the theoretical guarantees on all testing graphs and can find better or competitive solutions compared with the 2-approximation algorithm. Our proposed heuristics can further improve the solution quality of our PTAS implementation. We think these results show that separator-based PTASes are promising to be used in practical applications.

Although we avoid expensive dynamic programming step in our PTAS, we still need an exact algorithm for small subgraphs after the decomposition of kernel. Our experiments show that the running time of this exact algorithm is the bottleneck of the total runtime of our PTAS. Because of this, our PTAS needs much more time than the simple 2-approximation algorithm. One method to speed up our PTAS is to apply the exact algorithm in a parallel way to solve multiple subgraphs simultaneously. However, we find in our experiment that there exist subgraphs with 120 vertices that our exact algorithm needs more than one hour to solve the problem. So a faster exact algorithm will be very helpful not only in essentially reducing the runtime of our PTAS, but also in obtaining better solutions.

References

1. Abu-Khzam, F.N., Bou Khuzam, M.: An improved kernel for the undirected planar feedback vertex set problem. In: Thilikos, D.M., Woeginger, G.J. (eds.) IPEC 2012. LNCS, vol. 7535, pp. 264–273. Springer, Heidelberg (2012). https://doi.org/10.1007/978-3-642-33293-7_25
2. Aleksandrov, L., Djidjev, H., Guo, H., Maheshwari, A.: Partitioning planar graphs with costs and weights. J. Exp. Algorithm. (JEA) **11**, 1–5 (2007)
3. Alon, N., Seymour, P., Thomas, R.: A separator theorem for nonplanar graphs. J. Am. Math. Soc. **3**(4), 801–808 (1990)
4. Arora, S., Grigni, M., Karger, D.R., Klein, P.N., Woloszyn, A.: A polynomial-time approximation scheme for weighted planar graph TSP. In: SODA, vol. 98, pp. 33–41 (1998)
5. Baker, B.S.: Approximation algorithms for NP-complete problems on planar graphs. J. ACM (JACM) **41**(1), 153–180 (1994)
6. Bar-Yehuda, R., Geiger, D., Naor, J., Roth, R.M.: Approximation algorithms for the feedback vertex set problem with applications to constraint satisfaction and Bayesian inference. SIAM J. Comput. **27**(4), 942–959 (1998)
7. Bateni, M.H., Hajiaghayi, M.T., Marx, D.: Approximation schemes for Steiner forest on planar graphs and graphs of bounded treewidth. J. ACM **58**(5), 21 (2011)
8. Becker, A., Fox-Epstein, E., Klein, P.N., Meierfrankenfeld, D.: Engineering an approximation scheme for traveling salesman in planar graphs. In: LIPIcs-Leibniz International Proceedings in Informatics, vol. 75 (2017)
9. Becker, A., Geiger, D.: Optimization of Pearl's method of conditioning and greedy-like approximation algorithms for the vertex feedback set problem. Artif. Intell. **83**(1), 167–188 (1996)
10. Bodlaender, H.L., Fomin, F.V., Lokshtanov, D., Penninkx, E., Saurabh, S., Thilikos, D.M.: (Meta) kernelization. J. ACM (JACM) **63**(5), 44 (2016)
11. Bonamy, M., Kowalik, Ł.: A 13k-kernel for planar feedback vertex set via region decomposition. Theoret. Comput. Sci. **645**, 25–40 (2016)
12. Borradaile, G., Klein, P., Mathieu, C.: An $O(n \log n)$ approximation scheme for Steiner tree in planar graphs. ACM Trans. Algorithms (TALG) **5**(3), 1–31 (2009)
13. Boyer, J.M., Myrvold, W.J.: On the cutting edge: simplified $O(n)$ planarity by edge addition. J. Graph Algorithms Appl. **8**(2), 241–273 (2004)
14. Chiba, N., Nishizeki, T., Saito, N.: Applications of the Lipton and Tarjan's planar separator theorem. J. Inf. Process. **4**(4), 203–207 (1981)
15. Chiba, N., Nishizeki, T., Saito, N.: An approximation algorithm for the maximum independent set problem on planar graphs. SIAM J. Comput. **11**(4), 663–675 (1982)
16. Cohen-Addad, V., de Verdière, É.C., Klein, P.N., Mathieu, C., Meierfrankenfeld, D.: Approximating connectivity domination in weighted bounded-genus graphs. In: Proceedings of the 48th Annual ACM SIGACT Symposium on Theory of Computing, pp. 584–597. ACM (2016)
17. Demaine, E.D., Hajiaghayi, M.T., Kawarabayashi, K.-i.: Algorithmic graph minor theory: decomposition, approximation, and coloring. In: Proceedings of the 46th Annual IEEE Symposium on Foundations of Computer Science, pp. 637–646 (2005)
18. Demaine, E.D., Hajiaghayi, M.T.: Bidimensionality: new connections between FPT algorithms and PTASs. In: Proceedings of the Sixteenth Annual ACM-SIAM Symposium on Discrete Algorithms, SODA 2005, pp. 590–601 (2005)

19. Demetrescu, C., Goldberg, A., Johnson, D.: Implementation challenge for shortest paths. In: Kao, M.Y. (ed.) Encyclopedia of Algorithms. Springer, Boston (2008). https://doi.org/10.1007/978-0-387-30162-4
20. Eisenstat, D., Klein, P., Mathieu, C.: An efficient polynomial-time approximation scheme for Steiner forest in planar graphs. In: Proceedings of the Twenty-Third Annual ACM-SIAM Symposium on Discrete Algorithms, pp. 626–638. SIAM (2012)
21. Fomin, F.V., Lokshtanov, D., Saurabh, S., Thilikos, D.M.: Bidimensionality and kernels. In: Proceedings of the Twenty-First Annual ACM-SIAM Symposium on Discrete Algorithms, pp. 503–510 (2010)
22. Fomin, F.V., Lokshtanov, D., Saurabh, S., Thilikos, D.M.: Linear kernels for (connected) dominating set on H-minor-free graphs. In: Proceedings of the Twenty-Third Annual ACM-SIAM Symposium on Discrete Algorithms, pp. 82–93 (2012)
23. Fox-Epstein, E., Mozes, S., Phothilimthana, P.M., Sommer, C.: Short and simple cycle separators in planar graphs. J. Exp. Algorithm. (JEA) **21**, 2 (2016)
24. Grohe, M.: Local tree-width, excluded minors, and approximation algorithms. Combinatorica **23**(4), 613–632 (2003)
25. Holzer, M., Schulz, F., Wagner, D., Prasinos, G., Zaroliagis, C.: Engineering planar separator algorithms. J. Exp. Algorithm. (JEA) **14**, 5 (2009)
26. Iwata, Y.: Linear-time kernelization for feedback vertex set. CoRR (2016)
27. Iwata, Y., Wahlström, M., Yoshida, Y.: Half-integrality, LP-branching, and FPT algorithms. SIAM J. Comput. **45**(4), 1377–1411 (2016)
28. Karp, R.M.: Reducibility among combinatorial problems. In: Miller, R.E., Thatcher, J.W., Bohlinger, J.D. (eds.) Complexity of Computer Computations. The IBM Research Symposia Series. Springer, Boston (1972). https://doi.org/10.1007/978-1-4684-2001-2_9
29. Klein, P.N.: A linear-time approximation scheme for TSP in undirected planar graphs with edge-weights. SIAM J. Comput. **37**(6), 1926–1952 (2008)
30. Kozen, D.C.: The Design and Analysis of Algorithms. Springer, Heidelberg (2012)
31. Le, H., Zheng, B.: Local search is a PTAS for feedback vertex set in minor-free graphs. CoRR, abs/1804.06428 (2018)
32. Lipton, R.J., Tarjan, R.E.: A separator theorem for planar graphs. SIAM J. Appl. Math. **36**(2), 177–189 (1979)
33. Lipton, R.J., Tarjan, R.E.: Applications of a planar separator theorem. SIAM J. Comput. **9**(3), 615–627 (1980)
34. Marzban, M., Qian-Ping, G.: Computational study on a PTAS for planar dominating set problem. Algorithms **6**(1), 43–59 (2013)
35. Mehlhorn, K., Näher, S., Uhrig, C.: The LEDA platform for combinatorial and geometric computing. In: Degano, P., Gorrieri, R., Marchetti-Spaccamela, A. (eds.) ICALP 1997. LNCS, vol. 1256, pp. 7–16. Springer, Heidelberg (1997). https://doi.org/10.1007/3-540-63165-8_161
36. Tazari, S., Müller-Hannemann, M.: Dealing with large hidden constants: engineering a planar Steiner tree PTAS. J. Exp. Algorithm. (JEA) **16**, 3–6 (2011)

On New Rebalancing Algorithm

Koba Gelashvili[1], Nikoloz Grdzelidze[2], and Mikheil Tutberidze[3](✉)

[1] School of Business, Computing and Social Sciences, St. Andrew the First-Called Georgian University of the Patriarchate of Georgia, 53a Chavchavadze Ave., Tbilisi 0179, Georgia
koba.gelashvili@sangu.edu.ge

[2] Department of Computer Sciences, Faculty of Exact and Natural Sciences, Tbilisi State University, 2, University St., Tbilisi 0143, Georgia
nikolozgrdzelidze@gmail.com

[3] Institute of Applied Physics, Ilia State University, 3/5 Kakutsa Cholokashvili Ave., Tbilisi 0162, Georgia
mtutberidze@gmail.com

Abstract. This paper proposes a new algorithm for rebalancing a binary search tree, called qBalance, by a certain analogy. The running time of this algorithm is proportional to the number of keys in the tree. To compare qBalance with existing rebalancing algorithms, several versions of the algorithm are implemented. We compare qBalance with the well-known DSW algorithm on binary search trees (BSTs), with the Sedgewick algorithm on an ordered tree, with the modified Sedgewick algorithm on Red-Black (RB) trees with the specific structure of the node. For RB trees whose nodes have a standard (typical) structure, qBalance is implemented using asynchronous mode. This version of the algorithm is considerably complicated, but it is twice as fast as the serial implementation. The results of numerical experiments confirm the advantage of the new algorithm compared to DSW. The Sedgewick algorithm and its modification retain advantage by 30% in terms of running time on ordered trees. On red-black trees, the advantage of the new algorithm is significant. This especially applies to the asynchronous version of the algorithm.

Keywords: Binary search tree · RB-tree · DSW algorithm · Sedgewick balancing algorithm

1 Introduction

The issue of balancing of binary search trees is quite an old one for computer sciences, albeit few publications are devoted to this subject. Day (see [1]) in 1976 proposed an algorithm that balances the binary search tree in linear time without using the extra memory. Theoretically, Day's algorithm is the fastest balancing algorithm but its variant - the DSW algorithm is more refined because it creates a complete tree (see [2]) - in which every level, except possibly the last, is completely filled, and the bottommost tree level is filled from left to right.

I. Kotsireas et al. (Eds.): SEA² 2019, LNCS 11544, pp. 114–124, 2019.
https://doi.org/10.1007/978-3-030-34029-2_8

There is also an alternative approach proposed by Sedgewick (see [3]). Even though his recursive algorithm has the worst-case time complexity $n \log(n)$ and requires the node structure augmentation via adding the subtree size, in practical applications, it is more effective than DSW algorithm. The tree with nodes having subtree sizes as attributes are called ordered search tree or simply OST.

In 2016, the modified Sedgewick's algorithm (MSA) was proposed (see [4]). It works on OSTs, but when using the color field as the subtree size, it can rebalance Red-Black trees (in short RB-tree). After rebalancing, the tree should be colored again, and the properties of RB-tree will be restored. Despite some advantages, the Sedgewick algorithm and its modification are not applicable to general BSTs.

RB-trees (see [5]) are wide-spread. A lot of data structures used in high-level programming languages or in computational geometry, and the Completely Fair Scheduler used in current Linux kernels, are based on RB trees. RB trees belong to the category of balanced trees, so the rebalancing algorithms on RB trees are not of great importance in general. But, the situation has changed recently for the following reasons:

- The usage of greater data became necessary;
- In realistic tasks the height of the RB tree is often very close to its theoretical upper bound;
- Computers are not switched in the passive state immediately (e.g. hibernate) and naturally appears the possibility to rebalance the certain structure;
- The rebalancing algorithms might be faster when using multithreading. For example, typical laptops can ensure 4–12 threads or even more;
- The "boost" library uses RB trees to store large sparse matrices.

In Sect. 2, we introduce new rebalancing algorithm, which, like DSW, works on BSTs. We call it qBalance (because of its similarity to qSort algorithm). qBalance has similar to the DSW algorithm theoretical characteristics, it works in linear time. Numerical results show that in all benchmarking scenarios qBalance outperforms DSW algorithm.

In Sect. 3, we compare our qBalance algorithm to the Sedgewick rebalancing algorithm on OSTs. The latter is approximately 30% faster than the former. The reason is explained.

Like the original and modified Sedgewick algorithms, qBalance can be easily programmed on RB trees with double structured nodes. In Sect. 4 we compare qBalance and the modified Sedgewick algorithms. The new algorithm keeps a stable advantage during all possible scenarios.

In Sect. 5 qBalance is implemented using two threads in case of RB tree with standard node. As a result, the execution time is decreased two times in comparison to the serial qBalance.

Note that qBalance algorithms implemented over RB trees do not use the notion of rotation at all.

Projects and other materials used in numerical experiments were uploaded to GitHub and are available at address https://github.com/kobage. Four folders are created there. In the first folder "BST" the implementation of BST three class is stored which includes DSW and qBalance algorithms. In the second folder "OST" the implementation of OST tree is stored, which includes Sedgewick's modified and qBalance algorithms. In the third folder "RB" the implementation of RB- tree with the double structure of the node is stored, which includes modified Sedgewick's and qBalance algorithms. The latter is arranged so that as a result a complete tree is created. In the fourth folder "RB-asynchronous", the implementation of RB tree with nodes having the standard structure is stored. The implementation includes qBalance serial and asynchronous versions. For simplicity purposes we stay within rebalancing algorithms, therefore we have not implemented a node deletion algorithm in these classes.

For the children nodes, we use an array of pointers. Such approach significantly shortens the code. It is possible to use an alternative representation. The representation used here is without an alternative for the asynchronous rebalancing of RB-trees.

For the sake of simplicity, we assume that the keys of the trees have some T type, for which the binary operation "$\leq$" of comparison is defined. For more generalized implementation it is necessary to pass the comparison binary predicate to the class template as a parameter, which is not difficult but complicates notations and code.

2 qBalance Algorithm Over Binary Search Tree. The Comparison with DSW Algorithm

DSW algorithm works in the general case of BST tree, the node structure of which contains a minimal number of members. In our implementations, nodes have the following structure

```
template<typename T>
struct Node{
      T key;
      Node* child[2];
      Node();
      Node(T keyValue);
};
```

Which is often used in practice (see [5, p. 32]).

DSW is a two-stage algorithm. In the first stage, the tree is transformed into a particular simple form. The authors call this transformation by treeToWine, because of the analogy with the visual image of the tree obtained as a result. On the second stage, from this kind of tree, the balanced tree is constructed. We don't display algorithms here; their codes can be accessed on GitHub.

The treeToWine algorithm is useful for other purposes as well. For example, treeToWine is effective for coding the tree class destructor. BST may not be balanced and in such case tree traversing recursive algorithms will cause a stack overflow. The standard solution of the problem is to use a stack in the destructor code. But the destructor will be faster when using treeToWine.

In the case of BST, the treeToWine algorithm is used by qBalance algorithm. The better alternative for balanced trees will be considered in further paragraphs.

Let us describe qBalance algorithm.

Unless otherwise specified, we assume that the templated class $Tree < T >$ is created without using of the fictive node for leaves. If any node does not have left or right child, the address *nullptr* is written in the appropriate field. The attribute "root" in the tree class defines the address of the root.

The public method that balances the tree is as follows:

```
template<typename T>
void Tree<T>::qBalance()
{
    Node<T>* pseudo_root = new Node<T>();
    pseudo_root->child[1] = root;
    tree_to_vine(pseudo_root);

    root = qBalance(pseudo_root->child[1], size);
    delete pseudo_root;
}
```

It starts working after the tree to be balanced is transformed to the linked list by the treeToWine algorithm. The obtained list consists of tree nodes and in this list, any node (of the tree) stores the address of its next element in pointer $x->$ *child*[1]. Balancing is performed by the private recursive method, which will receive the address of some node of the list and the number of nodes to be balanced as parameters.

```
template<typename T>
Node<T>* Tree<T>::qBalance(Node<T>*& node, int m)
{
     if (m == 2)
     {
          Node<T>* tmp = node->child[1];
          tmp->child[0] = node;
          node->child[0] = node->child[1] = nullptr;
          node = tmp->child[1];
          tmp->child[1] = nullptr;
          return tmp;
     }
     if (m == 1)
     {
          Node<T>* tmp = node;
          node = node->child[1];
          tmp->child[0] = tmp->child[1] = nullptr;
          return tmp;
     }
     int q = m / 2;
```

```
        Node<T>* a = qBalance(node, q);
        Node<T>* b = node;
        b->child[0] = a;
        node = node->child[1];
        a = qBalance(node, m - q - 1);
        b->child[1] = a;
        return b;
}
```

Each node of the list (created by treeToWine algorithm) is visited only once by the qBalance algorithm. Therefore the algorithm works in time proportional to the number of nodes (linear time). the algorithm is recursive and starts with checking the stopping conditions. If it is invoked for two nodes or one node, it constructs the corresponding subtree and returns its root. It is obvious that any tree with one or two nodes is balanced. Note that when the algorithm (function) exits, the address of the node passed to function by reference stores the address of the successor node to the node containing maximal key in the subtree.

If the number of nodes is more than 2, then it is divided into two parts. After balancing the first part the next address of the node containing the maximal key of the balanced subtree is placed in the parameter "node". This node will become the root of the rebalanced tree and its address will be returned by algorithm qBalance. Before this, the address returned after rebalancing second subtree (excluding future root of this subtree) will become the right child of the root (which is returned by algorithm qBalance). Finally, algorithm will return the root of the balanced subtree and the address of the node which is the successor of the node containing the maximal key of the balanced subtree will be written in the "node".

We don't think that rebalancing general BSTs has actual practical importance, therefore we do not complicate the algorithm in order to receive complete tree as a result.

Results of numerical experiments show that qBalance is 30% faster than DSW.

3 qBalance Algorithm over Ordered Search Tree (OST). The Comparison with Sedgewick Algorithm

The node in our simple implementation of OST has the following structure:

```
template<typename T>
struct Node
{
        T key;
        Node* child[2];
        int bf;
        Node();
        Node(T);
};
```

For a given node x, the attribute $x\!-\!> bf$ contains the number of nodes in the subtree rooted at x. In some cases it is more convenient to use a function $N(x)$, which coincides with $x\!-\!> bf$, if x is an address of any existing node, or 0 if x is *nullptr*.

Sedgewick algorithm is well known, and in notations of this section, it is described in [4]. Therefore, we will not stay at this.

The two versions of qBalance, developed for OST and BST trees, slightly differs from each other. The public method that balances the tree is the same,

```
template<typename T>
Node<T>* Tree<T>::qBalance(Node<T>*& node, int m)
```

whilst in the private method the three additional rows ensure that the "bf" field of the node of the rebalanced tree reflects the size of the subtree, rooted at this node.

To this end, before returning the root of the two-noded balanced tree we are pointing that $tmp\!-\!> bf = 2$; before returning the root of the one-noded tree we are pointing that $tmp\!-\!> bf = 1$; finally, before returning the root of the subtree balanced via method qBalance (Node$< T >$ $*\&$ node, int m) we calculate its size:

```
b->bf = N(b->child[0]) + N(b->child[1]) + 1;
```

This algorithm works in the linear time but it needs to traverse the tree twice to make it balanced. Because of this Sedgewick algorithm is 30% faster than qBalance in experiments conducted on OST with random data. When data is coming in increasing or decreasing order, it is difficult to notice the difference since the large-sized tree is constructed in square time, very slowly. Besides, this case has only theoretic interest.

It is also easy to develop versions of qBalance for OST trees, resulting in a complete tree. However, we will do it only for the standard version of RedBlack tree because otherwise we will not be able to create the asynchronous rebalancing version of qBalance.

4 qBalance Algorithm Over Doublestructured RB Tree. The Comparison with the Modified Sedgewick Algorithm

As well as in [4] let us consider RB tree with nodes having the following structure:

```
template<typename T>
struct Node
{
      T key;
      Node* p;
      Node* child[2];
      int bf;
```

```
    Node();
    Node(T);
};
```

This allows us to consider the field (data member) "bf" as either size or color of subtree, depending on which properties are satisfied by tree – OST or RB. The tree having such node structure is double structured in the certain sense - it is not difficult to convert OST into RB and vice versa. The modified Sedgewick algorithm for the Double Structured RB tree is little faster than basic algorithm, therefore it is used for comparisons.

```
template<typename T>
void Tree<T>::balanceMod()
{
    updateSizes(root);
    root = balanceMod(root);
    int maxHeight = (int)log2(size);
    updateColors(root, maxHeight);
}
```

It is seen that together with the code being executed directly by rebalancing algorithm (over OST tree and which is faster than qBalance in this case) it is necessary to double traverse of the tree - the first one transforms RB tree to OST tree and after rebalancing the second traverse refreshes colors in order to restore RB tree properties. These procedures are described in [4] and are uploaded to GitHub.

qBalance algorithm has no need to take into account OST tree specifics, therefore this algorithm does only one traverse on the tree which determines its advantage in the sense of run time. qBalance algorithm should convert the tree into a list and select colors to obtain RB- tree again after rebalancing. These two tasks are done simultaneously within one traverse by the following private method:

```
template<typename T>
void Tree<T>::treeToList(Node<T>* x);
```

Which is later used by rebalancing algorithm:

```
template<typename T>
void Tree<T>::qBalanceComplete()
{
    int height = (int)log2(size);
    mateInt = pow(2, height) - 1;
    mateCounter = 0;
    treeToList(root);
    root = qBalanceComplete(0, size - 1);
}
```

Numerical experiments show that qBalance works approximately 30% faster for the RB tree with double structure than modified Sedgwick algorithm.

We developed this version of the qBalance algorithm to obtain a complete picture of the efficiency of qBalance for RB trees. Hence, its code is not completely processed in the sense of optimization and the design. These tasks are solved in the case of the standard structure of RB-trees.

5 qBalane Algorithm for Standard RB Tree. The Comparison Between Sequential and Parallel (Two Threads) Implementations

In the structure of the tree node is only one difference. Instead of the field

```
int bf;
```

we have

```
char color;
```

Hence we can consider only DSW and qBalance algorithms and their modifications. We consider an asynchronous modification of qBalance algorithm which uses two threads. According to Amdahl's law (see [7]) there is no reason for using more threads. Indeed, qBalance algorithm traverses the whole tree twice to rebalance it. In any case, it is necessary to traverse the tree at least once. It is clear that a tree traversal is loaded by various tasks of processing the nodes (see code), but the main part of the execution time corresponds with the traverse of the tree.

There is another issue. In the case of the standardly structured node, we do not know any algorithm which simultaneously will part tree nodes into more than two linked lists. The size of lists must be approximately equal and lists must be ordered linearly in the following sense: for any two lists either the key of arbitrary node of the first list is less or equal to the key of the arbitrary node of the second list or the key of arbitrary node of the first list is greater or equal to the key of the arbitrary node of the second list.

Describe briefly the rebalancing asynchronous algorithm

```
template<typename T>
void Tree<T>::qBalanceAsync();
```

The core difference from other variants of the qBalance algorithm is that on the first stage from the tree to be rebalanced two singly linked lists are created. Lists are created at the same time asynchronously. To do this the two generic algorithms are prepared in advance:

```
template<typename T>
Node<T>* Tree<T>::extremum(Node<T>* x, bool direction);
```

and

```
template<typename T>
Node<T>* Tree<T>::next(Node<T>* x, bool direction);
```

An invocation extremum($x, true$); will return the address of the node of the subtree with x root containing maximal key and extremum($x, false$); will return the address of the node of the subtree with x root containing the minimal key.

Similarly, any invocation of next($x, true$); will return address of the node that is the successor of the x node and an invocation next($x, false$); will return address of the node that is the predecessor of the x node.

The algorithm that transforms the tree into the list has the declaration

```
template<typename T>
void Tree<T>::treeToList_Colored
(
     Node<T>*& head,
     int NUmberOfNodes,
     const bool direction,
     const int height
)
```

First it defines the head of the list by the statement

```
    head = extremum(root, !direction);
```

the head represents the address of the node of the tree containing the maximal or minimal key. Then the algorithm starts construction of the list and determines colors of the nodes taking into account their final position. Any invocation containing "true" as an argument will give singly linked list with the head "head". In this list the address of node next to x node is stored in field $x->$ *child*[0]. Similarly, when in an invocation of the algorithm participates argument "false" then the address of the node next to x node is stored in field $x->$ *child*[1]. In the process of the transformation to the list, the determination of the color of the node is done considering the position which will be occupied by the node in the rebalanced tree. But this is a simple technical aspect and we will not pay much attention to it.

The idea of the second part of qBalanceAsync is simple. The tree is already divided into two parts (lists), and these parts will be rebalanced asynchronously (simultaneously) excepting the middle node, which will become the root after executing specific necessary procedures. The rebalancing algorithm which runs on two parts of the tree simultaneously principally is the same as considered in the previous section, but technically it is considerably complicated as two lists obtained from the tree store in two different fields addresses of next nodes.

```
template<typename T>
void Tree<T>::qBalancePublic(const bool direction)
```

Any invocation containing argument "true" creates a complete tree and fills in the bottom level from left to right (see left part of Fig. 1). On the other hand,

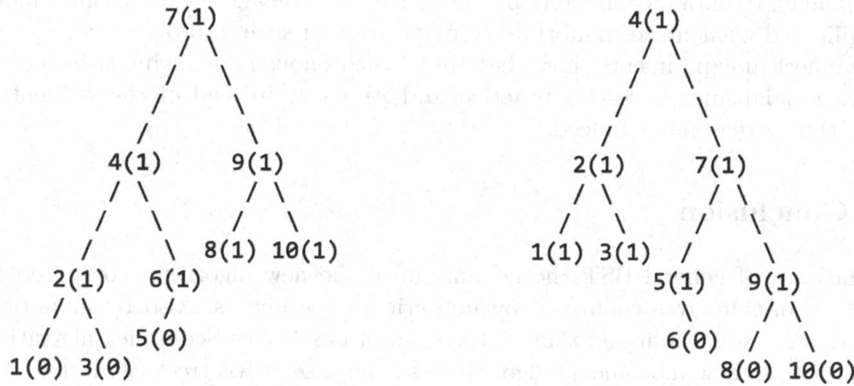

Fig. 1. Two different rebalancing scenarios, used in asynchronous mode.

any invocation containing argument "false" fills in bottom level from right to left (see right part of Fig. 1). Such flexibility is necessary for asynchronous rebalancing

```
template<typename T>
void Tree<T>::qBalanceAsync();
```

to consider all possible scenarios. First of all when heights of right and left subtrees are equal, then first thread works on left part of the tree through the scenario of the left part of Fig. 1 whilst second thread works through the scenario of the right part of the same figure (see left part of Fig. 2). When the heights of the subtrees are different, then qBalanceAsync algorithm paints the right subtree only in black. further improvement is possible (to improve) of asynchronous

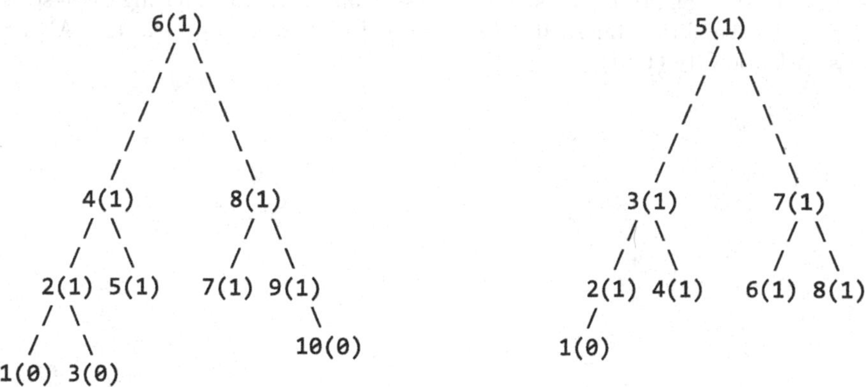

Fig. 2. Two different painting scenarios, used in asynchronous mode.

rebalancing to obtain completely balanced tree but the algorithm becomes more complicated what makes doubtful the motivation of such improvement.

Numerical experiments show that the asynchronous rebalancing twice accelerates rebalancing process in practice and its usage instead of the sequential algorithm is reasonable indeed.

6 Conclusion

In the case of general BST the advantage of the new algorithm compared to DSW algorithms was confirmed by numerical experiments. Also by numerical experiments was confirmed that in the case of OST trees Sedgewick algorithm is faster than new rebalancing algorithm. In the case of RB trees (with different structures of nodes) advantage of new algorithm is obvious, especially if its asynchronous version is used. However, the code of the latter is considerably complicated in comparison to other versions of qBalance. In the case when the new algorithm is not faster than its alternatives, the perspective of speeding-up of qBalance exists.

References

1. Day, C.: Comput. J. **XIX**, 360–361 (1976)
2. Stout, Q.F., Warren, B.L.: Commun. ACM **29**(9), 902–908 (1986)
3. Sedgewick, R.: Algorithms in C, Parts 1–5. Addison-Wesley Professional, Boston (2001)
4. Gelashvili, K., Grdzelidze, N., Shvelidze, G.: The modification of the Sedgewick's balancing algorithm. Bull. Georgian Acad. Sci. **10**(3), 60–67 (2016)
5. Pfaff, B.: An Introduction to Binary Search Trees and Balanced Trees. Libavl Binary Search Tree Library, vo.1.1: Source Code Ver. 2.0.2 (2004)
6. Rolfe, J.T.: One-time binary search tree balancing: the Day/Stout/Warren (DSW) algorithm. SIGCSE Bull. **34**, 85–88 (2002)
7. Amdahl, G.M.: Validity of the single processor approach to achieving large-scale computing capabilities. In: AFIPS Conference Proceedings, pp. 483–485. AFIPS Press, Atlantic City (1967)

Colorful Frontier-Based Search: Implicit Enumeration of Chordal and Interval Subgraphs

Jun Kawahara[1(✉)], Toshiki Saitoh[2], Hirofumi Suzuki[3], and Ryo Yoshinaka[4]

[1] Nara Institute of Science and Technology, Ikoma, Japan
jkawahara@is.naist.jp
[2] Kyushu Institute of Technology, Iizuka, Japan
toshikis@ces.kyutech.ac.jp
[3] Hokkaido University, Sapporo, Japan
h-suzuki@ist.hokudai.ac.jp
[4] Tohoku University, Sendai, Japan
ryoshinaka@tohoku.ac.jp

Abstract. This paper considers enumeration of specific subgraphs of a given graph by using a data structure called a zero-suppressed binary decision diagram (ZDD). A ZDD can represent the set of solutions quite compactly. Recent studies have demonstrated that a technique generically called frontier-based search (FBS) is a powerful framework for using ZDDs to enumerate various yet rather simple types of subgraphs. We in this paper, propose colorful FBS, an enhancement of FBS, which enables us to enumerate more complex types of subgraphs than existing FBS techniques do. On the basis of colorful FBS, we design methods that construct ZDDs representing the sets of chordal and interval subgraphs from an input graph. Computer experiments show that the proposed methods run faster than reverse search based algorithms.

Keywords: Graph algorithm · Graph enumeration · Decision diagram · Frontier-based search · Interval graph · Chordal graph

1 Introduction

Enumeration problems are fundamental in computer science and have many applications in areas of bioinformatics, operations research, high performance computing, and so on. Various kinds of enumeration problems and algorithms for them have been studied so far [25]. One of the most powerful frameworks for enumeration algorithms is *reverse search*, proposed by Avis and Fukuda [1], which is the basis for a vast number of enumeration algorithms thanks to its simplicity. However, since reverse search outputs solutions one by one, its computation time is proportional to at least the number of solutions to enumerate, which may be exponentially large in input and unacceptable for many applications.

I. Kotsireas et al. (Eds.): SEA^2 2019, LNCS 11544, pp. 125–141, 2019.
https://doi.org/10.1007/978-3-030-34029-2_9

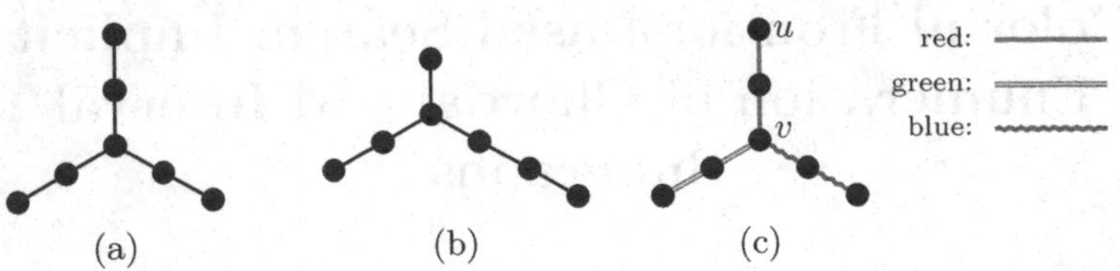

Fig. 1. Variants of T_2. (Color figure online)

In this paper, we consider enumeration of subgraphs of a given graph. For subgraph enumeration, another framework uses zero-suppressed binary decision diagrams (ZDDs) [21]. A ZDD is a data structure that compactly represents and efficiently manipulates a family of sets. A set of subgraphs of a graph is represented by a ZDD by regarding a subgraph as an edge set and a set of subgraphs as a family of edge sets, while disregarding isolated vertices. The size of the ZDD is usually much, often exponentially, smaller than the cardinality of the set of the subgraphs. Sekine et al. [24] and Knuth [16] proposed algorithms that construct decision diagrams representing all the spanning trees and all the paths of a given graph, respectively. Kawahara et al. [11] generalized their algorithms for many kinds of subgraphs, which are specified by degrees of vertices, connectivity of subgraphs and existence of cycles (for example, a path is a connected subgraph having two vertices with degree one and others with degree two), and called the resulting framework *frontier-based search* (FBS). The running time of FBS algorithms depends more on the size of the ZDD than the number of subgraphs. FBS has been used for many applications such as smart grid [6], political districting [10], hotspot detection [7], (a variant of) the longest path problem [12], and influence diffusion in WWW [19]. However, the existing FBS framework can handle only graphs that have the rather simple specifications mentioned above. More complex graphs such as *interval graphs* and *chordal graphs* are out of its range.

Graphs with geometric representations are important in areas of graph algorithms and computational geometry. For example, interval graphs have many applications in areas of bioinformatics, scheduling, and so on [5]. Chordal graphs also have such geometric representations and have many applications such as matrix computation and relational databases [2,3]. Despite their complex geometric representations, many of those classes are characterized by some relatively simple forbidden induced subgraphs. For example, chordal graphs have no induced cycle with length at least four. Actually it is easy to enumerate cycles of size at least four by an FBS algorithm.

In this paper, we propose a novel technique that enhances FBS, which we call *colorful frontier-based search* (colorful FBS), for subgraph enumeration problems, specifically enumeration of chordal and interval subgraphs. The idea of colorful FBS is to "colorize" edges of a subgraph. The set of subgraphs with colored edges is represented by a multiple-valued decision diagram (MDD) [20]. For example, the graph T_2, shown in Fig. 1(a) cannot be uniquely determined only by the FBS specification: three vertices with degree one, three vertices with degree two, and a vertex with degree three, because the one in Fig. 1(b) also satisfies this specification. Our colorful FBS employs "colored degree" of vertices. For example, the vertex u in the graph of Fig. 1(c) has red degree one, green degree zero, and blue degree zero, and the vertex v has red degree one, green degree one, and blue degree one, and so on. No other graphs satisfy the same colored degree specification. By "decolorizing" the graph in Fig. 1(c), we obtain the one in Fig. 1(a). By colorful FBS and decolorization, we can treat many more kinds of graph classes including $\mathcal{T}_2$, $\mathcal{X}_{31}$, $\mathcal{XF}_2^{\geq 1}$, and $\mathcal{XF}_3^{\geq 0}$, whose members are shown in Fig. 2 (we borrow the names of the five classes from [23]). Furthermore, we develop a technique that enables us to handle graphs characterized by forbidden induced subgraphs as an important application of colorful FBS, like chordal and interval graphs. We construct characterizing graphs by adding induced edges as colored edges to given forbidden subgraphs and then enumerate desired subgraphs characterized by forbidden induced subgraphs.

The contributions of the paper are as follows:

- We propose colorful FBS, with which one can construct ZDDs for more graph classes than the existing FBS techniques.
- We also show that a colorful FBS algorithm can construct ZDDs for subgraphs containing no subgraph in a given graph class as an induced subgraph.
- Using these algorithms, we propose methods that construct the ZDDs representing chordal subgraphs and interval subgraphs.
- By numerical experiments, we show that our methods surpass existing reverse-search algorithms for enumerating chordal and interval subgraphs.

The rest of the paper is organized as follows. We provide some preliminaries in Sect. 2. Section 3 describes colorful FBS for some graph classes. We propose a decolorization algorithm that converts an MDD constructed by colorful FBS into an ordinary ZDD in Sect. 3.3. In Sect. 4, we develop a method for constructing the ZDD representing the set of subgraphs that contain no subgraphs, given as a ZDD, as an induced subgraph. Section 5 compares the proposed methods and existing algorithms. Finally, we conclude this paper in Sect. 6.

2 Preliminaries

2.1 Forbidden Induced Subgraphs and Chordal and Interval Graphs

A graph is a tuple $G = (V, E)$, where V is a vertex set and $E \subseteq \{\{v, w\} \mid v, w \in V\}$ is an edge set, respectively. The problems we consider in this paper are that for some graph classes $\mathcal{H}$ in concern, we enumerate all the subgraphs H of an input graph G such that $H \in \mathcal{H}$. We assume that the input graph is simple (i.e., the graph has no self-loop or parallel edges) and has at least two vertices. Therefore, each edge is identified with a set of exactly two distinct vertices.

For any vertex subset $U \subseteq V$, $E[U]$ denotes the set of edges whose end points are both in U, i.e., $E[U] = \{\, e \in E \mid e \subseteq U \,\}$, called *induced edges* (*by* U). For any edge subset $D \subseteq E, \bigcup D$ denotes the set of the end points of each edge in D, i.e., $\bigcup D = \bigcup_{\{u,v\} \in D} \{u, v\}$, called *induced vertices* (*by* D). We call $(U, E[U])$ the (*vertex*) *induced subgraph* (*by* U). Let $G[D] = (\bigcup D, D)$, called the *edge*

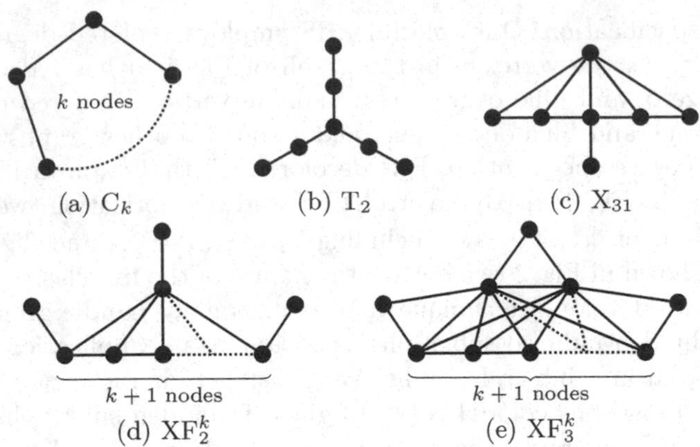

Fig. 2. Graphs C_k, T_2, X_{31}, XF_2^k and XF_3^k. C_k belongs to the graph class $\mathcal{C}_{\geq 4}$ if $k \geq 4$, T_2 belongs to $\mathcal{T}_2$, X_{31} belongs to $\mathcal{X}_{31}$, XF_2^k belongs to $\mathcal{XF}_2^{\geq 1}$ if $k \geq 1$, and XF_3^k belongs to $\mathcal{XF}_3^{\geq 0}$ if $k \geq 0$, respectively.

induced subgraph (by D). This paper often identifies an edge induced subgraph $G' = (\bigcup D, D)$ with the edge set D, ignoring isolated vertices.

Some graph classes are characterized by forbidden induced subgraphs. We say that a graph class $\mathcal{H}$ is *FIS-characterized by a graph class $\mathcal{F}$* if $\mathcal{H}$ consists of graphs $G = (V, E)$ such that no vertex subset of V induces a graph belonging to $\mathcal{F}$, i.e.,

$$(V, E) \in \mathcal{H} \iff \forall U \subseteq V, (U, E[U]) \notin \mathcal{F}.$$

A *chord* of a cycle is an edge such that both endpoints of the edge are contained in the cycle but the edge itself is not. A graph is *chordal* if any cycle with size at least four of the graph has a chord. Thus, the class of chordal graphs is FIS-characterized by the class of cycles of size at least four.

For a graph $G = (V, E)$, a set of intervals $\mathcal{I} = \{I_1, \ldots, I_n\}$ with $I_i = [a_i, b_i]$ and $a_i, b_i \in \mathbb{R}$ is called an *interval model of G* if there is a one-to-one correspondence $f : V \to \mathcal{I}$ such that $\{v, w\} \in E$ holds if and only if $f(v) \cap f(w) \neq \emptyset$. An *interval graph* is a graph that has an interval model. A *proper interval graph* is an interval graph that has an interval model whose intervals are not contained in any other intervals in the model.

The class of interval graphs is known to be FIS-characterized by the five graph classes, $\mathcal{C}_{\geq 4}$, $\mathcal{T}_2$, $\mathcal{X}_{31}$, $\mathcal{XF}_2^{\geq 1}$, and $\mathcal{XF}_3^{\geq 0}$ [18], whose members are shown in Fig. 2. The class of proper interval graphs is FIS-characterized by the five graph classes and $\mathcal{K}_{1,3}$ [9], where $\mathcal{K}_{1,3}$ is the class of graphs isomorphic to $\mathrm{K}_{1,3} = (\{0, 1, 2, 3\}, \{\{0, 1\}, \{0, 2\}, \{0, 3\}\})$.

2.2 Multi-valued Decision Diagrams

A *c-colored subset* of a finite set E is a c-tuple $\vec{D} = (D_1, \ldots, D_c)$ of subsets $D_i \subseteq E$ such that $D_i \cap D_j = \emptyset$ for any distinct i and j. An element $e \in E$ is *colored in i* if $e \in D_i$. To represent and manipulate sets of c-colored subsets, we use $(c+1)$*-valued decision diagrams* ($(c+1)$*-DDs*), which are special types of multiple-valued decision diagrams [20].

A $(c+1)$-DD over a finite set $E = \{e_1, \ldots, e_m\}$ is a labeled rooted directed acyclic graph $\mathbf{Z} = (N, A, \ell)$ with a node set N, an arc set A and a labeling function $\ell : N \to \{1, \ldots, m\}$. We use the terms *nodes* and *arcs* for members of N and A, while the terms *vertices* and *edges* are reserved for constituents of the input graph. The node set N has exactly one root node, denoted by $\rho_{\mathbf{Z}}$, and exactly two terminal nodes $\bot$ and $\top$. Each non-terminal node $\alpha \in N \setminus \{\top, \bot\}$ has a label $\ell(\alpha) \in \{1, \ldots, m\}$ and has exactly $c+1$ outgoing arcs called 0-arc, 1-arc, ..., and c-arc. The node pointed at by the j-arc of α is called the j-child and denoted by α_j for each $j \in \{0, 1, \ldots, c\}$. It is satisfied that $\ell(\alpha_j) = \ell(\alpha) + 1$ if α_j is not a terminal.

Each path π in a $(c+1)$-DD represents a c-colored subset $[\![\pi]\!] = (D_1, \ldots, D_c)$ of E defined by

$$D_j = \{\, e_{\ell(\beta)} \mid \pi \text{ includes the } j\text{-arc of a node } \beta \,\}$$

for $j \in \{1, \ldots, c\}$. The $(c+1)$-DD $\mathbf{Z}$ itself represents a set of c-colored subsets

$$[\![\mathbf{Z}]\!] = \{\, [\![\pi]\!] \mid \pi \text{ is a path from } \rho_{\mathbf{Z}} \text{ to the terminal } \top \,\}.$$

We call a $(c+1)$-DD *reduced* if there are no distinct nodes α and β such that $\ell(\alpha) = \ell(\beta)$ and $\alpha_j = \beta_j$ for all $j \in \{0, \ldots, c\}$. If a $(c+1)$-DD has nodes that violate this condition, those nodes can be merged repeatedly until the $(c+1)$-DD becomes reduced. This reduction does not change the semantics of the $(c+1)$-DD.

We remark that $(c+1)$-DDs, 2-DDs, and 3-DDs are almost identical to MDDs, binary decision diagrams as well as zero-suppressed binary decision diagrams, and ternary decision diagrams, respectively, except a reduction rule that eliminates nodes so that the obtained data structure will be more compact. Our algorithms with slight modification can handle "zero-suppressed" $(c+1)$-DDs, where a node can be eliminated if all the j-children for $1 \leq j \leq c$ point at the terminal $\bot$. However, for simplicity, we have defined $(c+1)$-DDs without employing such a reduction rule, where the label of a child node is always bigger than the parent's by one.

A (c-)colored graph is a tuple $H = (U, (D_1, \ldots, D_c))$, where U is the vertex set and each D_i is a set of edges called *c-colored edges* such that $D_i \cap D_j = \emptyset$ for distinct i and j. The *i-degree* of $u \in U$ is the number of edges in D_i incident to u and the *colored degree* of u is a sequence $\vec{\delta} = (\delta_1, \ldots, \delta_c)$, where δ_i is the i-degree of u. The *colored degree multiset* of H is a multiset s consisting of the colored degrees of all the vertices. The multiplicity of a colored degree $\vec{\delta} \in s$ is denoted by $s(\vec{\delta})$. The *decolorization of* H is the graph $\widetilde{H} = (U, \bigcup_{i=1,\ldots,c} D_i)$ and H is a *c-coloring* of $\widetilde{H}$. A *(c-)colored subgraph* of a graph G is a (c-)colored graph whose decolorization is a subgraph of G. By identifying a colored subgraph with its c-colored edge set, the set of c-colored subgraphs can be represented by a $(c+1)$-DD. Throughout the paper, we use red, green, and blue to represent the first, second, and third colors, respectively. Figure 3 shows the 3-DD representing the 2-colored subgraphs of some three edge graph that have no more red edges than green ones.

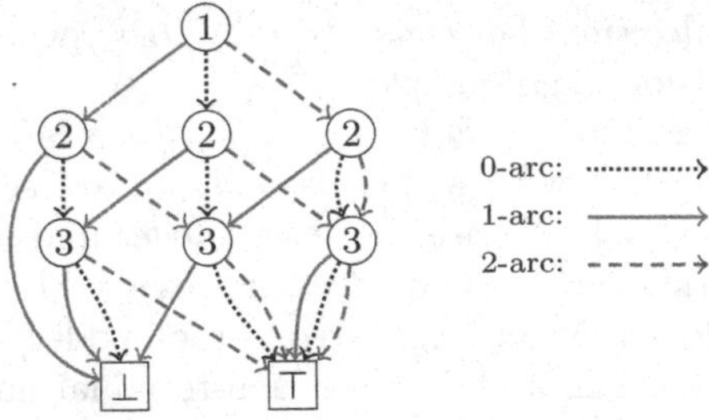

Fig. 3. 3-DD for 2-colored subgraphs of a three edge graph that have no more red edges than green ones. (Color figure online)

2.3 Frontier-Based Search

In this subsection, we explain FBS for subgraph enumeration problems [11,16, 24]. For a fixed input graph $G = (V, E)$, by identifying a subgraph with its edge set, a set of subgraphs can be represented by a 2-DD. In FBS, we order the edges of G and write them as $e_1, \ldots, e_m$. An FBS algorithm constructs the 2-DD representing the set of all the subgraphs in a concerned graph class $\mathcal{H}$. Once we obtain the 2-DD, it is easy to output all the subgraphs one by one. Therefore, in this paper, we describe how to construct the 2-DD instead of explicitly outputting the subgraphs.

One can see FBS as a framework to construct a 2-DD representing the computation of a dynamic programming algorithm searching for all the target subgraphs of an input graph. The dynamic programming algorithm processes edges $e_1, \ldots, e_m$ one by one and, accordingly, an FBS algorithm constructs the 2-DD in a breadth-first and top-down manner. The dynamic programming algorithm must involve a small data structure called *configuration* defined on each tentative decision of edge use, which corresponds to a path π from the root of the 2-DD to a node and is denoted as $\mathsf{config}(\pi)$ in the FBS algorithm. For notational convenience, let us represent a path π as a binary sequence and let $|\pi|$ denote the length of π. The configuration $\mathsf{config}(\pi)$ must express a characteristic of the subgraph $[\![\pi]\!]$ so that

- $\mathsf{config}(\pi \cdot a)$ can be computed from $\mathsf{config}(\pi)$, $|\pi|$, and $a \in \{0, 1\}$ for $|\pi| < m$,
- $\mathsf{config}(\pi) = \top$ if $[\![\pi]\!] \in \mathcal{H}$ and $\mathsf{config}(\pi) = \bot$ if $[\![\pi]\!] \notin \mathcal{H}$ for $|\pi| = m$,

where we identify the edge set $[\![\pi]\!]$ with the subgraph $(\bigcup[\![\pi]\!], [\![\pi]\!])$. This entails

if $|\pi_1| = |\pi_2|$ and $\mathsf{config}(\pi_1) = \mathsf{config}(\pi_2)$, then $[\![\pi_1 \cdot \pi']\!] \in \mathcal{H} \iff [\![\pi_2 \cdot \pi']\!] \in \mathcal{H}$ for any π' of length $m - |\pi_1|$.

Therefore, we may make the two paths having the same configuration reach the same node, while two paths with different configurations reach different nodes. In other words, each node can be identified with a configuration. The design of a configuration in FBS significantly affects the efficiency of the algorithm.

It is desirable to design config so that computation of config$(\pi \cdot a)$ from config(π) is cheap and as many as possible paths join to the same node.

In what follows, as an example, we consider the case in which $\mathcal{H}$ is the class of graphs each of which consists of any number of distinct cycles. In this case, the configuration of a path π is comprised of the degrees of the vertices in $F_{|\pi|+1}$ of the subgraph $[\![\pi]\!]$, where $F_i = (\bigcup_{j=1,\dots,i-1} e_j) \cap (\bigcup_{j=i,\dots,m} e_j)$, that is, the set of vertices incident to both processed and unprocessed edges. Then, config$(\pi \cdot 0)$ is identical to config(π) modulo the difference of the domains $F_{|\pi|+1}$ and $F_{|\pi|+2}$, and config$(\pi \cdot 1)$ simply increments the degrees of both vertices of $e_{|\pi|+1}$. The configuration is also used for *pruning*. No vertex of a graph in $\mathcal{H}$ has degree 3 or more. If the degree of a vertex $u \in e_{|\pi|+1}$ has already been two in config(π), then using $e_{|\pi|+1}$ makes the degree three. In this case we must perform pruning and let the path $\pi \cdot 1$ directly connect the terminal $\bot$. Moreover, no vertex of a graph in $\mathcal{H}$ has degree 1. If the degree of some vertex $u \in e_{|\pi|+1}$ is one in config(π) and $u \notin F_{|\pi|+2}$, then $e_{|\pi|+1}$ is the last edge that can make the degree of u two. Therefore, we connect the path $\pi \cdot 0$ to the terminal $\bot$, since afterwards the degree of u will be determined to one. For π, we call a vertex $u \notin \bigcup_{i=|\pi|+1}^{m} e_i$ *forgotten.*

FBS can be used for classes of *degree specified graphs* [12], whose members consist of connected graphs having a specified degree sequence. For example, the graph $K_{1,3}$ has the degree sequence $(3, 1, 1, 1)$, and conversely, any graph with the degree sequence $(3, 1, 1, 1)$ must be isomorphic to $K_{1,3}$. Therefore, the graph class $\mathcal{K}_{1,3}$ is a class of degree specified graphs. To construct the 2-DD for degree specified graphs by FBS, in addition to the degree of each vertex, we store the number of forgotten vertices having each degree into each 2-DD node, and a partition of the frontier to maintain connected components as a configuration.

3 Graph Classes Constructed by Colorful Frontier-Based Search

We construct a 2-DD for interval subgraphs of an input graph based on its FIS-characterization by the following three steps:

Step 1. We construct the 2-DDs for the subgraphs belonging to classes $\mathcal{C}_{\geq 4}$, $\mathcal{T}_2$, $\mathcal{X}_{31}$, $\mathcal{XF}_{2}^{\geq 1}$ and $\mathcal{XF}_{3}^{\geq 0}$.

Step 2. We make the union of those 2-DDs made by Step 1.

Step 3. Processing the 2-DD obtained by Step 2, taking induced edges into account, we construct the goal 2-DD.

The second step is no more than a standard operation over 2-DDs. This section is concerned with the first step, and the third step is described in Sect. 4. Both steps involve our proposed framework, *colorful FBS*, which constructs a $(c+1)$-DD representing c-colored subgraphs in a breadth-first and top-down manner.

3.1 Colorful FBS for Colored Degree Specified Graphs

Recall that a graph H is in the class $\mathcal{C}_{\geq 4}$ if and only if H is a connected graph with at least four vertices all of which have degree 2. Such a condition is a typical characterization that FBS can handle. The 2-DD for $\mathcal{C}_{\geq 4}$ can be constructed by an FBS algorithm. However, the other classes, $\mathcal{T}_2$, $\mathcal{X}_{31}$, $\mathcal{XF}_2^{\geq 1}$ and $\mathcal{XF}_3^{\geq 0}$, are not easy to treat for conventional FBS. For example, the graph $T_2 \in \mathcal{T}_2$ shown in Fig. 1(a) and the one in Fig. 1(b) have the same degree sequence $(3, 2, 2, 2, 1, 1, 1)$, but they are not isomorphic. In ordinary FBS, it is difficult to distinguish two subgraphs, although it would be possible if we far enrich the configuration stored in each node of a 2-DD, which should spoil the efficiency of FBS. To overcome the difficulty, we use colored graphs. It is easy to see that all 3-colored graphs with the same colored degree multiset $\{(2,0,0),(1,1,1),(1,0,0),(0,2,0),(0,1,0),(0,0,2),(0,0,1)\}$ are isomorphic to the one in Fig. 1(c). Enumeration of "colored degree specified graphs" is a typical application of colorful FBS.

Step 1 for constructing the 2-DD for subgraphs belonging to each of $\mathcal{T}_2$, $\mathcal{X}_{31}$, $\mathcal{XF}_2^{\geq 1}$ and $\mathcal{XF}_3^{\geq 0}$ consists of further two smaller steps. Step 1-1 constructs a $(c+1)$-DD for colored subgraphs, whose decolorizations belong to the concerned class, and Step 1–2 converts the $(c+1)$-DD into the desired 2-DD by decolorizing represented subgraphs.

Step 1-1 for $\mathcal{T}_2$ is only a special case of enumerating colored degree specified graphs over $(c+1)$-DDs. We first create the root node with label 1, and for $i = 1, \ldots, m$, and for $j = 1, \ldots, c$, we create the j-arc and the j-child of each node with label i. The j-arc of a node with label i means that we use e_i and give the j-th color to the edge e_i. Its 0-arc means that we do not use e_i.

We describe what to be stored into each $(c+1)$-DD node as a configuration. We define the configuration for the colored degree specified graphs by the colored degrees of the vertices in the frontier in the colored subgraph, and the number of forgotten vertices having each colored degree. In this way, one can enumerate all colored subgraphs of the input graph whose decolorizations are in $\mathcal{T}_2$. Pseudocode is shown in the full version of the paper.

Step 1-2, the decolorization of those obtained colored graphs, is explained in Sect. 3.3.

3.2 Coloring Graphs in $\mathcal{X}_{31}$, $\mathcal{XF}_2^{\geq 1}$ and $\mathcal{XF}_3^{\geq 0}$

The remained classes $\mathcal{X}_{31}$, $\mathcal{XF}_2^{\geq 1}$ and $\mathcal{XF}_3^{\geq 0}$ are not obtained by decolorizing colored degree specified graphs, but a little more additional conditions will suffice for characterizing those classes via 2-colored graphs. Those conditions are easily handled by colorful FBS.

Class $\mathcal{X}_{31}$. We give two colors to X_{31} as shown in Fig. 4(a). We can show that a graph is in $\mathcal{X}_{31}$ if and only if it admits a 2-coloring such that

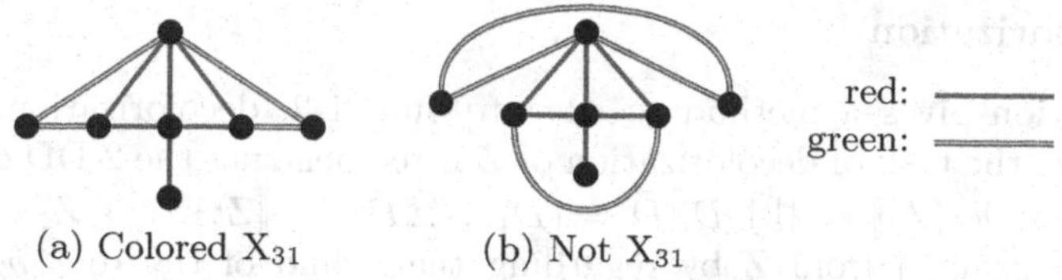

(a) Colored X31 (b) Not X31

Fig. 4. Two graphs with colored degree multiset $\{(4,0),(3,2),(2,1)^2,(1,0),(0,2)^2\}$.

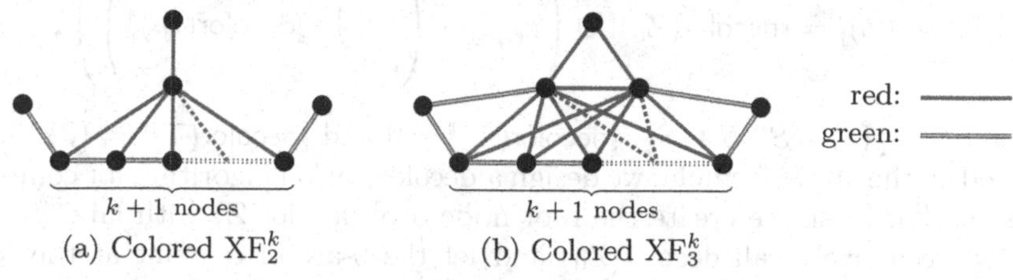

(a) Colored XF_2^k (b) Colored XF_3^k.

Fig. 5. Coloring XF_2^k and XF_3^k.

- The colored degree multiset is $\{(4,0),(3,2),(2,1)^2,(1,0),(0,2)^2\}$, where $\vec{\delta}^2$ means that the multiplicity of $\vec{\delta}$ is 2.
- The subgraph edge-induced by all the green edges is connected.

Note that we use the assumption that the input graph contains no self-loop or parallel edges. The second condition is necessary to exclude the graph shown in Fig. 4(b). (The subgraph edge-induced by all the red edges in a graph satisfying the first condition is always connected.) To impose the second condition, we can use the technique used in ordinary FBS that makes (monochrome) subgraphs represented by the 2-DD connected [11].

Class $\mathcal{XF}_2^{\geq 1}$. For $k \geq 1$, we give two colors to XF_2^k as shown in Fig. 5(a). We can show that a graph is in $\mathcal{XF}_2^{\geq 1}$ if and only if it admits a 2-coloring that satisfies the following two conditions:

- The colored degree multiset is $\{(k+2,0),(1,2)^{k+1},(1,0),(0,1)^2\}$ for some $k \geq 1$,
- The subgraph edge-induced by all the edges of each color is connected.

Class $\mathcal{XF}_3^{\geq 0}$. For $k \geq 0$, we give two colors to XF_3^k as shown in Fig. 5(b). We can show that a graph is in $\mathcal{XF}_3^{\geq 0}$ if and only if it admits a 2-coloring such that the following two conditions hold:

- The colored degree multiset is $\{(k+3,1)^2,(2,2)^{k+1},(2,0),(0,2)^2\}$ for some $k \geq 0$,
- The subgraph edge-induced by all the edges of each color is connected.

3.3 Decolorization

This subsection gives a method to execute Step 1-2, decolorization. Given a $(c+1)$-DD $\mathbf{Z}$, the task of decolorization of $\mathbf{Z}$ is to construct the 2-DD $\mathsf{decolor}(\mathbf{Z})$ such that $[\![\mathsf{decolor}(\mathbf{Z})]\!] = \{\bigcup_i D_i | \vec{D} = (D_1, \dots, D_c) \in [\![\mathbf{Z}]\!]\}$. Let $\mathbf{Z}_i$ denote the $(c+1)$-DD obtained from $\mathbf{Z}$ by regarding the i-child of the root $\rho_{\mathbf{Z}}$ of $\mathbf{Z}$ as the root $\rho_{\mathbf{Z}_i}$ and inheriting the nodes and the arcs of $\mathbf{Z}$ reachable from $\rho_{\mathbf{Z}_i}$. Our decolorization algorithm is based on the following recursive formula. If $\mathbf{Z} \neq \bot, \top$,

$$[\![\mathsf{decolor}(\mathbf{Z})]\!] = [\![\mathsf{decolor}(\mathbf{Z}_0)]\!] \cup \left(e_{\ell(\rho_{\mathbf{Z}})} * \left(\bigcup_{i=1,\dots,c} [\![\mathsf{decolor}(\mathbf{Z}_i)]\!] \right) \right),$$

where $e * \mathcal{S} = \{\{e\} \cup S \mid S \in \mathcal{S}\}$, $[\![\mathsf{decolor}(\bot)]\!] = \emptyset$ and $[\![\mathsf{decolor}(\top)]\!] = \{\emptyset\}$.

Based on the above formula, we design a decolorization algorithm for computing $\mathsf{decolor}(\mathbf{Z})$. First, we create the root node α of $\mathsf{decolor}(\mathbf{Z})$ with label $\ell(\rho_{\mathbf{Z}})$. Then, we recursively call $\mathsf{decolor}(\mathbf{Z}_0)$ and let the 0-arc of α point at the root of the resulting 2-DD. Next, we call $\mathsf{decolor}(\mathbf{Z}_1), \dots, \mathsf{decolor}(\mathbf{Z}_c)$ and compute $\mathsf{decolor}(\mathbf{Z}_1) \cup \cdots \cup \mathsf{decolor}(\mathbf{Z}_c)$ by the conventional binary operation '$\cup$' [4,21], where for 2-DDs $\mathbf{Z}_A$ and $\mathbf{Z}_B$, we define $\mathbf{Z}_A \cup \mathbf{Z}_B$ by the 2-DD representing $[\![\mathbf{Z}_A]\!] \cup [\![\mathbf{Z}_B]\!]$. We let the 1-arc of α point at the root of the resulting 2-DD.

4 Induced Subgraph and FIS-Characterization

This section describes Step 3, which constructs the 2-DD representing the subgraphs of the input graph that are FIS-characterized by $\mathcal{F}$ represented by a 2-DD $\mathbf{F}$. We have three parts:

3-1. By colorful FBS, we construct the 3-DD, say $\mathbf{Z}^3$, representing the set of colored subgraphs each of which is obtained by giving red to the edges of each member of $\mathcal{F}$ and green to the new edges induced by those red edges.
3-2. By using a recursive algorithm on $\mathbf{Z}^3$, we construct the 2-DD, say $\mathbf{Z}^2$, representing the set of (monochrome) subgraphs each of which contains a member of $\mathcal{F}$ as an induced subgraph.
3-3. We construct the desired 2-DD by computing $\mathbf{Z}^{\text{all}} \setminus \mathbf{Z}^2$ by the conventional binary operation '$\setminus$' [4,21], where $\mathbf{Z}^{\text{all}}$ is the 2-DD for all the subgraphs of the input graph and $\mathbf{Z}^{\text{all}} \setminus \mathbf{Z}^2$ is the 2-DD for $[\![\mathbf{Z}^{\text{all}}]\!] \setminus [\![\mathbf{Z}^2]\!]$.

Since the 2-DD for $\mathcal{C}_{\geq 4}$ can be obtained by ordinary FBS, we immediately obtain the 2-DD representing the set of all the chordal subgraphs by our method. In Step 3-3, it is easy to construct $\mathbf{Z}^{\text{all}}$. We explain Steps 3-1 and 3-2 in the rest of this section. Then, we also describe a more efficient way of simultaneously carrying out Steps 3-2 and 3-3.

4.1 Colorful FBS for Edge-Inducing

We design a colorful FBS method for $\mathbf{Z}^3$ for Step 3-1. Suppose that $\mathcal{F}$ is represented by a 2-DD $\mathbf{F}$. First, we describe how to store which edges are colored in red as a configuration of FBS. Recall that a path from the root to a node α in $\mathbf{F}$ corresponds to a subgraph consisting of edges in $\{e_1, \ldots, e_{\ell(\alpha)-1}\}$. To maintain which edges are colored in red, we store a node of $\mathbf{F}$ into each node of the 3-DD $\mathbf{Z}^3$, which we are constructing. We store the root of $\mathbf{F}$ into the root of $\mathbf{Z}^3$. Then, when we create the 1-child of a node α storing β, we store β_1 into the 1-child, where β_j is the j-child of β for $j = 0, 1$. If $\beta_1 = \bot$, we make the 1-arc of α point at $\bot$. When we create the 0-child or 2-child of α, we store β_0 into the child. If $\beta_0 = \bot$, we make the 0-arc or 2-arc of α point at $\bot$.

Next, we describe how to add green edges so that the coloring condition required in Step 3-1 holds. Let us observe some cases. Consider creating the children of a node α. Let $e_{\ell(\alpha)} = \{v, w\}$ (recall that $e_{\ell(\alpha)}$ is the $\ell(\alpha)$-th edge). If both v and w are incident to (distinct) red edges, a red edge or green edge must connect v and w. In this case, we make the 0-arc of α point at $\bot$. If v is incident to a red edge but w is not, connecting v and w by a green edge does not immediately violate the coloring condition, but w must be incident to a red edge in the future. Therefore, the 2-child of α must have the condition that a red edge must be connected to w in the future. Conversely, in the same situation, if we decide that v and w are connected by neither a red nor green edge, although it does not immediately violate the coloring condition, we cannot connect any red edge to w in the future. Therefore, the 0-child of α must remember that no red edge must be connected to w in the future.

On the basis of the above observation, we store the value $f_\alpha(u)$ for each vertex u on the frontier into each node α of $\mathbf{Z}^3$ as (a part of) a configuration:

- $f_\alpha(u) = 0$ means that there is no condition about u,
- $f_\alpha(u) = -1$ means that u must not be incident to any red edge,
- $f_\alpha(u) = 1$ means that u is not currently incident to any red edge, but will have to be incident to a red edge in the future,
- $f_\alpha(u) = 2$ means that u is incident to a red edge.

We can update the values of f in our colorful FBS and prune nodes so that no contradiction occurs.

4.2 Recursive Algorithm

We describe a recursive algorithm for Step 3-2. Let $G = (V, E)$ be an input graph. The task for Step 3-2 is to construct the 2-DD $\mathbf{Z}^2$, which we denote by $\mathsf{ind}(\mathbf{Z}^3)$, representing $[\![\mathbf{Z}^2]\!] = \{F \cup E' \mid F \in [\![\mathbf{F}]\!],\ E' \subseteq V \setminus E[\bigcup F]\} = \{F \cup E' \mid (F, E'') \in [\![\mathbf{Z}^3]\!], E' \subseteq V \setminus (F \cup E'')\}$ from the input 2-DD $\mathbf{F}$.

We use the following recursive structure to compute ind:

$$[\![\mathsf{ind}(\mathbf{Z}^3)]\!] = ([\![\mathsf{ind}(\mathbf{Z}^3_0)]\!] \cup [\![\mathsf{ind}(\mathbf{Z}^3_2)]\!]) \cup (e_z * ([\![\mathsf{ind}(\mathbf{Z}^3_0)]\!] \cup [\![\mathsf{ind}(\mathbf{Z}^3_1)]\!]))$$

where $z = \ell(\rho_{\mathbf{Z}^3})$. Let us describe an intuition on this formula. We decompose $[\![\mathsf{ind}(\mathbf{Z}^3)]\!]$ as $[\![\mathsf{ind}(\mathbf{Z}^3)]\!] = [\![\mathsf{ind}(\mathbf{Z}^3)_0]\!] \cup (e_z * [\![\mathsf{ind}(\mathbf{Z}^3)_1]\!])$, where $\mathsf{ind}(\mathbf{Z}^3)_j$ denotes the 3-DD whose root is the j-child of the root of $\mathsf{ind}(\mathbf{Z}^3)$. We also decompose $[\![\mathbf{Z}^3]\!]$ into the three groups of subgraphs: (i) not having e_z, (ii) having e_z with red, and (iii) having e_z with green. All the members of group (ii) must be included in $[\![\mathsf{ind}(\mathbf{Z}^3)_1]\!]$, and all the members of group (iii) must be included in $[\![\mathsf{ind}(\mathbf{Z}^3)_0]\!]$. All the members of group (i) must be included both in $[\![\mathsf{ind}(\mathbf{Z}^3)_0]\!]$ and $[\![\mathsf{ind}(\mathbf{Z}^3)_1]\!]$ because not using e_z in the 3-DD means that e_z is allowed to be used or not in the 2-DD. Therefore, we obtain $[\![\mathsf{ind}(\mathbf{Z}^3)_0]\!] = [\![\mathsf{ind}(\mathbf{Z}^3_0)]\!] \cup [\![\mathsf{ind}(\mathbf{Z}^3_2)]\!]$ and $[\![\mathsf{ind}(\mathbf{Z}^3)_1]\!] = [\![\mathsf{ind}(\mathbf{Z}^3_0)]\!] \cup [\![\mathsf{ind}(\mathbf{Z}^3_1)]\!]$.

We can integrate Steps 3-2 and 3-3 as follows. For a 2-DD $\mathbf{Z}$, let $[\![\mathbf{Z}]\!]^{\mathrm{c}}$ and $\mathbf{Z}^{\mathrm{c}}$ denote the complement of $[\![\mathbf{Z}]\!]$, i.e., $[\![\mathbf{Z}^{\mathrm{all}} \setminus \mathbf{Z}]\!]$, and the 2-DD for $[\![\mathbf{Z}]\!]^{\mathrm{c}}$, respectively. Then,

$$\begin{aligned}[\![\mathsf{ind}(\mathbf{Z}^3)]\!]^{\mathrm{c}} &= (([\![\mathsf{ind}(\mathbf{Z}^3_0)]\!] \cup [\![\mathsf{ind}(\mathbf{Z}^3_2)]\!]) \cup (e_z * ([\![\mathsf{ind}(\mathbf{Z}^3_0)]\!] \cup [\![\mathsf{ind}(\mathbf{Z}^3_1)]\!])))^{\mathrm{c}} \\ &= ([\![\mathsf{ind}(\mathbf{Z}^3_0)]\!]^{\mathrm{c}} \cap [\![\mathsf{ind}(\mathbf{Z}^3_2)]\!]^{\mathrm{c}}) \cup (e_z * ([\![\mathsf{ind}(\mathbf{Z}^3_0)]\!]^{\mathrm{c}} \cap [\![\mathsf{ind}(\mathbf{Z}^3_1)]\!]^{\mathrm{c}}))\end{aligned}$$

Hence, we can directly and recursively compute $\mathsf{ind}(\cdot)^{\mathrm{c}}$ on the basis of the above equation.

4.3 2-DDs for Interval and Proper Interval Subgraphs

We show how to construct the 2-DD for interval subgraphs. For a graph class $\mathcal{H}$, let $\mathbf{Z}(\mathcal{H})$ denote the 2-DD for subgraphs of the input graph belonging to $\mathcal{H}$. We construct the 2-DD $\mathbf{Z}(\mathcal{C}_{\geq 4})$ by conventional FBS. We also construct the 4-DD $\mathbf{Z}(\mathcal{T}_2)$ by colorful FBS in Sect. 3.1, and the 3-DDs $\mathbf{Z}(\mathcal{X}_{31})$, $\mathbf{Z}(\mathcal{XF}_2^{\geq 1})$, and $\mathbf{Z}(\mathcal{XF}_3^{\geq 0})$ by colorful FBS in Sect. 3.2, respectively. Then, we carry out the decolorization algorithm for the four DDs, that is, we compute $\mathsf{decolor}(\mathbf{Z}(\mathcal{T}_2))$, $\mathsf{decolor}(\mathbf{Z}(\mathcal{X}_{31}))$, $\mathsf{decolor}(\mathbf{Z}(\mathcal{XF}_2^{\geq 1}))$, and $\mathsf{decolor}(\mathbf{Z}(\mathcal{XF}_3^{\geq 0}))$. We take the union of the five DDs as $\mathbf{Z}(\mathcal{C}_{\geq 4}) \cup \mathsf{decolor}(\mathbf{Z}(\mathcal{T}_2)) \cup \mathsf{decolor}(\mathbf{Z}(\mathcal{X}_{31})) \cup \mathsf{decolor}(\mathbf{Z}(\mathcal{XF}_2^{\geq 1})) \cup \mathsf{decolor}(\mathbf{Z}(\mathcal{XF}_3^{\geq 0}))$, which we denote $\hat{\mathbf{Z}}$. Next, we compute the inducing 3-DD, say $\hat{\mathbf{Z}}^3$, for $\hat{\mathbf{Z}}$ described in Sect. 4. Finally, we compute $\mathsf{ind}(\hat{\mathbf{Z}}^3)^{\mathrm{c}}$ described in Sect. 4.2. The obtained 2-DD represents the set of all the interval subgraphs of the input graph.

Next, we construct the 2-DD for proper interval subgraphs. A proper interval subgraph is an interval subgraph that contains no member of $\mathcal{K}_{1,3}$ as an induced subgraph. Therefore, we construct the 2-DD, say $\mathbf{Z}(\mathcal{K}_{1,3})$ for $\mathcal{K}_{1,3}$ and when taking the union above, we compute $\mathbf{Z}(\mathcal{C}_{\geq 4}) \cup \mathsf{decolor}(\mathbf{Z}(\mathcal{T}_2)) \cup \mathsf{decolor}(\mathbf{Z}(\mathcal{X}_{31})) \cup \mathsf{decolor}(\mathbf{Z}(\mathcal{XF}_2^{\geq 1})) \cup \mathsf{decolor}(\mathbf{Z}(\mathcal{XF}_3^{\geq 0})) \cup \mathbf{Z}(\mathcal{K}_{1,3})$. The rest is the same as above. The obtained 2-DD represents the set of all the proper interval subgraphs of the input graph.

5 Experiments

In this section, we evaluate the performance of our methods by conducting experiments. Input graphs we use are complete graphs with n vertices (K_n) and graphs provided by the Internet Topology Zoo (TZ) [15], KONECT (KO) [17], and javaAwtComponent (JC) [22]. The Internet Topology Zoo provides benchmark graphs representing communication networks, KONECT contains network datasets for network science and related fields, and javaAwt-Component includes chordal interference graphs obtained from the compilation of `java.awt.Component`. We compare our methods with existing ones based on reverse search. The authors implemented our methods, a reverse search based algorithm for chordal subgraphs (RS-c) [14], and one for interval subgraphs (RS-i) [13] in the C++ language. For handling DDs, we used the SAPPOROBDD library, which has not been officially published yet but is available at https://github.com/takemaru/graphillion, and for implementing colored FBS, we used the TdZdd library [8]. The implementations were compiled by `g++` with the `-O3` optimization option and run on a machine with Intel Xeon E5-2630 (2.30 GHz) CPU and 128 GB memory (Linux Centos 7.6).

RS-c and RS-i do not actually output chordal and interval subgraphs, respectively, but only count the number of them. The proposed methods construct 2-DDs but do not explicitly output subgraphs. The cardinality of a family

Table 1. Comparison of methods for chordal graphs. "RS-c time" and "Ours time" indicate the running time of RS-c and the proposed method (in seconds), respectively. "<0.01" means that the time is less than 0.01 s. "# graphs" means the number of output chordal subgraphs.

Graph	$\lvert V\rvert$	$\lvert E\rvert$	RS-c time	Ours time	# graphs
K_2	2	1	<0.01	<0.01	2
K_3	3	3	<0.01	<0.01	8
K_4	4	6	<0.01	<0.01	61
K_5	5	10	<0.01	<0.01	822
K_6	6	15	0.04	<0.01	18154
K_7	7	21	0.68	0.02	617675
K_8	8	28	28.52	0.32	30888596
K_9	9	36	1892.48	14.31	2192816760
K_{10}	10	45	T/O	680.91	215488096587
K_{11}	11	55	T/O	M/O	–
TZ Darkstrand	28	31	30762.83	<0.01	2108348424
TZ Sunet	26	32	41792.31	<0.01	3523488768
TZ TataNld	145	186	T/O	1.96	4.1×10^{55}
TZ UsCarrier	158	189	T/O	2.96	4.3×10^{56}
TZ Kdl	754	895	T/O	M/O	–
KO Southern-women-2	10	14	0.04	<0.01	11822
KO South-African-Companies	11	13	0.03	<0.01	6432
KO American-Revolution	141	160	T/O	0.01	9.4×10^{46}
KO PDZBase	212	242	T/O	1.27	3.3×10^{70}
JC createBufferStrategy	37	145	T/O	104.33	1.3×10^{37}
JC getListeners	113	342	T/O	1.13	4.3×10^{96}
JC dispatchMouseWheelToAncestor	43	198	T/O	M/O	–

represented by a DD can be easily computed by a simple dynamic programming-based algorithm [16] in time proportional to the number of nodes in the DD.

Table 1 shows the comparison of RS-c and the proposed method of constructing the 2-DD for chordal subgraphs. We pick up input graphs with the largest edge set for which both algorithms succeeded in completing, and only the proposed method succeeded in, and pick up input graphs with the smallest edge set for which both algorithms failed to complete. "T/O" and "M/O" mean that the program failed due to timeout (the computation time exceeds 100,000 s) and out of memory (used up 128 GB memory), respectively. Both methods output the same number of subgraphs for all input graphs shown in the table. Table 2 shows the comparison of RS-i and the proposed method of constructing the 2-DD for interval subgraphs. The tables indicate that the proposed methods work for larger graphs than RS-c and RS-i.

Table 2. Comparison of RS-i and the proposed method for interval graphs. "# graphs" means the number of output interval subgraphs.

Graph	$\lvert V\rvert$	$\lvert E\rvert$	RS-i time	Ours time	# graphs
K_2	2	1	0.03	0.00	2
K_3	3	3	0.01	0.00	8
K_4	4	6	0.02	0.01	61
K_5	5	10	0.14	0.01	822
K_6	6	15	4.08	0.05	17914
K_7	7	21	223.03	0.54	571475
K_8	8	28	T/O	8.65	24566756
K_9	9	36	T/O	228.66	1346167320
K_{10}	10	45	T/O	M/O	–
TZ Nextgen	17	19	171.22	0.01	456375
TZ VisionNet	24	23	4427.51	0.02	8004608
TZ Interoute	110	146	T/O	2.89	5.2×10^{42}
TZ DialtelecomCz	138	151	T/O	4.65	9.5×10^{44}
TZ Ion	125	146	T/O	M/O	–
KO South-African-Companies	11	13	1.31	0.01	6184
KO Southern-women-2	10	14	2.50	0.02	11178
KO Chicago	1467	1298	T/O	1.38	4.4×10^{378}
KO Facebook-NIPS	2888	2981	T/O	120.95	1.3×10^{894}
KO Highland-Tribes	16	58	T/O	M/O	–
JC enableEvents	38	96	T/O	19.83	1.6×10^{24}
JC readObject	66	112	T/O	0.60	1.5×10^{31}
JC checkImage	13	46	T/O	M/O	–

We show the details of the computation time for interval subgraphs for K_9 as follows. FBS for $\mathcal{C}_{\geq 4}$ and colorful FBS for $\mathcal{T}_2$, $\mathcal{X}_{31}$, $\mathcal{XF}_2^{\gtreqless 1}$, and $\mathcal{XF}_3^{\gtreqless 0}$ took 0.02, 0.73, 5.64, 22.31, and 109.09 s, respectively. The decolorization for $\mathcal{T}_2$, $\mathcal{X}_{31}$, $\mathcal{XF}_2^{\gtreqless 1}$, and $\mathcal{XF}_3^{\gtreqless 0}$ took 0.06, 0.12, 1.00, and 10.58 s, respectively. The union of the 2-DDs for the five classes took less than 0.01 s. The colorful FBS method for Step 3-1 and ind operations took 0.26 and 94.08 s, respectively. The bottlenecks of our method for K_9 were colorful FBS for $\mathcal{XF}_3^{\gtreqless 0}$ and the ind operation. For graphs "KO Facebook-NIPS" and "JC enableEvents," colorful FBS for $\mathcal{XF}_2^{\gtreqless 1}$ took 90.9 and 2.89 s, that for $\mathcal{XF}_3^{\gtreqless 0}$ took 11.5 and 0.94 s, and ind took 1.28 and 5.26 s, respectively. The bottleneck of our method for the two graphs was colorful FBS for $\mathcal{XF}_2^{\gtreqless 1}$.

6 Conclusion

We proposed algorithms to construct the 2-DDs for the sets of all the chordal and interval subgraphs of a given graph. Our algorithms employ colorful frontier-based search for two different purposes. One is for enumerating forbidden subgraphs and the other is for inducing edges from those forbidden subgraphs. We also presented different recursive methods converting a $(c+1)$-DD into a 2-DD: one simply decolorizes subgraphs and the other produces subgraphs that have no forbidden induced subgraphs, when the edges of the forbidden graphs and their inducing edges have different colors in the input. Those demonstrate the potential of our colorful FBS framework. Future directions of this research include determining how many colors are needed for graph classes and theoretically investigating relations of graph classes and colored degrees.

Acknowledgment. This work was supported in part by JSPS KAKENHI Grant Numbers JP15H05711, JP18K04610, JP16K16006, JP18H04091 and JP19K12098, and NAIST Big Data Project.

References

1. Avis, D., Fukuda, K.: Reverse search for enumeration. Discrete Appl. Math. **65**(1–3), 21–46 (1996). https://doi.org/10.1016/0166-218X(95)00026-N
2. Beeri, C., Fagin, R., Maier, D., Yannakakis, M.: On the desirability of acyclic database schemes. J. ACM **30**(3), 479–513 (1983). https://doi.org/10.1145/2402.322389
3. Blair, J.R.S., Peyton, B.: An introduction to chordal graphs and clique trees. In: George, A., Gilbert, J.R., Liu, J.W.H. (eds.) Graph Theory and Sparse Matrix Computation. The IMA Volumes in Mathematics and its Applications, vol. 56, pp. 1–29. Springer, New York, New York, NY (1993). https://doi.org/10.1007/978-1-4613-8369-7_1
4. Bryant, R.E.: Graph-based algorithms for Boolean function manipulation. IEEE Trans. Comput. **C–35**(8), 677–691 (1986). https://doi.org/10.1109/TC.1986.1676819

5. Golumbic, M.C.: Algorithmic Graph Theory and Perfect Graphs (Annals of Discrete Mathematics, Vol. 57). North-Holland Publishing Co., Amsterdam (2004)
6. Inoue, T., et al.: Distribution loss minimization with guaranteed error bound. IEEE Trans. Smart Grid **5**(1), 102–111 (2014). https://doi.org/10.1109/TSG.2013.2288976
7. Ishioka, F., Kawahara, J., Mizuta, M., Minato, S., Kurihara, K.: Evaluation of hotspot cluster detection using spatial scan statistic based on exact counting. Jpn. J. Stat. Data Sci. **2**(1), 1–15 (2019). https://doi.org/10.1007/s42081-018-0030-6
8. Iwashita, H., Minato, S.: Efficient top-down ZDD construction techniques using recursive specifications. TCS Technical Reports TCS-TR-A-13-69 (2013)
9. Jackowski, Z.: A new characterization of proper interval graphs. Discrete Math. **105**(1), 103–109 (1992). https://doi.org/10.1016/0012-365X(92)90135-3
10. Kawahara, J., Horiyama, T., Hotta, K., Minato, S.: Generating all patterns of graph partitions within a disparity bound. In: Poon, S.-H., Rahman, M.S., Yen, H.-C. (eds.) WALCOM 2017. LNCS, vol. 10167, pp. 119–131. Springer, Cham (2017). https://doi.org/10.1007/978-3-319-53925-6_10
11. Kawahara, J., Inoue, T., Iwashita, H., Minato, S.: Frontier-based search for enumerating all constrained subgraphs with compressed representation. IEICE Trans. Inf. Syst. **E100–A**(9), 1773–1784 (2017)
12. Kawahara, J., Saitoh, T., Suzuki, H., Yoshinaka, R.: Solving the longest oneway-ticket problem and enumerating letter graphs by augmenting the two representative approaches with ZDDs. In: Phon-Amnuaisuk, S., Au, T.-W., Omar, S. (eds.) CIIS 2016. AISC, vol. 532, pp. 294–305. Springer, Cham (2017). https://doi.org/10.1007/978-3-319-48517-1_26
13. Kiyomi, M., Kijima, S., Uno, T.: Listing chordal graphs and interval graphs. In: Fomin, F.V. (ed.) WG 2006. LNCS, vol. 4271, pp. 68–77. Springer, Heidelberg (2006). https://doi.org/10.1007/11917496_7
14. Kiyomi, M., Uno, T.: Generating chordal graphs included in given graphs. IEICE Trans. Inf. Syst. **E89–D**(2), 763–770 (2006). https://doi.org/10.1093/ietisy/e89-d.2.763
15. Knight, S., Nguyen, H.X., Falkner, N., Bowden, R., Roughan, M.: The internet topology zoo. IEEE J. Sel. Areas Commun. **29**(9), 1765–1775 (2011). https://doi.org/10.1109/JSAC.2011.111002
16. Knuth, D.E.: The Art of Computer Programming. Combinatorial Algorithms, Part 1, vol. 4A. Addison-Wesley, Upper Saddle River (2011)
17. Kunegis, J.: KONECT: the Koblenz network collection. In: Proceedings of the 22nd International Conference on World Wide Web, pp. 1343–1350 (2013). https://doi.org/10.1145/2487788.2488173
18. Lekkerkerker, C., Boland, J.: Representation of a finite graph by a set of intervals on the real line. Fundamenta Mathematicae **51**(1), 45–64 (1962)
19. Maehara, T., Suzuki, H., Ishihata, M.: Exact computation of influence spread by binary decision diagrams. In: Proceedings of the 26th International Conference on World Wide Web, pp. 947–956 (2017). https://doi.org/10.1145/3038912.3052567
20. Miller, D.: Multiple-valued logic design tools. In: Proceedings of the 23rd International Symposium on Multiple-Valued Logic, pp. 2–11 (1993). https://doi.org/10.1109/ISMVL.1993.289589
21. Minato, S.: Zero-suppressed BDDs for set manipulation in combinatorial problems. In: Proceedings of the 30th ACM/IEEE Design Automation Conference, pp. 272–277 (1993). https://doi.org/10.1145/157485.164890

22. Pereira, F.M.Q., Palsberg, J.: Register allocation via coloring of chordal graphs. In: Yi, K. (ed.) APLAS 2005. LNCS, vol. 3780, pp. 315–329. Springer, Heidelberg (2005). https://doi.org/10.1007/11575467_21
23. de Ridder, H.N., et al.: Information System on Graph Classes and their Inclusions (ISGCI). http://www.graphclasses.org
24. Sekine, K., Imai, H., Tani, S.: Computing the Tutte polynomial of a graph of moderate size. In: Staples, J., Eades, P., Katoh, N., Moffat, A. (eds.) ISAAC 1995. LNCS, vol. 1004, pp. 224–233. Springer, Heidelberg (1995). https://doi.org/10.1007/BFb0015427
25. Wasa, K.: Enumeration of enumeration algorithms. CoRR abs/1605.05102 (2016)

Unit Disk Cover for Massive Point Sets

Anirban Ghosh(✉), Brian Hicks, and Ronald Shevchenko

School of Computing, University of North Florida, Jacksonville, FL, USA
{anirban.ghosh,n00133251,n01385011}@unf.edu

Abstract. Given a set of points in the plane, the Unit Disk Cover (UDC) problem asks to compute the minimum number of unit disks required to cover the points, along with a placement of the disks. The problem is NP-Hard and several approximation algorithms have been designed over the last three decades.

In this paper, we experimentally compare practical performances of some of these algorithms on massive point sets. The goal is to investigate which algorithms run fast and give good approximation in practice.

We present an elementary online 7-approximation algorithm for UDC which runs in $\mathcal{O}(n)$ time on average and is easy to implement. In our experiments with both synthetic and real-world massive point sets, we have observed that this algorithm is up to 61.63 times and at least 2.9 times faster than the existing algorithms implemented in this paper. It gave 2.7-approximation in practice for the point sets used in our experiments. In our knowledge, this is the first work which experimentally compares the existing algorithms for UDC.

Keywords: Geometric covering · Unit disks · Clustering

1 Introduction

Geometric covering is a well-researched family of problems in computational geometry and have been studied for decades. The Unit Disk Cover (UDC) problem has turned out to be one of the fundamental geometric covering problems. Given a set P of n points $p_1, \ldots, p_n$ in the Euclidean plane, the UDC problem asks to compute the minimum number of possibly intersecting unit disks (closed disks of unit radius) required to cover the points in P, along with a placement of the disks. See Fig. 1 for an example. The algorithms for UDC can be easily scaled for covering points using disks of any fixed radius $r > 0$. In this paper, we use $r = 1$.

The UDC problem has applications in wireless networking, facility location, robotics, image processing, and machine learning. For instance, P can be perceived as a set of clients or locations of interest seeking service from service providers which can be modelled using fixed-radius disks. The goal is to provide service or cover these locations using the minimum number of service providers.

Research supported by the University of North Florida start-up fund.

I. Kotsireas et al. (Eds.): SEA[2] 2019, LNCS 11544, pp. 142–157, 2019.
https://doi.org/10.1007/978-3-030-34029-2_10

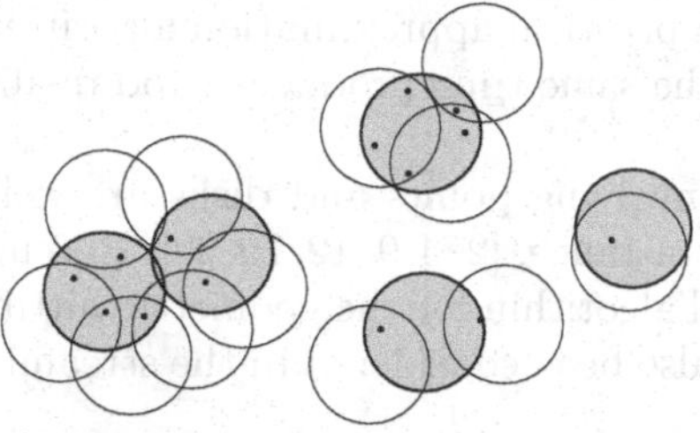

Fig. 1. Any optimal solution for this point set contains exactly 5 disks; an optimal solution is shown using gray disks. The solution shown in this figure used 12 disks.

The UDC problem has a long history. Back in 1981, UDC was shown to be NP-Hard by Fowler [14]. The first known approximation algorithm for UDC is a PTAS designed by Hochbaum and Maass [18], which runs in $\mathcal{O}(\ell^4(2n)^{4\ell^2+1})$ time having an approximation factor of $(1+\frac{1}{\ell})^2$, for any integer $\ell \geq 1$. Gonzalez [17] presented two approximation algorithms; a $2(1+\frac{1}{\ell})$-approximation algorithm which runs in $\mathcal{O}(\ell^2 n^7)$ time, where ℓ is a positive integer and another constant-factor[1] approximation algorithm with runtime of $\mathcal{O}(n \log |\mathrm{OPT}|)$, where $|\mathrm{OPT}|$ is the number of disks in an optimal cover. Charikar, Chekuri, Feder, and Motwani [8] devised a 7-approximation algorithm for the UDC problem (the authors used the name Dual Clustering for this problem). A $\mathcal{O}(1)$-approximation algorithm with runtime of $\mathcal{O}(n^3 \log n)$ is presented by Brönnimann and Goodrich [6]. Franceschetti, Cook, and Bruck [15] developed an algorithm with an approximation factor of $3(1+\frac{1}{\ell})^2$ having a runtime of $\mathcal{O}(Kn)$, where ℓ is a positive integer and K is a constant which depends on ℓ. A 2.8334-approximation algorithm is designed by Fu, Chen, and Abdelguerfi [16] which runs in $\mathcal{O}(n(\log n \log\log n)^2)$ time. Liu and Lu [22] designed a 25/4-approximation algorithm having a runtime of $\mathcal{O}(n \log n)$. Biniaz, Liu, Maheshwari, and Smid [5] devised a 4-approximation algorithm which has runtime of $\mathcal{O}(n \log n)$. Recently, Dumitrescu, Ghosh, and Tóth [13] have designed an online 5-approximation[2] algorithm for the problem, but no comment about its asymptotic runtime is made.

In the era of Big Data, the sizes of data sets are growing exponentially. Finding good solutions efficiently for NP-Hard geometric optimization problems has posed a great challenge. Due to the practical importance of the UDC problem, it is worthwhile to investigate which algorithms designed for UDC are fit for processing massive point sets in practice. It is no surprise that algorithms having runtime worse than $\mathcal{O}(n \log n)$ are unfit for practical uses. Moreover, the constants in their asymptotic runtimes should be small enough to be tolerated for practical purposes. This motivated us to implement some of the existing algorithms to find out which ones among these are efficient in practice. In particular,

[1] In [17], the author claims this constant to be 4, whereas in [5,15,22] the authors claims it to be 8. Unfortunately, in all these papers the claims appear unjustified.

[2] In the literature of online algorithms, the term *competitive ratio* is used instead of approximation factor.

our objective is to find a practical approximation algorithm for UDC which runs fast in practice and at the same time produces good quality solution with some theoretical guarantee.

Covering problems involving points and disks are well-studied in computational geometry; see for instance, [2–4,9–12,19,20]. Recently, Bus, Mustafa, and Ray designed a practical algorithm for the geometric hitting set problem; see [7]. The UDC problem has also been considered in the streaming setup by Liaw, Liu, and Reiss [21].

Our Contributions. For our experiments, we have implemented the following algorithms; appropriate abbreviations are used for naming purposes.

1. HM-1985 by Hochbaum and Mass (1985) [18]
2. G-1991 by Gonzalez (1991) [17]
3. CCFM-1997 by Charikar, Chekuri, Feder, and Motwani (1997) [8]
4. FCB-2001 by Franceschetti, Cook, and Bruck (2001) [15]
5. LL-2014 by Liu and Lu (2014) [22]
6. BLMS-2017 by Biniaz, Liu, Maheshwari, and Smid (2017) [5]
7. DGT-2018 by Dumitrescu, Ghosh, and Tóth (2018) [13]

We have refrained from implementing the algorithms from [6] and [16] since their runtimes are worse than $\mathcal{O}(n \log n)$. However, we have implemented HM-1985 [18] despite its poor runtime, primarily due to its historical importance in covering problems.

We have designed a simple online 7-approximation algorithm named GHS which runs in $\mathcal{O}(n)$ time on average; see Sect. 2.8. The solutions generated by GHS in our experiments is competitive with the ones generated by the above sophisticated algorithms. A simple proof is presented to show that GHS has an approximation factor of 7. Our experiments with 36 synthetic and 11 real-world massive point sets show that GHS gives 2.7-approximation in practice and is remarkably faster than the other implemented algorithms. In fact, we have observed up to 61.63 times and at least 2.9 times speedup while maintaining 2.7-approximation. The largest point set used in the experiments has ≈10.8 million points. The algorithms are implemented in C++17 using the CGAL 4.13 library [23]. For some algorithms, we have used some built-in algorithms from CGAL for high precision and efficient practical performance.

In Sect. 2, we discuss the algorithms implemented in this paper along with the GHS algorithm. In Sect. 3, we present our experimental results.

Notations. We denote a point $p \in \mathbf{R}^2$ using a pair of real numbers (a, b). By p_x and p_y, we denote its x and y-coordinates, respectively.

2 Algorithms

2.1 HM-1985: Hochbaum and Mass (1985)

HM-1985 [18] is a PTAS designed for covering points in d-space using unit balls. It has an approximation factor of $(1 + \frac{1}{\ell})^d$ and runs in $\mathcal{O}(\ell^d(\ell\sqrt{d})^d(2n)^{d(\ell\sqrt{d})^d+1})$

time, for any integer $\ell \geq 1$. In 2-space, it runs in $\mathcal{O}(\ell^4(2n)^{4\ell^2+1})$ time and has an approximation factor of $(1+\frac{1}{\ell})^2$. For $\ell = 1, 2, 3$, the approximation factors are $4, 2.25, \approx 1.77$ and the running times are $\mathcal{O}(n^5), \mathcal{O}(n^{17}), \mathcal{O}(n^{37})$, respectively. Although the approximation factors look attractive, this algorithm is unusable for massive point sets. Even with small point sets, the algorithm can be embarrassingly slow. However, from theoretical perspective this algorithm has a historical importance due to the *shifting strategy* introduced for geometric covering. Also, a few algorithms which came after this used the shifting strategy as a basis. Nevertheless, we have implemented this algorithm in order to gain further insight.

The shifting strategy is a divide and conquer approach. Let P be enclosed between two vertical lines L_1 and L_2. The space between these two lines are divided into vertical strips of width 2. A *group* is a collection of ℓ consecutive strips. Thus, a group has width 2ℓ. There are ℓ different ways of partitioning the region between L_1, L_2 into ℓ groups; let the partitions be $S := S_1, \ldots, S_\ell$, such that every partition in this sequence can be obtained from the previous one by shifting it 2 units to the right. If we repeat the shift ℓ times on S_1, the partition S_1 itself is obtained.

Algorithm 1 : HM-1985(P)

Obtain the partitions $S := S_1, \ldots, S_\ell$ by applying the shifting strategy in x;
for every $S_i \in S$ **do**
    Divide S_i into groups $G_1, G_2, \ldots$ of width 2ℓ;
    for every group G_i **do**
        apply the shifting strategy in y-dimension and obtain the best solution for G_i among ℓ solutions using the algorithm for finding optimal covers in $2\ell \times 2\ell$ squares;
    end for
    The solution C_i for S_i is the union of the solutions obtained for the G_is;
end for
return $C \in \{C_1, \ldots, C_\ell\}$ having the minimum number of disks;

For every partition $S_i \in S$, we compute a cover C_i in the following way. For every group in S_i, we compute a cover. The solution for S_i is the union of these covers obtained for these groups. The authors have showed that if we use a c-approximation algorithm for the groups, then the approximation factor of the global algorithm is $c(1+\frac{1}{\ell})$. Interestingly, the shifting strategy can also be applied to every group independently. In this case, the global algorithm has an approximation factor of $(1+\frac{1}{\ell})^2$. Now observe that for a particular group of width 2ℓ, at each iteration of the shifting strategy, we obtain $2\ell \times 2\ell$ squares. The author showed that one can find the optimal solutions for these squares. Assume that such a square contains m points. The authors argued that $2\ell^2$ disks are enough to cover these points. The candidate disks are the ones which have at least two points on their boundaries. It is easy to see that there are at most $2\binom{m}{2}$ such disks. Hence, we need to check $\mathcal{O}(m^{4\ell^2})$ arrangements of disks, starting with arrangements of size 1 and continuing till arrangements of size $2\ell^2$, until an arrangement is found which covers the m points. The final solution returned by the global algorithm is a cover $C \in \{C_1, \ldots, C_\ell\}$ having the minimum number of disks. Refer to Algorithm 1, for a high-level description of HM-1985.

The main slowdown comes from the time taken to find optimal solutions in the $2\ell \times 2\ell$ squares. Although HM-1985 appears to be unusable in practice, it finishes within a tolerable amount of time for point sets where the number of points in these $2\ell \times 2\ell$ squares is low, say less than 10.

2.2 G-1991: Gonzalez (1991)

Gonzalez designed two algorithms for UDC in d-space; refer to [17]. One of these algorithms is a PTAS which uses the shifting strategy introduced in HM-1985. This PTAS has an approximation factor of $2(1+\frac{1}{l})^{d-1}$ and runtime of $\mathcal{O}(\ell^{d-1}d(2\sqrt{d})(\ell\sqrt{d})^{d-1}n^{d(2\sqrt{d})^{d-1}+1})$, for every integer $\ell \geq 1$. In the plane, this algorithm has an approximation factor of $2(1+\frac{1}{l})$ and runs in $\mathcal{O}(\ell^2 n^{4\sqrt{2}+1})$ time. Although this algorithm is asymptotically faster than HM-1985, but in practice, it can be embarrassingly slow for massive point sets. Hence, we did not implement this algorithm.

Algorithm 2 : G-1991(P)

1: Let $P_1 := \{p \in P | i_y(p) \text{ is odd}\}$ and $P_2 := \{p \in P | i_y(p) \text{ is even}\}$. Execute the lines 2-12 independently on P_1 and P_2. The final solution is the union of these two solutions;

2: Sort P_i (P_1 or P_2) w.r.t $i_x(p)$ into sets $S := S_1, \ldots, S_k$;
3: $R \leftarrow S_1 \cup S_2$; $j \leftarrow 2$;
4: **while** $R \neq \emptyset$ **do**
5:     $q \leftarrow \min\{p_x \mid p \in R\}$;
6:     Let Q be the set of points in R at a distance $\leq \sqrt{2}$ (w.r.t x only) from q, $R \leftarrow R \setminus Q$;
7:     Output the $\sqrt{2} \times \sqrt{2}$ square whose left boundary includes q and whose top boundary coincides with the top boundary of the slab;
8:     **while** $j < k$ **and** R contains elements from at most one of the sets in S **do**
9:         $j \leftarrow j + 1$; $R \leftarrow R \cup S_j$;
10:     **end while**
11: **end while**
12: For every $\sqrt{2} \times \sqrt{2}$ square in the solution, place a unit disk at its center;

The other algorithm G-1991 has an approximation factor of $2^{d-1}(\sqrt{d})^d$ and runtime of $\mathcal{O}(dn + n \log |\text{OPT}|)$, where $|\text{OPT}|$ is the number of disks in an optimal cover. In the plane, G-1991 is a 4-approximation algorithm and runs in

$\mathcal{O}(n \log |\text{OPT}|)$ time. Let $p \in P$, then $i_x(p) = \lfloor p_x/\sqrt{2} \rfloor$ and $i_y(p) = \lfloor p_y/\sqrt{2} \rfloor$. This notation is used in the high-level description of G-1991; see Algorithm 2.

The author presented this algorithm for covering points using axis-parallel squares of fixed size and claimed that the same can be used for UDC. In our implementation we have used squares of length $\sqrt{2}$ and then placed a unit disk at the center of every square. Since a square of length $\sqrt{2}$ can be inscribed inside a unit disk, every point is covered in this approach.

Note that G-1991 does not use any comparison-based sorting algorithm. Refer to [17] for more details. Our experiments show that this algorithm is fast in practice and gives good quality solution.

2.3 CCFM-1997: Charikar, Chekuri, Feder, and Motwani (1997)

The algorithm by Charikar et al. [8] was originally designed for the online version of UDC, which the authors named DUAL CLUSTERING. In d-space, CCFM-1991 has an approximation factor of $\mathcal{O}(2^d d \log d)$. In 2-space, CCFM-1997 has an approximation factor of 7; refer to Algorithm 3. However, no comment was made about its runtime or implementation. So, have used the standard Delaunay triangulation from CGAL for nearest neighbour queries.

Algorithm 3 : CCFM-1997(P)

Let `active-centers` and `inactive-centers` be two empty Delaunay triangulations;
for $p \in P$ **do**
  if the distance to the nearest disk center in `active-centers` > 1 **then**
    if the distance to the nearest disk center q in `inactive-centers` ≤ 1 **then**
      Delete q from `inactive-centers` and add q to `active-centers`;
    else
      Add p to `active-centers` and add the following points to `inactive-centers`: $(p_x + \sqrt{3}, p_y), (p_x + \sqrt{3}/2, p_y + 1.5), (p_x + \sqrt{3}/2, p_y - 1.5), (p_x - \sqrt{3}/2, p_y + 1.5), (p_x - \sqrt{3}, p_y), (p_x - \sqrt{3}/2, p_y - 1.5)$;
    end if
  end if
end for
return `active-centers`;

2.4 FCB-2001: Franceschetti, Cook, and Bruck (2001)

FCB-2001 [15] is based on the shifting strategy from HM-1985. The main slowdown in HM-1985 was the exhaustive checking with the disk arrangements inside $2\ell \times 2\ell$ squares. FCB-2001 tries to address this issue using the following strategy thereby improving the overall asymptotic runtime.

Recall that in HM-1985 we need to check $\mathcal{O}(m^{4\ell^2})$ arrangements of disks to find the optimal solution for a square, where m is the number of points in a square. But in FCB-2001, a square grid of certain size is overlayed on the point set. Now consider any $2\ell \times 2\ell$ square. Find the grid points which belong to this square. Let the number of such points be p. These p points are the candidate disk centers for covering the m points in the square. A sub-optimal covering is found by checking all possible arrangements containing at most $2\ell^2 - 1$ disks. If all these arrangements to cover the points, a compact covering arrangement of $2\ell^2$ disks is chosen. Now the question remains that what should be the size of the grid. According to the authors, to achieve the best possible approximation, one can choose the grid size to be $4/5\sqrt{2} \approx 0.57$ (scaling applied for unit disks). This strategy achieves an approximation factor of $3(1+\frac{1}{\ell})^2$ having a runtime of Kn, where $K = \ell^2 \sum_{i=1}^{\lceil \ell\sqrt{2}\rceil^2-1} \binom{p}{i} i$. The pseudocode of FCB-2001 is similar to that of HM-1985 and hence is not presented.

For UDC, it is natural to consider the greedy strategy where at each iteration a disk placed which covers the maximum number of uncovered points. The authors showed that this greedy strategy does not give constant factor approximation.

2.5 LL-2014: Liu and Lu (2014)

In LL-2014 [22], the plane is divided into vertical strips of width $\sqrt{3}$ each. Inside each strip, we obtain an approximate solution by sorting the points in non-increasing order according to y-coordinate. The next uncovered point is covered by placing a disk as low as possible. The final solution is obtained by taking the union of all the solutions obtained for the strips. This technique is applied by shifting the strip system five times to the right. The final solution is the best out of these six solutions. Refer to Algorithm 4. The authors show that this algorithm has an approximation factor of $25/6 \approx 4.17$ and runs in $\mathcal{O}(n \log n)$ time.

The algorithm can be quite slow in practice since the algorithm has six expensive iterations. However, it gives high quality solution in practice. Also, this algorithm is quite easy to implement, not requiring any complicated data-structure and/or advanced algorithmic techniques. In our implementation, we have assumed that every strip is left closed and right open.

2.6 BLMS-2017: Biniaz, Liu, Maheshwari, and Smid (2017)

The algorithm by Biniaz et al. [5] is a 4-approximation algorithm which runs in $\mathcal{O}(n \log n)$ time. Refer to Algorithm 5 for a high-level description of the algorithm. We have implemented DISK-CENTERS in the algorithm using Delaunay triangulation from CGAL.

Although the algorithm has a good approximation factor, placing four disks in advance sometimes yields poor quality solution compared to the other algorithms. For instance, if the distance between any two points in P is greater than 2, BLMS-2017 places exactly 4 times the optimal number of disks. In comparison, other algorithms such as, G-1991 or DGT-2018 places an optimal number of disks.

Algorithm 4 : LL-2014(P)

```
DISK-CENTERS ← ∅, min ← n + 1;
Sort P w.r.t x-coordinate using an optimal sorting algorithm;
for i ∈ {0, 1, 2, 3, 4, 5} do
    current ← 1, C ← ∅, right ← P[1]_x + i√3/6;
    while current ≤ n do
        index ← current;
        while P[current]_x < right and current ≤ n do
            current ← current + 1;
        end while
        x-of-restriction-line ← right − √3/2, segments ← ∅;
        for j ← index to current−1 do
            d ← P[j]_x − x-of-restriction-line, y ← √(1 − d²);
            Create a segment s with the endpoints (x-of-restriction-line, P[j]_y + y)
            and (x-of-restriction-line, P[j]_y − y), insert s into segments;
        end for
        Sort segments in non-ascending order based on y coordinate of the top and
        greedily stab them by choosing the stabbing point as low as possible, while
        still stabbing the topmost unstabbed segment. Put the stabbing points (the
        disk centers) in C;
        Increment right by a multiple of √3 such that P[current] − right ≤ √3;
    end while
    if |C| < min then
        DISK-CENTERS ← C, min ← |C|;
    end if
end for
return DISK-CENTERS;
```

Algorithm 5 : BLMS-2017(P)

```
C ← ∅, DISK-CENTERS ← ∅;
Sort P from left to right using an optimal sorting algorithm;
for p := (x, y) ∈ P do
    if the nearest point in C is more than 2 units away from p then
        Place four disks centered at (x, y), (x + √3, y), (x + √3/2, y + 1.5), (x + √3/2, y − 1.5)
        as add these four points to DISK-CENTERS;
        C ← C ∪ p;
    end if
end for
return DISK-CENTERS;
```

2.7 DGT-2018: Dumitrescu, Ghosh, and Tóth (2018)

DGT-2018 is a simple online algorithm which gives 5-approximation in the plane. In d-space, the algorithm has an approximation factor of $\mathcal{O}(1.321^d)$ which is an improvement over CCFM-1997. Being an online algorithm like CCFM-1997, these two algorithms do not require any expensive pre-processing, such as, sorting. We have used Delaunay triangulation for nearest-neighbour queries. Refer to Algorithm 6 for a high-level description of this algorithm.

Algorithm 6 : DGT-2018(P)

```
DISK-CENTERS ← ∅;
for p ∈ P do
    if the distance from p to the nearest point in DISK-CENTERS is > 1 then
        DISK-CENTERS ← DISK-CENTERS ∪ p;
    end if
end for
return DISK-CENTERS;
```

2.8 GHS: A Fast 7-Approximation Algorithm

For massive point sets, any kind of pre-processing which requires $\omega(n)$ time is expensive. For instance, sorting n elements using any optimal sorting algorithm, takes a considerable amount of time. In this section, we present a simple algorithm GHS which does not require any pre-processing and is an online algorithm.

We use a square grid of size $\sqrt{2}$ as shown in Fig. 2(left). Every square in this grid is inscribed within a unit disk. If a point is already covered by one of the disks previously placed, then no action is required, else place a disk at the center of the square in which the point lies. We call such a disk, a *grid disk*. GHS is presented with all the technical details in Algorithm 7.

In our implementation, for fast searching we have used a hash table for storing the centers of the disks placed previously. In order to avoid floating-point inaccuracies, we have used a pair of integers to denote a disk center; see Fig. 2(left). The actual coordinates can be easily obtained by multiplying these integers by $\sqrt{2}$ and then adding $\sqrt{2}/2 = 1/\sqrt{2}$ to each of them. Given a point $p \in P$, the pair of integers used to represent the square and also the circumscribed disk in which it lies are $v_p := \lfloor p_x/\sqrt{2} \rfloor$ and $h_p := \lfloor p_y/\sqrt{2} \rfloor$. The exact location of the disk is $(\sqrt{2}v_p + \frac{1}{\sqrt{2}}, \sqrt{2}h_p + \frac{1}{\sqrt{2}})$. In GHS, for every point we perform at most five searches in the table; refer to Algorithm 7. Hence, GHS runs in $\mathcal{O}(n)$ time on average.

It is computationally expensive to check if a point is covered by a disk placed before since it involves floating-point distance calculations. To avoid these heavy calculations in some cases, we have used the following heuristic. Refer to Fig. 2(right). If a point lies in the gray square, then one can safely conclude that this point is not covered any other disks. This checking does not require any distance calculation and hence can be executed fast.

In the following, we show that GHS is an 7-approximation algorithm.

Theorem 1. *GHS is an 7-approximation algorithm. For every integer $n \geq 1$, there exists an $7n$-element point-set for which GHS places seven times the optimal number of disks. GHS requires $\mathcal{O}(s)$ additional runtime space, where s denotes the size of the solution.*

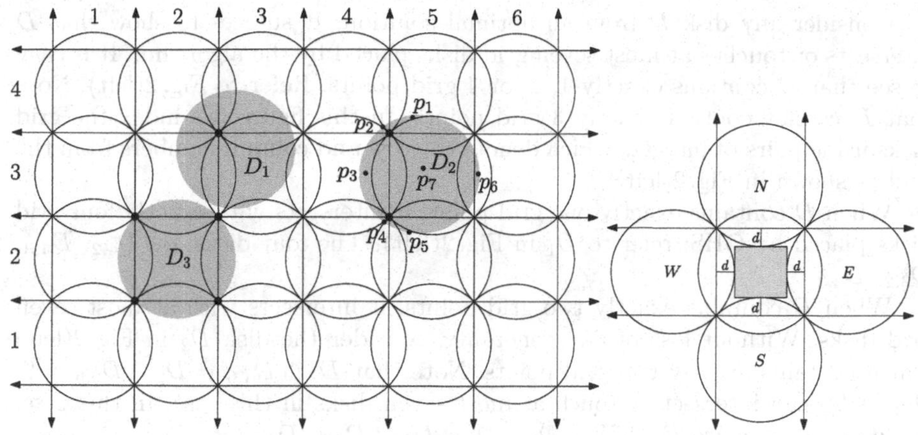

Fig. 2. Left: Illustration of the GHS algorithm. Right: If a point in a grid square belongs to the inner gray square, then the point is not covered by any disk placed before; it follows from rudimentary geometry that $d = 1 - \frac{\sqrt{2}}{2} \approx 0.293$.

Algorithm 7 : GHS(P)

$\mathcal{H} \leftarrow \emptyset$, DISK-CENTERS $\leftarrow \emptyset$;
for $p \in P$ **do**
  $v \leftarrow \lfloor p_x/\sqrt{2} \rfloor$, $h \leftarrow \lfloor p_y/\sqrt{2} \rfloor$;
  if $(v, h) \in \mathcal{H}$ **then**
    no action is required; {p is covered by a disk placed before}
  else if $p_x \geq \sqrt{2}(v + 1.5) - 1$ **and** $(v + 1, h) \in \mathcal{H}$ **and**
    `distance`$(p, (\sqrt{2}(v + 1) + \frac{1}{\sqrt{2}}, \sqrt{2}h + \frac{1}{\sqrt{2}})) \leq 1$ **then**
    do nothing; {p is covered by the disk E placed before; see Fig. 2(right)}
  else if $p_x \leq \sqrt{2}(v - 0.5) + 1$ **and** $(v - 1, h) \in \mathcal{H}$ **and**
    `distance`$(p, (\sqrt{2}(v - 1) + \frac{1}{\sqrt{2}}, \sqrt{2}h + \frac{1}{\sqrt{2}})) \leq 1$ **then**
    do nothing; {p is covered by the disk W placed before; see Fig. 2(right)}
  else if $p_y \geq \sqrt{2}(h + 1.5) - 1$ **and** $(v, h + 1) \in \mathcal{H}$ **and**
    `distance`$(p, (\sqrt{2}v + \frac{1}{\sqrt{2}}, \sqrt{2}(h + 1) + \frac{1}{\sqrt{2}})) \leq 1$ **then**
    do nothing; {p is covered by the disk N placed before; see Fig. 2(right)}
  else if $p_y \leq \sqrt{2}(h - 0.5) + 1$ **and** $(v, h - 1) \in \mathcal{H}$ **and**
    `distance`$(p, (\sqrt{2}v + \frac{1}{\sqrt{2}}, \sqrt{2}(h - 1) + \frac{1}{\sqrt{2}})) \leq 1$ **then**
    do nothing; {p is covered by the disk S placed before; see Fig. 2(right)}
  else
    Insert (v, h) into $\mathcal{H}$ and $(\sqrt{2}v + \frac{1}{\sqrt{2}}, \sqrt{2}h + \frac{1}{\sqrt{2}})$ into DISK-CENTERS;
  end if
end for
return DISK-CENTERS;

Proof. The union of grid disks gives $\mathbf{R}^2$. Hence, it is enough to consider the grid disks to cover the points in P.

Consider any disk D from an optimal solution. It suffices to show that D intersects or touches at most seven grid disks placed by the algorithm. It is easy to see that D contains exactly 1, 2, or 4 grid points. Refer to Fig. 2(left). Note that D cannot contain exactly 3 grid points. In this figure, we index the grid disks using pairs of integers which denote the row and column numbers from the grid as shown in Fig. 2(left).

When D contains exactly one grid point, it intersects with exactly four grid disks placed by GHS; refer to D_1 in Fig. 2(left). The four disks are $D_{4,2}$, $D_{4,3}$, $D_{3,2}$, $D_{3,3}$.

When D contains exactly two grid points, it intersects with at most seven grid disks. Without loss of any generality, consider the disk D_2 in Fig. 2(left) which contains exactly two grid points. Note that $D_2 \cap D_{4,6} = D_2 \cap D_{2,6} = \emptyset$. Hence D_2 can intersect or touch at most seven disks in this case. In this case, these seven disks are $D_{4,4}, D_{4,5}, D_{3,4}, D_{3,5}, D_{3,6}, D_{2,4}, D_{2,5}$.

When D contains exactly four grid points, it intersects or touches six grid disks; refer to D_3 in Fig. 2(left). The six disks are $D_{3,2}, D_{3,3}, D_{2,1}, D_{2,2}, D_{2,3}, D_{1,2}$. Observe that in this case D_4 is also a grid disk. Hence, we have proved that GHS is an 7-approximation algorithm.

Now observe that if the points $p_1, \ldots, p_7$ (see Fig. 2(left)) are present in the input in this sequence, our algorithm places the seven disks $D_{4,4}$, $D_{4,5}$, $D_{3,4}$, $D_{3,5}$, $D_{3,6}$, $D_{2,4}$, $D_{2,5}$. These seven points can be covered by a single disk D_2. This shows that our analysis is tight.

Let us call $p_1, \ldots, p_7$, a *cluster*. Given an integer $n \geq 1$, place n such clusters sufficiently apart, say at least 3 units away from each other. In this case, any optimal solution will contain n disks, but our algorithm will place $7n$ disks.

Note that we insert a disk into the hash-table only when it belongs to the solution. Hence, additional space required by GHS is $\mathcal{O}(s)$, where s denotes the number of disks in the solution.

Remark. If the bounding box of P is known in advance and sufficient space is available, GHS can be implemented to run exactly in $\mathcal{O}(n)$ time using a matrix for storing the disk centers. In our experiments, we have assumed that the bounding box is unknown.

3 Experimental Results

We have implemented these algorithms in C++17 using the CGAL library [23]. The machine used for experiments is equipped with Intel Core i7-6700 (3.40 GHz) and 24 GB of main memory, running Ubuntu Linux 18.04 LTS. The `g++` compiler was used with `-O3` flag. Native C++ containers `list` and `vector` are used wherever possible. For sorting, the standard `std::sort` is utilized. From CGAL, the `Exact_predicates_exact_constructions_kernel` is used. We have tried our best to tune our codes to run faster. For instance, for constant expressions used in the code, we pre-calculated those and stored them in variables to avoid repeated calculations.

Execution time of many UDC algorithms vastly depends on the spread of point sets. If the points are within a small bounding box, covers have small size. This enhances speed of many algorithms, such as, DGT-2018. For a point p, DGT-2018 checks a low number of disks to see if p is covered by a disk place previously. To demonstrate this, we have used point sets which have bounding boxes of varied sizes.

We have experimented the algorithms with synthetic and real-world point sets. For synthetic point sets, we have used three built-in point generators from CGAL: `Random_points_in_square_2` (see Fig. 3), `Random_points_in_disc_2` (see Fig. 4), and `random_convex_set_2` (see Fig. 5). We also have used point sets drawn from an annulus (see Fig. 6). For each of these four generators, we have varied the size of the domain from which the points are drawn randomly: small, medium, and large. Exact sizes are stated in the respective tables. Using each of these three domain sizes, we have generated point sets of three different sizes: 0.1 million, 0.5 million, and 1 million to see the changes in performances with the increase in n. In total, we have experimented with 36 synthetic point sets having different geometry.

We have used the following eleven real-world point sets for our experiments. Some of these are used in [7] for experiments and few others are obtained from Kaggle. See Fig. 7 for the experimental results obtained for these point sets.

1. **`birch3`** [7]: An 100,000-element point set representing random sized clusters in random locations.
2. **`monalisa`** [1]: A 100,000-city TSP instance representing a continuous-line drawing of the Mona Lisa.
3. **`usa`** [1]: A 115,475-city TSP instance representing (nearly) all towns, villages, and cities in the United States.
4. **`KDDCU2D`** [7]: An 145,751-element point set representing the first two dimensions of a protein data-set.
5. **`europe`** [7]: An 169,308-element point set representing differential coordinates of the map of Europe.
6. **`wildfires`**[3]: An 1,880,465-element point set representing wildfire locations in USA.
7. **`world`** [1]: A 1,904,711-city TSP instance consisting of all locations in the world that are registered as populated cities or towns, as well as several research bases in Antarctica.
8. **`china`** [7]: An 1,636,613-element point set representing locations in China.
9. **`nyctaxi`**[4]: An 2,917,288-element point set representing NYC taxi pickup and drop-off locations.
10. **`uber`**[5]: An 4,534,327-element point set representing Uber pickup locations in New York City.
11. **`hail2015`**[6]: An 10,824,080-element point set representing hail storm cell locations based on NEXRAD radar data obtained in 2015.

[3] https://www.kaggle.com/rtatman/188-million-us-wildfires/home.

[4] https://www.kaggle.com/wikunia/nyc-taxis-combined-with-dimacs/home.

[5] https://www.kaggle.com/fivethirtyeight/uber-pickups-in-new-york-city.

[6] https://www.kaggle.com/noaa/severe-weather-data-inventory.

In the tables, α stands for the approximation obtained for GHS, i.e., quality of the solution generated for a specific point set. We measure this using the size of solution obtained using BLMS-2017. Assume that for a particular point set, GHS placed m disks and BLMS-2017 placed n disks, then $\alpha = 4m/n$, since BLMS-2017 has an approximation factor of 4. Note that we could do this since in our experiments, the sizes of the solutions generated by GHS is always less than those generated by BLMS-2017. For a given point set, we define β to be the ratio of the running time of the fastest algorithm (other than GHS) to that of GHS. For every table, we report the highest and lowest observed β. We have truncated the running times and the values for α, β in the tables to 2 decimal places. Every point set was run 3 times and the average running time is reported.

In our experiments, we have found that the shifting-strategy based algorithms HM-1985 and FCB-2001 are embarrassingly slow (as expected from their asymptotic runtimes). However, for point sets where the size of the bounding square is large (say, 5 times n), HM-1985 can produce result within a tolerable amount of time. This implies that the number of points in every square (refer to HM-1985 from Sect. 2) must be very low, less than 10. High number of points in a square slows down HM-1985 because of expensive checking with exponential number of disk arrangements. For instance, for a 10,000-element point set generated inside a square of size 50,000 and $\ell = 2$, HM-1985 gives the answer in 25.3 s. But when $\ell = 2$, for a 60-element point set generated randomly within a square of size 20, HM-1985 ran in 88.86 s. FCB-2001 is also slow in practice since for every square, we consider every lattice point inside it as a candidate disk center. For instance, when $\ell = 2$, for a set of 60 points drawn from a square of size 20, FCB-2001 took 34.61 s. Since we are concerned with massive point sets, we do not include any further experimental data for them as they are unusable in practice.

n	G-1991	CCFM-1997	LL-2014	BLMS-2017	DGT-2018	GHS	α
0.1M	238, 0.16	321, 1.5	224, 1.18	444, 1.88	409, 2.25	256, 0.02	2.31
0.5M	5068, 3.1	6565, 12.65	4783, 6.12	9860, 12.27	9067, 13.75	5184, 0.09	2.11
1M	19746, 11.58	25350, 41.55	17832, 12.34	37896, 29.01	33984, 35.16	20164, 0.19	2.13

n	G-1991	CCFM-1997	LL-2014	BLMS-2017	DGT-2018	GHS	α
0.1M	16805, 1.1	19707, 6.17	12847, 1	30544, 3.03	22978, 4.08	20013, 0.04	2.63
0.5M	250203, 18.26	279990, 201.59	203893, 5.31	535872, 46.61	287632, 71.13	316321, 0.48	2.37
1M	666359, 51.78	720744, 840.98	576779, 10.82	1648108, 162.85	724938, 224.25	787077, 1.33	1.92

n	G-1991	CCFM-1997	LL-2014	BLMS-2017	DGT-2018	GHS	α
0.1M	99952, 0.31	99961, 46.17	99923, 0.63	399388, 11.54	99961, 15.57	99973, 0.08	1.01
0.5M	499944, 1.6	499962, 519.72	499923, 3.3	1999388, 121.57	499962, 122.49	499974, 0.52	1.01
1M	999951, 3.25	999960, 1482.72	999925, 6.79	3999356, 325.12	999960, 319.25	999978, 1.13	1.01

Fig. 3. Points drawn from square of varied size; Top: $2n/10^4$, Middle: $2n/10^3$, Bottom: $2n/10$. A pair in a cell denotes the solution size followed by the time taken in seconds. Highest β observed: 61.63 (Top, $n = 1M$); lowest β observed: 2.9 (Bottom, $n = 1M$).

n	G-1991	CCFM-1997	LL-2014	BLMS-2017	DGT-2018	GHS	α
$0.1M$	176, 0.19	247, 1.44	186, 1.21	352, 1.82	325, 2.13	180, 0.02	2.05
$0.5M$	3995, 2.8	5251, 11.47	3877, 6.15	7840, 11.31	7246, 19.11	4051, 0.09	2.07
$1M$	15637, 10.22	19964, 35.98	14419, 12.37	30248, 39.46	27283, 32.63	15945, 0.19	2.11

n	G-1991	CCFM-1997	LL-2014	BLMS-2017	DGT-2018	GHS	α
$0.1M$	13665, 0.97	16059, 5.17	10541, 1	24908, 2.86	19353, 3.77	15836, 0.03	2.55
$0.5M$	220153, 16.96	248765, 149.43	176313, 5.25	452400, 45.25	258850, 64.07	282912, 0.43	2.51
$1M$	611288, 50.29	669191, 620.83	518130, 10.8	1433020, 139.23	676477, 222.79	739930, 1.28	2.07

n	G-1991	CCFM-1997	LL-2014	BLMS-2017	DGT-2018	GHS	α
$0.1M$	99933, 0.31	99950, 54.78	99906, 0.68	399264, 11.59	99950, 13.43	99967, 0.08	1.01
$0.5M$	499932, 1.63	499956, 522.37	499895, 3.33	1999224, 120.58	499956, 114.94	499969, 0.52	1.01
$1M$	999933, 3.3	999953, 1475.6	999901, 6.66	3999256, 310.35	999953, 308.74	999969, 1.1	1.01

Fig. 4. Points drawn from disk of varied radius; Top: $n/10^4$, Middle: $n/10^3$, Bottom: $n/10$. A pair in a cell denotes the solution size followed by the time taken in seconds. Highest β observed: 55.53 (Top, $n = 1M$); lowest β observed: 3.01 (Bottom, $n = 1M$).

n	G-1991	CCFM-1997	LL-2014	BLMS-2017	DGT-2018	GHS	α
$0.1M$	44, 0.1	50, 0.88	42, 1.19	120, 1.24	61, 1.23	60, 0.02	2
$0.5M$	219, 0.58	246, 4.92	213, 6.1	608, 7.3	305, 10.17	284, 0.09	1.87
$1M$	436, 1.25	491, 15.57	430, 12.28	1220, 19.94	611, 40.45	564, 0.18	1.85

n	G-1991	CCFM-1997	LL-2014	BLMS-2017	DGT-2018	GHS	α
$0.1M$	435, 0.12	492, 1.28	431, 0.98	1216, 2	608, 3.35	563, 0.02	1.86
$0.5M$	2181, 0.69	2453, 17.4	2148, 4.97	6100, 35.61	3045, 48.82	2818, 0.09	1.85
$1M$	4361, 1.46	4910, 60.03	4304, 10.01	12208, 120.42	6095, 155.75	5643, 0.18	1.85

n	G-1991	CCFM-1997	LL-2014	BLMS-2017	DGT-2018	GHS	α
$0.1M$	36894, 0.24	43078, 26.69	32296, 0.57	103996, 26.19	44555, 32.05	48430, 0.05	1.87
$0.5M$	183342, 1.31	216910, 126.67	161822, 2.94	516988, 115.13	221835, 137.88	242351, 0.27	1.88
$1M$	366295, 2.72	434110, 360.19	323789, 6	1033880, 343.91	444418, 411.39	485557, 0.58	1.88

Fig. 5. Convex sets drawn from square of varied size; Top: $2n/10^4$, Middle: $2n/10^3$, Bottom: $2n/10$. A pair in a cell denotes the solution size followed by the time taken in seconds. Highest β observed: 8.35 (Middle, $n = 1M$); lowest β observed: 4.75 (Bottom, $n = 1M$).

n	G-1991	CCFM-1997	LL-2014	BLMS-2017	DGT-2018	GHS	α
$0.1M$	60992, 0.78	65147, 35.86	51606, 0.85	142732, 3.72	65874, 13.2	73732, 0.07	2.07
$0.5M$	118545, 5.86	129636, 112.55	91171, 4.86	221368, 15.62	137501, 76.15	149376, 0.29	2.7
$1M$	134750, 12.43	146795, 194.47	104288, 10.37	245564, 44.37	159973, 157.73	156022, 0.51	2.55

n	G-1991	CCFM-1997	LL-2014	BLMS-2017	DGT-2018	GHS	α
$0.1M$	87190, 1	89567, 110.03	82194, 0.9	277588, 8.45	89603, 20.7	93080, 0.09	1.35
$0.5M$	289255, 12.89	311483, 438.85	242137, 5.03	660384, 37.91	315135, 168.83	355083, 0.69	2.16
$1M$	407291, 33.03	445555, 732.36	323103, 10.63	821488, 80.36	457964, 383.71	526408, 1.64	2.57

n	G-1991	CCFM-1997	LL-2014	BLMS-2017	DGT-2018	GHS	α
$0.1M$	91934, 1.04	93503, 151.04	88442, 0.88	316020, 9.9	93544, 24.11	95762, 0.09	1.22
$0.5M$	348200, 15.6	368498, 715.05	305286, 5.07	892436, 55.68	370709, 216.63	405084, 0.79	1.82
$1M$	536250, 44.06	580054, 1246.16	443492, 10.68	1187956, 109.78	588665, 520.77	667366, 2.1	2.25

Fig. 6. Point sets drawn from an annulus of varied width, in this case, the radius r_1 of the inner circle is varied. Radius r_2 of the outer circle is fixed to 10^3; Top: $r_1 = 0.95r_2$, Middle: $r_1 = 0.75r_2$, Bottom: $r_1 = 0.5r_2$. A pair in a cell denotes the solution size followed by the time taken in seconds. Highest β observed: 20.56 (Top, $n = 1M$); lowest β observed: 5.1 (Bottom, $n = 1M$).

	G-1991	CCFM-1997	LL-2014	BLMS-2017	DGT-2018	GHS	α
birch3	99990, 0.33	99993, 204.66	99989, 0.76	399912, 10.94	99993, 19.86	99996, 0.08	1.01
monalisa	100000, 0.32	100000, 283.85	100000, 0.65	400000, 15.74	100000, 28.98	100000, 0.08	1
usa	115475, 0.39	115475, 269.99	115475, 0.78	461900, 20.57	115475, 33.4	115475, 0.1	1
KDDCU2D	1288, 0.34	1459, 5.29	1147, 1.59	2652, 2.97	1626, 5.12	1418, 0.03	2.14
europe	168087, 0.59	168079, 546.16	168253, 1.19	670524, 40.11	168069, 66.72	168333, 0.14	1.01
wildfires	623, 3.52	710, 65.24	622, 26.84	1256, 26.72	787, 69.42	663, 0.35	2.12
world	7098, 17.3	8597, 69.98	6667, 26.01	14188, 70.42	10980, 74.01	7874, 0.34	2.22
china	562, 4.72	639, 40.81	565, 24.12	1124, 26.9	723, 44.35	599, 0.29	2.14
nyctaxi	31, 3.31	29, 28.7	25, 53.36	84, 38.68	31, 30.86	34, 0.55	1.62
uber	4, 3.53	5, 41.39	3, 79.5	8, 8.51	5, 29.16	5, 0.78	2.5
hail2015	867, 29.14	998, 399.54	888, 174.67	1724, 234.64	1128, 410.03	901, 1.91	2.1

Fig. 7. Performances of the algorithms on real-world point sets. A pair in a cell denotes the solution size followed by the time taken in seconds. Highest β observed: 51.51 (world); lowest β observed: 4.17 (usa).

Why GHS is Efficient in Practice? GHS uses a hash-table to store the disks. Every point in $\mathbf{R}^2$ belongs to a constant number of grid disks. Since hash-tables look-ups take $\mathcal{O}(1)$ time on average, these checks can be done quickly resulting in fast running time in practice. Also, no expensive pre-processing is required such as sorting. The heuristic used to minimize the number of distance calculations speeds up GHS further.

In the proof of Theorem 1, it is shown that GHS places 7 times the optimal number of disks only under certain circumstances. In real-world point sets, these patterns of point clusters occur rarely. This gives good quality solution in practice.

4 Conclusion

The algorithms HM-1985 and FCB-2001 cannot be recommended for practical purposes. If quality of solution is of more interest than real-life running time, we recommend using LL-2014 or G-1991. If running time is important, we recommend GHS.

Interestingly, GHS not only runs fast but also computes good quality solutions on diverse point distributions, as seen in our experimental results; highest observed value of α is 2.7. For every point set used in our experiments, GHS ran faster than the other algorithms; highest and lowest observed values of β are 61.63 and 2.9, respectively. Moreover, even for the point sets for which highest speedups are obtained, GHS produced competitive solutions.

For future work, a natural direction is to design fast practical heuristics to improve the quality of the solutions generated by GHS. Another possible direction is to investigate which algorithms are efficient in practice for higher dimensions and also in other norms such as, L_1 and L_∞.

References

1. www.math.uwaterloo.ca/tsp/
2. Agarwal, P.K., Pan, J.: Near-linear algorithms for geometric hitting sets and set covers. In: Proceedings of the Thirtieth Annual Symposium on Computational Geometry, p. 271. ACM (2014)
3. Aloupis, G., Hearn, R.A., Iwasawa, H., Uehara, R.: Covering points with disjoint unit disks. In: CCCG, pp. 41–46 (2012)
4. Bar-Yehuda, R., Rawitz, D.: A note on multicovering with disks. Comput. Geom. **46**(3), 394–399 (2013)
5. Biniaz, A., Liu, P., Maheshwari, A., Smid, M.: Approximation algorithms for the unit disk cover problem in 2D and 3D. Comput. Geom. **60**, 8–18 (2017)
6. Brönnimann, H., Goodrich, M.T.: Almost optimal set covers in finite VC-dimension. Discrete Comput. Geom. **14**(4), 463–479 (1995)
7. Bus, N., Mustafa, N.H., Ray, S.: Practical and efficient algorithms for the geometric hitting set problem. Discrete Appl. Math. **240**, 25–32 (2018)
8. Charikar, M., Chekuri, C., Feder, T., Motwani, R.: Incremental clustering and dynamic information retrieval. SIAM J. Comput. **33**(6), 1417–1440 (2004)
9. Chazelle, B.M., Lee, D.T.: On a circle placement problem. Computing **36**(1–2), 1–16 (1986)
10. Das, G.K., Fraser, R., Lóopez-Ortiz, A., Nickerson, B.G.: On the discrete unit disk cover problem. Int. J. Comput. Geom. Appl. **22**(05), 407–419 (2012)
11. De Berg, M., Cabello, S., Har-Peled, S.: Covering many or few points with unit disks. Theory Comput. Syst. **45**(3), 446–469 (2009)
12. Dumitrescu, A.: Computational geometry column 68. ACM SIGACT News **49**(4), 46–54 (2018)
13. Dumitrescu, A., Ghosh, A., Tóth, C.D.: Online unit covering in euclidean space. In: Kim, D., Uma, R.N., Zelikovsky, A. (eds.) COCOA 2018. LNCS, vol. 11346, pp. 609–623. Springer, Cham (2018). https://doi.org/10.1007/978-3-030-04651-4_41
14. Fowler, R.J.: Optimal packing and covering in the plane are NP-complete. Inf. Process. Lett. **12**(3), 133–137 (1981)
15. Franceschetti, M., Cook, M., Bruck, J.: A geometric theorem for approximate disk covering algorithms (2001)
16. Fu, B., Chen, Z., Abdelguerfi, M.: An almost linear time 2.8334-approximation algorithm for the disc covering problem. In: Kao, M.-Y., Li, X.-Y. (eds.) AAIM 2007. LNCS, vol. 4508, pp. 317–326. Springer, Heidelberg (2007). https://doi.org/10.1007/978-3-540-72870-2_30
17. Gonzalez, T.F.: Covering a set of points in multidimensional space. Inf. Process. Lett. **40**(4), 181–188 (1991)
18. Hochbaum, D.S., Maass, W.: Approximation schemes for covering and packing problems in image processing and VLSI. J. ACM (JACM) **32**(1), 130–136 (1985)
19. Kaplan, H., Katz, M.J., Morgenstern, G., Sharir, M.: Optimal cover of points by disks in a simple polygon. SIAM J. Comput. **40**(6), 1647–1661 (2011)
20. Liao, C., Hu, S.: Polynomial time approximation schemes for minimum disk cover problems. J. Comb. Optim. **20**(4), 399–412 (2010)
21. Liaw, C., Liu, P., Reiss, R.: Approximation schemes for covering and packing in the streaming model. In: Canadian Conference on Computational Geometry (2018)
22. Liu, P., Lu, D.: A fast 25/6-approximation for the minimum unit disk cover problem. arXiv preprint arXiv:1406.3838 (2014)
23. The CGAL Project: CGAL User and Reference Manual. CGAL Editorial Board, 4.13 edn. (2018). https://doc.cgal.org/4.13/Manual/packages.html

Improved Contraction Hierarchy Queries via Perfect Stalling

Stefan Funke and Thomas Mendel(✉)

Universität Stuttgart, Stuttgart, Germany
mendel@fmi.uni-stuttgart.de

Abstract. Contraction Hierarchies (CH) are one of the most relevant techniques for accelerating shortest path-queries on road networks in practice. We reconsider the CH query routine and devise an additional preprocessing step which gathers auxiliary information such that CH queries can be answered even faster than before. Compared to the standard CH query, response times decrease by more than 70%; compared to a well-known refined CH query routine with so-called *stall-on-demand*, response times still decrease by more than 33% on average. While faster speed-up schemes like hub labels incur a serious space overhead, our precomputed auxiliary information takes less space than the graph representation itself.

Keywords: Shortest path · Contraction Hierarchies

1 Introduction

While the problem of computing shortest paths in general graphs with non-negative edge weights seems to have been well understood already decades ago, the last 10–15 years have seen tremendous progress when it comes to the specific problem of efficiently computing shortest paths in *real-world road networks*. Here the main idea is to spend some time in a preprocessing step where auxiliary information about the network is computed and stored, such that subsequent queries can be answered much faster than via standard Dijkstra's algorithm. One might classify most of the employed techniques into two classes: ones that are based on *pruned graph search* and such that are based on *distance lookups*. Most approaches fall into the former class, e.g., reach-based methods [11,12], highway hierarchies [13], arc-flags-based methods [6], or contraction hierarchies (CH) [10]. Here, basically Dijkstra's algorithm is given a hand to ignore some vertices or edges during the graph search. The achievable speed-up compared to plain Dijkstra's algorithm ranges from one magnitude [12] up to three orders of magnitudes [10]. In practice, this means that a query on a country-sized network like that of Germany (around 20 million nodes) can be answered in less than a *millisecond* compared to few seconds of Dijkstra's algorithm. While these methods directly yield the actual shortest path, the latter class is primarily concerned with the computation of the (exact) distance between given source

I. Kotsireas et al. (Eds.): SEA[2] 2019, LNCS 11544, pp. 158–166, 2019.
https://doi.org/10.1007/978-3-030-34029-2_11

and target queries – recovering the actual path often requires some additional effort. Examples for such distance-lookup-based methods are transit nodes [4,5] and hub labels [2]. They allow for the answering of *distance queries* another one or two orders of magnitudes faster.

In spite of their inferior query times, the methods based on pruned graph search are more popular in practice, because most of the time the actual shortest paths are in fact needed, and the methods based on distance lookups typically incur quite a considerable space overhead. For example, for a network of around 20 million nodes, the hub labelling scheme [2] requires to store for each node in the order of hundreds distance labels to allow for quick query answering. Hence the space consumption of the precomputed auxiliary information by far exceeds the space consumption of the original graph itself. For most methods based on pruned graph search, the space consumption of the precomputed auxiliary information is very moderate compared to the original graph itself. See [3] for a comprehensive survey on the topic.

1.1 Contribution and Outline

In this note we propose a refinement of the standard contraction hierarchies (CH) query procedure which decreases the average number of settled nodes during a query by 65% and the average query times by around 70% compared to the standard CH query procedure on a country-sized road network. Compared to a known 'stall-on-demand' query refinement our approach still decreases the query times by around 33%. In that sense it brings the CH query times closer to the distance-lookup-based methods, yet incurs only a very modest space overhead.

In Sect. 2 we first recapitulate basics of the contraction hierarchy (CH) scheme as well as the CH-based hierarchical hub labelling scheme. Then in Sect. 3 we show how to generate perfect stalling information for CH queries based on hub labels. Finally, we conclude with some experimental results and future work.

2 Preliminaries

2.1 Contraction Hierarchies

The contraction hierarchies approach [10] computes an overlay graph in which so-called shortcut edges span large sections of the shortest path. This reduces the hop length of optimal paths and therefore allows a variant of Dijkstra's algorithm to answer queries more efficiently.

The preprocessing is based on the so-called *node contraction* operation. Here, a node v as well as its adjacent edges are removed from the graph. In order not to affect shortest path distances between the remaining nodes, shortcut edges are inserted between all neighbors u, w of v, if and only if uvw was a shortest path (which can easily be checked via a Dijkstra run). The cost of the new shortcut edge (u, w) is set to the summed costs of (u, v) and (v, w). In the preprocessing phase all nodes are contracted one-by-one in some order. The rank of the node in this contraction order is also called the *level* of the node.

After having contracted all nodes, a new graph $G^+(V, E^+)$ is constructed, containing all original edges of G as well as all shortcuts that were inserted in the contraction process. An edge $e = (v, w)$ – original or shortcut – is called upwards, if the level of v is smaller than the level of w, and downwards otherwise. By construction, the following property holds: For every pair of nodes $s, t \in V$, there exists a shortest path in G^+, which first only consist of upwards edges, and then exclusively of downwards edges. This property allows to search for the optimal path with a bidirectional Dijkstra only considering upwards edges in the search starting at s, and only downwards edges in the reverse search starting in t. This reduces the search space significantly and allows for answering of shortest path queries within the *milliseconds* range compared to *seconds* on a country-sized road network.

2.2 CH-Based Hub Labels

Hub Labelling is a scheme to answer shortest path distance queries which differs fundamentally from graph search based methods. Here the idea is to compute for every $v \in V$ a *label* $L(v)$ such that for given $s, t \in V$ the distance between s and t can be determined by just inspecting the labels $L(s)$ and $L(t)$. All the labels are determined in a preprocessing step (based on the graph G), later on, the graph G can even be thrown away. There have been different approaches to compute such labels (even in theory); we will be concerned with labels that work well for road networks and are based on CH again, following the ideas in [2]. To be more concrete, the labels we are constructing have the following form:

$$L(v) = \{(w, d(v, w)) : w \in H(v)\}$$

Here we call $H(v)$ a set of *hubs* – important nodes – for v. The hubs should be chosen such that for any s and t, the shortest path from s to t intersects $L(s) \cap L(t)$.

If such label sets could be computed, the computation of the shortest path distance between s and t boils down to determining the node $w \in L(s) \cap L(t)$ minimizing the summed distance. If the labels $L(.)$ are stored lexicographically sorted, this can be done in a very cache-efficient manner in time $O(|L(s)|+|L(t)|)$.

Knowing about CH, there is a natural way of computing such labels: simply run an upward Dijkstra from each node v and let the label $L(v)$ be the settled nodes with their respective distances. Clearly, this yields valid labels since CH answers queries exactly. The drawback is that the space requirement is quite large; depending on the metric and the CH construction, one can expect labels consisting of several hundreds to thousands node-distance pairs. It turns out, though, that many of the labels created in such a manner are useless as they do not represent shortest-path distance (as we restricted ourselves to a search in the upgraph only); pruning out those reduces the number of labels by a factor of 4. A source target distance query can then be answered in the *microseconds* range.

3 Perfect Stalling

As we already observed in the construction of hub labels, the exploration of the upgraph during the CH query phase might visit nodes with non-shortest path distances. Obviously, none of the nodes settled with non-shortest path distance are relevant for answering a shortest path query, yet they contribute to the query time. In the original CH paper [10] a technique which they call *stall-on-demand* was suggested which identifies some of the nodes with non-shortest path distances in the exploration of the upgraphs. Note though, that there is a trade-off between the decrease in query time due to the reduced number of nodes to consider and the effort to identify nodes with non-shortest path distance.

In its simplest form, the stall-on-demand strategy from [10] works as follows: Consider the upgraph search from the source s (the reverse upgraph search from the target t works analogously). When a node v is pulled from the priority queue with distance label $d(v)$, one inspects all *incoming* neighbors w with (v, w) and $level(w) > level(v)$. Clearly, if $d(w) + c(w, v) < d(v)$, $d(v)$ cannot be the shortest path distance from s to v and hence the exploration (in particular relaxation of outgoing edges) of v can be 'stalled'. Of course, this procedure does not necessarily identify all nodes with non-shortest path distances, yet it is easy to implement and still prunes the search considerably. More involved stall-on-demand strategies explore a larger neighborhood to conclude for even more nodes to bear non-shortest-path distances. Yet, the additional effort at query time is not rewarded by a respective more reduced search space.

3.1 Precomputing Perfect Stalling Decisions

The contribution of this paper is the idea of precomputing perfect stalling decisions. In that way, we can benefit from a maximally reduced search space during the CH search without incurring a runtime penalty for performing a stall-on-demand computation at query time. It turns out that this can be done with moderate space overhead.

The first idea that comes to mind is to store for each node of the upgraph of s (and analogously for t) a bit whether it is reachable within the upgraph with shortest path distance. There are some disadvantages of this idea: First, we might store information for nodes in the upgraph that would never be encountered during the search because all immediate predecessors have already been stalled. Second, since the desired information varies for different sources s, we have to store for each s and each v in the upgraph of s whether v is reachable from s on a shortest path within the upgraph. While the actual information is only a single bit, storing the identity of each v is quite costly, e.g., a node ID is typically 64 bits. Storing several hundreds or thousands of such items results in several kilobytes additional memory for each node in the graph.

Note that if CH-based hub labels are available, the decision whether a node v just pulled from the priority queue with distance label $d(v)$ in the upgraph search from s can be made by using hub labels to look up the correct shortest path distance and comparing with d. Clearly, these decisions are perfect in a sense that we stall exactly those nodes that are not reachable on a shortest path within the upgraph. Yet, the requirement of having precomputed hub labels in the background just to speed up CH queries is prohibitive in practice due to the considerable space consumption.

Furthermore we want to remark that the notion of 'stalled nodes' also appears in theoretical analyses of contraction hierarchies, e.g., [7] or [9]; there, the so-called 'direct search space' $DSS(v)$ of a node v refers to all nodes in the upgraph of v with true shortest path distances, whereas the actual 'search space' $SS(v)$ comprises also nodes which are not reachable on shortest paths within the upgraph. Similar to the practical implementations, the latter nodes create some nuisance in the theoretical analysis.

Stalling Traces. Now the main idea to enjoy the benefits of stalling at query time without the runtime penalty of stall-on-demand or the space overhead of hub labels is to simulate the upgraph searches with perfect stalling during a preprocessing step (with the help of precomputed CH-based hub labels) and only record the respective decisions as a bit stream – which we call *stalling trace* – in the order they are taken. For each node v we store two stalling traces – one for the upgraph search where v acts as source, one for the reverse upgraph search where v acts as target. Apart from representing perfect stalling decisions, this approach not necessarily requires a bit for every node of the upgraph of v; if for a node w in the upgraph of v, all immediate predecessors are not reachable on shortest paths from v, w will never be pulled from the priority queue (due to the perfect stalling decisions) and hence does not require a bit in the stalling trace.

In summary, the preprocessing phase of our method looks as follows:

```
PREPROCESSING(G)
1. Construct CH
2. Construct CH-based hub labels HL
3. for each node:
   a. simulate upgraph searches
   b. store stalling traces
4. discard HL and only keep CH and stalling traces
```

At query time we simply run an ordinary CH-query but use the stalling traces to have perfect stalling decisions during the exploration of the upgraphs. Algorithms 1 and 2 illustrate the precomputation of a trace as well as the query for an unidirectional dijkstra run. In the bi-directional case we simply use the traces of the source as well as the target node.

We will see in the next section that this strategy pays off; queries are considerably accelerated without incurring a major space overhead compared to pure CH representation.

```
Data: Node s, UpGraph G
Result: Boolean Vector "trace"
trace = Vector[Bool];
distance = Vector[Integer](default=INF);
pq = MinHeap;
pq.push(0, s);
distance[s] = 0;
while pq not empty do
    settle next node from pq;
    if distance is correct then
        trace.push(TRUE);
        relax outgoing edges;
    else
        trace.push(FALSE);
    end
end
```

Algorithm 1: Computing the trace for a single node. The decision whether a distance is correct is made via precomputed hub labels.

```
Data: Node s, UpGraph G, Trace t
Result: Distance Vector "distance"
distance = Vector[Integer](default=INF);
pq = MinHeap;
pq.push(0, s);
distance[s] = 0;
while pq not empty do
    settle next node from pq;
    if t.next() == TRUE then
        relax outgoing edges;
    end
end
```

Algorithm 2: Using the trace during upgraph search.

4 Experiments

In the following we report on our experiments with the improved query scheme for contraction hierarchies. All implementations were compiled using `g++ 7.3.0` and executed on a single core of a Ubuntu Linux 18.04 system with an Intel Xeon E3-1225v3, 3.2 GHz and 32 GB of RAM.

4.1 Data Sets, CH and HL Precomputation

We consider three data sets that were extracted from the OpenStreetMap project [1]. For all three data sets contraction hierarchies as well as CH-based hub labels were computed using the standard approaches in [10] and [2]. See Table 1 for the resulting characteristics of our data sets. Note that both graphs and metrics differ from the ones used in [10] or [3], hence in particular the number of shortcuts as well as the average label sizes differ.

Table 1. Data sets for benchmarking

	STGT	BW	GER
Nodes	1.16 M	3.67 M	24.61 M
Edges (original + shortcuts)	4.32 M	13.76 M	91.61 M
CH shortcuts	1.97 M	6.37 M	41.82 M
HL avg. label size	81.2	109.5	236.28

4.2 Stalling Trace Construction

Since our machine did not suffice to store the hub labels for the largest instance in main memory, we had the hub labels stored in an external array on hard disk. The mapping was realized via STXXL [8], which can be used as a caching mechanism. Page and block sizes were chosen to fit into 16 GB of Memory to mimic a simple standard Desktop/Laptop. Furthermore the block-size was chosen such that most of the labels would fit into a single block. While the access of the labels is mostly sporadic, we always read the whole label into memory. Due to this blockwise access the overhead of disk-based storage was around a factor of 4 compared to a purely RAM-based implementation (given enough memory).

Table 2 summarizes the result of our preprocessing step. Preprocessing times were 5 min/31 min/19:21 h for our three data sets. While the first two instances at some point simply resided in main memory, the largest instance required frequent disk accesses. Note that our implementation is single-threaded; we expect considerable speedup by parallelization since any number of vertices can be processed in parallel. Observe that we require less than the graph representation itself of additional memory to store the stalling traces.

Table 2. Stalling trace construction

	STGT	BW	GER
avg. trace length (Bits/node)	130.2	202.6	489.4
max. trace length (Bits/node)	233	332	851
total trace space (MBytes)	36.0	177.3	2871.5
Graph size (incl. CH) (MBytes)	182.5	580.9	3870.2

4.3 Queries

Now let us analyze the effect on query times. We report on average query times for random source target queries for plain Dijkstra (**Dijk**), plain CH without stall-on-demand (**CH**), CH with standard stall-on-demand (**CHso**), and CH with perfect stalling (**CHps**), stating both number of settled nodes as well as the actual query times. All numbers are averaged over 100,000 queries except for plain Dijkstra, where we only made 100 queries due to time constraints. See Table 3 for the results.

Table 3. Query benchmarks: measuring average number of settled nodes as well as query times.

	STGT				BW				GER			
	Dijk	CH	CHso	CHps	Dijk	CH	CHso	CHps	Dijk	CH	CHso	CHps
# settled												
avg	627 k	313	206	201	1.65 M	586	321	311	12.9 M	2212	782	761
max	1.14 M	634	390	377	3.59 M	1183	585	577	24.4 M	4161	1388	1297
query-time												
avg in μs	155 ms	85	66	47	439 ms	200	124	83	4.4 s	1159	457	301
max in μs	298 ms	133	95	70	951 ms	313	175	119	8.6 s	1874	644	427

As to be expected all CH variants are at least a factor of 1,000 more efficient than plain Dijkstra, both in terms of number of settled nodes as well as actual query time (both max and average). Comparing the CH variants for the largest graph, the CHso variant on average settles only around 35% of the nodes, the query times are around 39% of the standard CH query. Somewhat to our surprise, CHps only settles slightly less nodes than CHso, yet the average query times improve to just 25% of the standard CH query. So one-hop stall-on-demand is almost perfect in detecting nodes with non-shortest-path distances; the additional improvement of CHps in query time is due to not having to perform stall-on-demand and simply use the precomputed stalling trace. We also observe that both stalling variants become more effective the larger the graphs get. The respective maximum values behave similar to the averages.

5 Conclusion

We have proposed a conceptually very simple technique to precompute and space-efficiently store perfect stalling decisions for contraction hierarchy based shortest path queries. While incurring only very moderate additional space (less than the graph and CH itself), considerable speedup in terms of number of touched nodes as well as actual query times can be achieved. Our preprocessing times still leave some room for improvement, yet straightforward parallelization should result in considerable speed-up of the preprocessing phase. We were quite surprised to see simple one-hop stall-on-demand to be so effective in detecting nodes with non-shortest-path distances; only very few such nodes survived the simple one-hop stall-on-demand. Our approach is obvlious to the concrete CH construction and hence could be combined with more sophisticated CH construction techniques as well.

References

1. OpenStreetMap. https://www.openstreetmap.org/
2. Abraham, I., Delling, D., Goldberg, A.V., Werneck, R.F.: Hierarchical hub labelings for shortest paths. In: Epstein, L., Ferragina, P. (eds.) ESA 2012. LNCS, vol. 7501, pp. 24–35. Springer, Heidelberg (2012). https://doi.org/10.1007/978-3-642-33090-2_4
3. Bast, H., et al.: Route planning in transportation networks. CoRR, abs/1504.05140 (2015)
4. Bast, H., Funke, S., Matijevic, D., Sanders, P., Schultes, D.: In transit to constant time shortest-path queries in road networks. In: ALENEX. SIAM (2007)
5. Bast, H., Funke, S., Sanders, P., Schultes, D.: Fast routing in road networks with transit nodes. Science **316**(5824), 566 (2007)
6. Bauer, R., Delling, D.: SHARC: fast and robust unidirectional routing. In: ALENEX, pp. 13–26. SIAM (2008)
7. Blum, J., Funke, S., Storandt, S.: Sublinear search spaces for shortest path planning in grid and road networks. In: AAAI, pp. 6119–6126. AAAI Press (2018)
8. Dementiev, R., Kettner, L., Sanders, P.: STXXL: standard template library for XXL data sets. Softw. Pract. Exper. **38**(6), 589–637 (2008)
9. Funke, S., Storandt, S.: Provable efficiency of contraction hierarchies with randomized preprocessing. In: Elbassioni, K., Makino, K. (eds.) ISAAC 2015. LNCS, vol. 9472, pp. 479–490. Springer, Heidelberg (2015). https://doi.org/10.1007/978-3-662-48971-0_41
10. Geisberger, R., Sanders, P., Schultes, D., Vetter, C.: Exact routing in large road networks using contraction hierarchies. Transp. Sci. **46**(3), 388–404 (2012)
11. Goldberg, A.V., Kaplan, H., Werneck, R.F.: Reach for a*: efficient point-to-point shortest path algorithms. In: ALENEX, pp. 129–143. SIAM (2006)
12. Gutman, R.J.: Reach-based routing: a new approach to shortest path algorithms optimized for road networks. In: ALENEX/ANALCO, pp. 100–111. SIAM (2004)
13. Sanders, P., Schultes, D.: Engineering highway hierarchies. ACM J. Exp. Algorithmics, **17**(1) (2012)

Constraint Generation Algorithm for the Minimum Connectivity Inference Problem

Édouard Bonnet[1], Diana-Elena Fălămaş[1,2], and Rémi Watrigant[1(✉)]

[1] Univ Lyon, CNRS, ENS de Lyon, Université Claude Bernard Lyon 1, LIP, 69342 Lyon Cedex 07, France
{edouard.bonnet,remi.watrigant}@ens-lyon.fr, falamasd@yahoo.com
[2] Technical University of Cluj-Napoca, Cluj-Napoca, Romania

Abstract. Given a hypergraph H, the Minimum Connectivity Inference problem asks for a graph on the same vertex set as H with the minimum number of edges such that the subgraph induced by every hyperedge of H is connected. This problem has received a lot of attention these recent years, both from a theoretical and practical perspective, leading to several implemented approximation, greedy and heuristic algorithms. Concerning exact algorithms, only Mixed Integer Linear Programming (MILP) formulations have been experimented, all representing connectivity constraints by the means of graph flows. In this work, we investigate the efficiency of a *constraint generation algorithm*, where we iteratively add cut constraints to a simple ILP until a feasible (and optimal) solution is found. It turns out that our method is faster than the previous best flow-based MILP algorithm on random generated instances, which suggests that a constraint generation approach might be also useful for other optimization problems dealing with connectivity constraints. At last, we present the results of an enumeration algorithm for the problem.

Keywords: Hypergraph · Constraint generation algorithm · Connectivity problem

1 Introduction and Related Work

We study the problem where one wants to infer a binary relation over a set of items V (that is, a graph), where the input consists of some subsets of those items which are known to be connected in the solution we are looking for. In other words, the input can be represented by a hypergraph $H = (V, \mathcal{E})$, and we are looking for an underlying undirected graph $G = (V, E)$ such that for every hyperedge $S \in \mathcal{E}$, the subgraph induced by S, denoted by $G[S]$, is connected (such a graph G will be called a *feasible solution* in the sequel). Observe that it is easy to construct trivial feasible solutions to this problem: consider for instance the graph $K(H)$ having vertex set V and an edge uv iff u and v belong to a same hyperedge. Since these solutions are unlikely to be of great interest in

I. Kotsireas et al. (Eds.): SEA[2] 2019, LNCS 11544, pp. 167–183, 2019.
https://doi.org/10.1007/978-3-030-34029-2_12

practice, it makes sense to add an optimization criteria. In this paper, we focus on minimizing the number of edges of the solution. More formally, we study the following problem:

Minimum Connectivity Inference (MCI)
Input: a hypergraph $H = (V, \mathcal{E})$
Output: a graph $G = (V, E)$ such that $G[S]$ is connected $\forall S \in \mathcal{E}$
Goal: minimize $|E(G)|$

This optimization problem is NP-hard [11], and was first introduced for the design of vacuum systems [12]. It has then be studied independently in several different contexts, mainly dealing with network design: computer networks [13], social networks [3] (more precisely modeling the *publish/subscribe* communication paradigm [7,15,19]), but also other fields, such as auction systems [8] and structural biology [1,2]. Finally, we can mention the issue of hypergraph drawing, where, in addition to the connectivity constraints, one usually looks for graphs with additional properties (*e.g.* planarity, having a tree-like structure... *etc.*) [5,16–18]. This plethora of applications explains why this problem is known under different names, such as Subset Interconnected Design, Minimum Topic Overlay or Interconnection Design. For a comprehensive survey of the theoretical work done on this problem, see [6] and the references therein.

Concerning the implementation of algorithms, previous works mainly focused on approximation, greedy and other heuristic techniques [19]. To the best of our knowledge, the first exact algorithm was designed by Agarwal *et al.* [1, 2] in the context of structural biology, where the sought graph represents the contact relations between proteins of a macro-molecule, which has to be inferred from a hypergraph constructed by chemical experiments and mass spectrometry. In this work, the authors define a Mixed Integer Linear Programming (MILP) formulation of the problem, representing the connectivity constraints by flows. They also provide an enumeration method using their algorithm as a black box, by iteratively adding constraints to the MILP in order to forbid already found solutions. Both their optimization and enumeration algorithms were tested on some real-life (from a structural biology perspective) instances for which the contact graph was already known.

This MILP model was then improved recently by Dar *et al.* [10], who mainly reduced the number of variables and constraints of the formulation, but still representing the connectivity constraints by the means of flows. In addition, they also presented and implemented a number of (already known and new) reduction rules. This new MILP formulation together with the reduction rules were then compared to the algorithm of Agarwal *et al.* on randomly-generated instances. For every kind of tested hypergraphs (different number and sizes of hyperedges), they observed a drastic improvement of both the execution time and the maximum size of instances that could be solved.

In this paper we initiate a different approach for this problem, by defining a simple constraint generation algorithm relying on a cut-based ILP. This method can be seen as an application of *Benders' decomposition* [4], where one wants to solve a (generally large) ILP called *master problem* by decomposing it into a smaller (and easier to solve) one, adding new constraints from the master problem when the obtained solution is infeasible (this approach is sometimes known as *row generation*, because new constraints are added throughout the resolution). We first present different approaches for the addition of new constraints and compare their efficiency on random instances. We then evaluate the performance of our method by comparing it to the MILP formulation of Dar *et al.* on randomly generated instances (using the same random generator).

Finally, we present an algorithm for enumerating all optimal solutions of an instance, which we compare to the approach developed by Agarwal *et al.*

Organization of the Paper. In the next section, we introduce our constraint generation algorithm. In Sect. 3, we recall the random generator of Dar *et al.* and present the results of the comparison between our constraint generation algorithm and the flow-based MILP formulation. Finally, Sect. 4 is devoted to our enumeration algorithm.

2 Constraint Generation Algorithm for MCI

2.1 Presentation

Rather than defining a single (M)ILP model whose optimal solutions coincide with optimal solutions of the MCI problem, our approach is a *constraint generation* algorithm which starts with a simple ILP whose optimal solutions do not necessarily correspond to feasible solutions for MCI. Then, some constraints are added to the model which is solved again. This process is repeated until we reach a feasible solution.

Let us define more formally our approach. In the sequel, $H = (V, \mathcal{E})$ will always denote our input hypergraph, and n and m will always denote the number of vertices and hyperedges of H, respectively. Recall that $K(H)$ denotes the graph with vertex set V having an edge uv iff u and v belong to a same hyperedge. Let us first define our starting ILP model. It has one binary variable x_e for every possible edge e of $K(H)$, which takes value 1 iff the corresponding edge is in the solution. In the following, we will thus make no distinction between solutions of our ILP and graphs with vertex set V.

The constraints that will be added are defined by *cuts* $(X_1, X_2, \dots, X_r)$, $r \geq 2$, where $X_i \subseteq V$, $X_i \neq \emptyset$ and $X_i \cap X_j = \emptyset$ for every $i, j \in \{1, \dots, r\}$, $i \neq j$. Given a cut $C := (X_1, \dots, X_r)$, we define its corresponding set of edges $E(C) := \{xy \in E(G), x \in X_i, y \in X_j, i \neq j\}$. Given a set of cuts $\mathcal{C}$, let $\mathcal{M}(\mathcal{C})$ be the following ILP:

$$\begin{array}{lll}
\text{Minimize} & \sum_{e\in K(H)} x_e & \\
\text{subject to:} & & \\
& \sum_{u,v\in S} x_{uv} \geq |S|-1 & \forall S \in \mathcal{E}\ (1)\\
& \sum_{e\in E(C)} x_e \geq r-1 & \forall C := (X_1,\ldots,X_r) \in \mathcal{C}\ (2)\\
& x_e \in \{0,1\} & \forall e \in K(H)
\end{array}$$

Constraints (1) forces the solution to contain at least $|S|-1$ edges within every hyperedge. Although this constraints is not sufficient to guarantee the connectivity in every hyperedge (for instance, two disjoint cycles also satisfy this constraint), its purpose is mainly to speed-up the resolution.

The purpose of constraints (2) is to forbid $X_1, \ldots, X_r$ to be connected components in the solution: it forces the quotient graph[1] *w.r.t.* $X_1, \ldots, X_r$ to contain at least $r-1$ edges. Notice that if $r=2$, then it forces the solution to have an edge between the two parts X_1 and X_2.

For a set $S \subseteq V$, define $\mathcal{B}_S := \{(X, S\backslash X) : X \subseteq S, X \notin \{\emptyset, S\}\}$ the set of cuts constructed from all non-trivial[2] bipartitions of S, and $\mathcal{P}_S = \{(X_1, \ldots, X_r) : r \geq 2, X_i \subseteq S, X_i \neq \emptyset, \cup_{i=1}^{r} X_i = S \text{ and } X_i \cap X_j = \emptyset \text{ for all } i,j \in \{1,\ldots,r\}, i \neq j\}$ the set of cuts constructed from all non-trivial partitions of S. Moreover let $\mathcal{B}_H := \bigcup_{S\in\mathcal{E}} \mathcal{B}_S$ and $\mathcal{P}_H := \bigcup_{S\in\mathcal{E}} \mathcal{P}_S$. We have the following:

Proposition 1. *Optimal solutions of $\mathcal{M}(\mathcal{P}_H)$ are in one-to-one correspondence with optimal solutions of $\mathcal{M}(\mathcal{B}_H)$ which are themselves in one-to-one correspondence with optimal solutions of the* MCI *instance.*

Proof. We have $\mathcal{B}_H \subseteq \mathcal{P}_H$, hence a feasible solution of $\mathcal{M}(\mathcal{P}_H)$ is also a feasible solution of $\mathcal{M}(\mathcal{B}_H)$. A feasible solution of $\mathcal{M}(\mathcal{B}_H)$ is also a feasible solution of MCI, since otherwise Constraint (2) would not be satisfied for some bipartition of some hyperedge. Finally a feasible solution of MCI is a feasible solution of $\mathcal{M}(\mathcal{P}_H)$, otherwise a hyperedge would not induce a connected subgraph. □

By the previous proposition, it would be sufficient to solve $\mathcal{M}(\mathcal{B}_H)$ or $\mathcal{M}(\mathcal{P}_H)$. However, we have $|\mathcal{P}_H| = \sum_{S\in\mathcal{E}} 2^{|S|} - 1$ and $|\mathcal{B}_H| = \sum_{S\in\mathcal{E}} 2^{|S|-1} - 1$, which makes these naive ILPs inefficient from a practical point of view. Fortunately, it turns out that for many instances in practice, only a small number of cuts among $\mathcal{B}_H$ (resp. $\mathcal{P}_H$) is actually needed in order to ensure connectivity in every hyperedge. This idea is the basis of our constraint generation algorithm described below.

[1] The *quotient graph w.r.t.* $X_1, \ldots, X_r$ has r vertices $v_1, \ldots, v_r$, and an edge $v_i v_j$ whenever there is an edge between a vertex of X_i and a vertex of X_j, $i \neq j$.

[2] A non-trivial partition of a set V is a partition where each set is different from $\emptyset$ and V.

Algorithm 1: constraint generation algorithm for MCI

Input: a hypergraph $H = (V, \mathcal{E})$
Output: a solution $G = (V, E)$
$\mathcal{C} \leftarrow \mathcal{C}_{init}(H)$
$G \leftarrow solve(\mathcal{M}(\mathcal{C}))$
while *G is not feasible* **do**
  $\mathcal{C} \leftarrow \mathcal{C} \cup newCuts(G)$
  $G \leftarrow solve(\mathcal{M}(\mathcal{C}))$
end

Our strategy is specified by a set of initial cuts of the input hypergraph $\mathcal{C}_{init}(H)$, and a routine $newCuts(G)$ which takes a non-feasible solution G as input, and outputs a set of cuts. If the $newCuts(.)$ routine always returns cuts from $\mathcal{B}_H$ (resp. $\mathcal{P}_H$) that were not considered before, then the algorithm clearly outputs a feasible optimal solution for the problem, since it only stops when a feasible solution is found and, in the worst case, it ends by solving $\mathcal{M}(\mathcal{B}_H)$ (resp. $\mathcal{M}(\mathcal{P}_H)$). This proves that Algorithm 1 always terminates and returns an optimal solution for MCI, provided that the $newCuts(.)$ routine satisfies the property described above. The choices of the initial set of cuts and this routine are described in the next sub-section.

2.2 Choice of Cuts

The choice of cuts is a crucial feature of our algorithm. The main challenge is to find the policies that will lead to a right balance of the number of added constraints: if too few constraints are added in each iteration, then the number of these iterations will increase, which will then result in a lack of efficiency. On the opposite, if too many constraints are added at the beginning and/or in each iteration, then the size of the ILP will increase too quickly, which will slow down the solver, and then result in a lack of efficiency once again. Here we present a set of initial set of cuts, and three possible $newCuts(.)$ routines. We then conducted an empirical evaluation of these strategies (using the initial set of cuts or not, followed by one of the three $newCuts(.)$ routine, thus defining six possible strategies).

Initial Set of Cuts. For every hyperedge $S \in \mathcal{E}$, and every vertex $v \in S$, the idea is to add the cut $(\{v\}, S \setminus \{v\})$. This set of cuts forbids solutions with isolated vertices in every hyperedge. One could also consider cuts $(X, X \setminus S)$ formed from every subset $X \subseteq S$ of a fixed size q. However, for $q = 2$ already, we noticed a drop of efficiency, mainly caused by the large number of constraints it creates. Hence, we shall initialize $\mathcal{C}_{init}$ with the cuts formed by singletons only. In the sequel, this initial set of cuts will sometimes be called *singleton cuts*.

The newCuts(.) Routine. Given a non-feasible solution G of MCI, recall that we shall add, for every hyperedge S such that $G[S]$ is disconnected, a set of cuts. Let S be such a hyperedge. Notice that the objective is not to guarantee connectivity in the *very next* iteration of the algorithm, but to constrain the model more and more. Let $S_1, \ldots, S_p$ be the connected components of $G[S]$, with $p \geq 2$. We considered three natural ideas for the set of new cuts corresponding to S in this situation:

- **Routine 1:** add only one cut (A, B) corresponding to a balanced bipartition of the connected components, that is, $A \cup B = S$, $A \cap B = \emptyset$ and $S_i \subseteq A$ or $S_i \subseteq B$ for every $i \in \{1, \ldots, p\}$, and the absolute value of $|A| - |B|$ is as minimum as possible. Since the problem of finding a balanced bipartition of a given set of numbers is an NP-hard problem, the computation of the bipartition was done using a polynomial greedy algorithm which considers connected components in decreasing order *w.r.t.* their sizes, and iteratively adds each of them to A (resp. B) whenever $|A| < |B|$ (resp. $|A| \geq |B|$). Notice that this algorithm provides a $\frac{7}{6}$-approximation of an optimal bipartition, and runs in $O(p \log p)$ time [14].
- **Routine 2:** add the cut $(S_i, \cup_{j \neq i} S_j)$, for every $i \in \{1, \ldots, p\}$. This idea forbids S_i to be disconnected from the rest of S in the next iteration.
- **Routine 3:** add the cut $(S_1, \ldots, S_p)$. Here, we simply forbid $G[S]$ to have the exact same connected components in the next round.

Observe that the first two strategies return cuts from the set $\mathcal{B}_H$ defined previously, while the third one returns a cut which belongs to $\mathcal{P}_H$. In all three cases, the routine returns cuts which were not in the model, hence guaranteeing the optimality and termination of our algorithms, as seen previously.

Combining the above choices, it gives six different strategies:

- Strategy 1: initial set of cuts: none; *newCuts*(.): Routine 1
- Strategy 2: initial set of cuts: none; *newCuts*(.): Routine 2
- Strategy 3: initial set of cuts: none; *newCuts*(.): Routine 3
- Strategy 4: initial set of cuts: singleton cuts; *newCuts*(.): Routine 1
- Strategy 5: initial set of cuts: singleton cuts; *newCuts*(.): Routine 2
- Strategy 6: initial set of cuts: singleton cuts; *newCuts*(.): Routine 3

After an empirical evaluation of the above strategies for different kind of instances, we observed a similar behaviour for all of them, with a high deviation for seemingly similar instances. Nevertheless, we could observe that on average, strategies 4, 5, and 6 were more efficient than strategies 1, 2 and 3, especially for instances with a high number of vertices, which suggests that using a non-empty set of initial set of cuts should always be better. The closeness of the results for the three routines can be explained by the fact that in practice (in our random instances, all having less than 25 vertices), the number of connected components of every hyperedge of non-feasible solution is usually small (frequently 2 or 3, and often smaller than 5), which leads to similar ILP models to be solved (for instance, when there are only two connected components, all three routines output exactly the same set of cuts).

Our first empirical results suggest that a more fine-grained comparison should be performed in order to better understand which hypergraph parameters influence the efficiency of our different strategies. This approach could then be used in a more general algorithm which would first analyze the instance to solve, and then choose the right strategy to use. Another option would be to run all strategies in parallel in order to obtain the least running time for every instance.

In the sequel, we decided to effectively use the singleton cuts as initial set of cuts, and to use Routine 1 as *newCuts*(.) (that is, it corresponds to strategy 4 described above).

3 Experimental Evaluation

3.1 Generation of Instances

Our random generator of instances follows the same rules as in the experiment conducted by Dar *et al.* [10]. A given scenario depends on the following features:

- **Number of vertices** n of the hypergraph.
- **Density of the hypergraph** $d = \frac{m}{n}$. As in [10], we used the following values: $d \in \{1, 3, 5\}$.
- **Hyperedge size bounds and distributions.** For this parameter, we used the four types defined by [10] plus a new fifth type. For the first four, a size is chosen uniformly at random for each hyperedge among prescribed upper and lower bounds:
  - Type 1: sizes of hyperedges between 2 and n
  - Type 2: sizes of hyperedges between 2 and $\lceil n/2 \rceil$
  - Type 3: sizes of hyperedges between $\lceil n/4 \rceil$ and n
  - Type 4: sizes of hyperedges between $\lceil n/4 \rceil$ and $\lceil n/2 \rceil$.

  Then, for each hyperedge, vertices are chosen uniformly at random until the desired size is reached. For the fifth type, hyperedges are chosen uniformly at random among all possible hyperedges. To do so, for each hyperedge, each vertex is added with probability 1/2 until the desired number of distinct hyperedges is reached. Hence, the sizes of hyperedges follow a uniform distribution for the first four types, and a gaussian distribution (centered at $\frac{n}{2}$) for the fifth one.

In the following, a *scenario* corresponds to a triple $(n, d, Type)$. In all experiments conducted in this paper, 50 instances were generated for each scenario. Moreover, a time limit of 900 s (15 min) was set for each instance.

3.2 Comparison with the Flow-Based MILP Formulation

In this sub-section, we present the results of the comparison between our constraint generation algorithm and the best state-of-the-art exact algorithm for MCI, which is the improved flow-based MILP model of Dar *et al.* [10]. As explained in the introduction, this algorithm is itself an improvement of a previous algorithm of Agarwal *et al.* [1]. Although both algorithms rely on a flow-based MILP formulation of the problem, the improvement of Dar *et al.* can be summarized as follows:

- The MILP formulation of Dar *et al.* contains less variables and constraints, mainly because of a factoring of several linearly-dependent constraints in the previous formulation. They also added some new constraints in order to speed-up the resolution.
- The algorithm of Dar *et al.* also contains several pre-processing rules whose purpose is to reduce the number of vertices and hyperedges of the input instance, and thus reduce the size of the MILP formulation. These reduction rules rely on some observations of the problem, dealing with parts of the instances where the structure of an optimal solution can be inferred in polynomial time (*e.g.* when a set of vertices belong to a same set of hyperedges of a large size). Notice that Dar *et al.* conducted an experimental evaluation of their reduction rules in [9].

For the sake of completeness, we provide the MILP formulation of Dar *et al.* To this end, let us first introduce some notions and definitions. For every hyperedge $S \in \mathcal{E}$ they choose an arbitrary vertex $r_S \in S$ to be the source of the flow which will ensures connectivity. Hence, they define a complete digraph $A(S)$ with vertex set S and, in addition to a variable x_e for every edge of $K(H)$, their model has also a variable f_a^S for every arc a of $A(S)$. For a vertex $v \in S$, $A_S^-(v)$ (resp. $A_S^+(v)$) denotes the set of arcs of $A(S)$ entering v (resp. leaving v). The model is the following:

$$
\begin{array}{lll}
\text{Minimize} & \sum_{e \in K(H)} x_e & \\
\text{subject to:} & & \\
& \sum_{u,v \in S} x_{uv} \geq |S| - 1 & \forall S \in \mathcal{E} \\
& \sum_{a \in A_S^-(v)} f_a^S - \sum_{a \in A_S^+(v)} f_a^S = -1 & \forall S \in \mathcal{E},\ \forall v \in S \setminus r_S \\
& f_{uv}^S + f_{vu}^S \leq (|S| - 1) x_e & \forall S \in \mathcal{E},\ \forall u, v \in S \\
& f_a^S \geq 0 & \forall S \in \mathcal{E},\ \forall a \in A(S) \\
& x_e \in \{0, 1\} & \forall e \in K(H)
\end{array}
$$

Since our goal was mainly to compare the performance of our constraint generation algorithm to a simple (M)ILP formulation, the reduction rules of Dar *et al.* were not used for both algorithms. In the sequel, the algorithm of Dar *et al.* will be denoted by FLOW-MILP, and our constraint generation algorithm by CGA.

All experiments were conducted on a computer equipped with an Intel® Xeon® E5620 processor (64 bits) at 2.4 GHz, 24 GB of RAM and a Linux system (Ubuntu version 18.04.1 LTS). The implementation of our constraint generation algorithm (Strategy 4 described above) was written and run in SageMath version

8.2 (release date 05/05/2018). The algorithm of Dar *et al.* was written[3] and run in MATLAB® Released R2016b. The MILP solver used in both algorithms was CPLEX® version 12.8 from IBM®. All algorithms (including all MILP resolutions) were conducted sequentially, *i.e.* not exploiting multi-threading. Notice that the measured time of the algorithm of Dar *et al.* only consists in the resolution of the MILP model (the purpose of the MATLAB® code is thus only to construct the MILP model from the instance), hence the difference of programming languages does not matter for the comparison.

For each scenario $(n, d, Type)$, a set of 50 instances were generated and given to both FLOW-MILP and CGA. As said previously, for each instance, a time limit of 900 s was set. Tables 1, 2 and 3 represent the results of the comparison for densities 1, 3 and 5, respectively, where the running time is the average running time of all instances solved within the time limit, and the number in brackets indicates the number of instances (out of 50) effectively solved within this limit in the case this number was different from 50. The tables also show the average number of constraints in the MILP formulation of both algorithms: for FLOW-MILP it corresponds directly to the number of constraints of the MILP model, while for CGA it corresponds to the number of constraints it had to add in order to be able to solve the instance (hence, it corresponds to the number of constraints in the last ILP solved).

As we can see in the results, our approach has a much lower average running time compared to the previous algorithm in every scenario. Indeed, on average (for all instances of all scenarios) CGA has a running time more than 13 times smaller than FLOW-MILP. As we could expect, the newly introduced type 5 of instances is the most difficult for both algorithms, certainly because these instances contain much less small hyperedges than the others. This also explains why type 2 instances are often the easiest to solve for both algorithms. These results also highlights the fact that our algorithm is able to solve larger instances than previously. When considering types 1, 2, 3, and 4 only:

- For $m = n$ and $n = 26$ for instance, our algorithm is able to solve 100% of instances within the time limit, while FLOW-MILP can only solve less than 85% of them.
- For $m = 3n$, $n = 20$, CGA is able to solve 90% of instances, while FLOW-MILP can only solve 66%.
- For $m = 5n$ and $n = 18$, CGA is able to solve 98% of instances while FLOW-MILP can only solve 82% of them.

Observe also that our algorithm generate much smaller MILP models. Indeed, firstly the number of variables is always smaller, since our models do not contain any flow variables. Secondly, as we can observe in the results, the number of added constraints is roughly 6 times smaller than in the flow-based MILP model. Despite the fact that for each instance our algorithm needs to call the MILP

[3] We used the implementation of [10] provided by their authors.

Table 1. Comparison of running times and number of constraints between FLOW-MILP and CGA for density 1. Columns labeled with (sec) (resp. (con) represent the average running time (resp. number of constraints).

n	Type	FLOW-MILP (sec)	FLOW-MILP (con)	CGA (sec)	CGA (con)
14	1	0.40	598.56	0.05	130.54
	2	0.10	195.26	0.02	78.42
	3	0.40	636.28	0.05	135.02
	4	0.12	226.58	0.03	85.70
	5	0.31	428.56	0.04	115.12
16	1	0.84	830.22	0.07	162.08
	2	0.16	277.54	0.04	99.18
	3	1.05	958.52	0.08	178.28
	4	0.25	358.30	0.06	115.60
	5	1.75	618.86	0.18	152.50
18	1	2.58	1163.16	0.11	201.56
	2	0.27	372.92	0.07	123.00
	3	8.51	1263.40	0.19	219.42
	4	0.40	466.40	0.09	139.08
	5	3.72	826.74	0.37	187.18
20	1	24.17	1569.76	0.34	253.22
	2	0.52	471.60	0.11	145.18
	3	35.33	1799.16	0.58	282.62
	4	1.88	672.04	0.54	182.14
	5	35.64	1158.72	2.97	250.80
22	1	60.53	2023.32	0.85	307.94
	2	1.05	630.04	0.27	177.80
	3	107.17 [49]	2386.94	1.08	342.64
	4	6.31	838.70	1.14	216.18
	5	137.17 [47]	1504.98	12.02	315.64
24	1	119.20 [49]	2566.84	2.15	365.82
	2	3.76	807.26	0.54	211.76
	3	314.43 [39]	3147.10	8.58	443.44
	4	42.49 [49]	1140.20	4.29	272.38
	5	344.12 [28]	1929.75	122.41 [49]	404.24
26	1	194.88 [38]	3342.79	28.25	448.26
	2	19.83	1019.42	4.28	253.16
	3	365.14 [33]	3733.33	8.39	498.22
	4	101.67 [48]	1338.85	16.92	318.62
	5	606.45 [8]	2382.62	285.70 [31]	478.58

Table 2. Comparison of running times and number of constraints between FLOW-MILP and CGA for density 3. Columns labeled with (sec) (resp. (con)) represent the average running time (resp. number of constraints).

n	Type	FLOW-MILP (sec)	FLOW-MILP (con)	CGA (sec)	CGA (con)
12	1	0.90	1056.10	0.08	275.24
	2	0.19	399.16	0.05	181.20
	3	0.78	1153.72	0.08	291.50
	4	0.30	469.44	0.06	198.74
	5	0.99	814.18	0.13	253.84
14	1	2.40	1645.76	0.15	366.24
	2	0.38	586.10	0.08	233.46
	3	2.75	1707.58	0.19	377.32
	4	0.59	665.06	0.14	251.90
	5	6.14	1254.08	0.69	340.64
16	1	9.67	2424.56	0.42	469.52
	2	0.97	827.48	0.21	293.36
	3	31.98	2773.78	1.38	518.96
	4	8.25	1056.66	1.60	340.16
	5	155.75 [49]	1857.67	27.56	454.78
18	1	35.95	3269.58	1.72	578.04
	2	3.55	1107.86	0.66	356.14
	3	187.60 [49]	3773.76	15.92	645.78
	4	69.12	1411.38	12.30	417.76
	5	393.61 [11]	2458.09	241.37 [33]	566.21
20	1	178.20 [44]	4413.09	11.05	712.24
	2	21.53	1454.96	4.73	432.20
	3	418.32 [22]	5395.00	115.07 [48]	829.40
	4	367.51 [16]	1943.69	274.18 [33]	532.66
	5	−1.00 [0]	0.00	888.23 [1]	685.00
22	1	330.37 [29]	5896.55	76.89	872.12
	2	105.96 [49]	1901.41	23.52 [49]	519.82
	3	544.25 [2]	6935.50	210.25 [35]	993.80
	4	−1.00 [0]	0.00	689.20 [8]	623.13
	5	−1.00 [0]	0.00	−1.00 [0]	0.00

solver several times, calling it on much smaller MILP models offers a better overall running time.

We also generated instances with hyperedges sizes bounded by a (small) constant, in order to see how far we could increase the number of vertices for both algorithms. More precisely, we generated instances with hyperedges of size 7, and

Table 3. Comparison of running times and number of constraints between FLOW-MILP and CGA for density 5. Columns labeled with (sec) (resp. (con)) represent the average running time (resp. number of constraints).

n	Type	FLOW-MILP (sec)	FLOW-MILP (con)	CGA (sec)	CGA (con)
10	1	0.44	1009.94	0.06	322.38
	2	0.19	432.96	0.05	226.90
	3	0.47	1023.98	0.07	324.68
	4	0.19	438.26	0.05	228.50
	5	0.53	816.82	0.07	301.58
12	1	1.28	1717.92	0.14	452.32
	2	0.36	674.92	0.07	303.98
	3	2.37	1862.66	0.19	479.42
	4	0.76	786.34	0.13	331.62
	5	2.63	1346.06	0.27	420.58
14	1	6.54	2684.80	0.28	601.18
	2	0.82	989.18	0.17	390.82
	3	10.00	2809.78	0.42	624.30
	4	1.79	1112.46	0.31	419.54
	5	57.82	2108.38	5.90	568.06
16	1	27.61	3892.42	0.95	765.86
	2	2.35	1375.52	0.45	485.30
	3	146.77 [49]	4414.90	6.91	843.04
	4	73.50 [46]	1757.11	42.70	567.12
	5	546.07 [20]	2976.35	176.13 [38]	728.76
18	1	91.41 [48]	5527.35	3.31	963.82
	2	15.35	1884.84	1.61	597.76
	3	381.68 [30]	6082.20	55.23	1050.14
	4	357.08 [36]	2313.11	104.60 [46]	684.29
	5	440.86 [1]	4244.00	283.08 [2]	911.00
20	1	178.73 [35]	7266.37	17.41	1169.50
	2	65.76	2430.20	6.06	708.82
	3	588.01 [1]	9003.00	347.20 [21]	1322.05
	4	−1.00 [0]	0.00	−1.00 [0]	0.00
	5	−1.00 [0]	0.00	−1.00 [0]	0.00

density $d \in \{1, 3\}$ (for density 5, the maximum number of vertices for which our algorithm was able to solve 100% of the instances was only 300).

The differences of running time is even more significant in this experiment (see Table 4). The algorithm of Dar *et al.* fails to solve 100% of the instances within the time limit for 200 vertices already (density 3). Moreover, for density

1, there is a huge lack of efficiency between 750 and 1000 vertices for the flow-based MILP algorithm, going from 100% of instances solved to 8%. Overall, we can observe that our approach allows to solve instances or a much larger size than the previous algorithm.

4 Enumeration Algorithm

In this section, we describe an approach to enumerate all optimal solutions of an instance of MCI. When solving an optimization problem using an MILP formulation in which the solution is represented by 0–1 variables, a natural way to obtain an enumeration algorithm consists in adding new constraints in order to forbid previously found solutions. More formally, if the objective of the MILP is

$$Minimize \sum_{i=1}^{n} x_i$$

where each x_i is a 0–1 variable, then one can forbid a given solution $S \subseteq \{1, \dots, n\}$ represented by the indices of all variables set to 1 by adding the following constraint:

Table 4. Results for instances with hyperedges of size 7. Columns labeled with (sec) (resp. (con)) represent the average running time (resp. number of constraints).

n	d	FLOW-MILP (sec)	FLOW-MILP (con)	CGA (sec)	CGA (con)
30	1	0.31	407.04	0.12	169.58
30	3	5.14	1259.36	2.03	512.14
50	1	0.47	699.32	0.25	286.58
50	3	12.85	2066.28	2.89	856.66
100	1	1.33	1396.66	0.62	572.82
100	3	39.35	4176.84	4.85	1752.02
200	1	5.38	2809.72	0.79	1132.54
200	3	106.51 [46]	8269.80	4.76	3436.34
300	1	17.20	4209.62	1.40	1691.58
300	3	148.21 [33]	12415.18	8.04	5117.94
400	1	41.62	5596.66	1.61	2239.86
400	3	220.36 [37]	16584.57	23.73	7033.42
500	1	83.89	7002.62	2.29	2792.82
500	3	369.10 [46]	20739.85	94.24	8969.68
750	1	296.43	10521.40	15.07	4265.36
750	3	−1.00 [0]	0.00	266.82	13454.44
1000	1	627.295 [4]	14018.50	34.54	5645.48
1000	3	−1.00 [0]	0.00	666.70 [33]	17785.06

$$\sum_{i \in S} x_i < |S|$$

Hence, forbidding a set of solutions $\mathcal{A}$ can be done by adding $|\mathcal{A}|$ new constraints to the model. This idea was used by Agarwal *et al.* [1] in order to obtain an algorithm enumerating all optimal solutions of an instance of MCI. This strategy, although being easy to implement, becomes much less efficient when the number of solutions of the instances increases, because the size of the MILP model becomes too large for the solver. We propose a new method for the enumeration of solutions, which, in a nutshell, consists in forbidding the solutions "chunk by chunk". To this end, we iteratively accumulate optimal solutions by exploring the neighborhood of a solution found (the way we explore this neighborhood will be explained later). Once this exploration is done, we forbid all optimal solutions found at the same time. A pseudo-code of this approach is presented in Algorithm 2.

Algorithm 2: Enumeration algorithm for MCI

```
   Input: a hypergraph H = (V, E)
   Output: A: the set of all optimal solutions of H
1  A ← ∅
2  c* ← cost of an optimal solution of H
3  while there exists a solution S of cost c* which does not belong to A do
4  |   N ← neighborhood of S
5  |   A ← A ∪ N
6  end
7  return A
```

Naturally, we use our constraint generation algorithm described previously in order to find new optimal solutions. Notice that once we have found one optimal solution of cost c^*, we shall add a new constraint to our ILP in order to find new solutions of size exactly c^* in the next rounds, which usually speeds up the resolution. We now describe the way we explore the neighborhood of a solution (which corresponds to Line 4 of Algorithm 2). This step is done by forbidding an arbitrary edge e of the previously found solution, by simply adding a new constraint to our ILP forcing the corresponding variable x_e to 0. We thus iteratively accumulate new optimal solutions until the solver returns that the obtained ILP does not admit a solution of the desired cost, which means that the exploration of the neighborhood is done. We then remove the newly added constraints used in this routine for the next loop in Line 3 of Algorithm 2.

We evaluated the performance of our approach by comparing its running time to the natural approach of forbidding each new optimal found in the ILP described at the beginning of this section (still using our exact algorithm as a black box for finding new solutions). To this end, we generated a set of 1000 random instances of type 1 with a density of $\frac{m}{n} = 2$, and $n = 10$. These settings were chosen because they allow the random generation to produce instances of various different structures. In particular, we observed a quite fair

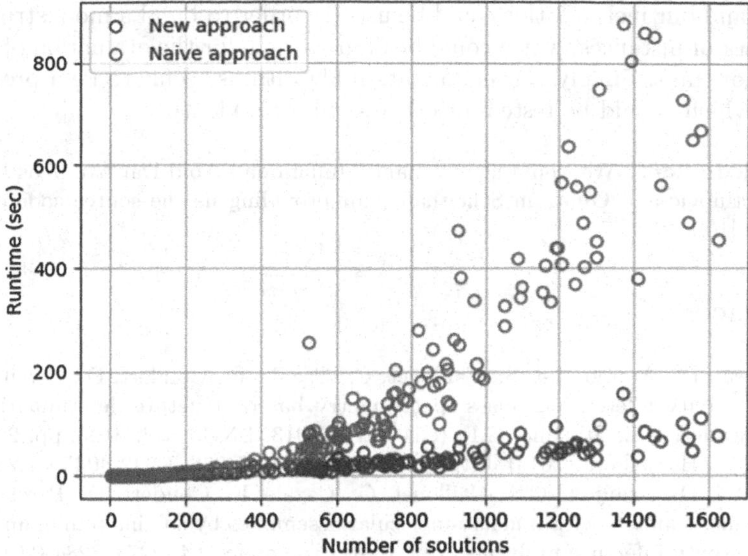

Fig. 1. Comparison of running times between the naive enumeration algorithm and our new approach, as a function of the number of solutions of the instances.

distribution of the numbers of solutions, which seemed to be a meaningful parameter for the comparison of the two approaches. Figure 1 presents the result of these experiments. As we can see, our new method offers a great improvement when the number of solutions is high, by reducing by more than 8 times the running time in our generated instances. These results suggest that our algorithm has a running time which is linear in the number of solutions in practice.

5 Conclusion

In this paper we presented and evaluated an exact algorithm for the Minimum Connectivity Inference problem, based on a constraint generation strategy in order to ensure connectivity. Our experiments, conducted on various randomly generated instances, demonstrated that our method outperforms the best previously known exact algorithm for this problem, relying on a flow-based MILP formulation. Since connectivity constraints appear very often in practical situations which are usually solved by the means of MILP, our results suggest that a constraint generation strategy can sometimes be much more efficient. As a further research, it would be interesting to apply this technique to other optimization problems in which connectivity plays an important role. It should be noted that during the empirical evaluation of the different sub-routines for our algorithm, we noticed high standard deviations in the running times. It would be thus interesting to understand which hypergraph parameters influence the complexity of our strategies. Apart from providing useful information about the problem and our method, this could be used in order to build a more structured benchmark of instances, which could be of great help for the evaluation of future exact algorithms. Finally, our enumeration algorithms seems to be a promising method which should be tested for other similar problems.

Acknowledgment. We would like to thank Muhammad Abid Dar, Andreas Fischer, John Martinovic and Guntram Scheithauer for providing us the source code of their algorithm [10].

References

1. Agarwal, D., Araujo, J.-C.S., Caillouet, C., Cazals, F., Coudert, D., Pérennes, S.: Connectivity inference in mass spectrometry based structure determination. In: Bodlaender, H.L., Italiano, G.F. (eds.) ESA 2013. LNCS, vol. 8125, pp. 289–300. Springer, Heidelberg (2013). https://doi.org/10.1007/978-3-642-40450-4_25
2. Agarwal, D., Araújo, J.C.S., Caillouet, C., Cazals, F., Coudert, D., Pérennes, S.: Unveiling contacts within macro-molecular assemblies by solving minimum weight connectivity inference problems. Mol. Cell. Proteomics **14**, 2274–2284 (2015)
3. Angluin, D., Aspnes, J., Reyzin, L.: Inferring social networks from outbreaks. In: Hutter, M., Stephan, F., Vovk, V., Zeugmann, T. (eds.) ALT 2010. LNCS (LNAI), vol. 6331, pp. 104–118. Springer, Heidelberg (2010). https://doi.org/10.1007/978-3-642-16108-7_12
4. Benders, J.F.: Partitioning procedures for solving mixed-variables programming problems. Numerische Mathematik **4**(1), 238–252 (1962)
5. Brandes, U., Cornelsen, S., Pampel, B., Sallaberry, A.: Path-based supports for hypergraphs. J. Discrete Algorithms **14**, 248–261 (2012). Proceedings of the 21st International Workshop on Combinatorial Algorithms (IWOCA 2010)
6. Chen, J., Komusiewicz, C., Niedermeier, R., Sorge, M., Suchý, O., Weller, M.: Polynomial-time data reduction for the subset interconnection design problem. SIAM J. Discrete Math. **29**(1), 1–25 (2015)
7. Chockler, G., Melamed, R., Tock, Y., Vitenberg, R.: Constructing scalable overlays for pub-sub with many topics. In: Proceedings of the 26th Annual ACM Symposium on Principles of Distributed Computing (PODC 2007), pp. 109–118 (2007)
8. Conitzer, V., Derryberry, J., Sandholm, T.: Combinatorial auctions with structured item graphs. In: Proceedings of the 19th National Conference on Artifical Intelligence, AAAI 2004, pp. 212–218 (2004)
9. Dar, M.A., Fischer, A., Martinovic, J., Scheithauer, G.: A computational study of reduction techniques for the minimum connectivity inference problem. In: Singh, V.K., Gao, D., Fischer, A. (eds.) Advances in Mathematical Methods and High Performance Computing. AMM, vol. 41, pp. 135–148. Springer, Cham (2019). https://doi.org/10.1007/978-3-030-02487-1_7
10. Dar, M.A., Fischer, A., Martinovic, J., Scheithauer, G.: An improved flow-based formulation and reduction principles for the minimum connectivity inference problem. Optimization **0**(0), 1–21 (2018)
11. Du, D.Z., Miller, Z.: Matroids and subset interconnection design. SIAM J. Discrete Math. **1**(4), 416–424 (1988)
12. Du, D.Z., Miller, Z.: On complexity of subset interconnection designs. J. Global Optim. **6**(2), 193–205 (1995)

13. Fan, H., Hundt, C., Wu, Y.-L., Ernst, J.: Algorithms and implementation for interconnection graph problem. In: Yang, B., Du, D.-Z., Wang, C.A. (eds.) COCOA 2008. LNCS, vol. 5165, pp. 201–210. Springer, Heidelberg (2008). https://doi.org/10.1007/978-3-540-85097-7_19
14. Graham, R.L.: Bounds on multiprocessing timing anomalies. SIAM J. Appl. Math. **17**(2), 416–429 (1969)
15. Hosoda, J., Hromkovič, J., Izumi, T., Ono, H., Steinová, M., Wada, K.: On the approximability and hardness of minimum topic connected overlay and its special instances. Theoret. Comput. Sci. **429**, 144–154 (2012)
16. Johnson, D.S., Pollak, H.O.: Hypergraph planarity and the complexity of drawing venn diagrams. J. Graph Theory **11**(3), 309–325 (1987)
17. Klemz, B., Mchedlidze, T., Nöllenburg, M.: Minimum tree supports for hypergraphs and low-concurrency Euler diagrams. In: Ravi, R., Gørtz, I.L. (eds.) SWAT 2014. LNCS, vol. 8503, pp. 265–276. Springer, Cham (2014). https://doi.org/10.1007/978-3-319-08404-6_23
18. Korach, E., Stern, M.: The clustering matroid and the optimal clustering tree. Math. Program. **98**(1), 385–414 (2003)
19. Onus, M., Richa, A.W.: Minimum maximum degree publish-subscribe overlay network design. In: IEEE INFOCOM 2009, pp. 882–890 (2009)

Efficient Split-Radix and Radix-4 DCT Algorithms and Applications

Sirani M. Perera[(✉)], Daniel Silverio, and Austin Ogle

Embry-Riddle Aeronautical University, Daytona Beach, USA
pereras2@erau.edu
{silverid,oglea1}@my.erau.edu
https://faculty.erau.edu/Sirani.Perera

Abstract. This paper proposes efficient split-radix and radix-4 Discrete Cosine Transform (DCT) of types II/III algorithms. The proposed fast split-radix and radix-4 algorithms extend the previous work on the lowest multiplication complexity, self-recursive, radix-2 DCT II/III algorithms. The paper also addresses the self-recursive and stable aspects of split-radix and radix-4 DCT II/III algorithms having simple, sparse, and scaled orthogonal factors. Moreover, the proposed split-radix and radix-4 algorithms attain the lowest theoretical multiplication complexity and arithmetic complexity for 8-point DCT II/III matrices. The factorization corresponding to the proposed DCT algorithms contains sparse and scaled orthogonal matrices. Numerical results are presented for the arithmetic complexity comparison of the proposed algorithms with the known fast and stable DCT algorithms. Execution time of the proposed algorithms is presented while verifying the connection to the order of the arithmetic complexity. Moreover, we will show that the execution time of the proposed split-radix and radix-4 algorithms are more efficient than the radix-2 DCT algorithms. Finally, the implementations of the proposed DCT algorithms are stated using signal-flow graphs.

Keywords: Discrete cosine transforms · Sparse and orthogonal factors · Split-radix and radix-4 algorithms · Self recursive algorithms · Complexity and performance of algorithms · Signal flow graphs

1 Introduction

The Discrete Fourier Transform (DFT) has a plethora of applications in applied mathematics and electrical engineering. Discrete Cosine Transform (DCT) is a real-arithmetic analogue of DFT. DCTs with orthogonal trigonometric transforms have been especially popular in recent decades due to their applications in digital video technology and high efficiency video coding; see, e.g., [2,29,34,39].

This work was funded by the Faculty Research Development Program, the Office of Undergraduate Research, and the Office of Provost at Embry-Riddle Aeronautical University.

I. Kotsireas et al. (Eds.): SEA^2 2019, LNCS 11544, pp. 184–201, 2019.
https://doi.org/10.1007/978-3-030-34029-2_13

Due to the savings in the transformed data, DCTs can be considered the building blocks for video and image compression algorithms, especially in High Efficiency Video Coding (HEVC) and H.264/H.265 video compression [26]. Also, DCT II is accepted as the best suboptimal transform since its performance is very close to that of the optimal Karhunen-Loeve Transform; see, e.g., [2,29,30]. One can say that DCT is the key transform in image processing, signal processing, finger print enhancement, quick response code (QR code), multi-mode interface, etc. [1,3,7,9,12–15,17,24,40].

Fast algorithms can be derived to compute DCT and its inverse efficiently. In this paper, we discuss self-recursive split-radix and radix-4 DCT and inverse DCT algorithms with reduction of the execution time. Our DCT algorithms are based on the recently published lowest multiplication complexity, self-recursive, radix-2 DCT algorithms [24]. We have considered simple, sparse, and scaled orthogonal factorization for the DCT matrices and reduced multiplication complexity while scaling at the end of the computation. The arithmetic complexity of the proposed DCT algorithms are the same as in [24]. The proposed algorithms have shown favorable results with the execution time as opposed to the radix-2 DCT algorithms. One should recall that algorithms are most beneficial when the hardware can take advantage. Based on the VLSI designs, the proposed DCT algorithms are neat [24]. The DCTs have main types varying from I to IV based on Dirichlet and Neumann boundary conditions. In this paper we have considered orthogonal DCT matrices of type II and III which we call DCT and inverse DCT, respectively, as follows:

$$C_n^{II} = \sqrt{\frac{2}{n}}\left[\epsilon_n(j)\cos\frac{j(2k+1)\pi}{2n}\right], C_n^{III} = \sqrt{\frac{2}{n}}\left[\epsilon_n(k)\cos\frac{(2j+1)k\pi}{2n}\right],$$

where $\epsilon_n(0) = \epsilon_n(n) = \frac{1}{\sqrt{2}}$, $\epsilon_n(j) = 1$ for $j \in \{1, 2, \cdots, n-1\}$, $n \geq 2$ is an even integer, and $j, k = 0, 1, \cdots, n-1$.

Since the 1960's, split-radix DCT and Discrete Sine Transform (DST) algorithms have been obtained as a special case of DFT algorithms [5,6,11,31,41,45]. Simply put, split-radix DCT and DST algorithms are computed using DFT algorithms. The crucial work on the reduced flop count of the split-radix DCT and DST algorithms in [11,31] is also based on DFT algorithms. The authors in [11,31] had reduced redundant operations using even symmetry of the input data. When DCT is computed using DFT we are using complex arithmetic as opposed to real arithmetic. Thus, one has to pay attention for the practical implementation of the DCT algorithms which utilize DFTs. Moreover, when DCTs are computed using DFT with several routines, this effects significantly the execution time and memory hierarchy. Thus, as opposed to the split-radix DCT algorithms in the literature, the proposed split-radix DCT algorithms are self-recursive and contain simple, sparse, and scaled orthogonal matrices. One could recall here that, if the factorizations do not preserve orthogonality, the resulting DCT and DST algorithms can lead to inferior numerical stability [37]. Apart from split-radix DCT computation through the DFTs, one can find fast split-radix DCT II algorithms utilizing real arithmetic in [19]. Unlike split-radix

DCT algorithms, radix-4 DCT algorithms have not quite been studied. But one can design a converged processor for 64-point DCT using radix-4 FFT as in [38]. Also, a fast radix-q and split-radix type-IV DCT algorithm, which was obtained using DCT-II, can be seen in [10]. Fast radix-p DCT and inverse DCT algorithms can be derived using a divide-and-conquer method and Chebyshev polynomials as in [33]. In this paper, we have proposed radix-4 DCT and inverse DCT algorithms with simple, sparse, and scaled orthogonal matrices while leading to self-recursive algorithms as opposed to the existing radix-4 DCT algorithms.

It is natural to compute fast (in the sense of using $\mathcal{O}(n \log n)$ complexity) DCT and DST algorithms using the polynomial arithmetic technique (using the divide and conquer technique to reduce the degree of the polynomial; see, e.g., [27,28,32,42]) and the matrix factorization technique (direct factorization of the DCT/DST matrix into the product of real and sparse matrices; see, e.g., [19–25,43,44]). Apart from these main techniques, one can use polynomial division based on comrade matrices to derive fast DCT and DST algorithms as in [18].

This paper addresses efficient split-radix and radix-4 DCT and inverse DCT algorithms by using real arithmetic with simple, sparse, and scaled orthogonal matrices. Moreover, we have presented self-recursive algorithms with real arithmetic as opposed to complex arithmetic, and compared arithmetic complexity results of the proposed algorithms with the existing fast and stable DCT algorithms. We have also shown optimized C code for comparing the execution time of the proposed DCT algorithm with existing radix-2 DCT algorithms. The proposed algorithms are also implemented to demonstrate signal flow graphs.

In Sect. 2, we utilize the factorization proposed in [24] to obtain self/completely recursive split-radix and radix-4 DCT II/III algorithms having sparse and scaled orthogonal matrices. Next, in Sect. 3 we derive addition and multiplication counts of the proposed split-radix and radix-4 algorithms. Within Subsect. 3.2, we will compare the proposed split-radix and radix-4 DCT algorithms with the known fast and stable DCT algorithms. Moreover, in Subsect. 3.3, we will utilize optimized C code to show the connection between the order of arithmetic complexity and execution time of the proposed algorithms. We have also presented the execution time of the proposed algorithms with existing radix-2 algorithms. Finally, in Sect. 4 we illustrate the signal flow graphs of the proposed algorithms for a 16-point case.

2 Simple, Self-recursive, Split-Radix and Radix-4 DCT Algorithms

In this section we introduce self-contained factorizations for DCT II/III having sparse and orthogonal matrices. The proposed factorizations are presented based on the lowest multiplication complexity, self-recursive, radix-2 DCT algorithms introduced in [24]. Once the factorizations are established we will state the corresponding simple, self-recursive, split-radix and radix-4 DCT II/III algorithms. The purpose of presenting these novel algorithms is to introduce self-recursive, fast split-radix and radix-4 algorithms having sparse and scaled orthogonal matrices. Let us first introduce all notations before discussing the factorizations of DCT matrices.

2.1 Frequently Use Notations

Here we introduce notations for sparse and scaled orthogonal matrices which will frequently be used in this paper. For a given vector $\mathbf{x} = [x_0, x_1, \cdots, x_{n-1}] \in R^n$, let us introduce an even-odd permutation matrix P_n $(n \geq 3)$ by

$$P_n \, \mathbf{x} = \begin{cases} [x_0, x_2, \cdots, x_{n-2}, x_1, x_3, \cdots, x_{n-1}]^T & \text{even } n, \\ [x_0, x_2, \cdots, x_{n-1}, x_1, x_3, \cdots, x_{n-2}]^T & \text{odd } n. \end{cases}$$

We also introduce a scaled orthogonal matrix $H_n = \begin{bmatrix} I_{\frac{n}{2}} & \tilde{I}_{\frac{n}{2}} \\ I_{\frac{n}{2}} & -\tilde{I}_{\frac{n}{2}} \end{bmatrix}$, where I_n is the identity matrix and $\tilde{I}_n$ is the counter-identity matrix, an orthogonal matrix $\bar{H}_n = \frac{1}{\sqrt{2}} H_n$, a bidiagonal matrix $B_{\frac{n}{2}} = \begin{bmatrix} \sqrt{2} & 1 & & & \\ & 1 & 1 & & \\ & & \ddots & \ddots & \\ & & & & 1 \\ & & & & 1 \end{bmatrix}$, a diagonal matrix $W_{\frac{n}{2}} = \text{diag} \left[\frac{\sec(\frac{(2k-1)\pi}{2n})}{2} \right]_{k=1}^{\frac{n}{2}}$, and a block diagonal matrix of say A and B as blkdiag (A, B).

2.2 Self-enclosed, Sparse, and Scaled Orthogonal Factors for DCT II/III

We will first show that the DCT II matrix admits a self-enclosed factorization with sparse and orthogonal matrices with the help of the DCT factorization in [24]. Next, we will use the factorization for DCT II to state the factorization for DCT III. Let us first recall the self-contained, sparse, orthogonal DCT II factorization proposed in [24].

Lemma 1. *For an even integer $n \geq 4$, the matrix C_n^{II} can be factored in the form*

$$C_n^{II} = P_n^T \begin{bmatrix} I_{\frac{n}{2}} & 0 \\ 0 & B_{\frac{n}{2}} \end{bmatrix} \begin{bmatrix} C_{\frac{n}{2}}^{II} & 0 \\ 0 & C_{\frac{n}{2}}^{II} \end{bmatrix} \begin{bmatrix} I_{\frac{n}{2}} & 0 \\ 0 & W_{\frac{n}{2}} \end{bmatrix} \bar{H}_n. \tag{1}$$

Proof. See [24].

The followings are immediate results of Lemma 1.

Corollary 1. *For a given* $n = 2^t (t \geq 2)$, *the split factorization for the matrix* C_n^{II} *can be stated in the form*

$$C_n^{II} = P_n^T \begin{bmatrix} P_{\frac{n}{2}}^T & \\ & I_{\frac{n}{2}} \end{bmatrix} \begin{bmatrix} I_{\frac{n}{4}} & & \\ & B_{\frac{n}{4}} & \\ & & B_{\frac{n}{2}} \end{bmatrix} \begin{bmatrix} C_{\frac{n}{4}}^{II} & & \\ & C_{\frac{n}{4}}^{II} & \\ & & C_{\frac{n}{2}}^{II} \end{bmatrix} \begin{bmatrix} I_{\frac{n}{4}} & & \\ & W_{\frac{n}{4}} & \\ & & W_{\frac{n}{2}} \end{bmatrix} \begin{bmatrix} \bar{H}_{\frac{n}{2}} & \\ & I_{\frac{n}{2}} \end{bmatrix} \bar{H}_n. \quad (2)$$

Proof. This can easily be seen by factoring $C_{\frac{n}{2}}^{II}$ in the top half of Eq. (1).

Corollary 2. *For a given* $n = 2^t (t \geq 2)$, *the split factorization for the matrix* C_n^{III} *can be stated in the form*

$$C_n^{III} = \bar{H}_n^T \begin{bmatrix} \bar{H}_{\frac{n}{2}}^T & \\ & I_{\frac{n}{2}} \end{bmatrix} \begin{bmatrix} I_{\frac{n}{4}} & & \\ & W_{\frac{n}{4}} & \\ & & W_{\frac{n}{2}} \end{bmatrix} \begin{bmatrix} C_{\frac{n}{4}}^{III} & & \\ & C_{\frac{n}{4}}^{III} & \\ & & C_{\frac{n}{2}}^{III} \end{bmatrix} \begin{bmatrix} I_{\frac{n}{4}} & & \\ & B_{\frac{n}{4}}^T & \\ & & B_{\frac{n}{2}}^T \end{bmatrix} \begin{bmatrix} P_{\frac{n}{2}} & \\ & I_{\frac{n}{2}} \end{bmatrix} P_n. \quad (3)$$

Proof. This is trivial by Corollary 1 as $C_n^{III} = \left[C_n^{II}\right]^T$.

Corollary 3. *For a given* $m = 4^t (t \geq 1)$, *the factorization for the matrix* C_m^{II} *can be stated in the form*

$$\begin{aligned} C_m^{II} = P_m^T \begin{bmatrix} I_{\frac{m}{2}} & \\ & B_{\frac{m}{2}} \end{bmatrix} \begin{bmatrix} P_{\frac{m}{2}}^T & \\ & P_{\frac{m}{2}}^T \end{bmatrix} & \begin{bmatrix} I_{\frac{m}{4}} & & & \\ & B_{\frac{m}{4}} & & \\ & & I_{\frac{m}{4}} & \\ & & & B_{\frac{m}{4}} \end{bmatrix} \begin{bmatrix} C_{\frac{m}{4}}^{II} & & & \\ & C_{\frac{m}{4}}^{II} & & \\ & & C_{\frac{m}{4}}^{II} & \\ & & & C_{\frac{m}{4}}^{II} \end{bmatrix} \\ & \begin{bmatrix} I_{\frac{m}{4}} & & & \\ & W_{\frac{m}{4}} & & \\ & & I_{\frac{m}{4}} & \\ & & & W_{\frac{m}{4}} \end{bmatrix} \begin{bmatrix} \bar{H}_{\frac{m}{2}} & \\ & \bar{H}_{\frac{m}{2}} \end{bmatrix} \begin{bmatrix} I_{\frac{m}{2}} & \\ & W_{\frac{m}{2}} \end{bmatrix} \bar{H}_m. \end{aligned} \quad (4)$$

Proof. This can easily be seen by factoring each $C_{\frac{n}{2}}^{II}$ in Eq. (1).

Corollary 4. *For a given* $m = 4^t (t \geq 1)$, *the factorization for the matrix* C_m^{III} *can be stated in the form*

$$\begin{aligned} C_m^{III} = \bar{H}_m^T \begin{bmatrix} I_{\frac{m}{2}} & \\ & W_{\frac{m}{2}} \end{bmatrix} \begin{bmatrix} \bar{H}_{\frac{m}{2}}^T & \\ & \bar{H}_{\frac{m}{2}}^T \end{bmatrix} & \begin{bmatrix} I_{\frac{m}{4}} & & & \\ & W_{\frac{m}{4}} & & \\ & & I_{\frac{m}{4}} & \\ & & & W_{\frac{m}{4}} \end{bmatrix} \begin{bmatrix} C_{\frac{m}{4}}^{III} & & & \\ & C_{\frac{m}{4}}^{III} & & \\ & & C_{\frac{m}{4}}^{III} & \\ & & & C_{\frac{m}{4}}^{III} \end{bmatrix} \\ & \begin{bmatrix} I_{\frac{m}{4}} & & & \\ & B_{\frac{m}{4}}^T & & \\ & & I_{\frac{m}{4}} & \\ & & & B_{\frac{m}{4}}^T \end{bmatrix} \begin{bmatrix} P_{\frac{m}{2}} & \\ & P_{\frac{m}{2}} \end{bmatrix} \begin{bmatrix} I_{\frac{m}{2}} & \\ & B_{\frac{m}{2}}^T \end{bmatrix} P_m. \end{aligned} \quad (5)$$

Proof. This is trivial by Corollary 3 as $C_m^{III} = \left[C_m^{II}\right]^T$.

2.3 Self Recursive Split-Radix and Radix-4 DCT II/III Algorithms

The factorization of DCT II/III matrices established in Sect. 2.2 leads to self-recursive split-radix and radix-4 algorithms for computing $\mathbf{y} = C^{II}\mathbf{x}$ and $\mathbf{y} = C^{III}\mathbf{x}$ for a given $\mathbf{x}$. To further reduce the number of multiplications, we can move the factor $\frac{1}{\sqrt{2}}$ in matrix $\bar{H}$ (i.e. using scaled orthogonal matrix H) to the end of the calculation to compute $\mathbf{y} = \sqrt{n}C_n^{II}\mathbf{x}$, $\mathbf{y} = \sqrt{n}C_n^{III}\mathbf{x}$, $\mathbf{y} = \sqrt{m}C_m^{II}\mathbf{x}$, and $\mathbf{y} = \sqrt{m}C_m^{III}\mathbf{x}$. Let us present the self-recursive split-radix and radix-4 DCT II/III algorithms to compute **cos2sr(x, n)**, **cos3sr(x, n)**, **cos2r4(x, n)**, and **cos3r4(x, n)**, respectively.

Before stating the self recursive split-radix and radix-4 algorithms, let's use the following notations to denote diagonal and bidiagonal matrices which will be used hereafter for $n \geq 4$

$$\tilde{W}_n = \begin{bmatrix} I_{\frac{n}{2}} & 0 \\ 0 & W_{\frac{n}{2}} \end{bmatrix} \text{ and } \tilde{B}_n = \begin{bmatrix} I_{\frac{n}{2}} & 0 \\ 0 & B_{\frac{n}{2}} \end{bmatrix}. \qquad (6)$$

Split-radix DCT-II Algorithm i.e. cos2sr(x, n)

Input: $n = 2^t (t \geq 1)$, $n_1 = \frac{n}{2}$, $n_2 = \frac{n}{4}$, $\mathbf{x} \in \mathbb{R}^n$.

1. If $n = 2$, then
$$\mathbf{y} := \begin{bmatrix} 1 & 1 \\ 1 & -1 \end{bmatrix} \mathbf{x}.$$
2. If $n = 4$, then
$$\mathbf{y} := P_4^T \begin{bmatrix} I_2 & \\ & B_2 \end{bmatrix} \begin{bmatrix} I_2 & & \\ & 1 & 1 \\ & 1 & -1 \end{bmatrix} \begin{bmatrix} I_2 & \\ & W_2 \end{bmatrix} \begin{bmatrix} H_2 & \\ & I_2 \end{bmatrix} H_4\, \mathbf{x}.$$
3. If $n \geq 8$, then
$$\begin{aligned} \mathbf{u} &:= H_n\, \mathbf{x}, \\ \mathbf{v} &:= [\text{blkdiag}(H_{n_1}, I_{n_1})]\, \mathbf{u}, \\ [w_j]_{j=0}^{n-1} &:= [\text{blkdiag}(\tilde{W}_{n_1}, W_{n_1})]\, \mathbf{v}, \\ \mathbf{z1} &:= \mathbf{cos2sr}\left([w_j]_{j=0}^{n_2-1}, n_2\right), \\ \mathbf{z2} &:= \mathbf{cos2sr}\left([w_j]_{j=n_2}^{n_1-1}, n_2\right), \\ \mathbf{z3} &:= \mathbf{cos2sr}\left([w_j]_{j=n_1}^{n-1}, n_1\right), \\ \mathbf{q} &:= [\text{blkdiag}(\tilde{B}_{n_1}, B_{n_1})]\left(\mathbf{z1}^T, \mathbf{z2}^T, \mathbf{z3}^T\right)^T, \\ \mathbf{y} &:= P_n^T\, [\text{blkdiag}(P_{n_1}^T, I_{n_1})]\mathbf{q}. \end{aligned}$$

Output: $\mathbf{y} = \sqrt{n}C_n^{II}\mathbf{x}$.

Split-radix DCT-III Algorithm i.e. cos3sr(x, n)

Input: $n = 2^t (t \geq 1)$, $n_1 = \frac{n}{2}$, $n_2 = \frac{n}{4}$, $\mathbf{x} \in \mathbb{R}^n$.

1. If $n = 2$, then
$$\mathbf{y} := \begin{bmatrix} 1 & 1 \\ 1 & -1 \end{bmatrix} \mathbf{x}.$$

2. If $n = 4$, then
$$\mathbf{y} := H_4^T \begin{bmatrix} H_2 & \\ & I_2 \end{bmatrix} \begin{bmatrix} I_2 & \\ & W_2 \end{bmatrix} \begin{bmatrix} I_2 & & \\ & 1 & 1 \\ & 1 & -1 \end{bmatrix} \begin{bmatrix} I_2 & \\ & B_2^T \end{bmatrix} P_4\, \mathbf{x}.$$
3. If $n \geq 8$, then
$$\mathbf{u} := P_n\, \mathbf{x},$$
$$\mathbf{v} := [\text{blkdiag}(P_{n_1}, I_{n_1})]\, \mathbf{u},$$
$$[w_j]_{j=0}^{n-1} := [\text{blkdiag}(\tilde{B}_{n_1}^T, B_{n_1}^T)\, \mathbf{v},$$
$$\mathbf{z1} := \mathbf{cos3sr}\left([w_j]_{j=0}^{n_2-1}, n_2\right),$$
$$\mathbf{z2} := \mathbf{cos3sr}\left([w_j]_{j=n_2}^{n_1-1}, n_2\right),$$
$$\mathbf{z3} := \mathbf{cos3sr}\left([w_j]_{j=n_1}^{n-1}, n_1\right),$$
$$\mathbf{q} := [\text{blkdiag}(\tilde{W}_{n_1}, W_{n_1})]\left(\mathbf{z1}^T, \mathbf{z2}^T, \mathbf{z3}^T\right)^T,$$
$$\mathbf{r} := [\text{blkdiag}(H_{n_1}^T, I_{n_1})]\mathbf{q},$$
$$\mathbf{y} := H_n^T\, \mathbf{r}.$$

Output: $\mathbf{y} = \sqrt{n} C_n^{III} \mathbf{x}$.

Radix-4 DCT-II Algorithm i.e. cos2r4(x, n)

Input: $m = 4^t (t \geq 1)$, $m_1 = \frac{m}{2}$, $m_2 = \frac{m}{4}$, $\mathbf{x} \in \mathbb{R}^m$.

1. If $m = 4$, then
$$\mathbf{y} := P_4^T \begin{bmatrix} I_2 & \\ & B_2 \end{bmatrix} \begin{bmatrix} H_2 & \\ & H_2 \end{bmatrix} \begin{bmatrix} I_2 & \\ & W_2 \end{bmatrix} H_4\, \mathbf{x}.$$
2. If $m \geq 16$, then
$$\mathbf{u1} := H_m\, \mathbf{x},$$
$$\mathbf{u2} := \tilde{W}_m \mathbf{u1},$$
$$\mathbf{v} := [\text{blkdiag}(H_{m_1}, H_{m_1})]\, \mathbf{u2},$$
$$[w_j]_{j=0}^{m-1} := [\text{blkdiag}(\tilde{W}_{m_1}, \tilde{W}_{m_1})]\, \mathbf{v},$$
$$\mathbf{z1} := \mathbf{cos2r4}\left([w_j]_{j=0}^{m_2-1}, m_2\right),$$
$$\mathbf{z2} := \mathbf{cos2r4}\left([w_j]_{j=m_2}^{m_1-1}, m_2\right),$$
$$\mathbf{z3} := \mathbf{cos2r4}\left([w_j]_{j=m_1}^{3m_2-1}, m_2\right),$$
$$\mathbf{z4} := \mathbf{cos2r4}\left([w_j]_{j=3m_2}^{m-1}, m_2\right),$$
$$\mathbf{q} := [\text{blkdiag}(\tilde{B}_{m_1}, \tilde{B}_{m_1})]\ \left(\mathbf{z1}^T, \mathbf{z2}^T, \mathbf{z3}^T, \mathbf{z4}^T\right)^T,$$
$$\mathbf{y} := P_m^T\, \tilde{B}_m\, [\text{blkdiag}(P_{m_1}^T, P_{m_1}^T)]\mathbf{q}$$

Output: $\mathbf{y} = \sqrt{m} C_m^{II} \mathbf{x}$.

Radix-4 DCT-III Algorithm i.e. cos3r4(x, n)

Input: $m = 4^t (t \geq 1)$, $m_1 = \frac{m}{2}$, $m_2 = \frac{m}{4}$, $\mathbf{x} \in \mathbb{R}^m$.

1. If $m = 4$, then
$$\mathbf{y} := H_4^T \begin{bmatrix} I_2 & \\ & W_2 \end{bmatrix} \begin{bmatrix} H_2 & \\ & H_2 \end{bmatrix} \begin{bmatrix} I_2 & \\ & B_2^T \end{bmatrix} P_4 \, \mathbf{x}.$$
2. If $m \geq 16$, then
$$\mathbf{u} := [\text{blkdiag}(P_{m_1}, P_{m_1})] \, \tilde{B}_m^T \, P_m \, \mathbf{x},$$
$$[w_j]_{j=0}^{m-1} := [\text{blkdiag}(\tilde{B}_{m_1}^T, \tilde{B}_{m_1}^T)] \, \mathbf{u},$$
$$\mathbf{z1} := \mathbf{cos3r4}\left([w_j]_{j=0}^{m_2-1}, m_2\right),$$
$$\mathbf{z2} := \mathbf{cos3r4}\left([w_j]_{j=m_2}^{m_1-1}, m_2\right),$$
$$\mathbf{z3} := \mathbf{cos3r4}\left([w_j]_{j=m_1}^{3m_2-1}, m_2\right),$$
$$\mathbf{z4} := \mathbf{cos3r4}\left([w_j]_{j=3m_2}^{m-1}, m_2\right),$$
$$\mathbf{q} := [\text{blkdiag}(\tilde{W}_{m_1}, \tilde{W}_{m_1})] \, \left(\mathbf{z1}^T, \mathbf{z2}^T, \mathbf{z3}^T, \mathbf{z4}^T\right)^T,$$
$$\mathbf{r} := [\text{blkdiag}(H_{m_1}^T, H_{m_1}^T)] \, \mathbf{q},$$
$$\mathbf{s} := \tilde{W}_m \, \mathbf{r}$$
$$\mathbf{y} := H_m^T \, \mathbf{s}$$

Output: $\mathbf{y} = \sqrt{m} C_m^{III} \mathbf{x}$.

3 Complexity of the Proposed DCT II/III Algorithms

We will analyze explicitly the number of additions and multiplications required to execute the proposed split-radix and radix-4 DCT algorithms. If we compute an order $n \times n$ DCT matrix by a vector in the usual way, it requires $\mathcal{O}(n^2)$ arithmetic operations. But using these self recursive radix-2 DCT II/III algorithms, it is possible to compute $\mathbf{y}$ with significant multiplication complexity reduction. Once the arithmetic complexity results are presented, we compare the complexity results of the proposed algorithms with the well-known fast and stable DCT algorithms. Finally, we will utilize optimized C code to show the execution time of the proposed DCT algorithms with existing radix-2 DCT algorithms and also to verify the order of the complexity.

3.1 Arithmetic Complexity of Self-recursive Split-Radix and Radix-4 DCT II/III Algorithms

In this section, we compute the number of additions (say α) and multiplications (say β) required to produce $\mathbf{y} = \sqrt{n} C^{II/III} \, \mathbf{x}$ for $n = 2^t (t \geq 1)$ and $\mathbf{y} = \sqrt{m} C^{II/III} \, \mathbf{x}$ for $m = 4^t (t \geq 1)$. Note that we do not count multiplication by ± 1.

Lemma 2. *Let $n = 2^t$ $(t \geq 2)$ be given. If the split-radix DCT II algorithm is computed using the algorithm* **cos2sr(x, n)**, *then the arithmetic complexity is* *given by*

$$\alpha\,(DCT\ II, n) = \frac{3}{2}nt - n + 1,$$
$$\beta\,(DCT\ II, n) = \frac{1}{2}nt - 1. \tag{7}$$

Proof. Referring to the split-radix DCT II algorithm **cos2sr(x, n)**, we get

$$\begin{aligned}\beta(\text{DCT II}, n) = 2\cdot\beta\left(\text{DCT II}, \frac{n}{4}\right) + \beta\left(\text{DCT II}, \frac{n}{2}\right) + \beta\,(H_n) + \beta\,(H_{\frac{n}{2}})\\ + \beta\left(\tilde{W}_{\frac{n}{2}}\right) + \beta\,(W_{\frac{n}{2}}) + \beta\left(\tilde{B}_{\frac{n}{2}}\right) + \beta\,(B_{\frac{n}{2}})\end{aligned} \tag{8}$$

Following the structures of $H_n, \tilde{W}_n, W_{\frac{n}{2}}, \tilde{B}_n$, and $B_{\frac{n}{2}}$, we have

$$\begin{aligned}\alpha\,(H_n) &= n, & \beta\,(H_n) &= 0,\\ \alpha\left(\tilde{W}_n\right) &= \alpha\,(W_{\frac{n}{2}}) = 0, & \beta\left(\tilde{W}_n\right) &= \beta\,(W_{\frac{n}{2}}) = \tfrac{n}{2},\\ \alpha\left(\tilde{B}_n\right) &= \alpha\,(B_{\frac{n}{2}}) = \tfrac{n}{2} - 1, & \beta\left(\tilde{B}_n\right) &= \beta\,(B_{\frac{n}{2}}) = 1.\end{aligned} \tag{9}$$

Using the above result, we can rewrite (8) as

$$\beta(\text{DCT II}, n) = 2\cdot\beta\left(\text{DCT II}, \frac{n}{4}\right) + \beta\left(\text{DCT II}, \frac{n}{2}\right) + \frac{3}{4}n + 2.$$

Since $n = 2^t$, the above simplifies to the second order linear difference equation with respect to $t \geq 2$

$$\beta(\text{DCT II}, 2^t) - \beta\,(\text{DCT II}, 2^{t-1}) - 2\cdot\beta\,(\text{DCT II}, 2^{t-2}) = 3\cdot 2^{t-2} + 2.$$

Solving the above second order linear difference equation using the initial conditions β (DCT II, 2) = 0 and β (DCT II, 4) = 3, we can obtain

$$\beta(\text{DCT II}, 2^t) = \frac{1}{2}nt - 1$$

Similarly, using the initial conditions $\alpha\,(\text{DCT II}, 2) = 2$ and $\alpha\,(\text{DCT II}, 4) = 9$, one can derive the analogous result for the number of multiplications as shown in (7).

Corollary 5. *Let $n = 2^t$ $(t \geq 2)$ be given. If the split-radix DCT III algorithm is computed using the algorithm* **cos3sr(x, n)**, *then the arithmetic complexity is given by*

$$\alpha\,(DCT\ III, n) = \frac{3}{2}nt - n + 1,$$
$$\beta\,(DCT\ III, n) = \frac{1}{2}nt - 1. \tag{10}$$

Proof. This is trivial as the factorization for DCT III is obtained using the factorization of DCT II with the help of the transpose property.

Lemma 3. *Let $m = 4^t$ $(t \geq 1)$ be given. If the radix-4 DCT II algorithm is computed using the algorithm* **cos2r4(x, m)**, *then the arithmetic complexity is given by*

$$\begin{aligned}\alpha\,(DCT\ II, m) &= 3mt - m + 1,\\ \beta\,(DCT\ II, m) &= mt - 1.\end{aligned} \tag{11}$$

Proof. Referring to the radix-4 DCT II algorithm **cos2r4(x, m)**, we get

$$\begin{aligned}\beta(\text{DCT II}, m) =& 4 \cdot \beta\left(\text{DCT II}, \frac{m}{4}\right) + \beta\left(H_m\right) + 2 \cdot \beta\left(H_{\frac{m}{2}}\right)\\ &+ \beta\left(\tilde{W}_m\right) + 2 \cdot \beta\left(\tilde{W}_{\frac{m}{2}}\right) + 2 \cdot \beta\left(\tilde{B}_{\frac{m}{2}}\right) + \beta\left(\tilde{B}_m\right)\end{aligned} \tag{12}$$

Following the structures of H_n, $\tilde{W}_n$, and $\tilde{B}_n$ i.e. using (9), we can rewrite (12) as

$$\beta(\text{DCT II}, m) = 4 \cdot \beta\left(\text{DCT II}, \frac{m}{4}\right) + m + 3.$$

Since $m = 4^t$, the above simplifies to the first order linear difference equation with respect to $t \geq 1$

$$\beta(\text{DCT II}, 4^t) - 4 \cdot \beta\left(\text{DCT II}, 4^{t-1}\right) = 4^t + 3.$$

Solving the above first order linear difference equation using the initial condition β (DCT II, 4) = 3, we can obtain

$$\beta(\text{DCT II}, 4^t) = mt - 1$$

Similarly, using the initial condition α (DCT II, 4) = 9, one can derive the analogous result for the number of multiplications as shown in (10).

Corollary 6. *Let $m = 4^t$ $(t \geq 1)$ be given. If the radix-4 DCT III algorithm is computed using the algorithm* **cos3r4(x, m)**, *then the arithmetic complexity is given by*

$$\begin{aligned}\alpha\,(DCT\ III, m) &= 3mt - m + 1,\\ \beta\,(DCT\ III, m) &= mt - 1.\end{aligned} \tag{13}$$

Proof. This is trivial as the factorization for DCT III is obtained using the factorization of DCT II with the help of the transpose property.

3.2 Complexity Comparison of DCT II/III Algorithms

We provide addition and multiplication complexity comparisons of the proposed split-radix and radix-4 DCT II/III algorithms with the fast and stable DCT algorithms in [4,16,19,24,28,31,35,36,41,43]. The numerical results are presented for the matrix size varying from 8 × 8 to 4096 × 4096 in Tables 1, 2, 3 and 4.

Although 8-point DCT is proposed in [36] with 22 multiplications and 28 additions, we did not include that in the Tables 1 and 2, as that paper established

Table 1. Number of multiplications required to compute DCT II algorithms

n	**cos2sr**(x,n)	**cos2r4**(x,m)	[24]	[28,41]	[19]	[4]	[43]	[35]	[16]
8	11	–	11	12	14	16	13	13	11
16	31	31	31	32	44	44	35	33	31
32	79	–	79	80	118	116	91	81	–
64	191	191	191	192	300	292	227	193	–
128	447	–	447	448	726	708	547	449	–
256	1023	1023	1023	1024	1708	1668	1283	1025	–
512	2303	–	2303	2304	3926	3844	2947	2305	–
1024	5119	5119	5119	5120	8876	8708	6659	5121	–
2048	11263	–	11263	11264	19798	19460	14851	11265	–
4096	24575	24575	24575	24576	43692	43012	32771	24577	–

DCT II approximation, not the exact same as we discuss here. We also did not include the multiplication and addition counts in [25] as it is based on the stability of DCT algorithms.

As shown in Table 1, the multiplication complexity of the proposed split-radix and radix-4 DCT II algorithms is closer to that of [35]. But there is no formula for the factorization for the DCT II matrix in [35] (only the arithmetic complexity of DCT II). Simply put, the authors in [35] state that a new DCT II algorithm can be obtained by applying the method specified in [44] for their proposed DCT IV matrix factorization (without derivation or stating a DCT II matrix factorization explicitly).

Table 2. Number of additions required to compute DCT II algorithms

n	**cos2sr**(x,n)	**cos2r4**(x,m)	[24,28,41]	[19]	[4]	[43]	[35]	[16]
8	29	–	29	26	26	29	29	29
16	81	81	81	72	74	83	81	81
32	209	–	209	186	194	219	209	–
64	513	513	513	456	482	547	513	–
128	1217	–	1217	1082	1154	1315	1217	–
256	2817	2817	2817	2504	2690	3075	2817	–
512	6401	–	6401	5690	6146	7043	6401	–
1024	14337	14337	14337	12744	13826	15875	14337	–
2048	31745	–	31745	28218	30722	35331	31745	–
4096	69633	69633	69633	61896	67586	77827	69633	–

We can see from Table 1 that the DCT II algorithm given by [16] has the same number of multiplications as our split-radix and radix-4 DCT II algorithms for $n = 8, 16$. Yet, [16] does not have any result for $n \geq 32$. Note that the proposed split-radix and radix-4 DCT algorithms and DCT algorithms in [24] have the lowest multiplication complexity compared to all existing algorithms for $n \geq 32$.

Table 3. Number of multiplications required to compute DCT III algorithms

n	**cos3sr**(x,n)	**cos3r4**(x,m)	[24]	[28]	[19]	[43]
8	11	–	11	12	28	13
16	31	31	31	32	74	35
32	79	–	79	80	180	91
64	191	191	191	192	426	227
128	447	–	447	448	980	547
256	1023	1023	1023	1024	2218	1283
512	2303	–	2303	2304	4948	2947
1024	5119	5119	5119	5120	10922	6659
2048	11263	–	11263	11264	23892	14851
4096	24575	24575	24575	24576	51882	32771

Table 4. Number of additions required to compute DCT III algorithms

n	**cos3sr**(x,n)	**cos3r4**(x,m)	[24, 28]	[19]	[43]
8	29	–	29	26	29
16	81	81	81	72	83
32	209	–	209	186	219
64	513	513	513	456	547
128	1217	–	1217	1082	1315
256	2817	2817	2817	2504	3075
512	6401	–	6401	5690	7043
1024	14337	14337	14337	12744	15875
2048	31745	–	31745	28218	35331
4096	69633	69633	69633	61896	77827

There is no result for the comparison of DCT III algorithms in [4,16,35]. Thus, in Tables 3 and 4, we only compare the computational complexity results of the new DCT III algorithm with the DCT III algorithms in [19,24,28,43]. One can observe from Tables 1 and 3 that the multiplication complexity of the proposed DCT II/III algorithms is closer to that of [28]. But the derivation for the factorizations of DCT II/III matrices in [28] is more tedious. Moreover, the DCT II matrix factorization in [28] is similar to the traditional DCT II matrix factorization where the DCT II matrix is expressed using half the matrix size of itself and half the matrix size of DCT IV. But the proposed split-radix and radix-4 DCT II/III algorithms are expressed using itself together with simple, sparse, and scaled orthogonal matrices. Multiplication count of the proposed split-radix DCT II algorithm is closer to the multiplication count of the split-radix DCT II algorithm in [41].

But the split-radix DCT-II algorithm in [41] was computed through a DFT and neither contain orthogonal nor scaled orthogonal factors. Hence, such algorithms affect the stability, performance, and implementations [8,11,37].

The reduced flop counts (with 5.6% reduction) for split-radix DCT II/III algorithms were presented in [31]. We didn't include the results in that paper in Tables 1, 2, 3 and 4, as that paper has the flop count but not the explicit multiplication and addition counts. Based on the comparison of the proposed split-radix DCT II/III algorithms with [31], the proposed ones have the lowest multiplication complexity and also the arithmetic complexity for 8-point DCT II/III matrices. When the size of the matrices are getting higher than the 32-point DCT II/III, the arithmetic complexity of split-radix DCT II/III algorithms in [31] are better than the proposed split-radix DCT II/III algorithms. But the split-radix DCT II/III algorithms in [31] neither being orthogonal nor scaled orthogonal, affects the stability of the DCT algorithms [37]. Moreover, practical implementation of the split-radix algorithms proposed in [31] have to be modified significantly because several subdivisions are performed at once within the execution and hence affect the modern memory hierarchy [8,11].

3.3 Performance and Execution Time of the Split-Radix and Radix-4 DCT Algorithms

In this section we will compare the execution time of the proposed split-radix and radix-4 DCT algorithms with the DCT-II algorithms in [2,19,24]. These numerical results are presented using C codes with CPU running at 1.80 GHz and Intel Core i7-8550U processor. The execution time of the DCT algorithms is presented for the matrix of size varying from 8×8 to 1024×1024 in Fig. 1. We have denoted SRP and R4P as the proposed split-radix and radix-4 DCT II algorithms respectively, R2P as the radix-2 DCT II algorithm in [24], and R2T as the radix-2 DCT II algorithms in [2,19]. In these results, we have stored matrices as vectors and not considered time required for active memory access and multiplication by 0 and ± 1.

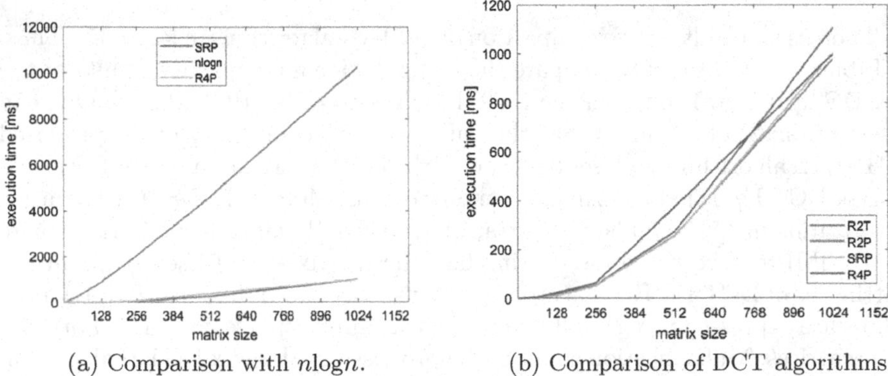

(a) Comparison with $n\log n$. (b) Comparison of DCT algorithms.

Fig. 1. Execution time of DCT II algorithms.

We have included $n\log n$ to observe the relationship between the execution time and the order of the arithmetic complexity of DCT algorithms. As shown in Fig. 1(a), the execution time of the proposed split-radix and radix-4 DCT algorithms is proportional to the order of the arithmetic complexity i.e. $\mathcal{O}(n\log n)$. Moreover, the proposed DCT algorithms are faster than $n\log n$ with speed improvement factor (ratio between $n\log n$ and the execution time of the proposed algorithms) of 10 even for 1024×1024 matrices. As shown in Fig. 1(b), the execution time of the proposed split-radix DCT algorithm is better than the radix-4 DCT algorithms. Also, the execution time of the proposed split-radix DCT algorithm is better than that of the radix-2 DCT algorithms. Finally, the execution time of the radix-2 DCT algorithm in [24] is better than that of the radix-2 DCT algorithms in [2,19].

4 Signal Flow Graphs for Split-Radix and Radix-4 DCT II/III Algorithms

Signal flow graphs can be utilized to realize a given system as an integrated circuit. In this section, we will present signal flow graphs to establish the connection

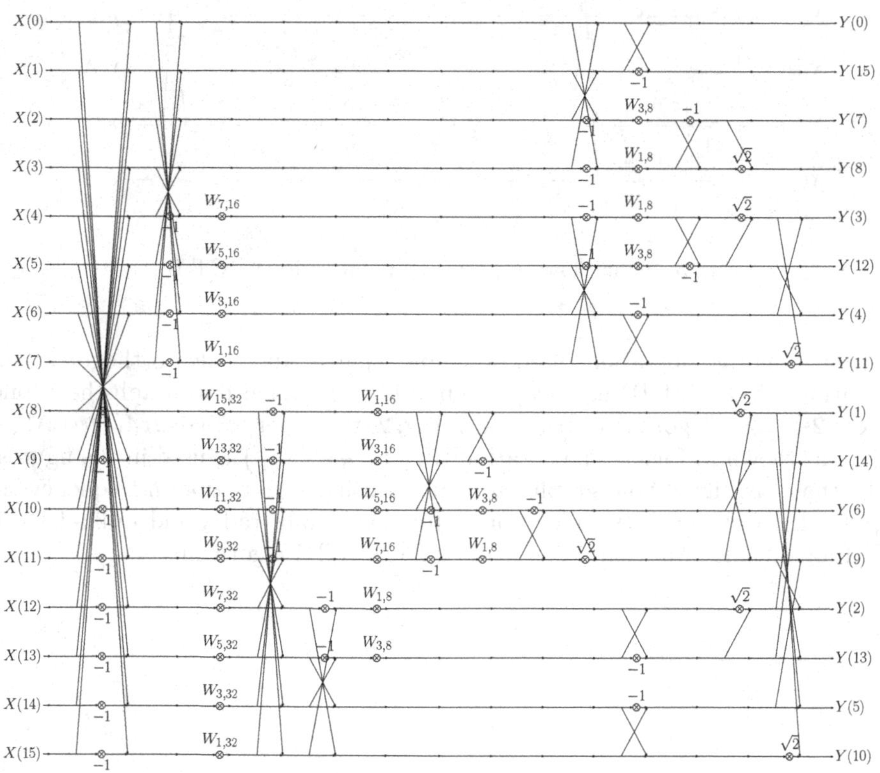

Fig. 2. Signal flow graph for 16-point split-radix DCT II.

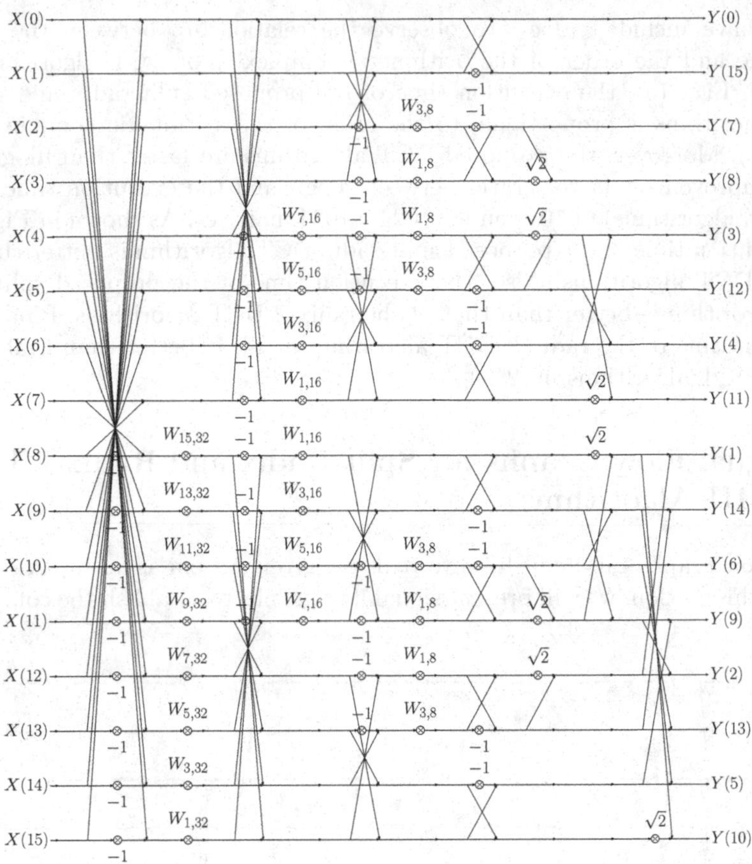

Fig. 3. Signal flow graph for 16-point radix-4 DCT II.

between algebraic operations used in simple, sparse, and scaled orthogonal factorizations of DCT II/III matrices. We provide two signal flow graphs here, one for **cos2sr**(x, n) Algorithm, as shown in Fig. 2, the other for **cos2r4**(x, n) Algorithm, as shown in Fig. 3. The notation $W_{j,k} := \frac{1}{2} \sec\left(\frac{j\pi}{2k}\right)$ is used in the figures. Note that the signal-flow graphs are drawn using decimation-in-frequency. It is possible to convert the decimation-in-frequency split-radix and radix-4 DCT II/III algorithms into decimation-in-time DCT II/III algorithms.

5 Conclusion

This paper extended the lowest complexity, self-recursive, radix-2 DCT II/III algorithms into efficient, self-recursive, split-radix and radix-4 Discrete Cosine Transform II/III algorithms. The proposed DCT algorithms have sparse and scaled orthogonal factors. The presented algorithms represent self-recursive and stable split-radix and radix-4 DCT II/III algorithms. The proposed algorithms have attained the lowest theoretical multiplication complexity and also the arithmetic complexity for 8-point DCT II/III matrices. Arithmetic complexity comparison of the proposed algorithms has been discussed including the existing fast and stable DCT algorithms. The optimized C code has shown favorable results on the execution time of the proposed algorithms as compared to radix-2 algorithms. This code has verified the order of arithmetic complexity. Finally, for the integrated circuits design and to show the simplicity of the proposed algorithms, signal flow graphs have been shown in decimation-of-frequency.

References

1. Britanak, V.: New generalized conversion method of the MDCT and MDST coefficients in the frequency domain for arbitrary symmetric windowing function. Digit. Sig. Proc. **23**, 1783–1797 (2013)
2. Britanak, V., Yip, P.C., Rao, K.R.: Discrete Cosine and Sine Transforms: General Properties Fast Algorithms and Integer Approximations. Academic Press, Great Britain (2007)
3. Chakraborty, S., Rao, K.R.: Fingerprint enhancement by directional filtering. In: 2012 9th International Conference on Electrical Engineering/Electronics, Computer, Telecommunications and Information Technology (ECTI-CON), Phetchaburi, pp. 1–4, May 2012. https://doi.org/10.1109/ECTICon.2012.6254113
4. Chen, W.H., Smith, C.H., Fralick, S.: A fast computational algorithm for the discrete cosine transform. IEEE Trans. Commun. **25**(9), 1004–1009 (1977)
5. Duhamel, P.: Implementation of split-radix FFT algorithms for complex, real, and real-symmetric data. IEEE Trans. Acoust. Speech Sig. Process. ASSP **34**(2), 285–295 (1986)
6. Duhamel, P., Vetterli, M.: Fast Fourier transforms: a tutorial review and a state of the art. Sig. Process. **19**(4), 259–299 (1990)
7. Fan, D., et al.: Optical identity authentication scheme based on elliptic curve digital signature algorithm and phase retrieval algorithm. Appl. Opt. **52**(23), 5645–5652 (2013)
8. Frigo, M., Johnson, S.G.: The design and implementation of FFTW3. Proc. IEEE **93**(2), 216–231 (2005)
9. Han, J., Saxena, A., Melkote, V., Rose, K.: Towards jointly optimal spatial prediction and adaptive transform in video/image coding. IEEE Trans. Image Process. **21**(4), 1874–1884 (2012)
10. Hsu, H.-W., Liu, C.-M.: Fast radix-q and mixed radix algorithms for type-IV DCT. IEEE Sig. Process. Lett. **15**, 910–913 (2008)
11. Johnson, S.G., Frigo, M.: A modified split-radix FFT with fewer arithmetic operations. IEEE Trans. Sig. Process. **55**(1), 111–119 (2007)

12. Kekre, H.B., Sarode, T.K., Save, J.K.: Column transform based feature generation for classification of image database. Int. J. Appl. Innov. Eng. Manag. (IJAIEM) **3**(7), 172–181 (2014)
13. Kekre, H.B., Sarode, T., Natu, P.: Performance comparison of hybrid wavelet transform formed by combination of different base transforms with DCT on image compression. Int. J. Image Graph. Sig. Process. **6**(4), 39–45 (2014)
14. Kekre, H.B., Solanki, J.K.: Comparative performance of various trigonometric unitary transforms for transform image coding. Int. J. Electron. **44**, 305–315 (1978)
15. Lee, M.H., Khan, M.H.A., Kim, K.J., Park, D.: A fast hybrid jacket-hadamard matrix based diagonal block-wise transform. Sig. Process. Image Commun. **29**(1), 49–65 (2014)
16. Loeffler, C., Ligtenberg, A., Moschytz, G.S.: Practical fast 1-D DCT algorithms with 11 multiplications. In: 1989 International Conference on Acoustics, Speech, and Signal Processing (ICASSP-1989), vol. 2, pp. 988–991 (1989). https://doi.org/10.1109/ICASSP.1989.266596
17. Ma, J., Plonka, G., Hussaini, M.Y.: Compressive video sampling with approximate message passing decoding. IEEE Trans. Circuits Syst. Video Technol. **22**(9), 1354–1364 (2012)
18. Olshevsky, A., Olshevsky, V., Wang, J.: A comrade-matrix-based derivation of the eight versions of fast cosine and sine transforms. In: Olshevsky, V. (ed.) Fast Algorithms for Structured Matrices: Theory and Applications, CONM, vol. 323, pp. 119–150. AMS Publications, Providence (2003)
19. Plonka, G., Tasche, M.: Fast and numerically stable algorithms for discrete cosine transforms. Linear Algebra Appl. **394**, 309–345 (2005)
20. Perera, S.M., Olshevsky, V.: Fast and stable algorithms for discrete sine transformations having orthogonal factors. In: Cojocaru, M.G., Kotsireas, I.S., Makarov, R.N., Melnik, R.V.N., Shodiev, H. (eds.) Interdisciplinary Topics in Applied Mathematics, Modeling and Computational Science, vol. 117, pp. 347–354. Springer, Basel (2015). https://doi.org/10.1007/978-3-319-12307-3_50
21. Perera, S.M.: Signal flow graph approach to efficient and forward stable DST algorithms. In: Proceedings of the 20th ILAS Conference, Leuven, Belgium (2016). Linear Algebra Appl. **542**, 360–390 (2017)
22. Perera, S.M., Madanayake, A., Dornback, N., Udayanga, N.: Design and digital implementation of fast and recursive DCT II-IV algorithms. Circuits Syst. Sig. Process. **38**(2), 529–555 (2018). https://doi.org/10.1007/s00034-018-0891-8
23. Perera, S.M., Olshevsky, V.: Stable, recursive and fast algorithms for discrete sine transformations having orthogonal factors. J. Coupled Syst. Multiscale Dyn. **1**(3), 358–371 (2013)
24. Perera, S.M., Liu, J.: Lowest complexity self recursive radix-2 DCT II/III algorithms. SIAM J. Matrix Anal. Appl. **39**(2), 664–682 (2018)
25. Perera, S.M.: Signal processing based on stable radix-2 DCT I-IV algorithms having orthogonal factors. Electron. J. Linear Algebra **31**, 362–380 (2016)
26. Pourazad, M.T., Doutre, C., Azimi, M., Nasiopoulos, P.: The new gold standard for video compression: how does HEVC compare with H.264/AVC? IEEE Consum. Electron. Mag. **1**(3), 36–46 (2012). https://doi.org/10.1109/MCE.2012.2192754
27. Püschel, M., Moura, J.M.F.: The algebraic approach to the discrete cosine and sine transforms and their fast algorithms. SIAM J. Comput. **32**, 1280–1316 (2003)
28. Püschel, M., Moura, J.M.F.: Algebraic signal processing theory: Cooley-Tukey type algorithms for DCTs and DSTs. IEEE Trans. Sig. Process. **56**(4), 1502–1521 (2008)
29. Rao, K.R., Kim, D.N., Hwang, J.J.: Fast Fourier Transform: Algorithm and Applications. Springer, New York (2010). https://doi.org/10.1007/978-1-4020-6629-0

30. Rao, K.R., Yip, P.: Discrete Cosine Transform: Algorithms, Advantages, Applications. Academic Press, San Diego (1990)
31. Shao, X., Johnson, S.G.: Type-II/III DCT/DST algorithms with reduced number of arithmetic operations. Sig. Process. **88**, 1553–1564 (2008)
32. Steidl, G., Tasche, M.: A polynomial approach to fast algorithms for discrete Fourier-cosine and Fourier-sine transforms. Math. Comput. **56**, 281–296 (1991)
33. Steidl, G.: Fast radix-p discrete cosine transforms. Appl. Algebra Eng. Commun. Comput. **3**(1), 39–46 (1992)
34. Strang, G.: The discrete cosine transform. SIAM Rev. **41**, 135–147 (1999)
35. Suehiro, N., Hatori, M.: Fast algorithms for the DFT and other sinusoidal transforms. IEEE Trans. Acoust. Speech Sig. Process. **34**(3), 642–644 (1986)
36. Tablada, C.J., Bayer, F.M., Cintra, R.J.: A class of DCT approximations based on the Feig-Winograd algorithm. Sig. Process. **113**, 38–51 (2015)
37. Tasche, M., Zeuner, H.: Roundoff error analysis for fast trigonometric transforms. In: Anastassiou, G. (ed.) Handbook of Analytic-Computational Methods in Applied Mathematics, pp. 357–406. Chapman and Hall/CRC Press, Boca Raton (2000)
38. Tell, E., Seger, O., Liu, D.: A converged hardware solution for FFT, DCT and Walsh transform. In: Proceedings of Seventh International Symposium on Signal Processing and Its Applications, pp. 609–612. IEEE (2003)
39. Van Loan, C.: Computational Frameworks for the Fast Fourier Transform. SIAM Publications, Philadelphia (1992)
40. Veerla, R., Zhang, Z., Rao, K.R.: Advanced image coding and its comparison with various still image codecs. Am. J. Sig. Process. **2**(5), 113–121 (2012)
41. Vetterli, M., Nussabaumer, H.J.: Simple FFT and DCT algorithms with reduced number of operations. Sig. Process. **6**, 267–278 (1984)
42. Voronenko, Y., Püschel, M.: Algebraic signal processing theory: Cooley-Tukey type algorithms for real DFTs. Trans. Sig. Process. **57**(1), 1–19 (2009)
43. Wang, Z.: Fast algorithms for the discrete W transform and the discrete Fourier transform. IEEE Trans. Acoust. Speech Sig. Process. **32**, 803–816 (1984)
44. Wang, Z.: On computing the Fourier and cosine transforms. IEEE Trans. Acoust. Speech Sig. Process. **33**, 1341–1344 (1985)
45. Yavne, R.: An economical method for calculating the discrete Fourier transform. In: Proceedings of the AFIPS Fall Joint Computer Conference, San Francisco, vol. 33, pp. 115–125, December 1968

Analysis of Max-Min Ant System with Local Search Applied to the Asymmetric and Dynamic Travelling Salesman Problem with Moving Vehicle

João P. Schmitt(✉), Rafael S. Parpinelli(✉), and Fabiano Baldo(✉)

Graduate Program in Applied Computing,
Santa Catarina State University (UDESC), Joinville, SC, Brazil
schmittjoaopedro@gmail.com, {rafael.parpinelli,fabiano.baldo}@udesc.br

Abstract. Vehicle routing problems require efficient computational solutions to reduce operational costs. Therefore, this paper presents a benchmark analysis of Max-Min Ant System (MMAS) combined with local search applied to the Asymmetric and Dynamic Travelling Salesman Problem with Moving Vehicle (ADTSPMV). Different from the well known ADTSP, in the moving vehicle scenario the optimization algorithm continues to improve the TSP solution while the vehicle is visiting the clients. The challenge of this scenario is mainly concerned with the fulfilment of hard time restrictions. In this study we evaluate how MMAS performs combined with US local search, 3-opt local search, and a memory mechanism. Besides that, we demonstrate how to model the moving vehicle restrictions under the MMAS algorithm. To perform the benchmark analysis instances from TSBLIB were selected. The dynamism was emulated by means of changes in traffic factors. The results indicate that for ADTSP the MMAS-US is the best algorithm while for ADTSPMV the MMAS-3opt is the most suitable.

Keywords: Computational optimization · Swarm intelligence · Hybrid methods · Dynamic problems

1 Introduction

The travelling salesman problem (TSP) is a well know problem in computer science, in which a salesman needs to visit a set of clients passing each one only once at the minimum cost. The TSP is important because it simulates problems from our daily live, like courier services. The TSP is still a rich field of study due to the complexity of finding exact solution methods [4]. By nature, TSP have exponential time execution and can not be optimally solved in a suitable time by exact methods, at least for scenarios with large amount of vertices.

Besides the classical TSP formulation, we can add other real-world features, like dynamism. Under dynamic problems there are studies that focus on TSP [12] and others that address the vehicle routing problem (VRP) [10,18]. The VRP

I. Kotsireas et al. (Eds.): SEA^2 2019, LNCS 11544, pp. 202–218, 2019.
https://doi.org/10.1007/978-3-030-34029-2_14

formulation is applied to the TSP because the VRP is considered a generalisation of the TSP [10].

Dynamic problems require algorithms to deal with unexpected events during the optimization, in contrast to classical TSP problems where the optimization happens before the salesman starts to visit the clients. One example of dynamism happens when a vehicle travelling across the clients needs to adapt its route regarding the traffic conditions changes or new request arrives [10]. Such scenario increases the complexity of finding good solutions.

Due to the intrinsic complexity of TSP and VRP problems, meta-heuristics are considered powerful techniques to approach them. The survey presented by [2] states that around 70% of the evaluated studies apply meta-heuristics, while only 17% of the studies consider exact methods. In this context, the Ant Colony Optimization (ACO) is seen as a suitable meta-heuristic to tackle the TSP [6]. The ACO is a bio-inspired meta-heuristic that imitates the behaviour of real ants in finding good paths between their next and the food sources. The literature shows that ACO is applicable for dealing with both static [20] and dynamic scenarios [12]. Some of them focus on the use of *Max-Min* Ant System (MMAS) as the ACO solving approach due to its characteristic of managing the pheromone, as well as, its easy integration with daemon procedures, as local search algorithms [12,23] and memory mechanisms [13–16,19]. Such daemon procedures are mainly used to improve the algorithm performance.

The performance of MMAS applied to the asymmetric and dynamic travelling salesman problem (ADTSP) was already evaluated in [12]. However, there is another class of TSP that adds a new feature to the problem, the moving vehicle constraint, named ADTSPMV. This constraint is particularly important to the logistic companies because they are concerned about the traffic dynamics along the working day to improve their vehicle routes. To the best of our knowledge, this problem was first evaluated in [19] applying the MMAS with a memory mechanism, and there are no studies comparing the MMAS with local search operators. Therefore, we propose a benchmark comparison of the MMAS combined with different local search operators and a memory mechanism. The objective is to identify the most suitable combination of MMAS with local search operators [12,21] and memory mechanism [19] for the ADTSPMV.

The remaining of this paper is organised as follows. Section 2 details the literature review. Section 3 states the problem and the algorithms modelling. Section 4 presents the protocol of experiments, results and analysis. Finally, Sect. 5 highlights the conclusion and points out some future works.

2 Related Work

Many Ant Colony algorithms were already proposed to deal with dynamic scenarios. Besides that, the literature indicates that hybrid algorithms are some of the best strategies to find suitable solutions [3,12,23]. This section presents some related works concerning with solution methods for the ADTSP. In [13–15,19] are proposed variations of the Ant Colony Algorithms (ACO), combined with memory mechanisms for the ADTSP, where the traffic dynamics is performed by modifying the graph edges costs along the time. These studies compare different implementations of the ACO with a build-in memory. The results indicate decreasing in the solutions' cost due to memory application.

Concerning the ADTSPMV, [19] presents a memory-based MMAS algorithm that deals with dynamic traffic factors while the vehicle is visiting the clients. Therefore, the algorithm keeps calculating the best-so-far solution even though the vehicle positioning is changing. The results figured out that the MMAS with memory outperform the canonical MMAS and the EIACO [16] algorithms.

Besides the memory mechanisms, the ACO also allows easy integration with local search operators. In [23] is presented a statistical comparison between implementations of local search heuristics for the static version of TSP. The study evaluate how Evolutionary and Ant-based Algorithms perform combined with different local search operator. The authors concluded that ACO algorithms with local search present better results than the one without such operators. Related to the application of local search for ADTSP, [12] presents a comparison between variants of MMAS with the 3-opt, 3-opt-res, 2-opt and Unstringing-Stringing (US) local search operators. The results present that MMAS with US is the best combination for ADTSP. Another study that presents a hybrid algorithm for solving the ADTSP is presented in [3]. They proposed an ACO algorithm combined with a Adaptive Large Neighborhood Search (ALNS). The analysis presents that ACO with ALNS reached better results compared with R-ACO and P-ACO variants.

The reviewed works some up the main found approaches of hybrid ACO in the literature. As can be seen, the use of ACO combined with operators are commonly applied to speed up its optimization. Given the benefits of this hybrid approaches for ADTSPMV, the next section presents the problem statement.

3 Problem Formulation and Modeling

This section presents a review and formulation of each one of the subjects addressed by this work. It starts formalizing ADTSPMV in Sect. 3.1. Then, it is outlined the MMAS modelling in Sect. 3.2. After than, the MMAS with Local Search Operator is stated in Sect. 3.3. Finally, the MMAS combined with a memory mechanism is formalized in Sect. 3.4.

3.1 The Asymmetric and Dynamic Travelling Salesman Problem with Moving Vehicle (ADTSPMV)

The travelling salesman problem (TSP) is formally defined by a fully connected graph $G = (N, A)$, where $N = \{v_1, v_2, ..., v_n\}$ represents the vertices set and $A = \{(v_i, v_j)|v_i, v_j \in N, i \neq j\}$ represents the edges sets. In the classic TSP formulation the vertices represent the cities fully connected with each other by a edge, except to itself. Besides that, each edge (v_i, v_j) is associated with a traversing cost information $c_{ij} > 0$. The objective of the solution approach is to build a route to salesman that starts in the depot, visits all cities passing in each one only once, and then returns back to the depot. The route must be minimized to have the smaller possible cost.

The TSP is said symmetric if for all edges the cost is the same in both directions, it means $c_{ij} = c_{ji}, \forall i, j \in N$. If the symmetric restriction is broken it is said that the TSP is asymmetric (ATSP). ATSP does not fulfil the triangular inequality $(d_{ij} + d_{jk} \geq d_{ik})$, therefore the problem gets harder because solution methods must be concerned about this information during optimization tasks. In general, the ATSP solutions used to have more complex data structures to represent the graph with the same amount of vertices than the symmetric ones [8].

Another characteristic that adds complexity in the TSP is the dynamism. Dynamic problems are well know and largely addressed in literature [3,10,12–16,18,19]. Its most common types are: dynamic requests, dynamic traffic factors and stochastic demands [10]. In this work we focus on the dynamic traffic factors, where the edges cost change along the time. This kind of dynamism, in real world, is caused by traffic jams, accidents or any other environmental conditions. To deal with such dynamic problems there are two strategies, the *a-priori* and the *real-time* methods [10]. The *a-priori* methods consider probabilistic information about the future, while *real-time* methods consider new information while the vehicle is on the road. In this work, we tackle the *real-time* approach, considering that the vehicle can change their route when traffic factors are modified.

To simulate the dynamic traffic factors, we apply a dynamic benchmark generator, the same used in [12]. We apply this simulator over static TSP instances, that works by modifying the edges costs applying (1).

$$c'_{ij} = c_{ij} \times t_{ij} \tag{1}$$

Where t_{ij} is a dynamic traffic factor, c_{ij} is the original cost and the c'_{ij} is the new cost. The t_{ij} traffic factor is generated by (2).

$$t_{ij} = \begin{cases} t_{ij} \leftarrow 1 + r \in [F_L, F_U], & \text{if } q < m \\ t_{ij} \leftarrow 1, & \text{otherwise} \end{cases} \tag{2}$$

where r is randomly distributed in the $[F_L, F_U]$ bounds, q is a randomly distributed number between $[0, 1]$ and m defines the magnitude of the change, respecting $0 < m \leq 1$. The role of r is to define the traffic jam scale for a given edge (street). When this value is near of F_L means less traffic jam, but when this value is close to F_U means more traffic jam [12]. Besides that, m represents the magnitude of changes, that is, it states how many edges will be modified in the dynamism simulation.

In general terms, the dynamism is performed by the application of (1) for every edge of the graph G within the frequency f [12]. The frequency defines the number of iterations that the algorithm will wait to apply a dynamism. Therefore, the frequency of dynamism is synchronized with the algorithm execution. So, assuming the current iteration as i, for each $i \mod f \equiv 0$ a dynamic modification is applied over all edges of the graph G using (1). According this reasoning, when f gets lower the modifications happen more often, on the other way around, when f gets higher the modifications happen less frequently.

The last aspect to be considered concerning ADTSPMV is the moving vehicle benchmark simulator [19]. Such mechanism is useful to simulate real world conditions meanwhile the vehicle is visiting the clients during the algorithm optimization. Formally, given a problem instance with N vertices, and an optimization algorithm that executes for i_{max} iterations, the vehicle will advance one client at each $\kappa = i_{max}/N$ intervals. Besides that, when a client is visited, it is ignored by the algorithm in the next optimization iterations. Therefore, for each κ number of iterations the set of vertices to be analysed reduces one vertex. Such scenario increases the problem complexity, because if the algorithm is not able to adapt the solution accordingly to the dynamic changes, the vehicle will have more probability to choose not suitable clients to visit, and then the final solution cost will get worse.

As performance indicator for the ADTSPMV, it is needed to compute the last iteration cost. This is necessary because it indicates the final solution cost at the end of the vehicle route, when all clients were already visited.

3.2 Max-Min Ant System (MMAS)

The MMAS [20] is an Ant Colony Optimization (ACO) derived from the Ant System (AS) [6]. This algorithm works based on pheromone and heuristic information. Pheromone is an information deposited by ants to propagate knowledge about the optimization process for future generations, and heuristic is an information about the problem itself. A very common scenario of MMAS application is the TSP problem, where it is used a population of ants to build routes that visit all clients. In MMAS, the first step is the solution construction, where all ants are positioned in a given start vertex and at each construction iteration the next node to be visited is selected based (3).

$$p_{ij}^k = \frac{[\tau_{ij}]^\alpha [\eta_{ij}]^\beta}{\sum_{l \in N_i^k} [\tau_{il}]^\alpha [\eta_{il}]^\beta}, \text{if } j \in N_i^k \tag{3}$$

Where $\eta_{ij} = 1/d_{ij}$ is the heuristic information related to the TSP problem, τ_{ij} is the pheromone information, α and β are two parameters to control the influence of heuristic and pheromone, and N_i^k is the neighborhood list of not visited clients from node i. Basically, it is a roulette-wheel selection, where nodes with better heuristics, pheromone or both, will have more probability to be selected.

After the construction step, the second step executes the pheromone evaporation. This step is useful to ignore not suitable nodes in the next constructions, reducing the algorithm exploitation bounds. The pheromone evaporation is given by (4).

$$\tau_{ij} = (1 - \rho)\tau_{ij}, \forall (i, j) \tag{4}$$

Where ρ is the evaporation rate defined in the $0 < \rho < 1$ interval, and τ_{ij} is the current pheromone value of edge (i, j).

The third step is the pheromone deposit, where the best so far ant deposits pheromone accordingly (5).

$$\tau_{ij} = \tau_{ij} + \triangle\tau_{ij}^{\text{best}}, \forall (i,j) \in R^{best} \tag{5}$$

Where $\triangle\tau_{ij}^{\text{best}} = 1/C^{best}$ is the amount of pheromone to be deposited, given by the inverse cost C^{best} of the best route R^{best}.

The MMAS is different from other AS variations because it defines a τ_{min} and a τ_{max} values, that limit the pheromone evaporation and deposit, respectively, to balance the intensification and the exploration process, and a $\overline{\lambda}$-branching to identify stagnation of the search process. Besides that, the MMAS allows the integration with local search operators, that is useful to exploits candidate solutions.

In this work, we adapted the MMAS construction step for better dealing with the ADTSPMV by considering $P = [v_0, v_1, v_2, ..., v_l]$ as the moving vehicle (salesman) partial route. It contains the visited clients sorted by the visited order, where l is the length of the partial route P. Besides that, R^k is the route of each ant $k \in K$. Before the next iteration of the construction step, we first copy the partial route P for each route R_k using (6), where i represents the route index. After that, we start the transition rule (3) to fill the remaining vertices (ignoring the vertices contained in P).

$$\forall k \in K, i \in [1, 2, ..., l], R_i^k = P_i \tag{6}$$

3.3 MMAS with Local Search

Local search operators are routines that receive a constructed route (in this case built by MMAS) and try to optimized it executing a sequence of predefined procedures [23]. Some of the most known heuristics for the TSP are Lin-Kernighan [11], Unstringing-String (US) [7] and the λ-opt. In this work we are focusing in US and λ-opt. The integration of MMAS (presented in Sect. 3.2) with local search operators is straightforward [12]. It is executed every time a best-so-far solution is found by the AS. So, the new best-so-far ant is passed as input to the local search operator that updates the ant route if it found a better solution.

The VRP is an asymmetric problem because it deals with real route networks that are asymmetric by nature. So, one of the challenges to solve TSP over a VRP is that some of the local search operators support only symmetric problems. The US version proposed by [12] is prepared to deal with asymmetric problems, however, the λ-opt operators are not. To overcome this limitation we need to convert the asymmetric graph to a symmetric one. It is necessary because λ-opt not deals with the triangular inequality and can stuck in a infinity loop. To convert the asymmetric graph we apply (1) the distance matrix conversion proposed in [9], and (2) adapt the route with the new symmetric nodes. After that, we execute the local search and convert its results to the original route format. Before to update the best so far route, we compare the cost improvement. This is necessary for λ-opt because the additional nodes inserted by the matrix conversion can present improvements in the symmetric matrix but not in the asymmetric matrix. Therefore, we only update the best so far ant route if the cost was decreased.

Due to the characteristic of ADTSPMV that optimizes the route while the vehicle is moving, when a client is visited it no longer needs to be considered in the next iterations. Therefore, to deal with this restriction we create a sub-graph and a sub-route which contain only the active clients (not yet visited). Formally, we create a sub-graph $G' = (N', A')$ where $N' = \forall v_i \notin P \cup v_0, i = [0, 1, 2, ..., n]$, where P is the moving vehicle partial route, and n is the total number of clients. After that, we copy all valid edges $A' = \forall(i, j) \in N'$. In G' we ignore the node v_0 because it represents the depot position, that is the first visited node before the vehicle starts to move. About the ant route, we create a sub-route $R'^k = \forall v_i \notin P \cup v_0, i \in R^k$. Given that, the first client (v_0) of route R'^k represents the depot position, but this depot is not the original problem depot (that was removed), we need to adjust last node costs to represent the link with the original depot. This adjustment is necessary because after the local search execution we replace in the original route with the new partial optimized route.

3.4 MMAS with Memory

The MMAS with memory is an algorithm that uses a long-term or short-term memory as a daemon process to help the algorithm during the optimization [15]. The idea is that, during the dynamic changes the memory will transfer knowledge from the previous environment to the next environment. In this study we are using the MMAS-MEM algorithm proposed by [19]. This algorithm selects the elitist best ants from the previous environment and update the memory if the new ants have better cost than the current ants. After that, the ants of this memory are used to update the pheromone trails. Finally, to increase the algorithm diversity the memory receives immigrants that are generated using an process to randomize the route in a predefined scale.

4 Protocol of Experiments, Results and Analysis

This section describes the methodology used to evaluate the MMAS variations under the ADTSP [12] and the ADTSPMV [19] scenarios. The test instances selected were KroA100, KroA150 and KroA200 [22]. These instances were extracted from previous studies that had considered the same dynamic characteristics [12,19]. To generate the dynamism (traffic factors), we used the dynamic benchmark generator proposed in [12]. To apply the moving vehicle, we used the moving vehicle simulator proposed in [19]. Besides that, we combined each instance with different levels of dynamism, defining different frequencies and magnitudes to be used by the dynamic benchmark generator.

In both scenarios we applied the canonical MMAS [1], the MMAS with 3-opt local search (MMAS-3OPT), the MMAS with unstringing-stringing local search (MMAS-US), the MMAS with memory (MMAS-MEM), and the MMAS with memory and US local search (MMAS-MEM-US). We focused on compare the performance of these MMAS versions.

There are some parameters to be tuned for the dynamic benchmark generator. They are the magnitude, frequency, lower and upper bounds (F_L, F_U). The values applied were extracted from the literature [12]. Concerning the magnitude, it is defined with $m = \{0.1, 0.5, 0.75\}$ indicating small, medium and high number of vertices suffering dynamics, respectively. Frequency is varied with $f = \{10, 100\}$ indicating low and high number of dynamics, respectively. Finally, lower and upped bounds are defined as $F_L = 0$ and $F_U = 2$. Such values generate 18 test instances (3 test instances $\times$ 2 frequencies $\times$ 3 magnitudes).

Also, parameters for the optimization algorithms were defined based on literature. For all algorithms the number of ants was defined as 50, $\alpha = 1$ and $\beta = 5$ [12]. The ρ parameter was defined as 0.02 for MMAS [20], and for MMAS-US, MMAS-3OPT, MMAS-MEM and MMAS-MEM-US it was defined as 0.8 [12, 19].

All algorithms were run under a Core i7-7700HQ CPU 2.8 GHz with 8 GB of RAM. All algorithms were coded in Java programming language and compiled using the JDK 1.8. Also, 30 independent runs were performed for each test instance. For each different run a different seed was provided for the dynamic benchmark generator. Therefore, each algorithm receives the same seed for the same run. Hence, all algorithms can run over the same initial conditions. The collected statistic was the P_{OFF}, given by (7).

$$P_{OFF} = \frac{1}{I}\sum_{i=1}^{I}(\frac{1}{E}\sum_{j=1}^{E} P^*_{ij}) \tag{7}$$

Where I is the total number of iterations, E is the number of trials, and P^*_{ij} is the best-so-far solution cost of that iteration. All algorithms were executed using an amount of 1,000 iterations. It means that for problem instances with frequency of 10 it is performed 100 environmental changes, while problems with frequency of 100 it is performed 10 environmental changes. For the moving vehicle benchmark simulator (ADTSPMV) the vehicle considers the number of vertices and iterations to calculate the time interval used to visit clients. It means that, using KroA100 instance for example, with the algorithm executing over 1000 iterations, the vehicle visits one client at each 10 iterations. It is worth remembering that, for the moving vehicle restriction of ADTSPMV, we need to analyse the last iteration cost as performance indicator.

We organised the results assessment in two sections, the Sect. 4.1 presents the ADTSP evaluation and the Sect. 4.2 presents the ADTSPMV evaluation.

4.1 ADTSP Results

Table 1 presents the P_{OFF} and its standard deviation for the ADTSP. The table is organized as follows: columns "Inst.", "Freq." and "Mag." present the test instances configuration, and the column "Algorithms" presents the algorithms' results. Besides that, all bold results indicate the best result with statistical significance of 5%. The statistical test applied was the Wilcoxon Rank Sum test. In the two bottom rows of Table 1 we present the overall average, and the number of best/worst/same results obtained by each algorithm compared to results obtained by MMAS-US algorithm. MMAS-US was selected because it presented the best results in most instances. Related to best/worst/same, best and worst are related to results in which statistical significance of 5% is achieved and by same we mean results that do not achieved statistical significance of 5%.

Table 1. P_{OFF} results comparing algorithms execution under the asymmetric and dynamic traveling salesman problem (ADTSP). Best/worst/same compare the algorithms results against the MMAS-US algorithm.

ADTSP			Algorithms				
Inst.	Freq.	Mag.	MMAS	$MMAS_{MEM}$	$MMAS_{MEM_{US}}$	$MMAS_{3OPT}$	$MMAS_{US}$
KroA100	10	0.1	24016 ± 600	24028 ± 576	23344 ± 543	22601 ± 472	**22204 ± 373**
		0.5	30672 ± 988	30448 ± 979	28642 ± 1158	27656 ± 1053	**27154 ± 998**
		0.75	35822 ± 1167	35582 ± 1138	33236 ± 1393	32463 ± 1435	**31803 ± 1275**
	100	0.1	23644 ± 567	23197 ± 459	22933 ± 491	22464 ± 364	**22152 ± 314**
		0.5	29385 ± 967	28774 ± 929	27639 ± 979	26819 ± 822	**26427 ± 799**
		0.75	34263 ± 1206	33297 ± 1076	32071 ± 1275	31112 ± 1047	**30804 ± 1050**
KroA150	10	0.1	31351 ± 704	31475 ± 645	30689 ± 706	28585 ± 551	**28008 ± 477**
		0.5	39122 ± 957	38789 ± 942	36840 ± 1320	34724 ± 1268	**34463 ± 1140**
		0.75	45568 ± 1147	45191 ± 1143	42632 ± 1605	40803 ± 1805	**40359 ± 1405**
	100	0.1	30520 ± 671	29890 ± 679	29598 ± 666	28315 ± 476	**27958 ± 399**
		0.5	37861 ± 994	36842 ± 920	35469 ± 1138	33709 ± 882	**33454 ± 933**
		0.75	43904 ± 1202	42783 ± 1219	41130 ± 1469	39231 ± 1221	**38972 ± 1228**
KroA200	10	0.1	34679 ± 753	35135 ± 744	34058 ± 747	31737 ± 720	**31143 ± 591**
		0.5	44091 ± 983	43832 ± 979	41805 ± 1338	38955 ± 1598	**38851 ± 1246**
		0.75	51414 ± 1154	51068 ± 1175	48338 ± 1654	46044 ± 2179	**45574 ± 1543**
	100	0.1	34065 ± 790	33476 ± 806	33079 ± 756	31393 ± 517	**30997 ± 502**
		0.5	42869 ± 1100	41955 ± 1001	40450 ± 1265	38060 ± 1026	**37775 ± 1148**
		0.75	50033 ± 1224	48704 ± 1210	46671 ± 1598	44368 ± 1317	**44196 ± 1514**
Overall average			36849 ± 954	36359 ± 923	34923 ± 1117	33280 ± 1042	**32905 ± 941**
Best/worst/same			0/18/0	0/18/0	0/18/0	0/18/0	18/0/-

From Table 1, a quantitative analysis of the P_{OFF} results are performed. The analysis shows that the best algorithm for this kind of scenario is the MMAS-US, also verified in [12], because it achieved the lowest average cost among all algorithms (32905.2) with the lowest values in all test instances. A qualitative analysis indicates that MMAS-US reached the best cost probably due to the usage of US local search. It is explained because US routine executes more types of swaps compared to 3-opt routine, and it is not influenced by the memory, that is used to maintain the diversity in higher levels. Besides that, we can see that the canonical version of MMAS has the higher overall average, this is because it does not use any local search nor memory operator to improve the optimization.

Analysing the MMAS with memory, in Table 1 we can see that MMAS is better than MMAS-MEM only in the scenarios with frequency of 10 and magnitude of 0.1. It indicates that, without local search, the usage of memory is more suitable for the ADTSP. Analysing the memory with local search, the MMAS-MEM-US is better than MMAS-MEM in all cases. It indicates that local search plays an essential role during the optimization. Finally, comparing memory-local search against local search only, we see that MMAS-MEM-US is worst than MMAS-US and MMAS-3OPT in all test cases.

In Fig. 1 we present the algorithms cost function for the KroA150 scenario. In this graphic, we can observe that MMAS-US converges faster than other algorithms due to US operator. Besides that, analysing the MMAS only, we observe that it takes a long time to start to converge, while other algorithms in the first iterations have achieved better solutions.

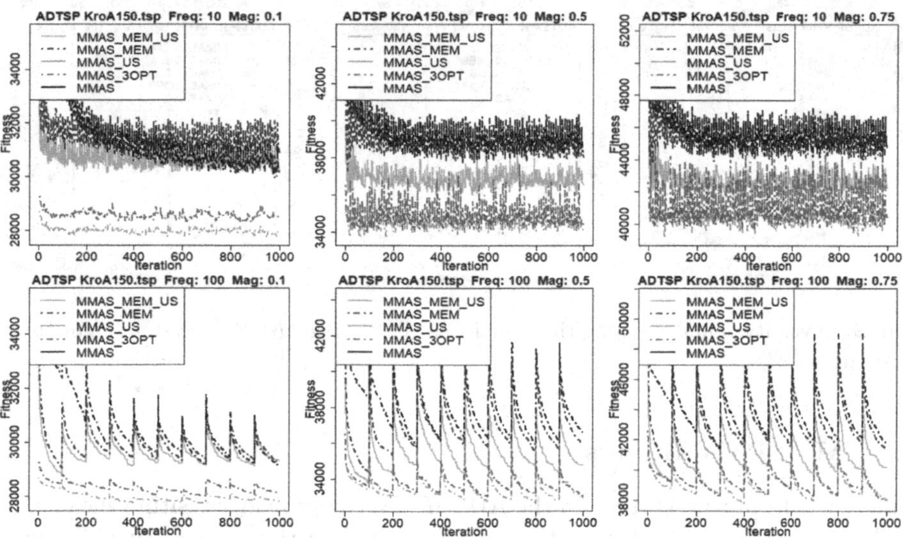

Fig. 1. Cost function along the time in ADTSP, *Max-Min* Ant System algorithms comparison.

In Fig. 2 we present the diversity function, that is the mean number of different edges found in the constructed routes in each iteration. It helps us to understand the convergence curves of Fig. 1. There we can see the MMAS-MEM-US and MMAS-MEM keep high diversity levels, while MMAS keeps a midterm, and MMAS-US and MMAS-3OPT keep the lowest levels. Analysing MMAS-US and MMAS-3OPT we see that both have the highest amplitude variation in diversity when dynamic changes occur. Such variation is caused by pheromone evaporation rate, memory and local search procedures. The diversity increasing in dynamic changes is healthful for the convergence process, as it allows the algorithm to explore the search space and adapts to the new environment. Therefore, for the ADTSP scenarios, we conclude that high convergence rates caused by abrupt reduction in diversity combined with diversity increasing after a dynamic change present a suitable analysis to select algorithms for the ADTSP. Besides that, MMAS combined with local search operators only, presents to be the best approach for the ADTSP.

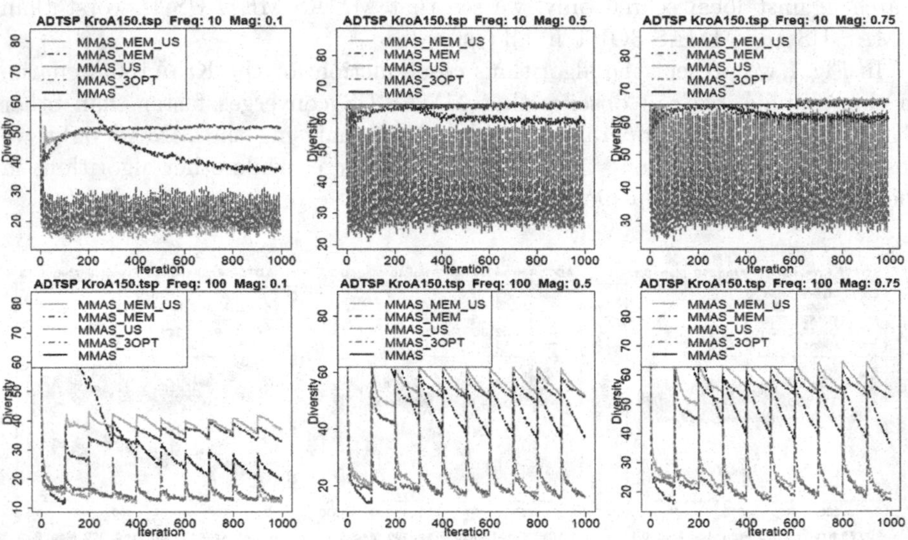

Fig. 2. Diversity function along the time in ADTSP, *Max-Min* Ant System algorithms comparison.

4.2 ADTSPMV Results

Differently from the ADTSP, in the ADTSPMV the vehicle is visiting the clients of the best-so-far solution while the optimization algorithm is executing. To analyse such scenario, Table 2 presents the results obtained. The best/worst/same line computes the results obtained by each algorithm compared to results obtained by MMAS-3OPT algorithm, chosen because it presented the best results in most instances.

A quantitative analysis indicates that MMAS-3OPT is the most suitable algorithm for this problem, because in most of the instances it presents the best results, except for KroA100 with frequency of 100. We also see that the best overall average result and the higher number of best results hit were achieved by MMAS-3OPT.

A qualitative analysis of the ADTSPMV requires from the algorithms to keep robust solutions along the time, because as the vehicle is visiting the clients during the optimization, there is a limited time budget to optimize the route before select the next client. Therefore, if the solution is not improved, the vehicle will have more probability to selected a poor next client and increase the final solution cost. Related to algorithms, we observed that memory mechanisms disturb in high scales the intensification of local searches. Hence, the 3-OPT is probably better than US due to the asymmetric to symmetric matrix conversion that allows search space exploration.

Table 2. P_{OFF} results comparing algorithms execution under the asymmetric and dynamic traveling salesman problem with moving vehicle (ADTSPM).

ADTSPMV			Algorithms				
Inst.	Freq.	Mag.	MMAS	MMAS$_{MEM}$	MMAS$_{MEM_{US}}$	MMAS$_{3OPT}$	MMAS$_{US}$
KroA100	10	0.1	24441 ± 739	24218 ± 811	23758 ± 867	**22998 ± 596**	23554 ± 763
		0.5	31418 ± 1570	31449 ± 1566	31176 ± 1581	**29860 ± 1318**	30797 ± 1465
		0.75	37025 ± 1780	36303 ± 1789	36702 ± 2111	**35834 ± 1833**	36370 ± 2127
	100	0.1	23985 ± 544	23702 ± 646	22863 ± 515	22880 ± 517	**22862 ± 518**
		0.5	29806 ± 1177	29620 ± 1206	27706 ± 1047	27856 ± 1090	27669 ± 1072
		0.75	34808 ± 1373	34260 ± 1368	33013 ± 1614	32775 ± 1171	**32333 ± 1373**
KroA150	10	0.1	31352 ± 888	31098 ± 887	30654 ± 977	**29298 ± 568**	30380 ± 1018
		0.5	40250 ± 1619	39784 ± 1629	39463 ± 1833	**37579 ± 1682**	39443 ± 2055
		0.75	46915 ± 1994	46245 ± 1968	46692 ± 2451	**44629 ± 1825**	45421 ± 1798
	100	0.1	30865 ± 735	30028 ± 709	29156 ± 671	**28789 ± 550**	28949 ± 610
		0.5	38094 ± 1430	37297 ± 1115	35826 ± 1347	**35044 ± 1156**	35264 ± 1053
		0.75	44209 ± 1447	43144 ± 1350	41758 ± 1537	**40954 ± 1413**	41108 ± 1535
KroA200	10	0.1	35067 ± 995	34796 ± 825	34459 ± 1297	**32736 ± 931**	34008 ± 1209
		0.5	44643 ± 1603	44418 ± 1632	45106 ± 2069	**42061 ± 1654**	44876 ± 2023
		0.75	52342 ± 1824	52122 ± 2042	52791 ± 2561	**50620 ± 2418**	52470 ± 2608
	100	0.1	34537 ± 813	33967 ± 851	32454 ± 832	**32017 ± 530**	32215 ± 850
		0.5	43171 ± 1363	42219 ± 1341	40665 ± 1661	**39168 ± 1235**	39610 ± 1279
		0.75	50077 ± 1426	49277 ± 1387	47175 ± 1846	**46482 ± 1587**	46709 ± 1864
Overall average			37389 ± 1295	36886 ± 1285	36190 ± 1490	**35088 ± 1226**	35780 ± 1401
Best/worst/same			0/18/0	0/18/0	3/15/0	15/3/-	3/15/0

The results from ADTSPMV tends to be greater than results from ADTSP, because in the ADTSPMV the algorithms should give fast response and such characteristic tends to increase the final solution cost if the time restriction is not satisfied. In Fig. 3 we see that algorithms with high variation tends to have worst final solutions. In this analysis, the MMAS-3OPT is the algorithm that suffers less variation compared with the others. Therefore, the solutions tends to be better.

In the ADTSPMV, as in the ADTSP, Fig. 4 shows that algorithms with local search presents the highest diversity variation during the environment changes indicating more chance to self adaptation in new environments. Besides that, for the ADTSPMV, the diversity decreases along the iterations, because the number of clients decrease as long as the vehicle is visiting them, so less different solutions can be constructed.

In the ADTSPMV, the P_{OFF} only give us a direction about the best algorithm. However, in this kind of problem is important to analyse the solution cost average of the last algorithm iteration. This value indicates the total vehicle navigation cost along the route, and is a better performance indicator. In Table 3 we see that the MMAS-3OPT seems to be the best algorithm, because it achieves the best results for nine test instances, while MMAS-US achieves the best solution for five and MMAS and MMAS-MEM achieved the best results for some instances. However, in general, we conclude that MMAS-3OPT is the most suitable algorithm for the ADTSPMV (Fig. 4).

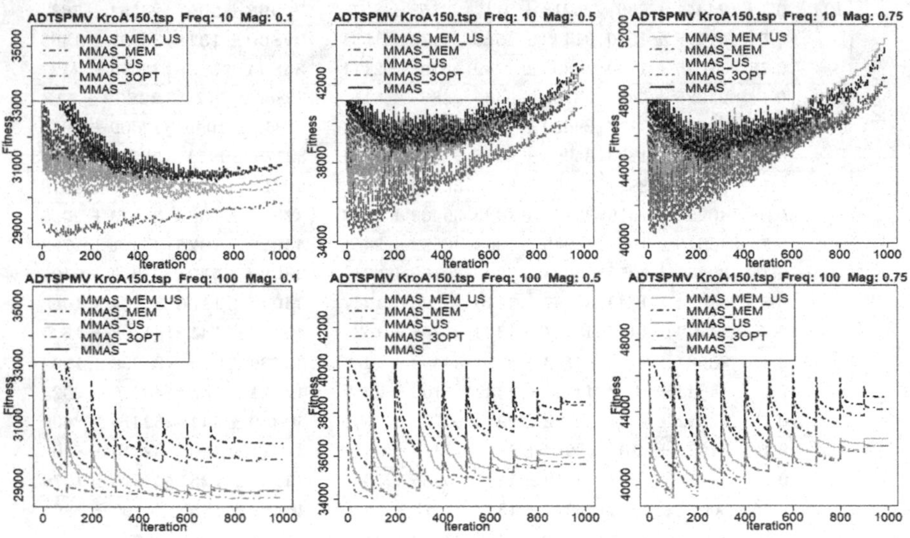

Fig. 3. Cost function along the time in ADTSPM, *Max-Min* Ant System algorithms comparison.

Table 3. Last iteration solution cost algorithms comparison in the asymmetric and dynamic travelling salesman problem with moving vehicle (ADTSPM).

ADTSPMV			Algorithms				
Inst.	Freq.	Mag.	MMAS	MMAS_{MEM}	$\text{MMAS}_{MEM_{US}}$	MMAS_{3OPT}	MMAS_{US}
KroA100	10	0.1	24252 ± 883	24280 ± 896	24089 ± 980	$\mathbf{23416 \pm 687}$	23877 ± 810
		0.5	32732 ± 2485	33332 ± 2975	33758 ± 2152	$\mathbf{32656 \pm 2207}$	33626 ± 1724
		0.75	40373 ± 3289	$\mathbf{39026 \pm 3254}$	40498 ± 2689	39616 ± 2742	40634 ± 2833
	100	0.1	23696 ± 493	23724 ± 695	$\mathbf{22721 \pm 515}$	22992 ± 572	22791 ± 585
		0.5	30348 ± 1455	30294 ± 1556	$\mathbf{28018 \pm 1046}$	28416 ± 1217	28203 ± 1138
		0.75	35536 ± 2009	35078 ± 1869	34000 ± 1796	34013 ± 1432	$\mathbf{33347 \pm 1749}$
KroA150	10	0.1	31048 ± 1215	31050 ± 987	30669 ± 1149	$\mathbf{29787 \pm 708}$	30445 ± 996
		0.5	42912 ± 2676	41879 ± 2925	42691 ± 2531	$\mathbf{40828 \pm 2681}$	42711 ± 3024
		0.75	51026 ± 4503	49315 ± 3774	51562 ± 3883	48826 ± 2262	$\mathbf{48767 \pm 2167}$
	100	0.1	30390 ± 830	29849 ± 757	28815 ± 643	28793 ± 523	$\mathbf{28585 \pm 563}$
		0.5	38487 ± 1901	38324 ± 1535	36184 ± 1338	$\mathbf{35595 \pm 1299}$	35894 ± 954
		0.75	44811 ± 1897	44133 ± 1767	42517 ± 1603	$\mathbf{42117 \pm 2014}$	42129 ± 1659
KroA200	10	0.1	34449 ± 1170	34459 ± 798	34281 ± 1313	$\mathbf{33388 \pm 1140}$	33526 ± 1161
		0.5	46708 ± 2853	46419 ± 2606	47741 ± 2493	$\mathbf{45071 \pm 1996}$	47655 ± 2531
		0.75	55845 ± 4013	$\mathbf{55234 \pm 3630}$	57696 ± 3504	55630 ± 3055	57126 ± 3684
	100	0.1	33900 ± 917	33764 ± 837	31774 ± 669	32172 ± 521	$\mathbf{31666 \pm 881}$
		0.5	43753 ± 1651	43147 ± 1868	40860 ± 1750	$\mathbf{39945 \pm 1348}$	40116 ± 1322
		0.75	50840 ± 1864	50307 ± 1647	47757 ± 1744	47899 ± 1611	$\mathbf{47743 \pm 1892}$
Overall average			38395 ± 2006	37979 ± 1910	37535 ± 1767	$\mathbf{36731 \pm 1556}$	37158 ± 1648
Best/worst			0/18	2/16	5/13	**9/9**	7/11

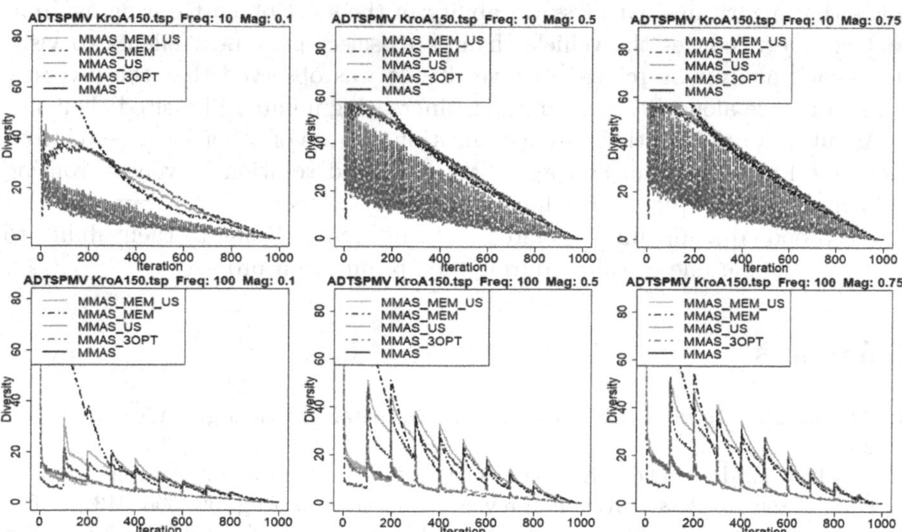

Fig. 4. Diversity function along the time in ADTSPM, *Max-Min* Ant System algorithms comparison.

5 Conclusions and Future Work

In this paper we evaluated the MMAS with local search and memory for the asymmetric and dynamic travelling salesman problem with moving vehicle (ADTSPMV). To evaluate the MMAS, we selected its following variations: MMAS, MMAS-MEM, MMAS-US, MMAS-3OPT, and MMAS-MEM-US. The analysed metrics were the convergence curves, diversity curves and the P_{OFF}, for both ADTSP and ADTSPMV scenarios.

For the ADTSP, a quantitative analysis indicates that MMAS-US is the best algorithm, because in all test instances it achieved the best results with statistical significance. These results are probably caused because US executes more types of swaps compared to 3-opt, and it is not influenced by the memory, that is used to maintain the diversity in higher levels. Besides that, we evaluated that local search plays an essential role in the ADTSP, as it can intensify the search in changed environments quickly.

For the ADTSPMV, the results indicate that MMAS-3OPT is the most suitable algorithm for this scenario. Besides that, differently from ADTSP where the P_{OFF} statistic is suitable to execute the analysis, for the ADTSPMV we evaluated the last algorithm iteration cost, that carry the total solution cost of the moving vehicle. Analysing the statistics, the MMAS-3OPT achieved best results in 9 out of 18 test instances. Hence, among the local search procedures, the most suitable is the 3-opt that achieves the overall best result.

For the ADTSPMV problem, two conclusions can be drawn. First, the final solution cost is impacted by the variability of the convergence process. This means that algorithms with less variability in the solution cost during environment changes, reduces the vehicle chances to select poor next clients to visit. The second analysis is related to diversity. It was observed that the diversity tends to reduce along the time given the increasing number of visited clients.

As future work directions, we can mention the use of other local search routines, like Lin-Kernighan, and apply the proposed solution in vehicle routing problems. Also, strategies of online adjustment and control of parameters [17] for the proposed algorithm is pointed as future research due to their ability to self-adapt the parameter values during the optimization process.

References

1. Ant Colony Optimization. http://www.aco-metaheuristic.org/. Accessed 8 Feb 2019
2. Braekers, K., Ramaekers, K., Van Nieuwenhuyse, I.: The vehicle routing problem: state of the art classification and review. Comput. Ind. Eng. **99**, 300–313 (2016)
3. Chowdhury, S., Marufuzzaman, M., Tunc, H., Bian, L., Bullington, W.: A modified ant colony optimization algorithm to solve a dynamic traveling salesman problem: a case study with drones for wildlife surveillance. J. Comput. Des. Eng. **6**, 368–386 (2018)
4. Cormen, T.: Introduction to Algorithms. MIT Press, Cambridge (2009)

5. Dorigo, M., Stutzle, T.: Ant Colony Optimization. MIT Press, Cambridge (2004)
6. Dorigo, M., Maniezzo, V., Colorni, A.: Ant system: optimization by a colony of cooperating agents. IEEE Trans. Syst. Man Cybern. Part B (Cybern.) **26**, 29–41 (1996)
7. Gendreau, M., Hertz, A., Laporte, G.: New insertion and postoptimization procedures for the traveling salesman problem. Oper. Res. **40**, 1086–1094 (1992)
8. Johnson, D., Gutin, G., McGeoch, L., Yeo, A., Zhang, W., Zverovitch, A.: Experimental analysis of heuristics for the ATSP. In: Gutin, G., Punnen, A.P. (eds.) The Traveling Salesman Problem and Its Variations, Combinatorial Optimization. Combinatorial Optimization, vol. 12, pp. 445–487. Springer, Boston (2007). https://doi.org/10.1007/0-306-48213-4_10
9. Jonker, R., Volgenant, T.: Transforming asymmetric into symmetric traveling salesman problems. Oper. Res. Lett. **2**, 161–163 (1983)
10. Larsen, A.: The dynamic vehicle routing problem. Ph.D. thesis, Technical University of Denmark, Kongens, Lyngby, Denmark (2000)
11. Lin, S., Kernighan, B.: An effective heuristic algorithm for the traveling-salesman problem. Oper. Res. **21**, 498–516 (1973)
12. Mavrovouniotis, M., Muller, F., Yang, S.: Ant colony optimization with local search for dynamic traveling salesman problems. IEEE Trans. Cybern. **47**, 1743–1756 (2017)
13. Mavrovouniotis, M., Yang, S.: Memory-based immigrants for ant colony optimization in changing environments. In: Di Chio, C., et al. (eds.) EvoApplications 2011. LNCS, vol. 6624, pp. 324–333. Springer, Heidelberg (2011). https://doi.org/10.1007/978-3-642-20525-5_33
14. Mavrovouniotis, M., Yang, S.: Interactive and non-interactive hybrid immigrants schemes for ant algorithms in dynamic environments. In: 2014 IEEE Congress on Evolutionary Computation (CEC), pp. 1542–1549 (2014)
15. Mavrovouniotis, M., Yang, S.: Ant colony optimization with immigrants schemes for the dynamic travelling salesman problem with traffic factors. Appl. Soft Comput. **13**, 4023–4037 (2013)
16. Mavrovouniotis, M., Yang, S.: Ant colony optimization with immigrants schemes in dynamic environments. In: Schaefer, R., Cotta, C., Kołodziej, J., Rudolph, G. (eds.) PPSN 2010. LNCS, vol. 6239, pp. 371–380. Springer, Heidelberg (2010). https://doi.org/10.1007/978-3-642-15871-1_38
17. Parpinelli, R.S., Plichoski, G.F., Silva, R., Narloch, P.H.: A review of techniques for on-line control of parameters in swarm intelligence and evolutionary computation algorithms. Int. J. Bio-Inspired Comput. **13**(1), 1–20 (2019)
18. Pillac, V., Gendreau, M., Guéret, C., Medaglia, A.: A review of dynamic vehicle routing problems. Eur. J. Oper. Res. **225**, 1–11 (2013)
19. Pedro Schmitt, J., Baldo, F., Stubs Parpinelli, R.: A MAX-MIN ant system with short-term memory applied to the dynamic and asymmetric traveling salesman problem. In: 2018 7th Brazilian Conference on Intelligent Systems (BRACIS), pp. 1–6 (2018)
20. Stützle, T., Hoos, H.: Max-min - ant system. Future Gener. Comput. Syst. **16**, 889–914 (2000)
21. Stutzle, T., Hoos, H.: MAX-MIN ant system and local search for the traveling salesman problem. In: Proceedings of 1997 IEEE International Conference on Evolutionary Computation (ICEC 1997), pp. 309–314 (1997)

22. TSPLIB. https://www.iwr.uni-heidelberg.de/groups/comopt/software/TSPLIB95/. Accessed 8 Feb 2019
23. Wu, Y., Weise, T., Chiong, R.: Local search for the traveling salesman problem: a comparative study. In: 2015 IEEE 14th International Conference on Cognitive Informatics & Cognitive Computing (ICC&ICC), pp. 213–220 (2015)

Computing Treewidth via Exact and Heuristic Lists of Minimal Separators

Hisao Tamaki(✉)

Meiji University, Kawasaki 214-8571, Japan
tamaki@cs.meiji.ac.jp

Abstract. We develop practically efficient algorithms for computing the treewidth $\mathrm{tw}(G)$ of a graph G. The core of our approach is a new dynamic programming algorithm which, given a graph G, a positive integer k, and a set Δ of minimal separators of G, decides if G has a tree-decomposition of width at most k of a certain canonical form that uses minimal separators only from Δ, in the sense that the intersection of every pair of adjacent bags belongs to Δ. This algorithm is used to show a lower bound of $k + 1$ on $\mathrm{tw}(G)$, setting Δ to be the set of all minimal separators of cardinality at most k and to show an upper bound of k on $\mathrm{tw}(G)$, setting Δ to be some, hopefully rich, set of such minimal separators. Combining this algorithm with new algorithms for exact and heuristic listing of minimal separators, we obtain exact algorithms for treewidth which overwhelmingly outperform previously implemented algorithms.

1 Introduction

Treewidth is a graph parameter introduced by Robertson and Seymour [14] which not only plays an essential role in their graph minor theory [15] but also serves as a powerful tool for designing efficient algorithms for graph problems (see, for example, a survey [5]). Computing treewidth is NP-complete [1] but is fixed-parameter tractable [3,15]. The currently fastest non-parameterized algorithm [7] uses the dynamic programming algorithm due to Bouchitté and Todinca [6] (BT algorithm henceforth) based on the notions of minimal separators and potential maximal cliques.

For practical computation of treewidth, the dominant approaches had been based on the perfect elimination order (PEO) of minimal chordal completions of the given graph [4,8] until recently. The landscape changed when the algorithm implementation challenges PACE2016 [12] and PACE2017 [13] featured exact treewidth tracks, where the winning implementations were not based on PEO but on a new mode, called positive-instance driven (PID), of executing other types of dynamic programming algorithms. In particular, a PID variant of the BT-algorithm due to the present author [17] performed extremely well on large instances, which were completely out of reach for PEO-based exact algorithms.

In this paper, we present an alternative to the PID approach. Roughly speaking, it lies between the original BT algorithm, which lists all minimal separators and all potential maximal cliques before the main dynamic programming iteration, and the PID algorithm which generates relevant instances of those

I. Kotsireas et al. (Eds.): SEA² 2019, LNCS 11544, pp. 219–236, 2019.
https://doi.org/10.1007/978-3-030-34029-2_15

combinatorial objects on the fly during the dynamic programming iteration. In our approach, we start by listing the minimal separators but not the potential maximal cliques. Then, in the dynamic programming iteration, we scan the components separated by those minimal separators in the ascending order of cardinality and decide if each component admits a partial tree-decomposition within the target width. This decision procedure is based on the ideas underlying the BT algorithm but do not explicitly involve potential maximal cliques, either in the algorithm or in the correctness proof.

To be more specific, let us say that a tree-decomposition T *uses separator* S if S is the intersection of some two adjacent bags of T. Our dynamic programming algorithm, which we call MSDP, accepts a set Δ of minimal separators of G in addition to a graph G and a positive integer k and decides if G has a tree-decomposition of width at most k that uses minimal separators only from Δ and is well-formed in the sense defined in Sect. 2.

To use MSDP to decide if $\text{tw}(G) \leq k$, we first need to list all minimal separators of cardinality at most k. Although an efficient algorithm for generating *all* minimal separators of G is known [2], which runs in $O(n^3)$ time on average per each generated instance, no similar result is known for the generation of minimal separators of cardinality at most k. We develop a practically efficient algorithm LIST-EXACT for generating this set. Experiments show that the combination of LIST-EXACT and MSDP clearly outperform the PID algorithm on large instances: many graph instances that cannot be solved by the PID algorithm with 6 h time-out can be solved by the combination of LIST-EXACT and MSDP in less than half an hour. See Sect. 6.

The most time consuming parts in the exact treewidth computation using a decision algorithm for $\text{tw}(G) \leq k$ are the computations for $k = \text{tw}(G) - 1$ and $k = \text{tw}(G)$, as they involve the largest numbers of minimal separators over all values $k \leq \text{tw}(G)$. Although the exact computation for $k = \text{tw}(G) - 1$ is unavoidable, we may use a heuristic algorithm for $k = \text{tw}(G)$ since finding a single tree-decomposition of width $\text{tw}(G)$ suffices. Our heuristic algorithm for listing minimal separators, which we call LIST-HEURISTIC, is designed for this purpose. The combination of LIST-HEURISTIC and MSDP can be used to produce a descending sequence of upper bounds, often ending with $\text{tw}(G)$, in time much smaller than the exact computation for $k = \text{tw}(G)$. Experiments show that our exact treewidth algorithms based on this idea overwhelmingly outperform the PID algorithm on large graphs.

The rest of this paper is organized as follows. In Sect. 2, we give preliminaries of the paper. In Sect. 3, we describe our dynamic programming algorithm MSDP. In Sect. 4, we describe our algorithms LIST-EXACT and LIST-HEURISTIC for listing minimal separators. In Sect. 5, we describe three algorithms for treewidth computation which consist of the component algorithms described in the previous sections. Finally in Sect. 6, we present experimental results.

2 Preliminaries

In this paper, all graphs are simple, that is, without self loops or parallel edges. Let G be a graph. We denote by $V(G)$ the vertex set of G and by $E(G)$ the edge set of G. The subgraph of G induced by $U \subseteq V(G)$ is denoted by $G[U]$. We sometimes use an abbreviation $G \setminus U$ to stand for $G[V(G) \setminus U]$. A vertex set $C \subseteq V(G)$ is a *clique* of G if $G[C]$ is a complete graph. For each $v \in V(G)$, $N_G(v)$ denotes the set of neighbors of v in G: $N_G(v) = \{u \in V(G) \mid \{u, v\} \in E(G)\}$. For $U \subseteq V(G)$, the *open neighborhood of U in G*, denoted by $N_G(U)$, is the set of vertices adjacent to some vertex in U but not belonging to U itself: $N_G(U) = (\bigcup_{v \in U} N_G(v)) \setminus U$. The *closed neighborhood of U in G*, denoted by $N_G[U]$, is defined by $N_G[U] = U \cup N_G(U)$. We also write $N_G[v]$ for $N_G[\{v\}] = N_G(v) \cup \{v\}$. In the above notation, as well as in the notation further introduced below, we will often drop the subscript G when the graph is clear from the context.

We say that vertex set $C \subseteq V(G)$ is *connected in G* if, for every pair of vertices $u, v \in C$, there is a path in $G[C]$ between u and v. It is a *connected component* or simply a *component* of G if it is connected and is inclusion-wise maximal subject to this condition. A vertex set $S \subseteq V(G)$ is a *separator* of G if $G \setminus S$ has more than one components. Note that the empty set is a separator in this definition if G is disconnected. We call each component C of $G \setminus S$ a *component associated with S*; we call it a *full component associated with S* if moreover $N(C) = S$. For $a, b \in V(G)$, a separator S is an *a-b separator* if there is no path between a and b in $G \setminus S$; it is a *minimal a-b separator* if it is an a-b separator and no proper subset of S is an a-b separator. A separator is a *minimal separator* if it is a minimal a-b separator for some $a, b \in V(G)$. Observe that S is a minimal separator if and only if there are at least two full components associated with S. Indeed, if A and B are two full components associated with S, then S is an a-b minimal separator for every pair $a \in A$ and $b \in B$. We generalize these notions by generalizing vertices a and b to disjoint connected vertex sets A and B. A separator S of G is an *A-B separator* if there are two distinct components of $G \setminus S$ each containing A and B; it is a *minimal A-B separator* if those two components are full components associated with S. When one of A and B is a singleton, we use notation such as A-b and a-B separators.

We denote by $\Delta(G)$ the set of all minimal separators of G and by $\Delta_k(G)$ the set of all minimal separators of G with cardinality at most k.

Graph H is *chordal* if every induced cycle of H has length exactly three. H is a *minimal triangulation of G* if it is chordal, $V(H) = V(G)$, $E(G) \subseteq E(H)$, and for every proper subset F of $E(H) \setminus E(G)$, the graph on $V(G)$ with edge set $E(G) \cup F$ is not chordal. A vertex set $\Omega \subseteq V(G)$ is a *potential maximal clique* of G, if Ω is a clique in some minimal triangulation of G.

A *tree-decomposition* of G is a pair $(T, \mathcal{X})$ where T is a tree and $\mathcal{X}$ is a family $\{X_i\}_{i \in V(T)}$ of vertex sets of G such that the following three conditions are satisfied. We call members of $V(T)$ *nodes* of T and each X_i the *bag* at node i.

1. $\bigcup_{i \in V(T)} X_i = V(G)$.
2. For each edge $\{u, v\} \in E(G)$, there is some $i \in V(T)$ such that $u, v \in X_i$.

3. The set of nodes $I_v = \{i \in V(T) \mid v \in X_i\} \subseteq V(T)$ induces a connected subtree of T.

The *width* of this tree-decomposition is $\max_{i \in V(T)} |X_i| - 1$. The *treewidth* of G, denoted by $\text{tw}(G)$ is the minimum width of all tree-decompositions of G. We may assume that the bags X_i and X_j are distinct from each other for $i \neq j$ and, under this assumption, we will regard a tree-decomposition as a tree T in which each node is a bag.

The subproblems arising in the BT dynamic programming algorithm are formulated in [17] as deciding the "feasibility" of connected sets of G with respect to the target treewidth k. We generalize this feasibility notion making it dependent on the set Δ of available separators.

Fix a graph G and a positive integer k. Let Δ be some set of minimal separators of G. We say that a connected set $C \subseteq V(G)$ is *feasible with respect to* Δ if $G[N[C]]$ has a tree-decomposition T of width at most k such that T uses separators only from Δ and has a bag containing $N(C)$.

We also need a slightly stronger version of feasibility. We say that a vertex set U of G is *baggy* if there is no connected set C such that $N(C) = U$ and moreover, for every non-empty $X \subseteq U$, there is a connected set C containing X such that $N(C) = U \setminus X$. We remark that a potential maximal clique is always baggy but the converse does not hold. We say that the tree-decomposition T of G is *well-formed* if every bag of T is baggy and, for every connected vertex set C of G such that C is a component of $G \setminus X$ for some bag X of T, there is a subtree T' of T and a bag Y in T' such that

1. T' is a tree-decomposition of $G[N[C]]$,
2. Y is adjacent to a bag of T, say Z, not in T' with $Y \cap Z = N(C)$.

It is well-known (see a survey [9] on minimal triangulations, for example) that a minimal triangulation H of G naturally gives rise to a tree-decomposition T of H, and hence of G, such that the bags of T are maximal cliques of H and every separator T uses is a minimal separator of G. Moreover, an optimal tree-decomposition of G can be obtained in this manner. Since it is straightforward to verify that such a tree-decomposition is well-formed, we have the following, which allows us to focus on well-formed tree-decompositions in deciding the treewidth.

Proposition 1. *Every graph G has a well-formed tree-decomposition of width* $\text{tw}(G)$.

We say that a connected vertex set C of G is *well-feasible* with respect to Δ if $G[N[C]]$ has a well-formed tree-decomposition T of width at most k such that T has a bag containing $N(C)$ and every separator used by T belongs to Δ.

3 Dynamic Programming for Tree-Decompositions with a Given Set Of available Minimal Separators

In this section, we formulate a dynamic programming algorithm MSDP which, given a graph G, a positive integer k, and a set Δ of minimal separators, decides if G has a tree-decomposition of width at most k that is well-formed and uses separators only from Δ. We fix graph G and positive integer k in the rest of this section. We assume that G is connected. Under this assumption, our problem is equivalent to the well-feasibility of $V(G)$ with respect to Δ.

Following [17], we orient minimal separators as follows. We assume a total order $<$ on $V(G)$ and, for $U \subseteq V(G)$, denote by $\min(U)$ the smallest vertex in U. We say that a connected set C is *inbound* if there is some full component D associated with $N(C)$ such that $\min(D) < \min(C)$; otherwise, it is *outbound*. Observe that if $N(C)$ is not a minimal separator then C is necessarily outbound, since there is no full component $D \neq C$ associated with $N(C)$.

Now we describe our dynamic programming algorithm MSDP. In the description, we fix $\Delta \subseteq \Delta_k(G)$ in addition to G and k. The main iteration of MSDP is given as Algorithm 1.

Algorithm 1. Main iteration of MSDP: decides the feasibility of each inbound connected set with respect to $\Delta \subseteq \Delta_k(G)$

Let L be the list of all inbound connected sets C with $N(C) \in \Delta$
Sort L in the increasing order of cardinality
for all C in L **do**
    if ISFEASIBLE(C) **then** mark C as feasible
end for

The procedure ISFEASIBLE used in the iteration is defined as Algorithm 2, together with an auxiliary procedure ALLFEASIBLE. We remark that the argument C of ISFEASIBLE is not restricted to be inbound: indeed, when the call is made from ALLFEASIBLE, $N(C)$ is not a minimal separator. When ISFEASIBLE(C) holds, however, marking C as feasible is done only if $N(C) \in \Delta$ and C is inbound. In that sense, our dynamic programming table, that maintains the markings, is partial. Consequently, for C such that $N(C)$ is not a minimal separator, it is possible that the call ISFEASIBLE(C) is made more than once.

Theorem 1. *Let $C \subseteq V(G)$ be connected with $|N(C)| \leq k$. If, during the execution of our dynamic programming algorithm, call* ISFEASIBLE*(C) is made and returns* **true** *then C is feasible with respect to Δ. On the other hand, if C is inbound with $N(C) \in \Delta$ and moreover is well-feasible with respect to Δ, then the algorithm marks C as feasible.*

The two statements of this theorem are immediate consequences of the following two lemmas, respectively.

Lemma 1. *Let $C \subseteq V(G)$ be connected with $|N(C)| \leq k$. Suppose that every proper subset C' of C that has been marked feasible is indeed feasible with respect to Δ. If* ISFEASIBLE*(C) is called in this situation and returns* **true**, *then C is feasible with respect to Δ.*

Algorithm 2. Feasibility procedures

```
procedure isFeasible(C)
    if N(C) ∈ Δ and |N[C]| ≤ k + 1 then return true
    for all inbound D with min(C) ∈ D marked feasible do
        if allFeasible(N(C) ∪ N(D), C) then return true
    end for
    return allFeasible(N(C) ∪ {min(C)}, C)
end procedure
procedure allFeasible(S, C)
    if |S| > k + 1 then return false
    for all component D associated with S such that D ⊆ C do
        if N(D) = S then
            if not isFeasible(D) then return false
        else
            assert D is inbound
            if D is not marked as feasible then return false
        end if
    end for
    return true
end procedure
```

Proof. The proof is by induction on the cardinality of C.

If isFeasible(C) returns **true** in the first **if** statement, then C is trivially feasible with respect to Δ. Suppose isFeasible(C) returns **true** within or after the **for** statement. Then, for some superset S of $N(C)$ such that $S \cap C \neq \emptyset$, allFeasible(S, C) returns **true**. From the description of allFeasible we see that, for every component D associated with S such that $D \subset C$, either isFeasible(D) returns **true** or D is marked as feasible. By the induction hypothesis, in the former case, and by the assumption in the Lemma on previous markings, in the latter case, each such D is feasible with respect to Δ. For each such D, let T_D be a tree-decomposition of $G[N[D]]$ of width at most k that has a bag X_D containing $N(D)$. Combining T_D for all D with a new bag S, making each X_D adjacent to S, we obtain a tree-decomposition T of $G[N(C)]$. Note that T indeed satisfies the connectivity condition for tree-decompositions, since $N(D) \subseteq S$ for each D. Since $|S| \leq k+1$ is ensured in the first **if** statement of allFeasible, the width of T is at most k and hence C is feasible as claimed. □

Lemma 2. *Let C be a connected set and suppose that C is well-feasible with respect to Δ. Then, if* isFeasible*(C) is called during the execution of Algorithm 1, it returns* **true**.

Proof. The proof is by induction on the cardinality of C.

Let T be a well-formed tree-decomposition of $G[N[C]]$ that attests the well-feasibility of C with respect to Δ: T has a bag X_C that contains $N(C)$ and every separator used by T belongs to Δ. If T consists of the single bag X_C, then the call isFeasible(C) returns **true** in the first **if** statement.

Next suppose that T has two or more bags. Let $\mathcal{C}$ be the set of components of $G \setminus X_C$ that are contained in C. Let $D \in \mathcal{C}$ be arbitrary. Since T is well-formed, there is a subtree T_D of T that attests the well-feasibility of D with respect to Δ, such that the intersection of the vertices in the bags T_D and the vertices in the bags not in T_D equals $N(D)$. As T uses $N(D)$, $N(D)$ belongs to Δ. Moreover, $N(D)$ is a proper subset of X_C since X_C is baggy. We also claim that D is inbound. To see this, let A be the connected set containing $X_C \setminus N(D)$ such that $N(A) = N(D)$, which exists since X_C is baggy. Then, A is a full component associated with $N(D)$. Since C is inbound, there is a full component B associated with $N(C)$ such that $\min(B) < \min(C)$. Since D is a proper subset of C and $N(D)$ is a minimal separator, $N[D]$ is a proper subset of $N[C]$ and $N(C) \setminus N(D)$ is non-empty. Therefore, $X_C \setminus N(D)$ intersects $N(C)$ and hence A contains A'. We have $\min(A) \leq \min(A') \leq \min(C) \leq \min(D)$ and hence D is inbound. We have shown that each $D \in \mathcal{C}$ is well-feasible with respect to Δ, is a proper subset of C, and is inbound. Therefore, each $D \in \mathcal{C}$ must be marked as feasible since call isFeasible(D) must have been made in the main iteration in Algorithm 1 and, by the induction hypothesis, must have returned **true**.

We are to show that isFeasible(C) returns **true**. Let $v_0 = \min(C)$ and suppose first that $v_0 \notin C \cap X_C$. Then, v_0 must belong to some member, say D_0, of $\mathcal{C}$. Then, at some point in the **for** iteration allFeasible, call allFeasible($S, \mathcal{C}$) is made with $S = N(C) \cup N(D_0)$. Let $\mathcal{C}_1 = \{D \in \mathcal{C} | N(D) \subseteq S\}$. If $\mathcal{C}_1 = \mathcal{C}$ then we are done, since the **else** branch is always taken in the **if** statement within the **for** iteration of allFeasible and hence allFeasible($S, \mathcal{C}$) returns true. So, suppose $\mathcal{C}_2 = \mathcal{C} \setminus \mathcal{C}_1$ is non-empty. Then, $X_C \setminus S$ is non-empty and, since X_C is baggy, there is a connected set D_{full} containing $X_C \setminus S$ such that $N(D_{\text{full}}) = S$. Observe that D_{full} is well-feasible with respect to Δ: the attesting tree-decomposition can be constructed by collecting the well-formed tree-decomposition attesting the feasibility of each $D \in \mathcal{C}_2$ and combining them with a new bag S. Therefore, the call isFeasible(D_{full}) in allFeasible returns **true** and hence allFeasible($S, \mathcal{C}$) returns **true** in this case as well.

The case where $v_0 \in C \cap X_C$ is similar: we consider the call allFeasible($S, \mathcal{C}$) with $S = N(C) \cup \{v_0\}$ and show it returns **true**. $\square$

4 Listing Minimal Separators

We continue to fix G and k and assume G is connected in this section. The goal of this section is to develop a practically efficient algorithm for generating $\Delta_k(G)$.

4.1 A Minimal *a*-*b* Separator Algorithm

Let $a, b \in V(G)$ be distinct and non-adjacent vertices, $A \subseteq V(G)$ a connected set such that $a \in A$, $b \in V(G) \setminus N[A]$, and $N(A) \in \Delta(G)$, and F a subset of $N(A)$. Given these parameters, we define a set $\mathcal{S}_{a,b}(A, F)$ which, informally, collects all minimal a-b separators S such that $|S| \leq k$, $F \subseteq S$, and S does not separate A. We add some more conditions on S for the purpose of efficiency: a minimal a-b separator S belongs to $\mathcal{S}_{a,b}(A, F)$ if and only if the following conditions are satisfied. Let C_a and C_b denote the components of $G \backslash S$ to which a and b belongs, respectively.

1. $|S| \leq k$
2. $F \subseteq S$
3. $A \subseteq C_a$ and $C_b \cap N[A] = \emptyset$
4. $a = \min(C_a)$ and $b = \min(C_b)$
5. $|C_a| \leq |C_b|$

The following proposition is straightforward.

Proposition 2. *We have* $\Delta_k(G) = \bigcup_{a,b} \mathcal{S}_{a,b}(\{a\}, N(a) \cap N(b))$.

We remark that the conditions 4 and 5 above are intended to make the sets $\mathcal{S}_{a,b}(\{a\}, N(a) \cap N(b))$ and $\mathcal{S}_{a',b'}(\{a'\}, N(a') \cap N(b'))$ almost disjoint to each other, for distinct pairs (a, b) and (a', b'): the exception may occur only when $(a', b') = (b, a)$ and the condition 5 is satisfied with $|C_a| = |C_b|$.

The following observations, which provide the base cases in our recursive computation of $\mathcal{S}_{a,b}(A, F)$, are also straightforward.

Proposition 3. *Let C_b denote the component of $G \setminus N[A]$ to which b belongs. In each of the following cases, $\mathcal{S}_{a,b}(A, F)$ is empty.*

1. $|F| > k$
2. $\min(A) \neq a$ *or* $\min(C_b) \neq b$
3. F *is not a subset of* $N(C_b)$
4. $|F| = k$ *and* $N(A) \neq F$
5. $|A| > |C_b|$
6. $|N(A)| > k$ *and* $|A| + (|N(A)| - k) > \min\{|C_b|, (|V(G)| - k)/2\}$

Moreover, if $|F| = k$, $N(A) = F$, $N(C_b) = F$, and $|A| \leq |C_b|$ then $\mathcal{S}_{a,b}(A, F) = \{N(A)\}$.

For the recursive steps, the notion of *close separators* [2,11,16] used in the literature for generating $\Delta(G)$ is essential. Let $A \subseteq V(G)$ be connected and $b \in V(G) \setminus N[A]$. Let B be the component of $G \setminus N(A)$ to which b belongs. Then, $N(B)$ is a minimal A-B separator and, indeed, is the unique minimal A-b separator contained in $N(A)$. We call $N(B)$ the *minimal A-b separator close to A*. We also say that a minimal separator is *close to A* if it is an A-b minimal separator close to A for some $b \in V(G) \setminus N[A]$.

The following recurrence is an adaptation of the one used by Takata [16] for generating $\Delta(G)$.

Lemma 3. *Let a, b, A, and F be as in the definition of $\mathcal{S}_{a,b}(A, F)$. Suppose none of the conditions in Proposition 3 for $\mathcal{S}_{a,b}(A, F)$ to be empty applies and moreover, $N(A) \setminus F \neq \emptyset$. Let $v_0 \in N(A) \setminus F$ be arbitrary. Then, we have*

$$\mathcal{S}_{a,b}(A, F) = \mathcal{S}_{a,b}(A, F \cup \{v_0\}) \cup \mathcal{S}_{a,b}(A', F)$$

where A' is the component associated with the minimal $(A \cup \{v_0\})$-b separator close to $A \cup \{v_0\}$, that contains A.

Our basic algorithm for listing minimal a-b separators recursively evaluates the recurrence in Lemma 3, using the base cases in Propositions 3. The following pruning rule provides a quite effective speed-up.

Lemma 4. *Let a, b, A, and F be as in the definition of $\mathcal{S}_{a,b}(A,F)$ and suppose that $N(A) > k$ and $N(A) \setminus F \neq \emptyset$. Let C_b denote the component of $G \setminus N[A]$ that contains b. Let $\mathcal{P}$ be a set of vertex disjoint paths of G such that each $P \in \mathcal{P}$ has one end in $N(A) \setminus F$ and all other vertices in $V(G) \setminus N[A]$. Let d be the number of vertices in the $(|\mathcal{P}| - (k - |F|))$ shortest paths in $\mathcal{P}$. If $|A| + d > \min\{C_b, (|V(G)| - k)/2\}$, then $\mathcal{S}_{a,b}(A,F)$ is empty.*

Proof. Let C be an arbitrary superset of A such that $F \subseteq N(C)$ and $|N(C)| \leq k$. Then, $N(C)$ contains at most $k - |F|$ vertices that are contained in the paths in $\mathcal{P}$. Since each path in $\mathcal{P}$ has one end in $N(A)$ and all other vertices outside of $N[A]$, at least $(|\mathcal{P}| - (k - |F|))$ paths in $\mathcal{P}$ are entirely contained in C. We have $|C| \geq |A| + d$ and the claim follows. □

To obtain the vertex disjoint paths $\mathcal{P}$, we use a greedy procedure. We first initialize the set $\mathcal{P}$ to consist of a single vertex path containing v for every v in $N(A) \setminus F$. At each iteration step, we scan the paths in $\mathcal{P}$ and extend each path, if possible, by a vertex neither in $N[A]$ nor in any path of $\mathcal{P}$, including the extensions made in this iteration. We stop when no more extensions are possible. As applying this pruning rule involves a non-trivial overhead, we invoke it only when $|N(A) \setminus F|$ is large and the chances of successful pruning appears plausible. More specifically, we have a threshold parameter r_{prune} and invoke the rule only when $|A| + r_{\mathrm{prune}}(|N(A) \setminus F| - (k - |F|)) \geq |V(G) - k|/2$ holds.

4.2 Nibble and Conquer

To generate $\Delta_k(G)$, rather than naively iterating the a-b minimal separator algorithm for all (a, b) pairs, we adopt the following recursive approach. Given a separator X of G, we divide the task of generating $\Delta_k(G)$ into the task of generating those separators in $\Delta_k(G)$ "crossing" X and the task of generating those "local to" each component of $G \setminus X$. We formalize this idea below.

Let $X \subseteq V(G)$ and S a minimal separator of G. We say S *crosses* X if S is a minimal a-b separator for some distinct two vertices a and b in X. For a component C of $G \setminus X$, we say that S *is local to* C *with respect to* X if there is a full component D associated with S such that $N[D] \subseteq X \cup C$.

Lemma 5. *Let $X \subseteq V(G)$ and S a minimal separator of G. Then, either S crosses X or there is some component C of $G \setminus X$ such that S is local to C with respect to X.*

Proof. Suppose that S is not local to any component of $G \setminus X$. Let D be an arbitrary full component associated with S. From the definition of locality, there are two distinct components C_1 and C_2 of $G \setminus X$ such that $N[D]$ intersects both C_1 and C_2. Therefore, at least one vertex of D must belong to X. Since there are at least two full components associated with S, S is a minimal a-b separator for some $a, b \in X$. □

Let U be a vertex set of graph G. The *localization* of G to U, denoted by local(G, U) is the graph obtained from $G[U]$ by filling $N(C)$ into a clique for every component C of $G \setminus U$.

Lemma 6. *Let $X \subseteq V(G)$ and C a component of $G \setminus X$. Suppose S is a minimal separator of G that is local to C with respect to X. Then, either S is a minimal separator of* local$(G, X \cup C)$ *or there is some component $C' \neq C$ of $G \setminus X$ such that $N(C') = S$.*

Proof. If there is a full component C' associated with S that is disjoint from $X \cup C$ then there is nothing to show. So suppose that every full component associated with S intersects $X \cup C$. We show that S is a minimal separator of local$(G, X \cup C)$. Let D_1 and D_2 be two distinct full components associated with S and let $D'_i = D_i \cap (X \cup C)$ for $i = 1, 2$. Then, the edges added to $G[X \cup C]$ to form local$(G, X \cup C)$ ensure that D'_i, $i = 1, 2$, is connected in local$(G, X \cup C)$. Those edges also ensure that the open neighborhood of D'_i in local$(G, X \cup C)$ equals $N_G(D_i) = S$, for $i = 1, 2$. Therefore, S is a minimal separator in local$(G, X \cup C)$. □

Given these two lemmas, it is tempting to adopt the following divide-and-conquer approach for generating $\Delta_k(G)$: find a balanced separator X, list separators in $\Delta_k(G)$ that cross X, and then recurse into local$(G, X \cup C)$ for every component C of $G \setminus X$. Experiments reveal, however, that listing separators crossing X is the dominant part of this computation if X is large: having small X, or, more precisely, X with small number of non-adjacent *a*-*b* pairs, is more important than balancing the sizes of components of $G \setminus X$. Since a small separator is often the neighborhood of a single vertex, we adopt the following nibbling approach listed as Algorithm 3. Here, the "fill in" of vertex v in G is the set of unordered pairs of distinct vertices a and b in $N(v)$ such that $\{a, b\}$ is not an edge of G.

Algorithm 3. Nibble and conquer to generate $\Delta_k(G)$

---

procedure LIST-EXACT(G, k)
    Let v be a vertex of the smallest fill-in in G
    Let $X = N(v)$
    Return the union of the followings:
    – The set of all minimal separators of G, each of cardinality at most k, that cross X
    – $\Delta_k(H)$, where $H = $ local$(G, X \cup C)$, for each component C of $G \setminus X$
end procedure

One of the components associated with $X = N(v)$ in the algorithm description is $\{v\}$ and for this component, the enumeration of $\Delta_k(H)$ for $H =$ local$(G, X \cup \{v\})$ is trivially handled. So, we recurse into $H = $ local$(G, X \cup C)$ for other components C. Typically, there is only one such component, namely $V(G) \setminus N[v]$, the result of nibbling v away. Compared to naively iterating the *a*-*b* separator algorithm for all pairs (a, b), the nibbling approach has the advantage of smaller graph size for *a*-*b* pairs handled deep in recursion. Experiments show that the this advantage indeed leads to substantial speed-ups. See Sect. 6.

4.3 Heuristic Listing of Minimal Separators

To compute upper bounds of treewidth, we may apply our dynamic programming algorithm with an arbitrary (though reasonably rich) subset Δ of $\Delta_k(G)$. In this subsection, we develop a heuristic algorithm LIST-HEURISTIC which incrementally generates larger and larger subsets of $\Delta_k(G)$.

Consider the digraph $\Lambda(G)$ on $\Delta(G)$, in which $(R, S) \in \Delta(G) \times \Delta(G)$ is an edge if and only if S is close to $A \cup \{v\}$ for some full component A associated with R and some $v \in R$. We denote by $\Lambda_k(G)$ the subdigraph of $\Lambda(G)$ induced by the vertex set $\Delta_k(G)$. It is shown in [2] that all members of $\Delta(G)$ are reachable in $\Lambda(G)$ from the set of minimal separators close to singleton vertex sets. Unfortunately, the subdigraph $\Lambda_k(G)$ does not seem to have such a nice reachability property.

In our heuristic generation algorithm LIST-HEURISTIC, we assume that a tree-decomposition T of width $k+1$ or greater is available and construct an initial set which contains, for each bag X of T, a minimal separator S of cardinality at most k if either $S = N(C)$ for some component of $G \setminus X$ or S is a minimal separator of local(G, X). When T is obtained by a greedy heuristic such as MINDEG or MINFILL, this initial set is often sufficient to yield an improved upper bound.

Suppose we have $\Delta \subseteq \Delta_k(G)$ after some iteration steps. Then, in the next iteration step, we expand Δ into Δ' by adding, for each $S \in \Delta$, all successors of S in Λ_k. This is repeated until either MSDP with Δ produces an improved upper bound or Δ has no external successors in Λ_k.

5 Treewidth Algorithms

We have the following three algorithms for computing treewidth: ASCEND, DESCEND, and ALTERNATE.

ASCEND: The combination of MSDP and LIST-EXACT gives an algorithm for deciding if $\mathrm{tw}(G) \leq k$ for given G and k. Algorithm ASCEND calls this procedure for ascending values of k, starting from the trivial lower-bound, the minimum vertex degree of G, and ending with $\mathrm{tw}(G)$. This ascending flow of computation is also employed in the PID algorithm in [17]. As long as we do the optimization based solely on an exact decision algorithm, this seems to be the only reasonable choice, since exactly deciding if $\mathrm{tw}(G) \leq k$ for $k > \mathrm{tw}(G)$ usually requires a cost much larger than that for $k \leq \mathrm{tw}(G)$ and should be avoided. We observe, however, that the only exact decision indispensable in demonstrating the correct value of $\mathrm{tw}(G)$ is that of the question $\mathrm{tw}(G) \leq k$ for $k = \mathrm{tw}(G) - 1$. The answers for this question for $k < \mathrm{tw}(G) - 1$ can be inferred from the answer for $k = \mathrm{tw}(G) - 1$ and, for $k = \mathrm{tw}(G)$, we only need the tree-decomposition of width k which may be found in a heuristic manner. This observation motivates the second algorithm.

DESCEND: In this Algorithm, we start from a greedily computed upper bound and try to lower the upper bound one by one. At each such improvement step, we first try a small list of minimal separators and gradually enlarge the list, using LIST-HEURISTIC, until either an improvement is found or further heuristic expansion of the list is impossible. When the latter happens, then it is the time to start the effort on showing the lower bound. Suppose the current upper bound is $k+1$ and the final heuristic subset of $\Delta_k(G)$ used in the failed improvement step is Δ. To show the lower bound, we first generate $\Delta_k(G)$ using LIST-EXACT and first test if $\Delta = \Delta_k(G)$. If the result is **true**, we can immediately conclude that the upper bound of $k+1$ is tight. Otherwise we apply MSDP with this exact set of minimal separators and, if necessary, repeat the exact decision procedure for smaller values of k.

For all instances in the experiment for which DESCEND successfully computes the exact treewidth, it invokes the exact decision procedure only with $k = \text{tw}(G) - 1$: the decisions for smaller values of k are dispensed with and the decision for $k = \text{tw}(G)$ is replaced by a heuristic upper bound computation. Experiments show that the saving in computation time is often dramatic. See Sect. 6.

ALTERNATE: For hard instances for which we do not have enough resource to compute the exact treewidth, we may want to have a pair of reasonably good upper and lower bounds. As ASCEND provides only a lower bound and DESCEND only an upper bound for those instances, it is natural to combine those algorithms. We may execute each once, allocating a fixed amount of time to each algorithm. In general, however, it is not easy to estimate the right amount of time in advance and avoid wasting the time for unsuccessful improvement effort: the bound on the other side may have been easier to improve.

In ALTERNATE, we alternate between upper and lower bound computations trying to balance the invested time so that the improvements on both sides of the bounds get reasonable chances. This balancing is done through the number of minimal separators involved, in the following manner. Suppose a lower bound of l on $\text{tw}(G)$ is established using the list $\Delta_{l-1}(G)$ during the computation. If l is smaller than the current upper bound, then we turn to improving the upper bound. In this improvement step, the size of the heuristic list of minimal separators is bounded roughly by $2|\Delta_{l-1}(G)|$. When the size of this list is exceeded, then the upper bound computation is suspended and the lower bound of $l+1$ is tried. If this lower bound computation is successful, then the list $\Delta_l(G)$ used is larger than the previous one and the upper bound computation is allowed to use a larger heuristic list. In this manner, the upper and lower bounds alternately approach the exact treewidth as the number of minimal separators allowed increases.

6 Experimental Results

This section describes the results of computational experiments. The computing environment for the experiment is as follows. CPU: Intel Core i7-6700 (4 cores), 3.40 GHz, 8192 KB cache; RAM: 32 GB; Operating system: Ubuntu 18.04.1 LTS; Programming language: Java 1.8; JVM: jre1.8.0_111; The maximum heap space size: 28 GB. The implementation is single threaded, except that multiple threads may be invoked for garbage collection by JVM. The time measured is the elapsed time. To minimize the influence of system processes, the computer is detached from the network and the graphic user interface is disabled.

6.1 Graph Instances

We used two sets of graph instances in our experiments. One consists of random graphs generated by the $G(n, m)$ model: given n and m, a graph is drawn uniformly at random from the set of all labeled graphs of n vertices and m edges. The number of vertices n is in $\{40, 50, 60, 70, 80, 90\}$ and, for each n, the number of edges m is in $\{in \mid 3 \leq i \leq 10\}$. For each pair (n, m), we used a single instance generated, fixing the pseudorandom number seed. Table 1 lists all of these instances. For each graph G, the following characteristic values are listed: $|V(G)|$, $|E(G)|$, $\mathrm{tw}(G)$, $|\Delta_k(G)|$ for $k = \mathrm{tw}(G) - 1$, the number of feasible inbound sets for $k = \mathrm{tw}(G) - 1$, $|\Delta_k(G)|$ for $k = \mathrm{tw}(G)$, and the number of feasible inbound sets for $k = \mathrm{tw}(G)$.

Observe that for each n, the number of relevant minimal separators is the largest at $m = 4n$ or $5n$. As we will see in the performance results, this number, together with the number of vertices and the treewidth, is a good indicator for the hardness of an instance for treewidth computation. Thus, we have a partial explanation of why the computation becomes easier as the graph gets denser after a threshold, despite the increase in the treewidth.

Observe also that the number of feasible inbound sets are, perhaps unexpectedly, quite close to the number of minimal separators. This shows that the advantage of the PID approach over conventional dynamic programming algorithms does not lie in the small number of positive (feasible) subproblem instances compared to all subproblem instances. It lies in that PID approach led to a new method of generating those subproblem instances.

The second set consists of a few hard instances from the DIMACS graph-coloring benchmark set [10]. Table 2 lists all of these instances with their characteristic values.

6.2 Minimal Separator Listing Algorithms

We compared the performances of our three algorithms for exactly listing minimal separators: (1) the basic algorithm based on the recurrence in Lemma 3 with naive iteration on *a*-*b* pairs; (2) with the pruning rule of Lemma 4; (3) with the pruning rule and the nibbling approach in Subsect. 4.2. The threshold r_{prune} is set to 10 in this experiment. In addition, we included our heuristic listing algorithm in the comparison. Table 3 shows the results for random graphs with 70 and 80 vertices, where the task for graph G is to generate $\Delta_k(G)$ for $k = \mathrm{tw}(G) - 1$, the task inevitable in establishing the exact bound on $\mathrm{tw}(G)$. The basic algorithm is able to complete listing within the 1-h timeout for only 4 out of 16 instances, the algorithm with the pruning rule succeeds on 13 instances,

Table 1. Random graph instances used in our experiments: columns FI1 and FI show the number of feasible inbound sets for $k = \text{tw}(G) - 1$ and $k = \text{tw}(G)$, respectively; an empty field means unsuccessful computation within reasonable amount of resource

$\|V\|$	$\|E\|$	tw	$\|\Delta_{tw-1}\|$	FI1	$\|\Delta_{tw}\|$	FI	$\|V\|$	$\|E\|$	tw	$\|\Delta_{tw-1}\|$	FI1	$\|\Delta_{tw}\|$	FI
40	120	14	1021	912	2356	2080	70	210	22	299681	227030	786777	602892
40	160	18	1640	1344	3952	3289	70	280	28	498944	412612	1137482	930417
40	200	20	875	735	1790	1502	70	350	33	590136	464161	1291834	1006981
40	240	22	812	667	1861	1615	70	420	37	472728	386375	991158	797858
40	280	24	631	518	1275	1103	70	490	38	106296	85958	203148	161982
40	320	25	342	296	626	521	70	560	42	150427	122423	293595	235726
40	360	27	292	246	579	474	70	630	45	150442	117591	304528	233298
40	400	28	232	203	469	405	70	700	47	101673	79286	205276	158689
50	150	16	3895	3565	9152	8099	80	240	25	1621664	1424712	4081263	3503941
50	200	20	3772	3377	7878	6956	80	320	31	2284149	1936667	5189162	4362765
50	250	24	5127	4397	10555	8949	80	400	35	988166	827068	2065839	1710869
50	300	26	2788	2299	5345	4417	80	480	39	751344	622050	1481223	1208020
50	350	29	3437	2682	6685	5293	80	560	42	458205	373407	872193	700485
50	400	31	2302	1766	4512	3586	80	640	46	608006	489690	1181883	934389
50	450	32	1163	945	2089	1656	80	720	49	471433	379049	896693	706639
50	500	34	1048	889	1987	1638	80	800	52	438636	355371	846794	671334
60	180	18	11698	9238	26313	22416	90	270	27	7947239	5585295	19521897	13560016
60	240	22	12743	10540	27052	21984	90	360	35	30498292	25231339	71039889	–
60	300	27	27359	20595	56991	41584	90	450	40	24205797	18839873	51925771	–
60	360	30	17956	13829	34793	26356	90	540	45	19877659	15311306	41166209	31119888
60	420	33	17281	13843	33755	26586	90	630	49	11958408	9812327	23932551	–
60	480	34	5862	4789	10320	8248	90	720	52	7106240	5573022	13888202	10716600
60	540	38	11693	9746	22610	18241	90	810	55	4770228	3805194	9237122	7228454
60	600	40	9612	7931	19319	15958	90	900	57	2115790	1721063	3980250	3176283

Table 2. A few hard instances from DIMACS graph coloring benchmark set: FI stands for feasible inbounds; [†]these lower bounds are $|\Delta_{48}(G)|$, $|\Delta_{83}(G)|$, and $|\Delta_{221}(G)|$, for each G

name	$\|V\|$	$\|E\|$	tw	$\|\Delta_{tw-1}\|$	FIs for $k = tw - 1$	$\|\Delta_{tw}\|$	FIs for $k = tw$
myciel6	95	755	35	2639	2583	3938	3848
myciel7	191	2360	66	223317	219381	316296	309735
queen10_10	100	1470	72	2442357	1523527	4199412	2633702
queen11_11	121	1980	87	22351589	13793133	36424473	22429873
DSJC125.1	125	736	[48, 65]	–	–	$\geq^{\dagger}$23302449	–
DSJC125.5	125	3891	108	190816	158478	347012	280655
DSJC250.1	250	3218	[83, 177]	–	–	$\geq^{\dagger}$1248182	–
DSJC250.5	250	15668	[221,230]	–	–	$\geq^{\dagger}$1882525	–

Table 3. Performances of our minimal separator listing algorithms with 1-h timeout

$\|V\|$	$\|E\|$	tw − 1	$\|\Delta_{\text{tw}-1}\|$	time (secs) for generating $\Delta_{\text{tw}-1}$				$\|$heuristic list$\|$ (no. of missings)
				Basic	Pruning	Pruning+Nibbling	Heuristic	
70	210	21	299681	–	876	34	17.4	299681(0)
70	280	27	498944	–	1262	102	33.5	498944(0)
70	350	32	590136	–	1348	117	42.6	590136(0)
70	420	36	472728	–	970	134	27.2	472728(0)
70	490	37	106296	932	212	33	3.84	106296(0)
70	560	41	150427	766	234	47	5.41	150427(0)
70	630	44	150442	551	221	42	5.39	150442(0)
70	700	46	101673	280	128	30	3.45	101673(0)
80	240	24	1621664	–	–	259	68.5	1621664(0)
80	320	30	2284149	–	–	409	99.8	2284149(0)
80	400	34	988166	–	–	231	42.0	988166(0)
80	480	38	751344	–	2498	239	31.2	751344(0)
80	560	41	458205	–	1372	163	18.7	458205(0)
80	640	45	608006	–	1453	201	25.5	608006(0)
80	720	48	471433	–	1106	186	19.5	471433(0)
80	800	51	438636	–	843	166	18.0	438636(0)

and the one with pruning and nibbling succeeds on all of the 16 instances within 7 min each.

The heuristic algorithm is even faster and the list it generates is complete for all the tested instances listed here, as can be confirmed by the last column of the table. We cannot, however, expect this phenomenon to always happen: there do exist instances on which the heuristic algorithm fails to generate the complete list.

6.3 Treewidth Algorithms

We compared the performances of our algorithms with the PID algorithm in [17]. The implementation of the PID algorithm used in this experiment is essentially the same as the one reported in [17] except in the following two points. (1) The implementation used here does not incorporate the safe separator preprocessing, which was an essential part of the implementation described in [17]. This is, however, inessential in our experiments where the intention is to measure the solution ability of the basic "engine" and the instances used do not have useful safe separators. (2) We use a new implementation of the "block sieve" data structure [17], which is the same as the one used in the new algorithms.

Table 4 shows the performances of the PID and our algorithms on the random graph instances.

The advantages of our algorithms are clear. For instances with up to 80 vertices, all of our algorithms computed the exact treewidth, while the PID

Table 4. Performances of PID and our algorithms on random graphs with 6-h timeout: UB and LB are the upper and lower bounds on the treewidth, respectively, computed by the respective method; the time listed is that of the last improvement

$\|V\|$	$\|E\|$	tw	PID			ASCEND			DESCEND			ALTERNATE		
			UB	LB	time (secs)	UB	LB	time (secs)	UB	LB	time (secs)	UB	LB	time (secs)
40	120	14	14	14	0.530	14	14	0.505	14	14	0.325	14	14	0.335
40	160	18	18	18	0.558	18	18	0.992	18	18	0.483	18	18	0.621
40	200	20	20	20	0.287	20	20	0.664	20	20	0.308	20	20	0.617
40	240	22	22	22	0.203	22	22	0.536	22	22	0.232	22	22	0.379
40	280	24	24	24	0.122	24	24	0.429	24	24	0.221	24	24	0.300
40	320	25	25	25	0.068	25	25	0.330	25	25	0.155	25	25	0.279
40	360	27	27	27	0.058	27	27	0.297	27	27	0.166	27	27	0.267
40	400	28	28	28	0.049	28	28	0.276	28	28	0.125	28	28	0.244
50	150	16	16	16	9.24	16	16	1.97	16	16	1.67	16	16	2.34
50	200	20	20	20	6.09	20	20	2.52	20	20	1.48	20	20	3.46
50	250	24	24	24	6.28	24	24	3.54	24	24	1.71	24	24	2.15
50	300	26	26	26	1.26	26	26	2.51	26	26	1.04	26	26	2.12
50	350	29	29	29	0.974	29	29	2.95	29	29	1.38	29	29	2.04
50	400	31	31	31	0.590	31	31	2.27	31	31	0.949	31	31	1.34
50	450	32	32	32	0.224	32	32	1.51	32	32	0.528	32	32	1.09
50	500	34	34	34	0.198	34	34	1.23	34	34	0.615	34	34	1.06
60	180	18	18	18	190	18	18	5.90	18	18	4.08	18	18	4.10
60	240	22	22	22	168	22	22	7.61	22	22	5.19	22	22	5.75
60	300	27	27	27	263	27	27	18.2	27	27	10.5	27	27	16.8
60	360	30	30	30	124	30	30	15.0	30	30	8.26	30	30	14.6
60	420	33	33	33	75.7	33	33	15.2	33	33	7.70	33	33	10.8
60	480	34	34	34	2.96	34	34	7.47	34	34	3.32	34	34	6.87
60	540	38	38	38	7.19	38	38	10.9	38	38	5.44	38	38	10.5
60	600	40	40	40	3.85	40	40	8.84	40	40	4.19	40	40	5.89
70	210	22	–	22	17808	22	22	255	22	22	129	22	22	166
70	280	28	–	28	16959	28	28	749	28	28	802	28	28	1225
70	350	33	–	33	16839	33	33	806	33	33	342	33	33	432
70	420	37	–	37	8173	37	37	733	37	37	656	37	37	870
70	490	38	38	38	2278	38	38	151	38	38	66.0	38	38	204
70	560	42	42	42	3211	42	42	214	42	42	151	42	42	225
70	630	45	45	45	3052	45	45	185	45	45	137	45	45	122
70	700	47	47	47	1178	47	47	131	47	47	56.5	47	47	116
80	240	25	–	21	4812	25	25	3283	25	25	1396	25	25	1548
80	320	31	–	27	7095	31	31	6516	31	31	3167	31	31	3632
80	400	35	–	33	15647	35	35	1573	35	35	682	35	35	1781
80	480	39	–	38	15458	39	39	1377	39	39	691	39	39	1578
80	560	42	–	42	16767	42	42	813	42	42	339	42	42	650
80	640	46	–	46	19764	46	46	1150	46	46	734	46	46	1339
80	720	49	49	49	20958	49	49	934	49	49	362	49	49	1099
80	800	52	52	52	17740	52	52	842	52	52	296	52	52	559
90	270	27	–	21	6210	–	26	8755	27	–	1357	30	26	16522
90	360	35	–	27	12283	–	32	6785	35	–	109	35	32	10972
90	450	40	–	33	19743	–	37	7398	40	–	252	40	37	9870
90	540	45	–	38	14458	–	42	7300	46	–	14896	47	42	10458
90	630	49	–	43	17118	–	47	10983	50	–	17.9	50	47	12893
90	720	52	–	47	16713	–	51	10294	52	–	128	53	51	12485
90	810	55	–	51	18279	–	55	15370	55	55	15284	55	55	19642
90	900	57	–	54	11509	57	57	8684	57	57	11450	57	57	3864

algorithm timed out (6 h) for most of the instances with 80 vertices. Even for many instances with 70 vertices, it only computed the tight lower bound but failed to produce the matching upper bound. For instances with 90 vertices, the lower bounds computed by the PID algorithm is much weaker than those computed by ASCEND.

The results for those large instances that are not exactly solved by our algorithms show the respective advantages and disadvantages of our three algorithms: ASCEND gives only lower bounds, DESCEND only upper bounds, and, though ALTERNATE gives both, the bounds are sometimes weaker than computed by ASCEND or DESCEND. It is worth noting that DESCEND often produces very quickly an upper bound which coincides with the exact treewidth, although it times out in its effort to show the tightness.

The results for the DIMACS instances are listed in Table 5. For the first four instances, the advantage of our algorithms is clear. The result for "myciel7" is particularly dramatic: PID can compute a very weak lower bound of 39 with 6-h time-out, while all of our three algorithms compute the exact treewidth of 66 in 10 or a few more minutes. Although the computation for "queen11_11" times out with all the algorithms, DESCEND discovers the tight upper bound in 3 h and, though not shown in the table, establishes the matching lower bound in 30 more hours. For these two instances, the exact treewidth was previously unknown. We should note, however, that our algorithms are as helpless as PID for the last two large instances, which are random graphs with 250 vertices and are far from being solvable with the current state of the art.

Overall, the experimental results show the effectiveness of our approach that combines MSDP with exact and heuristic listing of minimal separators.

Table 5. Performances of PID and our algorithms on some DIMACS graph coloring instances with 6-h timeout: time (in seconds) is that of the last improvement

name	$\|V\|$	$\|E\|$	tw	PID			ASCEND			DESCEND			ALTERNATE		
				UB	LB	time	UB	LB	time	UB	LB	time	UB	LB	time
myciel6	95	755	35	35	35	617	35	35	3.90	35	35	1.89	35	35	3.09
myciel7	191	2360	66	–	39	10583	66	66	779	66	66	652	66	66	584
queen10_10	100	1470	72	–	70	10792	72	72	12363	72	72	3267	72	72	8224
queen11_11	121	1980	87	–	79	14199	–	81	19525	87	–	9114	88	81	20216
DSJC125.1	125	736	–	–	38	16140	–	45	15259	65	–	0.025	65	37	12758
DSJC125.5	125	3891	108	108	108	1275	108	108	581	108	108	176	108	108	590
DSJC250.1	250	3218	–	–	71	15859	–	72	19840	177	–	0.235	177	51	10598
DSJC250.5	250	15668	–	–	215	20218	–	212	20732	230	–	48.8	231	211	19167

Acknowledgments. A preliminary part of this work was reported and discussed at NWO-JSPS joint seminar "Computation on Networks with a Tree-Structure: From Theory to Practice", held in September 2018. The author thanks Hans Bodlaender and Yota Otachi for organizing this seminar.

References

1. Arnborg, S., Corneil, D.G., Proskurowski, A.: Complexity of finding embeddings in a k-tree. SIAM J. Algebr. Discret. Methods **8**, 277–284 (1987)
2. Berry, A., Bordat, J.-P., Cogis, O.: Generating all the minimal separators of a graph. Int. J. Found. Comput. Sci. **11**(03), 397–403 (2000)
3. Bodlaender, H.L.: A linear-time algorithm for finding tree-decompositions of small treewidth. SIAM J. Comput. **25**(6), 1305–1317 (1996)
4. Bodlaender, H.L., Fomin, F.V., Koster, A.M.C.A., Kratsch, D., Thilikos, D.M.: On exact algorithms for treewidth. ACM Trans. Algorithms **9**(1), 12 (2012)
5. Bodlaender, H.L., Koster, A.M.C.A.: Combinatorial optimization on graphs of bounded treewidth. Comput. J. **51**(3), 255–269 (2008)
6. Bouchitté, V., Todinca, I.: Treewidth and minimum fill-in: grouping the minimal separators. SIAM J. Comput. **31**(1), 212–232 (2001)
7. Fomin, F., Villanger, Y.: Treewidth computation and extremal combinatorics. Combinatorica **32**(3), 289–308 (2012)
8. Gogate, V., Dechter, R.: A complete anytime algorithm for treewidth. In: Proceedings of the 20th Conference on Uncertainty in Artificial Intelligence. AUAI Press (2004)
9. Heggernes, P.: Minimal triangulations of graphs: a survey. Discret. Math. **306**(3), 297–317 (2006)
10. Johnson, D.S., Trick, M.A. (eds.): Cliques, coloring, and satisfiability: second DIMACS implementation challenge. Series in Discrete Mathematics and Theoretical Computer Science, American Mathematical Society, vol. 26. American Mathematical Society (1996)
11. Kloks, T., Kratsch, D.: Listing all minimal separators of a graph. SIAM J. Comput. **27**(3), 605–613 (1998)
12. Dell, H., Husfeldt, T., Jansen, B.M.P., Kaski, P., Komusiewicz, C., Rosamond, F.: The first parameterized algorithms and computational experiments challenge. In: Proceedings of the 11th International Symposium on Parameterized and Exact Computation (IPEC 2016), pp. 30:1–30:9 (2017)
13. Dell, H., Komusiewicz, C., Talmon, N., Weller, M.: The PACE 2017 parameterized algorithms and computational experiments challenge: the second iteration. In: Proceedings of the 12th International Symposium on Parameterized and Exact Computation (IPEC 2017), pp. 30:1–30:12 (2018)
14. Robertson, N., Seymour, P.D.: Graph minors. II. Algorithmic aspects of tree-width. J. Algorithms **7**, 309–322 (1986)
15. Robertson, N., Seymour, P.D.: Graph minors. XX Wagner's conjecture. J. Comb. Theory Series B **92**(2), 325–357 (2004)
16. Takata, K.: Space-optimal, backtracking algorithms to list the minimal vertex separators of a graph. Discret. Appl. Math. **158**(15), 1660–1667 (2010)
17. Tamaki, H.: Positive-instance driven dynamic programming for treewidth. J. Comb. Optim., October 2018. https://doi.org/10.1007/s10878-018-0353-z

Fast Public Transit Routing with Unrestricted Walking Through Hub Labeling

Duc-Minh Phan[1] and Laurent Viennot[2(✉)]

[1] Got It Vietnam, Hanoi, Vietnam
minhp@got-it.ai

[2] Inria – Paris University, Paris, France
laurent.viennot@inria.fr

Abstract. We propose a novel technique for answering routing queries in public transportation networks that allows unrestricted walking. We consider several types of queries: earliest arrival time, Pareto-optimal journeys regarding arrival time, number of transfers and walking time, and profile, i.e. finding all Pareto-optimal journeys regarding travel time and arrival time in a given time interval. Our techniques uses hub labeling to represent unlimited foot transfers and can be adapted to both classical algorithms RAPTOR and CSA. We obtain significant speedup compared to the state-of-the-art approach based on contraction hierarchies. A research report version is deposited on HAL with number hal-02161283.

Keywords: Route planning · Public transportation · Hub labeling

1 Introduction

Despite remarkable progress of route planning algorithms in road networks [4], public transit routing still requires specific algorithms due to its temporal nature. Various efficient methods were proposed such as CSA [12], RAPTOR [11], Transfer Pattern [3,5], PTL [8]. They all consider a graph with two types of edges: the connections that correspond to a vehicle traveling from a stop to the next one, and the transfers that correspond to walking from a stop to another nearby stop. While each connection is scanned only once per query, transfer edges from a stop are considered each time an event is detected at the stop. Efficiency of such techniques thus relies on the sparsity of the transfer graph. Additionally, they all share the requirement that the graph resulting from walking transfers is transitively closed and are generally experimented with a sparse transfer graph by restricting transfers to very short distances only. Allowing unrestricted transfers, that is walking from a stop to any other stop, is indeed out of reach with

D.-M. Phan—This work was mostly performed while this author was at Irif – Paris University, CNRS, France.

L. Viennot—Supported by Irif laboratory from CNRS and Paris University, and ANR project Multimod (ANR-17-CE22-0016).

I. Kotsireas et al. (Eds.): SEA[2] 2019, LNCS 11544, pp. 237–247, 2019.
https://doi.org/10.1007/978-3-030-34029-2_16

these methods although it would allow to find better answers. Indeed, recent work [19] shows the benefit of using unrestricted walking over sparse transfers by measuring that it can reduce travel time by hours in Switzerland and Germany networks.

This paper is devoted to enable unrestricted walking in efficient public transit routing. The motivation for considering unrestricted walking goes beyond the gain of quality in the answers. It is indeed a fundamental step towards computing multimodal journeys as it is considered as a main bottleneck in [7]. Note that bicycle or taxi transfers can be handled similarly as walking transfers with different speed and cost. Techniques developed for unrestricted walking can thus generalize to other modes of transportation.

A first step towards unrestricted walking was made by MCR [7] and UCCH [13] algorithms that both use a contracted version of the full walking graph, inspired by contraction hierarchies [14], for representing the full walking graph which is much bigger. However, this contracted graph is not transitively closed and has to be globally scanned several times during a query. Accelerating such computations with unrestricted walking is still challenging as multi-criteria or profile queries require seconds to be performed on practical networks with these methods.

In static graphs such as road networks, hub labeling [1,10] (also called 2-hop labeling [6]) is a remarkable technique that achieves state-of-art response-time to shortest path queries. It consists in selecting for each node a small set of access nodes called hubs such that any shortest path can be described as a two hop travel through a common hub of the extremities. Intersecting the two lists of hubs of a source and a destination indeed allows to find efficiently the shortest path between them. Such technique was used in PTL [8] on the time expanded graph representation of a network to obtain fast transit routing. A similar approach is followed by TTL [20] which revisits hierarchical hub labeling [2] in the context of public transit networks. However, these approaches still assume sparse transfers. Note that the time expanded graph representation duplicates transfer edges from a stop for all events at that stop, and its size can blow up with dense transfer graphs.

In this work, we propose a new approach for handling unrestricted walking in public transit routing based on a different usage of hub labeling. It basically consists in decomposing walking transfers into two consecutive hops. We use hub labeling in the classical setting of a static graph but in a novel manner compared to distance or shortest path queries: we scan hub lists to propagate reachability information. Interestingly, the technique can easily be adapted to both RAPTOR and CSA based algorithms which are the two main classical approaches with restricted transfers. HLRaptor, our variant of RAPTOR obtains significant speedup compared to MCR. HLCSA, our variant of CSA obtains competitive running times for earliest arrival time and profile queries.

The paper is organized as follows. Section 2 defines public transit networks, describes briefly RAPTOR and CSA algorithms, and introduces hub labeling. Sections 3 and 4 present HLRaptor and HLCSA respectively. We describe in

Sect. 5 public transit data used to evaluate our algorithms. The results of our experiments are presented in Sect. 6.

2 Preliminaries

We define a public transit network with a triple (S, T, R) representing trips of vehicles (buses, trains, etc.): S is the set of *stops* where passengers can enter or disembark from a vehicle, T is the set of trips made by vehicles and are grouped into routes represented by the set R. More precisely, a *trip* t is given by a sequence of stops served by a vehicle and for each stop u in the sequence, an arrival time $\tau_{arr}(t, u)$ of the vehicle at stop u and a departure time $\tau_{dep}(t, u)$ of the vehicle from stop u. A *route* r consists in a set of trips with same stop sequence. This set of trips can be represented by a two-dimensional timetable where each line lists the arrival and departure times $\tau_{arr}(t, u), \tau_{dep}(t, u)$ of a trip t for all stops u in the sequence. Note that the sequence of times listed in a line is non-decreasing. Similarly to RAPTOR authors [11], we assume that no trip of a route can overtake another trip of the same route. In other words, the lines of the timetable can be sorted so that each column is non-decreasing. This property can easily be enforced by splitting the set of trips with same stop sequence into smaller subsets of trips if necessary.

The public transit network is complemented by a weighted footpath graph $G = (V, E, \tau_w)$ with $S \subseteq V$, and $\tau_w(u, v)$ denotes the time needed to walk from a node u to a node v. In the unrestricted walking setting, the graph is not assumed to be transitively closed and we expect that V is much larger than S. Each edge $(u, v) \in E$ typically corresponds to a segment of street than can be traversed by walking. We let $d_G(u, v)$ denote the length of a shortest path in G from u to v, i.e., the minimum walking time from u to v.

We consider several journey computation problems. Given a source stop s and a target stop t, a *journey* from s to t is an alternating sequence of trips and footpaths in the public transit network, which starts with s and ends with t. The goal of public transit planning is to compute journeys from s to t optimizing one or several criteria. Given a departure time τ, an *earliest arrival time* query consists in computing a journey with minimum arrival time at t that departs from s at τ or later. In a *multi-criteria* query, we are additionally interested in the number of transfers and the overall walking time of the journey, and ask for all Pareto-optimal journeys. Recall that a journey is Pareto-optimal if no other journey is better on one criterion and at least as good on all criteria. In a *profile* (or *range*) query, we ask for journeys whose departure time falls within a given time interval while optimizing both departure time and arrival time (where later departure time is considered better). The required answer again consists in all Pareto-optimal journeys within the given time interval.

The RAPTOR algorithm and its variants [7,11] compute journeys starting from a given stop at a given time in rounds, where each round extends partial journeys by one trip. More precisely, each round consists of two phases: the first phase explores each route of the public transit network and extends partial journeys arriving at stops served by a route using the first trip arriving at each stop.

In the second phase, each partial journey arriving at a stop is extended by walking paths from that stop. In the regular RAPTOR version [11], single-edge paths only are considered and the footpath graph is assumed to be transitively closed. In the unrestricted walking setting [7], a multi-source Dijkstra is performed on a contracted version of the footpath graph in order to find all stops whose arrival times can be improved by walking from the stops that were scanned during the first phase.

The Connection Scan Algorithm (CSA) [12] breaks each trip into consecutive *connections*, which represent a vehicle traveling from a stop to the next one in the stop sequence of the trip. All connections are sorted by departure times in a pre-computation step. The algorithm scans all the connections and transfers to update the earliest arrival time at each reachable stops. More precisely, for each connection c in increasing order of departure time, we need to check whether a passenger can travel on c or not: either the trip containing c has been reached earlier, or we can arrive at the departure stop of c before its departure time. Then we update the arrival time at the arrival stop of c if necessary, and scan the footpath transfers from the arrival stop of c. Similarly to RAPTOR, CSA also requires the footpath graph to be transitively closed.

Two-hop labeling [6], or equivalently, *hub labeling* [1,10], for a (weighted, directed) graph G consists in assigning two subsets of nodes $H_-(u)$ and $H_+(v)$ to each node u. Nodes in $H_-(u)$ (resp. $H_+(u)$) are called *in-hubs* (resp. *out-hubs*) and serve as intermediate nodes to reach u (resp. to leave u). The following *two-hop* property is required: for any pair u, v of nodes, there must exist a common hub $h \in H_+(u) \cap H_-(v)$ lying on a shortest path from u to v, i.e., satisfying $d_G(u,h) + d_G(h,v) = d_G(u,v)$. Equivalently, H_+ (resp. H_-) can be seen as a graph with vertex set V and edges (u,v) with weight $d_G(u,v)$ for every pair u, v such that $v \in H_+(u)$ (resp. $u \in H_-(v)$). The two-hop property can then be stated as $H_+ \cdot H_- = G^*$, where G^* denotes the transitive closure of G and $\cdot$ denotes the graph product resulting from the (min, +)-matrix product of adjacency matrices (the weight of an edge (u,w) in $H_+ \cdot H_-$ is $\min_{v \in H_+(u) \cap H_-(w)} d_G(u,v) + d_G(v,w)$). In other words, any shortest path in G corresponds to a two-hop path in $H_+ \cup H_-$. The interest for such representation comes from the fact that it is possible to compute very small hub sets (less than 100 nodes on average) in large road networks and footpath graphs [9], and thereby obtain the fastest known practical oracles for computing distances and shortest paths in such networks [4].

3 HLRaptor: RAPTOR with Two-Hop Transfers

Using a hub labeling H_-, H_+ of the footpath graph G, we propose the following modification of RAPTOR that we call HLRaptor. We replace the second phase of a round by two sub-phases: in the first sub-phase we scan every stop u for which arrival time τ_u was improved in the regular first phase of the round, and update arrival time at its out-hubs $h \in H_+(u)$ to $\min\{\tau_h, \tau_u + \tau_w(u,h)\}$. In the second sub-phase, we scan every hub h whose arrival time was improved in the first sub-phase and update arrival time at nodes v such that $h \in H_-(v)$ to $\min\{\tau_v, \tau_h + \tau_w(h,v)\}$.

The correctness of HLRaptor comes from the two-hop property of the hub labeling that ensures $H_+ \cdot H_- = G^*$. Our two sub-phases using H_+ and H_- are thus equivalent to the second phase of the regular RAPTOR algorithm using the transitive closure G^* of G. However, its performance depends on the out-degrees of H_+ and H_- rather than that of G^*.

Target Pruning Optimization. The lists $H_-^{-1}(h) = \{v \mid h \in H_-(v)\}$ and $H_+(u)$ can be pre-computed for all $u, h \in V$. Additionally, these lists can be sorted according to walking time from u (resp. h) in non-decreasing order. This enables a target pruning optimization where we stop scanning a list as soon as the arrival time computed for a node in the list exceeds the best arrival time known at the target.

HLprRaptor: Profile Queries with HLRaptor. We can follow the same approach as [19] to compute all Pareto-optimal journeys with respect to departure time and arrival time in a given interval of time. The difference is that we use HLRaptor instead of MCR. The idea is to use HLRaptor to compute the best arrival time τ_a when starting at a given time τ. Then we use a reverse version of HLRaptor (or simply a reversed version of the transit data) to compute the last departure time τ_d such that arrival at τ_a is still possible. We then repeat this procedure for departure time $\tau_d + \varepsilon$ for sufficiently small ε (we simply use $\varepsilon = 1$ s, which is the time unit in our datasets). We iterate this until all Pareto-optimal journeys in the given time interval have been found.

HLmcRaptor: HLRaptor with Multiple Criteria. To deal with more criteria than arrival time and number of transfers, we can keep multiple non-dominating labels for each stop u in round k in a bag structure similarly to McRAPTOR [11]. For each route r with a stop improved in the previous round, we scan the first trip departing after any improved arrival time at a stop u of the route and update bags accordingly at the stops served by the trip after u. In the second phase of the round, each newly inserted label is first propagated along out-hubs links and then newly inserted labels at hubs are propagated along in-hubs links similarly. We can adapt local and target pruning as in McRAPTOR. We can also adapt our target pruning optimization specific to HLRaptor to stop scanning hub lists as soon as the propagated label is dominated by the destination bag.

4 HLCSA: Connection Scan with Two-Hop Transfers

Given a hub labeling H_+, H_- of the walking graph G, we propose the following modification of CSA. For an earliest arrival time query from s to t, we first scan out-hubs $H_+(s)$ and update arrival time to them by walking from s. Similarly to CSA, we then scan connections by non-decreasing departure time. When considering a connection c, we first scan the in-hubs $H_-(u)$ of its departure stop

u and update the arrival time at u through walking from a hub. The connection can be boarded if the trip has been marked as boarded or if the arrival time at u plus the minimum transfer time at u is no later than the departure time of c. In that case, we update the arrival time at the arrival stop v of c and scan its out-hubs $H_+(v)$ to update their arrival times through walking from v. Finally, we scan the in-hubs $H_-(t)$ of the destination t and update the arrival time at t by walking from any of them.

The correctness of the algorithm comes again from the two-hop property of hubs. For any possible transfer from a connection c to another connection d in a journey, c must be considered before d. Let h denote a common hub for the arrival stop u of c and the departure stop v of d such that $d_G(u,v) = d_G(u,h) + d_G(h,v)$ according to the two-hop property. After c is considered, arrival time at h is thus no more than $\tau + d_G(u,h)$, where τ is the arrival time of c at u and $d_G(u,h)$ is the walking time from u to h. When d is then considered, arrival time to v is updated to $\tau + d_G(u,h) + d_G(h,v) = \tau + d_G(u,v)$ as if a transfer from u to v had been considered. A similar reasoning applies for a journey starting with a walk from s or ending with a walk to t. HLCSA thus behaves as in a regular CSA execution where all transitive transfers in G^* would be considered.

Optimization. In addition to all CSA classical optimizations, we can again sort out-hub lists in non-decreasing order of walking time, and apply target pruning similarly as in HLRaptor. In addition, we scan the in-hub list of the departure stop of a connection when the trip is not marked as boarded. Again, this list can be sorted by non-decreasing walking time and we stop scanning the list as soon as the walking time from the hub exceeds the estimated travel time to the departure stop (local pruning).

HLprCSA: Profile Queries with HLCSA. Similarly to the original extension of CSA to solve the profile problem [12], we store for each stop a bag containing Pareto-optimal pairs of departure time at stop with associated arrival time at destination. We also store such information for hubs. We also consider connections in non-increasing order of departure time. When scanning a connection c, we use the bags of the out-hubs of its arrival stop to obtain the best arrival time through walking after c. If the arrival time of the trip of c is improved, we then update the bags of the in-hubs of the departure stop of c for that arrival time with departure time corresponding to walking from the hub for boarding right in time the connection. We also scan in-hubs of the destination at the beginning of the procedure and out-hubs of the source at end in order to take care of walking from source and to destination. The correctness of the modification follows similar lines as for HLCSA.

5 Public Transit Data

To evaluate the algorithms, we use datasets from three locations: London, Paris, and Switzerland. The dataset for London was obtained from Transport for London [18]. The dataset for Paris was obtained from Open Data RATP [16]. And the dataset for Switzerland was provided by Wagner and Zündorf [19][1]. The extracted dates are 2015-11-06 for London and 2018-03-30 for Paris.

The public transit data of Paris already has transfers between stops, we simply need to make the transfer graph transitively closed for appropriate use with RAPTOR and CSA. However, the dataset of London does not have transfers, thus we have created transfers by linking any pair of stops separated by 75 m of walk one from another. This threshold was chosen to obtain a transitively closed transfer graph with similar size as in previous works. The graph obtained by transitive closure of restricted transfers is called *transfer graph* in the sequel.

The footpath graphs for London and Paris were extracted from Geofabrik's data [15], which is itself extracted from OpenStreetMap's data [17]. We call *walking graph* the union of this unrestricted footpath graph and the transfer graph. The method to merge a stop of the public transit network into the walking graph is the following. For each stop p, we find the closest node v in the walking graph. If the distance between p and v is less than 5 m, we identify p and v, connecting p with the in- and out-neighbors of v using the same weights. Otherwise, we find the 5 closest nodes of p in the walking graph, and connect p with those at distance 100 m at most. If there are no nodes in the walking graph within the radius of 100 m from p, then p is isolated. Walking times are computed according to a walking speed of 4 km/h. Table 1 provides statistics concerning the datasets. The columns stops and transfers provide the number of nodes and edges in the transitively closed restricted walking graph, while the last two columns give the numbers of vertices and edges in the unrestricted walking graphs.

Table 1. Dataset statistics

	Routes	Trips	Events	Stops	Transfers	Vertices	Edges
London	1622	122593	4695285	19746	46566	281167	840880
Paris	1973	78757	1915253	23519	362291	533470	1666386
Switzerland	13930	369744	4740869	25427	38265	604230	1882551

We computed hub labelings of the walking graphs using the sampling-based algorithm by Delling et al. [9] (1–2 h of pre-computation per graph). Table 2 provides statistics on the degrees of transfer graphs vs. in-hubs and out-hubs graphs: $\delta^+(Tr)$ and $\Delta^+(Tr)$ designate the average and maximum out-degree resp. of the transfer graph Tr, $\delta^+(H_+)$ and $\Delta^+(H_+)$ designate the average and maximum out-degree resp. of the out-hub graph H_+, $\delta^-(H_-)$ and $\Delta^-(H_-)$

[1] https://i11www.iti.kit.edu/PublicTransitData/Switzerland/.

designate the average and maximum in-degree resp. of the in-hub graph H_-. We let $|H_+|$ and $|H_-|$ designate the number of edges in H_+ and H_- resp. while $|V(H)|$ designates the number of hubs (including stops). We note that the size of hub lists is comparable to the number of events and their storage do not increase too much space requirements.

Table 2. Transfers, out-hubs and in-hubs degrees.

	$\delta^+(Tr)$	$\Delta^+(Tr)$	$\delta^+(H_+)$	$\Delta^+(H_+)$	$\delta^-(H_-)$	$\Delta^-(H_-)$	$\|V(H)\|$	$\|H_+\|$	$\|H_-\|$
London	2.36	20	70	150	71	142	65059	1393759	1395024
Paris	15.4	205	118	196	118	210	60519	2770336	2798315
Switzerland	1.5	26	78	229	79	230	117793	2005312	2005312

We also prepared two sets of roughly 1000 queries for each dataset. In the first one, source and destinations are selected independently uniformly at random among all stops similarly to experiments in [7,11,12]. In the second one, we select sources and destinations similarly to [19]: one hundred sources are selected uniformly at random. For each source, we order the destinations by increasing walking distance and select a random one uniformly among those with rank in $[2^i, 2^{i+1}]$ for $i = 2 \ldots 14$. For Switzerland, we use exactly the same pairs as in [19] where sources are selected with probability proportional the number of trips serving them. In both sets, we additionally selected uniformly at random a departure time in $[0, 24 \times 3600]$ for each source-destination pair. We will reference the two sets of queries as "uniform" and "rank" respectively. Note that most of the uniform queries (those in the uniform set) correspond to high rank pairs (2^{12} or higher) while the rank set of queries has a strong bias towards low rank pairs.

The datasets are made publicly available[2].

6 Experiments

Our algorithms were implemented in C++ and compiled with GCC version 7.2.0 (with flag `-O3`). Experiments were conducted on one core of a dual 10-core Intel Xeon E5-2670-v2 with with 25 MiB of L3 cache and 64 GiB of DDR3-1866 RAM. The code is made available[3].

Table 3 presents the average running times in milliseconds of HLRaptor and HLCSA variants on the three datasets. We indicate for each algorithm which criteria are optimized: arrival time (Arr.), number of transfers (Nb. tr.), overall walking time (Walk), and whether the query spans a range of departure times (Range).

In the restricted walking setting, our algorithms are equivalent to the corresponding Raptor or CSA based version. On the London instance with restricted walking and uniform queries, we obtain similar results as Raptor [11] for earliest arrival, multi-criteria and 2 h range queries: 5.1 ms vs. 7.3 ms, 87.3 ms vs.

[2] https://files.inria.fr/gang/graphs/public_transport/.

[3] https://github.com/lviennot/hl-csa-raptor.

Table 3. Average running times of HLRaptor and HLCSA.

					London				Paris			
					Restricted		Unrestr.		Restricted		Unrestr.	
Algorithm	Range	Arr.	Nb. tr.	Walk	Unif.	Rank	Unif.	Rank	Unif.	Rank	Unif.	Rank
HLRaptor	○	●	●	○	5.1	1.9	26.4	8.7	3.0	1.3	19.5	5.5
HLCSA	○	●	○	○	2.2	1.1	33.1	16.8	1.0	0.5	13.8	6.5
HLmcRaptor	○	●	●	●	87.3	33.0	417	140	60.0	25.9	248	85.4
HLprRaptor (2h)	●	●	●	○	76.5	31.1	685	237	53.0	23.1	652	205
HLprCSA (2h)	●	●	○	○	47.1	28.4	1012	539	60.9	35.4	628	330
HLprRaptor (24h)	●	●	●	○	805	322	7522	2524	567	262	7441	2511
HLprCSA (24h)	●	●	○	○	312	217	11644	8453	404	298	9523	7902

					Switzerland			
					Restricted		Unrestr.	
Algorithm	Range	Arr.	Nb. tr.	Walk	Unif.	Rank	Unif.	Rank
HLRaptor	○	●	●	○	13.4	4.0	59.4	7.6
HLCSA	○	●	○	○	6.6	2.9	54.2	19.4
HLmcRaptor	○	●	●	●	150	62	854	229
HLprRaptor (2h)	●	●	●	○	47.7	16.4	402	83.9
HLprCSA (2h)	●	●	○	○	51.9	32.1	563	240
HLprRaptor (24h)	●	●	●	○	293	111	3461	751
HLprCSA (24h)	●	●	○	○	128	96	4173	3076

107 ms, and 76.5 ms vs. 87 ms, respectively (we compare times reported in Table 3 to times reported in [11]). Our running times are 15–30% faster, probably due to the use of more recent hardware. We also obtain similar results as CSA [12] for earliest arrival and 24 h range queries: 2.2 ms vs. 1.2 ms and 312 vs. 107 ms. Our running times are 2–3 times slower than those reported in [12], probably due to less optimized code.

In the unrestricted walk setting, our algorithms are significantly faster than previous works. On the London instance with uniform queries and unrestricted walking, HLmcRaptor is 3.4 times faster than times reported for MCR in [7] (417 ms vs. 1438 ms) and HLRaptor is 1.7 times faster than the MR-∞ variant of MCR (26.4 ms vs. 44.4 ms). On the Switzerland instance with ranked based queries and unrestricted walking, HLprRaptor computes profile queries roughly 7 times faster than the profile variant of MCR proposed in [19]: 751 ms vs. 5.5 s approximately. Most uniform queries have high rank, and HLprRaptor obtains their profiles in roughly 3.5 s compared to 20 s approximately as reported in [19].

Interestingly, our hub-labeling-based versions of CSA obtain rather good performances with respect to Raptor based versions in the unrestricted walk setting: they are nearly as fast or even faster on uniform queries, and at most 2–3 times slower on rank queries. (Note that on low rank queries, Raptor-based solutions benefit from target pruning.)

Table 4. Average/median gain of unrestricted walking on travel time compared to restricted walking.

	Unif.	Unif. 6 h–20 h	Rank	Rank 6 h-20 h
London	12%/5.8%	6.9%/2.9%	24%/13%	16%/5.0%
Paris	22%/15%	15%/13%	31%/21%	22%/17%
Switzerland	47%/46%	37%/39%	47%/47%	35%/37%

Gain of Unrestricted Walking. We confirm the results of [19] showing the benefit of considering unrestricted walking. Table 4 presents the percentage of time gained by using a journey with unrestricted walking compared to the travel time with restricted walking. The average gain ranges from 12% to 47% on uniform queries depending on the dataset. City networks (especially London) seem to benefit less from unrestricted walking than the train network of Switzerland. As observed in [19], the gain is less important during daytime that is queries with departure time in the range 6 h-20 h here. We observe a higher gain on low rank queries. The median gain ranges from 13% to 47% for them. More precisely, the gain is at least 13% on half of the low rank queries for London, 21% for Paris and 47% for Switzerland.

7 Conclusion

We have demonstrated the efficiency of using a two-hop representation of unrestricted walk transfers in conjunction with CSA and RAPTOR algorithm. This shows that is possible to enable unrestricted walking in practical public transit routing engines and opens new perspectives for allowing complex multimodal scenarios. We also want to further investigate how this approach could be integrated in other efficient public transit routing algorithms.

References

1. Abraham, I., Delling, D., Goldberg, A.V., Werneck, R.F.: A hub-based labeling algorithm for shortest paths in road networks. In: Pardalos, P.M., Rebennack, S. (eds.) SEA 2011. LNCS, vol. 6630, pp. 230–241. Springer, Heidelberg (2011). https://doi.org/10.1007/978-3-642-20662-7_20
2. Abraham, I., Delling, D., Goldberg, A.V., Werneck, R.F.: Hierarchical hub labelings for shortest paths. In: Epstein, L., Ferragina, P. (eds.) ESA 2012. LNCS, vol. 7501, pp. 24–35. Springer, Heidelberg (2012). https://doi.org/10.1007/978-3-642-33090-2_4
3. Bast, H., et al.: Fast routing in very large public transportation networks using transfer patterns. In: de Berg, M., Meyer, U. (eds.) ESA 2010. LNCS, vol. 6346, pp. 290–301. Springer, Heidelberg (2010). https://doi.org/10.1007/978-3-642-15775-2_25

4. Bast, H., et al.: Route planning in transportation networks. In: Kliemann, L., Sanders, P. (eds.) Algorithm Engineering. LNCS, vol. 9220, pp. 19–80. Springer, Cham (2016). https://doi.org/10.1007/978-3-319-49487-6_2
5. Bast, H., Hertel, M., Storandt, S.: Scalable transfer patterns. In: Proceedings of the Eighteenth Workshop on Algorithm Engineering and Experiments, ALENEX 2016, pp. 15–29. SIAM (2016). https://doi.org/10.1137/1.9781611974317.2
6. Cohen, E., Halperin, E., Kaplan, H., Zwick, U.: Reachability and distance queries via 2-hop labels. SIAM J. Comput. **32**(5), 1338–1355 (2003). https://doi.org/10.1137/S0097539702403098
7. Delling, D., Dibbelt, J., Pajor, T., Wagner, D., Werneck, R.F.: Computing multimodal journeys in practice. In: Bonifaci, V., Demetrescu, C., Marchetti-Spaccamela, A. (eds.) SEA 2013. LNCS, vol. 7933, pp. 260–271. Springer, Heidelberg (2013). https://doi.org/10.1007/978-3-642-38527-8_24
8. Delling, D., Dibbelt, J., Pajor, T., Werneck, R.F.: Public transit labeling. In: Bampis, E. (ed.) SEA 2015. LNCS, vol. 9125, pp. 273–285. Springer, Cham (2015). https://doi.org/10.1007/978-3-319-20086-6_21
9. Delling, D., Goldberg, A.V., Pajor, T., Werneck, R.F.: Robust distance queries on massive networks. In: Schulz, A.S., Wagner, D. (eds.) ESA 2014. LNCS, vol. 8737, pp. 321–333. Springer, Heidelberg (2014). https://doi.org/10.1007/978-3-662-44777-2_27
10. Delling, D., Goldberg, A.V., Werneck, R.F.: Hub labeling (2-hop labeling). In: Kao, M.Y. (ed.) Encyclopedia of Algorithms, pp. 932–938. Springer, Heidelberg (2016). https://doi.org/10.1007/978-1-4939-2864-4_580
11. Delling, D., Pajor, T., Werneck, R.F.: Round-based public transit routing. Transp. Sci. **49**(3), 591–604 (2015). https://doi.org/10.1287/trsc.2014.0534
12. Dibbelt, J., Pajor, T., Strasser, B., Wagner, D.: Connection scan algorithm. ACM J. Exp. Algorithmics **23**, 17 (2018). https://dl.acm.org/citation.cfm?id=3274661
13. Dibbelt, J., Pajor, T., Wagner, D.: User-constrained multimodal route planning. ACM J. Exp. Algorithmics **19**(1), (2014). https://doi.org/10.1145/2699886
14. Geisberger, R., Sanders, P., Schultes, D., Vetter, C.: Exact routing in large road networks using contraction hierarchies. Transp. Sci. **46**(3), 388–404 (2012). https://doi.org/10.1287/trsc.1110.0401
15. Geofabrik. http://download.geofabrik.de/
16. Open Data RATP. https://data.ratp.fr/
17. OpenStreetMap. https://www.openstreetmap.org/
18. Transport for London Unified API. https://api.tfl.gov.uk/
19. Wagner, D., Zündorf, T.: Public transit routing with unrestricted walking. In: 17th Workshop on Algorithmic Approaches for Transportation Modelling, Optimization, and Systems, ATMOS OASICS, vol. 59, pp. 7:1–7:14 (2017). https://doi.org/10.4230/OASIcs.ATMOS.2017.7
20. Wang, S., Lin, W., Yang, Y., Xiao, X., Zhou, S.: Efficient route planning on public transportation networks: A labelling approach. In: Proceedings of the 2015 ACM SIGMOD International Conference on Management of Data, SIGMOD 2015, pp. 967–982. ACM, New York (2015). https://doi.org/10.1145/2723372.2749456

Effective Heuristics for Matchings in Hypergraphs

Fanny Dufossé[1], Kamer Kaya[2], Ioannis Panagiotas[3(✉)], and Bora Uçar[3,4]

[1] Inria Grenoble Rhône-Alpes, Montbonnot-Saint-Martin, France
fanny.dufosse@inria.fr

[2] Sabanci University, Istanbul, Turkey
kaya@sabanciuniv.edu

[3] ENS Lyon, Lyon, France
{ioannis.panagiotas,bora.ucar}@ens-lyon.fr

[4] CNRS and LIP (UMR5668, CNRS - ENS Lyon - UCB Lyon 1 - INRIA), Lyon, France

Abstract. The problem of finding a maximum cardinality matching in a d-partite, d-uniform hypergraph is an important problem in combinatorial optimization and has been theoretically analyzed. We first generalize some graph matching heuristics for this problem. We then propose a novel heuristic based on tensor scaling to extend the matching via judicious hyperedge selections. Experiments on random, synthetic and real-life hypergraphs show that this new heuristic is highly practical and superior to the others on finding a matching with large cardinality.

Keywords: d-dimensional matching · Tensor scaling · Matching in hypergraphs · Karp-Sipser heuristic

1 Introduction

A hypergraph $H = (V, E)$ consists of a finite set V and a collection E of subsets of V. The set V is called vertices, and the collection E is called hyperedges. A hypergraph is called *d-partite* and *d-uniform*, if $V = \bigcup_{i=1}^{d} V_i$ with disjoint V_is and every hyperedge contains a single vertex from each V_i. A matching in a hypergraph is a set of disjoint hyperedges. In this paper, we investigate effective heuristics for finding large matchings in d-partite, d-uniform hypergraphs.

Finding a maximum cardinality matching in a d-partite, d-uniform hypergraph for $d \geq 3$ is NP-Complete; the 3-partite case is called the Max-3-DM problem [27]. This problem has been studied mostly in the context of local search algorithms [24], and the best known algorithm is due to Cygan [8] who provides $((d+1+\varepsilon)/3)$-approximation, building on previous work [9,21]. It is NP-Hard to approximate Max-3-DM within 98/97 [3]. Similar bounds exist for higher dimensions: the hardness of approximation for $d = 4, 5$ and 6 are shown to be $54/53 - \varepsilon$, $30/29 - \varepsilon$, and $23/22 - \varepsilon$, respectively [22].

Finding a maximum cardinality matching in a d-partite, d-uniform hypergraph is a special case of the d-Set-Packing problem [23]. It has been shown that d-Set-Packing is hard to approximate within a factor of $\mathcal{O}(d/\log d)$ [23].

I. Kotsireas et al. (Eds.): SEA² 2019, LNCS 11544, pp. 248–264, 2019.
https://doi.org/10.1007/978-3-030-34029-2_17

The maximum/perfect set packing problem has many applications, including combinatorial auctions [20] and personnel scheduling [18]. Such a matching can also be used in the coarsening phase of multilevel hypergraph partitioning tools [6], when the input is d-uniform and d-partite, such as those used in modeling and partitioning tensors [28].

Our contributions in this paper are as follows. We propose five heuristics. The first two are adaptations of the well-known greedy [15] and Karp-Sipser [26] heuristics widely used for finding matchings in bipartite graphs. We use Greedy^g and Karp-Sipser^g to refer to these heuristics, and Greedy and Karp-Sipser for the proposed generalizations. Greedy traverses the hyperedge list in random order and adds a hyperedge to the matching whenever possible. Karp-Sipser introduces certain rules to Greedy to improve the cardinality. The third heuristic is inspired by a recent scaling-based approach proposed for the maximum cardinality matching problem on graphs [11–13]. The fourth heuristic is a modification of the third one that allows for faster execution time. The last one finds a matching for a reduced, $(d-1)$-dimensional problem and exploits it for d dimensions. This heuristic uses an exact algorithm for the bipartite matching problem. We perform experiments to evaluate the performance of these heuristics on special classes of random hypergraphs and real-life data.

Another way to tackle the problem at hand is to create the line graph G for a given hypergraph H. The line graph is created by identifying each hyperedge of H with a vertex in G, and connecting two vertices of G with an edge, iff the corresponding hyperedges share a common vertex in H. Then, successful heuristics for computing large independent sets in graphs, e.g., KaMIS [29], can be used to compute large matchings in hypergraphs. This approach, although promising quality-wise, can be impractical. This is so, since building G from H requires quadratic run time and storage (in terms of the number of hyperedges) in the worst case. While this can be acceptable in some instances, in others it is not. We have such instances in the experiments.

The rest of the paper is organized as follows. Section 2 introduces the notation and summarizes the background material. The proposed heuristics are summarized in Sect. 3. Section 4 presents the experimental results and Sect. 5 concludes the paper.

2 Background and Notation

Tensors are multidimensional arrays, generalizing matrices to higher orders. Let $\mathbf{T}$ be a d-dimensional tensor whose size is $n_1 \times \cdots \times n_d$. The elements of $\mathbf{T}$ are shown with $\mathbf{T}_{i_1,\ldots,i_d}$, where $i_j \in \{1,\ldots,n_j\}$. A marginal is a $(d-1)$-dimensional section of a d-dimensional tensor, obtained by fixing one of its indices. A d-dimensional tensor where the entries in each of its marginals sum to one is called d-stochastic. In a d-stochastic tensor, all dimensions necessarily have the same size n. A d-stochastic tensor where each marginal contains exactly one nonzero entry (equal to one) is called a permutation tensor. Franklin and Lorenz [16] show that if a nonnegative tensor $\mathbf{T}$ has the same zero-pattern as a d-stochastic tensor $\mathbf{B}$, then one can find a set of d vectors $x^{(1)}, x^{(2)}, \ldots, x^{(d)}$ such that $\mathbf{T}_{i_1,\ldots,i_d} \cdot$

$x^{(1)}_{i_1} \cdots \cdot x^{(d)}_{i_d} = \mathbf{B}_{i_1,\ldots,i_d}$ for all $i_1, \ldots, i_d \in \{1, \ldots, n\}$. In fact, a multidimensional version of the algorithm for doubly-stochastic scaling (of matrices) by Sinkhorn and Knopp [32] can be used to obtain these d vectors.

A d-partite, d-uniform hypergraph $H = (V_1 \cup \cdots \cup V_d, E)$ can be naturally represented by a d-dimensional tensor. This is done by associating each tensor dimension with a vertex class. Let $|V_i| = n_i$, and the tensor $\mathbf{T} \in \{0,1\}^{n_1 \times \cdots \times n_d}$ have a nonzero element $\mathbf{T}_{v_1,\ldots,v_d}$ iff $(v_1, \ldots, v_d)$ is a hyperedge of H. Then, $\mathbf{T}$ is called the adjacency tensor of H. In H, if a vertex is a member of only a single hyperedge we call it a degree-1 vertex. Similarly, if it is a member of only two, we call it a degree-2 vertex.

In the *k-out random hypergraph model*, given V, each vertex $u \in V$ selects k hyperedges from the set $E_u = \{e : e \subseteq V, u \in e\}$ in a uniformly random fashion and the union of these hyperedges forms E. We are interested in the d-partite, d-uniform case, and hence $E_u = \{e : |e \cap V_i| = 1 \text{ for } 1 \leq i \leq d, u \in e\}$. This model generalizes random k-out bipartite graphs [34]. Devlin and Kahn [10] investigate fractional matchings in these hypergraphs, and mention in passing that k should be exponential in d to ensure that a perfect matching exists.

3 Heuristics for Maximum d-Dimensional Matching

A matching which cannot be extended with more hyperedges is called *maximal*. In this work, we propose heuristics for finding maximal matchings on d-partite, d-uniform hypergraphs. For such hypergraphs, any maximal matching is a d-approximate matching. The bound is tight and can be verified for $d = 3$. Let H be a 3-partite $3 \times 3 \times 3$ hypergraph with the following hyperedges $e_1 = (1,1,1), e_2 = (2,2,2), e_3 = (3,3,3)$ and $e_4 = (1,2,3)$. The maximum matching is $\{e_1, e_2, e_3\}$, and the hyperedge $\{e_4\}$ alone forms a maximal matching.

3.1 A Greedy Heuristic for Max-d-DM

There exist two variants of Greedy^g in the literature. The first one [15] randomly visits the edges and adds the current edge to the matching if both endpoints are available. The second one randomly visits the vertices [30], and matches the vertex with the first available neighbor, if any, visited in a random order. We adapt the first variant to our problem and call it Greedy. It traverses the hyperedges in random order and adds the current hyperedge to the matching whenever possible. Since any maximal matching is possible as its output, Greedy is a d-approximation heuristic. It obtains matchings of varying quality, depending upon the order in which the hyperedges are processed.

3.2 Karp-Sipser for Max-d-DM

A widely-used heuristic to obtain large matchings in graphs is Karp-Sipser^g [26]. On a graph, the heuristic iteratively adds a random edge to the matching and reduces the graph by removing its endpoints, as well as their edges. Whenever possible, Karp-Sipser^g does not apply a random selection but reduces the problem size, i.e., number of vertices in the graph by one via two rules:

- At any time during the heuristic, if a degree-1 vertex appears it is matched with its only neighbor.
- If a degree-2 vertex u appears with neighbors $\{v, w\}$ and no degree-1 vertex exists, u (and its edges) is removed from the current graph, and v and w are merged to create a new vertex vw whose set of neighbors is the union of those of v and w (except u). A maximum cardinality matching for the reduced graph can be extended to obtain one for the current graph by matching u with either v or w depending on vw's match.

Both rules are optimal in the sense that they do not reduce the cardinality of a maximum matching in the current graph they are applied on. We now propose an adaptation of Karp-Sipser^g for d-partite, d-uniform hypergraphs, and call this heuristic Karp-Sipser. Similar to the original one, Karp-Sipser iteratively adds a random hyperedge to the matching, remove its d endpoints and their hyperedges. However, the random selection is not applied whenever hyperedges defined by the lemmas below appear.

Lemma 1. *During the heuristic, if a hyperedge e with at least $d-1$ degree-1 endpoints appears, there exists a maximum cardinality matching in the current hypergraph containing e.*

Proof. Let H' be the current hypergraph at hand and $e = (u_1, \ldots, u_d)$ be a hyperedge in H' whose first $d-1$ endpoints are degree-1 vertices. Let M' be a maximum cardinality matching in H'. If $e \in M'$, we are done. Otherwise, assume that u_d is the endpoint matched by a hyperedge $e' \in M'$ (note that if u_d is not matched M' can be extended with e). Since u_i, $1 \le i < d$, are not matched in M', $M' \setminus \{e'\} \cup \{e\}$ defines a valid maximum cardinality matching for H'. □

We note that it is not possible to relax the condition by using a hyperedge e with less than $d-1$ endpoints of degree-1; in M', two of e's higher degree endpoints could potentially be matched with two different hyperedges, in which case the substitution as done in the proof of the lemma is not valid.

Lemma 2. *During the heuristic, let $e = (u_1, \ldots, u_d)$ and $e' = (u'_1, \ldots, u'_d)$ be two hyperedges sharing at least one endpoint where for an index set $\mathcal{I} \subset \{1, \ldots, d\}$ of cardinality $d-1$, the vertices u_i, u'_i for all $i \in \mathcal{I}$ only touch e and/or e'. That is for each $i \in \mathcal{I}$, either $u_i = u'_i$ is a degree-2 vertex or $u_i \neq u'_i$ and they are both degree-1 vertices. For $j \notin \mathcal{I}, u_j$ and u'_j are arbitrary vertices. Then, in the current hypergraph, there exists a maximum cardinality matching having either e or e'.*

Proof. Let H' be the current hypergraph at hand and $j \notin \mathcal{I}$ be the remaining index. Let M' be a maximum cardinality matching in H'. If either $e \in M'$ or $e' \in M'$, we are done. Otherwise, u_i and u'_i for all $i \in \mathcal{I}$ are unmatched by M'. Furthermore, since M' is maximal, u_j must be matched by M' (otherwise, M' can be extended by e). Let $e'' \in M'$ be the hyperedge matching u_j. Then $M' \setminus \{e''\} \cup \{e\}$ defines a valid maximum cardinality matching for H'. □

Whenever such hyperedges appear, the rules below are applied in the same order:

- **Rule-1**: At any time during the heuristic, if a hyperedge e with at least $d-1$ degree-1 endpoints appears, instead of a random hyperedge, e is added to the matching and removed from the hypergraph.
- **Rule-2**: Otherwise, if two hyperedges e and e' as defined in Lemma 2 appear, they are removed from the current hypergraph with the endpoints u_i, u'_i for all $i \in \mathcal{I}$. Then, we consider u_j and u'_j. If u_j and u'_j are distinct, they are merged to create a new vertex $u_j u'_j$, whose hyperedge list is defined as the union of u_j's and u'_j's hyperedge lists. If u_j and u'_j are identical, we only rename u_j as $u_j u'_j$. After obtaining a maximal matching on the reduced hypergraph, depending on the hyperedge matching $u_j u'_j$, either e or e' can be used to obtain a larger matching in the current hypergraph.

When Rule-2 is applied, the two hyperedges identified in Lemma 2 are removed from the hypergraph, and only the hyperedges containing u_j and/or u'_j have an update in their vertex list. Since the original hypergraph is d-partite and d-uniform, that update is just a renaming of a vertex in the concerned hyperedges (hence the resulting hypergraph is also d-partite and d-uniform).

Although the two rules usually lead to improved results in comparison to Greedy, Karp-Sipser still adheres to the d-approximation bound of maximal matchings. To see this, we use the toy example given as a worst-case for Greedy. For the example given at the beginning of Sect. 3, Karp-Sipser generates a maximum cardinality matching by applying Rule-1. However, if $e_5 = (2, 1, 3)$ and $e_6 = (3, 1, 3)$ are added to the example, neither of the two rules can be applied. As before, if e_4 is randomly selected, it forms a maximal matching.

3.3 Karp-Sipser-scaling for Max-d-DM

Karp-Sipser can be modified for better decisions in case neither of the two rules applies. In this variant, called Karp-Sipser-scaling, instead of a random selection, we first scale the adjacency tensor of H and obtain an approximate d-stochastic tensor $\mathbf{T}$. We then augment the matching by adding the hyperedge which corresponds to the largest value in $\mathbf{T}$. The modified heuristic is summarized in Algorithm 1.

Our inspiration comes from the $d = 2$ case and more specifically from the relation between scaling and matching. It is known due to Birkhoff [4] that the polytope of $n \times n$ doubly stochastic matrices is the convex hull of the $n \times n$ permutation matrices. A nonnegative matrix $\mathbf{A}$ where all entries participate in some perfect matching can be scaled with two positive diagonal matrices $\mathbf{R}$ and $\mathbf{C}$ such that $\mathbf{RAC}$ is doubly stochastic. Otherwise, provided that $\mathbf{A}$ has a perfect matching, it can still be scaled to a doubly stochastic form asymptotically. In this case, the entries not participating in any perfect matching tend to zero in the scaled matrix. This fact is exploited to design randomized approximation algorithms for the maximum cardinality matching problem in graphs [12,13]. By scaling the adjacency matrix in a preprocess and choosing edges with a

Algorithm 1: Karp-Sipser-scaling

Input: A d-partite, d-uniform $n_1 \times \cdots \times n_d$ hypergraph $H = (V, E)$
Output: A maximal matching M of H

$M \leftarrow \emptyset$ ▶ Initially M is empty
$S \leftarrow \emptyset$ ▶ Stack for the merges for Rule-2
while H is not empty **do**
  Remove the isolated vertices from H
  if $\exists e = (u_1, \dots, u_d)$ as in Rule-1 **then**
    $M \leftarrow M \cup \{e\}$ ▶ Add e to the matching
    Apply the reduction for Rule-1 on H
  else if $\exists e = (u_1, \dots, u_d), e' = (u'_1, \dots, u'_d)$ and $\mathcal{I}$ as in Rule-2 **then**
    Let j be the part index where $j \notin \mathcal{I}$
    Apply the reduction for Rule-2 on H by introducing the vertex $u_j u'_j$
    $E' = \{(v_1, \dots, u_j u'_j, \dots, v_d) : \text{for all } (v_1, \dots, u_j, \dots, v_d) \in E\}$
      ▶ memorize the hyperedges of u_j
    S.push(e, e', $u_j u'_j$, E') ▶ Store the current merge
  else
    $\mathbf{T} \leftarrow \textsc{Scale}(adj(H))$ ▶ Scale the adjacency tensor of H
    $e \leftarrow \arg\max_{(u_1, \dots, u_d)} (\mathbf{T}_{u_1, \dots, u_d})$ ▶ Find the maximum entry in $\mathbf{T}$
    $M \leftarrow M \cup \{e\}$ ▶ Add e to the matching
    Remove all hyperedges of $u_1, \dots, u_d$ from E
    $V \leftarrow V \setminus \{u_1, \dots, u_d\}$
while $S \neq \emptyset$ **do**
  $(e, e', u_j u'_j, E') \leftarrow S$.pop() ▶ Get the most recent merge
  if $u_j u'_j$ is not matched by M **then**
    $M \leftarrow M \cup \{e\}$
  else
    Let $e'' \in M$ be the hyperedge matching $u_j u'_j$
    if $e'' \in E'$ **then**
      Replace $u_j u'_j$ in e'' with u'_j
      $M \leftarrow M \cup \{e'\}$
    else
      Replace $u_j u'_j$ in e'' with u_j
      $M \leftarrow M \cup \{e\}$

probability corresponding to the scaled value of the associated matrix entry, the edges which are not included in a perfect matching become less likely to be chosen. The current algorithm differs from these approaches by selecting a single hyperedge at each step and applying scaling again before the next selection.

For $d \geq 3$, there is no equivalent of Birkhoff's theorem as demonstrated by the following lemma.

Lemma 3. *For $d \geq 3$, there exist extreme points in the set of d-stochastic tensors which are not permutations tensors.*

The proof can be found in the accompanying technical report [14], where we give extreme points that are not permutation tensors. Due to the lemma above, we do not have the theoretical foundation to imply that hyperedges corresponding to the large entries in the scaled tensor must necessarily participate in a perfect matching. Nonetheless, the entries not in any perfect matching tend to become zero (not guaranteed for all though). For the worst case example of Karp-Sipser described above, the scaling indeed helps the entries corresponding to e_4, e_5 and e_6 to become zero. Additionally even if the heuristic selects an entry in the non-zero pattern of an extreme point without a perfect matching, we do not necessarily reduce our chances of obtaining a good matching (see the discussion following the proof of Lemma 3 in the technical report).

On a d-partite, d-uniform hypergraph $H = (V, E)$, the Sinkhorn-Knopp algorithm used for scaling operates in iterations, each of which requires $\mathcal{O}(|E| \times d)$ time. In practice, only a few iterations (e.g., 10–20) can be performed. Since we can match at most $|V|/d$ hyperedges, the overall run time of scaling is $\mathcal{O}(|V| \times |E|)$. A straightforward implementation of the second rule can require quadratic time in the case of a large number of repetitive merges with a given vertex. In practice, more of a linear time behavior should be observed.

3.4 Hypergraph Matching via Pseudo Scaling

In Algorithm 1, applying scaling at every step can be very costly. Here we propose an alternative idea inspired by the specifics of the Sinkhorn-Knopp algorithm to reduce the overall cost.

The Sinkhorn-Knopp algorithm scales a d-dimensional tensor $\mathbf{T}$ in a series of iterations by updating the set of vectors $x^{(1)}, \ldots, x^{(d)}$ where initially all values in all vectors are equal to 1. During an iteration, the coefficient vector $x^{(j)}$ for a given dimension j is updated by using

$$x_{i_j}^{(j)} = \frac{x_{i_j}^{(j)}}{\sum_{i_1,\ldots,i_{j-1},i_{j+1},\ldots i_d} \left(\mathbf{T}_{i_1,\ldots,i_j,\ldots,i_d} \prod_{k=1}^{d} x_{i_k}^{(k)} \right)}, \text{for all } i_j \in \{1, \ldots, n_j\}\,. \quad (1)$$

These updates are done in a sequential order and for simplicity we assume that they happen in the dimension order: $1, \ldots, d$. Each vector entry $x_{i_j}^{(j)}$ corresponds to a vertex in the hypergraph. Let λ_{i_j} denote the degree of the vertex i_j from jth part. For the first iteration of (1), each $x_{i_1}^{(1)}$ is set to $\frac{1}{\lambda_{i_1}}$ since all values in the vectors are one. The pseudo scaling approach applies d parallel executions of updates (1) and sets each $x_{i_j}^{(j)} = \frac{1}{\lambda_{i_j}}$ for all $j \in \{1, \ldots, d\}$ and $i_j \in \{1, \ldots, n_j\}$. That is, each vertex gets a value inversely proportional to its degree. This avoids 10–20 iterations of Sinkhorn-Knopp and the $O(|E|)$ cost for each. However, as the name of the approach implies, this scaling is not exact.

With this approach each hyperedge $\{i_1, \ldots, i_d\}$ is associated with a value $\frac{1}{\prod_{j=1}^{d} \lambda_{i_j}}$. The selection procedure is the same as that of Algorithm 1, i.e., the hyperedge with the maximum value is added to the matching set. We refer to this algorithm as Karp-Sipser-mindegree, as it selects a hyperedge based on a function of the degrees of the vertices. With a straightforward implementation, finding this hyperedge takes $O(|E|)$ time.

3.5 Reduction to Bipartite Graph Matching

A perfect matching in a d-partite, d-uniform hypergraph H remains perfect when projected on a $(d-1)$-partite, $(d-1)$-uniform hypergraph obtained by removing one of H's vertex parts. Matchability in $(d-1)$-partite sub-hypergraphs has been investigated [1] to provide an equivalent of Hall's Theorem for d-partite hypergraphs. These observations lead us to propose a heuristic called Bipartite-reduction. This heuristic tackles the d-partite, d-uniform case by recursively asking for matchings in $(d-1)$-partite, $(d-1)$-uniform hypergraphs and so on, until $d = 2$.

Let us start with $d = 3$. Let $G = (V_G, E_G)$ be the bipartite graph where the vertex set $V_G = V_1 \cup V_2$ is obtained by deleting V_3 from a 3-partite, 3-regular hypergraph $H = (V, E)$. The edge $(u, v) \in E_G$ iff there exists a hyperedge $(u, v, z) \in E$. One can assign weights to the edges during this step, e.g., $w(u, v) = |\{z : (u, v, z) \in E\}|$. A maximum weighted matching algorithm can be used to obtain a matching M_G on G. A second bipartite graph $G' = (V_{G'}, E_{G'})$ is then created with $V_{G'} = (V_1 \times V_2) \cup V_3$ and $E_{G'} = \{(uv, z) : (u, v) \in M_G, (u, v, z) \in H\}$. Any matching in G' corresponds a valid matching in H. Furthermore, if the weight function defined above is used with a maximum weighted matching algorithm, the number of edges surviving for G' is maximized.

For d-dimensional matching, a similar process is followed. First, an ordering $i_1, i_2, \ldots, i_d$ of the dimensions is defined. At the jth bipartite reduction step, the matching is found between the dimension cluster $i_1 i_2 \cdots i_j$ and the dimension i_{j+1} by similarly solving a bipartite matching problem, where the edge $(u_1 \cdots u_j, v)$ exists in the bipartite graph iff the vertices $u_1, \ldots, u_j$ were matched previously, and there exists a hyperedge $(u_1, \ldots, u_j, v, z_{j+2}, \ldots, z_d)$ in H. Unlike the previous heuristics, Bipartite-reduction does not have any approximation guarantee, as stated in the following lemma (the proof is in the accompanying technical report [14], where we describe a family of hypergraphs for which Bipartite-reduction yields $\frac{5}{n}$-approximation for $n \geq 5$).

Lemma 4. *If algorithms for the maximum cardinality or the maximum weighted matching (with the suggested edge weights) problems are used, then Bipartite-reduction has a worst-case approximation ratio of $\Omega(n)$.*

3.6 Performing Local Search

A local search heuristic is proposed by Hurkens and Schrijver [24]. It starts from a feasible maximal matching M and performs a series of swaps until it is no longer possible. In a swap, k hyperedges of M are replaced with at least $k+1$ new hyperedges from $E \setminus M$ so that the cardinality of M increases by at least one. These k hyperedges from M can be replaced with at most $d \times k$ new edges. Hence, these hyperedges can be found by a polynomial algorithm enumerating all the possibilities. The approximation guarantee improves with higher k values. Local search algorithms are limited in practice due to their high time complexity. The algorithm might have to examine all $\binom{|M|}{k}$ subsets of M to find a feasible swap at each step. The algorithm by Cygan [8] which achieves $\left(\frac{d+1+\varepsilon}{3}\right)$-approximation is based on a different swap scheme but is also not suited for large hypergraphs.

4 Experiments

To understand the relative performance of the proposed heuristics, we conducted a wide variety of experiments with both synthetic and real-life data. The experiments were performed on a computer equipped with intel Core i7-7600 CPU and 16 GB RAM. For $d = 3$, we also implemented a local search heuristic [24], called Local-Search, which replaces one hyperedge from a maximal matching M with at least two hyperedges from $E \setminus M$ to increase the cardinality of M. We did not consider local search schemes for higher dimensions or with better approximation ratios as they are computationally too expensive. For each hypergraph, we perform ten runs of Greedy and Karp-Sipser with different random decisions and take the maximum cardinality obtained. Since Karp-Sipser-scaling or Karp-Sipser-mindegree do not pick hyperedges randomly, we run them only once. We perform 20 steps of the scaling procedure in Karp-Sipser-scaling. We refer to quality of a matching M in a hypergraph H as the ratio of M's cardinality to the size of the smallest vertex partition of H.

4.1 Experiments on Random Hypergraphs

We perform experiments on two classes of d-partite, d-uniform random hypergraphs where each part has n vertices. The first class contains random k-out hypergraphs, and the second one contains sparse random hypergraphs.

Random k-out, d-partite, d-uniform hypergraphs
Here, we consider random k-out, d-partite, d-uniform hypergraphs described in Sect. 2. Hence (ignoring the duplicate ones), these hypergraphs have around $d \times k \times n$ hyperedges. These k-out, d-partite, d-uniform hypergraphs have been recently analyzed in the matching context by Devlin and Kahn [10]. They state in passing that k should be exponential in d for a perfect matching to exist with high probability. The bipartite graph variant of the same problem has been extensively studied in the literature [17,25,34]; a perfect matching almost always exists in a random 2-out bipartite graph [34].

We first investigate the existence of perfect matchings in random k-out, d-partite, d-uniform hypergraphs. For this purpose, we implemented the linear program of d-dimensional matching in CPLEX and found the maximum cardinality of a matching in these hypergraphs with $k \in \{d^{d-3}, d^{d-2}, d^{d-1}\}$ for $d \in \{2, \ldots, 5\}$ and $n \in \{10, 20, 30, 50\}$. For each (k, d, n) triple, we created five hypergraphs and computed their maximum cardinality matchings. For $k = d^{d-3}$, we encountered several hypergraphs with no perfect matching, especially for $d = 3$. The hypergraphs with $k = d^{d-2}$ were also lacking a perfect matching for $d = 2$. However, all the hypergraphs we created with $k = d^{d-1}$ had at least one. Based on these results, we experimentally confirm Devlin and Kahn's statement. We also conjecture that d^{d-1}-out random hypergraphs have perfect matchings almost surely. The average maximum matching cardinalities we obtained in this experiment are given in Table 1. In this table, we do not have results for $k = d^{d-3}$ for $d = 2$, and the cases marked with $*$ were not solved within 24 h.

Table 1. The average maximum matching cardinalities of five random instances over n on random k-out, d-partite, d-uniform hypergraphs for different k, d, and n. No runs for $k = d^{d-3}$ and $d = 2$; the problems marked with $*$ were not solved within 24 h.

	d	k				d	k		
		d^{d-3}	d^{d-2}	d^{d-1}			d^{d-3}	d^{d-2}	d^{d-1}
$n = 10$	2	-	0.87	1.00	$n = 30$	2	-	0.84	1.00
	3	0.80	1.00	1.00		3	0.88	1.00	1.00
	4	1.00	1.00	1.00		4	0.99	1.00	1.00
	5	1.00	1.00	1.00		5	$*$	1.00	1.00
$n = 20$	2	-	0.88	1.00	$n = 50$	2	-	0.87	1.00
	3	0.85	1.00	1.00		3	0.84	1.00	1.00
	4	1.00	1.00	1.00		4	$*$	1.00	1.00
	5	1.00	1.00	1.00		5	$*$	$*$	$*$

We now compare the performance of the proposed heuristics on random k-out, d-partite, d-uniform hypergraphs with $d \in \{3, 9\}$ and $n \in \{1000, 10000\}$. We tested with k being equal to powers of two, and $k \leq d \log d$. The results are summarized in Fig. 1. For each (k, d, n) triplet, we created ten random instances and present the average performance of the heuristics on them. Further figures for $d = 6$ can be found in the accompanying technical report [14]. The x-axis in each figure denotes k, and the y-axis reports the matching cardinality over n. As seen, Karp-Sipser-scaling and Karp-Sipser-mindegree have the best performance, comfortably beating the other alternatives. For $d = 3$ Karp-Sipser-scaling dominates Karp-Sipser-mindegree, but when $d > 3$ we see that Karp-Sipser-mindegree has the best performance. Karp-Sipser performs better than Greedy. However, their performances get closer as d increases. This is due to the fact that the conditions for Rule-1 and Rule-2 hold less often for larger d. Bipartite-reduction has worse performance than the others, and the gap in the performance grows as d increases. This happens, since at each step, we impose more and more conditions on the edges involved and there is no chance to recover from bad decisions.

Sparse random d-partite, d-uniform hypergraphs
Here, we consider a random d-partite, d-uniform hypergraph H_i that has $i \times n$ random hyperedges. The parameters used for these experiments are $i \in \{1, 3, 5, 7\}$, $n \in \{4000, 8000\}$, and $d \in \{3, 9\}$. Each H_i is created by choosing the vertices of a hyperedge uniformly at random for each dimension. We do not allow duplicate hyperedges. Another random hypergraph H_{i+M} is then obtained by planting a perfect matching to H_i. We again generate ten random instances for each parameter setting. We do not present results for Bipartite-reduction as it was always worse than the others, as before. The average quality of different heuristics on these instances is shown in Fig. 2 (the accompanying report [14] contains further results). The experiments confirm that Karp-Sipser performs consistently better than Greedy. Furthermore, Karp-Sipser-scaling performs significantly better than Karp-Sipser. Karp-Sipser-scaling works even better than the local search

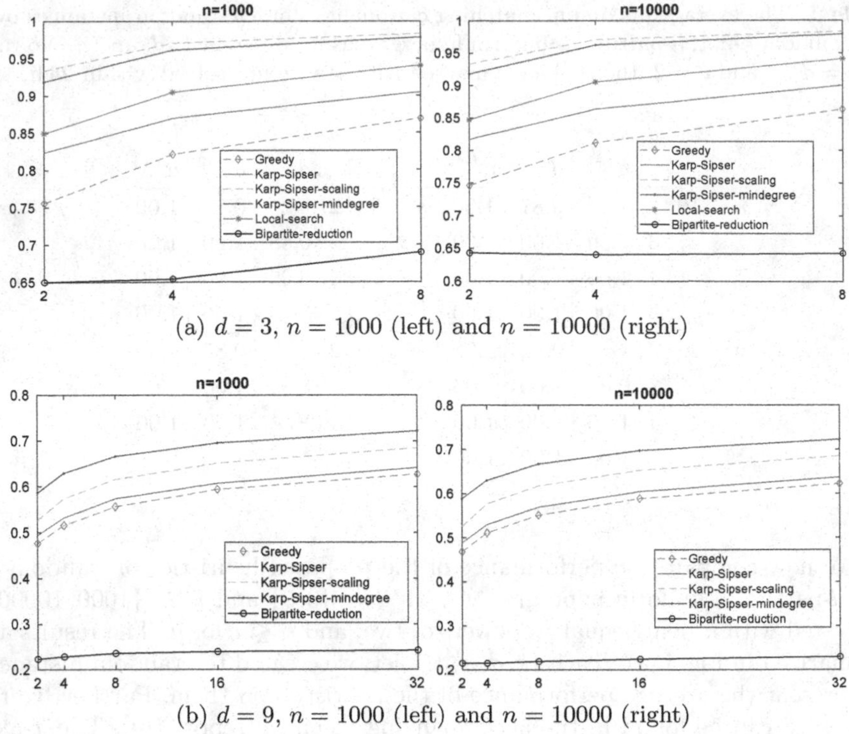

(a) $d = 3$, $n = 1000$ (left) and $n = 10000$ (right)

(b) $d = 9$, $n = 1000$ (left) and $n = 10000$ (right)

Fig. 1. The performance of the heuristics on k-out, d-partite, d-uniform hypergraphs with n vertices at each part. The y-axis is the ratio of matching cardinality to n whereas the x-axis is k. No Local-Search for $d = 9$.

	H_i: Random Hypergraph										H_{i+M}: Random Hypergraph with Perfect Matching									
	Greedy		Local Search		Karp-Sipser		Karp-Sipser-scaling		Karp-Sipser-minDegree		Greedy		Local Search		Karp-Sipser		Karp-Sipser-scaling		Karp-Sipser-minDegree	
i	4000	8000	4000	8000	4000	8000	4000	8000	4000	8000	4000	8000	4000	8000	4000	8000	4000	8000	4000	8000
1	0.43	0.42	0.47	0.47	0.49	0.48	0.49	0.48	0.49	0.48	0.75	0.75	0.93	0.93	1.00	1.00	1.00	1.00	1.00	1.00
3	0.63	0.63	0.71	0.71	0.73	0.72	0.76	0.76	0.78	0.77	0.72	0.71	0.82	0.81	0.81	0.81	0.99	0.99	0.92	0.92
5	0.70	0.70	0.80	0.80	0.78	0.78	0.86	0.86	0.88	0.88	0.75	0.74	0.84	0.84	0.82	0.82	0.94	0.94	0.92	0.92
7	0.75	0.75	0.84	0.84	0.81	0.81	0.94	0.94	0.93	0.93	0.77	0.77	0.87	0.87	0.83	0.83	0.96	0.96	0.94	0.94

(a) $d = 3$, without (left) and with (right) the planted matching

	H_i: Random Hypergraph								H_{i+M}: Random Hypergraph with Perfect Matching							
	Greedy		Karp-Sipser		Karp-Sipser-scaling		Karp-Sipser-minDegree		Greedy		Karp-Sipser		Karp-Sipser-scaling		Karp-Sipser-minDegree	
i	4000	8000	4000	8000	4000	8000	4000	8000	4000	8000	4000	8000	4000	8000	4000	8000
1	0.25	0.24	0.27	0.27	0.27	0.27	0.30	0.30	0.56	0.55	0.80	0.79	1.00	1.00	1.00	1.00
3	0.34	0.33	0.36	0.36	0.36	0.36	0.43	0.43	0.40	0.40	0.44	0.44	1.00	1.00	0.99	1.00
5	0.38	0.37	0.40	0.40	0.41	0.41	0.48	0.48	0.41	0.40	0.43	0.43	1.00	1.00	0.99	0.99
7	0.40	0.40	0.42	0.42	0.44	0.44	0.51	0.51	0.42	0.42	0.44	0.44	1.00	1.00	0.97	0.96

(b) $d = 9$, without (left) and with (right) the planted matching

Fig. 2. Performance comparisons on d-partite, d-uniform hypergraphs with $n = \{4000, 8000\}$. H_i contains $i \times n$ random hyperedges, and H_{i+M} contains an additional perfect matching.

heuristic, and it is the only heuristic that is capable of finding planted perfect matchings for a significant number of the runs. In particular when $d > 3$, it finds a perfect matching on H_{i+M}s in all cases shown. For $d = 3$, it finds a perfect matching only when $i = 1$ and attains a near perfect matching when $i = 3$. Interestingly Karp-Sipser-mindegree outperforms Karp-Sipser-scaling on H_is but is dominated on H_{i+M}s, where it is the second best performing heuristic.

4.2 Evaluating Algorithmic Choices

Here, we evaluate the use of scaling and the importance of Rule-1 and Rule-2.

Scaling vs No-Scaling

To evaluate and emphasize the contribution of scaling better, we compare the performance of the heuristics on a particular family of d-partite, d-uniform hypergraphs where their bipartite counterparts have been used before as challenging instances for the original Karp-Sipser^g heuristic [12].

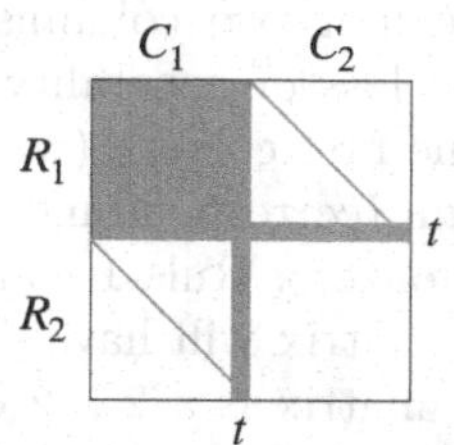

Fig. 3. $\mathbf{A}_{KS}$: A challenging instance for Karp-Sipser^g.

Table 2. Performance of the proposed heuristics on 3-partite, 3-uniform hypergraphs corresponding to $\mathbf{T}_{KS}$ with $n = 300$ vertices in each part.

t	Greedy	Local Search	Karp-Sipser	Karp-Sipser-scaling	Karp-Sipser-minDegree
2	0.53	0.99	0.53	1.00	1.00
4	0.53	0.99	0.53	1.00	1.00
8	0.54	0.99	0.55	1.00	1.00
16	0.55	0.99	0.56	1.00	1.00
32	0.59	0.99	0.59	1.00	1.00

Let $\mathbf{A}_{KS}$ be an $n \times n$ matrix. Let R_1 and C_1 be $\mathbf{A}_{KS}$'s first $n/2$ rows and columns, respectively, and R_2 and C_2 be the remaining $n/2$ rows and columns, respectively. Let the block $R_1 \times C_1$ be full and the block $R_2 \times C_2$ be empty. A perfect bipartite graph matching is hidden inside the blocks $R_1 \times C_2$ and $R_2 \times C_1$ by introducing a non-zero diagonal to each. In addition, a parameter t connects the last t rows of R_1 with all the columns in C_2. Similarly, the last t columns in C_1 are connected to all the rows in R_2. An instance from this family of matrices is depicted in Fig. 3. Karp-Sipser^g is impacted negatively when $t \geq 2$ whereas Greedy^g struggles even with $t = 0$ because random edge selections will almost always be from the dense $R_1 \times C_1$ block. To adapt this scheme to hypergraphs/tensors, we generate a 3-dimensional tensor $\mathbf{T}_{KS}$ such that the nonzero pattern of each marginal of the 3rd dimension is identical to that of $\mathbf{A}_{KS}$. Table 2 shows the performance of the heuristics (i.e., matching cardinality normalized with n) for 3-dimensional tensors with $n = 300$ and $t \in \{2, 4, 8, 16, 32\}$.

The use of scaling indeed reduces the influence of the misleading hyperedges in the dense block $R_1 \times C_1$, and the proposed Karp-Sipser-scaling heuristic always finds the perfect matching as does Karp-Sipser-mindegree. However, Greedy and Karp-Sipser perform significantly worse. Furthermore, Local-Search returns 0.99-approximation in every case because it ends up in a local optima.

Rule-1 vs Rule-2

We finish the discussion on the synthetic data by focusing on Karp-Sipser. Recall from Sect. 3.2 that Karp-Sipser has two rules. In the bipartite case, a variant of Karp-Sipser[g] in which Rule-2 is not applied received more attention than the original version, because it is simpler to implement and easier to analyze. This simpler variant has been shown to obtain good results both theoretically [26] and experimentally [12]. Recent work [2] shows that both rules are needed to obtain perfect matchings in random cubic graphs.

We present a family of hypergraphs to demonstrate that using Rule-2 can lead to better performance than using Rule-1 only. We use Karp-Sipser$_{R_1}$ to refer to Karp-Sipser without Rule-2. As before, we describe first the bipartite case. Let $\mathbf{A}_{RF}$ be a $n \times n$ matrix with $(\mathbf{A}_{RF})_{i,j} = 1$ for $1 \le i \le j \le n$, and $(\mathbf{A}_{RF})_{2,1} = (\mathbf{A}_{RF})_{n,n-1} = 1$. That is, $\mathbf{A}_{RF}$ is composed of an upper triangular matrix and two additional subdiagonal nonzeros. The first two columns and the last two rows have two nonzeros. Assume without loss of generality that the first two rows are merged by applying Rule-2 on the first column (which is discarded). Then in the reduced matrix, the first column (corresponding to the second column in the original matrix) will have one nonzero. Rule-1 can now be applied whereupon the first column in the reduced matrix will have degree one. The process continues similarly until the reduced matrix is a 2×2 dense block, where applying Rule-2 followed by Rule-1 yields a perfect matching. If only Rule-1 reductions are allowed, initially no reduction can be applied and randomly chosen edges will be matched, which negatively affects the quality of the returned matching.

For higher dimensions we proceed as follows. Let $\mathbf{T}_{RF}$ be a d-dimensional $n \times \cdots \times n$ tensor. We set $(\mathbf{T}_{RF})_{i,j,\ldots,j} = 1$ for $1 \le i \le j \le n$ and $(\mathbf{T}_{RF})_{1,2,\ldots,2} = (\mathbf{T}_{RF})_{n,n-1,\ldots,n-1} = 1$. By similar reasoning, we see that Karp-Sipser with both reduction rules will obtain a perfect matching, whereas Karp-Sipser$_{R_1}$ will strug-

Table 3. Quality of matching and the number r of the applications of Rule-1 over n in Karp-Sipser$_{R_1}$, for hypergraphs corresponding to $\mathbf{T}_{RF}$. Karp-Sipser obtains perfect matchings.

n	d					
	2		3		6	
	quality	$\frac{r}{n}$	quality	$\frac{r}{n}$	quality	$\frac{r}{n}$
1000	0.83	0.45	0.85	0.47	0.80	0.31
2000	0.86	0.53	0.87	0.56	0.80	0.30
4000	0.82	0.42	0.75	0.17	0.84	0.45

gle. We give some results in Table 3 that show the difference between the two. We test for $n \in \{1000, 2000, 4000\}$ and $d \in \{2, 3, 6\}$, and show the quality of Karp-Sipser$_{R_1}$ and the number of times that Rule-1 is applied over n. We present the best result over 10 runs.

As seen in Table 3, Karp-Sipser$_{R_1}$ obtains matchings that are about 13–25% worse than Karp-Sipser. Furthermore, the larger the number of Rule-1 applications is, the higher the quality is.

4.3 Experiments with Real-Life Tensor Data

We also evaluate the performance of the proposed heuristics on some real-life tensors selected from FROSTT library [33]. The descriptions of the tensors are given in Table 4. For `nips` and `Uber`, a dimension of size 17 and 24 is dropped respectively, as they restrict the size of the maximum cardinality matching. As described before, a d-partite, d-uniform hypergraph is obtained from a d-dimensional tensor by associating a vertex for each dimension index, and a hyperedge for each nonzero. Unlike the previous experiments, the parts of the hypergraphs obtained from real-life tensors in Table 4 do not have an equal number of vertices. In this case, the scaling algorithm works along the same lines. Let $n_i = |V_i|$ be the cardinality at the ith dimension, and $n_{max} = \max_{1 \leq i \leq d} n_i$ be the maximum one. By slightly modifying Sinkhorn-Knopp, for each iteration of Karp-Sipser-scaling, we scale the tensor such that the marginals in dimension i sum up to n_{max}/n_i instead of one. The results in Table 4 resemble those from previous sections; Karp-Sipser-scaling has the best performance and is slightly superior to Karp-Sipser-mindegree. Greedy and Karp-Sipser are close to each other and when it is feasible, Local-Search is better than them. In addition we see that in these instances Bipartite-reduction exhibits a good performance: its performance is at least as good as Karp-Sipser-scaling for the first three instances, but about 10% worse for the last one.

Table 4. The performance of the proposed heuristics on the hypergraphs corresponding to real-life tensors. No Local-Search for four dimensional tensor `Enron`.

Tensor	d	Dimensions	nnz	Greedy	Local-Search	Karp-Sipser	Karp-Sipser-minDegree	Karp-Sipser-scaling	Bipartite-Reduction
`Uber`	3	183 × 1140 × 1717	1,117,629	183	183	183	183	183	183
`nips` [19]	3	2,482 × 2,862 × 14,036	3,101,609	1,847	1,991	1,839	2005	2,007	2,007
`Nell-2` [5]	3	12,092 × 9,184 × 28,818	76,879,419	3,913	4,987	3,935	5,100	5,154	5,175
`Enron` [31]	4	6,066 × 5,699 × 244,268 × 1,176	54,202,099	875	-	875	988	1,001	898

4.4 Experiments with an Independent Set Solver

We compare Karp-Sipser-scaling and Karp-Sipser-mindegree with the idea of reducing MAX-d-DM to the problem of finding an independent set in the line graph of the given hypergraph. We show that this transformation can yield good results, but is restricted because line graphs can require too much space.

We use KaMIS [29] to find independent sets in graphs. KaMIS uses a plethora of reductions and a genetic algorithm in order to return high cardinality independent sets. We use the default settings of KaMIS (where execution time is limited to 600 s) and generate the line graphs with efficient sparse matrix–matrix multiplication routines. We run KaMIS, Greedy, Karp-Sipser-scaling, and Karp-Sipser-mindegree on a few hypergraphs from previous tests. The results are summarized in Table 5. The run time of Greedy was less than one second in all instances. KaMIS operates in rounds, and we give the quality and the run time of the first round and the final output. We note that KaMIS considers the time-limit only after the first round has been completed. As can be seen, while the quality of KaMIS is always good and in most cases superior to Karp-Sipser-scaling and Karp-Sipser-mindegree, it is also significantly slower (its principle is to deliver high quality results). We also observe that the pseudo scaling of Karp-Sipser-mindegree indeed helps to reduce the run time compared to Karp-Sipser-scaling.

Table 5. Run time (in seconds) and performance comparisons between KaMIS, Greedy, and Karp-Sipser-scaling. The time required to create the line graphs should be added to KaMIS's overall time.

Hypergraph	KaMIS					Greedy	Karp-Sipser-scaling		Karp-Sipser-mindegree	
	Line graph gen. time	Round 1		Output						
		Quality	Time	Quality	Time	Quality	Quality	Time	Quality	Time
8-out, $n = 1000$, $d = 3$	10	0.98	80	0.99	600	0.86	0.98	1	0.98	1
8-out, $n = 10000$, $d = 3$	112	0.98	507	0.99	600	0.86	0.98	197	0.98	1
8-out, $n = 1000$, $d = 9$	298	0.67	798	0.69	802	0.55	0.62	2	0.67	1
$n = 8000$, $d = 3$, H_3	1	0.77	16	0.81	602	0.63	0.76	5	0.77	1
$n = 8000$, $d = 3$, H_{3+M}	2	0.89	25	1.00	430	0.70	1.00	11	0.91	1

The line graphs of the real-life instances from Table 4 are too large to be handled. We estimated (using known techniques [7]) the number of edges in these graphs to range from 1.5×10^{10} to 4.7×10^{13}. The memory needed ranges from 126 GB to 380 TB if edges are stored twice (assuming 4 bytes per edge).

5 Conclusion and Future Work

We have proposed heuristics for the Max-d-DM problem by generalizing existing heuristics for the maximum cardinality matching in bipartite graphs. The experimental analyses on various hypergraphs/tensors show the effectiveness and efficiency of the proposed heuristics. As future work, we plan to investigate the stated conjecture that d^{d-1}-out random hypergraphs have perfect matchings almost always, and analyze the theoretical guarantees of the proposed algorithms.

References

1. Aharoni, R., Haxell, P.: Hall's theorem for hypergraphs. J. Graph Theory **35**(2), 83–88 (2000)
2. Anastos, M., Frieze, A.: Finding perfect matchings in random cubic graphs in linear time. arXiv preprint arXiv:1808.00825 (2018)
3. Berman, P., Karpinski, M.: Improved approximation lower bounds on small occurence optimization. ECCC Report (2003)
4. Birkhoff, G.: Tres observaciones sobre el algebra lineal. Univ. Nac. Tucuman, Ser. A **5**, 147–154 (1946)
5. Carlson, A., Betteridge, J., Kisiel, B., Settles, B., Hruschka Jr., E.R., Mitchell, T.M.: Toward an architecture for never-ending language learning. In: AAAI, vol. 5, p. 3 (2010)
6. Çatalyürek, Ü.V., Aykanat, C.: PaToH: A Multilevel Hypergraph Partitioning Tool, Version 3.0. Bilkent University, Department of Computer Engineering, Ankara, 06533 Turkey. https://www.cc.gatech.edu/~umit/software.html (1999)
7. Cohen, E.: Structure prediction and computation of sparse matrix products. J. Comb. Optim. **2**(4), 307–332 (1998)
8. Cygan, M.: Improved approximation for 3-dimensional matching via bounded pathwidth local search. In: 2013 IEEE 54th Annual Symposium on Foundations of Computer Science (FOCS), pp. 509–518. IEEE (2013)
9. Cygan, M., Grandoni, F., Mastrolilli, M.: How to sell hyperedges: the hypermatching assignment problem. In: Proceedings of the Twenty-Fourth Annual ACM-SIAM Symposium on Discrete Algorithms, pp. 342–351. SIAM (2013)
10. Devlin, P., Kahn, J.: Perfect fractional matchings in k-out hypergraphs. arXiv preprint arXiv:1703.03513 (2017)
11. Dufossé, F., Kaya, K., Panagiotas, I., Uçar, B.: Scaling matrices and counting perfect matchings in graphs. Technical Report RR-9161, Inria - Research Centre Grenoble - Rhône-Alpes (2018)
12. Dufossé, F., Kaya, K., Uçar, B.: Two approximation algorithms for bipartite matching on multicore architectures. J. Parallel Distr. Com. **85**, 62–78 (2015)
13. Dufossé, F., Kaya, K., Panagiotas, I., Uçar, B.: Approximation algorithms for maximum matchings in undirected graphs. In: Proceedings Seventh SIAM Workshop on Combinatorial Scientific Computing, pp. 56–65. SIAM, Bergen (2018)
14. Dufossé, F., Kaya, K., Panagiotas, I., Uçar, B.: Effective heuristics for matchings in hypergraphs. Research Report RR-9224, Inria Grenoble Rhône-Alpes, November 2018. https://hal.archives-ouvertes.fr/hal-01924180
15. Dyer, M., Frieze, A.: Randomized greedy matching. Random Struct. Algorithms **2**(1), 29–45 (1991)
16. Franklin, J., Lorenz, J.: On the scaling of multidimensional matrices. Linear Algebra Appl. **114**, 717–735 (1989)
17. Frieze, A.M.: Maximum matchings in a class of random graphs. J. Comb. Theory B **40**(2), 196–212 (1986)
18. Froger, A., Guyon, O., Pinson, E.: A set packing approach for scheduling passenger train drivers: the French experience. In: RailTokyo2015. Tokyo, Japan, March 2015. https://hal.archives-ouvertes.fr/hal-01138067
19. Globerson, A., Chechik, G., Pereira, F., Tishby, N.: Euclidean embedding of co-occurrence data. J. Mach. Learn. Res. **8**, 2265–2295 (2007)
20. Gottlob, G., Greco, G.: Decomposing combinatorial auctions and set packing problems. J. ACM **60**(4), 24:1–24:39 (2013)

21. Halldórsson, M.M.: Approximating discrete collections via local improvements. In: SODA, vol. 95, pp. 160–169 (1995)
22. Hazan, E., Safra, S., Schwartz, O.: On the complexity of approximating k-dimensional matching. In: Arora, S., Jansen, K., Rolim, J.D.P., Sahai, A. (eds.) APPROX/RANDOM -2003. LNCS, vol. 2764, pp. 83–97. Springer, Heidelberg (2003). https://doi.org/10.1007/978-3-540-45198-3_8
23. Hazan, E., Safra, S., Schwartz, O.: On the complexity of approximating k-set packing. Comput. Complex. **15**(1), 20–39 (2006)
24. Hurkens, C.A.J., Schrijver, A.: On the size of systems of sets every t of which have an SDR, with an application to the worst-case ratio of heuristics for packing problems. SIAM J. Discrete Math. **2**(1), 68–72 (1989)
25. Karoński, M., Pittel, B.: Existence of a perfect matching in a random $(1+e^{-1})$-out bipartite graph. J. Comb. Theory B **88**(1), 1–16 (2003)
26. Karp, R.M., Sipser, M.: Maximum matching in sparse random graphs. In: FOCS 1981, Nashville, TN, USA, pp. 364–375 (1981)
27. Karp, R.M.: Reducibility among combinatorial problems. In: Miller, R.E., Thatcher, J.W., Bohlinger, J.D. (eds.) Complexity of Computer Computations, pp. 85–103. Springer, Boston (1972). https://doi.org/10.1007/978-1-4684-2001-2_9
28. Kaya, O., Uçar, B.: Scalable sparse tensor decompositions in distributed memory systems. In: Proceedings of the International Conference for High Performance Computing, Networking, Storage and Analysis. SC 2015, pp. 77:1–77:11. ACM, Austin (2015)
29. Lamm, S., Sanders, P., Schulz, C., Strash, D., Werneck, R.F.: Finding Near-Optimal Independent Sets at Scale. In: Proceedings of the 16th Meeting on Algorithm Engineering and Exerpimentation (ALENEX'16) (2016)
30. Pothen, A., Fan, C.J.: Computing the block triangular form of a sparse matrix. ACM T. Math. Software **16**, 303–324 (1990)
31. Shetty, J., Adibi, J.: The enron email dataset database schema and brief statistical report. Information sciences institute technical report, University of Southern California 4 (2004)
32. Sinkhorn, R., Knopp, P.: Concerning nonnegative matrices and doubly stochastic matrices. Pacific J. Math. **21**, 343–348 (1967)
33. Smith, S., et al.: FROSTT: the formidable repository of open sparse tensors and tools (2017). http://frostt.io/
34. Walkup, D.W.: Matchings in random regular bipartite digraphs. Discrete Math. **31**(1), 59–64 (1980)

Approximated ZDD Construction Considering Inclusion Relations of Models

Kotaro Matsuda[1(✉)], Shuhei Denzumi[1], Kengo Nakamura[2], Masaaki Nishino[2], and Norihito Yasuda[2]

[1] Graduate School of Information Science and Technology, The University of Tokyo, Tokyo, Japan
{kotaro_matsuda,denzumi}@mist.i.u-tokyo.ac.jp

[2] NTT Communication Science Laboratories, NTT Corporation, Kyoto, Japan
{nakamura.kengo,nishino.masaaki,yasuda.n}@lab.ntt.co.jp

Abstract. Zero-suppressed binary decision diagrams (ZDDs) are data structures that can represent families of sets in compressed form. By using ZDDs, we can perform several useful operations over families of sets in time polynomial to ZDD size. However, a ZDD representing a large family of sets tends to consume a prohibitive amount of memory. In this paper, we attempt to reduce ZDD size by allowing them to have some false positive entries. Such inexact ZDDs are still useful for optimization and counting problems since they permit only false positive errors. We propose two algorithms that can reduce ZDD size without making any false negative errors. The first is a general algorithm that can be applied to any ZDD. The second one is a faster algorithm that can be used if the ZDD represents a monotone family of sets. Our algorithms find pairs of nodes in a ZDD that do not yield any false negatives if they are merged, and then merge those pairs in a greedy manner to reduce ZDD size. Furthermore, our algorithms can be easily combined with existing top-down ZDD construction methods to directly construct approximated ZDDs. We conduct experiments with representative benchmark datasets and empirically confirm that our proposed algorithms can construct ZDDs with 1,000 times fewer false positives than those made with baseline methods, when the ZDD sizes are halved from the original sizes.

Keywords: Zero-suppressed binary decision diagrams · Probabilistic data structure · Enumeration

1 Introduction

A zero-suppressed binary decision diagram (ZDD) [13], a variant of the binary decision diagram (BDD) [7], is a data structure that can represent a family of sets as a directed acyclic graph. A ZDD can represent a family of sets in a compressed form that permits several useful operations including computing the cardinality of the family or performing binary operations between sets of families, in time

I. Kotsireas et al. (Eds.): SEA[2] 2019, LNCS 11544, pp. 265–282, 2019.
https://doi.org/10.1007/978-3-030-34029-2_18

polynomial to ZDD size. Due to these benefits, ZDDs and BDDs are widely used for solving optimization and enumeration problems.

The effectiveness of any ZDD-based problem solving method strongly depends on how compactly the ZDD can represent the target family of sets. ZDDs can become prohibitively large if the inputs are large. Indeed, large ZDDs may exceed the memory capacity of the computer, which prevents ZDDs from being used to solve problems. In this paper, we propose an algorithm for constructing approximated ZDDs. Here we say a ZDD is approximated if it represents a family of sets $\mathcal{F}'$ that is close to the target family of sets $\mathcal{F}$. However, we may lose many of the benefits of ZDDs if $\mathcal{F}$ is approximated arbitrarily. Our solution is to consider a special type of approximated ZDDs with the constraint that $\mathcal{F} \subseteq \mathcal{F}'$. This means that while the approximated ZDD might have false positive entries, it will never have false negative entries. This type of ZDD is called *relaxed* ZDD [2].

Even though relaxed ZDDs cannot provide exact computational results, they are still useful since they can be applied in the following situations:

Probabilistic Membership Testing. ZDDs can answer membership queries to a family of sets in time linear to the size of the base set. Relaxed ZDDs can also answer membership queries, but they may cause false positive errors. However, such approximated membership queries are useful as Bloom filters [4], famous probabilistic data structures that can answer membership queries but sometimes make false positive errors. They are used in a wide range of application fields. Unlike Bloom filters, relaxed ZDDs might work well if the cardinality of input family is huge. This is because if we want a Bloom filter whose false positive rate is below a threshold value, then the size of a Bloom filter must be linear to the cardinality of the family, which can be prohibitively large.

Obtaining Upper Bound Scores in Optimization. If we represent a family of sets as a ZDD, we can, in time linear to ZDD size, find the set present within the family that maximizes a linear objective function. This property is lost if we relax ZDDs, but they do allow us to obtain an upper bound solution value of the problem. When solving optimization problems, obtaining good upper bound values is important in making the search procedure more efficient.

Approximated Counting with a Membership Oracle. ZDDs support uniform sampling of their entries. If we combine a membership oracle, which answers a membership query, with a relaxed ZDD, we can perform uniform random sampling by combining uniform sampling on a relaxed ZDD and using the oracle to judge membership. Moreover, estimation of the cardinality of the family also can be performed in the same manner.

In addition to these important applications, we should note that ZDDs obtained by applying union or intersection operations between relaxed ZDDs are still relaxed. Therefore, they retain flexibility of ZDDs.

Our goal is to construct a ZDD that is small with the fewest false positive entries possible. We propose two algorithms; the first one is general in the sense that any ZDD is accepted as input. The second is faster and consumes less memory but can be applied only if the input ZDD represents a monotone family of sets. Both algorithms consist of two steps; first, they find pairs of nodes in the given ZDD that can be merged without yielding false negatives; next, they merge those pairs of nodes one by one by a kind of greedy algorithm to reduce ZDD size. Let $|G|$ be the number of nodes in the original ZDD. For general cases, the first step takes $O(|G|^2)$ time and space; the second step takes $O(|G|^2)$ time and $O(|G|)$ space for each repetition. Overall, it is an $O(m|G|^2)$ time and $O(|G|^2)$ space algorithm where m is the number of merge operations. If the ZDD represents a monotone family of sets, the first step takes $O(|G|\alpha(|G|))$ time and $O(|G|)$ space where $\alpha(\cdot)$ is the inverse Ackermann function, and the second step takes $O(|G|)$ time and $O(|G|)$ space for each repetition. Overall, it is an $O(|G|(m+\alpha(|G|))$ time and $O(|G|)$ space algorithm. Furthermore, our algorithms can be easily combined with existing top-down ZDD construction methods to directly construct approximated ZDDs. Experiments empirically confirm that our algorithms are better than existing alternatives including heavy branch subsetting [14,18] and ZDD relaxation [3].

2 Related Work

Making small decision diagrams has long been considered to be important. Therefore, most BDD or ZDD processing software available today implement methods for reducing the sizes of decision diagrams. One of the most popular methods is variable reordering. The size of ordered BDDs and ZDDs strongly depends on the variable order used. Since finding an optimal variable ordering is known to be NP-hard [5], efficient heuristic algorithms like sifting [16] are used in BDD and ZDD packages. Negative edges [6] can also be used for obtaining small BDDs. However, since these methods are exact, i.e., they do not change the target family of sets being represented, their reduction ability is limited. It is known that there are Boolean functions such that a BDD representing them becomes exponentially larger under any variable order (e.g., [7]). Exact methods cannot cope with these functions.

Many methods for approximating decision diagrams have been proposed over the years. Approximation methods can be divided into two classes. Methods in one class try to reduce the size of a given decision diagram. Previous methods in this class reduce the size of decision diagrams by applying some reduction rules. The proposed method lies in this class. Heavy branch subsetting [14,18] reduces ZDD size by removing a node and then making the edges pointing to that node point to a terminal node. Unlike the proposed method, heavy branch subsetting is applicable only in limited situations. Another approach close to ours is [14]. Their method, called Remap, is similar in that they use inclusion relationships between two children nodes of a BDD to decide which node to delete. However, compared with ours, the situation wherein two nodes can be merged is limited. Moreover, the reduction rule applied in Remap cannot reduce ZDD size, while our method can be used with both BDDs and ZDDs.

The other class of approximation methods directly construct DDs without first making exact DDs [3,9]. These methods are useful in the situations where an exact DD is hard to construct due to its size. However, these methods are specialized for solving optimization problems. Therefore, their performance is not so high when we want to solve other than optimization problems. Moreover, these methods must be carefully designed to suit the target family of sets. In contrast, our methods can be used in combination with existing top-down construction algorithms without any modification.

3 Preliminaries

Let $C = \{1, \ldots, c\}$ be the universal set. Throughout this paper, sets are subset of C. The empty set is denoted by $\emptyset$. For set $S = \{a_1, \ldots, a_s\}$ $(\subseteq C)$, $s \geq 0$, we denote the *size* of S by $|S| = s$. A *family* is a subset of the power set of C. We say a family of sets $\mathcal{F}$ is monotone if $\mathcal{F}$ satisfies $S \in \mathcal{F} \Rightarrow \forall k \in S, S \setminus \{k\} \in \mathcal{F}$ or $S \in \mathcal{F} \Rightarrow \forall k \in C, S \cup \{k\} \in \mathcal{F}$. We describe families of sets that satisfy the former condition as monotone decreasing and families of sets that satisfy the latter condition as monotone increasing.

3.1 Zero-Suppressed Binary Decision Diagrams

A *zero-suppressed binary decision diagram* (ZDD) [13] is a data structure for manipulating finite families of sets. A ZDD is a single-rooted directed acyclic graph written as $G = (V, E)$ that satisfies the following properties. A ZDD has two types of nodes: branch nodes and terminal nodes. There are exactly two *terminal nodes*: $\bot$ and $\top$. Terminal nodes have no outgoing edges. *Branch node* v has an integer $\ell(v) \in \{1, \ldots, c\}$ as the *label* of v. v has exactly two distinguishable edges called the *0-edge* and the *1-edge* of v. We call the destination node of 0-edge $zero(v)$ (1-edge $one(v)$ resp.) *0-child* (*1-child* resp.) of v. If there is no confusion, we simply write the 0-child and 1-child of v as v_0 and v_1, respectively. We say a ZDD is ordered if $\ell(v) < \ell(v_0)$ and $\ell(v) < \ell(v_1)$ holds for any branch node v. For convenience, we define $\ell(v) = c + 1$ for terminal node v. In this paper, we consider only ordered ZDDs. Let $L_1, \ldots, L_{c+1}$ be layers. Each layer L_i is the set of nodes whose label is i for $i = 1, \ldots, c+1$. Note that there is no edge from a node in L_i to a node in L_j if $i \geq j$. In the figures in the paper, terminal and branch nodes are drawn as squares and circles, respectively, and 0-edges and 1-edges are drawn as dashed and solid arrows, respectively. We define the *size* of ZDD G as the number of its nodes and denote it by $|G|$. We define families of sets represented by ZDDs as follows:

Definition 1 (a family of sets represented by a ZDD). *Let v be a ZDD node. Then, a finite family of sets $\mathcal{F}_v$ is defined recursively as follows: (1) If v is a terminal node: $\mathcal{F}_v = \{\emptyset\}$ if $v = \top$, and $\mathcal{F}_v = \emptyset$ if $v = \bot$. (2) If v is a branch node: $\mathcal{F}_v = \{S \cup \{\ell(v)\} \mid S \in \mathcal{F}_{v_1}\} \cup \mathcal{F}_{v_0}$.*

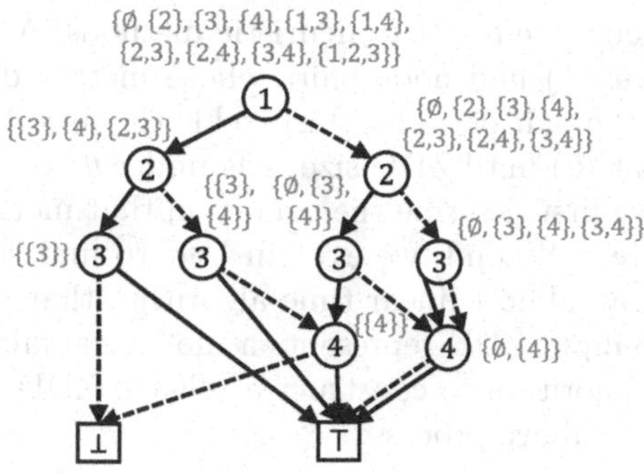

Fig. 1. Example of a ZDD.

If r is the root of ZDD G, then $\mathcal{F}_r$ represents the family of sets that G corresponds to. We also write this family of sets as $\mathcal{F}_G$. The example in Fig. 1 represents the following family of sets $\{\emptyset, \{2\}, \{3\}, \{4\}, \{1,3\}, \{1,4\}, \{2,3\}, \{2,4\}, \{3,4\}, \{1,2,3\}\}$. Set $S = \{a_1, \ldots, a_s\}$ corresponds to a path in ZDD G starting from root r in the following way: At each branch node with label b, the path continues to the 0-child if $b \notin S$ and to the 1-child if $b \in S$; the path eventually reaches $\top$ ($\perp$ resp.), indicating that $S \in \mathcal{F}_r$ ($S \notin \mathcal{F}_r$ resp.). If a ZDD represents a monotone decreasing (increasing resp.) family of sets, then for any branch node v in the ZDD, $\mathcal{F}_{v_0} \supset \mathcal{F}_{v_1}$ ($\mathcal{F}_{v_0} \subset \mathcal{F}_{v_1}$ resp.) holds.

3.2 False Positives/False Negatives

Our objective in this paper is to construct an approximated ZDD that represents the family of sets $\mathcal{F}'$ that is similar to the given sets family $\mathcal{F}$. We say $S \subseteq C$ is a *false positive* if $S \in \mathcal{F}' \setminus \mathcal{F}$. Similarly, we say S is a *false negative* if $S \in \mathcal{F} \setminus \mathcal{F}'$. For given ZDD G that represents $\mathcal{F}$, we say a ZDD that represents $\mathcal{F}'$ is a *relaxed ZDD* of G if $\mathcal{F}'$ does not have any false negative entries, that is, $\mathcal{F} \subseteq \mathcal{F}'$. Conversely, if $\mathcal{F}'$ has no false positive entries, then the ZDD representing $\mathcal{F}'$ is said to be a *restricted ZDD* of G.

4 Approximation of a Given ZDD

In this section, we propose an algorithm that constructs a relaxed ZDD G' from a given ZDD G. Smaller relaxed ZDDs that have fewer false positives are preferred. Therefore, we formulate the problem of finding a good relaxed ZDD as the following combinatorial optimization problem:

$$\text{Minimize } |\mathcal{F}_{G'} \setminus \mathcal{F}_G|$$
$$\text{Subject to } |G'| < \theta, \quad \mathcal{F}_G \subseteq \mathcal{F}_{G'},$$

where $|\mathcal{F}_{G'} \setminus \mathcal{F}_G|$ represents the number of false positives and θ is the parameter that limits the size of the relaxed ZDD. Since finding an optimal solution of the above problem is difficult, we resort to heuristic methods. A rough sketch of our algorithm is as follows: (1) find node pairs whose merger does not cause false negatives; (2) choose the pair of nodes that yields the fewest false positives and merge them; (3) repeat (2) until ZDD size falls under θ.

In the following, we first describe the condition that merging a pair of nodes causes no false negatives. Second, we explain how to find such node pairs efficiently. We also show an almost linear time algorithm that can find node pairs, particularly when the input ZDD represents a monotone family of sets. Finally, we propose a greedy algorithm to construct a relaxed ZDD that uses the information obtained by the above process.

4.1 Node Merging Without False Negatives

In our algorithms, we make the relaxed ZDD by repeatedly merging pairs of ZDD nodes. We merge two nodes u and v by first deleting u and redirecting to v all incoming edges to u. Then, we delete descendants of u that cannot be reached from the root after deleting u. We call this manipulation *incorporation of u into v*. Generally speaking, incorporating nodes may cause both false positive and negative errors. However, our condition of incorporation does not yield any false negative errors.

Theorem 1. *Given ZDD nodes u and v, if $\mathcal{F}_u \subseteq \mathcal{F}_v$ and $\ell(u) = \ell(v)$, then incorporating u into v does not cause false negatives.*

From Definition 1, for node p that has u as its 1-child, the family of sets represented by p does not contain false negatives in comparison to the original $\mathcal{F}_p$ after incorporation of u into v because $\{S \cup \{\ell(p)\} \mid S \in \mathcal{F}_u\} \cup \mathcal{F}_{p_0} \subseteq \{S \cup \{\ell(p)\} \mid S \in \mathcal{F}_v\} \cup \mathcal{F}_{p_0}$. The above statement holds true for subsequent ancestors and nodes having u as their 0-child. Finally, we can prove Theorem 1. An example of incorporation that does not cause false negatives is shown in Fig. 2. Since $\mathcal{F}_u = \{\{3\}\}$ and $\mathcal{F}_v = \{\emptyset, \{3\}\}$, $\mathcal{F}_u \subseteq \mathcal{F}_v$ holds. Incorporating u into v causes new false positive sets $\emptyset$ and $\{1\}$, but no false negative sets.

Readers may notice that an inclusion relation such as $\mathcal{F}_p \subseteq \mathcal{F}_q$ may be broken by incorporating descendant nodes, and thus it seems that we cannot incorporate p into q after that. However, even in this situation we can incorporate p into q. Let $\mathcal{F}_p^*$ and $\mathcal{F}_q^*$ be the family of sets p and q initially represents, respectively, and assume $\mathcal{F}_p^* \subseteq \mathcal{F}_q^*$. Then after incorporating descendant nodes, $\mathcal{F}_p \supseteq \mathcal{F}_p^*$ and $\mathcal{F}_q \supseteq \mathcal{F}_q^*$ hold. Since $\mathcal{F}_p^* \subseteq \mathcal{F}_q^* \subseteq \mathcal{F}_q$, incorporating p into q causes some false positives but no false negatives compared to the original set family $\mathcal{F}_p^*$. These arguments suggest that we can continuously use the inclusion relations computed with the initial ZDD. Therefore we compute the inclusion relations only once (at first). From the above, we first need to find the pairs of nodes that can be incorporated to obtain a relaxed ZDD by incorporation.

4.2 Finding All Inclusion Relations in Each Layer

Here, we show how to find inclusion relations among ZDD nodes. To identify the node pairs that can be incorporated, we search the inclusion relations in each

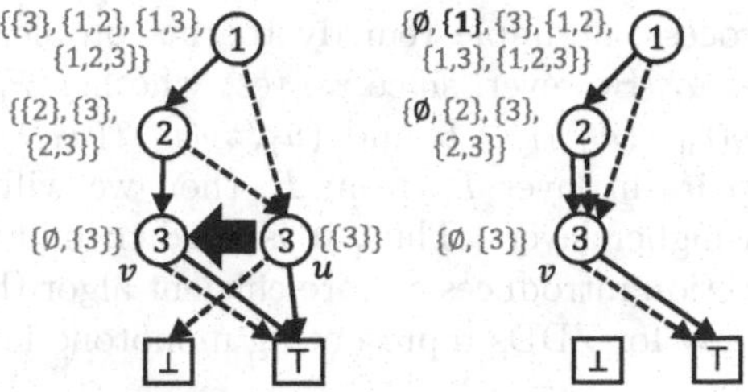

Fig. 2. An example showing that incorporation does not cause false negative errors. Bold arrow from u to v indicates that u is included in v.

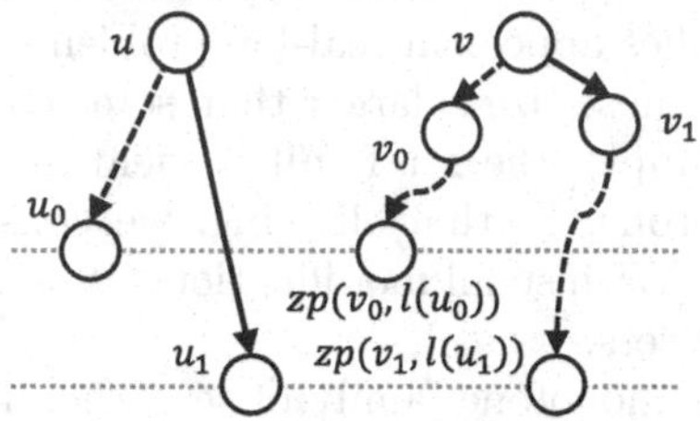

Fig. 3. How to decide whether $(u, v) \in R$. Curved dashed lines denote paths consisting of only 0-edges.

layer. Let R be the set of all node pairs that can be incorporated, i.e., $(u, v) \in R$ if and only if $\mathcal{F}_u \subseteq \mathcal{F}_v$ and $\ell(u) = \ell(v)$. Note that, for each node u, we define $(u, u) \in R$. Since $\mathcal{F}_u \subseteq \mathcal{F}_v$ is equivalent to $\mathcal{F}_{u_0} \subseteq \mathcal{F}_{v_0}$ and $\mathcal{F}_{u_1} \subseteq \mathcal{F}_{v_1}$ when $\ell(u) = \ell(v)$, we can compute R in a bottom-up order from layer L_{c+1} to layer L_1. We define $zp(u, k)$ as follows: if there is node v of label k that can be reached from u by traversing only 0-edges, $zp(u, k) = v$; if such a node does not exist, $zp(u, k) = null$. We should note that such v is uniquely determined if exists. For a pair of distinct nodes (u, v) in L_k, we can decide whether u can be incorporated into v or not by the following process: For $k = c+1$, it is trivial that $(\bot, \top) \in R$ and $(\top, \bot) \notin R$. For $k = c, c-1, \ldots, 2$, we check whether $(u, v) \in R$ or not by seeing whether both $\mathcal{F}_{u_0} \subseteq \mathcal{F}_{v_0}$ and $\mathcal{F}_{u_1} \subseteq \mathcal{F}_{v_1}$ are satisfied or not. We use the following fact.

Theorem 2. *For any branch nodes u, v, $(u, v) \in R$ iff $(u_0, zp(v_0, \ell(u_0))) \in R$ and $(u_1, zp(v_1, \ell(u_1))) \in R$ are satisfied.*

Figure 3 shows the condition. The above theorem says that if we have $zp(u, k)$ for every node u and $1 \leq k \leq c+1$ and we know all pairs in R whose label is larger than $\ell(u)$, then we can check the inclusion relations of node pairs in constant time. $zp(u, k)$ can be computed by bottom-up dynamic programming in $O(|G|c)$ time and we can store the results in a table of size $O(|G|c)$. As a result, our bottom-up algorithm runs in $O(|G|^2)$ space and $O(|G|^2)$ time.

The size of R obtained by the above procedure is $O(|G|^2)$, which can be hard to store in memory if the input ZDD is large. Since we do not need all elements of R in the later process, a simple remedy to this problem is to give up on finding all elements of R. However, since we test whether $\mathcal{F}_u \subset \mathcal{F}_v$ by checking whether both $(u_0, zp(v_0, \ell(u_0))) \in R$ and $(u_1, zp(v_1, \ell(u_1))) \in R$ are satisfied, if we remove some pairs in layer L_k from R, then we will overlook inclusion relations of nodes in higher layers. Thus, it is hard to find only valuable node pairs. The next subsection introduces a more efficient algorithm to find inclusion relations that work only for ZDDs representing monotone families of sets.

4.3 Finding Inclusion Relations Under Monotonicity

Next method can be used for ZDDs that represent monotone families of sets. Many monotone set families appear in real-life problems. For example, a family of sets consisting of sets whose size is larger than some constant, k, is monotone. As a more practical example, the set families yielded by frequent set mining problems are also monotone. In the following, we consider only a monotone decreasing family of sets. With small modifications, it can be applied to a monotone increasing family of sets.

If a ZDD represents a monotone family of sets, then finding mergeable node pairs becomes easier than the general case as we can use the following fact.

Theorem 3. *Suppose that $\mathcal{F}_G$ is monotone decreasing. Then $\mathcal{F}_{u_1} \subseteq \mathcal{F}_v$ holds for any branch node u, where $v = zp(u, \ell(u_1))$ (Cond. (b)).*

Different from the general case, we can find inclusion relations if we know some inclusion relations of parent nodes.

Theorem 4. *Suppose that $\mathcal{F}_G$ is monotone decreasing. If two branch nodes p, q satisfy $\ell(p) = \ell(q)$ and $\mathcal{F}_p \subseteq \mathcal{F}_q$, then $\mathcal{F}_{p_0} \subseteq \mathcal{F}_v$ (Cond. (a)) and $\mathcal{F}_{p_1} \subseteq \mathcal{F}_w$ (Cond. (c)), where $v = zp(q, \ell(p_0))$ and $w = zp(q_1, \ell(p_1))$.*

The above three conditions are shown in Fig. 4. Furthermore, the following inclusion relation also holds.

Theorem 5. *If $\mathcal{F}_G$ is monotone, for each node u other than the special node in each layer, there is at least one node pair (u, v) such that $(u, v) \in R$ and $u \neq v$.*

Here we say a node is *special* if it can be reached from the root by using only 0-edges. All these facts suggest that we can easily find mergeable node pairs if $\mathcal{F}_G$ is monotone. We introduce a heuristic method that processes ZDD nodes in order from the top layer to the bottom layer to find at most one mergeable node pair (u, v) for every branch node u. In the following, we use $sup(u)$ to represent the node v that satisfies $(u, v) \in R'$, where $R' \subseteq R$ is the set of mergeable node pairs found by the algorithm introduced in this section. Since the size of R' is $O(|G|)$, memory requirements are reasonable. Clearly, our algorithm may overlook important node pairs. However, by selecting a "good" pair (u, v) for every node u by using some criterion, our algorithm is practical and works well.

Algorithm 1 shows the algorithm that finds mergeable node pairs of ZDD G that represents a monotone decreasing family of sets. The algorithm proceeds

Algorithm 1. *FindInclMono*: Compute some node pairs in the same layer that hold inclusion relations under monotonicity.

Input: a ZDD $G = (V, E)$

1: **for** $k = 1, \ldots, c$ **do**
2:     **for** $u \in L_k$ that is not special **do**
3:         $P_0 \leftarrow \{v \mid v \in V, u = v_0\}$
4:         $P_1 \leftarrow \{v \mid v \in V, u = v_1\}$
5:         **for all** $p \in P_0 \cup P_1$ **do** ▷ Checking Cond.(a), Cond. (b)
6:             $q \leftarrow p$
7:             **while** $u = zp(q, \ell(u))$ **do**
8:                 $q \leftarrow sup(q)$
9:             insert $zp(q, \ell(u))$ to the candidates of $sup(u)$
10:        **for all** $p \in P_1$ **do** ▷ Checking Cond.(c)
11:            $q \leftarrow p$
12:            **while** $u = zp(q_1, \ell(u))$ and q is not special **do**
13:                $q \leftarrow sup(q)$
14:            **if** $u \neq zp(q_1, \ell(u))$ **then**
15:                insert $zp(q_1, \ell(u))$ to the candidates of $sup(u)$
16:        $sup(u) \leftarrow \arg\min_v\{|\mathcal{F}_v| \mid v \in \{\text{candidates of } sup(u)\}\}$

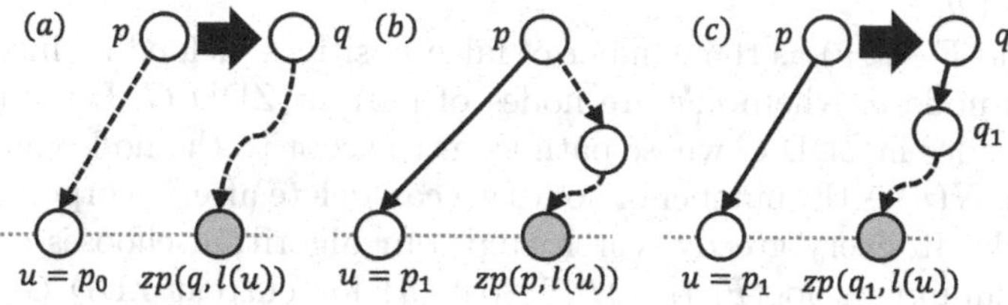

Fig. 4. Possible candidates of $sup(u)$ are shown in gray. Curved dashed lines denote paths consisting of only 0-edges. Bold arrow from p to q represents that $\mathcal{F}_p \subseteq \mathcal{F}_q$.

from the top layer to the bottom layer. For each layer, (1) it finds candidates of $sup(u)$ for every node u in the layer (lines 3 to 14), and (2) it sets $sup(u) = v$ where $|\mathcal{F}_v|$ is the smallest among the candidates (line 15). Here we use $|\mathcal{F}_v|$ as the criterion since the number of false positives caused by incorporating u into v equals $|\mathcal{F}_v \setminus \mathcal{F}_u|$. To find $sup(u)$ candidates, we use the three conditions listed in the above theorems.

Time Complexity. If we simply implement this algorithm, it runs in $\Theta(|G|^2)$ time in the worst case. However, we can reduce this worst case time complexity to $O(|G|\alpha(|G|))$ by using the disjoint-set data structure [8,19], where $\alpha(\cdot)$ is an inverse Ackermann function (Details are shown in Appendix A).

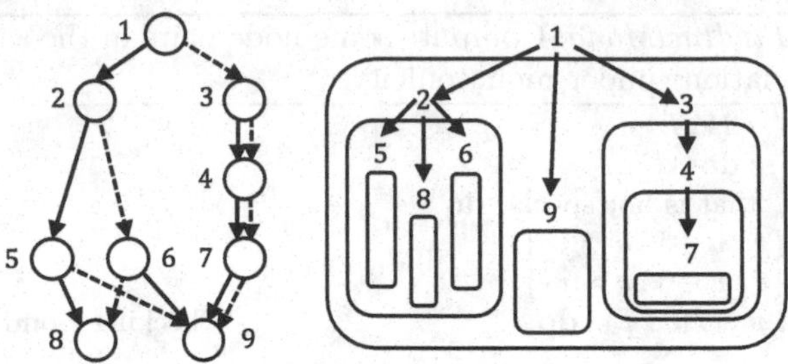

Fig. 5. An example of a dominator tree for a given graph.

4.4 Greedy Reduction by Selecting Nodes to Be Incorporated

If nodes u and v satisfy $\mathcal{F}_u \subset \mathcal{F}_v$, we can incorporate u into v without causing false negatives. For good approximation, we need to choose node pairs that yield fewer false positives. In this subsection, we introduce an algorithm for determining pairs of nodes to be merged. Since finding pairs of nodes that minimize ZDD size is difficult, we propose a greedy algorithm that repeatedly selects the node pairs to be incorporated until the size of the ZDD becomes smaller than the threshold θ.

We define $E_G(u, v)$ as the number of false positives yielded by incorporating node u into node v, where u, v are nodes of current ZDD G. Let $N(u)$ be the number of nodes in ZDD G whose path from the root to the node contains v. In other words, $N(u)$ is the number of nodes we can delete after incorporating u into another node. In every greedy search step, our algorithm chooses a node pair (u, v) that minimizes $\log(E_G(u, v)/|\mathcal{F}_G|)/N(u)$ for current ZDD G. We chose this objective function because we want the least possible additional error rate $\log(E_G(u, v)/|\mathcal{F}_G|)$ as well as more node deletion $N(u)$. We also tried other objective functions, but this function achieved the best results in preliminary experiments.

To execute the greedy algorithm efficiently, we have to rapidly calculate $E_G(u, v)$, $|\mathcal{F}_G|$ and $N(u)$, for every greedy search step.

Calculating $E_G(u, v)$ **and** $|\mathcal{F}_G|$**:** Since we know $\mathcal{F}_u \subset \mathcal{F}_v$, $E_G(u, v)$ is calculated simply by $(|\mathcal{F}_v| - |\mathcal{F}_u|)$ times the number of paths from the root to u. For every node u, we can calculate $|\mathcal{F}_u|$ and the number of paths from the root by executing dynamic programming [11] in $O(|G|)$ time. Note that $|\mathcal{F}_G|$ equals $|\mathcal{F}_r|$ where r is the root of G.

Calculating $N(u)$**:** Here, we introduce a concept called *dominator*. For a directed graph, if there exists a distinguished start node r such that the other nodes are reachable from r, the graph is called a flow-graph. A single-rooted ZDD is a flow-graph whose distinguished start node is the root node. In a flow-graph, if and only if there is node v on every path from the start node r to node u, we define node v as a dominator of u. It is known that this relationship among all nodes can be represented by a tree structure, which is called the dominator tree of the graph. An example of a dominator tree is shown in Fig. 5. In a dominator tree, node u is an ancestor of node v if and only if u dominates v in the given flow-graph. A dominator tree can be constructed in time linear to the number of nodes of a given graph [1,12]. Remember that $N(u)$ is the number of nodes that are reachable only via u. Thus, $N(u)$ equals the number of descendants of u in the dominator tree. $N(u)$ is calculated by bottom-up computation on the dominator tree in linear time.

5 On-the-Fly Approximation in the Construction of a ZDD

In Sect. 4, we described an algorithm to approximate a given ZDD. However, the discussion implicitly assumed that the size of the given ZDD is within the main memory limits. Indeed, approximation techniques are quite useful for huge ZDDs that exceed main memory. In this section, we show how to apply our algorithms for constructing ZDDs from scratch.

5.1 ZDD Construction by Depth-First Dynamic Programming

First, we briefly review an existing ZDD construction algorithm called the frontier-based search (FBS) method [11,17]. FBS is a kind of dynamic programming that can construct ZDDs that represent sets of all solutions of some problems, including the knapsack problem and enumeration problems of subgraphs satisfying certain conditions. FBS creates ZDD nodes in the order from the root to the bottom, i.e., it first creates a ZDD node and then creates its child nodes. If we naively create nodes in this way, layer L_k might have $O(2^k)$ nodes. To avoid such explosion, FBS methods apply the signature called *mate* to each subset of C. Mate is designed to ensure that if two sets S_1, S_2 with the same maximal element have the same mate signature value, then $\{S \setminus S_1 \mid S \in \mathcal{F}_G, S_1 \subseteq S\} = \{S \setminus S_2 \mid S \in \mathcal{F}_G, S_2 \subseteq S\}$ for the family of solution sets $\mathcal{F}_G$. In other words, if the mate values of S_1 and S_2 are the same, we know that the resultant ZDDs constructed after choosing each of S_1 and S_2 are equivalent. By designing appropriate mate values, FBS methods can avoid redundant computation; they reuse the results for the sets with the same mate values and thus create ZDDs efficiently. In our relaxation method, we use $\phi(S)$ to represent the mate value of $S \in C$ and $M[\cdot]$ is the table that maps mate values to corresponding ZDD nodes. If $\phi(S) = null$, S cannot be included any solution set. Thus, we can stop further computation. Algorithm 2 provides the pseudo code of this algorithm. We run $DFS(\emptyset, 1)$ to obtain the root of the solution ZDD.

5.2 Approximated Construction in Depth-First Manner

By combining the depth-first FBS algorithm with the node merging method introduced in Sect. 4, we can design an on-the-fly approximation algorithm. We

Algorithm 2. $DFS(S, k)$: Depth-first construction algorithm of a ZDD that contains all solutions of a given problem.

Input: a set $S \subseteq C$ and a level $k \in C$
Output: a ZDD node

if $k = |C| + 1$ **then**
  if S is a solution of the problem **then return** $\top$
  else return $\bot$
if $\phi(S) = null$ **then return** $\bot$
if key $\phi(S)$ exists in M **then return** $M[\phi(S)]$
$v_0 \leftarrow DFS(S, k+1)$
$v_1 \leftarrow DFS(S \cup \{k\}, k+1)$
if $v_1 = \bot$ **then return** v_0
if node v with label k and having child nodes v_0 and v_1 exists **then**
  $M[\phi(S)] \leftarrow v$
  return v
else
  Create node v in the layer L_k whose 0-child is v_0 and 1-child is v_1
  $M[\phi(S)] \leftarrow v$
  return v

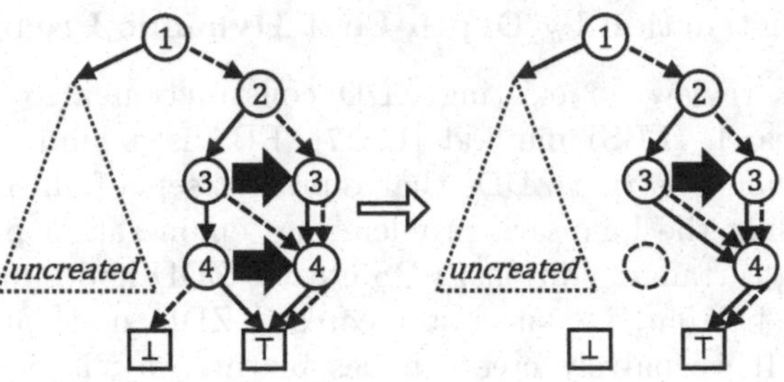

Fig. 6. Node incorporation in construction.

first set θ as the threshold of the allowed number of nodes before commencing ZDD construction. We then run the depth-first construction algorithm. During construction, we run our approximation algorithm when the number of created nodes exceeds threshold θ. An example of incorporation in construction is shown in Fig. 6. Since ZDDs are only partially constructed, we compute $E_G(u, v)$ or $N(u)$; uncreated parts are ignored.

When we incorporate node u into node v, we must update $M[x]$ to v, where x is the mate such that $M[x] = u$ holds before deleting u. However, storing all such relations may results in growing the size of M. We, therefore, set another threshold σ and divide M into c individual hashtables corresponding to each layer. If the size of a table for a layer reaches σ, we choose an element in the hashtable at random and delete it. Such forgetting weakens the benefit of memorization and may yield redundant repeated computation, but it does guarantee a bound on memory consumption. This idea of bounding the memory usage in each layer is also employed in [3].

6 Experiments and Results

6.1 Approximation of a Given ZDD

In this subsection, we show that the methods proposed in Sect. 4 cause fewer false positives than other algorithms. We implemented the algorithms proposed in Sect. 4 and existing approximation algorithms in C++ and compared them by measuring the numbers of nodes after approximation and the resulting sizes of set families including false positives.

- **proposed_method:** our algorithm (described in Sect. 4).
- **proposed_method_monotone:** our algorithm for ZDDs that represent monotone families of sets (described in Sect. 5).
- **random:** choose a pair of nodes uniformly at random from all node pairs (u, v) such that $\mathcal{F}_u \subseteq \mathcal{F}_v$ and incorporate u into v.
- **greedy_eraser:** choose node $u = \arg\max_w\{|N(w)| \mid w \in \{p \mid p, q \in V, \mathcal{F}_p \subseteq \mathcal{F}_q\}\}$ and incorporate u into node $v = \arg\min_w\{|\mathcal{F}_w| \mid w \in \{q \mid q \in V, \mathcal{F}_u \subseteq \mathcal{F}_q\}\}$.
- **heavy_branch:** algorithm proposed in [15,18]; it is called rounding up by heavy branch subsetting.[1]

Table 1. Detail of data sets and properties of their ZDDs.

Families of sets	$\lvert C\rvert$	$\lvert G\rvert$	$\lvert\mathcal{F}_G\rvert$
Matching edge sets of an 8×8 grid graph	112	4367	1.798×10^{17}
Matching edge sets in the network *"Interoute"*	146	11143	6.896×10^{24}
Frequent item sets in *"mushroom"* ($p = 0.001$)	117	26719	2.311×10^{9}
Frequent item sets in *"retail"* ($p = 0.00025$)	6053	12247	5.048×10^{4}
frequent item sets in *"T40I10D100K"* ($p = 0.005$)	838	47363	1.286×10^{6}

We use the ZDDs that represent the families of sets listed in Table 1. The top two are families of sets of edges in each graph that constitute matching in the graphs. The network *"Interoute"* is derived from [10][2]. The bottom three are the results of frequent set mining with support $p, 0 \leq p \leq 1$, on data sets that are taken from Frequent Itemset Mining Dataset Repository[3]. All of these data sets are monotonically decreasing.

The results are shown in Figs. 7 and 8. Note that we use semilog scale. Our proposed algorithms yield fewer false positives than the other algorithms; over

[1] Heavy branch subsetting was originally proposed for BDDs. We slightly modify it to suit ZDDs. In this method, when we want to delete node u with label k, we incorporate u into node p_k such that $\mathcal{F}_{p_k} = 2^{\{k,k+1,\cdots,c\}}$. There can be several node selection methods such as [15,18]. In our experiment, we decide nodes to delete similarly to our proposed method.

[2] http://www.topology-zoo.org/dataset.html.

[3] http://fimi.uantwerpen.be/data/.

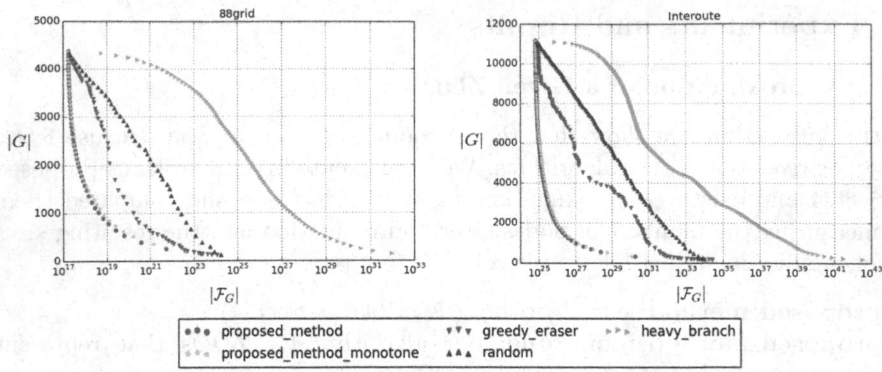

Fig. 7. Relations between size of ZDD G, and $\mathcal{F}_G$ for matching edge sets.

Table 2. Graphs used for on-the-fly approximation.

A graph $G = (V, E)$	$\lvert C\rvert$	#node $\lvert V\rvert$	#edge $\lvert E\rvert$
8×8 grid graph	112	64	112
A real communication network *"Interoute"*	146	110	146

1,000 times fewer false positives for a given ZDD than the algorithm that chooses $(x, y) \in R$ at random, when the ZDD compression rate is 50%. In all cases, our methods achieve the lowest false positive rate in almost all ranges. Our methods restrain the increase in the number of false positives much more than the other methods until they approach the limitation of approximation. When comparing both of our proposed methods, the performance of the method using monotonicity is not significantly inferior even though it does not compute all inclusion relations between nodes. Note that our method cannot delete nodes if no pair of nodes have the inclusion relation.

6.2 On-the-Fly Approximation

We implement our on-the-fly approximated ZDD construction algorithm and compare it with ZDD relaxation [3]. To decide which node is to be incorporated next, we employ the method called H_3 from [3]. In this experiment, we constructed approximated ZDDs that include the matching edge set of the graphs in Table 2. The variable ordering of ZDDs is determined by BFS order. We set various size thresholds to evaluate the number of false positives. If the number of nodes reaches a threshold during construction, we delete 300 nodes and continue construction. For ZDD relaxation, we conduct experiments with various upper limits on ZDD width because ZDD relaxation is a method that bounds the number of nodes in each layer. The results are shown in Fig. 9. Our proposed methods yield about 10–100 times fewer false positives than ZDD relaxation when the approximated ZDD size is halved.

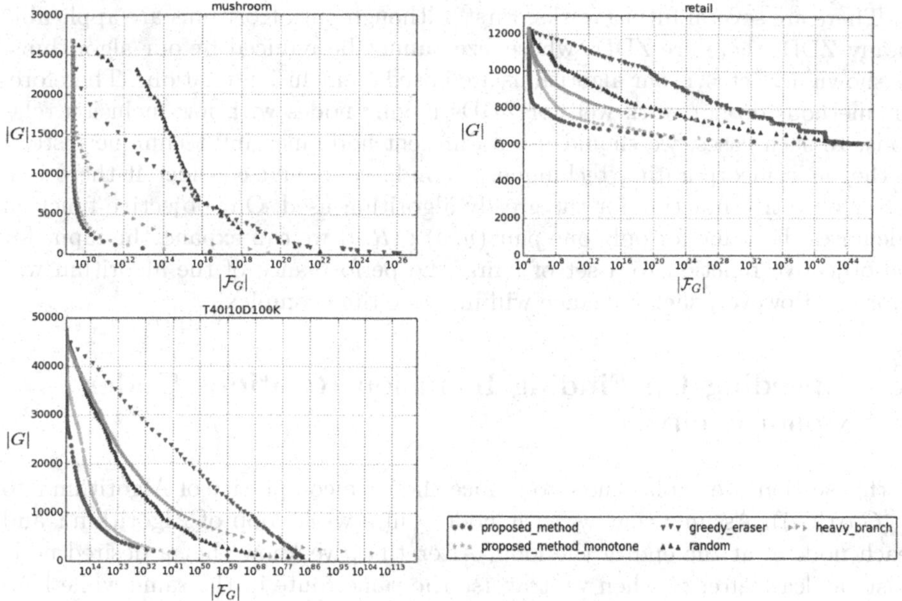

Fig. 8. Relations between size of ZDD G, and $\mathcal{F}_G$ for frequent item sets.

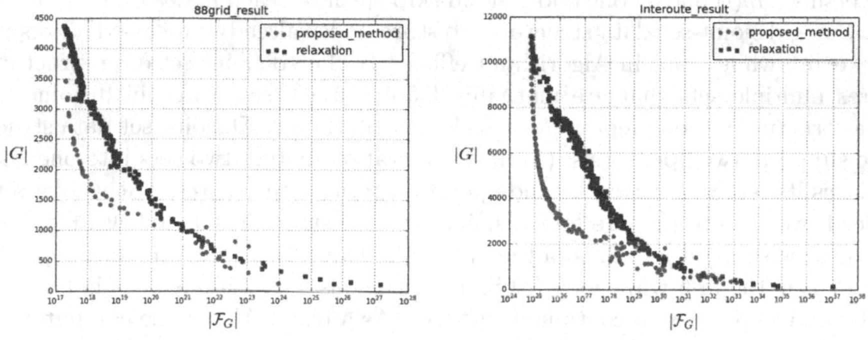

Fig. 9. Relations between size of ZDD, G, and $\mathcal{F}_G$ for on-the-fly approximation.

7 Conclusion

In this paper, we proposed two algorithms to construct approximated ZDDs that are allowed to have false positives. The algorithms compute inclusion relations between pairs of nodes in each layer, choose such a pair based on an objective function, and incorporate them one by one. Our experiments showed that our methods construct ZDDs that have smaller size and fewer errors than other methods.

There are several future works. First, although our algorithms are applicable to any ZDD, there are ZDDs whose size cannot be reduced by our algorithms. As shown in Sect. 4.1, our algorithms are based on inclusion relations. Therefore, our methods do not work well for ZDDs having nodes with few inclusion relations. In such cases, we should use a different node merging technique instead of the node incorporating technique. Second, we should consider if there is a better objective function for the greedy algorithm used. Our objective function calculates the value for only one pair $(u, v) \in R$. If we can expand the input for the objective function to a set of pairs, the performance of the algorithm will improve. However, such a change will increase time complexity.

A Speeding Up Finding Inclusion Relations Under Monotonicity

In this section, we explain how to reduce the time complexity of Algorithm 1 to $O(|G|\alpha(|G|))$. Assume that we visit node q in a while loop of Algorithm 1 and reach node q' at the end of the loop. After that, we know that a desired node exists at least after p' when we traverse the same route in the same while loop because of the loop conditions. Thus, traversing nodes one by one every time is redundant and can be avoided. For example, when we finished processing of nodes in layer L_k, we know that we can ignore the nodes with label smaller than k while computing $sup(\cdot)$ of nodes in lower layers. Therefore, we want to avoid traversing $sup(\cdot)$ one by one and instead skip them to reach nodes in desired layers. Using disjoint-set data structures to store nodes already processed allows us to execute while loops in Algorithm 1 efficiently. The disjoint-set data structure stores multiple sets that are mutually disjoint. Each set stored in disjoint-set data structure has one representative element in the set. Disjoint-set data structure supports two operations: (1) union operation merges two sets into one and updates its representative; (2) find operation returns the representative of a set. Indeed, we have to prepare four disjoint-set data structures as follows: (1) Skip continuous $sup(\cdot)$ if $p \in P_0$ is in the first while loop (2) Skip continuous $sup(\cdot)$ if $p \in P_1$ is in the first while loop (3) Skip continuous $sup(\cdot)$ if $p \in P_0$ is in the second while loop (4) Skip continuous 0-edges. As a result, the whole computation time of Algorithm 1 is $O(|G|\alpha(|G|))$ and its space complexity is $O(|G|)$.

We store node sets in the ZDD by using four disjoint-set data structures DS_1, DS_2, DS_3, DS_4. Each element in DS_i corresponds to a node in the ZDD. Let $ds_i(u)$ be the set that contains u in DS_i and $rep(ds_i(u))$ be the representative of $ds_i(u)$. For each node u in the ZDD, $ds_i(u) = \{u\}$ and $rep(ds_i(u)) = u$ as an initial value. First, we speed up the repeated updates $q \leftarrow sup(q)$ in two while loops by using three disjoint-set data structures DS_1, DS_2, DS_3. These are used in the following situations:

1. If $p \in P_0$ is in the first while loop, we use DS_1.
2. If $p \in P_1$ is in the first while loop, we use DS_2.
3. If $p \in P_0$ is in the second while loop, we use DS_3.

We explain only the case of $p \in P_0$ in the first loop (The other cases are dealt with in the same manner). For each node $p \in P_0$, let $p' = rep(ds_1(p))$ and $q' = rep(ds_1(sup(p')))$. While $u = zp(q', \ell(u))$, we compute the union of $ds(p')$ and $ds(q')$ and set $rep(ds(p')) = q'$. Then, node $zp(q', \ell(u))$ is the candidate of $sup(u)$. Since the union and find operations are executed at most $O(|G|)$ times, these operations run in $O(|G|\alpha(|G|))$ time. Second, we explain how to rapidly calculate $zp(q, \ell(u))$ and $zp(q_1, \ell(u))$. When we see only 0-edges, a ZDD can be considered as a tree whose root is the terminal node $\top$ as shown in Fig. 3. When we consider only the nodes included in $L_1, \dots, L_k$ $(1 \leq k \leq c+1)$, the induced subgraph of the tree is a forest. Note that each node $v \in L_1, \dots, L_k$ belongs to a tree of the forest. If T is a tree of the forest, calculating $zp(q, k)$ is equivalent to finding the node in L_k and an ancestor of q. We use the disjoint-set data structure DS_4 to represent the forest and update it dynamically alongside the processing of layers. When we compute $sup(\cdot)$ of nodes in L_k, for each 0-edge (p, p_0) such that $p_0 \in L_k$, we compute the union of $ds_4(p)$ and $ds_4(p_0)$ and set the representative of this new set to p_0. This process ensures that $zp(v, k) = rep(ds_4(u))$ holds. Therefore, we can compute $zp(q, k)$ by finding $rep(ds_4(q))$ in $O(\alpha(|G|))$ time. The union and find operations are executed $O(|G|)$ times because the number of the 0-edges is $|G|$. Moreover, $zp(\cdot, \cdot)$ is called $O(|G|)$ times. Finally, the whole computation time of this algorithm is $O(|G|\alpha(|G|))$.

References

1. Alstrup, S., Harel, D., Lauridsen, P.W., Thorup, M.: Dominators in linear time. SIAM J. Comput. **28**(6), 2117–2132 (1999)
2. Andersen, H.R., Hadzic, T., Hooker, J.N., Tiedemann, P.: A constraint store based on multivalued decision diagrams. In: Bessière, C. (ed.) CP 2007. LNCS, vol. 4741, pp. 118–132. Springer, Heidelberg (2007). https://doi.org/10.1007/978-3-540-74970-7_11
3. Bergman, D., van Hoeve, W.-J., Hooker, J.N.: Manipulating MDD relaxations for combinatorial optimization. In: Achterberg, T., Beck, J.C. (eds.) CPAIOR 2011. LNCS, vol. 6697, pp. 20–35. Springer, Heidelberg (2011). https://doi.org/10.1007/978-3-642-21311-3_5
4. Bloom, B.H.: Space/time trade-offs in hash coding with allowable errors. Commun. ACM **13**(7), 422–426 (1970)
5. Bollig, B., Wegener, I.: Improving the variable ordering of OBDDs is NP-complete. IEEE Trans. Comput. **45**(9), 993–1002 (1996)
6. Brace, K.S., Rudell, R.L., Bryant, R.E.: Efficient implementation of a BDD package. In: Proceedings of DAC 1990, pp. 40–45 (1990)
7. Bryant, R.E.: Graph-based algorithms for Boolean function manipulation. IEEE Trans. Comput. **35**, 677–691 (1986)
8. Galler, B.A., Fisher, M.J.: An improved equivalence algorithm. Commun. ACM **7**(5), 301–303 (1964)
9. Hadzic, T., Hooker, J.N., O'Sullivan, B., Tiedemann, P.: Approximate compilation of constraints into multivalued decision diagrams. In: Proceedings of CP 2008, pp. 448–462 (2008)
10. Knight, S., Nguyen, H.X., Falkner, N., Bowden, R., Roughan, M.: The internet topology zoo. IEEE J. Sel. Areas Commun. **29**(9), 1765–1775 (2011)

11. Knuth, D.E.: The Art of Computer Programming. Combinatorial Algorithms, Part 1, vol. 4A, 1st edn. Addison-Wesley Professional, Boston (2011)
12. Lengauer, T., Tarjan, R.E.: A fast algorithm for finding dominators in a flowgraph. ACM Trans. Prog. Lang. Syst. **1**(1), 121–141 (1979)
13. Minato, S.: Zero-suppressed BDDs for set manipulation in combinatorial problems. In: Proceedings of DAC 1993, pp. 272–277 (1993)
14. Ravi, K., McMillan, K.L., Shiple, T.R., Somenzi, F.: Approximation and decomposition of binary decision diagrams. In: Proceedings of DAC 1998, pp. 445–450 (1998)
15. Ravi, K., Somenzi, F.: High-density reachability analysis. In: Proceedings of ICCAD 1995, pp. 154–158 (1995)
16. Rudell, R.: Dynamic variable ordering for ordered binary decision diagrams. In: Proceedings of ICCAD 1993, pp. 42–47 (1993)
17. Sekine, K., Imai, H., Tani, S.: Computing the Tutte polynomial of a graph of moderate size. In: Proceedings of ISAAC 1995, pp. 224–233 (1995)
18. Soeken, M., Große, D., Chandrasekharan, A., Drechsler, R.: BDD minimization for approximate computing. In: Proceedings of ASP-DAC 2016, pp. 474–479 (2016)
19. Tarjan, R.E., van Leeuwen, J.: Worst-case analysis of set union algorithms. J. ACM **31**(2), 245–281 (1984)

Efficient Implementation of Color Coding Algorithm for Subgraph Isomorphism Problem

Josef Malík(✉), Ondřej Suchý, and Tomáš Valla

Department of Theoretical Computer Science, Faculty of Information Technology, Czech Technical University in Prague, Prague, Czech Republic
{josef.malik,ondrej.suchy,tomas.valla}@fit.cvut.cz

Abstract. We consider the subgraph isomorphism problem where, given two graphs G (source graph) and F (pattern graph), one is to decide whether there is a (not necessarily induced) subgraph of G isomorphic to F. While many practical heuristic algorithms have been developed for the problem, as pointed out by McCreesh et al. [JAIR 2018], for each of them there are rather small instances which they cannot cope. Therefore, developing an alternative approach that could possibly cope with these hard instances would be of interest.

A seminal paper by Alon, Yuster and Zwick [J. ACM 1995] introduced the color coding approach to solve the problem, where the main part is a dynamic programming over color subsets and partial mappings. As with many exponential-time dynamic programming algorithms, the memory requirements constitute the main limiting factor for its usage. Because these requirements grow exponentially with the treewidth of the pattern graph, all existing implementations based on the color coding principle restrict themselves to specific pattern graphs, e.g., paths or trees. In contrast, we provide an efficient implementation of the algorithm significantly reducing its memory requirements so that it can be used for pattern graphs of larger treewidth. Moreover, our implementation not only decides the existence of an isomorphic subgraph, but it also enumerates all such subgraphs (or given number of them).

We provide an extensive experimental comparison of our implementation to other available solvers for the problem.

Keywords: Subgraph isomorphism · Subgraph enumeration · Color coding · Tree decomposition · Treewidth

1 Introduction

Many real-world domains incorporate large and complex networks of interconnected units. Examples include social networks, the Internet, or biological and

J. Malík—Supported by grant 17-20065S of the Czech Science Foundation.
O. Suchý and T. Valla—The author acknowledges the support of the OP VVV MEYS funded project CZ.02.1.01/0.0/0.0/16_019/0000765 "Research Center for Informatics".

I. Kotsireas et al. (Eds.): SEA² 2019, LNCS 11544, pp. 283–299, 2019.
https://doi.org/10.1007/978-3-030-34029-2_19

chemical systems. These networks raise interesting questions regarding their structure. One of those questions asks whether a given network contains a particular pattern, which typically represents a specific behaviour of interest [1,4,12]. The problem of locating a particular pattern in the given network can be restated as a problem of locating a subgraph isomorphic to the given pattern graph in the network graph.

Formally, the SUBGRAPH ISOMORPHISM (SUBISO) problem is, given two undirected graphs G and F, to decide whether there is a (not necessarily induced) subgraph of G isomorphic to F. Or, in other words, whether there is an adjacency-preserving injective mapping from vertices of F to vertices of G. Since we do not require the subgraph to be induced (or the mapping to preserve non-adjacencies), some authors call this variant of the problem SUBGRAPH MONOMORPHISM.

For many applications it is not enough to just learn that the pattern does occur in the network, but it is necessary to actually obtain the location of an occurrence of the pattern or rather of all occurrences of the pattern [18,25]. Because of that, we aim to solve the problem of subgraph enumeration, in which it is required to output all subgraphs of the network graph isomorphic to the pattern graph. In SUBGRAPH ENUMERATION (SUBENUM), given again two graphs G and F, the goal is to enumerate all subgraphs of G isomorphic to F. Note, that SUBENUM is at least as hard as SUBISO. We call the variants, where the problem is required to be induced INDSUBISO and INDSUBENUM, respectively.

As CLIQUE, one of the problems on the Karp's original list of 21 NP-complete problems [15], is a special case of SUBISO, the problem is NP-complete. Nevertheless, there are many heuristic algorithms for SUBENUM, many of them based on ideas from constraint programming (see Sect. 1.1), which give results in reasonable time for most instances. However, for each of them there are rather small instances which they find genuinely hard, as pointed out by McCreesh et al. [23]. Therefore, developing an alternative approach that could possibly cope with these hard instances would be of interest.

In this paper we focus on the well known randomized color coding approach [2], which presumably has almost optimal worst case time complexity. Indeed, its time complexity is $\mathcal{O}\big(n_G^{\mathrm{TW}(F)+1}2^{\mathcal{O}(n_F)}\big)$ with memory requirements of $\mathcal{O}\big(n_G^{\mathrm{TW}(F)+1}\mathrm{TW}(F)n_F 2^{n_F}\big)$, where n_G and n_F denote the number of vertices in the network graph G and the pattern graph F, respectively, and $\mathrm{TW}(F)$ is the treewidth of graph F—a measure of tree-likeness (see Sect. 1.2 for exact definitions). Moreover, we presumably cannot avoid the factor exponential in treewidth in the worst case running time, as Marx [21] presented an ETH[1]-based lower bound for PARTITIONED SUBGRAPH ISOMORPHISM problem.

Proposition 1 (Marx [21]). *If there is a recursively enumerable class $\mathcal{F}$ of graphs with unbounded treewidth, an algorithm $\mathcal{A}$, and an arbitrary function f such that $\mathcal{A}$ correctly decides every instance of* PARTITIONED SUBGRAPH

[1] Exponential Time Hypothesis [14].

Isomorphism *with the smaller graph* F *in* $\mathcal{F}$ *in time* $f(F)n_G^{o(\text{TW}(F)/\log \text{TW}(F))}$, *then ETH fails.*

As the memory requirements of the color coding approach grow exponentially with treewidth of the pattern graph, existing implementations for subgraph enumeration based on this principle restrict themselves to paths [13] or trees [25], both having treewidth 1. As the real world applications might employ networks of possibly tens to hundreds of thousands of vertices and also pattern graphs with structure more complicated than trees, we need to significantly reduce the memory usage of the algorithm.

Using the principle of inclusion-exclusion, Amini et al. [3, Theorem 15] suggested a modification of the color coding algorithm, which can decide whether the pattern F occurs in the graph G in expected time $\mathcal{O}\big(n_G^{\text{TW}(F)+1}2^{\mathcal{O}(n_F)}\big)$ with memory requirements reduced to $\mathcal{O}\big(n_G^{\text{TW}(F)+1}\log n_F\big)$.[2] While single witnessing occurrence can be found by means of self-reduction (which is complicated in case of randomized algorithm), the inclusion-exclusion nature of the algorithm does not allow to find all occurrences of pattern in the graph, which is our main goal.

Therefore, our approach rather follows the paradigm of generating only those parts of a dynamic programming table that correspond to subproblems with a positive answer, recently called "positive instance driven" approach [28]. This further prohibits the use of the inclusion-exclusion approach of Amini et al. [3], since the inclusion-exclusion approach tends to use most of the table and the term $\mathcal{O}\big(n_G^{\text{TW}(F)+1}\big)$ is itself prohibitive in the memory requirements for $\text{TW}(F) \geq 2$.

Because of the time and memory requirements of the algorithm, for practical purposes we restrict ourselves to pattern graphs with at most 32 vertices.

Altogether, our main contribution is twofold:

- We provide a practical implementation of the color coding algorithm of Alon, Yuster, and Zwick [2] capable of processing large networks and (possibly disconnected) pattern graphs of small, yet not a priory bounded, treewidth.
- We supply a routine to extract the occurrences of the subgraphs found from a run of the algorithm.

It is important to note that all the modifications only improve the practical memory requirements and running time. The theoretical worst case time and space complexity remain the same as for the original color coding algorithm and the algorithm achieves these, e.g., if the network graph is complete. Also, in such a case, there are $n_G^{\Theta(n_F)}$ occurrences of the pattern graph in the network implying a lower bound on the running time of the enumeration part.

In Sect. 2 we describe our modifications to the algorithm and necessary tools used in the process. Then, in Sect. 3, we benchmark our algorithm on synthetic and realistic data and compare its performance with available existing implementations of algorithms for subgraph isomorphism and discuss the results obtained.

[2] While the formulation of Theorem 15 in [3] might suggest that the algorithm actually outputs a witnessing occurrence, the algorithm merely decides whether the number of occurrences is non-zero (see the proof of the theorem).

Section 4 presents future research directions. Parts of the paper not present in this extended abstract due to space restrictions, can be found it the ArXiv preprint [20] or in the full version of the paper.

1.1 Related Work

There are several algorithms tackling SubIso and its related variants. Some of them only solve the variant of subgraph counting, our main focus is however on algorithms actually solving SubEnum. Following Carletti et al. [7] and Kimmig et al. [16], we categorize the algorithms by the approach they use (see also Kotthoff et al. [17] for more detailed description of the algorithms). Many of the approaches can be used both for induced and non-induced variants of the problem, while some algorithms are applicable only for one of them.

Vast majority of known algorithms for the subgraph enumeration problem is based on the approach of representing the problem as a searching process. Usually, the state space is modelled as a tree and its nodes represent a state of a partial mapping. Finding a solution then typically resorts to the usage of DFS in order to find a path of mappings in the state space tree which is compliant with isomorphism requirements. The efficiency of those algorithms is largely based on early pruning of unprofitable paths in the state space. Indeed, McCreesh et al. [23] even measure the efficiency in the number of generated search tree nodes. The most prominent algorithms based on this idea are Ullmann's algorithm [29], VF algorithm and its variants [5,7,9,10] (the latest VF3 [5] only applies to IndSubEnum) and RI algorithm [4]. The differences between these algorithms are based both on employed pruning strategies and on the order in which the vertices of pattern graph are processed (i.e. in the shape of the state space tree).

Another approach is based on constraint programming, in which the problem is modelled as a set of variables (with respective domains) and constraints restricting simultaneous variable assignments. The solution is an assignment of values to variables in a way such that no constraint remains unsatisfied. In subgraph isomorphism, variables represent pattern graph vertices, their domain consists of target graph vertices to which they may be mapped and constraints ensure that the properties of isomorphism remain satisfied. Also in this approach, a state space of assignments is represented by a search tree, in which non-profitable branches are to be filtered. Typical algorithms in this category are LAD algorithm [26], Ullmann's bitvector algorithm [30], and Glasgow algorithm [22]. These algorithms differ in the constraints they use, the way they propagate constraints, and in the way they filter state space tree.

There are already some implementations based on the color coding paradigm, where the idea is to randomly color the input graph and search only for its subgraphs, isomorphic to the pattern graph, that are colored in distinct colors (see Sect. 2.1 for more detailed description). This approach is used in subgraph counting algorithms, e.g., in ParSE [31], FASCIA [24], and in [1], or in algorithms for path enumeration described in [25] or in [13]. Each of these algorithms, after the color coding step, tries to exploit the benefits offered by this technique in its own way; although usually a dynamic programming sees its use. Counting algorithms as ParSE and FASCIA make use of specifically partitioned pattern graphs, which allow to use combinatorial computation. Weighted path enumeration algorithms [13,25] describe a dynamic programming approach and try to optimize it in various ways. However, to the best of our knowledge there is no color coding algorithm capable of enumerating patterns of treewidth larger than 1.

Our aim is to make step towards competitive implementation of color coding based algorithm for SubEnum, in order to see, where this approach can be potentially beneficial against the existing algorithms. To this end, we extend the comparisons of SubEnum algorithms [6,17,23] to color coding based algorithms, including the one proposed in this paper.

1.2 Basic Definitions

All graphs in this paper are undirected and simple. For a graph G we denote $V(G)$ its vertex set, n_G the size of this set, $E(G)$ its edge set, and m_G the size of its edge set.

As already said, we use the color coding algorithm. The algorithm is based on a dynamic programming on a nice tree decomposition of the pattern graph. We first define a tree decomposition and then its nice counterpart.

Definition 1. *A* tree decomposition *of a graph F is a triple (T, β, r), where T is a tree rooted at node r and $\beta\colon V(T) \mapsto 2^{V(F)}$ is a mapping satisfying: (i) $\bigcup_{x \in V(T)} \beta(x) = V(F)$; (ii) $\forall \{u, v\} \in E(F)\ \exists x \in V(T)$, such that $u, v \in \beta(x)$; (iii) $\forall u \in V(F)$ the nodes $\{x \in V(T) \mid u \in \beta(x)\}$ form a connected subtree of T.*

We shall denote bag $\beta(x)$ as $\mathcal{V}_x$. The width of tree decomposition (T, β, r) is $\max_{x \in V(T)} |\mathcal{V}_x| - 1$. Treewidth $\text{TW}(F)$ of graph F is the minimal width of a tree decomposition of F over all such decompositions.

Definition 2. *A tree decomposition of a graph F is* nice *if $\deg_T(r) = 1$, $\mathcal{V}_r = \emptyset$, and each node $x \in V(T)$ is of one of the following four types:*

- Leaf node—*x has no children and $|\mathcal{V}_x| = 1$;*
- Introduce node—*x has exactly one child y and $\mathcal{V}_x = \mathcal{V}_y \cup \{u\}$ for some $u \in V(F) \setminus \mathcal{V}_y$;*
- Forget node—*x has exactly one child y and $\mathcal{V}_x = \mathcal{V}_y \setminus \{u\}$ for some $u \in \mathcal{V}_y$;*
- Join node—*x has exactly two children y, z and $\mathcal{V}_x = \mathcal{V}_y = \mathcal{V}_z$.*

Note that for practical purposes, we use a slightly modified definition of nice tree decomposition in this paper. As the algorithm starts the computation in a leaf node, using the standard definition with empty bags of leaves [11] would imply that the tables for leaves would be somewhat meaningless and redundant. Therefore, we make bags of leaf nodes contain a single vertex.

Definition 3. *For a tree decomposition (T, β, r), we denote by $\mathcal{V}^*_x$ the set of vertices in $\mathcal{V}_x$ and in $\mathcal{V}_y$ for all descendants y of x in T. Formally $\mathcal{V}^*_x = \mathcal{V}_x \cup \bigcup_{y \text{ is a descendant of } x \text{ in } T} \mathcal{V}_y$.*

Note that, by Definition 3, for the root r of T we have $\mathcal{V}^*_r = V(F)$ and $F[\mathcal{V}^*_r] = F$.

2 Algorithm Description

In this section we first briefly describe the idea of the original color coding algorithm [2], show, how to alter the computation in order to reduce its time and memory requirements, and describe implementation details and further optimizations of the algorithm. Due to space restrictions, the way to obtain a nice tree decomposition of the pattern and the reconstruction of results are deferred to the full version of the paper.

2.1 Idea of the Algorithm

The critical idea of color coding is to reduce the problem to its colorful version. For a graph G and a pattern graph F, we color the vertices of G with exactly n_F colors. We use the randomized version, i.e., we create a random coloring $\zeta\colon V(G) \mapsto \{1, 2, \ldots, n_F\}$. After the coloring, the algorithm considers as valid only subgraphs G' of G that are colorful copies of F as follows.

Definition 4. *Subgraph G' of a graph G is a* colorful copy *of F with respect to coloring $\zeta\colon V(G) \mapsto \{1, 2, \ldots, n_F\}$, if G' is isomorphic to F and all of its vertices are colored by distinct colors in ζ.*

As the output of the algorithm heavily depends on the chosen random coloring of G, in order to reach some predefined success rate of the algorithm, we need to repeat the process of coloring several times. The probability of a particular occurrence of pattern graph F becoming colorful with respect to the random coloring is $\frac{n_F!}{n_F^{n_F}}$, which tends to e^{-n_F} for large n_F. Therefore, by running the algorithm $e^{n_F \log \frac{1}{\varepsilon}}$ times, each time with a random coloring $\zeta\colon V(G) \mapsto \{1, 2, \ldots, n_F\}$, the probability that an existing occurrence of the pattern will be revealed in none of the runs is at most ε. While using more colors can reduce the number of iterations needed, it also significantly increases the memory requirements. Hence, we stick to n_F colors. Even though it is possible to derandomize such algorithms, e.g., by the approach shown in [11], in practice the randomized approach usually yields the results much quicker, as discussed in [25]. Moreover, we are not aware of any actual implementation of the derandomization methods.

The main computational part of the algorithm is a dynamic programming. The target is to create a graph isomorphism $\Phi\colon V(F) \mapsto V(G)$. We do so by traversing the nice tree decomposition (T, β, r) of the pattern graph F and at each node $x \in V(T)$ of the tree decomposition, we construct possible partial mappings $\varphi\colon \mathcal{V}_x^* \to V(G)$ with regard to required colorfulness of the copy. Combination of partial mappings consistent in colorings then forms a desired resulting mapping.

The semantics of the dynamic programming table is as follows. For any tree decomposition node $x \in V(T)$, any partial mapping $\varphi\colon \mathcal{V}_x \mapsto V(G)$ and any color subset $C \subseteq \{1, 2, \ldots, n_F\}$, we define $\mathcal{D}(x, \varphi, C) = 1$ if there is an isomorphism Φ of $F[\mathcal{V}_x^*]$ to a subgraph G' of G such that:

(i) for all $u \in \mathcal{V}_x, \Phi(u) = \varphi(u)$;
(ii) G' is a colorful copy of $F[\mathcal{V}_x^*]$ using exactly the colors in C, that is, $\zeta(\Phi(\mathcal{V}_x^*)) = C$ and ζ is injective on $\Phi(\mathcal{V}_x^*)$.

If there is no such isomorphism, then we let $\mathcal{D}(x, \varphi, C) = 0$. We denote all configurations (x, φ, C) for which $\mathcal{D}(x, \varphi, C) = 1$ as *nonzero* configurations.

The original version of the algorithm is based on top-down dynamic programming approach with memoization of already computed results. That immediately implies a big disadvantage of this approach—it requires the underlying dynamic programming table (which is used for memoization) to be fully available throughout the whole run of the algorithm. To avoid this inefficiency in our modification we aim to store only nonzero configurations, similarly to the recent "positive instance driven" dynamic programming approach [28].

2.2 Initial Algorithm Modification

In our implementation, we aim to store only nonzero configurations, therefore we need to be able to construct nonzero configurations of a parent node just from the list of nonzero configurations in its child/children.

We divide the dynamic programming table $\mathcal{D}$ into lists of nonzero configurations, where each nice tree decomposition node has a list of its own. Formally, for every node $x \in V(T)$, let us denote by $\mathcal{D}_x$ a list of all mappings φ with a list of their corresponding color sets C, for which $\mathcal{D}(x, \varphi, C) = 1$. The list $\mathcal{D}_x$ for all $x \in V(T)$ is, in terms of contained information, equivalent to maintaining the whole table $\mathcal{D}$—all configurations not present in the lists can be considered as configurations with a result equal to zero.

Dynamic Programming Description. We now describe how to compute the lists $\mathcal{D}(x, \varphi, C)$ for each type of a nice tree decomposition node.

For a *leaf* node $x \in T$, there is only a single vertex u in $\mathcal{V}_x^*$ to consider. We can thus map u to all possible vertices of G, and we obtain a list with n_G partial mappings φ, in which the color list for each mapping contains a single color set $\{\zeta(\varphi(u))\}$.

For an *introduce* node $x \in T$ and its child y in T, we denote by u the vertex being introduced in x, i.e., $\{u\} = \mathcal{V}_x \backslash \mathcal{V}_y$. For all nonzero combinations of a partial mapping and a color set (φ', C') in the list $\mathcal{D}_y$, we try to extend φ' by all possible mappings of the vertex u to the vertices of G. We denote one such a mapping as φ. We can consider mapping φ as correct, if (i) the new mapping $\varphi(u)$ of the vertex u extends the previous colorset C', that is, $C = C' \cup \{\zeta(\varphi(u))\} \neq C'$, and (ii) φ is *edge consistent*, that is, for all edges $\{v, w\} \in E(F)$ between currently mapped vertices, i.e., in our case $v, w \in \mathcal{V}_x$, there must be an edge $\{\varphi(v), \varphi(w)\} \in E(G)$. However, because φ' was by construction already edge consistent, it suffices to check the edge consistency only for all edges in $F[\mathcal{V}_x]$ with u as one of their endpoints, i.e., for all edges $\{u, w\} \in E(F[\mathcal{V}_x])$ with $w \in N_{F[\mathcal{V}_x]}(u)$. After checking those two conditions, we can add (φ, C) to $\mathcal{D}_x$.

Due to space restrictions, the computation in forget and join nodes is deferred to the full version of the paper.

Because we build the result from the leaves of the nice tree decomposition, we employ a recursive procedure on its root, in which we perform the computations in a way of a post-order traversal of a tree. From each visited node, we obtain a bottom-up dynamic programming list of nonzero configurations. After the whole nice tree decomposition is traversed, we obtain a list of configurations, that were valid in its root. Such configurations thus represent solutions found during the algorithm, from which we afterwards reconstruct results. Note that as we prepend a root with no vertices in its bag to the nice tree decomposition, there is a nonzero number of solutions if and only if, at the end of the algorithm, the list $\mathcal{D}_r$ contains a single empty mapping using all colors.

2.3 Further Implementation Optimizations

Representation of Mappings. For mapping representation, we suppose that the content of all bags of the nice tree decomposition stays in the same order during the whole algorithm. This natural and easily satisfied condition allows us to represent a mapping $\varphi\colon \mathcal{V}_x \mapsto V(G)$ in a nice tree decomposition node x simply by an ordered tuple of $|\mathcal{V}_x|$ vertices from G. From this, we can easily determine which vertex from F is mapped to which vertex in G. Also, for a mapping in an introduce or a forget node, we can describe a position in the mapping, on which the process of introducing/forgetting takes place.

Representation of Color Sets. We represent color sets as bitmasks, where the i-th bit states whether color i is contained in the set or not. For optimization purposes, we represent bitmasks with an integer number. As we use n_F colors in the algorithm and restricted ourselves to pattern graphs with at most 32 vertices, we represent a color set with a 32-bit number.

Compressing the Lists. Because we process the dynamic programming lists one mapping at a time, we store these lists in a compressed way and decompress them only on a mapping retrieval basis. Due to space restrictions, the exact way we serialize the records, the use of delta compression and a special library is deferred to the full version of the paper.

Masking Unprofitable Mappings. Our implementation supports an extended format of input graphs where one can specify for each vertex of the network, which vertices of the pattern can be mapped to it. This immediately yields a simple degree-based optimization. Before the run of the main algorithm, we perform a linear time preprocessing of input graphs and only allow a vertex $y \in V(F)$ to be mapped to a vertex $x \in V(G)$ if $\deg_G(x) \geq \deg_F(y)$.

Mapping Expansion Optimizations. The main "brute-force" work of the algorithm is performed in two types of nodes—leaf and introduce nodes, as we need to try all possible mappings of a particular vertex in a leaf node or all possible mappings of an introduced vertex in a introduce node to a vertex from G. We describe ways to optimize the work in introduce nodes in this paragraph.

Let x be an introduce node, u the vertex introduced and φ a mapping from a nonzero configuration for the child of x. We always need to check whether the new mapping of u is edge consistent with the mapping φ of the remaining vertices for the corresponding bag, i.e., whether all edges of F incident on u would be realized by an edge in G. Therefore, if u has any neighbors in $F[\mathcal{V}_x]$, then a vertex of G is a candidate for the mapping of u only if it is a neighbor of all vertices in the set $\varphi(N_{F[\mathcal{V}_x]}(u))$, i.e., the vertices of G, where the neighbors of u in F are mapped. Hence, we limit the number of candidates by using the adjacency lists of the already mapped vertices.

In the case $\deg_{F[\mathcal{V}_x]}(u) = 0$ we have to use different approach. The pattern graphs F tend to be smaller than the input graphs G by several orders of magnitude. Hence, if the introduced vertex is in the same connected component of F as some vertex already present in the bag, a partial mapping processed in an introduce node anchors the possible resulting component to a certain position in G. Due to space restrictions, the exact way to exploit that is deferred to the full version of the paper.

Only if there is no vertex in the bag sharing a connected component of F with u, we have to fall back to trying all possible mappings.

3 Experimental Results

The testing was performed on a 64-bit linux system with Intel Xeon CPU E3-1245v6@3.70GHz and 32 GB 1333 MHz DDR3 SDRAM memory. The module was compiled with `gcc` compiler (version 7.3.1) with `-O3` optimizations enabled. Implementation and instances utilized in the testing are available at http://users.fit.cvut.cz/malikjo1/subiso/. All results are an average of 5 independent measurements.

We evaluated our implementation in several ways. Firstly, we compare available implementations on two different real world source graphs and a set of more-or-less standard target graph patterns. Secondly, we compare available implementations on instances from ICPR2014 Contest on Graph Matching Algorithms for Pattern Search in Biological Databases [8] with suitably small patterns. We also adapt the idea of testing the algorithms on Erdős-Rényi random graphs [23].

3.1 Algorithm Properties and Performance

In the first two subsection we used two different graphs of various properties as target graph G. The first instance, Images, is built from an segmented image, and is a courtesy of [27]. It consists of 4838 vertices and 7067 edges. The second instance, TRANS, is a graph of transfers on bank accounts. It is a very sparse network, which consists of 45733 vertices and 44727 undirected edges. Due to space restrictions, the results on this dataset are deferred to the full version of the paper.

For the pattern graphs, we first use a standard set of basic graph patterns, as the treewidth of such graphs is well known and allows a clear interpretation of the results. In particular, we use paths, stars, cycles, an complete graphs on n vertices, denoted P_n, S_n, C_n, and K_n with treewidth 1, 1, 2, and $n-1$, respectively. We further used grids $G_{n,m}$ on $n \times m$ vertices, with treewidth $\min\{n, m\}$. Secondly, we use a special set of pattern graphs in order to demonstrate performance on various patterns. Patterns A, B, C, and D have 9, 7, 9, and 7 vertices, 8, 7, 12, 6 edges, and treewidth 1, 2, 2, and 2, respectively. Patterns A, B, and D appear in both dataset, pattern C in neither and pattern D is disconnected. Description of these pattern graphs is deferred to the full version of the paper.

Due to randomization, in order to achieve some preselected constant error rate, we need to repeat the computation more than once. The number of found results thus depends not only on the quality of the algorithm, but also on the choice of the number of its repetitions. Hence, it is logical to measure performance of the single run of the algorithm. Results from such a testing, however, should be still taken as a rough average, because the running time of a single run of the algorithm depends on many factors.

Therefore, we first present measurements, where we average the results of many single runs of the algorithm (Table 1). We average not only the time and space needed, but also the number of found subgraphs. To obtain the expected time needed to run the whole algorithm, it suffices to sum the time needed to create a nice tree decomposition and ℓ times the time required for a single run, if there are ℓ runs in total.

Table 1. Performance of a single run of the algorithm on IMAGES dataset.

Pattern	Comp. time [ms]	Comp. memory [MB]	Occurrences
$\mathcal{P}_5$	240	12.73	3488.21
$\mathcal{P}_{10}$	160	8.52	732.46
$\mathcal{P}_{15}$	90	10.54	76.18
$\mathcal{S}_5$	4	5.37	114.72
$\mathcal{C}_5$	20	7.24	239.17
$\mathcal{C}_{10}$	70	9.34	26.64
$\mathcal{K}_4$	5	6.46	0
$G_{3,3}$	90	13.42	0
Pattern A	80	9.14	292.48
Pattern B	10	7.17	6.85
Pattern C	10	5.30	0
Pattern D	40	10.14	426.76

3.2 Comparison on Real World Graphs and Fixed Graph Patterns

We compare our implementation to three other tools for subgraph enumeration: RI algorithm [4] (as implemented in [19]), LAD algorithm [26] and color coding algorithm for weighted path enumeration [13] (by setting, for comparison purposes, all weights of edges to be equal). The comparison is done on the instances from previous subsection and only on pattern graphs which occur at least once in a particular target graph.

In comparison, note the following specifics of measured algorithms. The RI algorithm does not support outputting solutions, which might positively affect its performance. LAD algorithm uses adjacency matrix to store input graphs, and thus yields potentially limited use for graphs of larger scale. Neither of RI or LAD algorithms supports enumeration of disconnected patterns.[3] Also we did not measure the running time of the weighted path algorithm on non-path queries and also on TRANS dataset, as its implementation is limited to graph sizes of at most 32 000.

We run our algorithm repeatedly to achieve an error rate of $\varepsilon = \frac{1}{e}$. In order to be able to measure the computation for larger networks with many occurrences of the pattern, we measure only the time required to retrieve no more than first 100 000 solutions and we also consider running time greater than 10 min (600 s) as a timeout. Since we study non-induced occurrences (and due to automorphisms) there might be several ways to map the pattern to the same set of vertices. Other measured algorithms do count all of them. Our algorithm can behave also like this, or can be switched to count only occurrences that differ in vertex sets. For the sake of equal measurement, we use the former version of our algorithm.

From Table 2, we can see that RI algorithm outperforms all other measured algorithms. We can also say our algorithm is on par with LAD algorithm, as the results of comparison of running times are similar, but vary instance from instance. Our algorithm nevertheless clearly outperforms another color coding algorithm, which on one hand solves more complicated problem of weighted paths, but on the another, is still limited only to paths. Also, our algorithm is the only algorithm capable of enumerating disconnected patterns.

The weak point of the color coding approach (or possibly only of our implementation) appears to be the search for a pattern of larger size with very few (or possibly zero) occurrences. To achieve the desired error rate, we need to repeatedly run the algorithm many times. Therefore our algorithm takes longer time to run on some instances (especially close to zero-occurrence ones), which are easily solved by the other algorithms.

---

[3] When dealing with disconnected patterns, one could find the components of the pattern one by one, omitting the vertices of the host graph used by the previous component. However, this would basically raise the running time of the algorithm to the power equal to the number of components of the pattern graph.

Table 2. Comparison of running time on IMAGES dataset (in seconds).

Pattern	Our algorithm	RI algorithm	LAD algorithm	Weighted path
$\mathcal{P}_5$	31.12	0.11	28.86	362.41
$\mathcal{P}_{10}$	53.17	1.25	13.63	> 600
$\mathcal{P}_{15}$	104.30	3.7	8.18	> 600
$\mathcal{S}_5$	0.94	0.07	0.43	–
$\mathcal{C}_5$	4.98	0.14	35.18	–
$\mathcal{C}_{10}$	151.25	3.44	174.27	–
Pattern A	43.11	0.82	36.60	–
Pattern B	91.93	0.41	0.83	–
Pattern D	23.54	–	–	–

3.3 ICPR2014 Contest Graphs

To fully benchmark our algorithm without limitations on time or number of occurrences found, we perform a test on ICPR2014 Contest on Graph Matching Algorithms for Pattern Search in Biological Databases [8].

In particular, we focus our attention on a MOLECULES dataset, containing 10,000 (target) graphs representing the chemical structures of different small organic compounds and on a PROTEINS dataset, which contains 300 (target) graphs representing the chemical structures of proteins and protein backbones. Target graphs in both datasets are sparse and up to 99 vertices or up 10,081 vertices for MOLECULES and PROTEINS, respectively.

In order to benchmark our algorithm without limiting its number of iterations, we focus on pattern graphs of small sizes, which offer reasonable number of iterations for an error rate of $\frac{1}{e}$. Both datasets contain 10 patterns for each of considered sizes constructed by randomly choosing connected subgraphs of the target graphs. We obtained an average matching time of all pattern graphs of a given size to all target graphs in a particular dataset.

Table 3. Comparison of average running time on ICPR2014 graphs

Targets	Pattern size	Our algorithm	LAD algorithm	RI algorithm
MOLECULES	4	0.01	0.01	0.01
MOLECULES	8	0.67	0.14	0.01
XSC PROTEINS	8	19.45	8.83	0.51

From the results in Table 3, we can see our algorithm being on par with LAD algorithm, while being outperformed by RI algorithm. However, we mainly include these results as a proof of versatility of our algorithm. As discussed in [23], benchmarks created by constructing subgraphs of target graphs do not necessarily encompass the possible hardness of some instances and might even present a distorted view on algorithms' general performance. Thus, in the following benchmark we opt to theoretically analyze our algorithm.

3.4 Erdős-Rényi Graph Setup

In order to precisely analyze the strong and weak points of our algorithm we measure its performance is a setting where both the pattern and the target are taken as an Erdős-Rényi random graph of fixed size with varying edge density and compare the performance of our algorithm with the analysis of McCreesh et al. [23], which focused on algorithms Glasgow, LAD, and VF2.

An Erdős-Rényi graph $G(n, p)$ is a random graph on n vertices where each edge is included in the graph independently at random with probability p. We measure the performance on target graph of 150 vertices and pattern graph of 10 vertices with variable edge probabilities. As our algorithm cannot be classified in terms of search nodes used (as in [23]), we measure the time needed to complete 10 iterations of our algorithm.

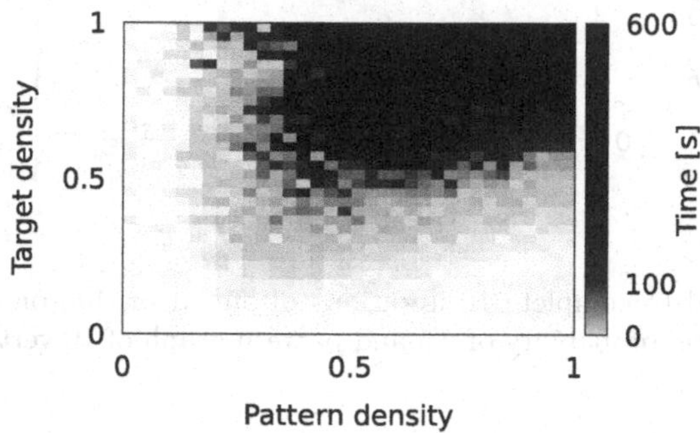

Fig. 1. Behavior for target graph of 150 vertices and pattern graph of 10 vertices. The x-axis is the pattern edge probability, the y-axis is the target edge probability, from 0 to 1 with step of 0.03. Graph shows the time required for our algorithm to complete 10 iterations (the darker, the more time is required). Black regions indicate instances on which a timeout of 600 s occurred.

From Fig. 3 we can see our algorithm indeed follows a well observed phase transition (transition between instances without occurrence of the pattern and with many occurrences of the pattern). If we compare our results from Fig. 1 to the results of [23], we can see that hard instances for our algorithm start to occur later (in terms of edge probabilities). However, due to the almost linear dependency of treewidth on edge probabilities (see Fig. 2), hard instances for our algorithm concentrate in the "upper right corner" of the diagram, which contains dense graphs with naturally large treewidth (Fig. 4).

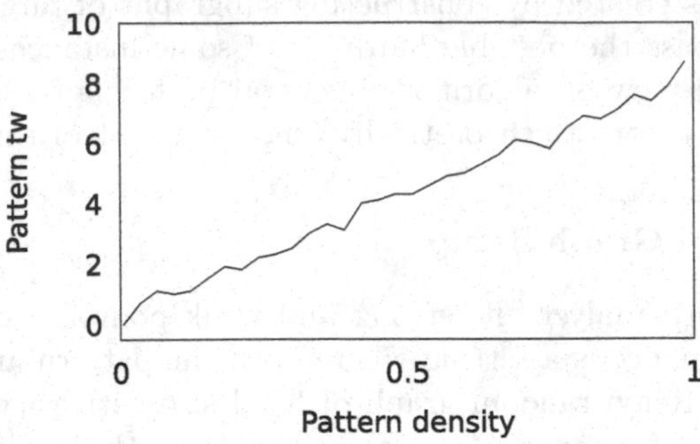

Fig. 2. Correspondence of treewidth to the edge probability of a pattern graph with 10 vertices.

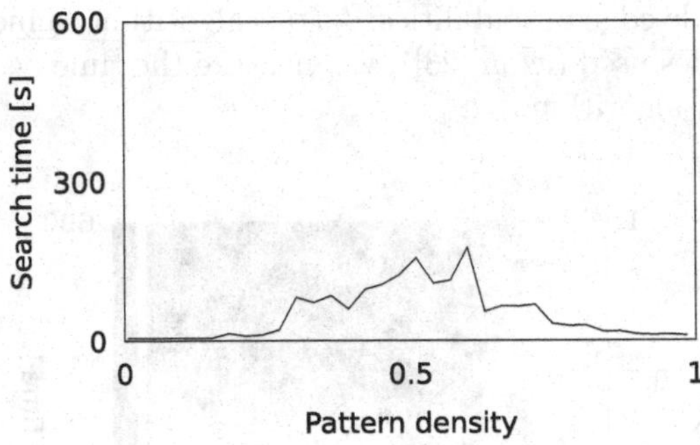

Fig. 3. Time needed to complete 10 iterations of our algorithm on a target graph of 150 vertices with edge probability of 0.5 and pattern graph of 10 vertices with variable edge probability.

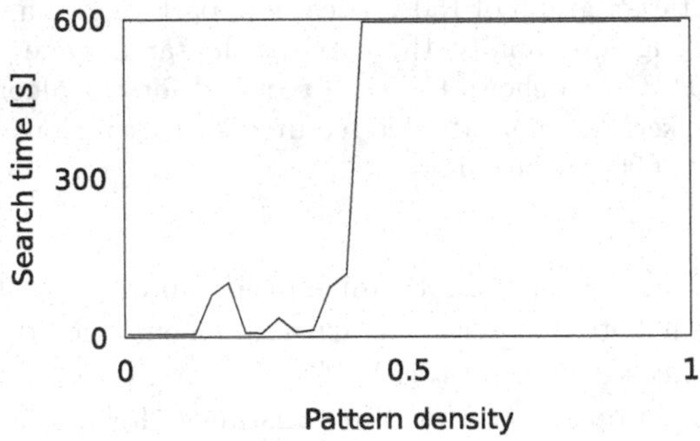

Fig. 4. Time needed to complete 10 iterations of our algorithm on a target graph of 150 vertices with edge probability of 0.8 and pattern graph of 10 vertices with variable edge probability.

Therefore, it seems that our algorithm complements the portfolio of algorithms studied by Kotthoff et al. [17] by an algorithm suitable just below the phase transition (in view of Fig. 1).

4 Conclusion

We described an efficient implementation of the well known color coding algorithm for the subgraph isomorphism problem. Our implementation is the first color-coding based algorithm capable of enumerating all occurrences of patterns of treewidth larger than one. Moreover, we have shown that our implementation is competitive with existing state-of-the-art solutions in the setting of locating small pattern graphs. As it exhibits significantly different behaviour than other solutions, it can be an interesting contribution to the portfolio of known algorithms [17,23].

As an obvious next step, the algorithm could be made to run in parallel. We also wonder whether the algorithm could be significantly optimized even further, possibly using some of the approaches based on constraint programming.

References

1. Alon, N., Dao, P., Hajirasouliha, I., Hormozdiari, F., Sahinalp, S.: Biomolecular network motif counting and discovery by color coding. Bioinformatics **24**, 241–249 (2008)
2. Alon, N., Yuster, R., Zwick, U.: Color-coding. J. ACM **42**(4), 844–856 (1995)
3. Amini, O., Fomin, F.V., Saurabh, S.: Counting subgraphs via homomorphisms. SIAM J. Discrete Math. **26**(2), 695–717 (2012)
4. Bonnici, V., Giugno, R., Pulvirenti, A., Shasha, D., Ferro, A.: A subgraph isomorphism algorithm and its application to biochemical data. BMC Bioinform. **14**, 1–13 (2013)
5. Carletti, V., Foggia, P., Saggese, A., Vento, M.: Introducing VF3: a new algorithm for subgraph isomorphism. In: Foggia, P., Liu, C.-L., Vento, M. (eds.) GbRPR 2017. LNCS, vol. 10310, pp. 128–139. Springer, Cham (2017). https://doi.org/10.1007/978-3-319-58961-9_12
6. Carletti, V., Foggia, P., Vento, M.: Performance comparison of five exact graph matching algorithms on biological databases. In: Petrosino, A., Maddalena, L., Pala, P. (eds.) ICIAP 2013. LNCS, vol. 8158, pp. 409–417. Springer, Heidelberg (2013). https://doi.org/10.1007/978-3-642-41190-8_44
7. Carletti, V., Foggia, P., Vento, M.: VF2 Plus: an improved version of VF2 for biological graphs. In: Liu, C.-L., Luo, B., Kropatsch, W.G., Cheng, J. (eds.) GbRPR 2015. LNCS, vol. 9069, pp. 168–177. Springer, Cham (2015). https://doi.org/10.1007/978-3-319-18224-7_17
8. Carletti, V., Foggia, P., Vento, M., Jiang, X.: Report on the first contest on graph matching algorithms for pattern search in biological databases. In: Liu, C.-L., Luo, B., Kropatsch, W.G., Cheng, J. (eds.) GbRPR 2015. LNCS, vol. 9069, pp. 178–187. Springer, Cham (2015). https://doi.org/10.1007/978-3-319-18224-7_18
9. Cordella, L.P., Foggia, P., Sansone, C., Vento, M.: Performance evaluation of the VF graph matching algorithm. In: 10th International Conference on Image Analysis and Processing, ICIAP 1999. pp. 1172–1177. IEEE Computer Society (1999)

10. Cordella, L.P., Foggia, P., Sansone, C., Vento, M.: A (sub)graph isomorphism algorithm for matching large graphs. IEEE Trans. Pattern Anal. Mach. Intell. **26**(10), 1367–1372 (2004)
11. Cygan, M., et al.: Parameterized Algorithms. Springer, Cham (2015). https://doi.org/10.1007/978-3-319-21275-3
12. Dahm, N., Bunke, H., Caelli, T., Gao, Y.: Efficient subgraph matching using topological node feature constraints. Pattern Recogn. **48**(2), 317–330 (2015)
13. Hüffner, F., Wernicke, S., Zichner, T.: Algorithm engineering for color-coding with applications to signaling pathway detection. Algorithmica **52**(2), 114–132 (2008)
14. Impagliazzo, R., Paturi, R.: On the complexity of k-SAT. J. Comput. Syst. Sci. **62**(2), 367–375 (2001)
15. Karp, R.M.: Reducibility among combinatorial problems. In: Symposium on the Complexity of Computer Computations, COCO 1972, The IBMResearch Symposia Series, pp. 85–103. Plenum Press, New York (1972)
16. Kimmig, R., Meyerhenke, H., Strash, D.: Shared memory parallel subgraph enumeration. In: 2017 IEEE International Parallel and Distributed Processing Symposium Workshops (IPDPSW), pp. 519–529. IEEE Computer Society (2017)
17. Kotthoff, L., McCreesh, C., Solnon, C.: Portfolios of subgraph isomorphism algorithms. In: Festa, P., Sellmann, M., Vanschoren, J. (eds.) LION 2016. LNCS, vol. 10079, pp. 107–122. Springer, Cham (2016). https://doi.org/10.1007/978-3-319-50349-3_8
18. Kuramochi, M., Karypis, G.: Frequent subgraph discovery. In: 2001 IEEE International Conference on Data Mining, pp. 313–320. IEEE Computer Society (2001)
19. Leskovec, J., Sosič, R.: Snap: a general-purpose network analysis and graph-mining library. ACM Trans. Intel. Syst. Technol. (TIST) **8**(1), 1 (2016)
20. Malík, J., Suchý, O., Valla, T.: Efficient implementation of color coding algorithm for subgraph isomorphism problem. CoRR abs/1908.11248 (2019)
21. Marx, D.: Can you beat treewidth? Theory Comput. **6**(1), 85–112 (2010)
22. McCreesh, C., Prosser, P.: A parallel, backjumping subgraph isomorphism algorithm using supplemental graphs. In: Pesant, G. (ed.) CP 2015. LNCS, vol. 9255, pp. 295–312. Springer, Cham (2015). https://doi.org/10.1007/978-3-319-23219-5_21
23. McCreesh, C., Prosser, P., Solnon, C., Trimble, J.: When subgraph isomorphism is really hard, and why this matters for graph databases. J. Artif. Intell. Res. **61**, 723–759 (2018)
24. Slota, G.M., Madduri, K.: Fast approximate subgraph counting and enumeration. In: ICPP 2013, pp. 210–219. IEEE Computer Society (2013)
25. Slota, G.M., Madduri, K.: Parallel color-coding. Parallel Comput. **47**, 51–69 (2015)
26. Solnon, C.: AllDifferent-based filtering for subgraph isomorphism. Artif. Intell. **174**(12–13), 850–864 (2010)
27. Solnon, C., Damiand, G., de la Higuera, C., Janodet, J.C.: On the complexity of submap isomorphism and maximum common submap problems. Pattern Recogn. **48**(2), 302–316 (2015)
28. Tamaki, H.: Positive-instance driven dynamic programming for treewidth. In: ESA 2017. LIPIcs, vol. 87, pp. 68:1–68:13. Schloss Dagstuhl (2017)
29. Ullmann, J.R.: An algorithm for subgraph isomorphism. J. ACM **23**(1), 31–42 (1976)
30. Ullmann, J.R.: Bit-vector algorithms for binary constraint satisfaction and subgraph isomorphism. J. Exp. Algorithmics **15**, 1.6:1.1–1.6:1.64 (2011)
31. Zhao, Z., Khan, M., Kumar, V.S.A., Marathe, M.V.: Subgraph enumeration in large social contact networks using parallel color coding and streaming. In: ICPP 2010, pp. 594–603. IEEE Computer Society (2010)

Quantum-Inspired Evolutionary Algorithms for Covering Arrays of Arbitrary Strength

Michael Wagner, Ludwig Kampel, and Dimitris E. Simos(✉)

SBA Research, 1040 Vienna, Austria
{mwagner,lkampel,dsimos}@sba-research.org

Abstract. The construction of covering arrays, the combinatorial structures underlying combinatorial test suites, is a highly researched topic. In previous works, various metaheuristic algorithms, such as Simulated Annealing and Tabu Search, were used to successfully construct covering arrays with a small number of rows. In this paper, we propose for the first time a quantum-inspired evolutionary algorithm for covering array generation. For this purpose, we introduce a simpler and more natural qubit representation as well as new rotation and mutation operators. We implemented different versions of our algorithm employing the different operators. We evaluate the different implementations against selected (optimal) covering array instances.

Keywords: Optimization · Covering arrays · Quantum algorithms

1 Introduction

Covering arrays (CAs) are discrete combinatorial structures that can be considered a generalization of orthogonal arrays and are most frequently represented as arrays, which columns fulfil certain *coverage criteria* regarding the appearance of *tuples* in submatrices. Their properties make CAs attractive for application in several fields, first and foremost in the field of automated software testing. The interested reader may have a look at [12]. For their application in testing it is generally desired to construct CAs with a small number of rows, while maintaining their defining coverage criteria. Resulting optimization problems are closely related to NP-hard problems, such as the ones presented in [4,15,17], suggesting that the problem of finding *optimal covering arrays* is a hard combinatorial optimization problem. However, the actual complexity of this problem remains unknown [10].

Aside from theoretical construction techniques, based on the theory of groups, finite fields or on combinatorial techniques (see [3] and references therein), there exist many algorithmic approaches dedicated to the construction of CAs. The latter include greedy heuristics [11,13], metaheuristics, see [20], as well as exact approaches as in [9]. For a survey of CA generation methods, the interested reader may also have a look in [19].

I. Kotsireas et al. (Eds.): SEA^2 2019, LNCS 11544, pp. 300–316, 2019.
https://doi.org/10.1007/978-3-030-34029-2_20

This paper proposes quantum-inspired evolutionary algorithms for CA generation. For this purpose we introduce a reduced qubit representation and a means to change the state of these qubits. In general, covering arrays can be defined over arbitrary alphabets (see for example [19]). In this work, however, we restrict our attention to CAs over binary alphabets as the 2-state nature of a qubit makes representing binary values straightforward.

We (informally) introduce CAs as follows: A binary $N \times k$ array $M = (\mathbf{m}_1, \dots, \mathbf{m}_k)$ is a *binary covering array* (CA), $\mathsf{CA}(N; t, k)$, if and only if M has the property that any array $(\mathbf{m}_{i_1}, \dots, \mathbf{m}_{i_t})$, with $\{i_1, \dots, i_t\} \subseteq \{1, \dots, k\}$, comprised of t columns of M has the property that each binary t-tuple in $\{0, 1\}^t$ appears at least once as a row.

The value t is referred to as the *strength* of a CA. As already mentioned previously, CAs with a small or the smallest number of rows are of particular interest. The smallest integer N, for which a $\mathsf{CA}(N; t, k)$ exists is called *covering array number* for t and k and is denoted as $\mathsf{CAN}(t, k)$. The proposed quantum-inspired evolutionary algorithm in this paper will take N,t and k, as input and attempts to find a $\mathsf{CA}(N; t, k)$. Thus for given (N, t, k), we speak of a *CA instance*.

The defining properties of CAs can be also expressed by means of *t-way interactions*. For given strength t and a number of columns k, a t-way interaction is a set of t pairs $\{(p_1, v_1), \dots, (p_t, v_t)\}$ with $1 \leq p_1 < p_2 < \dots < p_t \leq k$, and $v_i \in \{0, 1\}$ for all $i = 1, \dots, t$[1]. The value k is usually clear from the context and is omitted. We say the t-way interaction $\{(p_1, v_1), \dots, (p_t, v_t)\}$ is *covered* by an array A, if there exists a row in A that has the value v_i in position p_i for all $i = 1, \dots, t$. Then a $\mathsf{CA}(N; t, k)$ is characterized by covering all t-way interactions.

This paper is structured as follows. In Sect. 2 we provide the necessary preliminaries needed for this paper and cover related work. Furthermore, Sect. 3 introduces a quantum-inspired evolutionary algorithm for CA generation, which we will evaluate in Sect. 4. Finally, Sect. 5 concludes the paper and discusses future directions of work.

2 Evolutionary Algorithms and Quantum Computing

Evolutionary algorithms are nature-inspired, metaheuristic, stochastic search- and optimization algorithms based on a population of individuals, which is evolved by the concepts of *selection*, *recombination* and *mutation* [14]. Each individual represents a potential candidate solution to the problem instance. Operations like selection and recombination are used to select a part of the population and produce offspring by combining the selected individuals, generating a new generation of candidate solutions. In the selection process, an objective function, tailored to respective problem instance, is used to evaluate the individuals and selects them accordingly for reproduction. This process of creating

---

[1] In the literature t-way interactions are defined for arbitrary alphabets. However we restrict our attention to binary t-way interactions.

new generations of individuals, based on the fitness of the individuals, guides the search towards (local) maxima.

The concept of mutation is used to add variety to the individuals, which lets the algorithm explore different search spaces and provides means to escape local maxima. Depending on the problem, the algorithm generally gets terminated when either a sufficiently good solution is found or a certain number of generations is reached.

Quantum computing utilizes the quantum-mechanical phenomena of quantum entanglement and quantum interference to perform computational tasks. Entanglement allows one to encode data into superpositions of states and quantum interference can be used to evolve these quantum states [1]. Quantum algorithms make use of these superpositions of states and the resulting parallelism to perform certain tasks faster or more space efficient than classical algorithms can. Examples for such algorithms would be Grovers algorithm [5] for unstructured search and Shors algorithm [18] for factoring numbers.

The smallest unit of information in Quantum Computing is a qubit, which is a 2-state system consisting of the states $|0\rangle$ and $|1\rangle$. A qubit can either be in state $|0\rangle$, state $|1\rangle$ or in a superposition of the two. One way to fully specify the state of the qubit is

$$|\Psi\rangle = \alpha\,|0\rangle + \beta\,|1\rangle\,, \tag{1}$$

where $\alpha, \beta \in \mathbb{C}$ and $|\alpha|^2 + |\beta|^2 = 1$. The coefficients α and β are called the amplitudes of the qubit. Upon observation the qubit collapses into state $|0\rangle$ with probability $|\alpha|^2$ and into state $|1\rangle$ with probability $|\beta|^2$. Once a qubit has collapsed, without interference from outside, further measurement of the qubit will always result in the previously observed state.

Moreover, a qubit can also be represented with a so called Bloch Sphere, which we will make use of in Sect. 3 to represent the qubits for our binary CAs. A visualization of the Bloch Sphere representation is given in Fig. 1a.

Lemma 1. *The state of a 2-state system (qubit) can accurately be described by the Bloch Sphere representation:*

$$|\Psi\rangle = \cos\frac{\theta}{2}\,|0\rangle + e^{i\varphi}\sin\frac{\theta}{2}\,|1\rangle\,,\ \ \textit{with}\ \theta \in [0,\pi], \varphi \in [0, 2\pi).$$

Sketch of Proof. This qubit representation is well known [21] and can be directly derived from Eq. 1 by transforming the complex amplitudes α and β to polar coordinates and making use of the normalization criterion $|\alpha|^2 + |\beta|^2 = 1$. The angle φ represents the difference in complex phases of α and β, while $\cos\frac{\theta}{2}$ and $\sin\frac{\theta}{2}$ describe the radius of α and β respecting normalization. □

Quantum algorithms use quantum circuits to evolve the state of the system, where their smallest building blocks are quantum gates. Quantum gates can perform reversible operations on one or more qubits. The restriction of reversibility demands that a state obtained by applying a gate can be deterministically recreated by applying the gates inverse. Therefore an operation, where information gets lost, like for example in classical AND gates, has to be replaced by a reversible gate in quantum computing. For a more in-depth explanation and examples of common quantum gates, we refer the interested reader to [16].

Since the late 1990s, the combination of superpositions and the inherent parallelism of quantum computing with evolutionary algorithms has been explored. In contrast to approaches that try to implement evolutionary algorithms in a quantum computation environment [22] quantum-inspired evolutionary algorithms (QIEAs) are classical algorithms that take inspiration from concepts of quantum computing. In 2002 Han and Kim proposed a quantum-inspired binary observation evolutionary algorithm [6], which we briefly review below.

Like in most evolutionary algorithms, multiple individuals represent potential candidate solutions for the problem instance, but in addition, every individual also has a qubit representation, that stores two numbers α and β representing the amplitudes of a qubit. Generally these amplitudes both get initialized to $\frac{1}{\sqrt{2}}$, representing a uniform distribution where both states have the same probability of being measured. Instead of the concept of reproduction, in each generation new candidate solutions are created by observing the qubits. Observation of a qubit will return $|0\rangle$ with probability $|\alpha|^2$ and $|1\rangle$ otherwise, but unlike physical quantum states, the qubits do not collapse, but maintain their amplitudes.

To update these states, a novel gate was introduced, called rotation gate, which is used to adjust the probability of measuring a state while maintaining normalization. The proposed gate operator to rotate a qubit by the angle $\Delta\theta_i$ is defined as

$$U(\Delta\theta_i) = \begin{bmatrix} \cos(\Delta\theta_i) & -\sin(\Delta\theta_i) \\ \sin(\Delta\theta_i) & \cos(\Delta\theta_i) \end{bmatrix} \tag{2}$$

and acts on each qubit individually. The sign of the rotation angle $\Delta\theta$ for each qubit depends on the value it contributed to the best previously measured solution for the individual. By rotating the qubit towards this best observed value, the probability of measuring this value again in later generations increases. This guides the search towards the best found solution, while utilizing the random probability of observing the other state to explore the search space.

The original algorithm, as described in [6], used multiple individuals, each evolving towards their best found solutions, and included local and global migration conditions that allowed exchange of information between the individuals as a means to escape local maxima. The introduction of $H\epsilon$ gates [7] prevents the qubits from completely collapsing into one of the two states. Therefore, using a single individual proved sufficient for many different problems [8]. In this work, we have modified this algorithm and extended it for the case of CA generation which we describe in the next section.

3 A Quantum-Inspired Evolutionary Algorithm for CAs

In this section, first we introduce a qubit representation that is simpler and more efficient for our approach to the CA problem than the representation given in Eq. 1 and provides a more natural way of updating the states of the qubits than the rotation gates discussed in Sect. 2. Afterwards, we will propose a quantum-inspired evolutionary algorithm for covering array construction and explore different operators to guide the search.

3.1 Simplified Qubit-Representation

Considering that the quantum-inspired evolutionary algorithm proposed in [6] utilizes real numbers, generally $\frac{1}{\sqrt{2}}$, as starting amplitudes, applying the rotation gates defined in Eq. 2 will result in real amplitudes as well. Using the Bloch Sphere representation defined in Lemma 1 and the restraint of real starting amplitudes, hence no complex phase difference ($\varphi = 0$), it is sufficient to consider the following simpler qubit representation. The validity of the following corollary comes immediately from Lemma 1.

Corollary 1. *If the relative phase φ between the states $|0\rangle$ and $|1\rangle$ is 0, the state $|\Psi\rangle$ of a qubit can be fully described with*

$$|\Psi\rangle = \cos\Theta\,|0\rangle + \sin\Theta\,|1\rangle \qquad \Theta \in [0, \frac{\pi}{2}].$$

We call such a representation, the *circular representation* of a qubit. Compared to previous works, where only the rotation gates used to update the amplitudes of the qubits were depicted in polar coordinates [6], we fully describe the qubit state with a single real number, the angle Θ (see Fig. 1b). In the remaining work, whenever the term qubit is mentioned, we refer to this reduced representation.

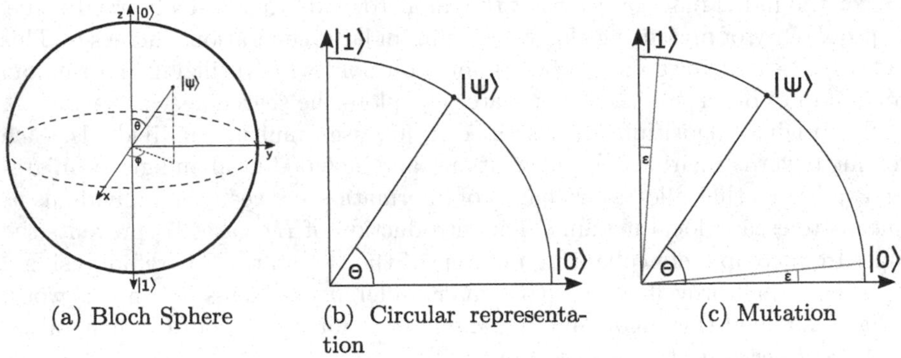

(a) Bloch Sphere (b) Circular representation (c) Mutation

Fig. 1. Derivation of the circular representation. The angle φ, drawn red in (a), gets set to zero and after dividing the angle by two, the representation reduces to (b). (c) visualizes the effect of mutation on the qubit representation (see Subsect. 3.2).

An observation of the state will result in $|0\rangle$ with probability $(\cos\Theta)^2$ and in state $|1\rangle$ otherwise. Moreover, we can now update the qubit states as follows.

Lemma 2. *Let the qubit representation be* $|\Psi\rangle = \cos\Theta\,|0\rangle + \sin\Theta\,|1\rangle$*, where* $\Theta \in [0, \frac{\pi}{2}]$*. Then applying a rotation gate to the qubit reduces to simple addition of the desired rotation amount:*

$$\begin{pmatrix} \cos\Delta\Theta & -\sin\Delta\Theta \\ \sin\Delta\Theta & \cos\Delta\Theta \end{pmatrix} \begin{pmatrix} \cos\Theta \\ \sin\Theta \end{pmatrix} = \begin{pmatrix} \cos(\Theta + \Delta\Theta) \\ \sin(\Theta + \Delta\Theta) \end{pmatrix} \tag{3}$$

Proof. Using the Ptolemy's theorem concerning trigonometric identities, the prove of the assertion is straight forward. □

3.2 Algorithmic Description

In this subsection we propose our algorithm QIEAforCA, provide the pseudocode with an explanation and also give some examples. The basic idea underlying our algorithm is to consider an $N \times k$ array Q of qubits, that serves as a source from which binary arrays are created by observing the qubits. In each generation this source gets updated according to the best found array B generated thus far, i.e. the array that covers the most t-way interactions. This update will happen based on a rotation operator, implemented by the procedure Rotation, that can be realized in different types (`universal rotation` and `individual rotation`) and influences each qubit of Q by changing its state, according to the so called `rotation speed`. To guarantee a certain possibility for mutation, we prevent the qubits in the source Q to go beyond a certain state and thus from collapsing into one of the states $|0\rangle$ or $|1\rangle$. This will happen based on an $H\epsilon$ operator, implemented by the procedure Mutation. The quantity for this possibility is described by the `mutation rate`. Also, for the procedure Mutation we will detail two different types (`universal mutation` and `individual mutation`) later in this subsection.

Our algorithm can be summarized as follows (see Algorithm 1). Initially a qubit representation $Q = (q_{ij})$ for each value in an $N \times k$ array gets created according to Corollary 1. The angle, describing the initial state, of each qubit is set to 45°, inducing a uniform distribution of the possible states $|0\rangle$ and $|1\rangle$. By measuring each qubit, a first candidate solution $C = (c_{ij})$, which is a binary-valued array, gets created. Measuring q_{ij} in state $|0\rangle$ results in an entry $c_{ij} = 0$, while measuring state $|1\rangle$ results in $c_{ij} = 1$ respectively (see Example 1). The initial best solution $B = (b_{ij})$ is set to the first candidate solution, $B(0) = C(0)$. Thereafter, the following steps get repeated until either a covering array is found or a specific number n of generations have passed. These criteria are implemented as terminating conditions by the procedure Termination. In each generation a new candidate solution $C(n)$ gets created by measuring the state of each qubit in $Q(n-1)$. In case the candidate solution has a higher *fitness*, i.e. it covers a higher number of t-way interactions than the current best solution $B(n-1)$, the best solution is updated to the candidate solution $B(n) = C(n)$. Depending on the best solution, the states of the qubits (q_{ij}) get updated yielding $Q(n)$. For this update, the direction in which the qubit q_{ij} gets rotated is defined by the value b_{ij} in the current best solution and the procedure Mutation. This direction is represented by the `target state` α_{ij}, which we explain detailed later in this section. The angles, by which the qubits get rotated in each generation, get determined by the procedure ROTATION.

Algorithm 1. QIEAFORCA(t, k, N)

```
Require: ROTATION, MUTATION, TERMINATION
n ← 0
Create Q(n) representing the N × k array
Create candidate solution C(n) by observing Q(n)
Evaluate C(n) based on the number of covered t-way interactions
B(n) ← C(n)
while (not TERMINATION(B(n), t) do
    n ← n + 1
    Create C(n) by observing Q(n − 1)
    Evaluate C(n)
    if C(n) is better than B(n − 1) then
        B(n) ← C(n)
    else
        B(n) ← B(n − 1)
    end if
    for all Qubits q_ij in Q(n) do
        α_ij ← MUTATION(b_ij)
        q_ij ← ROTATION(q_ij, α_ij)
    end for
end while
return B(n)

procedure TERMINATION(B(n), n)
Require: Termination number m
    if fitness of B(n) is 100% or n ≥ m then return true
    else
        return false
    end if
end procedure

procedure MUTATION(b_ij)
Require: ε_glob, MutationType
    mutation rate ε_ij ← 0
    if MutationType is universal mutation then
        ε_ij ← ε_glob
    else if MutationType is individual mutation then
        Calculate the relative mutation amount ε_ind based on the unique coverage of b_ij
        ε_ij ← ε_glob + ε_ind
    end if
    if b_ij is 0 then
        α_ij ← ε_ij
    else if b_ij is 1 then
        α_ij ← 90° − ε_ij
    end if
    return α_ij
end procedure

procedure ROTATION(q_ij, α_ij, b_ij)
Require: rotation speed s, RotationType
    s_ij ← 0
    if RotationType is universal mutation then
        s_ij ← s
    else if RotationType is individual mutation then
        Calculate rotation angle s_ij based on s and the unique coverage of b_ij
    end if
    Rotate qubit q_ij by s_ij towards target state α_ij
    return q_ij
end procedure
```

Example 1. A candidate solution is created by observation of the qubit angles. The majority of measurements return the expected value, while states closer to 45° have a higher chance of returning the opposite state. For example, the 30° state has a 25% probability of observing state $|1\rangle$.

$$\begin{pmatrix} \Psi_{02} \, \Psi_{12} \, \Psi_{22} \, \Psi_{32} \\ \Psi_{01} \, \Psi_{11} \, \Psi_{21} \, \Psi_{31} \\ \Psi_{00} \, \Psi_{10} \, \Psi_{20} \, \Psi_{30} \end{pmatrix}^T \longrightarrow \begin{pmatrix} 10^\circ & 60^\circ & 0^\circ & 90^\circ \\ 0^\circ & 30^\circ & 90^\circ & 60^\circ \\ 0^\circ & 90^\circ & 60^\circ & 20^\circ \end{pmatrix}^T \longrightarrow \begin{pmatrix} 0\,1\,0\,1 \\ 0\,1\,1\,1 \\ 0\,1\,1\,0 \end{pmatrix}^T \tag{4}$$

Mutation. Before we explain the procedure MUTATION in detail, we give a brief motivation. Whenever the state of a qubit q_{ij} reaches 0° or 90°, further measurements of this qubit will always return the same result, as the probability $(\cos\Theta)^2$ of measuring state $|0\rangle$ becomes 1 or 0 respectively. This locks the qubit into the corresponding state and can lead to premature convergence. To avoid this phenomenon, Han and Kim introduced $H\epsilon$ gates in [7] to keep the qubits from converging completely. Using our qubit representation, this concept is implemented in the procedure MUTATION as follows. The extremal states $|0\rangle$ and $|1\rangle$, represented by 0° and 90° respectively, are replaced by the states corresponding to $0° + \epsilon$ and $90° - \epsilon$ respectively, for a certain angle ϵ, called the `mutation rate` (see Fig. 1c). Depending on b_{ij}, the qubits now rotate towards one of these extremal states, which is why we refer to them as `target states` α_{ij}. For example with an ϵ of 5°, the possible `target states` are 5° and 85° (see Example 2). We call this procedure MUTATION, since it is very similar to the concept of mutation used in other evolutionary algorithms, as once a qubit converged to its `target state`, it will still maintain a small chance of measuring the opposite state.

We propose two different mutation types that are implemented as part of the procedure MUTATION:

- `universal mutation`: $\epsilon_{ij} = \epsilon_{glob}$. We call this first mutation type `universal mutation` as it assigns the same `mutation rate` to each qubit. Moreover, we denote the parameter specifying this angle, that is constant for all qubits, with ϵ_{glob}.
- `individual mutation`: $\epsilon_{ij} = \epsilon_{glob} + \epsilon_{ind}$. The second mutation type we introduce is called `individual mutation` as it assigns different `mutation rates` ϵ_{ij} to the individual qubits.

We elaborated further on how to make these rates for `individual mutation` specific to the CA generation problem, by proposing the property `unique coverage` below. For each entry b_{ij} in the current best solution $B(n)$, `unique coverage`(b_{ij}) is defined as the number of t-way interactions that are covered exactly once and involve b_{ij}. Removing b_{ij} from $B(n)$ would therefore reduce the total number of t-way interactions covered by the array exactly by `unique coverage`(b_{ij}). Thus, we use this measure to evaluate the significance of b_{ij} in the current best solution $B(n)$. In `individual mutation`, we make use of this property by calculating the angle ϵ_{ind}, that gets added to the `mutation rate` of qubit q_{ij}, indirectly proportional to `unique coverage`(b_{ij}). This approach penalizes entries q_{ij} of $Q(n)$, that correspond to entries b_{ij} that are deemed less significant for the number of t-way interactions covered by the current best solution $B(n)$. A visual representation of how `individual mutation` affects different qubits is given in Fig. 2b and d.

Example 2. `Target states` for all qubits get set dependent on the values of the best previous solution. If the value of the respective entry in the solution is 0, then the `target state` is set to 0°, else if the value is 1, the `target state` is set to 90°. The third array depicts how `universal mutation` with $\epsilon_{glob} = 5°$ affects the `target states`:

$$\begin{pmatrix} 0\,1\,0\,1 \\ 0\,1\,1\,0 \\ 0\,1\,1\,1 \end{pmatrix}^T \longrightarrow \begin{pmatrix} 0° & 90° & 0° & 90° \\ 0° & 90° & 90° & 0° \\ 0° & 90° & 90° & 90° \end{pmatrix}^T \longrightarrow \begin{pmatrix} 5° & 85° & 5° & 85° \\ 5° & 85° & 85° & 5° \\ 5° & 85° & 85° & 85° \end{pmatrix}^T . \quad (5)$$

Rotation. Another part of the algorithm that can be adjusted is the angle that the qubits get rotated by in each generation. Since this determines how quickly the qubits converge towards their respective `target states`, we call this the `rotation speed` s. Similar to mutation, we propose two different types of rotation:

- `universal rotation`: $s_{ij} = s$, where each qubit gets rotated by the same angle.
- `individual rotation`: $s_{ij} = s \cdot \omega(b_{ij})$, where each qubit gets rotated by an individual angle, where $\omega(b_{ij})$ is a weight based on b_{ij} with $0 < \omega(b_{ij}) < 1$.

In detail, during `universal rotation` every qubit q_{ij} gets rotated by the same predefined angle s towards it's respective `target state` α_{ij}. Furthermore, the introduced `individual rotation` is based on the `unique coverage` of b_{ij} in the best solution. In this rotation type, the angle by which a qubit q_{ij} gets rotated is proportional `unique coverage` of the respective value b_{ij} in the best solution. In other words, values that are contributing more to the fitness of the solution, i.e. the number of covered t-way interactions, get rotated faster towards their `target states` than others. Due to this, qubits q_{ij}, where the respective entry b_{ij} has a high `unique coverage`, quickly converge towards their `target state` α_{ij} and can serve as an anchor to guide the search faster towards a promising subset of the search space.

Combination of Rotation and Mutation. Figure 2 gives a visual representation of the four possible combinations of rotation and mutation types. In this example, for each combination two arbitrary qubits, q_1 and q_2, are depicted.

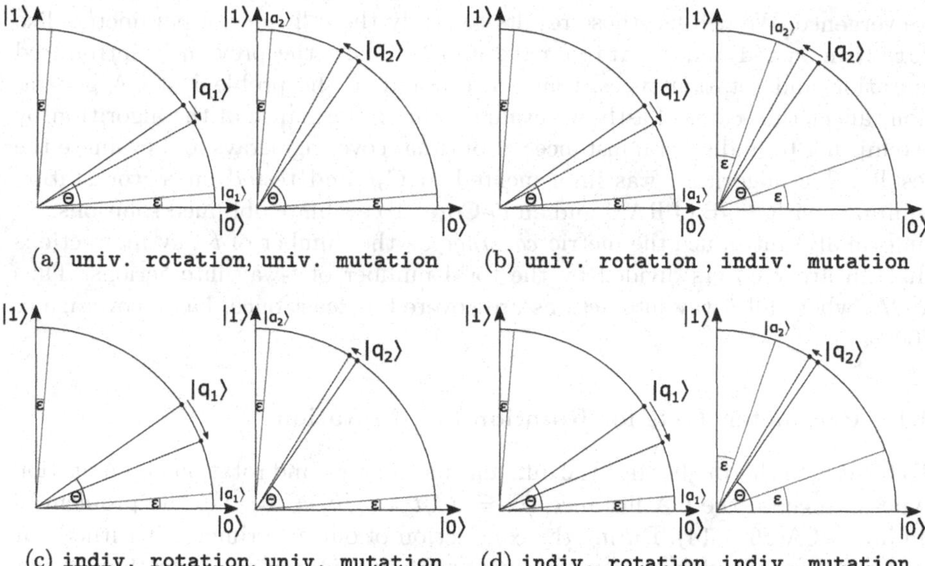

(a) **univ. rotation, univ. mutation** (b) **univ. rotation , indiv. mutation**

(c) **indiv. rotation, univ. mutation** (d) **indiv. rotation, indiv. mutation**

Fig. 2. All four different combinations of the proposed rotation and mutation types visually represented using two arbitrary qubits.

The `mutation rate` that gets applied to each qubit is represented by the angle ϵ_{ij} in red. The angle s_{ij}, by which the qubit q_{ij} gets rotated in every generation, is represented by the length of the arrow towards the respective `target state` α_{ij}. In this given example, the unique coverage of the value b_{ij} represented by qubit q_1 was higher than that represented by qubit q_2. In other words, the entry corresponding to q_1 was more significant to the best solution B. This is visible by the size of ϵ_{ij} and s_{ij} when `individual mutation` and `individual rotation` are used respectively. Figure 2a depicts a setup using `universal rotation` and `universal mutation`, where each qubit gets the same `mutation rate` ϵ and `rotation speed` s applied. In Fig. 2b, due to `individual mutation`, the `mutation rate` ϵ applied to q_1 is smaller, allowing it to converge close towards the extremal states $|0\rangle$ or $|1\rangle$. Similarly, qubit q_2 gets a higher `mutation rate` applied. Figure 2c illustrates how with `individual rotation`, the better qubit q_1 gets rotated by a larger angle per iteration, while the state of qubit q_2 does not change much. Lastly, Fig. 2d shows how when combining `individual mutation` and `individual rotation`, the angle by which qubits get rotated and the `mutation rate` of qubit q_1 are proportionally larger and smaller respectively than for the less significant qubit q_2.

4 Experimental Results

We evaluate our algorithm in two steps. First, we benchmark different configurations of our algorithm on the same problem instance and compare their convergence. We will use those results to study the influence of parameters like `rotation speed` and `mutation rate` and test how the previously introduced mutation and rotation types, that are tailored to the problem of CA generation, affect the search. Lastly we evaluate the performance of the algorithm by attempting to find several instances of optimal covering arrays and evaluate the results. The algorithm was implemented in C# and tested on a workstation equipped with 8 GB of RAM and an i5-Core. To evaluate obtained solutions, we (informally) introduce the metric *coverage* as the number of t-way interactions that an array covers divided by the total number of t-way interactions. Thus a CA, where all t-way interactions are covered at least once, has a coverage of 100%.

4.1 Parameter Tests for Rotation and Mutation

To be able to thoroughly test the different parameters and rotation and mutation types, we chose the CA instance ($N = 16, t = 3, k = 14$), i.e. the problem of finding a $\mathsf{CA}(16; 3, 14)$. During the conduction of our experiments it turned out that this is a difficult problem, that most configurations of our algorithm are not able to solve. We purposely chose this instance, as it allows for better comparison of the different configurations. For the experiments reported in this subsection each configuration was run 10 times with a limit of 500000 generations. For the plots in Figs. 3, 4 and 5 we recorded the average coverage of the best found solution after every 100 generations. The average run time of each experiment was approximately 10 min. The average coverage as well as the coverage of the best run for each configuration and `rotation speed` are reported in Table 1 at the end of this subsection.

Rotation Type Tests Without Mutation. First we evaluate how the rotation types `universal rotation` and `individual rotation` perform without any mutation at the `rotation speeds` $s = 0.001$, $s = 0.01$, $s = 0.1$ and $s = 1.0$ and compare the results in Fig. 3. The results very clearly show the premature convergence due to lack of escape mechanism without any type of mutation. Furthermore the graphs depict nicely that the faster the rotation speed, the less time the algorithm has to explore the solution space before fully converging. Interestingly, even with high `rotation speeds`, `individual rotation` found better arrays than any configuration using `universal rotation`. The quickest `rotation speed`, $s = 1.0$, converged to an array with 98.0% coverage after only 500 generations, while with `universal rotation`, the best array had 97.7% coverage and was found after around 100000 generations. We believe `individual rotation` performs better on this instance, as it rotates *better* qubits, where the respective b_{ij} had high `unique coverage`, quicker towards the best solution, hence restricting the search early towards a promising subset of the solution space. At the same time, the qubits representing weaker entries in the array are still relatively unbiased, allowing the algorithm to find the best solution around this restricted solution space.

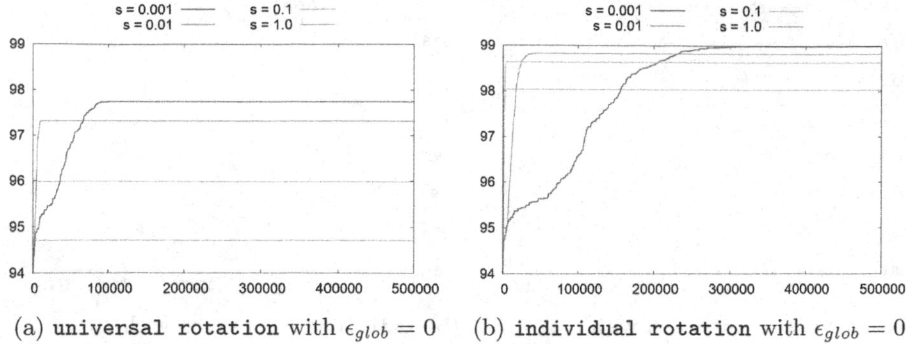

(a) `universal rotation` with $\epsilon_{glob} = 0$ (b) `individual rotation` with $\epsilon_{glob} = 0$

Fig. 3. Different `rotation speeds` ($s = 0.001$, $s = 0.01$, $s = 0.1$ and $s = 1.0$) for `universal rotation` and `individual rotation` without any mutation are compared side by side.

Rotation Type Tests with Universal Mutation. Next, we investigate how `universal mutation` affects the previously tested configurations. Figure 4 compares the convergence of the configurations with $\epsilon_{glob} = 5$. Adding `universal mutation` significantly improves the coverage of the best found arrays and no clear point of full convergence exists any more. Unlike in the previous tests without mutation, slow `rotation speeds` do not seem to provide any benefits and in fact result in noticeably worse solutions than fast `rotation speeds`. Furthermore, with $\epsilon_{glob} = 5$, `universal rotation` and `individual rotation` find solutions of similar quality, however `universal rotation` consistently performs slightly better than `individual rotation`. Again, the success of fast `rotation speeds` confirms the importance of quickly guiding the search to a promising subspace. The addition of `universal mutation` keeps the algorithm from fully converging and allows it to constantly explore the solution space around the current best candidate solution. If the `rotation speed` s is chosen too small, qubits can not adapt to a new solution quickly enough. We also tested configurations with $\epsilon_{glob} = 10$ (see Table 1), which produced worse results than a rate of $\epsilon_{glob} = 5$. This shows that if a `mutation rate` is too big, the system can become unstable, resulting in too much exploration of the search space with too little exploitation.

Comparison Tests of Mutation Types. To investigate the effect of `individual mutation`, we compare the results of four configurations using `universal rotation` with a fixed `rotation speed` of $s = 0.1$, see Fig. 5a. Compared to the configuration without any mutation, `individual mutation` with $\epsilon_{glob} = 0$ already leads to a significant improvement in solution quality. We can see that even though the majority of values now do get a mutation applied, some of the qubits still do not, which can lead to the algorithm getting stuck more easily at a local maximum. Additionally, a global mutation rate of $\epsilon_{glob} = 5$ increases the performance for both mutation types and they converge very similarly. At

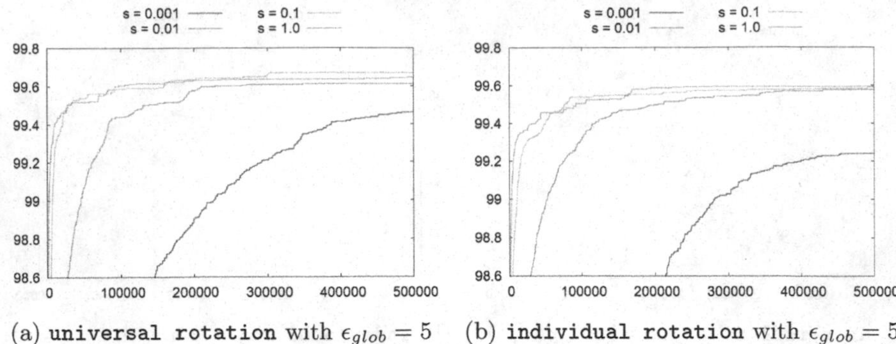

(a) **universal rotation** with $\epsilon_{glob} = 5$ (b) **individual rotation** with $\epsilon_{glob} = 5$

Fig. 4. Comparison between **universal rotation** and **individual rotation** with different **rotation speeds** ($s = 0.001$, $s = 0.01$, $s = 0.1$, $s = 1.0$) and **universal mutation** using $\epsilon_{glob} = 5$.

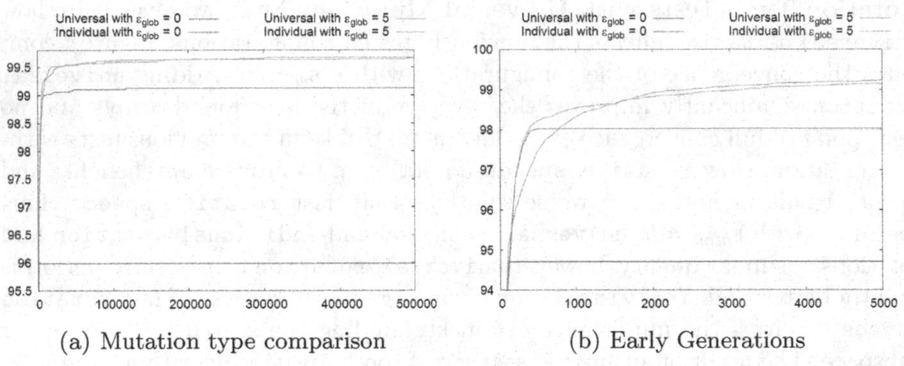

(a) Mutation type comparison (b) Early Generations

Fig. 5. (a) compares different mutation configurations using **universal rotation** with a **rotation speed** of $s = 0.1$. (b) depicts the convergence of selected configurations during the first 5000 generations at **rotation speed** $s = 1.0$.

the same time, in these experiments, **individual mutation** with $\epsilon_{glob} = 5$ is the only configuration tested that was able to actually find a CA for this instance (see also Table 1).

Influence of Individual Rotation in the Early Stages of the Algorithm. Lastly, in Fig. 5b we explore how selected configurations behave in the first 5000 generations. Since slower **rotation speeds** do not converge quickly enough for this evaluation, we only considered a **rotation speed** of $s = 1$. It is worth noting that the configurations using **individual rotation** performed significantly better during the first 1500 generations. In later generations it seems that the

Table 1. Results of parameter tests for $\mathsf{CA}(16, 3, 14)$.

Configuration			$s = 0.001$		$s = 0.01$		$s = 0.1$		$s = 1.0$	
RotType	MutType	MutRate	Average	Best	Average	Best	Average	Best	Average	Best
Univ	Univ	0	97.74	98.45	97.32	97.80	95.99	96.60	94.71	95.57
Ind	Univ	0	98.99	99.28	98.82	99.11	98.63	99.00	98.04	98.45
Univ	Univ	5	99.46	**99.73**	99.61	**99.79**	**99.67**	99.86	99.65	99.76
Ind	Univ	5	99.24	99.42	99.58	99.66	99.59	99.76	99.59	99.73
Univ	Univ	10	99.32	99.55	99.52	99.66	99.57	99.79	99.59	**99.79**
Ind	Univ	10	98.86	99.07	99.43	99.66	99.53	99.73	99.58	99.66
Univ	Ind	0	99.19	99.42	99.11	99.42	99.13	99.45	99.16	99.42
Univ	Ind	5	**99.55**	99.69	99.61	99.76	99.66	**100**	99.64	99.76
Ind	Ind	5	99.32	99.55	**99.67**	**99.79**	99.62	99.79	**99.66**	**99.79**

effect of `universal mutation` starts to negate this advantage and the configurations converge very similarly. We conducted these experiments to investigate which algorithm finds arrays with the highest coverage within a limited number of generations, which can serve as a foundation for future work.

4.2 Algorithm Evaluation

Having evaluated different combinations of rotation and mutation types in the previous section, we now analyze the performance of our quantum-inspired evolutionary algorithm. To that extend, we attempted to find CAs for some interesting CA instances. For many of those, CAs exist that are likely to be optimal, see [2]. Further, we consider two instances, for which the existence of a covering array is still unknown. The results of our experiments can be found in Table 2. In these experiments we use the configuration using `individual mutation` with $\epsilon_{glob} = 5$ and `individual rotation` with a `rotation speed` of $s = 0.1$, as this configuration seemed to perform the most consistent during our experiments. The upper bound for the number of generations was set to 3000000. To analyze the results, 30 runs were executed for each CA instance. We recorded the number of successful CA generations, the average and the best achieved coverage, as well as the average number of iterations and run time in seconds of the successful runs, or all 30 runs, if no solution was found.

For binary CAs of strength $t = 2$, the covering array numbers are known from theory, see [19] and references therein. It is hence possible to consider for a given number of rows N, the hardest CA instance, i.e. the instance with the largest k, such that a $\mathsf{CA}(N; 2, k)$ still exists. In our experiments we hence target the computation of $\mathsf{CA}(7; 2, 15)$, $\mathsf{CA}(8; 2, 35)$ and $\mathsf{CA}(9; 2, 56)$. For $t = 3$, we consider the CA instances $(14; 3, 12)$ and $(15; 3, 13)$, for which the existence of covering arrays is unknown, to the best of our knowledge.

As shown in Table 2, with the configuration specified above, the algorithm finds solutions for all CA instances in the given number of generations, with the exception of the CA instances $(14; 3, 12)$ and $(15; 3, 13)$. While all tested strength $t = 2$ instances were found, the results indicate that the algorithm has difficulties finding solutions to larger problem instances such as $(56; 2, 9)$. At the same time, the average coverage of the resulting arrays remains high (above 99.92%). For the strength $t = 3$ instances, in the best run, the returned array missed a total of two 3-way interactions to find a $\mathsf{CA}(14; 3, 12)$, covering 99.89% of the 3-way interactions, while the best array found for instance $(15; 3, 13)$ only had a coverage of 99.74%. However, adding one row, the algorithm finds solutions to the instances $(15; 3, 12)$ and $(16; 3, 13)$. Lastly, the optimal $\mathsf{CA}(24; 4, 12)$ is found 19 out of 30 times, demonstrating the viability of the algorithm also for higher strengths.

Table 2. Results for selected CA instances $(N; t, k)$. Row counts N marked with an asterisk are not yet confirmed to have a possible covering array solution.

CA instance	Found CAs	Average coverage	Best coverage	Average generations	Average runtime
(7; 2, 15)	30	100	100	641147	108
(8; 2, 35)	14	99.95	100	605462	1034
(9; 2, 52)	1	99.92	100	2790982	11619
(14*; 3, 12)	0	99.51	99.89	3000000	4626
(15; 3, 12)	6	99.71	100	1195933	1226
(15*; 3, 13)	0	99.46	99.74	3000000	6025
(16; 3, 13)	1	99.73	100	54113	61
(24; 4, 12)	19	98.47	100	345479	2280

5 Conclusion and Future Work

In this paper we introduced a quantum-inspired evolutionary algorithm for covering array generation. We used a simplified qubit representation and successfully encode properties of CAs in our algorithm. The algorithm manages to find many optimal CAs of various strengths and showed the ability to find good, albeit not always optimal, solutions in a small number of iterations. This property is desirable when our quantum-inspired evolutionary algorithm is used as a subroutine as part of greedy methods for CA construction (see [19]). Such hybrid algorithms are considered part of our future work. Furthermore, combining the quantum-inspired evolutionary algorithm with other metaheuristics, or subprocedures thereof, might provide additional escape mechanisms and is worth exploring. Lastly, in this work only binary arrays were considered. Expanding the algorithm to higher alphabets might require small modifications to the used representation and procedures.

Acknowledgements. This research was carried out partly in the context of the Austrian COMET K1 program and publicly funded by the Austrian Research Promotion Agency (FFG) and the Vienna Business Agency (WAW).

References

1. Cleve, R., Ekert, A., Macchiavello, C., Mosca, M.: Quantum algorithms revisited. In: Proceedings of the Royal Society A: Mathematical, Physical and Engineering Sciences, vol. 454, August 1997
2. Colbourn, C.J.: Covering Array Tables for t = 2, 3, 4, 5, 6. http://www.public.asu.edu/~ccolbou/src/tabby/catable.html. Accessed 26 Feb 2019
3. Colbourn, C.J.: Combinatorial aspects of covering arrays. Le Mathematiche **LIX**((I–II)), 125–172 (2004)
4. Danziger, P., Mendelsohn, E., Moura, L., Stevens, B.: Covering arrays avoiding forbidden edges. In: Yang, B., Du, D.-Z., Wang, C.A. (eds.) COCOA 2008. LNCS, vol. 5165, pp. 296–308. Springer, Heidelberg (2008). https://doi.org/10.1007/978-3-540-85097-7_28
5. Grover, L.K.: A fast quantum mechanical algorithm for database search. In: Proceedings of the Twenty-Eighth Annual ACM Symposium on Theory of Computing, STOC 1996, pp. 212–219. ACM, New York (1996)
6. Han, K.H., Kim, J.H.: Quantum-inspired evolutionary algorithm for a class of combinatorial optimization. IEEE Trans. Evol. Comput. **6**(6), 580–593 (2002)
7. Han, K.H., Kim, J.H.: Quantum-inspired evolutionary algorithms with a new termination criterion, h/sub /spl epsi// gate, and two-phase scheme. IEEE Trans. Evol. Comput. **8**(2), 156–169 (2004)
8. Han, K.H., Kim, J.H.: On the analysis of the quantum-inspired evolutionary algorithm with a single individual. In: 2006 IEEE International Conference on Evolutionary Computation, pp. 2622–2629, July 2006
9. Hnich, B., Prestwich, S.D., Selensky, E., Smith, B.M.: Constraint models for the covering test problem. Constraints **11**(2), 199–219 (2006)
10. Kampel, L., Simos, D.E.: A survey on the state of the art of complexity problems for covering arrays. Theor. Comput. Sci. (2019). https://doi.org/10.1016/j.tcs.2019.10.019
11. Kleine, K., Simos, D.E.: An efficient design and implementation of the in-parameter-order algorithm. Math. Comput. Sci. **12**(1), 51–67 (2018)
12. Kuhn, D., Kacker, R., Lei, Y.: Introduction to Combinatorial Testing. Chapman & Hall/CRC Innovations in Software Engineering and Software Development Series. Taylor & Francis, Routledge (2013)
13. Lei, Y., Tai, K.C.: In-parameter-order: a test generation strategy for pairwise testing. In: Proceedings Third IEEE International High-Assurance Systems Engineering Symposium (Cat. No. 98EX231), pp. 254–261, November 1998
14. Martí, R., Pardalos, P., Resende, M.: Handbook of Heuristics. Springer, Heidelberg (2018)
15. Nayeri, P., Colbourn, C.J., Konjevod, G.: Randomized post-optimization of covering arrays. Eur. J. Comb. **34**(1), 91–103 (2013)
16. Nielsen, M.A., Chuang, I.: Quantum Computation and Quantum Information. Cambridge University Press, New York (2002)
17. Seroussi, G., Bshouty, N.H.: Vector sets for exhaustive testing of logic circuits. IEEE Trans. Inf. Theory **34**(3), 513–522 (1988)

18. Shor, P.W.: Algorithms for quantum computation: discrete logarithms and factoring. In: Proceedings 35th Annual Symposium on Foundations of Computer Science, pp. 124–134, November 1994
19. Torres-Jimenez, J., Izquierdo-Marquez, I.: Survey of covering arrays. In: 2013 15th International Symposium on Symbolic and Numeric Algorithms for Scientific Computing, pp. 20–27, September 2013
20. Torres-Jimenez, J., Rodriguez-Tello, E.: New bounds for binary covering arrays using simulated annealing. Inf. Sci. **185**(1), 137–152 (2012)
21. Yanofsky, N.S., Mannucci, M.A., Mannucci, M.A.: Quantum Computing for Computer Scientists, vol. 20. Cambridge University Press, Cambridge (2008)
22. Zhang, G.: Quantum-inspired evolutionary algorithms: a survey and empirical study. J. Heuristics **17**(3), 303–351 (2011)

An Experimental Study of Algorithms for Geodesic Shortest Paths in the Constant-Workspace Model

Jonas Cleve(✉) and Wolfgang Mulzer

Institut für Informatik, Freie Universität Berlin, Takustr. 9, 14195 Berlin, Germany
{jonascleve,mulzer}@inf.fu-berlin.de

Abstract. We perform an experimental evaluation of algorithms for finding geodesic shortest paths between two points inside a simple polygon in the constant-workspace model. In this model, the input resides in a read-only array that can be accessed at random. In addition, the algorithm may use a constant number of words for reading and for writing. The constant-workspace model has been studied extensively in recent years, and algorithms for geodesic shortest paths have received particular attention.

We have implemented three such algorithms in Python, and we compare them to the classic algorithm by Lee and Preparata that uses linear time and linear space. We also clarify a few implementation details that were missing in the original description of the algorithms. Our experiments show that all algorithms perform as advertised in the original works and according to the theoretical guarantees. However, the constant factors in the running times turn out to be rather large for the algorithms to be fully useful in practice.

Keywords: Simple polygon · Geodesic shortest path · Constant workspace · Experimental evaluation

1 Introduction

In recent years, the *constant-workspace model* has enjoyed growing popularity in the computational geometry community [6]. Motivated by the increasing deployment of small devices with limited memory capacities, the goal is to develop simple and efficient algorithms for the situation where little workspace is available. The model posits that the input resides in a read-only array that can be accessed at random. In addition, the algorithm may use a constant number of memory words for reading and for writing. The output must be written to a write-only memory that cannot be accessed again for reading. Following the initial work by Asano *et al.* from 2011 [2], numerous results have been published for this model, leading to a solid theoretical foundation for dealing with geometric

Supported in part by DFG projects MU/3501-1 and RO/2338-6 and ERC StG 757609.

I. Kotsireas et al. (Eds.): SEA^2 2019, LNCS 11544, pp. 317–331, 2019.
https://doi.org/10.1007/978-3-030-34029-2_21

problems when the working memory is scarce. The recent survey by Banyassady *et al.* [6] gives an overview of the problems that have been considered and of the results that are available for them.

But how do these theoretical results measure up in practice, particularly in view of the original motivation? To investigate this question, we have implemented three different constant-workspace algorithms for computing geodesic shortest paths in simple polygons. This is one of the first problems to be studied in the constant-workspace model [2,3]. Given that the general shortest path problem is unlikely to be amenable to constant-workspace algorithms (it is NL-complete [18]), it may come as a surprise that a solution for the geodesic case exists at all. By now, several algorithms are known, both for constant workspace as well as in the *time-space-trade-off* regime, where the number of available cells of working memory may range from constant to linear [1,12].

Due to the wide variety of approaches and the fundamental nature of the problem, geodesic shortest paths are a natural candidate for a deeper experimental study. Our experiments show that all three constant-workspace algorithms work well in practice and live up to their theoretical guarantees. However, the large running times make them ill-suited for very large input sizes. During our implementation, we also noticed some missing details in the original publications, and we explain below how we have dealt with them.

As far as we know, our study constitutes the first large-scale comparative evaluation of geometric algorithms in the constant-workspace model. A previous implementation study, by Baffier *et al.* [5], focused on time-space trade-offs for stack-based algorithms and was centered on different applications of a powerful algorithmic technique. Given the practical motivation and wide applicability of constant-workspace algorithms for geometric problems, we hope that our work will lead to further experimental studies in this direction.

2 The Four Shortest-Path Algorithms

We provide a brief summary for each of the four algorithms in our implementation; further details can be found in the original papers [2,3,14]. In each case, we use P to denote a simple input polygon in the plane with n vertices. We consider P to be a closed, connected subset of the plane. Given two points $s, t \in P$, our goal is to compute a shortest path from s to t (with respect to the Euclidean length) that lies completely inside P.

2.1 The Classic Algorithm by Lee and Preparata

This is the classic linear-space algorithm for the geodesic shortest path problem that can be found in textbooks [11,14]. It works as follows: we triangulate P, and we find the triangle that contains s and the triangle that contains t. Next, we determine the unique path between these two triangles in the dual graph of the triangulation. The path is unique since the dual graph of a triangulation of

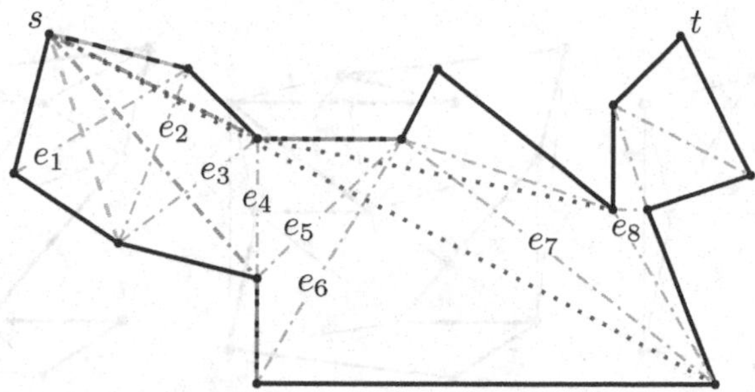

Fig. 1. Examples of three funnels during the algorithm for finding a shortest path from s to t. Each has cusp s and goes up to diagonals e_2 (green, dashed), e_6 (orange, dash dotted), and e_8 (purple, dotted). (Color figure online)

a simple polygon is a tree [7]. We obtain a sequence $e_1, \ldots, e_m$ of diagonals (incident to pairs of consecutive triangles on the dual path) crossed by the geodesic shortest path between s and t, in that order. The algorithm walks along these diagonals, while maintaining a *funnel*. The funnel consists of a *cusp* p, initialized to be s, and two concave *chains* from p to the two endpoints of the current diagonal e_i. An example of these funnels can be found in Fig. 1. In each step i of the algorithm, $i = 1, \ldots, m-1$, we update the funnel for e_i to the funnel for e_{i+1}. There are two cases: (i) if e_{i+1} remains visible from the cusp p, we update the appropriate concave chain, using a variant of Graham's scan; (ii) if e_{i+1} is no longer visible from p, we proceed along the appropriate chain until we find the cusp for the next funnel. We output the vertices encountered along the way as part of the shortest path.Implemented in the right way, this procedure takes linear time and space.[1]

2.2 Using Constrained Delaunay-Triangulations

The first constant-workspace-algorithm for geodesic shortest paths in simple polygons was presented by Asano *et al.* [3] in 2011. It is called *Delaunay*, and it constitutes a relatively direct adaptation of the method of Lee and Preparata to the constant-workspace model.

In the constant-workspace model, we cannot explicitly compute and store a triangulation of P. Instead, we use a uniquely defined implicit triangulation of P, namely the *constrained Delaunay triangulation* of P [9]. In this variant of the classic Delaunay triangulation, we prescribe the edges of P to be part of the desired triangulation. Then, the additional triangulation edges cannot cross the prescribed edges. Thus, unlike in the original Delaunay triangulation,

[1] If a triangulation of P is already available, the implementation is relatively straightforward. If not, a linear-time implementation of the triangulation procedure constitutes a significant challenge [8]. Simpler methods are available, albeit at the cost of a slightly increased running time of $O(n \log n)$ [7].

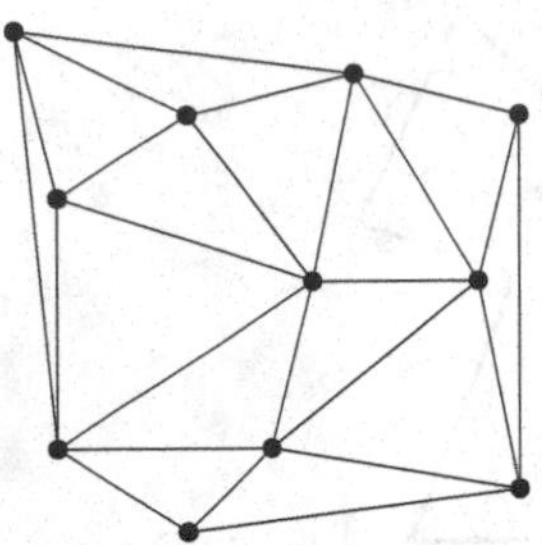

(a) The Delaunay triangulation of a point set.

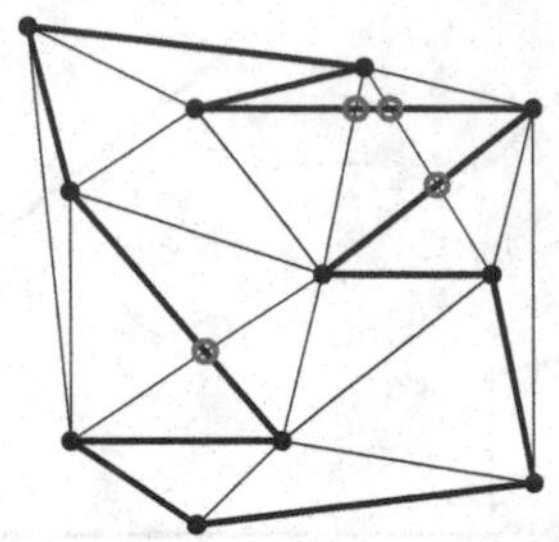

(b) A polygon on the point set which intersects the triangulation.

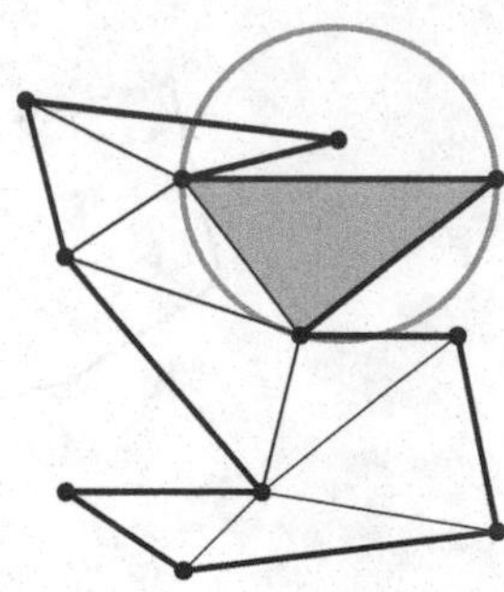

(c) The circumcircle may contain points not visible from some triangle vertices.

Fig. 2. An example of a constrained Delaunay triangulation of a simple polygon.

the circumcircle of a triangle may contain other vertices of P, as long as the line segment from a triangle endpoint to the vertex crosses a prescribed polygon edge, see Fig. 2 for an example.

The constrained Delaunay triangulation of P can be navigated efficiently using constant workspace: given a diagonal or a polygon edge, we can find the two incident triangles in $O(n^2)$ time [3]. Using an $O(n)$ time constant-workspace-algorithm for finding shortest paths in trees, also given by Asano *et al.* [3], we can thus enumerate all triangles in the dual path between the constrained Delaunay triangle that contains s and the constrained Delaunay triangle that contains t in $O(n^3)$ time.

As in the algorithm by Lee and Preparata, we need to maintain the visibility funnel while walking along the dual path of the constrained Delaunay triangulation. Instead of the complete chains, we store only the two line segments that define the current visibility cone (essentially the cusp together with the first vertex of each chain). We recompute the two chains whenever it becomes necessary. The total running time of the algorithm is $O(n^3)$. More details can be found in the paper by Asano *et al.* [3].

2.3 Using Trapezoidal Decompositions

This algorithm was also proposed by Asano *et al.* [3], as a faster alternative to the algorithm that uses constrained Delaunay triangulations. It is based on the same principle as *Delaunay*, but it uses the trapezoidal decomposition of P instead of the Delaunay triangulation [7]. See Fig. 3 for a depiction of the decomposition and the symbolic perturbation method to avoid a general position assumption. In the algorithm, we compute a trapezoidal decomposition of P, and we follow the dual path between the trapezoid that contains s and the trapezoid that contains t, while maintaining a funnel and outputting the new vertices of the geodesic shortest path as they are discovered. Assuming general position, we can find all

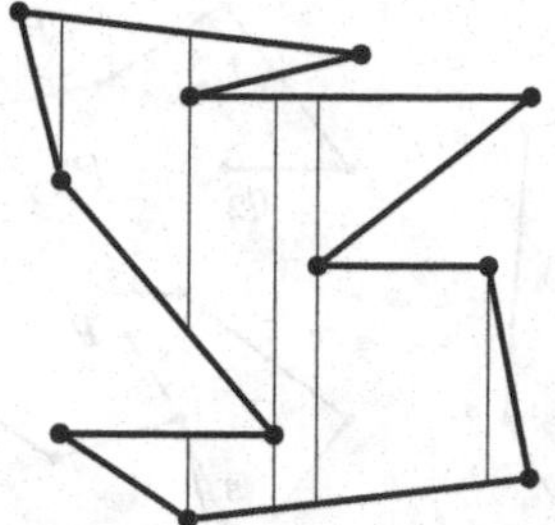

(a) The trapezoidal decomposition is obtained by shooting rays up and down at every vertex.

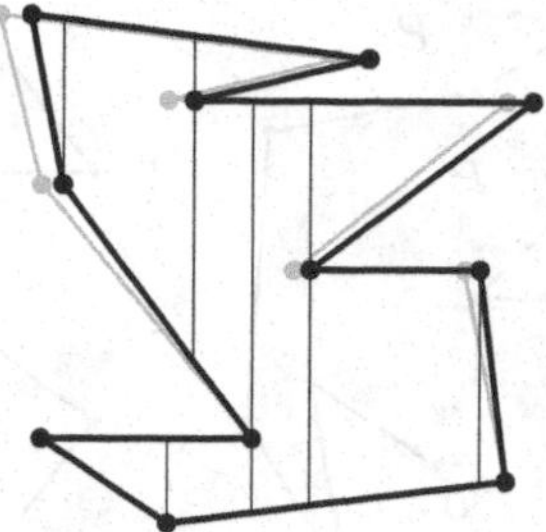

(b) Shifting all points to the right by $y\varepsilon$ makes sure no two share the same x-coordinate.

Fig. 3. The trapezoidal decomposition of a polygon. If the polygon is in general position (right) each trapezoid has at most four neighbors which can all be found in $O(n)$ time.

incident trapezoids of the current trapezoid and determine how to continue on the way to t in $O(n)$ time (instead of $O(n^2)$ time in the case of the *Delaunay* algorithm). Since there are still $O(n)$ steps, the running time improves to $O(n^2)$.

2.4 The Makestep Algorithm

This algorithm was presented by Asano *et al.* [2]. It uses a direct approach to the geodesic shortest path problem and unlike the two previous algorithms, it does not try to mimic on the algorithm by Lee and Preparata. In the traditional model, this approach would be deemed too inefficient, but in the constant-workspace world, its simplicity turns out to be beneficial. The main idea is as follows: we maintain a *current vertex* p of the geodesic shortest path, together with a *visibility cone*, defined by two points q_1 and q_2 on the boundary of P. The segments pq_1 and pq_2 cut off a subpolygon $P' \subseteq P$. We maintain the invariant that the target t lies in P'. In each step, we gradually shrink P' by advancing q_1 and q_2, sometimes also relocating p and outputting a new vertex of the geodesic shortest path. These steps are illustrated in Fig. 4. It is possible to realize the shrinking steps in such a way that there are only $O(n)$ of them. Each shrinking step takes $O(n)$ time, so the total running time of the MakeStep algorithm is $O(n^2)$.

3 Our Implementation

We have implemented the four algorithms from Sect. 2 in Python [15]. For graphical output and for plots, we use the `matplotlib` library [13]. Even though there are some packages for Python that provide geometric objects such as line segments, circles, etc., none of them seemed suitable for our needs. Thus, we decided to implement all geometric primitives on our own. The source code of the implementation is available online in a Git-repository.[2]

[2] https://github.com/jonasc/constant-workspace-algos.

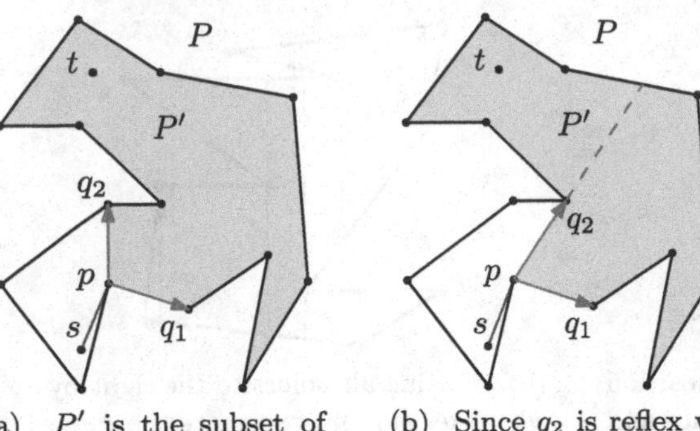

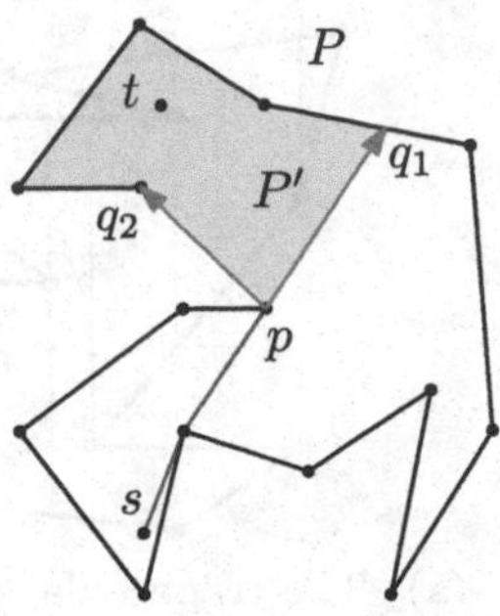

(a) P' is the subset of P cut off by the three points p, q_1, and q_2. Both points are convex, one is advanced.

(b) Since q_2 is reflex we shoot a ray until it hits the boundary. t lies to the left of this ray.

(c) p is relocated to the previous position of q_2 and P' is shrunk along the ray.

Fig. 4. An illustration of the steps in the *Makestep* algorithm.

In order to apply the algorithm *Lee-Preparata*, we must be able to triangulate the simple input polygon P efficiently. Since implementing an efficient polygon triangulation algorithm can be challenging and since this is not the main objective of our study, we relied for this on the *Python Triangle* library by Rufat [16], a Python wrapper for Shewchuk's *Triangle*, which was written in C [17]. We note that *Triangle* does not provide a linear-time triangulation algorithm, which would be needed to achieve the theoretically possible linear running time for the shortest path algorithm. Instead, it contains three different implementations, namely Fortune's sweep line algorithm, a randomized incremental construction, and a divide-and-conquer method. All three implementations give a running time of $O(n \log n)$. For our study, we used the divide-and conquer algorithm, the default choice. In the evaluation, we did not include the triangulation phase in the time and memory measurement for running the algorithm by Lee and Preparata.

3.1 General Implementation Details

All three constant-workspace algorithms have been presented with a general position assumption: *Delaunay* and *Makestep* assume that no three vertices lie on a line, while *Trapezoid* assumes that no two vertices have the same x-coordinate. Our implementations of *Delaunay* and *Makestep* also assume general position, but they throw exceptions if a non-recoverable general position violation is encountered. Most violations, however, can be dealt with easily in our code; e.g. when trying to find the constrained Delaunay triangle(s) for a diagonal, we can simply ignore points collinear to this diagonal. For the case of *Trapezoid*, Asano *et al.* [3] described how to enforce the general position assumption by changing the x-coordinate of every vertex to $x + \varepsilon y$ for some small enough $\varepsilon > 0$ such that the x-order of all vertices is maintained.

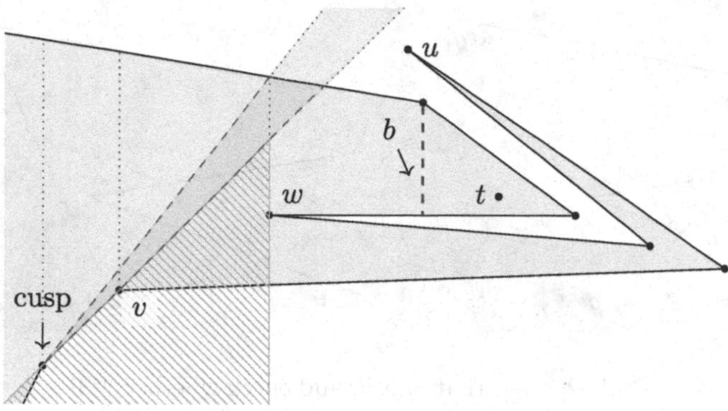

Fig. 5. During the gift wrapping from the cusp to the diagonal b, the vertices need to be restricted to the shaded area. Otherwise, u would be considered to be part of the geodesic shortest path, as it is to the left of vw. (Color figure online)

In our implementation, we apply this method to every polygon in which two vertices share the same x-coordinate.

The coordinates are stored as 64 bit IEEE 754 floats. In order to prevent problems with floating point precision or rounding, we take the following steps: first, we never explicitly calculate angles, but we rely on the usual three-point-orientation test, i.e., the computation of a determinant to find the position of a point c relative to the directed line through to points a and b [7]. Second, if an algorithm needs to place a point somewhere in the relative interior of a polygon edge, we store an additional edge reference to account for inaccuracies when calculating the new point's coordinates.

3.2 Implementing the Algorithm by Lee and Preparata

The algorithm by Lee and Preparata can be implemented easily, in a straightforward fashion. There are no particular edge cases or details that we need to take care of. Disregarding the code for the geometric primitives, the algorithm needs less then half as many lines of code than the other algorithms.

3.3 Implementing Delaunay and Trapezoid

In both constant-workspace adaptations of the algorithm by Lee and Preparata, we encounter the following problem: whenever the cusp of the current funnel changes, we need to find the cusp of the new funnel, and we need to find the piece of the geodesic shortest path that connects the former cusp to the new cusp. In their description of the algorithm, Asano *et al.* [3] only say that this should be done with an application of gift wrapping (Jarvis' march) [7]. While implementing these two algorithms, we noticed that a naive gift wrapping step that considers all the vertices on P between the cusp of the current funnel and the next diagonal might include vertices that are not visible inside the polygon.

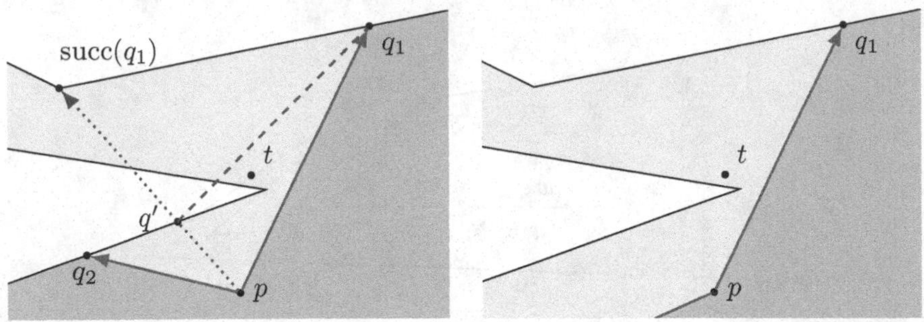

Fig. 6. Left: Asano *et al.* [2] state that one should check whether "t lies in the subpolygon from q' to q_1." This subpolygon, however, is not clearly defined as the line segment $q'q_1$ does not lie inside P. Considering pq' instead and using q_1pq' to shrink the cutoff region gives the correct result on the right.

Figure 5 shows an example: here b is the next diagonal, and naively we would look at all vertices along the polygon boundary between v and w. Hence, u would be considered as a gift wrapping candidate, and since it forms the largest angle with the cusp and v (in particular, an angle that is larger than the angle formed by w) it would be chosen as the next point, even though w should be the cusp of the next funnel. A simple fix for this problem would be an explicit check for visibility in each gift-wrapping step. Unfortunately, the resulting increase in the running time would be too expensive for a realistic implementation of the algorithms.

Our solution for *Trapezoid* is to consider only vertices whose x-coordinate is between the cusp of the current vertex and the point where the current visibility cone crosses the boundary of P for the first time. For ease of implementation, one can also limit it to the x-coordinate of the last trapezoid boundary visible from the cusp. Figure 5 shows this as the dotted green region. For *Delaunay*, a similar approach can be used. The only difference is that the triangle boundaries in general are not vertical lines.

3.4 Implementing Makestep

Our implementation of the *Makestep* algorithm is also relatively straightforward. Nonetheless, we would like to point out one interesting detail; see Fig. 6. The description by Asano *et al.* [2] says that to advance the visibility cone, we should check if "*t lies in the subpolygon from q' to q_1.*" If so, the visibility cone should be shrunk to $q'pq_1$, otherwise to q_2pq'.

However, the "*subpolygon from q' to q_1*" is not clearly defined for the case that the line segment $q'q_1$ is not contained in P. To avoid this difficulty, we instead consider the line segment pq'. This line segment is always contained in P, and it divides the cutoff region P' into two parts, a "subpolygon" between q' and q_1 and a "subpolygon" between q_2 and q'. Now we can easily choose the one containing t.

4 Experimental Setup

We now describe how we conducted the experimental evaluation of our four implementations for geodesic shortest path algorithms.

4.1 Generating the Test Instances

Our experimental approach is as follows: given a desired number of vertices n, we generate 4–10 (pseudo)random polygons with n vertices. For this, we use a tool developed in a software project carried out under the supervision of Günter Rote at the Institute of Computer Science at Freie Universität Berlin [10]. Among others, the tool provides an implementation of the *Space Partitioning* algorithm for generating random simple polygons presented by Auer and Held [4].

Since our main focus was in validating the theoretical guarantees that where published in the literature, we opted for pseudorandomly generated polygons as our test set. This allowed us to quickly produce large input sets of varying sizes. Of course, from a practical point of view, it would also be very interesting to test our implementations on real-world examples. We leave this as a topic for a future study.

Next, we generate the set S of desired endpoints for the geodesic shortest paths. This is done as follows: for each edge e of each generated polygon, we find the incident triangle t_e of e in the constrained Delaunay triangulation of the polygon. We add the barycenter of t_e to S. In the end, the set S will have between $\lfloor n/2 \rfloor$ and $n-2$ points. We will compute the geodesic shortest path for each pair of distinct points in S.

4.2 Executing the Tests

For each pair of points $s, t \in S$, we find the geodesic shortest path between s and t using each of the four implemented algorithms. Since the number of pairs grows quadratically in n, we restrict the tests to 1500 random pairs for all $n \geq 200$.

First, we run each algorithm once in order to assess the memory consumption. This is done by using the `get_traced_memory` function of the built-in `tracemalloc` module which returns the peak and current memory consumption—the difference tells us how much memory was used by the algorithm. Starting the memory tracing just before running the algorithm gives the correct values for the peak memory consumption. In order to obtain reproducible numbers we also disable Python's garbage collection functionality using the built-in `gc.disable` and `gc.enable` functions.

After that, we run the algorithm between 5 and 20 times, depending on how long it takes. We measure the processor time for each run with the `process_time` function of the `time` module which gives the time during which the process was active on the processor in user and in system mode. We then take the median of the times as a representative running time for this point pair.

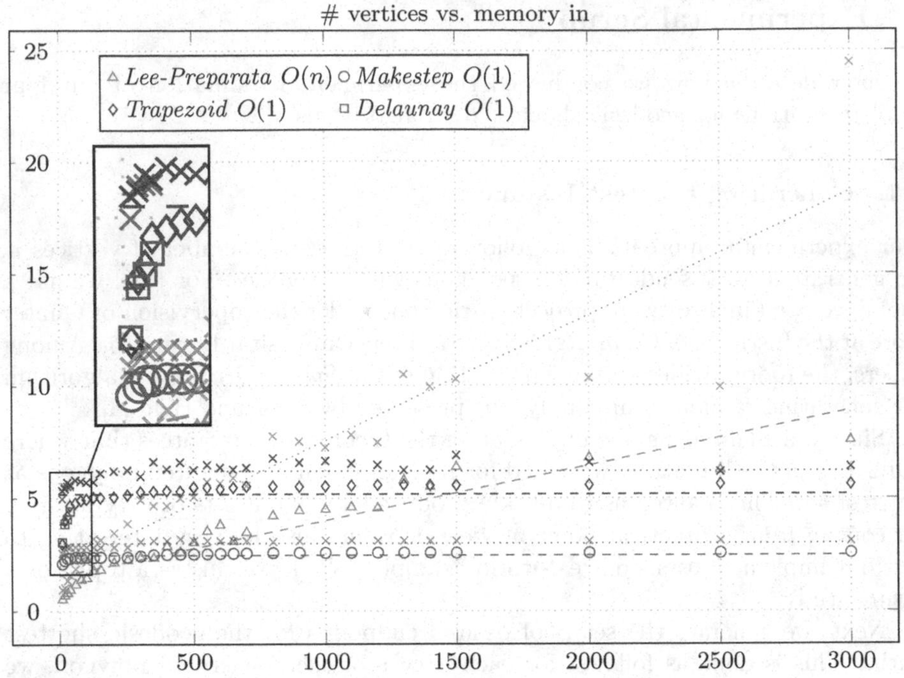

Fig. 7. Memory consumption for random instances. The outlined shapes are the median values; the semi-transparent crosses are maximum values.

4.3 Test Environment

Since we have a quadratic number of test cases for each instance, our experiments take a lot of time. Thus, the tests were distributed on multiple machines and on multiple cores. We had six computing machines at our disposal, each with two quad-core CPUs. Three machines had Intel Xeon E5430 CPUs with 2.67 GHz; the other three had AMD Opteron 2376 CPUs with 2.3 GHz. All machines had 32 GB RAM, even though, as can be seen in the next section, memory was never an issue. The operating system was a Debian 8 and we used version 3.5 of the Python interpreter to implement the algorithms and to execute the tests.

5 Experimental Results

The results of the experiments can be seen in the following plots. The plot in Fig. 7 shows the median and maximum memory consumption as solid shapes and transparent crosses, respectively, for each algorithm and for each input size. More precisely, the plot shows the median and the maximum over all polygons with a given size and over all pairs of points in each such polygon.

We observe that the memory consumption for *Trapezoid* and for *Makestep* is always smaller than a certain constant. At first glance, the shape of the median

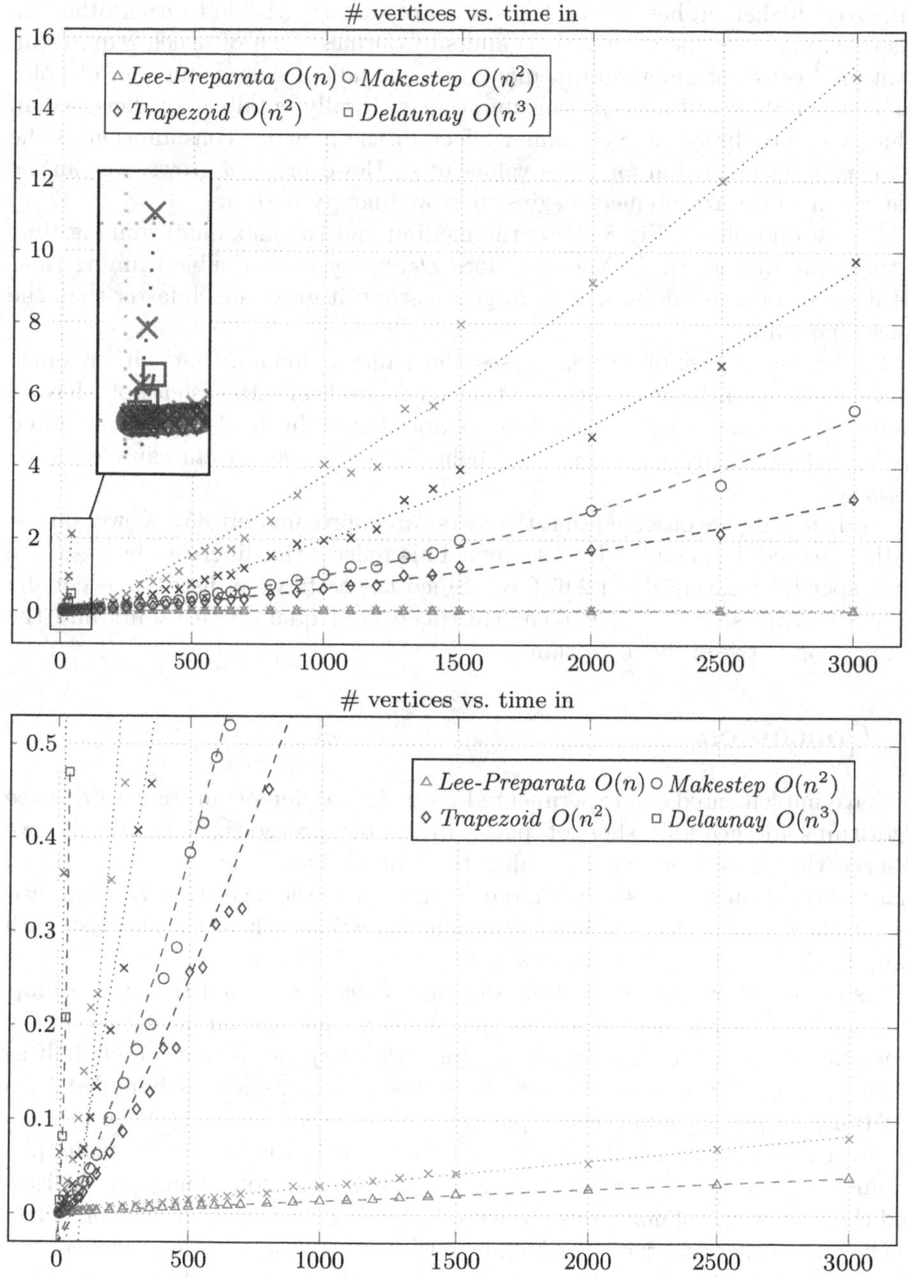

Fig. 8. Runtime for random instances. Outlined shapes are median values; semi-transparent crosses are maximum values. The bottom plot is a scaled version of the top.

values might suggest logarithmic growth. However, a smaller number of vertices leads to a higher probability that s and t are directly visible to each other. In this case, many geometric functions and subroutines, each of which requires an additional constant amount of memory, are not called. A large number of point pairs with only small memory consumption naturally entails a smaller median value. We can observe a very similar effect in the memory consumption of the *Lee-Preparata* algorithm for small values of n. However, as n grows, we can see that the memory requirement begins to grow linearly with n.

The second plot in Fig. 8 shows the median and the maximum running time in the same way as Fig. 7. Not only does *Delaunay* have a cubic running time, but it also seems to exhibit a quite large constant: it grows much faster than the other algorithms.

In the lower part of Fig. 8, we see the same x-domain, but with a much smaller y-domain. Here, we observe that *Trapezoid* and *Makestep* both have a quadratic running time; *Trapezoid* needs about two thirds of the time required by *Makestep*. Finally, the linear-time behavior of *Lee-Preparata* can clearly be discerned.

Additionally, we observed that the tests ran approximately 85% slower on the AMD machines than on the Intel servers. This reflects the difference between the clock speeds of 2.3 GHz and 2.67 GHz. Since the tests were distributed equally on the machines, this does not change the overall qualitative results and the comparison between the algorithms.

6 Conclusion

We have implemented and experimented with three different constant-workspace algorithms for geodesic shortest paths in simple polygons. Not only did we observe the cubic worst-case running time of *Delaunay*, but we also noticed that the constant factor is rather large. This renders the algorithm virtually useless already for polygons with a few hundred vertices, where the shortest path computation might, in the worst case, take several minutes.

As predicted by the theory, *Makestep* and *Trapezoid* exhibit the same asymptotic running time and space consumption. *Trapezoid* has an advantage in the constant factor of the running time, while *Makestep* needs only about half as much memory. Since in both cases the memory requirement is bounded by a constant, *Trapezoid* would be our preferred algorithm.

We chose Python for the implementation mostly due to our previous programming experience, good debugging facilities, fast prototyping possibilities, and the availability of numerous libraries. In hindsight, it might have been better to choose another programming language that allows for more low-level control of the underlying hardware. Python's memory profiling and tracking abilities are limited, so that we cannot easily get a detailed view of the used memory with all the variables. Furthermore, a more detailed control of the memory management could be useful for performing more detailed experiments.

A Tables of Experimental Results

Here we list the experimental results shown in Figs. 7 and 8 (Tables 1 and 2).

Table 1. The median and maximum memory usage in bytes for all runs with a specific number of vertices n.

n	*Delaunay*		*Lee-Preparata*		*Makestep*		*Trapezoid*	
	median	max	median	max	median	max	median	max
10	3048	4976	528	952	2096	2552	2976	5344
20	3864	5512	696	1032	2240	2776	3992	5432
30	4080	5536	808	1360	2344	2840	4208	5704
40	4416	5536	952	1592	2344	2840	4672	5616
60			1184	1872	2384	2840	4784	5808
80			1400	2264	2376	2840	4904	5752
100			1464	2200	2392	2840	4952	5704
125			1792	3216	2384	2840	5040	5720
150			1832	3160	2392	2840	5024	6104
200			2152	3472	2384	2840	5048	6200
250			2264	3880	2392	2840	5144	6240
300			2376	4928	2440	3072	5284	5964
350			2360	4672	2496	3120	5288	6608
400			2880	4672	2512	3120	5328	6284
450			2616	5008	2532	3200	5304	6588
500			3048	5064	2532	3120	5484	6248
550			3552	5736	2540	3120	5480	6476
600			3824	5680	2552	3120	5404	6360
650			3104	5904	2560	3120	5472	6276
700			3496	5568	2568	3120	5528	6352
800			4224	7752	2580	3072	5528	6768
900			4448	7512	2580	3112	5516	6576
1000			4504	7248	2580	3120	5528	6648
1100			4608	7808	2580	3120	5556	6532
1200			4560	7472	2588	3120	5588	6468
1300			5792	10480	2588	3120	5592	6240
1400			5512	9936	2588	3112	5572	6240
1500			6384	10264	2580	3112	5572	6896
2000			6792	10328	2580	3112	5584	6580
2500			6232	9912	2580	3120	5624	6168
3000			7560	24448	2580	3104	5616	6392

Table 2. The median and maximum running times in seconds for all runs with a specific number of vertices n.

n	*Delaunay*		*Lee-Preparata*		*Makestep*		*Trapezoid*	
	median	max	median	max	median	max	median	max
10	0.014019	0.064309	0.000367	0.000910	0.000519	0.004162	0.000820	0.002603
20	0.081347	0.361370	0.000616	0.001613	0.001156	0.011853	0.002010	0.007343
30	0.207516	0.943375	0.000830	0.002879	0.003319	0.026406	0.003655	0.014552
40	0.469530	2.112217	0.001045	0.002166	0.006867	0.033851	0.005716	0.020334
60			0.001399	0.002918	0.013428	0.056691	0.009516	0.030625
80			0.001756	0.003444	0.024055	0.100309	0.016658	0.061326
100			0.002056	0.004030	0.033428	0.150279	0.022560	0.068170
125			0.002501	0.005372	0.046976	0.217762	0.033954	0.101315
150			0.002888	0.005534	0.061505	0.232505	0.041352	0.133888
200			0.003576	0.007240	0.100989	0.354232	0.064532	0.193956
250			0.004321	0.008537	0.137829	0.458281	0.086141	0.260132
300			0.005073	0.009685	0.173749	0.739960	0.110249	0.407216
350			0.005579	0.010597	0.200256	0.808604	0.128425	0.457386
400			0.006372	0.011761	0.249399	0.887698	0.175070	0.589235
450			0.006710	0.013537	0.282497	1.096251	0.175412	0.587010
500			0.007469	0.015005	0.383682	1.541501	0.256470	0.746144
550			0.008528	0.016190	0.415579	1.666938	0.261306	0.859130
600			0.008899	0.017127	0.486157	1.660158	0.307261	0.969043
650			0.009350	0.018987	0.520370	1.707651	0.320547	0.991330
700			0.010033	0.021339	0.548668	2.018272	0.323926	1.187180
800			0.011583	0.021906	0.729638	2.502922	0.452032	1.526597
900			0.012974	0.029481	0.866354	3.218503	0.536978	1.866497
1000			0.014204	0.029770	1.019635	4.070813	0.623762	1.969603
1100			0.015121	0.032160	1.239577	3.846221	0.717239	2.040144
1200			0.016401	0.035842	1.251472	4.010515	0.733767	2.282640
1300			0.018357	0.039272	1.506918	5.627138	1.001474	3.095028
1400			0.019354	0.043886	1.641150	5.707774	1.026240	3.415236
1500			0.021279	0.043013	1.990088	7.978124	1.261539	3.941024
2000			0.026627	0.054653	2.821684	9.151548	1.731935	4.854338
2500			0.032861	0.070760	3.533656	12.003607	2.187277	6.840824
3000			0.039188	0.081773	5.616593	14.949720	3.159590	9.751315

References

1. Asano, T.: Memory-constrained algorithms for simple polygons. Comput. Geom. Theory Appl. **46**(8), 959–969 (2013)
2. Asano, T., Mulzer, W., Rote, G., Wang, Y.: Constant-work-space algorithms for geometric problems. J. Comput. Geom. **2**(1), 46–68 (2011)
3. Asano, T., Mulzer, W., Wang, Y.: Constant-work-space algorithms for shortest paths in trees and simple polygons. J. Graph Algorithms Appl. **15**(5), 569–586 (2011)
4. Auer, T., Held, M.: Heuristics for the generation of random polygons. In: Proceedings of 8th Canadian Conference on Computational Geometry (CCCG), pp. 38–43 (1996)

5. Baffier, J.F., Diez, Y., Korman, M.: Experimental study of compressed stack algorithms in limited memory environments. In: Proceedings of 17th International Symposium Experimental Algorithms (SEA), pp. 19:1–19:13 (2018)
6. Banyassady, B., Korman, M., Mulzer, W.: Computational geometry column 67. SIGACT News **49**(2), 77–94 (2018)
7. de Berg, M., Cheong, O., van Kreveld, M., Overmars, M.: Computational Geometry: Theory and Applications, 3rd edn. Springer, Heidelberg (2008). https://doi.org/10.1007/978-3-540-77974-2
8. Chazelle, B.: Triangulating a simple polygon in linear time. Discret. Comput. Geom. **6**, 485–524 (1991)
9. Chew, L.P.: Constrained Delaunay triangulations. Algorithmica **4**, 97–108 (1989)
10. Dierker, S., Ehrhardt, M., Ihrig, J., Rohde, M., Thobe, S., Tugan, K.: Abschlussbericht zum Softwareprojekt: Zufällige Polygone und kürzeste Wege. Institut für Informatik, Freie Universität, Berlin (2012). https://github.com/marehr/simple-polygon-generator
11. Ghosh, S.K.: Visibility Algorithms in the Plane. Cambridge University Press, New York (2007)
12. Har-Peled, S.: Shortest path in a polygon using sublinear space. J. Comput. Geom. **7**(2), 19–45 (2016)
13. Hunter, J.D.: Matplotlib: a 2D graphics environment. Comput. Sci. Eng. **9**(3), 90–95 (2007)
14. Lee, D.T., Preparata, F.P.: Euclidean shortest paths in the presence of rectilinear barriers. Networks **14**(3), 393–410 (1984)
15. Python Software Foundation: Python. https://www.python.org/. version 3.5
16. Rufat, D.: Python Triangle (2016). http://dzhelil.info/triangle/. version 20160203
17. Shewchuk, J.R.: Triangle: engineering a 2D quality mesh generator and Delaunay triangulator. In: Lin, M.C., Manocha, D. (eds.) WACG 1996. LNCS, vol. 1148, pp. 203–222. Springer, Heidelberg (1996). https://doi.org/10.1007/BFb0014497
18. Tantau, T.: Logspace optimization problems and their approximability properties. Theor. Comput. Sci. **41**(2), 327–350 (2007)

Searching for Best Karatsuba Recurrences

Çağdaş Çalık[1], Morris Dworkin[1], Nathan Dykas[2], and Rene Peralta[1(✉)]

[1] Computer Security Division, NIST, Gaithersburg, USA
{cagdas.calik,dworkin,peralta}@nist.gov
[2] Mathematics Department, University of Maryland, College Park, USA
ndykas@math.umd.edu

Abstract. Efficient circuits for multiplication of binary polynomials use what are known as Karatsuba recurrences. These methods divide the polynomials of size (i.e. number of terms) $k \cdot n$ into k pieces of size n. Multiplication is performed by treating the factors as degree-$(k-1)$ polynomials, with multiplication of the pieces of size n done recursively. This yields recurrences of the form $M(kn) \leq \alpha M(n) + \beta n + \gamma$, where $M(t)$ is the number of binary operations necessary and sufficient for multiplying two binary polynomials with t terms each. Efficiently determining the smallest achievable values of (in order) α, β, γ is an unsolved problem. We describe a search method that yields improvements to the best known Karatsuba recurrences for k = 6, 7 and 8. This yields improvements on the size of circuits for multiplication of binary polynomials in a range of practical interest.

1 Introduction

Polynomials over $\mathbb{F}_2$ are called *binary polynomials*. They have a number of applications, including in cryptography (see [2,5] and the references therein) and in error correcting codes. Let A, B be binary polynomials. We seek small circuits, over the basis $(\wedge, \oplus, 1)$ (that is, arithmetic over $\mathbb{F}_2$), that compute the polynomial $A \cdot B$. In addition to size, i.e. number of gates, we also consider the depth of such circuits, i.e. the length of critical paths.

Notation: We let $M(t)$ denote the number of gates necessary and sufficient to multiply two binary polynomials of size t.

Suppose the polynomials A, B are of odd degree $2n-1$. Karatsuba's algorithm [11] splits A, B into polynomials A_0, A_1 (B_0, B_1 resp.) of size n. Then it recursively computes the product $C = A \cdot B$ as shown in Fig. 1. Careful counting of operations leads to the *2-way Karatsuba recurrence* $M(2n) \leq 3M(n) + 7n - 3$ (see [9], equation (4)).

The product C is $A_0B_0 + X^n(A_0B_1 + A_1B_0) + X^{2n}A_1B_1$. The constant 3 in the 2-way Karatsuba recurrence comes from the fact that 3 multiplications are necessary and sufficient to calculate the three terms $A_0B_0, A_0B_1 + A_1B_0$, and A_1B_1 from A_0, A_1, B_0, B_1. The term $7n - 3$ counts the number of $\mathbb{F}_2$ additions necessary and sufficient to produce the term W and then combine the terms U, V, W into the result C (see [9]).

I. Kotsireas et al. (Eds.): SEA² 2019, LNCS 11544, pp. 332–342, 2019.
https://doi.org/10.1007/978-3-030-34029-2_22

$$
\begin{aligned}
A &= (a_0 + a_1X + \cdots a_{n-1}X^{n-1}) + X^n \cdot (a_n + a_{n+1}X + \cdots a_{2n-1}X^{n-1}) \\
A &= A_0 + X^n A_1 \\
B &= (b_0 + b_1X + \cdots b_{n-1}X^{n-1}) + X^n \cdot (b_n + b_{n+1}X + \cdots b_{2n-1}X^{n-1}) \\
B &= B_0 + X^n B_1 \\
U &\leftarrow A_0 \cdot B_0 \\
V &\leftarrow A_1 \cdot B_1 \\
W &\leftarrow (A_0 + A_1) \cdot (B_0 + B_1) + U + V \\
C &\leftarrow U + X^n W + X^{2n} V.
\end{aligned}
$$

Fig. 1. Karatsuba's algorithm

The generalized Karatsuba method takes two polynomials with kn terms, splits each into k pieces $A_0, \ldots, A_{k-1}$, $B_0, \ldots, B_{k-1}$, computes the polynomials

$$C_m = \sum_{m=i+j} A_i B_j$$

and finally combines the C_i's by summing the overlapping terms.

Karatsuba recurrences have been studied for some time. The paper [12] gives recurrences for the cases $n = 5, 6$, and 7. These recurrences have been improved over the years. The state of the art is [9].

The work [9] provides a unifying description of the generalized Karatsuba method, allowing for a systematic search for such recurrences. The steps in the search are outlined in Fig. 2. Steps 1 and 4 involve solving computationally hard problems. We rely on experimental methods to gain reasonable assurance that we have found the best Karatsuba recurrences in the defined search space.

1. find sets of bilinear forms of minimum size α from which the target C_i's can be computed via additions only.
2. as per [9], each set of bilinear forms determines three matrices T, R, E over $\mathbb{F}_2$.
3. the matrices T, R, E define linear maps L_T, L_R, L_E.
4. let the number of additions necessary for each of the maps be μ_T, μ_R, μ_E, respectively.
5. then the maps yield the recurrence

$$M(kn) \leq \alpha M(n) + \beta n + \gamma$$

with $\beta = 2\mu_T + \mu_E$ and $\gamma = \mu_R - \mu_E$.
6. pick the best recurrence.

Fig. 2. Methodology

2 Finding Minimum-Size Spanning Bilinear Forms

In this section we describe the method for computing (or finding upper bounds on) the constant α in the Karatsuba recurrence.

2.1 Description of the Problem

Consider the two n-term (degree $n-1$) binary polynomials

$$f(x) = \sum_{i=0}^{n-1} a_i x^i, \quad g(x) = \sum_{i=0}^{n-1} b_i x^i \quad \in \mathbb{F}_2[x]$$

with $(2n-1)$-term product

$$h(x) := (fg)(x) = \sum_{k=0}^{2n-2} c_k x^k = \sum_{k=0}^{2n-2} \sum_{i+j=k} a_i b_j x^k$$

We wish to describe the *target coefficients* $c_k = \sum_{i+j=k} a_i b_j$ as linear combinations of bilinear forms of the form

$$\Big(\sum_{i \in S} a_i\Big)\Big(\sum_{i \in S'} b_i\Big), \quad S, S' \subseteq [n-1] = \{0, 1, \ldots, n-1\}$$

Each such bilinear form represents one field multiplication, and the smallest number required to express the target coefficients equals the *multiplicative complexity* of the polynomial multiplication.

Finding these sets of bilinear forms involves searching a space that is doubly exponential in n. Because of this, we will mostly restrict our attention to the *symmetric bilinear forms*, those for which $S = S'$. Two justifications for this simplification are that heuristically they stand a good chance of efficiently generating the target coefficients, which are themselves symmetric, and also that in practice all known cases admit an optimal solution consisting solely of symmetric bilinear forms. However it should be noted that there do exist optimal solutions containing non-symmetric bilinear forms.

2.2 Method for Finding Spanning Sets of Bilinear Forms

Barbulescu et al. [1] published a method for finding minimum-size sets of bilinear forms that span a target set. Their method, which substantially reduces the search space, is described below in the context of Karatsuba recurrences.

The first step is to guess the size of the smallest set of symmetric bilinear forms that spans the target polynomials. Call this guess θ. If θ is too low, then no solution will be found. For the cases of $6, 7, 8$-terms θ is $17, 22, 26$, respectively.

We now assume that the target polynomials are contained in a space spanned by θ of the $(2^n-1)^2$ symmetric bilinear vectors. Checking all spanning sets of size θ is of complexity $\Omega\big(\binom{(2^n-1)^2}{\theta}\big)$, and even if we restrict attention to symmetric bilinear forms as explained above, this is of complexity $\Omega\left(\binom{2^n-1}{\theta}\right)$, which is still prohibitively large, even for $n = 7, \theta = 22$ (for $n = 6, \theta = 17$, this is about 2^{50} and thus close to the limit of what we can compute in practice).

The Barbulescu et al. method is as follows: Let $\mathcal{B}$ be the collection of $(2^n - 1)$ symmetric bilinear products and $\mathcal{T}$ the collection of $2n - 1$ target vectors. For a subset $\mathcal{S} \subset \mathcal{B}$ of size $\theta - (2n - 1)$, let $\mathcal{G} = \mathcal{T} \cup \mathcal{S}$ be a generating set of vectors of size θ and let $\mathbf{C}$ be the candidate subspace generated by $\mathcal{G}$.

We compute the intersection $\mathcal{B} \cap \mathbf{C}$ by applying the rank test to all B in $\mathcal{B}$:

$$B \in \mathbf{C} \iff \theta = \text{rank}(\mathbf{C}) = \text{rank}(\langle \mathbf{C}, B \rangle)$$

which can be computed efficiently via Gaussian elimination.

Now let $\mathbf{C}' := \langle \mathcal{B} \cap \mathbf{C} \rangle$ be the subspace spanned by the intersection. In order to determine $\mathcal{T} \cap \mathbf{C}'$, the collection of target vectors in $\mathbf{C}'$, we again apply a rank test to all T in $\mathcal{T}$:

$$T \in \mathbf{C}' \iff \text{rank}(\mathbf{C}') = \text{rank}(\langle \mathbf{C}', T \rangle)$$

If all the target vectors are spanned, i.e. if $\mathcal{T}' = \mathcal{T}$, then each set of θ independent vectors in $\mathcal{B} \cap \mathbf{C}$ is a solution.

We iterate through the different choices of $\mathcal{S}$ until a solution is found. This reduces the complexity to $O\left(2^n \binom{(2^n-1)}{\theta-(2n-1)}\right)$, which in the cases of $n = 6, 7, 8$ transforms the problem from computationally infeasible to feasible. For details, see [1].

This method generates a potentially large number of solutions with the target multiplicative complexity. Each such solution allows one to produce an arithmetic circuit that computes the product of two n-term polynomials. [9] Describes a way to translate this arithmetic circuit into three $\mathbb{F}_2$-matrices T, R, E, the *top*, *main*, and *extended* matrices. The additive complexities μ_T, μ_R, μ_E, respectively, of these matrices determine the parameters α, β, γ of a recursion (see Fig. 2). In the next section we describe our methods for bounding these additive complexities.

3 Finding Small Circuits for the Linear Maps Determined by each Bilinear Form

The problem is NP-hard and MAX-SNP hard [4], implying limits to its approximability. In practice, it is not currently possible to exactly solve this problem for matrices of the size that arise in this research. SAT-solvers have been used on small matrices, but at size about 8×20 the methods begin to fail (see [10]). The sizes of the matrices T, R, E in the method of [9] are given in Table 1.

For small-enough matrices (those with dimensions in written in **bold**) in Table 1, we used the heuristic of [4] (henceforth the *BMP* heuristic). For the larger matrices we used the randomized algorithm of [3]. More specifically, we used the RAND-GREEDY algorithm with generalized-Paar operation, allowing less than optimal choices in the greedy step (see [3], Sect. 3.4–3.6).

Table 1. Dimensions of linear optimization problems.

n	T	R	E
5	$\mathbf{13 \times 5}$	$\mathbf{9 \times 13}$	10×26
6	$\mathbf{17 \times 6}$	$\mathbf{11 \times 17}$	12×34
7	$\mathbf{22 \times 7}$	$\mathbf{13 \times 22}$	14×44
8	$\mathbf{26 \times 8}$	15×26	16×52

4 Experimental Results

We looked for recurrences for 6, 7, and 8-way Karatsuba. Only symmetric bilinear forms were considered. There exist spanning sets of bases, of optimal size, that contain one or more non-symmetric bilinear forms. However, it is believed, but has not been proven, that there always exists an optimal size spanning set containing only symmetric bilinear forms.

In the following subsections, we give the best T and R matrices found for $n = 6, 7$, and 8. In each case, the matrix E is defined as follows: letting R_i be the ith row of R, the matrix E is

$$E = \begin{pmatrix} R_1 & 0 \\ R_2 & R_1 \\ \vdots & \\ R_{2k-1} & R_{2k-2} \\ 0 & R_{2k-1} \end{pmatrix}.$$

4.1 6-way Split

The search included all symmetric bilinear forms. We searched but did not find solutions with 16 multiplications. We conjecture that the multiplicative complexity of multiplying two binary polynomials of size 6 is 17. 54 solutions with 17 multiplications were found. This matches results reported in [1]. For the matrices T and R, the BMP heuristic was used. For the E matrix, RAND-GREEDY was used. The best recurrence thus obtained was

$$M(6n) \leq 17M(n) + 83n - 26.$$

The best Karatsuba recurrence known before this work was ([9])

$$M(6n) \leq 17M(n) + 85n - 29.$$

The matrices are

$$T_6 = \begin{pmatrix} 1&0&0&0&0&0 \\ 0&1&0&0&0&0 \\ 1&1&0&0&0&0 \\ 0&1&1&0&0&0 \\ 1&1&1&0&0&0 \\ 0&0&0&1&0&0 \\ 1&0&1&0&1&0 \\ 0&1&1&0&1&0 \\ 1&1&0&1&1&0 \\ 0&0&0&0&0&1 \\ 0&1&1&0&0&1 \\ 0&1&0&1&0&1 \\ 1&0&1&1&0&1 \\ 0&1&0&0&1&1 \\ 0&1&1&0&1&1 \\ 0&1&0&1&1&1 \\ 1&1&1&1&1&1 \end{pmatrix} \quad R_6 = \begin{pmatrix} 1&0&0&0&0&0&0&0&0&0&0&0&0&0&0&0&0 \\ 1&1&1&0&0&0&0&0&0&0&0&0&0&0&0&0&0 \\ 0&0&1&1&1&0&0&0&0&0&0&0&0&0&0&0&0 \\ 1&1&0&1&0&1&0&1&1&1&0&0&1&1&0&0&1 \\ 1&0&1&1&0&0&0&0&1&1&1&1&1&1&1&1&1 \\ 0&1&1&1&0&1&1&0&1&1&0&0&1&0&1&1&0 \\ 0&1&0&0&0&1&0&0&0&1&1&0&0&1&1&0&0 \\ 0&1&0&1&0&0&0&0&0&0&1&1&0&1&0&1&0 \\ 0&1&0&0&1&1&1&0&0&1&0&0&1&0&0&1&1 \\ 0&0&0&1&0&0&0&1&0&0&1&0&0&0&1&0&0 \\ 0&0&0&0&0&0&0&0&0&1&0&0&0&0&0&0&0 \end{pmatrix}$$

4.2 7-way Split

The search included all symmetric bilinear forms. There are no solutions with 21 multiplications. This leads us to conjecture that the multiplicative complexity of multiplying two binary polynomials of size 7 is 22. 19550 solutions with 22 multiplications were found, which matches results reported in [1]. For the matrix T the BMP heuristic was used. For the R and E matrices, the RAND-GREEDY heuristic was used.

Both the BMP heuristic and the RAND-GREEDY are randomized algorithms. The way to use these algorithms is to run them many times and pick the best solution found. Since the linear optimization problem is NP-hard, we expect that at some value of n, we should no longer be confident that we can find the optimal solution. In practice, we aimed at running the algorithms about 100 thousand times. Since we wouldn't be able to do this for all 19550 sets of matrices, we proceeded in two rounds. In the first round, we ran the algorithms for 1000 times on each set of matrices. The results yielded four sets of matrices that implied values of the β parameter which were better than the rest. We then ran the algorithms for 100 thousand times on each of the four sets of matrices and picked the best.

The best recurrence thus obtained was

$$M(7n) \leq 22M(n) + 106n - 31.$$

The best Karatsuba recurrence known before this work was ([9])

$$M(7n) \leq 22M(n) + 107n - 33.$$

The matrices are

$$T_7 = \begin{pmatrix} 1&0&0&0&0&0&0 \\ 0&1&0&0&0&0&0 \\ 1&1&0&0&0&0&0 \\ 0&0&1&0&0&0&0 \\ 1&0&1&0&0&0&0 \\ 1&0&1&1&0&0&0 \\ 0&1&1&1&0&0&0 \\ 0&0&1&0&1&0&0 \\ 1&0&1&0&1&0&0 \\ 0&0&0&0&0&1&0 \\ 0&1&0&0&0&1&0 \\ 0&0&0&1&0&1&0 \\ 0&1&0&1&0&1&0 \\ 0&1&1&0&1&1&0 \\ 0&0&0&0&0&0&1 \\ 1&0&1&0&0&0&1 \\ 1&0&1&0&1&0&1 \\ 1&1&0&1&1&0&1 \\ 1&1&1&1&1&0&1 \\ 0&0&0&0&0&1&1 \\ 1&0&1&1&0&1&1 \\ 1&1&1&1&1&1&1 \end{pmatrix} \quad R_7 = \begin{pmatrix} 1&0&0&0&0&0&0&0&0&0&0&0&0&0&0&0&0&0&0&0&0&0 \\ 1&1&1&0&0&0&0&0&0&0&0&0&0&0&0&0&0&0&0&0&0&0 \\ 1&1&0&1&1&0&0&0&0&0&0&0&0&0&0&0&0&0&0&0&0&0 \\ 0&1&0&1&1&1&1&0&0&1&1&1&1&0&0&0&0&0&0&0&0&0 \\ 0&0&0&0&1&0&0&1&1&1&1&1&1&0&0&0&0&0&0&0&0&0 \\ 0&0&0&1&0&0&1&1&0&1&1&0&1&1&1&0&0&0&1&1&0&1 \\ 1&0&0&1&0&0&1&0&0&0&0&1&1&0&1&1&0&1&1&0&0&0 \\ 1&1&1&0&0&0&0&1&0&0&1&1&1&1&0&1&1&0&0&0&1&1 \\ 1&1&0&0&1&0&1&1&0&0&1&0&1&0&0&0&0&1&1&0&0&0 \\ 0&1&0&0&1&1&0&0&0&1&1&0&0&0&0&1&0&0&1&0&1&1 \\ 0&0&0&0&1&0&0&0&1&1&0&0&0&0&1&1&0&0&0&0&0&0 \\ 0&0&0&0&0&0&0&0&0&1&0&0&0&0&1&0&0&0&0&1&0&0 \\ 0&0&0&0&0&0&0&0&0&0&0&0&0&0&1&0&0&0&0&0&0&0 \end{pmatrix}$$

4.3 8-way Split

It is known that the multiplicative complexity of 8-term binary polynomials is at most 26 [8]. We were not able to improve on this, the search for solutions with multiplicative complexity 25 appears to require either a huge investment in computation time or an improvement in search methods.

For multiplicative complexity 26, we were not able to search the whole space of symmetric bilinear forms. We verified that there are no solutions with either 7 or 8 "singleton" bases (i.e. bases of the form $a_i b_i$), and there are exactly 77 solutions with 6 "singleton" bases. Additionally, we restricted the search space to sets of bases containing the bilinear forms $a_1 b_1$ and $(a_0+a_2+a_3+a_5+a_6)(b_0+b_2+b_3+b_5+b_6)$ and three among the following

$$\begin{aligned}
&(a_1 + a_3 + a_4 + a_5)(b_1 + b_3 + b_4 + b_5) \\
&(a_1 + a_2 + a_3 + a_6)(b_1 + b_2 + b_3 + b_6) \\
&(a_2 + a_4 + a_5 + a_6)(b_2 + b_4 + b_5 + b_6) \\
&(a_0 + a_2 + a_3 + a_4 + a_7)(b_0 + b_2 + b_3 + b_4 + b_7) \\
&(a_0 + a_1 + a_2 + a_5 + a_7)(b_0 + b_1 + b_2 + b_5 + b_7) \\
&(a_0 + a_1 + a_4 + a_6 + a_7)(b_0 + b_1 + b_4 + b_6 + b_7) \\
&(a_0 + a_3 + a_5 + a_6 + a_7)(b_0 + b_3 + b_5 + b_6 + b_7).
\end{aligned}$$

Our search yielded 2079 solutions, including 63 of the 77 solutions with 6 singletons. For the matrix T, the BMP heuristic was used. For the R and E matrices, RAND-GREEDY was used. Among these 2079 solutions, we found one for which T_8 could be computed with 24 gates, R_8 with 59 gates and E_8 with 99 gates.

The matrices are

$$T_8 = \begin{pmatrix}
1&0&0&0&0&0&0&0\\
0&1&0&0&0&0&0&0\\
1&1&0&0&0&0&0&0\\
0&0&1&0&0&0&0&0\\
1&0&1&0&0&0&0&0\\
0&1&0&1&1&1&0&0\\
0&0&0&0&0&0&1&0\\
0&1&1&1&0&0&1&0\\
0&0&1&1&1&0&1&0\\
1&0&1&0&0&1&1&0\\
1&0&1&1&0&1&1&0\\
0&1&0&0&1&1&1&0\\
0&0&1&0&1&1&1&0\\
0&0&0&0&0&0&0&1\\
1&0&1&0&0&0&0&1\\
1&1&1&0&1&0&0&1\\
1&0&1&1&1&0&0&1\\
1&1&1&0&0&1&0&1\\
0&1&1&0&1&1&0&1\\
1&0&0&1&1&1&0&1\\
0&0&0&0&0&0&1&1\\
1&1&0&1&0&0&1&1\\
1&1&0&0&1&0&1&1\\
1&1&0&1&1&0&1&1\\
1&0&1&0&0&1&1&1\\
1&1&1&1&1&1&1&1
\end{pmatrix}
\quad
R_8 = \begin{pmatrix}
1&0&0&0&0&0&0&0&0&0&0&0&0&0&0&0&0&0&0&0&0&0&0&0&0&0\\
1&1&1&0&0&0&0&0&0&0&0&0&0&0&0&0&0&0&0&0&0&0&0&0&0&0\\
1&1&0&1&1&0&0&0&0&0&0&0&0&0&0&0&0&0&0&0&0&0&0&0&0&0\\
0&1&0&1&0&1&1&0&0&1&1&0&1&0&1&0&0&0&0&1&0&1&1&1&0&1\\
0&0&1&1&0&0&0&1&1&1&0&0&1&0&1&1&0&1&1&0&1&1&1&1&0&0\\
0&0&0&0&1&1&0&0&1&1&0&1&1&1&1&0&1&1&0&0&1&0&0&0&0&1\\
0&0&0&0&0&0&1&1&0&0&0&1&1&1&0&1&0&1&0&1&1&0&1&0&1&1\\
1&0&0&0&0&0&0&0&0&0&0&0&0&1&0&0&1&1&0&1&0&1&1&0&1&1\\
1&1&1&0&0&1&0&0&1&0&0&0&1&0&0&1&0&0&0&1&0&0&1&0&0&1\\
1&1&0&1&1&1&0&1&0&0&0&1&0&0&0&1&0&1&0&0&0&1&0&0&0&1\\
0&1&0&1&0&0&1&1&1&1&1&0&1&0&1&1&1&0&0&1&0&0&0&1&1&0\\
0&0&1&1&0&0&0&0&0&1&0&1&0&0&1&1&1&1&1&1&1&0&1&1&0&1\\
0&0&0&0&1&0&0&0&0&1&0&0&0&1&1&0&0&0&0&0&1&0&0&0&1&0\\
0&0&0&0&0&0&1&0&0&0&0&0&0&1&0&0&0&0&0&0&1&0&0&0&0&0\\
0&0&0&0&0&0&0&0&0&0&0&0&0&1&0&0&0&0&0&0&0&0&0&0&0&0
\end{pmatrix}$$

This yields the recurrence

$$M(8n) \leq 26M(n) + 147n - 40.$$

The new recurrence for 8-way Karatsuba may be of practical interest. The smallest known Karatsuba-based circuit for multiplying two polynomials of size 96 has 7110 gates [9]. Using the new recurrence, along with $M(12) \leq 207$, yields

$$M(96) = M(8 \cdot 12) \leq 26 \cdot 207 + 147 \cdot 12 - 40 = 7106.$$

Table 2. New circuit sizes and depths for $n = 28$ to 99. Values of n for which we obtained and improvement in size are in **bold**.

n	Size in [9]	New size	Depth [9]	New depth	n	Size in [9]	New size	Depth in [9]	New depth
28	944	943	14	15	64	3673	3673	13	13
29	1009	1009	13	13	65	3920	3920	15	15
30	1038	1038	13	13	66	4041	4041	15	15
31	1113	1113	12	12	67	4152	4152	14	14
32	1156	1156	11	11	68	4220	4220	14	14
33	1271	1271	12	12	69	4353	4353	14	14
34	1333	1333	12	12	70	4417	4417	14	14
35	1392	1392	11	11	**71**	4478	4456	25	20
36	1428	1428	11	11	**72**	4510	4489	25	20
37	1552	1552	15	15	73	4782	4782	18	18
38	1604	1604	14	14	74	4815	4815	18	18
39	1669	1669	14	14	75	4847	4847	18	18
40	1703	1703	14	14	76	5075	5075	17	17
41	1806	1806	16	17	77	5198	5198	16	16
42	1862	1859	16	17	78	5255	5255	16	16
43	1982	1982	15	16	79	5329	5329	16	16
44	2036	2036	12	12	80	5366	5366	16	16
45	2105	2105	14	14	81	5593	5593	19	20
46	2179	2179	14	14	**82**	5702	5697	19	19
47	2228	2228	13	13	**83**	5769	5760	18	19
48	2259	2259	13	13	**84**	5804	5795	18	19
49	2436	2436	14	14	**85**	6118	6115	18	19
50	2523	2523	17	17	**86**	6224	6221	19	20
51	2663	2663	14	14	87	6344	6344	18	19
52	2725	2725	13	13	88	6413	6413	15	15
53	2841	2825	24	19	**89**	6516	6488	28	23
54	2878	2863	24	19	**90**	6550	6523	28	23
55	2987	2984	17	18	91	6776	6776	17	17
56	3022	3017	17	18	92	6842	6842	16	16
57	3145	3145	15	15	93	6929	6929	18	19
58	3212	3211	17	18	94	7010	7010	16	16
59	3273	3273	15	15	**95**	7073	7071	15	25
60	3306	3306	15	15	**96**	7110	7106	16	25
61	3472	3472	15	15	97	7465	7465	17	17
62	3553	3553	15	15	98	7636	7636	20	20
63	3626	3626	14	14	99	7801	7801	19	19

5 Implications for the Circuit Complexity of Binary Polynomial Multiplication

This work yielded three new Karatsuba recurrences:

$$M(6n) \leq 17M(n) + 83n - 26$$
$$M(7n) \leq 22M(n) + 106n - 31$$
$$M(8n) \leq 26M(n) + 147n - 40.$$

As per [9], the circuits for these recurrences can be leveraged into circuits for multiplication of binary polynomials of various sizes. Doing this, we found that the new recurrences improve known results for Karatsuba multiplication starting at size 28. The circuits were generated automatically from the circuits for each set of matrices for $n = 2, \ldots, 8$ (the cases $n = 6, 7, 8$ are reported in this work). We generated the circuits up to $n = 100$. The circuits were verified by generating and validating the algebraic normal form of each output. Table 2 compares the new circuit sizes and depths to the state of the art as reported in [9]. The table starts at the first size in which the new recurrences yield a smaller number of gates. The circuits have not been optimized for depth. The circuits will be posted at cs-www.cs.yale.edu/homes/peralta/CircuitStuff/CMT.html.

A different approach to gate-efficient circuits for binary polynomial multiplication is to use interpolation methods. These methods can yield smaller circuits than Karatsuba multiplication at the cost of higher depth (see, for example, [6,7]). An interesting open question is to characterize the depth/size tradeoff of Karatsuba versus interpolation methods for polynomials of sizes of practical interest. In elliptic curve cryptography, multiplication of binary polynomials with thousands of bits is used.

References

1. Barbulescu, R., Detrey, J., Estibals, N., Zimmermann, P.: Finding optimal formulae for bilinear maps. In: Özbudak, F., Rodríguez-Henríquez, F. (eds.) WAIFI 2012. LNCS, vol. 7369, pp. 168–186. Springer, Heidelberg (2012). https://doi.org/10.1007/978-3-642-31662-3_12
2. Bernstein, D.J.: Batch binary Edwards. In: Halevi, S. (ed.) CRYPTO 2009. LNCS, vol. 5677, pp. 317–336. Springer, Heidelberg (2009). https://doi.org/10.1007/978-3-642-03356-8_19
3. Boyar, J., Find, M.G., Peralta, R.: Small low-depth circuits for cryptographic applications. Crypt. Commun. **11**(1), 109–127 (2018). https://doi.org/10.1007/s12095-018-0296-3
4. Boyar, J., Matthews, P., Peralta, R.: Logic minimization techniques with applications to cryptology. J. Cryptol. **26**(2), 280–312 (2013)
5. Brent, R.P., Gaudry, P., Thomé, E., Zimmermann, P.: Faster multiplication in GF(2)[x]. In: van der Poorten, A.J., Stein, A. (eds.) ANTS 2008. LNCS, vol. 5011, pp. 153–166. Springer, Heidelberg (2008). https://doi.org/10.1007/978-3-540-79456-1_10

6. Cenk, M., Hasan, M.A.: Some new results on binary polynomial multiplication. J. Cryptogr. Eng. **5**, 289–303 (2015)
7. De Piccoli, A., Visconti, A., Rizzo, O.G.: Polynomial multiplication over binary finite fields: new upper bounds. J. Cryptogr. Eng. 1–14, April 2019. https://doi.org/10.1007/s13389-019-00210-w
8. Fan, H., Hasan, M.A.: Comments on five, six, and seven-term Karatsuba-like formulae. IEEE Trans. Comput. **56**(5), 716–717 (2007)
9. Find, M.G., Peralta, R.: Better circuits for binary polynomial multiplication. IEEE Trans. Comput. **68**(4), 624–630 (2018). https://doi.org/10.1109/TC.2018.2874662
10. Fuhs, C., Schneider-Kamp, P.: Optimizing the AES S-box using SAT. In: Proceedings International Workshop on Implementation of Logics (IWIL), pp. 64–70 (2010)
11. Karatsuba, A.A., Ofman, Y.: Multiplication of multidigit numbers on automata. Sov. Phys. Doklady **7**, 595–596 (1963)
12. Montgomery, P.L.: Five, six, and seven-term Karatsuba-like formulae. IEEE Trans. Comput. **54**(3), 362–369 (2005). https://doi.org/10.1109/TC.2005.49. http://doi.ieeecomputersociety.org/10.1109/TC.2005.49

Minimum and Maximum Category Constraints in the Orienteering Problem with Time Windows

Konstantinos Ameranis[1], Nikolaos Vathis[2(✉)], and Dimitris Fotakis[2]

[1] Boston University, Boston, MA 02215, USA
ameranis@bu.edu

[2] National Technical University of Athens,
9 Iroon Polytechniou Street, 15780 Zografou, Greece
nvathis@softlab.ntua.gr, fotakis@cs.ntua.gr

Abstract. We introduce a new variation of the Orienteering Problem (OP), the Minimum-Maximum Category Constraints Orienteering Problem with Time Windows. In the Orienteering Problem we seek to determine a path from node S to node T in a weighted graph where each node has a score. The total weight of the path must not exceed a predetermined budget and the goal is to maximize the total score. In this variation, each Activity is associated with a category and the final solution is required to contain at least a minimum and at most a maximum of specific categories. This variation better captures the problem of tourists visiting cities. For example, the tourists can decide to visit exactly one restaurant at a specific time window and at least one park. We present a Replace Local Search and an Iterated Local Search which utilizes Stochastic Gradient Descent to identify the tightness of the constraints. We perform exhaustive experimental evaluation of our results against state of the art implementations for the unconstrained problem and examine how it performs against increasingly more restricting settings.

Keywords: Orienteering · Local Search · Integer Programming · Heuristics · Category constraints

1 Introduction

When a tourist visits a city they want to get the most out of their time there. There is a plethora of places they can visit in various corners of the city. Defining the characteristics of a good itinerary is very hard, since there are various ideas of what the perfect solution is, that are difficult to formulate. For example, a tourist would expect to definitely visit iconic landmarks, to not visit mutually exclusive places such as multiple restaurants in the span of a small timeframe, and to not follow an itinerary that is saturated with similar POIs.

Our approach on honoring the aforementioned characteristics is to introduce a new kind of constraints, namely minimum and maximum category constraints.

I. Kotsireas et al. (Eds.): SEA[2] 2019, LNCS 11544, pp. 343–358, 2019.
https://doi.org/10.1007/978-3-030-34029-2_23

Using these constraints, compulsory nodes can be assigned to a category by themselves and request one node of that category. Nodes that are complimentary infeasible can be assigned in the same category whose maximum is set to one. We can also express composite constraints such as "visit at least two museums but no more than four, while visiting at least one park".

Almost all forms of the Orienteering Problem can be formulated as Shortest Path Problems [20]. To solve the Minimum-Maximum Category Constraints Orienteering Problem Team we propose two algorithms. The first is based on the approach described in [40], while the second is a local search approach focused on replacing POIs. We explore how to evaluate the tightness of these constraints while solving the problem instance and compare our results to the LP which acts as an upper limit and can be computed efficiently and note of how adding more constraints affects the results of our algorithm.

Our contributions include:

- We formulate an extension of the Orientation Problem
- Provide an IP formulation
- Propose two algorithms
- Compare with exact solutions in cases they could be found
- Compare our algorithms with the algorithm proposed in [40]
- Evaluate our approaches on increasingly more constrained settings

As it is widely known, IP is in the general case NP-hard and as a result any IP program attempting to solve an unconstrained might in the worst case take exponential time to the size of the input. Attempting to solve our instances in Sect. 4.4 using CPLEX requires too much time, otherwise the produced solution's score is only a fraction of what our algorithms compute.

On the other hand, our algorithms provide comparable results to previous works for unconstrained instances, while our two approaches provide high quality solutions for a wide spectrum of tightness of the category constraints.

2 Previous Work

The Orienteering Problem (OP) was first described in [17], but the term OP was first introduced in [39]. Subsequently [18] proved that the OP is NP-hard. OP has stemmed from the TSP and many papers have referred to it as the constrained TSP [15,22,38]. Since then a multitude of variations have appeared changing either the nature of the underlying graph, the constraints imposed on the solution, or the objective function.

In the OP we must determine a path from node S to node T on a weighted graph whose total weight does not exceed a certain budget and maximizes the score collected from the participating nodes. Variations which alter the graph include making either the node availability (OP with Time Windows) or the edge weights temporal (Time Dependent OP), making the score associated with a node stochastic (OP with Stochastic Profits) or the time needed to travel between nodes and/or to visit them (OP with Stochastic Travel and Service Times), or associating the score with edges instead of nodes (Arc OP). Variations on the constraints of the problem include having compulsory nodes, dropping by specific kinds of establishments (ATM, gas station, super market), limiting the amount of nodes belonging to certain categories and having multiple constraints such as money in addition to time. When the objective function is not linear we refer to that problem as the Generalized OP. Obviously some variations are orthogonal and therefore can be combined, such as the TDTOPTW.

For each one of these problems numerous exact, approximation and heuristic solutions have been proposed. While before the 2000s most proposed algorithms were exact and any heuristic algorithms were hand rolled, after the eve of the new millennium there are multiple papers implementing well known metaheuristics to various formats of the OP. The exact solutions include Branch and Bound [22,31], Branch and Cut [8,11,13,15], Branch and Price [4] and Cutting Plane [23] algorithms. Early heuristic approaches include Center of Gravity [18], the Four-phase heuristic [32] and the Five-step heuristic [6,7]. Newer attempts in heuristics include Tabu Search [2,16], several variants of Variable Neighborhood Search [2,24,41], a technique to which researches return to through the years and use to solve almost all variants of the OP. Among other metaheuristics used are Ant Colony Optimization [14,21], Iterated Local Search [19,40], Greedy Randomized Adaptive Search Procedure, also known as GRASP [5,28], Particle Swarm Optimization, also known as PSO [9,10,30,34,35], Simulated Annealing [42] and Genetic Algorithms [12].

Although there are multiple attempts at solving the problem for various maximum constraints [3,36,37], little to no research has been done for minimum category constraints. This paper aims to provide a first peek into this problem as it can be a useful tool while describing real life problems and applications, such as the Tourist Trip Design Problem (TTDP).

3 Problem Formalization

While most papers use the term POI (Point of Interest) for the nodes of the graph, we felt that this term does not fully capture their temporal nature in the case of OPTW. Therefore we will henceforth use the term Activity. In a single location we can have several concurrent Activities (e.g cinema) or different Activities in non overlapping Time Windows.

Let there be a set of N Activities, each associated with a profit p_i, a visiting time v_i and opening and closing times o_i, c_i Let there also be sets K_c such that $|\bigcup_c K_c| = N$ and $c_1 \neq c_2 \Leftrightarrow K_{c_1} \cap K_{c_2} = \emptyset$. Furthermore there is a traveling time t_{ij} between every pair of Activities i, j which respects the triangle inequality. Finally, let there be a starting location s, a finishing location f, a time budget T_{max}, and minimum and maximum category constraints m_c, M_c. We need to determine a path from s to f which maximizes the total profit, while the total time needed for travelling between and visiting the participating Activities does not exceed T_{max}. Additionally the participating Activities from each set K_c must be between m_c and M_c (inclusive).

Here we present the Integer Programming formulation of the problem. Let $x_{ij} = 1$ when a visit to Activity i is followed by a visit to Activity j. Let b_i be the time that the visit to Activity i begins. In that case, the problem can be written down thus:

$$\max \sum_{i=1}^{N-1} \sum_{j=2}^{N} p_i x_{ij} \tag{1}$$

$$\sum_{j \neq s} x_{sj} = \sum_{i \neq f} x_{if} = 1 \tag{2}$$

$$\sum_{i=1}^{N} x_{ik} = \sum_{j=1}^{N} x_{kj} \leq 1 \quad \forall k \neq s, f \tag{3}$$

$$b_i - b_j + x_{ij}(T_{max} + v_i + t_{ij}) \leq T_{max} \quad \forall i, j \tag{4}$$

$$o_i \leq b_i \leq c_i \quad \forall i \tag{5}$$

$$m_c \leq \sum_{i \in K_C} \sum_{j \neq i} x_{ij} \leq M_c \tag{6}$$

$$b_s = 0 \tag{7}$$

$$b_f \leq T_{max} \tag{8}$$

$$x_{ij} \in \{0,1\} \tag{9}$$

Equation 1 is the objective function, constraints 2 forces the route to start from s and finish at f, constraints 3 mean that each Activity will be visited at most once and if there is an incoming edge, there will also be an outgoing one. Constraints 4 make sure that the starting times will be correct and also serve as the Miller-Tucket-Zemlin (MTZ) subtour elimination constraints together with 7 and 8. Constraints 5 make sure that the Time Windows are respected. Constraints 6 express the minimum and maximum category constraints.

From the formulation it can be deduced that this is a shortest path problem seeking to move from s to f while optimizing the score collected along the path with the additional category constraints.

4 Algorithmic Approaches

4.1 Iterated Local Search

The Iterated Local Search [26] algorithm, as implemented by [40], is a rather simplistic variation of the Local Search metaheuristic. It repeatedly alternates between two phases. In the first phase the algorithm greedily selects one Activity at a time to be inserted in the current solution until it reaches a local optimum. For each Activity it determines the position that will take the least amount of additional time to visit, while respecting time constraints. After determining this quantity ($shift$) for all Activities it evaluates a metric similar to that of the continuous knapsack, $\frac{Profit^2}{shift}$ and selects the best one. Equation 10 shows how to calculate $shift$ when inserting Activity 3 between Activities 1 and 2.

Because of time windows inserting a new Activity might present no problem for the immediate next, but make an Activity unfeasible later in the path. Naively the algorithm could check every time until the end of the path that no such violations exist, but in an effort to optimize the authors introduce max_shift which describes how much later an Activity can start without violating the starting time of any of the Activities in the current solution. As a result after every insertion a pass forward from the inserted Activity to update starting times and max_shift and a pass backward to update max_shift is required.

$$shift_{12}^{3} = t_{13} + \max(o_3 - b_1 - t_{13}, 0) + v_3 + t_{32} - t_{12} \tag{10}$$

In the second phase, it deletes a contiguous sequence of Activities, chosen in a deterministic pseudorandom manner preparing for the next iteration. Through the alternation between these two phases the algorithm escapes local minima, while producing high quality solutions.

After 150 iterations of not improving the best solution found the algorithm terminates. The ILS heuristic offers characteristics that align with our approach. By focusing on repeated exploitation, it rapidly converges to good solutions. These solutions have proven in practice to be near-optimal for well-known benchmarks. With this in mind, we chose the ILS heuristic as a basis for our SLILS heuristic, as well as a building block in our Replace LS heuristic.

4.2 Supervised Learning Iterated Local Search

One characteristic of the MMCOP is that the difficulty of finding a feasible solution is dependent on the input, in a non obvious way. Depending on factors such as supply, demand, visiting time of individual Activities and existence of an Activity in the optimal solution, an instance of the MMCOP might be solvable as easily as the unconstrained OP instance, or it might be a very difficult problem. In fact, satisfiability of category constraints is NP-Hard in itself, as it is easy to craft an instance that reduces to Hamiltonian cycle: Just assign the same category to each activity and request a minimum constraint for that category equal to the number of Activities.

If the optimal solution satisfies the category constraints, we don't need to do anything more than solving the unconstrained OP instance. However, for difficult settings, we could perform a reweighting of the graph. The algorithm needs to include more points of the desired categories until the minimum constraints are covered. Therefore, all points of those categories need to have their ratios increased. This can be achieved by adding a multiplicative term to the ratio function when demand for this category is not satisfied.

$$WRatio = Ratio \times (1 + w_{cat}) \times \frac{demand_{cat}}{supply_{cat}}$$

$demand_{cat}$ is how many more POIs of this category our solution would require to be feasible. $supply_{cat}$ is how many candidates of this category we could possibly admit.

If an oracle could answer what this optimal $\overline{w}$ was, we could run the original ILS and it would provide us with multiple valid solutions to choose the best from. Our goal is to find this optimal $\overline{w}$. To arrive to this reweighting we use Stochastic Gradient Descent (SGD). Starting from some $\overline{w}$, we create a solution and alter the weights depending on how close we are to cover demands for each category. If we are far, we need to increase the corresponding w_{cat}. If the demands were met, then we can lower the value.

Since *WRatio* is the same as *Ratio* when demand (for the current solution) is zero, in tightly constrained settings the category weights will only increase. To avoid this count-to-infinity type scenario we introduced regularization. The update rule is shown in Eq. 11.

$$w_{cat}^{(i+1)} = w_{cat}^{(i)} + (demand_{cat}^{(i)} - \lambda w_{cat}^{(i)}) \times step \tag{11}$$

step is following an exponentially decreasing schedule to ensure convergence. λ, the regularization factor, is set to 0.10 in our experiments and yields superior results to the non-regularized variant when there are not too many constraints.

4.3 Replace-Based Local Search

One big shortcoming of SLILS is that it tries to find optimal and feasible solutions at the same time. As it blindly focuses on the best possible feasible solution, it might fail to find any feasible solution if constraints are really tight.

The second heuristic we employed to solve MMCOP is another algorithm based on local search, albeit one that works quite differently. Instead of trying to find solutions of maximal length, it tries to find solutions of all possible lengths. Thus, Replace-based Local Search (Replace LS for short) is able to find feasible solutions for very hard settings, since it first tries to find easy ways to satisfy the constraints. The downside is of course the running time and solution quality for less constrained settings.

Replace LS works in three phases. For each solution length `N`, it first generates an initial solution of length `N` in a controlled random fashion that is feasible regarding the time constraints, but not necessarily regarding the category constraints. Then, it alternates between satisfying the constraints and maximizing the profit.

In more detail, initially the goal is to generate any solution, irregardless of the total profit. In order to generate a solution that satisfies the time constraints, we try to generate the solution with the least duration. This is achieved by repeatedly inserting Activities that have the least impact on the duration of the solution.

Then, we set a target `dist(Solution, Constraints)` equal to `N`. Every time `dist(Solution, Constraints)` is above the target distance, the algorithm will replace one Activity with the first Activity that reduces the distance. On the contrary, if `dist(Solution, Constraints)` is below the target distance, the algorithm will replace one Activity with the first Activity that increases the total profit, while still respecting the category constraints.

```
function REPLACE LS
    while failure_count < max_failures do
        solution_length ← solution_length + 1
        for some iterations do
            generate an initial solution
            shuffle candidates
            calculate distance from feasible solution
            set target_distance to an initial value
            while target_distance > 0 do
                REPLACE STEP
                decrease target_distance
            end while
            if solution is feasible then
                keep solution if improves best solution
            else
                increase failure_count
            end if
        end for
    end while
end function
```

```
function REPLACE STEP
    for max_replaces do
        DELETE STEP
        try INSERT STEP
        if unable to insert then
            reinsert deleted element
        end if
    end for
end function
```

```
function DELETE STEP
    delete i-th candidate
    i ← (i + 1) mod solution length
end function
```

```
function INSERT STEP(i, j)
    for all candidates k after i do
        for all positions p after j do
            if REPLACE(k, i) has better profit and dist ≤ target then
                insert candidate
            end if
        end for
    end for
end function
```

4.4 IP Solutions

Before proposing any heuristics to solve the new problem we decided to try an exact algorithm. Even a slow exact solution will allow us to have a benchmark to compare any following algorithms. The problem was formulated using the python CPLEX API. Each problem ran for up to 30 min before reporting the best found integer solution.

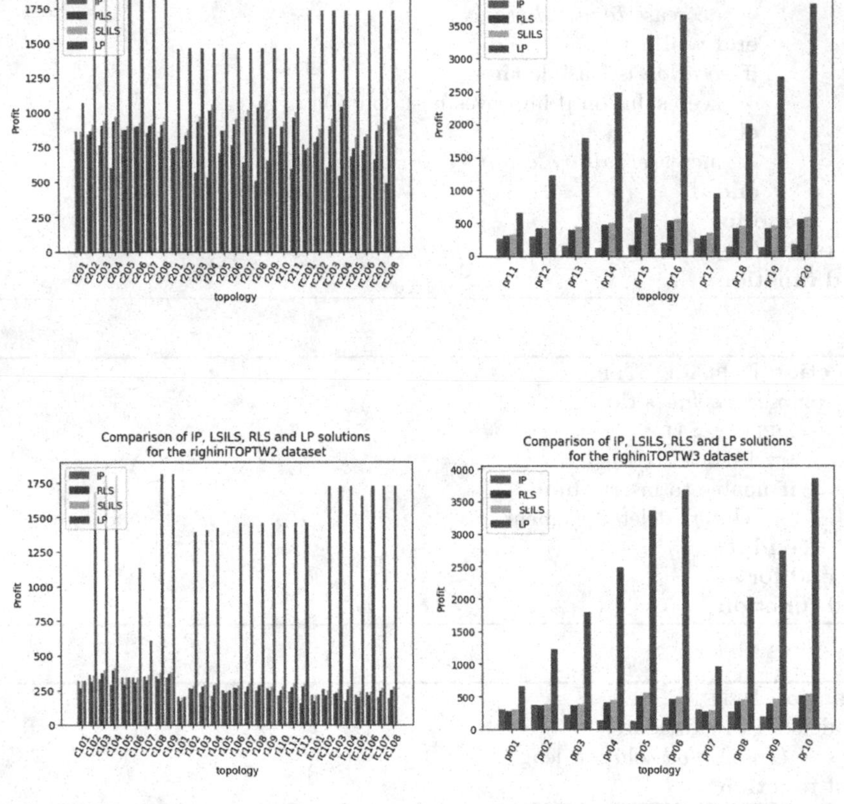

Fig. 1. Comparison of IP, SLILS, RLS and LP solutions

Since the problem is NP-hard (as a generalization of the orienteering problem) a polynomial time solution is beyond our grasp, but the LP solution provides an upper bound to the IP and can be found in polynomial time. Unfortunately, the results were disappointing with very high LP values (5 times larger than known results) and low values for the IP (comparable in small topologies and many times smaller in large topologies).

In 5 instances, the solver was able to find the optimum integer solution which was very close or equal to the SLILS solution, validating that our algorithm while not always optimum, is very close to that. Additionally, in the 8 cases where the IP algorithm found better results than SLILS, our algorithm ran for one sixth of a second, which compared to half an hour is 10.000 times faster. If we take into consideration that CPLEX was running on 16 cores versus the single core that SLILS used we can clearly see that without further optimizations, using an IP solver is not going to deliver a valuable, scalable solution.

Even letting CPLEX cheat and look into the solutions found by our algorithms or letting the solver toil for 12 h did not improve the quality of the solutions for some of the large topologies. In Fig. 1 the results can be seen graphically.

5 Experimental Results

5.1 Datasets

The datasets used in this paper are augmented versions of those in [40]; the Solomon, as well as the Cordeau, Gendrau and Laporte datasets, shown in Table 1. To the best of our knowledge there are no datasets with categories and therefore we had to create our own by augmenting each topology with a randomly chosen category according to the probabilities in Table 2. We chose to assign categories randomly in order not to insert bias in the best solution.

These datasets have been used in a plethora of related papers enabling us to establish a baseline for our algorithm and quantify our algorithms' performance in the unconstrained setting. Various papers have used different datasets, depending on the problem variation each paper was studying and data availability. A partial list of problems and datasets can be found following the link on [1].

Table 1. Summary of OPTW datasets

Name	Instance sets	Number of nodes	Reference
Solomon	c100, r100, rc100	100	[33]
Cordeau	pr01–pr10	48–288	
Solomon	c200, r200, rc200	100	[29]
Cordeau	pr11–pr20	48–288	

Table 2. Probability of category for each node

Category	1	2	3	4	5	6
Probability	30%	20%	20%	10%	10%	10%

Table 3. Aggregate results of the unconstrained setting for Solomon and Cordeau, Gendrau, and Laporte's test problems. Columns 1–3 show in how many instances SLILS performed worse, the same or better. Columns 4–6 show the average percentage gap. The formula is $100(score_{ILS} - score_{SLILS})/score_{ILS}$. A negative percentage gap means that SLILS out performs ILS, zero means equality and a positive percentage means ILS outperforms our algorithm.

Dataset	Worse	Same	Better	Max (%)	Min (%)	Average (%)
RighiniTOPTW2	12	7	10	7.07	−8.24	0.30
RighiniTOPTW3	9	0	1	8.36	−0.89	2.56
MontemanniTOPTW1	18	2	7	6.66	−2.38	1.62
MontemanniTOPTW2	6	1	3	7.62	−3.73	1.09
General	45	10	21	8.36	−8.24	1.17

Table 4. Aggregate results of the unconstrained setting for Solomon and Cordeau, Gendrau, and Laporte's test problems. Columns 1–3 show in how many instances RLS performed worse, the same or better. Columns 4–6 show the average percentage gap. The formula is $100(score_{ILS} - score_{RLS})/score_{ILS}$. A negative percentage gap means that RLS out performs ILS, zero means equality and a positive percentage means ILS outperforms our algorithm.

Dataset	Worse	Same	Better	Max (%)	Min (%)	Average (%)
RighiniTOPTW2	29	0	0	21.46	1.75	10.11
RighiniTOPTW3	10	0	0	18.00	6.12	10.66
MontemanniTOPTW1	27	0	0	12.50	0.88	5.57
MontemanniTOPTW2	10	0	0	17.64	2.90	9.22
General	76	0	0	21.46	0.88	8.46

5.2 Unconstrained Setting

The results of our algorithms in the unconstrained settings is listed in Tables 3 and 4. The results are aggregated over all topologies. Our algorithms contain randomness, therefore we are citing the rounded average profit of 100 runs which we believe is the fairest way to perform comparisons. SLILS is performing on par with ILS (as expected since it is a modified version) while RLS is behaving poorly in the unconstrained setting but makes up for it in very constrained settings.

In two out of seven topologies SLILS found on average better solutions, in some cases delivering an 8% increase. In one out of seven, SLILS and ILS came out tied, indicating that this might be the best feasible solution. Finally, in four out of seven topologies SLILS performed slightly worse than ILS. Both algorithms have similar execution times, so we decided against comparing execution times. The main difference of the algorithms is not computationally heavy, so that is to be expected. Both implementations were written by the authors and hence are equally optimized.

On the other hand, RLS is consistently performing worse in the unconstrained setting producing results up to 20% worse than already tested algorithms. This should be attributed to broader exploration, which means that the algorithm doesn't fully exploit the already found solutions and does not optimize them in order to achieve the best possible result. This is expected, as RLS was designed to perform in more constrained settings, where SLILS fails to find any solution.

As can be seen in Fig. 2 the spread of percentage gaps follows a normal distribution with different mean and standard deviation for each algorithm. Lowering these two values should be the aim of any algorithm. Having a lower mean gap from a tested algorithm means that the algorithm is performing better across all topologies, while a smaller standard deviation implies that there are fewer special cases where the algorithm is performing suboptimally.

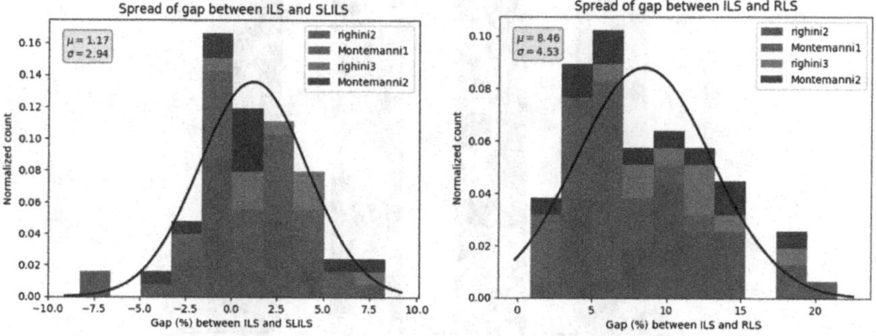

Fig. 2. Spread of percentage gap for SLILS and RLS

5.3 Constrained Setting

Our goal in formulating this problem was to be able to solve constrained topologies. There have been many papers focusing on constraining some sort of maximum. The original OP tries to find a solution with the most profit under the constraint of time. Other papers such as [36] deal with multiple maximum constraints. However little to no research is done on minimum constraints, other than mandatory visits [25, 27] therefore our results focus on exploring those settings.

Our experiments consist of running the algorithms with increasingly higher number of constraints, for our set of 76 topologies. The executions request a minimum of 0, 4, 8 and 12 POIs that belong to the first category. We compared plain ILS, SLILS and RLS. The expectation is that plain ILS will find feasible solutions in slightly constrained settings by chance, and will fail to find solutions in more constrained settings.

In almost all cases the profit decreases as the minimums increase. However, in many cases one of the constrained settings achieves a better result. There are two reasons for this behavior. First of all, Adding more constraints shrinks the search space. That allows for a more effective exploration, finding better solutions more consistently. Secondly, our algorithms are greedy heuristics and as such any changes may have slight positive or negative effect. In some cases, the reweighting due to the constraints has a positive effect and the final solution ends up being slightly better than the one in the unconstrained setting.

Also, in more constrained settings RLS is returning more feasible solutions and finding solutions in which SLILS is not able to. For 8 minimums RLS is able to solve 20 more instances than SLILS. For 12 minimums RLS solves 8 more instances and has better solutions in a further 8. In less constrained settings SLILS is handily beating RLS in most cases, with a few exceptions. The comparison between the two algorithms in the constrained setting can be found in Fig. 3.

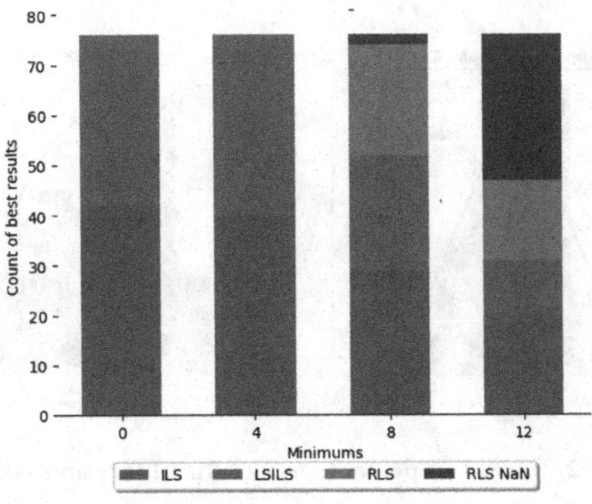

Fig. 3. Comparison of performance of ILS, SLILS and RLS in the constrained setting. The graph shows for each minimum in how many instances each algorithm produced the best result. No algorithm produced a solution when RLS failed. When no algorithm produced a solution (presumably because it does not exist) it is depicted with red. (Color figure online)

From the practical experimentation we can see that existing algorithms fail to always find solutions to the problem we proposed and our algorithms produce high quality solutions adapting themselves to the difficulty of the instance they are currently solving.

We tested three algorithms. The IP formulation on a consumer solver (CPLEX) which while it can provide an exact solution, might achieve that in exponential time, deeming it unfit for wide use. The second algorithm, SLILS, based on iterated local search is a very fast metaheuristic which achieves competitive results and adjusts on the tightness of the problem instance. Finally, RLS is more suitable for more constrained instances, as it is more guaranteed to return a solution. However RLS is more time consuming and performs worse when the constraints are relatively easy to satisfy and many solutions exist.

Using the described algorithms this new formulation of the Orienteering problem can be efficiently solved in a variety of cases. We hope that this paper is the start for more research into this field.

6 Conclusion and Future Work

In this paper we have presented a new variation on the Orienteering Problem and have proposed two algorithms for solving it. This variation is orthogonal to others and can be easily combined to produce more realistic problem formalizations. We began trying to solve it with integer programming which provided us with an upper bound to our solutions. In cases where the solver managed to find an exact solution it was very close to our own.

Our numeric experiments in the unconstrained setting showed that our approach is comparable with previous results. The experiments on the constrained setting showed that SLILS is faster and produces better results in the unconstrained setting, while RLS solves more easily the more demanding settings, but lacks the quality of SLILS in the less constrained ones.

There are many avenues to explore in the future. In this work each POI had only one category, but in reality a POI can fall under many categories. Having many categories changes substantially the way that this problem is solved. One question is how the different weights should be combined with these categories.

Another avenue to explore is how randomness should be utilized. The algorithm could start being deterministic to achieve high exploitation and over time relax to achieve better exploration. The schedule of this relaxation and how it relates to the size of the search space is another interesting question.

In this paper we have concerned ourselves with cold solving a single instance. However, solving multiple instances in the same topology could give us new ways to approach the problem. For instance, instead of using the SGD we could train a model to provide us with an *a priori* weight vector to use, taking into account all the input variables. Furthermore, solving many instances in parallel could pose its own challenges. Should instances be batched together, or having access to static information and each one being solved in a different CPU is the most efficient approach we can achieve?

As Moore's law has been ticking away for 40 years since the conception of the Orienteering Problem and as better and better heuristics have been devised to solve this problem, we have acquired the ability to solve instances for ever greater topologies. However all modern datasets are still around the hundreds or at best a few thousand nodes, a far cry from the real life datasets where there are a hundred POIs even in a small city and multiple thousands in the big metropolises around the globe. Being able to scale these solutions to the thousands and millions is crucial for the ability to transform this from a theoretical paper to an engineering solution which actually helps people make more informed decisions.

References

1. The orienteering problem: Test instances. https://www.mech.kuleuven.be/en/cib/op
2. Archetti, C., Hertz, A., Speranza, M.G.: Metaheuristics for the team orienteering problem. J. Heuristics **13**(1), 49–76 (2007)
3. Bolzoni, P., Helmer, S.: Hybrid best-first greedy search for orienteering with category constraints. In: Gertz, M., et al. (eds.) SSTD 2017. LNCS, vol. 10411, pp. 24–42. Springer, Cham (2017). https://doi.org/10.1007/978-3-319-64367-0_2
4. Boussier, S., Feillet, D., Gendreau, M.: An exact algorithm for team orienteering problems. 4OR **5**(3), 211–230 (2007)
5. Campos, V., Martí, R., Sánchez-Oro, J., Duarte, A.: Grasp with path relinking for the orienteering problem. J. Oper. Res. Soc. **65**(12), 1800–1813 (2014)
6. Chao, I.M., Golden, B.L., Wasil, E.A.: A fast and effective heuristic for the orienteering problem. Eur. J. Oper. Res. **88**(3), 475–489 (1996)
7. Chao, I.M., Golden, B.L., Wasil, E.A.: The team orienteering problem. Eur. J. Oper. Res. **88**(3), 464–474 (1996)
8. Dang, D.-C., El-Hajj, R., Moukrim, A.: A branch-and-cut algorithm for solving the team orienteering problem. In: Gomes, C., Sellmann, M. (eds.) CPAIOR 2013. LNCS, vol. 7874, pp. 332–339. Springer, Heidelberg (2013). https://doi.org/10.1007/978-3-642-38171-3_23
9. Dang, D.-C., Guibadj, R.N., Moukrim, A.: A PSO-based memetic algorithm for the team orienteering problem. In: Di Chio, C., et al. (eds.) EvoApplications 2011. LNCS, vol. 6625, pp. 471–480. Springer, Heidelberg (2011). https://doi.org/10.1007/978-3-642-20520-0_48
10. Dang, D.C., Guibadj, R.N., Moukrim, A.: An effective PSO-inspired algorithm for the team orienteering problem. Eur. J. Oper. Res. **229**(2), 332–344 (2013)
11. Feillet, D., Dejax, P., Gendreau, M.: Traveling salesman problems with profits. Transp. Sci. **39**(2), 188–205 (2005)
12. Ferreira, J., Quintas, A., Oliveira, J.A., Pereira, G.A.B., Dias, L.: Solving the team orienteering problem: developing a solution tool using a genetic algorithm approach. In: Snášel, V., Krömer, P., Köppen, M., Schaefer, G. (eds.) Soft Computing in Industrial Applications. AISC, vol. 223, pp. 365–375. Springer, Cham (2014). https://doi.org/10.1007/978-3-319-00930-8_32
13. Fischetti, M., Gonzalez, J.J.S., Toth, P.: Solving the orienteering problem through branch-and-cut. Inf. J. Comput. **10**(2), 133–148 (1998)
14. Gambardella, L.M., Montemanni, R., Weyland, D.: An enhanced ant colony system for the sequential ordering problem. In: Klatte, D., Lüthi, H.J., Schmedders, K. (eds.) Operations Research Proceedings 2011. Operations Research Proceedings (GOR (Gesellschaft für Operations Research e.V.)), pp. 355–360. Springer, Heidelberg (2012). https://doi.org/10.1007/978-3-642-29210-1_57
15. Gendreau, M., Laporte, G., Semet, F.: A branch-and-cut algorithm for the undirected selective traveling salesman problem. Networks **32**(4), 263–273 (1998)
16. Gendreau, M., Laporte, G., Semet, F.: A tabu search heuristic for the undirected selective travelling salesman problem. Eur. J. Oper. Res. **106**(2–3), 539–545 (1998)
17. Golden, B., Levy, L., Dahl, R.: Two generalizations of the traveling salesman problem. Omega **9**(4), 439–441 (1981)
18. Golden, B.L., Levy, L., Vohra, R.: The orienteering problem. Nav. Res. Logist. **34**(3), 307–318 (1987)

19. Gunawan, A., Lau, H.C., Lu, K.: An iterated local search algorithm for solving the orienteering problem with time windows. In: Ochoa, G., Chicano, F. (eds.) EvoCOP 2015. LNCS, vol. 9026, pp. 61–73. Springer, Cham (2015). https://doi.org/10.1007/978-3-319-16468-7_6
20. Irnich, S., Desaulniers, G.: Shortest path problems with resource constraints. In: Desaulniers, G., Desrosiers, J., Solomon, M.M. (eds.) Column Generation, pp. 33–65. Springer, Boston (2005). https://doi.org/10.1007/0-387-25486-2_2
21. Ke, L., Archetti, C., Feng, Z.: Ants can solve the team orienteering problem. Comput. Ind. Eng. **54**(3), 648–665 (2008)
22. Laporte, G., Martello, S.: The selective travelling salesman problem. Discret. Appl. Math. **26**(2–3), 193–207 (1990)
23. Leifer, A.C., Rosenwein, M.B.: Strong linear programming relaxations for the orienteering problem. Eur. J. Oper. Res. **73**(3), 517–523 (1994)
24. Liang, Y.C., Kulturel-Konak, S., Lo, M.H.: A multiple-level variable neighborhood search approach to the orienteering problem. J. Ind. Prod. Eng. **30**(4), 238–247 (2013)
25. Lin, S.W., Vincent, F.Y.: Solving the team orienteering problem with time windows and mandatory visits by multi-start simulated annealing. Comput. Ind. Eng. **114**, 195–205 (2017)
26. Lourenço, H.R., Martin, O.C., Stützle, T.: Iterated local search. In: Glover, F., Kochenberger, G.A. (eds.) Handbook of Metaheuristics, pp. 320–353. Springer, Boston (2003). https://doi.org/10.1007/0-306-48056-5_11
27. Lu, Y., Benlic, U., Wu, Q.: A memetic algorithm for the orienteering problem with mandatory visits and exclusionary constraints. Eur. J. Oper. Res. **268**(1), 54–69 (2018)
28. Marinakis, Y., Politis, M., Marinaki, M., Matsatsinis, N.: A memetic-GRASP algorithm for the solution of the orienteering problem. In: Le Thi, H.A., Pham Dinh, T., Nguyen, N.T. (eds.) Modelling, Computation and Optimization in Information Systems and Management Sciences. AISC, vol. 360, pp. 105–116. Springer, Cham (2015). https://doi.org/10.1007/978-3-319-18167-7_10
29. Montemanni, R., Gambardella, L.M.: An ant colony system for team orienteering problems with time windows. Found. Comput. Decis. Sci. **34**(4), 287 (2009)
30. Muthuswamy, S., Lam, S.S.: Discrete particle swarm optimization for the team orienteering problem. Memetic Comput. **3**(4), 287–303 (2011)
31. Ramesh, R., Yoon, Y.S., Karwan, M.H.: An optimal algorithm for the orienteering tour problem. ORSA J. Comput. **4**(2), 155–165 (1992)
32. Ramesh, R., Brown, K.M.: An efficient four-phase heuristic for the generalized orienteering problem. Comput. Oper. Res. **18**(2), 151–165 (1991)
33. Righini, G., Salani, M.: Decremental state space relaxation strategies and initialization heuristics for solving the orienteering problem with time windows with dynamic programming. Comput. Oper. Res. **36**(4), 1191–1203 (2009)
34. Şevkli, A.Z., Sevilgen, F.E.: StPSO: strengthened particle swarm optimization. Turk. J. Electr. Eng. Comput. Sci. **18**(6), 1095–1114 (2010)
35. Şevkli, Z., Sevilgen, F.E.: Discrete particle swarm optimization for the orienteering problem. In: 2010 IEEE Congress on Evolutionary Computation (CEC), pp. 1–8. IEEE (2010)
36. Souffriau, W., Vansteenwegen, P., Vanden Berghe, G., Van Oudheusden, D.: The multiconstraint team orienteering problem with multiple time windows. Transp. Sci. **47**(1), 53–63 (2013)

37. Sylejmani, K., Dorn, J., Musliu, N.: A tabu search approach for multi constrained team orienteering problem and its application in touristic trip planning. In: 2012 12th International Conference on Hybrid Intelligent Systems (HIS), pp. 300–305. IEEE (2012)
38. Thomadsen, T., Stidsen, T.K.: The quadratic selective travelling salesman problem. Technical report (2003)
39. Tsiligirides, T.: Heuristic methods applied to orienteering. J. Oper. Res. Soc. **35**, 797–809 (1984)
40. Vansteenwegen, P., Souffriau, W., Berghe, G.V., Van Oudheusden, D.: Iterated local search for the team orienteering problem with time windows. Comput. Oper. Res. **36**(12), 3281–3290 (2009)
41. Vansteenwegen, P., Souffriau, W., Berghe, G.V., Van Oudheusden, D.: Metaheuristics for tourist trip planning. In: Sörensen, K., Sevaux, M., Habenicht, W., Geiger, M. (eds.) Metaheuristics in the Service Industry, pp. 15–31. Springer, Heidelberg (2009). https://doi.org/10.1007/978-3-642-00939-6_2
42. Vincent, F.Y., Lin, S.W.: Multi-start simulated annealing heuristic for the location routing problem with simultaneous pickup and delivery. Appl. Soft Comput. **24**, 284–290 (2014)

Internal Versus External Balancing in the Evaluation of Graph-Based Number Types

Hanna Geppert and Martin Wilhelm(✉)

Otto-von-Guericke Universität, Magdeburg, Germany
martin.wilhelm@ovgu.de

Abstract. Number types for exact computation are usually based on directed acyclic graphs. A poor graph structure can impair the efficency of their evaluation. In such cases the performance of a number type can be drastically improved by restructuring the graph or by internally balancing error bounds with respect to the graph's structure. We compare advantages and disadvantages of these two concepts both theoretically and experimentally.

1 Introduction

Inexact computation causes many problems when algorithms are implemented, ranging from slightly wrong results to crashes or invalid program states. This is especially prevalent in the field of computational geometry, where real number computations and combinatorical properties intertwine [10]. In consequence, various exact number types have been developed [6,8,16]. It is an ongoing challenge to make these number types sufficiently efficient to be an acceptable alternative to floating-point primitives in practical applications. Number types based on the Exact Computation Paradigm recompute the value of complex expressions if the currently stored error bound is not sufficient for an exact decision [15]. Hence, they store the computation history of a value in a directed acyclic graph, which we call an *expression dag*. The structure of the stored graph is then determined by the order in which the program executes the operations. It lies in the nature of iterative programming that values are often computed step by step, resulting in list-like graph structures.

Re-evaluating expressions in an unbalanced graph is more expensive than in a balanced one [4,11]. We discuss two general approaches on reducing the impact of graph structure on the evaluation time. Prior to the evaluation, the expression dag can be restructured. Originally proposed by Yap [15], restructuring methods with varying degrees of invasiveness were developed [11,14]. Root-free expression trees can be restructured to reach optimal depth as shown by Brent [3]. In Sect. 2.1 we introduce a weighted version of Brent's algorithm applied on maximal subtrees inside an expression dag. Besides restructuring, which can be considered 'external' with respect to the evaluation process, we can make 'internal' adjustments during the evaluation to compensate for bad structure. Error bounds occuring during an evaluation can be balanced to better reflect

I. Kotsireas et al. (Eds.): SEA^2 2019, LNCS 11544, pp. 359–375, 2019.
https://doi.org/10.1007/978-3-030-34029-2_24

the structure of the graph [4]. Doing so requires a switch from an integer to a floating-point error bound representation, leading to numerical issues that need to be taken into consideration [7,13]. In Sect. 2.2 we show how error bounds can be balanced optimally in both the serial and the parallel case and compare several heuristics. Finally, in Sect. 3 we experimentally highlight strengths and weaknesses of each approach.

2 Concepts

An *expression dag* is a rooted ordered directed acyclic graph in which each node is either a floating-point number, a unary operation ($\sqrt[d]{\ }, -$) with one child, or a binary operation ($+, -, *, /$) with two children. We call an expression dag E', whose root is part of another expression dag E a *subexpression of* E. We write $v \in E$ to indicate that v is a node in E and we write $|E|$ to represent the number of operator nodes in E. In an *accuracy-driven evaluation* the goal is to evaluate the root node of an expression dag with absolute accuracy q, i.e., to compute an approximation $\tilde{x}$ for the value x of the represented expression, such that $|\tilde{x} - x| \leq 2^q$ (cf. [15]). To reach this goal, sufficiently small error bounds for the (up to two) child nodes and for the operation error are set and matching approximations are computed recursively for the children. Let $v \in E$ be a node with outgoing edges e_l to the left and e_r to the right child. Let $\mathrm{i}(e_l), \mathrm{i}(e_r)$ be the increase in accuracy for the left and the right child of v and $\mathrm{i}(v)$ be the increase in accuracy for the operation (i.e. the increase in precision) at v. Depending on the operation in v we assign constants $\mathrm{c}(e_l), \mathrm{c}(e_r)$ to its outgoing edges as depicted in Table 1. If the node v is known from the context, we shortly write i_v, i_l, i_r for the accuracy increases at v, e_l, e_r and c_l, c_r for the respective constants. To guarantee an accuracy of q at v, the choice of i_v, i_l, i_r must satisfy the inequality

$$2^{q+i_v} + c_l 2^{q+i_l} + c_r 2^{q+i_r} \leq 2^q \quad \text{or, equivalently,} \quad 2^{i_v} + c_l 2^{i_l} + c_r 2^{i_r} \leq 1 \quad (1)$$

Aside from this condition, the choice of i_v, i_l, i_r is arbitrary and usually done by a symmetric distribution of the error. In the exact number type `Real_algebraic` they are chosen such that $c_l 2^{i_l} \leq 0.25$, $c_r 2^{i_r} \leq 0.25$ and $2^{i_v} = 0.5$ (and adjusted accordingly for one or zero children). Let the *depth* of a node v in an expression dag be the length of the longest path from the root to v. In general, the precision p_v needed to evaluate a node v increases linearly with the depth of the node due to the steady increase through i_l, i_r. The approximated value of each node is stored in a multiple-precision floating-point type (bigfloat). The cost of evaluating a node is largely dominated by the cost of the bigfloat operation, which is linear in $|p_v|$ in case of addition and subtraction and linear up to a logarithmic factor in case of multiplication, division and roots. So the precision p_v is a good indicator for the total evaluation cost of a node (except for negations). Let E be an expression dag. We define the *cost of a node* $v \in E$ to be $|p_v|$ and the *cost of* E, denoted by $\mathrm{cost}(E)$, as the sum of the cost of all nodes in E. We set the *depth of* E to the maximum depth of all nodes in E. Let E_{list} be a list-like expression dag, i.e., an expression dag with depth $\Theta(n)$, where n is the number of its nodes and let E_{bal} be a balanced expression dag,

Table 1. Operation-dependent constants $c(e_l)$ and $c(e_r)$ for an accuracy-driven evaluation in `Real_algebraic`, with x_{high}, y_{high} upper bounds and x_{low}, y_{low} lower bounds on the child values.

	negation	add./sub.	multipl.	division	d-th root
$c(e_l)$	1	1	y_{high}	$\frac{1}{y_{low}}$	$\frac{1}{d}(x_{low})^{\frac{1-d}{d}}$
$c(e_r)$	0	1	x_{high}	$\frac{1}{{y_{low}}^2}$	0

i.e., an expression dag with depth $\Theta(\log(n))$. Since the precision increases linearly with the depth, we have $\text{cost}(E_{list}) = \Theta(n^2)$ and $\text{cost}(E_{bal}) = \Theta(n\log(n))$, assuming that the operation constants can be bounded (cf. [11]). In a parallel environment the cost of the evaluation is driven by dependencies between the nodes. For an expression dag E with n nodes let the *cost of a path* in E be the sum of the cost of the nodes along the path. Let $\text{cp}(E)$ be a path in E with the highest cost. We call $\text{cp}(E)$ a *critical path* in E. Then the cost of evaluating E in parallel is $\Theta(\text{cost}(\text{cp}(E)))$ with $O(n)$ processors. Let E_{list}, E_{bal} be defined as before. Then obviously $\text{cost}(\text{cp}(E_{list})) = \Theta(n)$ and $\text{cost}(\text{cp}(E_{bal})) = \Theta(\log n)$ (cf. [14]). So in both the serial and the parallel case, balanced graph structures are superior.

2.1 Graph Restructuring

By definition, exact number types that use accuracy-driven evaluation act lazy, i.e., expressions are not evaluated until a decision needs to be made. Before their first evaluation, underlying graph structures are lightweight and can be changed at low cost. Therefore graph restructuring algorithms ideally take place when the first decision is demanded. While it is not impossible to restructure graphs that have already been evaluated, it comes with several downsides. Since subexpressions will change during restructuring, all approximations and error bounds associated with these subexpressions are lost, although they could be reused in later evaluations. Since stored data may depend on data in subexpressions, the internal state of the whole expression dag may be invalidated. Those effects can make restructuring expensive if many decisions are requested without significant changes to the graph in between. Let E be an expression dag. We call a connected, rooted subgraph of E an *operator tree* if it consists solely of operator nodes, does not contain root operations and does not contain nodes with two or more parents (not necessarily in E), except for its root. We restructure each maximal operator tree in E according to a weighted version of Brent's algorithm. Let T be an operator tree in E. We call the children of the leaves of T the *operands* of T and associate a positive weight with each of those operands. We define a weight function, such that for each node $v \in T$ the weight of v is greater or equal than the weight of its children. The main difference between the original algorithm and the weighted variation lies in the choice of the split node. We give a brief outline of the algorithm.

The algorithm builds upon two operations, `compress` and `raise`. The operation `compress` takes an expression tree E and returns an expression tree of the form F/G and `raise` takes an expression tree E and a subtree X and returns an expression tree of the form $(AX + B)/(CX + D)$, where A, B, C, D, F, G are division-free expression trees with logarithmic depth.

Algorithm 1: The operations `compress` and `raise`.

Function `compress(`R`)`:
    if *R is not an operand* **then**
        X = `split(`R, $\frac{1}{2}$ weight(R)`)`;
        let X_1, X_2 be the children of X;
        `compress(`X_1`)`; `compress(`X_2`)`; `raise(`R,X`)`;
        substitute X in R;
    end

Function `raise(`R,X`)`:
    if $R \neq X$ **then**
        Y = `split(`R, $\frac{1}{2}$(weight(R) + weight(X))`)`;
        let Y_1, Y_2 be the children of Y, such that Y_1 contains X;
        `raise(`Y_1,X`)`; `compress(`Y_2`)`; `raise(`R,Y`)`;
        substitute Y in R;
    end

Let v_r be the root node of T. We choose v_s as a node with maximal weight in T such that both children have either weight $< \frac{1}{2}$ weight(v_r) or are operands. Note that this implies weight$(v_s) \geq \frac{1}{2}$ weight(v_r). We then recursively call `compress` on v_s and raise v_s to the root by repeating the following steps:

1. Search for a new split node v_s' on the path from v_r to v_s that splits at a weight of $\frac{1}{2}$(weight(v_r) + weight(v_s)).
2. Recursively raise v_s' to v_r and v_s to the respective child node in v_s'.
3. Substitute v_s' and its children into v_r by incorporating the operation at v_s'.

Let R be the expression at v_r, let Y be the expression at v_s' and let X be the expression at v_s. After the second step, $R = \frac{A'Y+B'}{C'Y+D'}$ and we have $Y = Y_L \circ Y_R$ with $Y_L = \frac{A''X+B''}{C''X+D''}$ and $Y_R = F''/G''$ or vice versa. Substituting Y (with respect to the operation $\circ$ at Y) then gives the desired $R = \frac{AX+B}{CX+D}$. Substituting $X = F'/G'$ finally leads to a balanced expression of the form $R = F/G$.

The new `split` operation is shown in Algorithm 2. If unit weight is chosen, there will never be an operand that does not satisfy the split condition. If furthermore the weight function is chosen as the number of operands in a subtree, satisfying the split condition implies having a bigger weight than the sibling. Therefore the algorithm is identical to Brent's original algorithm applied to subtrees of the expression dag and guarantees logarithmic depth for the new

Algorithm 2: The split operation.

```
Function split(X,w):
    if X.left is not operand and weight(X.left) ≥ w and
    weight(X.left) ≥ weight(X.right) then
        return split(X.left,w);
    else if X.right is not operand and weight(X.right) ≥ w then
        return split(X.right,w);
    else
        return X;
```

operator tree. Regarding the overall expression dag, nodes which contain root operations, have more than one parent or have been evaluated before are treated equally to the other operands in this case and therefore act as 'blocking nodes' for the balancing process. Let k be the number of these blocking nodes in E. If the number of incoming edges for each blocking node is bounded by a constant, the depth of E after applying the algorithm to each operator tree is in $O(k \log(\frac{n}{k}))$. This depth can be reduced by applying appropriate weights to the blocking nodes. From a conceptual perspective, a sensible choice for the weight of an operand (as well as for the weights of the inner nodes) would be the number of operator nodes in the subexpression rooted at the operand. Note that we are actually interested in the number of bigfloat operations. However, it is very expensive to compute the number of descendants for a node in a DAG, since one has to deal with duplicates [2]. Ignoring duplicates, we could choose the number of operators we would get by expanding the DAG to a tree. While computable in linear time, the number of operators can get exponential (cf. [4,11]) and therefore we cannot store the exact weight in an integer data type anymore. There are ways of managing such weights, as we discuss in Sect. 2.2, but they are imprecise and less efficient than relying on primitives. Both weight functions behave identical to the unit weight case when there are no blocking nodes present. If there are blocking nodes on the other hand, these nodes get weighted accordingly and expensive nodes are risen to the top of the operator tree. The depth after restructuring for k blocking nodes therefore becomes $O(k + \log n)$.

The weight functions described above are optimal, but hard to compute. Let the weight of both operators and operands be the depth of the subexpression rooted at the operand or operator in the underlying expression dag. Then the algorithm subsequently reduces the length of the longest paths in the expression dag. Note that this strategy does not necessarily lead to an optimal result. Nevertheless, computing the depth of a subexpression in an expression dag can be done fast and the depth can be represented efficiently. Therefore this strategy might prove to be a good heuristic to combine advantages of the unit weight algorithm and the weighted approach.

2.2 Error Bound Balancing

As described at the start of this section, the additional cost of unbalanced graph structures originates in the increase in accuracy associated with each node. A more careful choice of i_v, i_l, i_r in (1) may compensate for an unfavorable structure. If set correctly, linear depth still only leads to a logarithmic increase in accuracy aside from operation constants [4].

An increase in accuracy at an operator node only affects the operation itself. An increase in accuracy for a child node affects all operations in the subexpression of the child. We associate a non-negative *weight* $\mathrm{w}(e)$ with each edge e in an expression dag E, representing the impact a change in $\mathrm{i}(e)$ has on the total cost of E. For a node $v \in E$ with outgoing edges e_l, e_r let $w_l = \mathrm{w}(e_l)$ and $w_r = \mathrm{w}(e_r)$. We then say that, for an evaluation to accuracy q, the cost induced on E by the choice of parameters in v is given by

$$\mathrm{cost_i}(v) \doteq -(q + i_v + w_l i_l + w_r i_r) \tag{2}$$

whereas $\mathrm{cost}(E) = \sum_{v \in E} \mathrm{cost_i}(v)$. To minimize the total cost we want to minimize the cost induced by each node while maintaining the condition in (1). Let $z_l = c_l 2^{i_l}$, $z_r = c_r 2^{i_r}$ and let $w_{all} = 1 + w_l + w_r$. With an optimal choice of the parameters, (1) is an equality and we have $i_v = \log(1 - z_l - z_r)$. Substituting i_v into (2) and setting $\frac{\partial}{\partial i_l} \mathrm{cost_i}(v) = \frac{\partial}{\partial i_r} \mathrm{cost_i}(v) = 0$ we get

$$(1 + w_l) z_l + w_l z_r - w_l = 0 \tag{3}$$

$$(1 + w_r) z_r + w_r z_l - w_r = 0 \tag{4}$$

leading to $z_l = \frac{w_l}{w_{all}}$ and $z_r = \frac{w_r}{w_{all}}$. Resubstituting z_l and z_r, the optimal choice of the parameters for error bound distribution inside a node is

$$\begin{aligned} i_l &= \log(w_l) - \log(w_{all}) - \log(c_l) \\ i_r &= \log(w_r) - \log(w_{all}) - \log(c_r) \\ i_v &= -\log(w_{all}) \end{aligned} \tag{5}$$

We can show that this parameter choice makes the cost of the evaluation to some degree independent of the structure of the graph. For a node $v \in E$ we denote the set of paths between the root node of E and v by $\mathbf{P}(v)$. For a path $P \in \mathbf{P}(v)$ we write $e \in P$ to indicate that e is an edge along P. The precision requested at v along P can be expressed as $\mathrm{cost_r}(P) = \mathrm{cost_v}(P) + \mathrm{cost_f}(P)$ where

$$\mathrm{cost_v}(P) = -\textstyle\sum_{e \in P}(\mathrm{i}(e) + \log(\mathrm{c}(e))) - \mathrm{i}(v) \quad \text{and} \quad \mathrm{cost_f}(P) = \textstyle\sum_{e \in P} \log(\mathrm{c}(e))$$

denote the *variable cost* induced by the choice of i_l, i_r, i_v and the *fixed cost* induced by the operation constants along the path.

Theorem 1. *Let E be an expression dag consisting of n unevaluated operator nodes. Then the cost of evaluating E with accuracy $q \leq 0$ and with an optimal choice of parameters is*

$$\mathrm{cost}(E) = n \log(n) + \textstyle\sum_{v \in E} \log\left(\sum_{P \in \mathbf{P}(v)} 2^{\mathrm{cost_f}(P)}\right) - nq$$

Proof. We define weights for each node v and each edge e in E with respect to (2). Let $\mathrm{c_f}(v) = \sum_{P\in\mathbf{P}(v)} 2^{\mathrm{cost_f}(P)}$ and $\mathrm{c_f}(v,e) = \sum_{P_e\in\mathbf{P}(v),e\in P_e} 2^{\mathrm{cost_f}(P_e)}$. Then we set $\mathrm{w}(v) = w_{all} = 1 + w_l + w_r$ if v is an operator node and $\mathrm{w}(v) = 0$ otherwise. For an edge e leading to v we set

$$\mathrm{w}(e) = \frac{\mathrm{c_f}(v,e)}{\mathrm{c_f}(v)}\mathrm{w}(v) = \frac{\sum_{P_e\in\mathbf{P}(v),e\in P_e} 2^{\mathrm{cost_f}(P_e)}}{\sum_{P\in\mathbf{P}(v)} 2^{\mathrm{cost_f}(P)}}\mathrm{w}(v) \tag{6}$$

We show that choosing the parameters as in (5) with this weight function is optimal and that it leads to the desired total evaluation cost. For a node $v \in E$ let $P \in \mathbf{P}(v)$ be any path to v of the form $P = (v_0, e_0, ..., v_k, e_k, v_{k+1} = v)$, then

$$\begin{aligned}
&\mathrm{cost_r}(P) = \mathrm{cost_v}(P) + \mathrm{cost_f}(P)\\
&\quad = -\textstyle\sum_{e\in P}(\mathrm{i}(e) + \log(\mathrm{c}(e))) - \mathrm{i}(v) + \mathrm{cost_f}(P)\\
&\quad = -\textstyle\sum_{j=0}^{k}(\log(\mathrm{w}(e_j)) - \log(\mathrm{w}(v_j)) + \log(\mathrm{w}(v)) + \mathrm{cost_f}(P)\\
&\quad = \log(\mathrm{w}(v_0)) - \textstyle\sum_{j=0}^{k}(\log(\mathrm{c_f}(v_j)) + \log(\mathrm{c}(e_j)) - \log(\mathrm{c_f}(v_{j+1})) + \mathrm{cost_f}(P)\\
&\quad = \log(\mathrm{w}(v_0)) - \log(\mathrm{c_f}(v))
\end{aligned} \tag{7}$$

In particular, the precision requested at v along each path is the same. Assume that the parameter choice is not optimal. For an edge e let $\delta(e)$ be the difference in $\mathrm{i}(e)$ between the optimal value and the value resulting from (5) with weights as defined in (6) and let $\delta(v)$ be the respective difference in $\mathrm{i}(v)$ for a node v. Due to the optimization that led to (5), the slope of $\mathrm{i}(v)$ is $-\mathrm{w}(e_l)$ in direction of $\mathrm{i}(e_l)$ and $-\mathrm{w}(e_r)$ in direction of $\mathrm{i}(e_r)$ when keeping (1) equal. So the difference in $\mathrm{i}(v)$ can be bounded through

$$\delta(v) \leq -\delta(e_l)\,\mathrm{w}(e_l) - \delta(e_r)\,\mathrm{w}(e_r) = -\textstyle\sum_{v'\in E}\left(\delta(e_l)\frac{\mathrm{c_f}(v',e_l)}{\mathrm{c_f}(v')} + \delta(e_r)\frac{\mathrm{c_f}(v',e_r)}{\mathrm{c_f}(v')}\right)$$

Denote the difference in cost by preceeding it with Δ and let $\mathbf{E}(E)$ be the set of edges in E. For our parameter choice, the precision requested at a node v is the same along each path as shown in (7), so $\Delta\max_{P\in\mathbf{P}(v)}\mathrm{cost_r}(P) = \max_{P\in\mathbf{P}(v)}\Delta\,\mathrm{cost_r}(P)$. We then get

$$\begin{aligned}
\Delta\,\mathrm{cost}(E) &= \sum_{v\in E}\max_{P\in\mathbf{P}(v)}\Delta\,\mathrm{cost_r}(P)\\
&= -\sum_{v\in E}\min_{P\in\mathbf{P}(v)}\sum_{e\in P}\delta(e) - \sum_{v\in E}\delta(v)\\
&\geq -\sum_{v\in E}\min_{P\in\mathbf{P}(v)}\sum_{e\in P}\delta(e) + \sum_{v\in E}\sum_{e\in\mathbf{E}(E)}\delta(e)\frac{\mathrm{c_f}(v,e)}{\mathrm{c_f}(v)}\\
&= -\sum_{v\in E}\min_{P\in\mathbf{P}(v)}\sum_{e\in P}\delta(e) + \sum_{v\in E}\sum_{P\in\mathbf{P}(v)}\sum_{e\in P}\delta(e)\frac{2^{\mathrm{cost_f}(P)}}{\mathrm{c_f}(v)}\\
&\geq -\sum_{v\in E}\min_{P\in\mathbf{P}(v)}\sum_{e\in P}\delta(e) + \sum_{v\in E}\min_{P\in\mathbf{P}(v)}\sum_{e\in P}\delta(e) \;=\; 0
\end{aligned}$$

and therefore our parameter choice is optimal. It remains to calculate the total cost for evaluating E. Since $\mathrm{w}(v_0) = n$ and each path $P \in \mathbf{P}(v)$ leads to the same requested precision, the desired equation follows directly from (7) with

$$\mathrm{cost}(E) = \sum_{v \in E} \max_{P \in \mathbf{P}(v)} \mathrm{cost}_{\mathrm{r}}(P) - nq = \sum_{v \in E} (\log(n) - \log(c_{\mathrm{f}}(v))) - nq \qquad \square$$

Choosing the parameters as in (5) leads to an optimal distribution of error bounds under the assumption that the weights w_l, w_r accurately reflect the impact of an increase in i_l, i_r on the total cost. Computing the exact weight shown in (6) is hard since we have to know and to maintain the cost along all paths leading to a node. We discuss several heuristic approaches. From Theorem 1 we can immediately conclude:

Corollary 1. *Let T, $|T| = n$, be an expression tree, i.e., an expression dag where each node has at most one parent. Then the optimal weight choice for an edge leading to a node v is the number of operator nodes in the subexpression rooted at v and the cost of an evaluation of T to accuracy $q \leq 0$ is*

$$\mathrm{cost}(T) = n \log n + \sum_{v \in V} \mathrm{cost}_{\mathrm{f}}(\mathrm{path}(v)) - nq$$

where $\mathrm{path}(v)$ *denotes the unique path $P \in \boldsymbol{P}(v)$.* $\square$

So a natural choice for the weight of an edge is the number of operator nodes in the respective subexpression of the target node. Then the optimality condition holds for tree-like expression dags but fails when common subexpressions exist. Figure 1 shows a graph for which the optimal distribution (1a) differs from the distribution achieved through counting the operators (1b). In the example the weights for the middle node are $w_l = w_r = 1$. Since the lower addition is a common child of the left and the right path, it gets evaluated only once. The optimal weights would therefore be $w_l = w_r = 0.5$. When constants are present it may even occur that a common subexpression already needs to be evaluated at a much higher accuracy and therefore the weight can be set close to zero.

Computing the actual number of operators without duplicates in an expression dag is already a difficult task. As in Sect. 2.1, we can set the weight of an edge to the number of operators in the subexpression, counting duplicates, in which case we need to deal with a possible exponential increase in weight size. This leads to an additional loss in optimality (cf. Fig. 1c), but makes it algorithmically feasible to compute the weights. This approach is largely identical to the one of van der Hoeven, who defined the weights as the number of leaves in the left and right subexpression [4]. Regarding the exponential weight increase, van der Hoeven suggested the use of a floating-point representation. Effectively managing correct floating-point bounds can get expensive. We use a different approach. In the definition of i_v, i_l, i_r the actual value of the weights is never needed. This enables us to store the weight in a logarithmic representation from

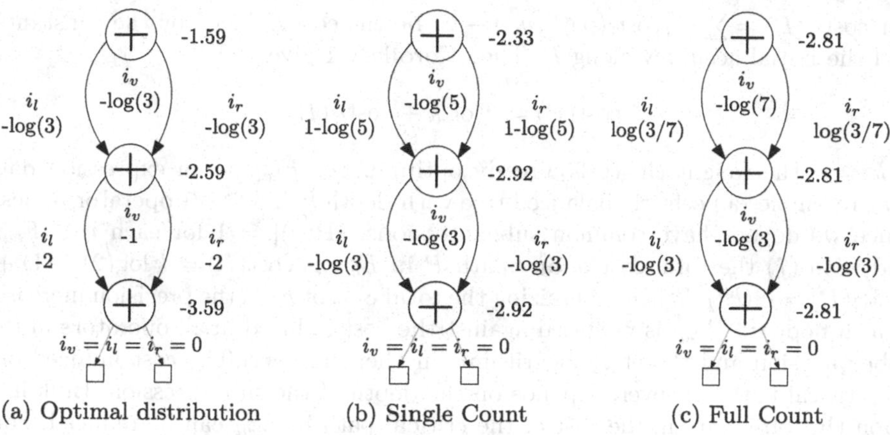

Fig. 1. Error bound distribution through different weight functions. The optimal distribution achieves a total cost of 7.77, while counting the operators with and without removing duplicates has total cost 8.17 and 8.43, respectively.

the start. The downside of this approach is that an exact computation of the weight is not possible even for small values. Note that an overestimation of the weights will never lead us to violate the condition in (1) and therefore maintains exact computation. When computing the weights, we need to compute terms of the form $\log(2^a+2^b)$. Let $a \geq b$, then we have $\log(2^a+2^b) = a+\log(1+2^{b-a})$ with $2^{b-a} \leq 1$. An upper bound on the logarithm can be obtained through repeated squaring [5]. For $a \gg b$ squaring $1 + 2^{b-a}$ is numerically unstable. In this case we can approximate the logarithm by linearization near 1. Then

$$\log(1+r) \leq \log(1) + r\tfrac{d}{dx}\log(x)|_1 = \tfrac{r}{\ln(2)} \tag{8}$$

and therefore $\log(2^a + 2^b) \leq a + \frac{1}{\ln(2)}2^{b-a}$. This approximation works well for a large difference between a and b. For small values of $a - b$ we can use repeated squaring. Otherwise we simply set the result to 1 for $a - b \leq \log(\ln(2))$. One way to efficiently compute an upper bound to the power term is to compute the product $2^{2^{d_1}} \cdots 2^{2^{d_k}}$ with $d_{min} \leq d_i \leq 0$ for $1 \leq i \leq k$ where $d_1, ..., d_k \in \mathbb{Z}$ are the digits set to one in the binary representation of $b - a$. Since the number of possible factors is finite, we can store upper bounds for them in a lookup table.

Error bound balancing does not alter the structure of the expression dag and therefore does not change its parallelizability. The maximum cost of a critical path is reduced from $\Theta(n^2)$ to $\Theta(n \log n)$, but multiple threads cannot be utilized effectively. If an arbitrary number of processors is available, the total cost of the evaluation reduces to the cost of evaluating a critical path. We can therefore choose the error bounds in such a way that the highest cost of a path from the root to a leaf is minimized. A lower bound on the cost of a critical path $P = (v_0, e_0, ..., e_{k-1}, v_k)$ with k operators can be obtained by isolating it, i.e., by assuming that each other edge in the expression dag leads to an operand.

Let $\text{cost}_\text{C}(P) = \sum_{i=0}^{k} \text{cost}_\text{f}(\text{path}(v_i)) - kq$ be the cost induced by the constants and the initial accuracy along P. Then Corollary 1 gives

$$\text{cost}(P) = k \log k + \text{cost}_\text{C}(P)$$

If $k = n$ the weight choice is already optimal. Let E_{bal} be an expression dag that resembles a perfectly balanced tree with depth k and $2^k - 1$ operator nodes. Since we do not have common subexpressions, $|\mathbf{P}(v)| = 1$ for each $v \in E_{bal}$ and with (7) the total cost of any path P in E_{bal} is $\text{cost}(P) = k \log(2^k - 1) + \text{cost}_\text{C}(P) = \Theta(k^2)$. When minimizing the total cost of E_{bal}, the precision increase i_v at a node $v \in E_{bal}$ is weighted against the cost induced in all operators in its subexpression and therefore logarithmic in their number. The cost induced on the critical path, however, depends on the depth of the subexpression. Building upon this observation, the cost of the critical path in E_{bal} can be reduced. For a node $v \in E_{bal}$ with subexpression depth j and outgoing edges e_l, e_r we set $i_v = -\log(j)$ and $i_l - \log(c_l) = i_r - \log(c_r) = \log(j-1) - \log(j) - 1$ (cf. (5)). Then the cost of the critical path P in E_{bal} is

$$\begin{aligned}\text{cost}(P) &= -\textstyle\sum_{j=2}^{k}(-\log(j) + (j-1)(\log(j-1) - \log(j) - 1)) + \text{cost}_\text{C}(P)\\ &= k \log k + \tfrac{k(k-1)}{2} + \text{cost}_\text{C}(P) \qquad (9)\end{aligned}$$

It can be shown that this parameter choice is optimal, aside from taking the operation constants into account. Although not an asymptotic improvement, the cost of the critical path was cut nearly in half. In the derivation of the chosen parameters, we made use of the symmetry of the expression. In general it is hard to compute the optimal parameters for minimizing the critical path. Let v be the root node of an expression dag X with outgoing edges e_l, e_r where the left subexpression L has depth $d_l \geq 1$ and the right subexpression R has depth $d_r \geq 1$. In an optimal parameter choice we have

$$\text{cost}(\text{cp}(L)) - d_l i_l = \text{cost}(\text{cp}(R)) - d_r i_r \qquad (10)$$

Otherwise, i_l or i_r could be decreased without increasing the cost of the critical path of X and i_v could be increased, reducing its cost. Let $d_f = \frac{d_l}{d_r}$, let $c_f = \frac{\text{cost}(\text{cp}(L)) - \text{cost}(\text{cp}(R))}{d_r}$ and let $c = 2^{c_f}$. Then $i_r = d_f i_l + c_f$ and with (1) and $z = 2^{i_l}$ we get $i_v = \log(1 - z - cz^{d_f})$. Due to (10) there is a critical path through e_l and therefore

$$\text{cost}(\text{cp}(X)) = \text{cost}(\text{cp}(L)) - d_l i_l - i_v$$

Substituting i_v and forming the derivative with respect to i_l we get

$$\frac{-z - c d_f z^{d_f}}{1 - z - cz^{d_f}} - d_l = 0 \iff c\frac{d_r - 1}{d_r} z^{d_f} + \frac{d_l - 1}{d_l} z - 1 = 0 \qquad (11)$$

Solving this equation yields an optimal choice for i_l (and hence with (10) and (1) for i_r and i_v). Note that for $d_f = 1$, $c_f = 0$ and $d_l = d_r$ we get the parameters used for E_{bal}. Unfortunately, there is no closed form for the solution of (11) for

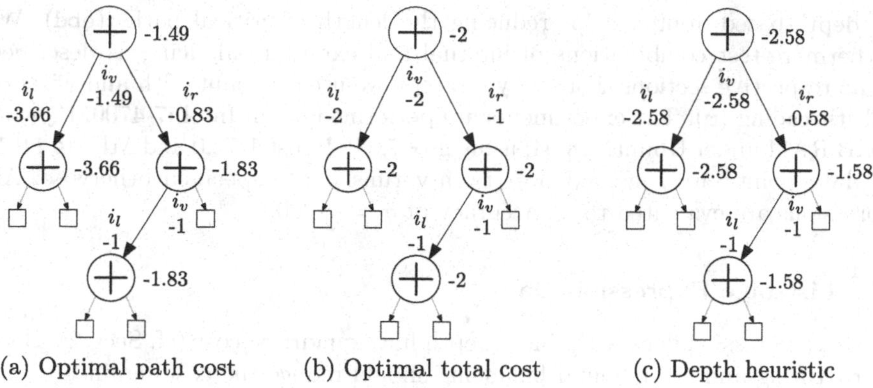

Fig. 2. Error bound distribution for a graph with two paths of different lengths. In the optimal case, both paths have cost 5.15. When minimizing total cost, the cost of the critical path is 6, which gets reduced to 5.74 with the depth heuristic.

arbitrary d_f. Thus, for an implementation a numerical or a heuristic approach is needed. The cost induced by operation constants and the initial accuracy usually increases with a higher depth. So it is plausible to assume for a node v that the child with the higher subexpression depth will contain a more expensive path in the evaluation, if the difference in accuracy increase at v is relatively small. We can use this observation in the following heuristic. We set

$$\begin{aligned} i_v &= i_r = -\log(d_l+1)-1, \quad i_l = \log(d_l)-\log(d_l+1), && \text{if } d_l > d_r \\ i_v &= i_l = -\log(d_r+1)-1, \quad i_r = \log(d_r)-\log(d_r+1), && \text{if } d_l < d_r \qquad (12) \\ i_v &= -\log(d_l+1), \quad i_l = i_r = \log(d_l)-\log(d_l+1)-1, && \text{if } d_l = d_r \end{aligned}$$

Figure 2 shows an example for the differences between the critical path optimization, total cost optimization and the depth heuristic. The heuristic reduces the weight of the critical path compared to the previous strategies.

3 Experiments

We present experiments to underline differences between restructuring (Sect. 2.1) and error bound balancing (Sect. 2.2). For the comparison, the policy-based exact-decisions number type `Real_algebraic` with multithreading is used [8, 12]. We compare several different strategies. In our default configuration for `Real_algebraic` we use `boost::interval` as floating-point filter and `mpfr_t` as bigfloat data type. Furthermore we always enable topological evaluation, bottom-up separation bound representation and error representation by exponents [9, 13]. We call the default strategy without balancing `def`. For restructuring we use the weighted version of Brent's algorithm with unit weights (`bru`) and with setting the weights to the expression depth (`brd`). For error bound balancing we use the weight function counting all operators without removing duplicates (`ebc`) and the depth-based approach for reducing the length of critical paths (`ebd`). We furthermore test combinations of internal and external balancing as described in the respective sections. For every strategy we use a variant with and without multithreading (`m`). The experiments are performed on an Intel i7-4700MQ with 16 GB RAM under Ubuntu 18.04, using g++ 7.3.0, Boost 1.62.0 and MPFR 4.0.1. All data points are averaged over twenty runs if not specified otherwise. All expressions are evaluated to an accuracy of $q = -10000$.

3.1 List-Like Expression Dags

List-like expression dags with linear depth have quadratic cost (cf. Sect. 2). Both restructuring and error bound balancing should reduce the cost significantly in this case. We build an expression dag E_{list} by computing $res := res \circ a_i$ in a simple loop starting with $res = a_0$, where $\circ \in \{+, -, *, /\}$ is chosen randomly and uniformly and a_i are operands $(0 \leq i \leq n)$. For the operands we choose random rationals, i.e., expressions of the form $a_i = d_{i,1}/d_{i,2}$ where $d_{i,j} \neq 0$ are random `double` numbers exponentially distributed around 1. By using exact divisions we assure that the operands have sufficient complexity for our experiments. To prevent them from being affected by restructuring, we assign an additional (external) reference to each operand. Figure 3 shows the results for evaluating E_{list}. Both balancing methods lead to a significant reduction in running time compared to the default configuration (note the logarithmic scale). For large numbers of operators, restructuring is superior to error bound balancing. While error bound balancing optimizes the variable precision increase, it does not reduce the cost associated with the operation constants. The precision increase due to operation constants affects more nodes in an unbalanced structure than in a balanced one, which gives restructuring an advantage. For small numbers of operators, error bound balancing leads to better results than restructuring, since the cost of evaluating additional operators created through restructuring becomes more relevant. The structure of E_{list} is highly detrimental to efficient parallelization.

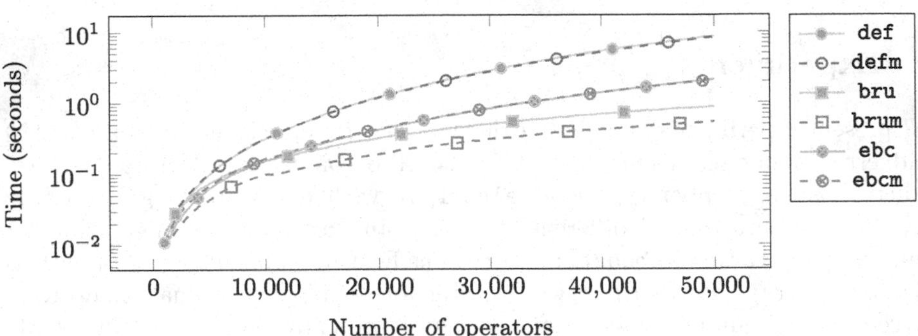

Fig. 3. Running times on a list-like expression dag. Restructuring reduces times by up to 90 % for single-threaded and by up to 94 % for multithreaded evaluation. Error bound balancing reduces the running time by up to 75 % in both cases.

Consequently, neither the default evaluation nor the error bound balanced evaluation show significant cost reduction when run on multiple processors. With Brent's algorithm a speedup of about 1.7, i.e., a runtime reduction of about 40%, can be observed. Since E_{list} does not contain any common subexpressions or other barriers, the results for other restructuring or error bound balancing strategies are indistinguishable from their counterparts. Interestingly, the evaluation does not benefit from a combination of both balancing strategies. Instead the results closely resemble the results obtained by using only restructuring and even get a bit worse in the multithreaded case. Since through restructuring a perfectly balanced dag is created, the default error bounds are already close to optimal (cf. Sect. 3.3).

3.2 Blocking Nodes

Restructuring gets difficult as soon as 'blocking nodes', such as nodes with multiple parents, occur in the expression dag (cf. Sect. 2.1). We repeat the experiment from Sect. 3.1, but randomly let about 30% of the operator nodes be blocking nodes by adding an additional parent (which is not part of our evaluation). Nodes with such a parent cannot be part of a restructuring process, since the subexpressions associated with them might be used somewhere else and therefore cannot be destroyed. Both the default and the internal balancing method are not affected by the change and thus show the same results as before. Restructuring on the other hand performs worse and falls back behind error bound balancing (cf. Fig. 4). The depth heuristic leads to fewer losses for both total and parallel running time. It reduces the running time by about 10% in single-threaded and about 20% in multithreaded execution compared to using unit weights. Combining internal and external balancing combines the advantages of both strategies in this case. Exemplarly, a combination of `brd` and `ebc`, named `cmb`, is shown in Fig. 4. For serial evaluation the running time of the combined approach mostly

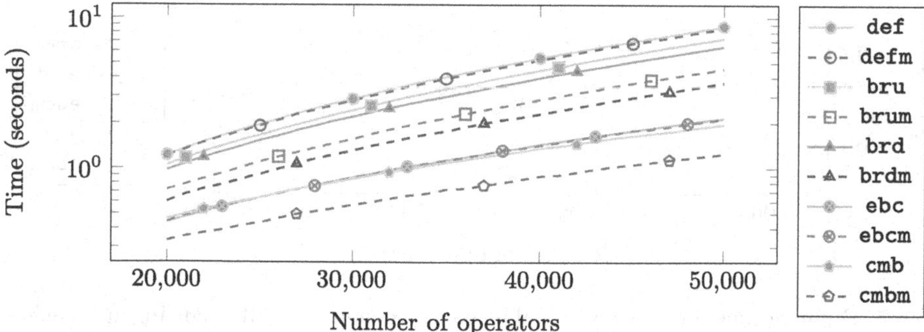

Fig. 4. Running times on a list-like expression dag where 30% of the operators have an additional reference. Error bound balancing is not affected by the references, restructuring performs much worse. Combining error bound balancing and restructuring leads to the best results for multithreading.

resembles the running time of error bound balancing, getting slightly faster for a large number of operators (about 9% for $N = 50000$). In parallel, however, it strongly increases parallelizability leading to a speedup of 1.6 and a total runtime reduction of up to 85% compared to the default strategy.

3.3 Balanced Expression Dags

When an expression dag is already balanced, there is not much to gain by either balancing method. In a perfectly balanced expression dag E_{bal}, restructuring cannot reduce the depth and therefore does not reduce its cost, neither in serial nor in parallel. Brent's algorithm still creates a normal form, which adds additional operations and might even increase the maximum depth. Error bound balancing on the other hand can potentially make a difference. For a balanced expression dag the total cost is strongly influenced by the operation constants, which is reflected in a high variance when choosing the operators at random. In the experiment shown in Fig. 5, we increase the number of test sets for each data point from 20 to 50 and use the same test data for each number type. The single data points lie in a range of about ±20% of the respective average. As expected, restructuring performs worse than the default number type, doubling the depth and replacing each division by, on average, two multiplications. Error bound balancing performs worse than not balancing as well. Neither the total operator count, nor the depth-based strategy have a significant impact on the running time of the bigfloat operations, since the cost decrease per operation is at most logarithmic in the number of operators. For the same reason and due to the limited number of processors, the expected cost reduction between `ebcm` and `ebdm` in the multithreaded case (cf. Sect. 2.2) can not be observed in the experimental data.

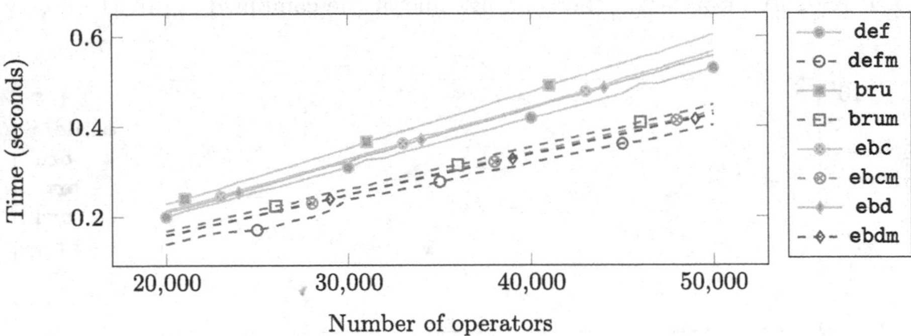

Fig. 5. Running times on a perfectly balanced expression dag. All balancing approaches lead to a performance loss. Restructuring increases the running time by about 15%, error bound balancing by 5% to 7% in the single-threaded case.

3.4 Common Subexpressions

Random expression dags, created by randomly applying operations on a forest of operands until it is reduced to a single DAG, tend to be balanced and therefore behave similarly to a perfectly balanced tree. This changes if common subexpressions are involved. With error bound balancing, common subexpressions can be recognized and the error bounds at the parent nodes can be adjusted, such that both request the same accuracies (cf. Theorem 1). The two implemented heuristics to some degree take common subexpressions into account, since they contribute the same weight to all of the subexpression's parents. We test the behavior of error bound balancing strategies by randomly reusing a certain percentage of subtrees during randomized bottom-up construction of the graph. To avoid zeros, ones, or an exponential explosion of the expression's value we only use additions if two subtrees are identical during construction. While error bound balancing still cannot outperform the default strategy due to the balanced nature, it moves on par with it. If 5% of the operations have more than one parent, the error bound balancing strategies improve the single-threaded running time by about 1% to 5%. In a parallel environment, it still performs worse with `ebdm` being slightly superior to `ebcm`.

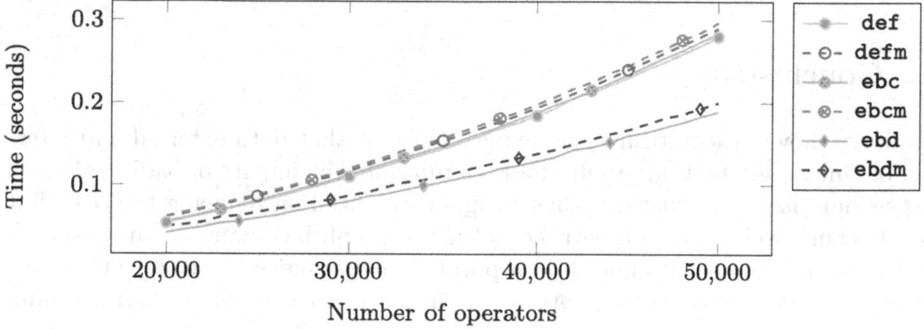

Fig. 6. Running times on a series of self-additions as depicted in Fig. 1. Counting operators without removing duplicates does not improve on the default running time. The depth heuristic reduces the default running time by up to 32%.

If common subexpressions lead to a large difference between the actual number of operators and the number of operators in a tree expansion, `ebc` significantly overestimates the optimal weight of its edges (cf. Fig. 1). Figure 6 shows results for evaluating a sequence of additions where the left and the right summand is the result of the previous addition. The full operator count heuristic does not reduce the running time and even performs worse than the default strategy for large numbers of operators, whereas the depth-based heuristic clearly outperforms the other strategies. Note that in this case, the depth-based heuristic leads to the optimal error distribution for both total and critical path cost.

3.5 A Note on Floating-Point Primitives

Error bound balancing requires the use of floating-point error bounds. While IEEE 754 requires that floating point computations must be exactly rounded, it is surprisingly difficult to find an adequate upper or lower bound to the result of such an operation. IEEE 754 specifies four rounding modes: Round to nearest, Round to positive/negative infinty and Round to zero [1]. For the last three modes, which are commonly referred to as directed rounding, it is easy to obtain a lower or upper bound by negating the operands adequately. Unfortunately, most systems implement Round to nearest. Switching the rounding mode is expensive. While `double` operations with appropriate negations for directed rounding are about two times slower, switching to an appropriate rounding mode can increase the running time of a single operation by a factor of 100. The same factor applies if we manually jump to the next (or previous) representable `double` value.

Handling floating-point primitives correctly can, depending on the architecture, be very expensive. In most cases, however, the computed error bounds massively overestimate the actual error. Moreover, for the actual bigfloats computations the error bounds are rounded up to the next integer. It is therefore almost impossible that floating-point rounding errors make an actual difference in any computation. For our experiments we refrained from handling those bounds correctly to make the results more meaningful and less architecture-dependent.

4 Conclusion

We have shown, theoretically and experimentally, that both external and internal balancing methods are useful tools to mitigate the impact of badly balanced expression dags. Restructuring has a higher potential on reducing the cost, but can become useless or even detrimental if the graph has many common subexpressions or is already balanced. In a parallel environment, restructuring is necessary to make use of multiple processors in an unbalanced graph. Error bound balancing is more widely applicable, but is limited in its effectivity. If the graph is small or already sufficiently balanced, neither of the methods has a significant positive impact on the evaluation cost. A general purpose number type should therefore always check whether the structure generally requires balancing before applying either of the algorithms. For both strategies we have described optimal weight functions. In both cases implementations require heuristics to be practicable. Our experiments show that carefully chosen heuristics are in most cases sufficient to increase the performance of exact number types.

References

1. IEEE standard for floating-point arithmetic: IEEE Std 754–2008, pp. 1–70 (2008)
2. Borassi, M.: A note on the complexity of computing the number of reachable vertices in a digraph. Inf. Process. Lett. **116**(10), 628–630 (2016). https://doi.org/10.1016/j.ipl.2016.05.002

3. Brent, R.P.: The parallel evaluation of general arithmetic expressions. J. ACM **21**(2), 201–206 (1974). https://doi.org/10.1145/321812.321815
4. van der Hoeven, J.: Computations with effective real numbers. Theor. Comput. Sci. **351**(1), 52–60 (2006). https://doi.org/10.1016/j.tcs.2005.09.060
5. Majithia, J.C., Levan, D.: A note on base-2 logarithm computations. Proc. IEEE **61**(10), 1519–1520 (1973). https://doi.org/10.1109/PROC.1973.9318
6. Mehlhorn, K., Näher, S.: LEDA a library of efficient data types and algorithms. In: Kreczmar, A., Mirkowska, G. (eds.) MFCS 1989. LNCS, vol. 379, pp. 88–106. Springer, Heidelberg (1989). https://doi.org/10.1007/3-540-51486-4_58
7. Monniaux, D.: The pitfalls of verifying floating-point computations. ACM Trans. Program. Lang. Syst. **30**(3), 12:1–12:41 (2008). https://doi.org/10.1145/1353445.1353446
8. Mörig, M., Rössling, I., Schirra, S.: On design and implementation of a generic number type for real algebraic number computations based on expression dags. Math. Comput. Sci. **4**(4), 539–556 (2010). https://doi.org/10.1007/s11786-011-0086-1
9. Mörig, M., Schirra, S.: Precision-driven computation in the evaluation of expression-dags with common subexpressions: problems and solutions. In: Kotsireas, I.S., Rump, S.M., Yap, C.K. (eds.) MACIS 2015. LNCS, vol. 9582, pp. 451–465. Springer, Cham (2016). https://doi.org/10.1007/978-3-319-32859-1_39
10. Schirra, S.: Robustness and precision issues in geometric computation. In: Handbook of Computational Geometry, pp. 597–632. Elsevier (2000)
11. Wilhelm, M.: Balancing expression dags for more efficient lazy adaptive evaluation. In: Blömer, J., Kotsireas, I.S., Kutsia, T., Simos, D.E. (eds.) MACIS 2017. LNCS, vol. 10693, pp. 19–33. Springer, Cham (2017). https://doi.org/10.1007/978-3-319-72453-9_2
12. Wilhelm, M.: Multithreading for the expression-dag-based number type Real_algebraic. Technical Report FIN-001-2018, Otto-von-Guericke-Universität, Magdeburg (2018)
13. Wilhelm, M.: On error representation in exact-decisions number types. In: Proceedings of the 30th Canadian Conference on Computational Geometry, CCCG, pp. 367–373 (2018)
14. Wilhelm, M.: Restructuring expression dags for efficient parallelization. In: 17th International Symposium on Experimental Algorithms, SEA, pp. 20:1–20:13 (2018). https://doi.org/10.4230/LIPIcs.SEA.2018.20
15. Yap, C.: Towards exact geometric computation. Comput. Geom. **7**, 3–23 (1997). https://doi.org/10.1016/0925-7721(95)00040-2
16. Yu, J., Yap, C., Du, Z., Pion, S., Brönnimann, H.: The design of core 2: a library for exact numeric computation in geometry and algebra. In: Fukuda, K., Hoeven, J., Joswig, M., Takayama, N. (eds.) ICMS 2010. LNCS, vol. 6327, pp. 121–141. Springer, Heidelberg (2010). https://doi.org/10.1007/978-3-642-15582-6_24

Hacker's Multiple-Precision Integer-Division Program in Close Scrutiny

Jyrki Katajainen[1,2(✉)]

[1] Department of Computer Science, University of Copenhagen, Universitetsparken 5, 2100 Copenhagen East, Denmark
jyrki@di.ku.dk

[2] Jyrki Katajainen and Company, 3390 Hundested, Denmark
http://hjemmesider.diku.dk/~jyrki/

Abstract. Before the era of ubiquitous computers, the long-division method was presented in primary schools as a paper-and-pencil technique to do whole-number division. In the book "Hacker's Delight" by Warren [2nd edition, 2013], an implementation of this algorithm was given using the C programming language. In this paper we will report our experiences when converting this program to a generic program-library routine.

The highlights of the paper are as follows: (1) We describe the long-division algorithm—this is done for educational purposes. (2) We outline its implementation—the goal is to show how to use modern C++ to achieve flexibility, portability, and efficiency. (3) We analyse its computational complexity by paying attention to how the digit width affects the running time. (4) We compare the practical performance of the library routine against Warren's original. It is pleasure to announce that the library routine is faster. (5) We release the developed routine as part of a software package that provides fixed-width integers of arbitrary length, e.g. a number of type `cphstl::ℕ<2019>` (editor's note: the non-transliterated form used in the code is `cphstl::bbbN<2019>`) has 2019 bits and it supports the same operations with the same semantics as a number of type **unsigned int**.

Keywords: Software library · Multiple-precision arithmetic · Algorithm · Long division · Description · Implementation · Meticulous analysis · Experimentation

1 Introduction

The algorithms for multiple-precision integer addition, subtraction, multiplication, and division are at the heart of algorithmics. In this paper we discuss the computer implementation of the long-division method introduced by Briggs

I. Kotsireas et al. (Eds.): SEA^2 2019, LNCS 11544, pp. 376–391, 2019.
https://doi.org/10.1007/978-3-030-34029-2_25

around 1600 A.D. [https://en.wikipedia.org/wiki/Long_division]. The underlying ideas are even older since the Chinese, Hindu, and Arabic division methods used before that show remarkable resemblance to it [8,13].

In a positional numeral system, a string $\langle d_{\ell-1}, d_{\ell-2}, \ldots, d_0 \rangle$ of *digits* d_i, $i \in \{0, 1, \ldots, \ell-1\}$, is used to represent an integer $\boldsymbol{d}$, ℓ being the length of the representation, $d_{\ell-1}$ the most-significant digit, and d_0 the least-significant digit. Let β, $\beta \geq 2$, denote the *base* of the numeral system. The individual digits are drawn from some bounded universe, the size of which is at least β, and the digit d_j has the *weight* β^j. In the decimal system, the digit universe is $\{0, 1, \ldots, 9\}$ and the weight of d_j is 10^j. For a general base β, the decimal *value* of $\boldsymbol{d}$ is $\sum_{j=0}^{\ell-1} d_j \cdot \beta^j$. As is customary, in this representation the leading zero digits (0) may be omitted, except when representing number zero ($\mathbf{0}$).

In the computer representation of a number, the digit width is often selected to be in harmony with the word size of the underlying hardware. We use `W` to denote the type of the digits and we assume that the width of `W` is a power of two. The numbers themselves are arrays of digits of type `W`. The length of these arrays can be specified at compile time (**std::array** in C++), or the length can be varying and may change at run time (**std::vector** in C++). The memory for these arrays can be allocated from the stack at run time (so-called C arrays) or from the heap relying on the memory-allocation and memory-deallocation methods provided by the operating system. Since memory management is not highly relevant for us, we will not discuss this issue here.

In the division problem for whole numbers (non-negative integers), the task is to find out how many times a number $\boldsymbol{y}$ (*divisor*) is contained in another number $\boldsymbol{x}$ (*dividend*). Throughout the paper, we use the division operator / to denote the whole-number division and we assume that $\boldsymbol{y} \neq \mathbf{0}$ since division by $\mathbf{0}$ has no meaning. That is, the output of $\lfloor \boldsymbol{x}/\boldsymbol{y} \rfloor$ is the largest whole number $\boldsymbol{q}$ (*quotient*) for which the inequality $\boldsymbol{q} * \boldsymbol{y} \leq \boldsymbol{x}$ holds. Throughout the paper, we use n to denote the *length* of the dividend (the number of its digits) and m the length of the divisor. After computing the quotient, the *remainder* $\boldsymbol{x} - \boldsymbol{q} * \boldsymbol{y}$ can be obtained by a single long multiplication and long subtraction. We ignore the computation of the remainder, but we acknowledge that a routine `divmod` that computes both the quotient and the remainder at the same time could be handy.

The main motivation for this study was the desire to implement a program package for the manipulation of multiple-precision integers. In our application (see [3]), we only needed addition, subtraction, and multiplication for numbers whose length was two or three words. When making the package complete and finishing the job, the implementation of the division algorithm turned out to be a non-trivial task. We are not the first to make this observation (see, e.g. [1]).

First, we reviewed the presentation of the long-division algorithm in "The Art of Computer Programming" (Volume 2) [6, Sect. 4.3.1]. Knuth described the algorithm (Algorithm D), proved its correctness (Theorem B), analysed its complexity, and gave an implementation (Program D) using his mythical MIX assembly language. The paper by Pope and Stein [11] was one of the significant sources used by him. Under reasonable assumptions, Knuth estimated that, in the average case, the program will execute about $30n \cdot m + O(n+m)$ MIX instructions. (Before reading Knuth's book, check the official errata available at [https://www-cs-faculty.stanford.edu/~knuth/taocp.html]—this can save you some troubles later.)

Next, we looked at the Pascal implementation described by Brinch Hansen [1] and the C implementation described by Warren in the book "Hacker's Delight" [12] (errata can be found at [https://www.hackersdelight.org/]). In particular, Warren carefully examined many implementation details so we decided to base our library implementation on his programs (the source code is available at [https://www.hackersdelight.org/]). We looked at other sources as well, but very quickly we got back to Algorithm D or some of its variants. When the numbers are not longer than a few thousand digits, the long-division algorithm should be good enough for most practical purposes. Although its asymptotic complexity is high $O(n \cdot m)$, the leading constant in the order notation is small.

In this write-up, we report our observations when implementing the division routine for the multiple-precision integers provided by the CPH STL [http://www.cphstl.dk/]. We put emphasis on the following issues:

Portability. In the old sources the digit universe is often fixed to be small. For example, in Warren's implementation the width of digits was set to 16 bits. For our implementation the digit width can be any power of two—it should just be specified at compile time. We wrote the programs using C++. This made it possible to hide some of the messy details inside some few subroutines called by the high-level code.

Analysis. Instead of the MIX cost used by Knuth or the RISC cost used by Warren, we analyse the Intel cost—the number of Intel assembler instructions—of the long-division routine as a function of the number of the bits in the inputs and the word size of the underlying computer. (Before reading any further, you should stop for a moment to think about what would be a good data type `W` for the digits when dividing an N-bit number with an $\frac{N}{2}$-bit number.)

Efficiency. We perform some experiments to check the validity of our back-of-the-envelope calculations in a real machine. The tests show unanimously that our program—with larger digit widths—is faster than Warren's program. And because of adaptability, it should be relatively easy to modify the code—if at all necessary—if the underlying hardware changes.

2 Long-Division Algorithm

Let $\odot$ be one of the operations supported by the C++ programming language for integers, e.g. $==$, $<$, $+$, $-$, $*$, $/$, $\%$, $>>$, $<<$, $\sim$ (**compl**), & (**bitand**), or || (**bitor**). To understand the beauty of the division algorithm, we use the notation $\boldsymbol{\odot(n, m)}$ to denote a subroutine that performs the $\odot$ operation when the first operand is an n-digit number and the second operand (if any) an m-digit number.

2.1 Software Stack

To perform the operation $/(n, m)$, the long-division algorithm needs the following subroutines:

$\odot(1)$, $\odot \in \{\sim, \mathbf{nlz}\}$. The primitive $\sim$ computes the bitwise complement of a digit and **nlz** the number of leading 0 bits in a digit. We assume that these primitives are available in hardware or provided by the environment.

$\odot(1, 1)$, $\odot \in \{==, <, /, \%, >>, <<, \&, ||\}$. We assume that these operations are also available in hardware or provided by the environment.

$+(1, 1)$. We assume that this operation is a built-in primitive. The overflow bit (*carry*) can be computed by checking whether the sum is smaller than one of the operands ($<(1, 1)$ operation) [12, Sect. 2-16].

$-(1, 1)$. We assume that this operation is a built-in primitive. The underflow bit (*borrow*) can be computed by checking whether the first operand is smaller than the second ($<(1, 1)$ operation) [12, Sect. 2-16].

$*(1, 1)$. We assume that this operation is a built-in primitive, but the output consists of two digits so the higher-order digit must be computed separately. A routine that computes the higher-order digit without overflows is described in [12, Fig. 8-2]. (It requires 16 RISC instructions.)

$+(2, 1)$. This operation involves two $+(1, 1)$ operations and one $<(1, 1)$ operation to forward the carry bit (if any) from the first position to the second. The operation is always used in a context where the overflow can be ignored.

$/(2, 1)$. The operation is only needed in a context where the output is one digit long. In principle, this operation implements the division tables which are the reversal of the multiplication tables we learnt at school. This operation is the most complicated subroutine; an implementation is given in [12, Fig. 9-4]. (According to Warren's analysis, for uniformly distributed random numbers, this operation executes about 52 RISC instructions.)

$*(n, 1)$. This operation can be accomplished in a single scan over the first operand by invoking n times the $*(1, 1)$ operation, forwarding the higher-order digit from the previous position and adding it to the result of the multiplication with a $+(2, 1)$ operation [1, Algorithm 2]. This form of multiplication is always used in a context where the overflow can be ignored.

$<(n, n)$. This operation is a simple scan over the digits starting from the most-significant end [1, Algorithm 6]. The first position where the digits differ is found (if any) using the $==(1, 1)$ operation and at the found position the $<(1, 1)$ operation is applied to get the answer.

$-(n, n)$. This operation can be accomplished in one scan by performing n $+(1, 1)$ operations, n $-(1, 1)$ operations, and $2n$ $<(1, 1)$ operations to handle the borrow from the previous position [1, Algorithm 7]. This form of subtraction is always used in a context where the underflow can be ignored.

Since the computational complexity of the long-division algorithm will be determined by the routines $*(n, 1)$, $<(n, n)$, and $-(n, n)$, we give them their own names `product`, `is_less`, and `difference`, respectively.

2.2 Algorithm Description

Let us consider how the division problem can be solved when the dividend is $\boldsymbol{x} = \langle x_{n-1}, x_{n-2}, \ldots, x_0 \rangle$ and the divisor $\boldsymbol{y} = \langle y_{m-1}, y_{m-2}, \ldots, y_0 \rangle$. We assume that n and m are the real lengths of the numbers so that $x_{n-1} \neq 0$ and $y_{m-1} \neq 0$. Recall that the digits are of type `W` and let w be the width of `W` in bits.

At a high level, the long-division algorithm is simple: it computes the quotient digits one at the time starting from the most-significant end. The basic complication is the need of a good estimate $\widehat{\mathrm{q}}$ for the next quotient digit. When this is available, the partial remainder can be updated and the computation can proceed to the next digit.

To get a reasonable estimate for the next quotient digit, the key algorithmic idea is *normalization* [11]: this means that the divisor is cast into the form where its most significant digit is higher than or equal to 2^{w-1}. One way to achieve this is to multiply both the dividend and the divisor with some factor f, which makes the most-significant digit of the divisor large enough. Let $\overline{\boldsymbol{x}} = f * \boldsymbol{x}$ and $\overline{\boldsymbol{y}} = f * \boldsymbol{y}$. Since $\lfloor \overline{\boldsymbol{x}} / \overline{\boldsymbol{y}} \rfloor = \lfloor \boldsymbol{x} / \boldsymbol{y} \rfloor$, the quotient for the normalized numbers is the same as that for the original numbers. Knuth used the factor $f = \lfloor 2^w / (y_{m-1} + 1) \rfloor$ (see the errata of [6, Sect. 4.3.1]). Warren [12, Fig. 9-4] used the factor $f = 2^\sigma$, where σ is the number of leading 0 bits in y_{m-1}.

During the execution of the algorithm, the *partial remainder* is maintained in $\boldsymbol{u} = \langle u_n, u_{n-1}, \ldots, u_0 \rangle$ which is initialized to contain the normalized dividend $f * \boldsymbol{x}$. The normalized divisor is maintained in $\boldsymbol{v} = \langle v_m, v_{m-1}, \ldots, v_0 \rangle$. In the main loop of the algorithm, the loop index j goes down from $n-m$ to 0. We call the subrange of length $m+1$ $\langle u_{j+m}, u_{j+m-1}, \ldots, u_j \rangle$ the *active part* of the partial remainder. Then the operation $\boldsymbol{/(2,1)}$ with the first two digits $\langle u_{j+m}, u_{j+m-1} \rangle$ of the active part and v_{m-1} is used to compute an estimate $\widehat{\mathrm{q}}$ for the next quotient digit. This estimate is the correct quotient digit, or it is one or two too high [6, Theorem B]. Collins and Musser [2] proved that for random numbers, with high probability, the estimate is correct or off by one.

These results have been improved in several ways: (1) Mifsud [9] (see an addendum in [10]) proved that with more aggressive normalization the estimate can be guaranteed to be correct or off by one. (2) Krishnamurthy and Nandi [7] obtained the same result by using the prefixes of 3 and 2 digits when calculating the estimate. So both of these approaches guarantee that not more than one correction is required to obtain the true quotient digit. (3) Also, people have tried to find conditions under which the normalization can be skipped (see, for example, [7,9]). Even if the use of $\boldsymbol{v}$ can be avoided, temporary storage is still needed to store the active part of the partial remainder $\boldsymbol{u}$ and the product $\boldsymbol{p}$ of $\widehat{\mathrm{q}}$ and the (normalized) divisor.

Now we can describe the algorithm in detail:

(1) If $\boldsymbol{x} < \boldsymbol{y}$, return $\boldsymbol{0}$ as the answer. This comparison is a generalization of the $\boldsymbol{<(n,n)}$ operation where the operands are not necessarily of the same length. It involves a synchronous scan over the digits starting from the end of the longer string. After this step we can be sure that $n \geq m$.

(2) Allocate space for the quotient $\boldsymbol{q} = \langle q_{n-m}, q_{n-m-1}, \ldots, q_0 \rangle$ and fill it with zeros.
(3) Allocate space for the partial remainder $\boldsymbol{u} = \langle u_n, u_{n-1}, \ldots, u_0 \rangle$ and copy $\boldsymbol{x}$ there; observe that $\boldsymbol{u}$ is one longer, so u_n is set to zero.
(4) Allocate space for the normalized divisor $\boldsymbol{v} = \langle v_m, v_{m-1}, \ldots, v_0 \rangle$ and copy $\boldsymbol{y}$ there; v_m is needed to make this string $m+1$ long, so v_m is set to zero.
(5) Compute the number of leading 0 bits in the digit y_{m-1}. Let this be σ.
(6) Shift the bits of $\boldsymbol{u}$ σ positions to the left. This operation is a special case of the $\boldsymbol{*(n+1,1)}$ operation where the multiplier is 2^σ. Since $\boldsymbol{u}$ is one digit longer than $\boldsymbol{x}$ and $\sigma < w$, no overflow is possible.
(7) Shift the bits of $\boldsymbol{v}$ σ positions to the left. Naturally, no overflow is possible and after this operation the leading bit of v_{m-1} is set as required.
(8) Compute now the digits of $\boldsymbol{q}$, one by one, by letting the loop index j go down from $n-m$ to 0.
   (a) Calculate an estimate $\widehat{\mathrm{q}}$ for the quotient digit by invoking the $\boldsymbol{/(2,1)}$ operation with the arguments $\langle u_{j+m}, u_{j+m-1} \rangle$ and v_{m-1}. However, if $u_{j+m} \geq v_{m-1}$, set $\widehat{\mathrm{q}}$ equal to $2^w - 1$ without performing the division.
   (b) Compute the product of $\boldsymbol{v}$ and $\widehat{\mathrm{q}}$ by invoking the $\boldsymbol{*(m+1,1)}$ operation. Keep the result temporarily in $\boldsymbol{p} = \langle p_m, p_{m-1}, \ldots, p_0 \rangle$.
   (c) Check if the estimate is too large by invoking the $\boldsymbol{<(m+1,m+1)}$ operation for the active part of the partial remainder $\langle u_{j+m}, u_{j+m-1}, \ldots, u_j \langle$ and $\boldsymbol{p}$.
   (d) If the estimate was too large, make it one smaller, subtract $\boldsymbol{v}$ from $\boldsymbol{p}$ by performing the $\boldsymbol{-(m+1,m+1)}$ operation, and go back to Step 8c.
   (e) Otherwise, set q_j equal to $\widehat{\mathrm{q}}$. Furthermore, update the partial remainder by subtracting the computed product $\boldsymbol{p}$ from the active part by invoking the $\boldsymbol{-(m+1,m+1)}$ operation. Hereafter we can proceed to the computation of the next quotient digit.
(9) Release the space allocated for $\boldsymbol{u}$, $\boldsymbol{v}$, and $\boldsymbol{p}$.
(10) Return $\boldsymbol{q}$ as the result of the computation.

2.3 Asymptotic Analysis

In this algorithm, Steps (1)–(7) all involve sequential scans over the digit strings. If the digits can be processed at unit cost, the amount of work done is $O(n+m)$. Most of the work is done in Step 8. Of the substeps, Step 8b calls the function `product`, Step 8c the function `is_less`, and Steps 8d and 8e the function `difference`. The arguments are of length $m+1$. Each of these operations involves a linear scan over the digits. Therefore, the asymptotic complexity of the algorithm is $O((n-m) \cdot m + n + m)$.

3 Implementation

In this section we describe our implementation of the long-division algorithm. The source code is extracted from the CPH STL so, unfortunately, it contains some noise that has to be explained first.

Standard library. A good documentation of the facilities available at the C++ standard library can be found at [https://en.cppreference.com/].

Constraints and concepts. In the code some requirements are specified for the template arguments to ensure that the components are used in a correct way. Here we rely on the features drafted in the upcoming C++2a standard, but some compilers support them already now.

Type functions. In the code some metaprogramming tools are used; these are taken from the CPH MPL (Copenhagen metaprogramming library) [4]. A *type function* maps a type to some value or to some type, and this computation is done at compile time. By convention, a type function, the name of which begins with **is_**, returns a Boolean value. As concrete examples, consider the following type functions specified for some type W:

(1) The built-in function **sizeof**(W) gives the size of the objects of type W, measured in bytes. Unfortunately, for this type function the syntax is not the same as that preferred in the CPH MPL.

(2) The type function **cphmpl::width**<W> returns the width of the objects of type W, measured in bits. In our test computer, the compiler will replace all occurrences of **cphmpl::width**<**int**> in the code with the number 32.

(3) The type function **cphmpl::twice_wider**<W> specifies an alias for the type, the width of which is twice as large as that of W. For example, **cphmpl::twice_wider**<**cphstl::ℕ**<512>> is an alias for **cphstl::ℕ**<1024>.

Ranges. The digit strings given for the programs can be stored in a **std::array**, in a **std::vector**, in a C array, or in any other container—or part of it—that supports (bidirectional) iterators. A *range* specifies such a sequence. To manipulate the digits, it must be possible to use a range as an argument for the functions **std::begin**, **std::cbegin**, **std::end**, **std::cend**, **std::size**, and **std::empty**. With this abstraction, the programs are independent of the representation of the digit strings.

Hidden details. The code for some functions is omitted on purpose. Many of the omitted functions defined inside the **namespace cphstl::detail** work for an arbitrary numeric type, but they are overloaded to work more efficiently for the standard integer types.

3.1 Function is_less

The implementation of function **is_less** is given in Listing 1. Starting from the most significant digit, the purpose is to find the first position where the two strings differ and then use the found digits to determine the answer. The critical inner loop is in lines 15–18.

Listing 1. Function **is_less** in C++.

```
1  template<typename L, typename R>
2  requires
3  /* 1 */ cphmpl::specifies_range<L> and
4  /* 2 */ cphmpl::specifies_range<R> and
5  /* 3 */ std::is_same_v<cphmpl::value<L>, cphmpl::value<R>>
```

```
bool is_less(L const& lhs, R const& rhs) {
  // check whether lhs < rhs or not
  assert(std::size(lhs) == std::size(rhs));
  assert(not std::empty(lhs));
  using I = cphmpl::const_iterator<L>;
  using J = cphmpl::const_iterator<R>;
  I p = std::cend(lhs);
  J q = std::cend(rhs);
  I first = std::cbegin(lhs);
  do {
    --p;
    --q;
  } while (p != first and *p == *q);
  return *p < *q;
}
```

We declared the digits to be of type **unsigned long long int**, the size of which was 8 bytes, and asked the compiler to generate the assembler code for the inner loop of `is_less`. The inner loop had 7 instructions. A micro-benchmark that was used to verify this count compared two equal numbers. The test revealed that, when the digits were of type **unsigned char**, the compiler could optimize the code so that the execution only required 0.17 instructions per digit. This optimization was not done for the other standard types.

3.2 Function `difference`

In long division, it is only necessary to do the subtraction $x - y$ when the two numbers have the same length and when $x \geq y$. Also, it is not necessary to keep the old value. Therefore, we implemented the operation x −= y in addition to the general subtraction. The C++ code for this is given in Listing 2. Here the inner loop is in lines 18–26.

Listing 2. Function `difference` in C++.

```
1  template<typename L, typename R>
2  requires
3  /* 1 */ cphmpl::specifies_range<L> and
4  /* 2 */ cphmpl::specifies_range<R> and
5  /* 3 */ std::is_same_v<cphmpl::value<L>, cphmpl::value<R>> and
6  /* 4 */ cphmpl::is_unsigned<cphmpl::value<L>>
7  void difference(L& minuend, R const& subtrahend) {
8    // compute minuend -= subtrahend
9    assert(std::size(minuend) == std::size(subtrahend));
10   assert(not std::empty(minuend));
11   using I = cphmpl::iterator<L>;
12   using J = cphmpl::const_iterator<R>;
13   using W = cphmpl::value<L>;
14   I p = std::begin(minuend);
15   J q = std::cbegin(subtrahend);
16   I past = std::end(minuend);
```

```
  bool borrow = 0;
  while (p ≠ past) {
    W t = *q + W(borrow);
    bool overflow = (t < *q);
    bool underflow = (*p < t);
    *p = *p − t;
    borrow = overflow or underflow;
    ++p;
    ++q;
  }
}
```

Again we let the compiler generate the assembly-language translation when the digits were of type **unsigned long long int**. The inner loop had 15 instructions. When the digits were of type **unsigned char**, the compiler could optimize the code so that the execution only required about 12 instructions per digit.

3.3 Function product

In long division, only a restricted form of multiplication is needed where a number $\boldsymbol{x}$ is multiplied by a single digit. Furthermore, it is not allowed to modify $\boldsymbol{x}$ so the result must be saved somewhere else. We assume that the caller has allocated space for the result. Listing 3 gives the C++ code that does this multiplication.

Listing 3. Function product in C++.

```
template<typename L, typename R, typename W>
requires
/* 1 */ cphmpl::specifies_range<L> and
/* 2 */ cphmpl::specifies_range<R> and
/* 3 */ cphmpl::is_unsigned<W> and
/* 4 */ std::is_same_v<cphmpl::value<L>, W> and
/* 5 */ std::is_same_v<cphmpl::value<R>, W>
void product(L& result, R const& multiplicand, W const& factor) {
  // compute result = multiplicand * factor
  assert(std::size(result) == std::size(multiplicand));
  using D = cphmpl::twice_wider<W>;
  using I = cphmpl::iterator<L>;
  using J = cphmpl::const_iterator<R>;
  J first = std::cbegin(multiplicand);
  J past = std::cend(multiplicand);
  W carry = W();
  I q = std::begin(result);
  for (J p = first; p ≠ past; ++p, ++q) {
    D t = cphstl::detail::multiply<D>(*p, factor);
    t = cphstl::detail::add(t, carry);
    *q = cphstl::detail::lower_half<W>(t);
    carry = cphstl::detail::upper_half<W>(t);
  }
}
```

Here `W` is the type of the digits and `D` is an alias for a type that is twice as wide as `W`. The function **cphstl::detail::multiply** performs the operation $*(\mathbf{1},\mathbf{1})$ and the function **cphstl::detail::add** the operation $+(\mathbf{2},\mathbf{1})$. Finally, the remaining functions **cphstl::detail::lower_half** and **cphstl::detail::upper_half** are used to get from a digit of type `D` its two halves of type `W`.

The inner loop is in lines 18–23. When **sizeof**(`W`) was 8 and `D` was an alias of **unsigned __int128**—an extension supported by the **g++** compiler, the assembly-language translation of this loop contained 10 instructions. On the other hand, when `D` was an alias of **std::array**<`W`, 2>, the inner loop contained 26 instructions. For **unsigned char** the instruction count was 9, and for **unsigned short** and **unsigned int** it was 10. For the digit widths 128 and 256, the instruction count dropped to around 6 which could be explained by the fact that the compiler had turned on the streaming SIMD extensions (SSE), allowing parallel operations on four values per instruction.

3.4 Main Loop

After these initial exercises, we can peek inside the long-division program. Its main loop is shown in Listing 4. The meaning of most functions should be clear by their names. The function **cphstl::detail::halves_together** concatenates two digits and the function **cphstl::detail::divide** performs the $/(\mathbf{2},\mathbf{1})$ operation. Because of the **if** test before this division operation, the output is always a single digit and the upper half can be discarded.

Listing 4. The main loop of the long-division program in C++; array **u** contains the partial remainder, array **v** the normalized divisor, and array **p** is for temporary use.

```
auto normalized_divisor = cphstl::range(&v[0], &v[m + 1]);
auto temporary = cphstl::range(&p[0], &p[m + 1]);
auto q = std::begin(quotient);
std::advance(q, n − m);

for (int j = n − m; j ≥ 0; −−j, −−q) {
  auto active_part = cphstl::range(&u[j], &u[j+m+1]);
  W q̂ = compl W(); // estimate for the quotient digit
  if (u[j+m] < v[m−1]) {
    D t = cphstl::detail::halves_together<D>(u[j+m−1], u[j+m]);
    t = cphstl::detail::divide(t, v[m−1]);
    q̂ = cphstl::detail::lower_half<W>(t);
  }
  cphstl::detail::product(temporary, normalized_divisor, q̂);
  while (cphstl::detail::is_less(active_part, temporary)) {
    −−q̂; // correction; estimate may be 1 or 2 too large
    cphstl::detail::difference(temporary, normalized_divisor);
  }
  *q = q̂;
  cphstl::detail::difference(active_part, temporary);
}
```

4 Meticulous Analysis

After describing the long-division program, we can analyse its performance. All the processing is sequential, so we are mainly interested in the number of instructions executed. In the analysis we keep the number of digits (n) fixed, but vary the width of the digits.

Table 1. Summary of the instruction counts (per digit) determined experimentally for the performance-critical functions.

Digit width	`is_less`	`difference`	`product`
8	0.17	12.30	9.17
16	7.16	14.28	10.16
32	7.16	14.24	10.15
64	7.18	15.24	26.17
128	10.40	33.51	6.39
256	16.68	71.88	6.67
512	29.26	148.62	2 874
1024	54.37	317.04	14 339

In Table 1, we summarize the instruction counts that were measured for the efficiency-determining functions. In the reported counts, the total number of instructions executed is divided by n. In the micro-benchmarks, (1) `is_less` compared two equal numbers; (2) `difference` processed two random numbers, except that the first was made larger by resetting the most significant digits; and (3) `product` multiplied a long random number with a random digit. These counts are approximations, but they are firmly linked to the generated assembler code.

Assume now that N is a power of two and that we want to divide an N-bit number with an $\frac{N}{2}$-bit number. When the word size is α, our theoretical analysis shows that the running time of the long-division program should be proportional to $\frac{N}{2\alpha} \cdot \frac{N}{2\alpha}$. This analysis is based on the assumption that digits can be processed at unit cost. The micro-benchmarks show that—in the test environment—this assumption is valid up to 64, or maybe all the way to 256.

Assuming that we rely on the more aggressive normalization proposed by Mifsud [9], the estimate is correct or off by one. Then, in the worst-case scenario, in each iteration of the main loop the functions `is_less` and `product` are called once, and `difference` is called twice. Thus, for the word size α ($\alpha \leq 64$), N-bit dividend, and $\frac{N}{2}$-bit divisor, the worst-case Intel cost of the long-division program is $\frac{1}{4} \cdot (7.18 + 2 \cdot 15.24 + 26.17) \cdot \left(\frac{N}{\alpha}\right)^2 + O\left(\frac{N}{\alpha}\right)$, which is $15.95 \cdot \left(\frac{N}{\alpha}\right)^2 + O\left(\frac{N}{\alpha}\right)$.

5 Integration with the Library

The class templates `cphstl::`$\mathbb{N}$ and `cphstl::`$\mathbb{Z}$ are designed to provide fixed-width integers of arbitrary length [5]. The number of bits (`b`) used in the representation is specified at compile time. Let us use `U` as a shorthand for the standard type **unsigned long long int** and let α = cphmpl::width<U>. The class template cphstl::$\mathbb{N}$ is written in two parts using constraint-based overloading.

(1) When $0 < \mathtt{b} \leq \alpha$, the classes cphstl::$\mathbb{N}$<b> are just thin wrappers around the standard unsigned integer types (Listing 5). If b is not a power of two, additional sanitation is needed to perform the calculations modulo $2^{\mathtt{b}}$.

Listing 5. An extract from the **private** part of cphstl::$\mathbb{N}$<b> for $0 < \mathtt{b} \leq \alpha$.

```
using uints = cphmpl::typelist<unsigned char, unsigned short int,
    ↪ unsigned int, unsigned long int, unsigned long long int>;
using W = uints::get<detail::first_wide_enough<uints, b>()>;

W data;
```

(2) When $\mathtt{b} > \alpha$, an integer is represented as a std::array<U, n>, where $\mathtt{n} = \lfloor(\mathtt{b} + \alpha - 1)/\alpha\rfloor$ (Listing 6). The long-division algorithm is in action first when the numbers are wider than α.

Listing 6. An extract from the **private** part of cphstl::$\mathbb{N}$<b> for $\mathtt{b} > \alpha$.

```
using U = unsigned long long int;

static constexpr std::size_t α = cphmpl::width<U>;
static constexpr std::size_t n = (b + α − 1) / α;

std::array<U, n> data;
```

In the first place, we needed the long-division program for the implementation of **operator**/ for the class templates cphstl::$\mathbb{N}$ and cphstl::$\mathbb{Z}$. Soon this program became an important test-bed for the whole library since the functions inside the library should work for these fixed-width integers themselves. In particular, in long division, it is now possible to choose the digits to be of type cphstl::$\mathbb{N}$<b> for arbitrary positive integer b that is a power of two.

To get some insight into the program transformations involved, when converting Warren's implementation [12, Chapter 9] into a generic library routine, look at the following code extracts taken from Hacker's Delight (Listing 7) and the CPH STL (Listing 8), respectively. When W is an alias of **unsigned int**, the assembler code generated by the compiler should be identical for both, but the latter works for any unsigned integer type and it can even be faster.

Listing 7. An extract from the function divlu in [12, Fig. 9-3].

```
// v is the divisor of type unsigned int

unsigned vn0, vn1;
int s;

s = nlz(v);            // 0 ≤ s ≤ 31
v = v << s;            // Normalize divisor.
vn1 = v >> 16;         // Break divisor into
vn0 = v & 0xFFFF;      // two 16-bit digits.
```

Listing 8. An extract from the function **divide_long_unsigned** in the CPH STL.

```
//W is a template parameter
// v is the divisor of type W

constexpr std::size_t w = cphmpl::width<W>;
constexpr W ooooffff = cphstl::some_trailing_ones<w / 2, W>;

std::size_t const s = cphstl::leading_zeros(v);
v = v << s;
W const vn1 = v >> (w / 2);
W const vn0 = v bitand ooooffff;
```

The functions `nlz` [12, Sect. 5-3] and `cphstl::leading_zeros` compute the number of leading 0 bits in the representation of a digit. In the CPH STL, this function is overloaded to work differently depending on the type of the argument. For the standard integer types, it can even call an intrinsic function that will be translated into a single hardware instruction[1]. There is also a **constexpr** form that computes the value at compile time if the argument is known at that time.

6 Benchmarking

In the following we will explain in more detail how we evaluated the quality of the division routines in the CPH STL.

6.1 Computing Environment

All the experiments were done on a personal computer that run Linux. The programs were written in C++ and the code was compiled using the `g++` compiler. The hardware and software specifications of the system were as follows.

Processor. Intel® Core™ i7-6600U CPU @ 2.6 GHz × 4
Word size. 64 bits
Operating system. Ubuntu 18.04.1 LTS
Linux kernel. 4.15.0-43-generic
Compiler. `g++` version 8.2.0—GNU project C++ compiler
Compiler options. `-O3 -Wall -Wextra -std=c++2a -fconcepts -DNDEBUG`
Profiler. `perf stat`—Performance analysis tool for Linux
Profiler options. `-e instructions`

6.2 Small Numbers

In our first experiment, we wanted to test how well the operations for the types `cphstl::`$\mathbb{N}$`<b>` perform. The benchmark was simple: For an array `x` of n digits and a digit `f`, execute the assignment `x[i] = x[i]` $\odot$ `f` for all `i` $\in \{0, 1, \ldots, n-1\}$. In the benchmark the number of instructions executed, divided by n, was measured for different digit types and operators $\odot \in \{+, -, *, /\}$.

[1] The Windows support of the bit tricks was programmed by Asger Bruun.

Table 2. The number of instructions executed (per operation) on an average when performing scalar-vector arithmetic for different types; in the implementation of `cphstl`::ℕ<128> the GNU extension **unsigned** `__int128` was not used.

Type	+	−	*	/
unsigned char	0.40	0.40	0.90	6.02
unsigned short int	0.46	0.46	0.46	4.53
unsigned int	0.89	0.89	2.39	4.52
unsigned long long int	1.77	1.77	5.77	4.53
unsigned `__int128`	5.53	5.53	7.04	19.55
`cphstl`::ℕ<8>	0.25	0.25	0.72	6.02
`cphstl`::ℕ<16>	0.46	0.46	0.46	7.02
`cphstl`::ℕ<24>	1.14	1.14	2.77	7.02
`cphstl`::ℕ<32>	0.89	0.89	2.39	7.02
`cphstl`::ℕ<48>	2.27	2.27	6.27	7.03
`cphstl`::ℕ<64>	1.77	1.77	5.77	7.03
`cphstl`::ℕ<128>	5.54	7.54	24.07	18.12
`cphstl`::ℕ<256>	18.06	27.06	161.9	49.37
`cphstl`::ℕ<512>	38.11	81.11	407.6	73.61
`cphstl`::ℕ<1024>	96.21	179.2	1396	129.3

The obtained instruction counts are reported in Table 2. Here the absolute values are not important due to loop overhead; one should look at the relative values instead. From these results, we make two conclusions:

(1) The `g++` compiler works well! When wrapping the standard types into a class, the abstraction penalty is surprisingly small. Division is somewhat slower due to the check if the divisor is zero, which is done to avoid undefined behaviour.

(2) On purpose, in the implementation of the type `cphstl`::ℕ<128>, we did not rely on the extension **unsigned** `__int128`. Instead the double-length arithmetic was implemented as explained in [12, Sect. 2-16, Sect. 8-2, and Fig. 9.3]. In particular, multiplication is slow compared to **unsigned** `__int128`.

6.3 Large Numbers

In our second experiment, we considered the special case where the dividend was an N-bit number and the divisor an $\frac{N}{2}$-bit number, and we run the long-division programs for different values of N and digit widths. In the benchmark, we generated two random numbers and measured the number of instructions executed. When reporting the results, we scaled the instruction counts using the scaling factor $\left(\frac{N}{64}\right)^2$. Here the rationale is that any program should be able to utilize the power of the words native in the underlying hardware.

Table 3. The performance of the long-division programs for different digit widths, measured in the number of instructions executed when processing two random numbers of N and $\frac{N}{2}$ bits. The values indicate the coefficient C in the formula $C \cdot \left(\frac{N}{64}\right)^2$.

Digit width	$N = 2^{12}$	$N = 2^{14}$	$N = 2^{16}$	$N = 2^{18}$	$N = 2^{20}$	$N = 2^{22}$
8	447.1	390.5	427.2	361.9	410.5	392.7
16	110.9	97.1	124.1	113.9	104.6	135.0
32	34.2	32.1	28.6	30.8	26.6	29.1
64	14.4	12.5	10.9	12.6	11.9	11.3
128	5.8	3.2	2.6	2.4	2.4	2.4
256	13.5	4.3	1.9	1.3	1.2	1.2
512	15.5	3.8	1.3	0.7	0.6	0.6
1024	36.1	18.5	14.7	13.9	13.7	13.6
16 [12, Fig. 9-3]	63.3	60.8	60.2	60.0	60.0	60.0

The test computer was a 64-bit machine. We fixed six measurement points: $N \in \{2^{12}, 2^{14}, 2^{16}, 2^{18}, 2^{20}, 2^{22}\}$. The obtained results are reported in Table 3.

We expected to get the best performance when the width of the digits matches the word size, but wider digits produced better results. As the instruction counts for Warren's program [12, Fig. 9-3] indicate, the choice 16 for the digit width is based on old technological assumptions. The figures for the width 16 also reveal that we have not followed the sources faithfully. (We have not used Knuth's optimization in Step D3 of Algorithm D [6, Sect. 4.3.1] and we have not fusioned the loops in `product` and `difference`.) The slowdown may also be due to abstraction overhead. Nonetheless, for larger digit widths, in these tests the library routine performed significantly better than Warren's program.

7 Final Remarks

The long-division program is based on the concept of digits. In a generic implementation, the type of digits is given as a template parameter so it will be fixed at compile time. If the lengths of the inputs are known beforehand, the code can be optimized to use the best possible digit width `b`. According to our experiments, for large values of N, the best performance is obtained for large values of `b`. The optimum depends on the operations supported by the hardware.

Seeing the program hierarchically, several levels of abstractions are visible:

User level. **operator**/ provided by the types `cphstl::ℕ<b>` and `cphstl::ℤ<b>` for any specific width `b`.

Implementation level. Operation $/(n, m)$ where n and m are the number of digits in the operands.

Efficiency-determining functions. Operations $<(n, n)$ (`is_less`), $-(n, n)$ (`difference`), and $*(n, 1)$ (`product`).

Intermediate level. Operation $+(\mathbf{2},\mathbf{1})$, which will not overflow, and operation $/(\mathbf{2},\mathbf{1})$, which just needs to work when the output is a single digit.

Overflowing primitives. Operations $+(\mathbf{1},\mathbf{1})$ and $-(\mathbf{1},\mathbf{1})$ can overflow or underflow by one bit, and operation $*(\mathbf{1},\mathbf{1})$ has a two-digit output.

Safe primitives. Operations $\odot(\mathbf{1},\mathbf{1})$, $\odot \in \{==,<,/,\%,>>,<<,\&,||\}$, must also be provided, but they cannot overflow.

Bit-manipulation primitives. Unary operations $\odot(\mathbf{1})$, $\odot \in \{\sim,\mathbf{nlz}\}$ are needed for division, but the library supports other bit tricks as well.

For many library functions, there exist several overloaded versions to get the best match with the instructions provided by the underlying hardware. Here the keywords are constraint-based function overloading and template specialization.

When dividing an N-bit number by an $\frac{N}{2}$-bit number, we determined a good digit width `b` experimentally. Instead of using this idea only once, one could use a divide-and-conquer approach where the digits are subdivided into subdigits and the method is applied recursively. As the results of our experiments suggest, for large values of N, this approach may have practical value. At least it would take the hacking to another level.

References

1. Brinch Hansen, P.: Multiple-length division revisited: a tour of the minefield. Report 9-1992, Syracuse University (1992). https://surface.syr.edu/eecs_techreports/166/
2. Collins, G.E., Musser, D.R.: Analysis of the Pope-Stein division algorithm. Inf. Process. Lett. **6**(5), 151–155 (1977). https://doi.org/10.1016/0020-0190(77)90012-6
3. Gamby, A.N., Katajainen, J.: Convex-hull algorithms: implementation, testing, and experimentation. Algorithms **11**(12) (2018). https://doi.org/10.3390/a11120195
4. Katajainen, J.: Pure compile-time functions and classes in the CPH MPL. CPH STL report 2017-2, Department of Computer Science, University of Copenhagen (2017). http://hjemmesider.diku.dk/~jyrki/Myris/Kat2017R.html
5. Katajainen, J.: Class templates `cphstl::`$\mathbb{N}$ and `cphstl::`$\mathbb{Z}$ for fixed-precision arithmetic. Work in progress (2017–2019)
6. Knuth, D.E.: Seminumerical Algorithms, The Art of Computer Programming, vol. 2, 3rd edn. Addison Wesley Longman, Boston (1998)
7. Krishnamurthy, E.V., Nandi, S.K.: On the normalization requirement of divisor in divide-and-correct methods. Commun. ACM **10**(12), 809–813 (1967). https://doi.org/10.1145/363848.363867
8. Lay-Yong, L.: On the Chinese origin of the galley method of arithmetical division. Br. J. Hist. Sci. **3**(1), 66–69 (1966). https://doi.org/10.1017/S0007087400000200
9. Mifsud, C.J.: A multiple-precision division algorithm. Commun. ACM **13**(11), 666–668 (1970). https://doi.org/10.1145/362790.362795
10. Mifsud, C.J., Bohlen, M.J.: Addendum to a multiple-precision division algorithm. Commun. ACM **16**(10), 628 (1973). https://doi.org/10.1145/362375.362400
11. Pope, D.A., Stein, M.L.: Multiple precision arithmetic. Commun. ACM **3**(12), 652–654 (1960). https://doi.org/10.1145/367487.367499
12. Warren Jr., H.S.: Hacker's Delight, 2nd edn. Pearson Education Inc., London (2013)
13. Yong, L.L.: The development of Hindu-Arabic and traditional Chinese arithmetic. Chin. Sci. **13**, 35–54 (1996). https://www.jstor.org/stable/43290379

Assessing Algorithm Parameter Importance Using Global Sensitivity Analysis

Alessio Greco[1], Salvatore Danilo Riccio[2,3], Jon Timmis[4], and Giuseppe Nicosia[3(✉)]

[1] Department of Mathematics and Computer Science, University of Catania, Catania, Italy
alessio.greco.it@gmail.com
[2] Politecnico di Milano, Milan, Italy
salvatore.riccio@mail.polimi.it
[3] Systems Biology Centre, University of Cambridge, Cambridge, UK
{sdr38,gn263}@cam.ac.uk
[4] Department of Electronic Engineering, University of York, York, UK
jon.timmis@york.ac.uk

Abstract. In general, biologically-inspired multi-objective optimization algorithms comprise several parameters which values have to be selected ahead of running the algorithm. In this paper we describe a global sensitivity analysis framework that enables a better understanding of the effects of parameters on algorithm performance. For this work, we tested NSGA-III and MOEA/D on multi-objective optimization testbeds, undertaking our proposed sensitivity analysis techniques on the relevant metrics, namely Generational Distance, Inverted Generational Distance, and Hypervolume. Experimental results show that both algorithms are most sensitive to the cardinality of the population. In all analyses, two clusters of parameter usually appear: (1) the population size (Pop) and (2) the Crossover Distribution Index, Crossover Probability, Mutation Distribution Index and Mutation Probability; where the first cluster, Pop, is the most important (sensitive) parameter with respect to the others. Choosing the correct population size for the tested algorithms has a significant impact on the solution accuracy and algorithm performance. It was already known how important the population of an evolutionary algorithm was, but it was not known its importance compared to the remaining parameters. The distance between the two clusters shows how crucial the size of the population is, compared to the other parameters. Detailed analysis clearly reveals a hierarchy of parameters: on the one hand the size of the population, on the other the remaining parameters that are always grouped together (in a single cluster) without a possible significant distinction. In fact, the other parameters all have the same importance, a secondary relevance for the performance of the algorithms, something which, to date, has not been observed in the evolutionary algorithm literature. The methodology designed in this paper can be adopted to evaluate the importance of the parameters of any algorithm.

I. Kotsireas et al. (Eds.): SEA[2] 2019, LNCS 11544, pp. 392–407, 2019.
https://doi.org/10.1007/978-3-030-34029-2_26

Keywords: NSGA-III · MOEA/D · Global Sensitivity Analysis · Elementary Effects · Sobol method · Variance Based Sensitivity Analysis

1 Introduction

Machine learning algorithms typically have many parameters, and the setting of these parameters is critical to the performance of the algorithm. Parameter setting is typically done by experts that are able, with their knowledge, to tune algorithm parameters. Ideally, this step should be automated, to save time and potentially improve performance. Trying each hyperparameter combination is usually unfeasible because of the high computational cost required to run this test. To address this problem, various approaches have been proposed. For example, Bergstra and Bengio demonstrated that random experiments are more efficient than grid search, suggesting that this property is due to the fact that hyperparameters are not equally important to tune [1]. Starting from this work, Wang et al. introduced an algorithm to deal with extremely high dimensions in Bayesian optimization problems [2]. Another possibility is to model algorithm performance as a sample from a Gaussian process [3]. Chapelle et al. exploited gradient descent to tackle the problem of tuning parameters in Support Vector Machines [4]. A similar approach that exploits a gradient-based method to choose hyperparameters for log-linear models is described in [5]. A first attempt to develop an automatic algorithm configuration framework for a large number of parameters can be found in [6]. Concerning Evolutionary Algorithms (EAs), a complete work describing parameter tuning techniques for EAs was proposed in [7]. Bartz-Beielstein et al. showed an algorithmic procedure to deal with parameter tuning, especially in the case of Genetic Algorithms (GAs) [8]. More recently, Wu et al. introduced an approach able to discover hidden variables that may affect search-based parameter tuning [9]. Conca et al. showed how to use sensitivity minimization to tune the parameter of a chosen algorithm [10].

One of the current main limitations is the lack of a tool that is able to determine dependencies between parameters. Furthermore, it is desirable to rank parameters with respect to how much each parameter affects the output. This paper presents a methodology based on Global Sensitivity Analysis to address the sensitivity of an algorithm with respect to a chosen subset of its parameters. In principle, our approach could be applied to any learning algorithm to produce a ranking of the chosen parameters.

The remainder of this paper is structured as follows. Section 2 introduces the main ideas and the general scheme to perform the analysis. Section 3 describes the tests done and shows some application scenarios. Section 4 concludes the work, giving some insights on the results obtained.

2 Experimental Setup

In this section we introduce the main concepts underlying the analysis. For work in this paper, we use Evolutionary Algorithms (EAs): population-based metaheuristic optimization algorithms [11] inspired by the mechanism of biological evolution, such as reproduction, mutation, recombination (or crossover), and selection [12]. The performance of any EA depends on parameters that have to be tuned.

We now briefly describe the algorithms, the metrics, and finally the Sensitivity Analysis techniques performed.

2.1 Non-dominated Sorting Genetic Algorithm III (NSGA-III)

Non-dominated Sorting Genetic Algorithm III [13] implements a niching strategy used to obtain a more well-distributed Pareto front. The algorithm is similar to its previous version, with a major change regarding the choice of points that have to be sent to the next generation (i.e. "least represented reference points" are chosen).

We choose "Polynomial Mutation" (PM) as the Mutation operator and "Simulated Binary Crossover" (SBX) as the Crossover operator. Each of them requires two parameters: (I) probability of applying the operator (from 0 to 1); (II) distribution index, a control parameter whose numeric value is inversely proportional to the amount of perturbation in the design variables. The last hyperparameter that has to be chosen is the population size (Pop). Thus, NSGA-III requires five hyperparameters: (1) Population, (2) PM Probability, (3) PM Distribution Index, (4) SBX Probability, (5) SBX Distribution Index. The Computational Complexity for a single generation is $max\{Pop^2\, log^{M-2}(Pop)\ ,\ Pop^2 M\}$ [13], where M is the number of objectives and *Pop* is the population size.

2.2 Multi-objective Evolutionary Algorithm Based on Decomposition (MOEA/D)

To experimentally show that our methodology has a more general applicability, we apply it to MOEA/D [14]. The decomposition strategy chosen in this work is Penalty-based Boundary Intersection Approach (PBI variant).

MOEA/D shares the same five hyperparameters previously described for NSGA-III. Since both algorithms have the same hyperparameters, we can exploit this feature to compare their results as it will be done in Sect. 3. The Computational Complexity for a single generation of MOEA/D is $M \cdot Pop \cdot T$ [15], where M is the number of objectives, *Pop* is the population size, and T is the number of weight vectors in the neighbourhood of each weight vector.

2.3 Metrics

To assess the performance of the algorithms, indices must be selected [16,17]. We choose three indices related to the Pareto front: Generational Distance (GD) [18,19], Inverted Generational Distance (IGD) [13], and Hypervolume (HV) [20].

2.4 Sensitivity Analysis Techniques

Sensitivity analysis is the study of how the uncertainty in the output of a mathematical model or system (numerical or otherwise) can be apportioned to different sources of uncertainty in its inputs [21–24].

In this work we experimentally show that sensitivity analysis can be used to study randomized many-objective learning algorithm, assessing how much its parameters affect the output uncertainty.

Among many sensitivity analysis techniques, we exploit two: Elementary Effects (EE) and Variance Based Sensitivity Analysis (VBSA).

Elementary Effects (EE). EE Test (EET, also known as Morris method) is a screening method with low computational cost. It produces two values: μ that represents the importance of an input factor on the model output, and σ that represents non-linear effects and interactions between variables.

Let r be the number of samples, m the number of inputs, and $Y(x)$ the output of the model linked to x. Then we can define two main design types:

- Trajectory Design: the $((i-1)m+1)$-th point represents the starting point of the i-th trajectory. Starting from that point, each input is varied by a certain amount d_j. Be the j-th input the one that is varying, and k the current point. Then, the Elemental Effect for the current sample i affected by the input variation j is

$$EE_{k,j} = \frac{Y(x_k) - Y(x_{k-1})}{d_j} \tag{1}$$

- Radial Design: The $((i-1)m+1)$-the point represents the i-th centre index c_i . From $((i-1)m+2)$ to $i \cdot m$, each input is varied starting from the first to the last one. As such, the Elementary Effect for the current sample i and the effect j, given the variation of the input d_j, is

$$EE_{i,j} = \frac{Y(x_{c_i+j}) - Y(x_{c_i})}{d_j} \tag{2}$$

The two main sampling strategy that we are going to use are Classical Morris [25] and Optimized strategy with Latin Hypercube Sampling (LHS) [26].

Variance Based Sensitivity Analysis (VBSA). We use Variance Based Sensitivity Analysis (VBSA, also known as Sobol method) [27] to compute two kinds of indices:

- First-Order Indices (Si), based on the variation of just one input
- Total-Order Indices (Sti), that take into account the variation of a single input, including all variance caused by its interactions with any other input variables.

Other methods can be chosen to estimate the variance, see [22] for further information. Sobol analysis is computationally expensive: it requires to explore a high number of points (i.e. $n(p+2)$, where n is the number of points, p is the number of inputs) to obtain good results.

2.5 The Problems

The following classical multi-objective problems have been used in the testbed: DTLZ1, DTLZ2, DTLZ3, DTLZ4 [28], CDTLZ2 [13], WFG6, WFG7 [29].

2.6 The Design Automation Framework

The route to find a good flow will be now discussed in detail. We exploit the existing Matlab platform for evolutionary multi-objective optimization PlatEMO [30]. We also use the workflows already available in the SAFE toolbox [31] as an inspiration to develop the combined EET/VBSA flow shown in Fig. 1.

The number of chosen variables is $Objectives + k - 1$, where k is 5 for DTLZ1, otherwise k is equal to 10. Table 1 summarizes the parameters values and ranges used for each algorithm.

Table 1. Algorithms and parameters chosen. Each parameter is associated with its range of feasible numeric values.

Algorithm	Population	SBX Probability	SBX Distr. Index	PM Probability	PM Distr. Index
NSGA-III	2–300	0–1	1–50	0–1	1–50
MOEA/D	2–300	0–1	1–50	0–1	1–50

The metrics used are the ones previously described, namely GD, IGD, and HV. Two *EET* are used for the analysis:

- Morris: sampling performed using the Classic Morris strategy, with *Trajectory* design as described in [25];
- Morris LHS: sampling performed using the LHS strategy, with *Radial* design. The distribution function used for the strategy is Uniform [26].

Latin hypercube sampling (LHS) is the sampling strategy used for VBSA.

3 Results

We now discuss the experiments undertaken. First, an appropriate value for $NRun$ has to be chosen. In this work we set $NRun = 20$ which is a common choice that can be found in literature [32–34]. In future works we will further exploit existing statistical technique packages like *spartan* [35] to understand how much a change in this parameter may affect sensitivity analysis overall performance. To ensure fairness in the evaluation of the metrics for low population cases that had a similarly low population in their resulting Pareto front, it was decided to consider in the computation of the metrics the Pareto front associated with all the points that were evaluated during the execution (instead of just the final one).

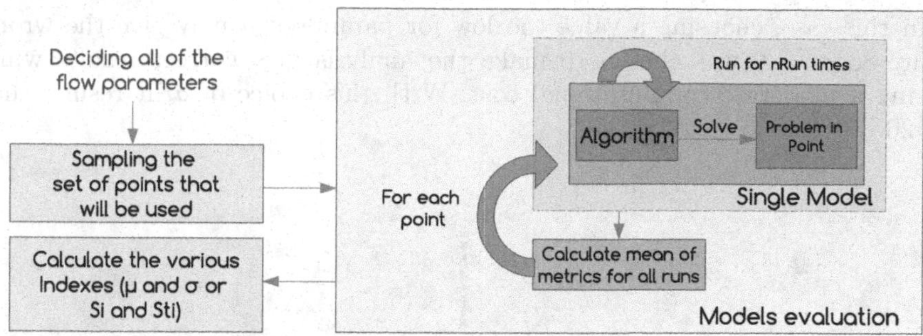

Fig. 1. Overall flow designed to perform sensitivity analysis.

Since both EET and VBSA provide a ranking of the parameters importance, we can use the normalized values of μ, σ, Si and Sti for every execution to make them comparable between the executions and between the metrics themselves.

Before running EET, we need to select the number of samples r. Thus, we first study whether the normalized values of the mean μ and variance σ converge. We solve DTLZ1 with 3 objectives using NSGA-III for increasing number of samples r, ranging from 2 to 200. We collect μ and σ values and we plot them together with a smoothed regression line obtained with LOESS, as shown in Fig. 2.

From the plots, it appears that the five hyperparameters can be divided into two clusters: the first one contains the Population (*Pop*), while the second one contains Simulated Binary Crossover Distribution Index (*SBX DI*), Simulated Binary Crossover Probability (*SBX P*), Polynomial Mutation Distribution Index (*PM DI*), and Polynomial Mutation Probability (*PM P*).

Analysing the second cluster only, see Fig. 3, a convergence can be seen in both μ and σ values as r increases. Although this property is more apparent for higher r values, it is still valid for lower r values. For this reason, $r = 5$ will be chosen for the following analyses.

Other than that, since there are 35 problems in the testbed, averaging the result of those analysis will still be *equal, in a certain sense,* to making the analysis with a higher r.

A similar analysis has to be applied to decide how many points n should be used for VBSA. Instead of trying the values of n directly, we define the variable a as the ratio $a = \frac{TotalPoints(VBSA)}{TotalPoints(EET)}$, where $TotalPoints(VBSA) = (M+2)n$ and $TotalPoints(EET) = (M+1)r$. Since $M = 5$, it results after few passages that $n = \frac{6}{7}ar$. To force this value to be an integer, we apply the ceiling function to the result.

The analysis was performed by varying a with values ranging from 1 to 20. The results obtained are similar to the plots in Fig. 2, with the same two clusters. Because of this similarity, we don't show the figures for sake of conciseness. Nonetheless, the second cluster is worth some attention and should be analysed, see Fig. 4.

In this case, choosing a value too low for parameter a may give the wrong result. So $a = 6$ was chosen to make the analysis give correct results while keeping a moderate computational cost. With this choice of a, it results that $n = 26$.

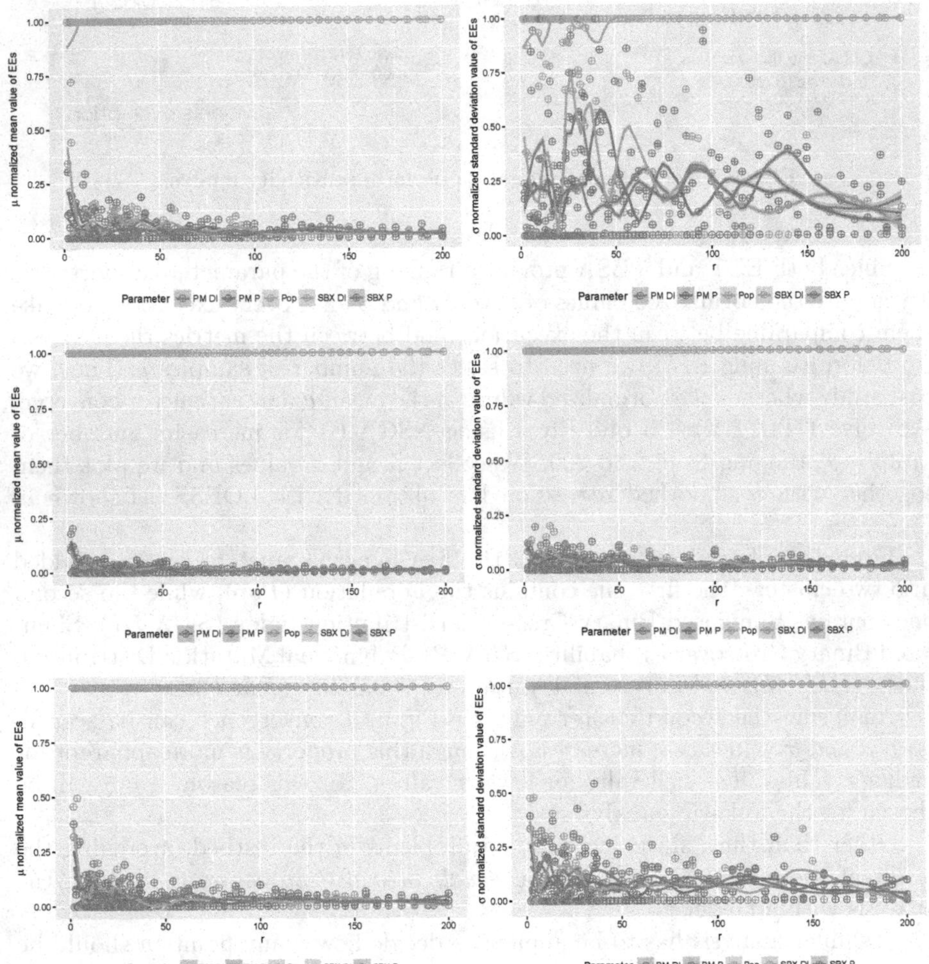

Fig. 2. These plots show the convergence of normalized metrics obtained as a function of the number of samples r. Data come from the solution of DTLZ1 with 3 objectives using NSGA-III. First row: μ, GD (left) and σ, GD (right). Second row: μ, IGD (left) and σ, IGD (right). Third row: μ, HV (left) and σ, HV (right). It can be seen that there are two clusters for every value of r.

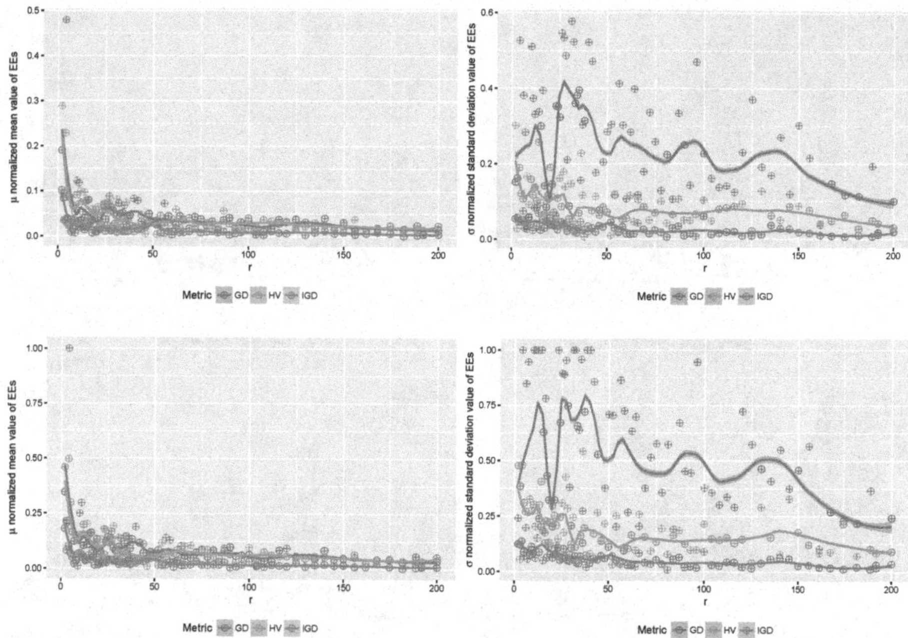

Fig. 3. These plots show metrics convergence dynamics of the second cluster as a function of number of samples r. First row: Mean μ (left) and Mean σ (right). Second row: Max μ (left) and Max σ (right). Although there is good convergence for high values of r, this may indeed lead to overfitting.

In the MOEA/D, the PBI method was chosen as the decomposition strategy ($kind = 1$ in PlatEMO). An error in the calculation of the HV using the PlatEMO library makes, in some cases, the result of the metric become 0, because all of the Pareto-Front are worse than the $RefPoint$. To avoid this issue, we choose $RefPoint$ over both the Reference Set and the Pareto Front.

We verify on a benchmark the time required to compute the metrics inserted in PlatEMO. In the first test we use NSGA-III on DTLZ1 with 3 objectives and 8 variables, see Fig. 5 for the results.

If the number of objectives is greater than 3, the PlatEMO library exploits Monte Carlo-like method for the estimation of HV. To assess how this change affects the time required, we tested NSGA-III on DTLZ1 with 5 objectives and 10 variables, see Fig. 6 for the results. We tried both values of $SampleNum$, namely 10^6 and 10^5, obtaining almost the same metrics value whereas with the second choice a sizeable speed up can be obtained. For this reason, we advise to change $SampleNum$ value to 10^5.

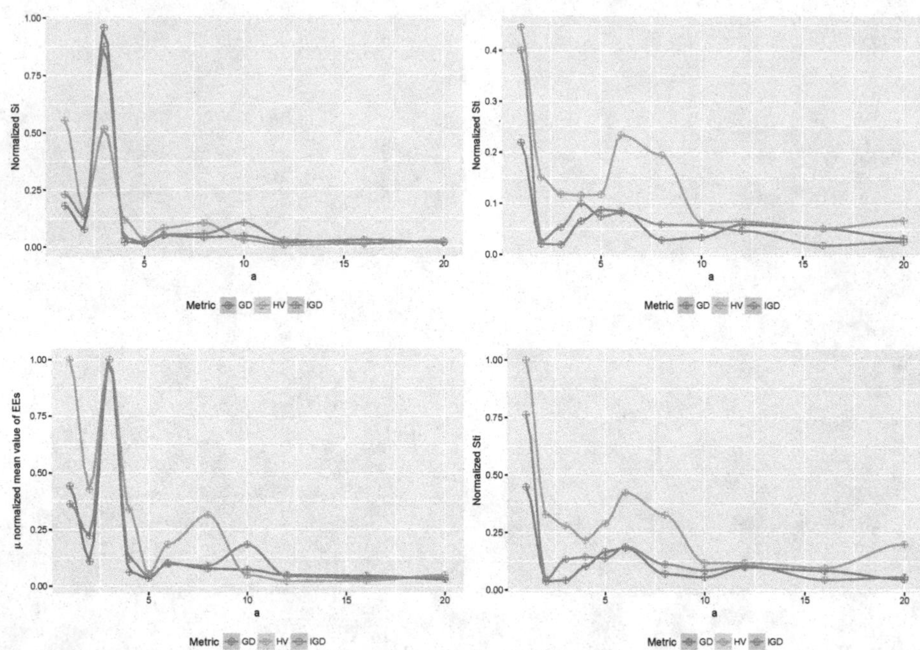

Fig. 4. These plots show metrics convergence dynamics of the second cluster as a function of number of samples a (ratio of total number of points considered in VSBA to EET). First row: Mean Si (left) and Mean Sti (right). Second row: Max Si (left) and Max Sti (right).

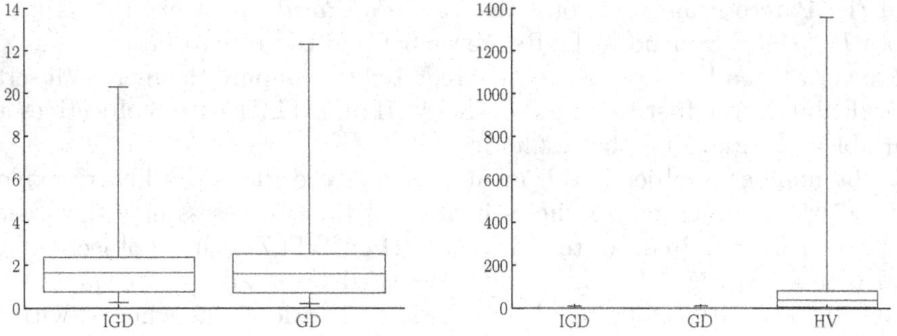

Fig. 5. Time (in milliseconds) required to compute the chosen metrics. Data come from the solution of DTLZ1 with 3 objectives and 8 variables using NSGA-III. Left: boxplot of the first two metrics, namely IGD and GD. Right: boxplot of all computed metrics; clearly, HV is much more time consuming than the others.

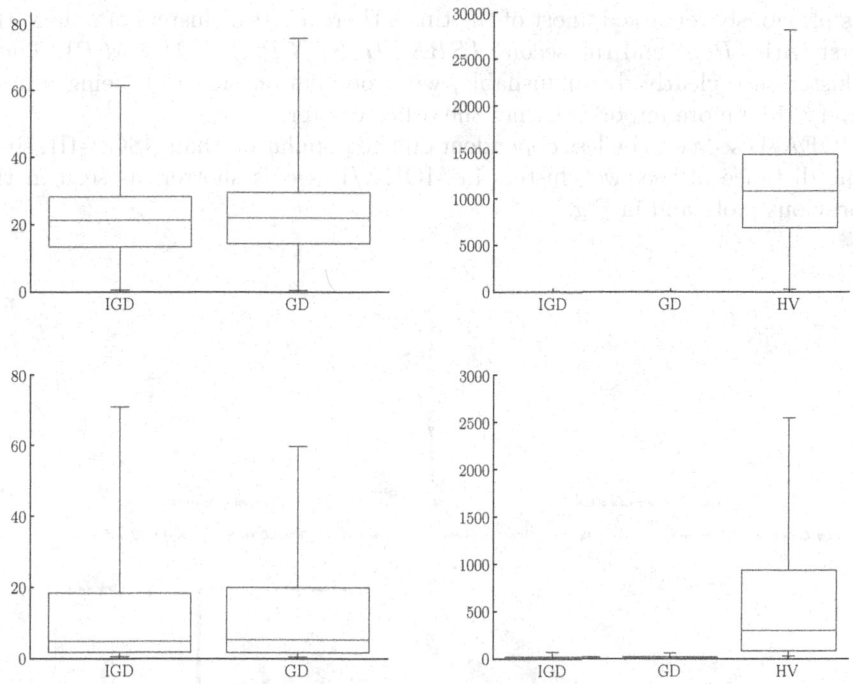

Fig. 6. Time (in milliseconds) required to compute the chosen metrics. Data come from the solution of DTLZ1 with 5 objectives and 10 variables using NSGA-III. $SampleNum = 10^6$ (first row), $SampleNum = 10^5$ (second row). Left: boxplot of the first two metrics, namely IGD and GD. Right: boxplot of all computed metrics; clearly, HV is much more time consuming than the others. It is also worth noting that using a lower value for *SampleNum* results in a dramatic decrease in the time required to perform indices computations, in particular for the HV index whose time decreases by a factor of 10.

Performing all the analyses, we can summarise as follows:

- the *Population* parameter usually has $\mu_{Normalized}$ and $\sigma_{Normalized}$ equal to 1. As such, it is the parameter with the highest effect, interaction with other parameters and with the highest non-linearity. The same behaviour can be found in VBSA where both normalized First and Total Order Indices are equal to 1;
- most of the *ranks of the parameters* are different between algorithms, especially in the case where a precise problem is analysed. Surprisingly enough in the EET, when the mean between the various cases is considered, they become more similar by either having the same ranking (with respect to μ) between the parameters or having a similar shape (see Fig. 7);

- as previously remarked, most of the times there are two clusters of points, the first with $\{Pop\}$ and the second $\{SBX\,DI, SBX\,P, PM\,DI, PM\,P\}$. These clusters are clearly distinguishable, with population size *Pop* being almost every time more important than the other cluster;
- MOEA/D seems to be less dependent on the population than NSGA-III, since the distance of the two clusters in MOEA/D seems shorter, as seen in the previous plots and in Fig. 7.

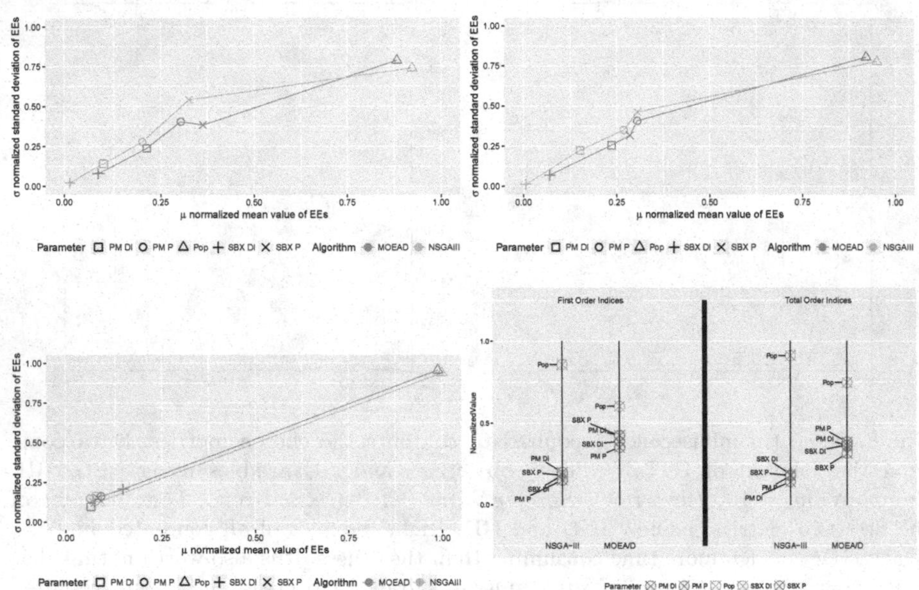

Fig. 7. In these plots we use metrics to compare the importance of the chosen hyperparameters for either NSGA-III and MOEA/D. First row: Morris LHS, NSGA-III Vs MOEA/D - HV (left), Morris LHS, NSGA-III Vs MOEA/D - Mean of all Metrics (right). Second row: Morris, NSGA-III Vs MOEA/D - GD (left), VBSA, NSGA-III Vs MOEA/D - Mean of all Metrics (right). It can be seen from the plots that Pop (Population size) is the most important hyperparameter with respect to the other ones.

Results obtained with Morris method show a different behaviour than the ones obtained with Morris LHS, as it can be seen in Figs. 8 and 9. Nonetheless, when the results of all multi-objective optimization problems are averaged, they start to exhibit similar behaviour (see Fig. 10).

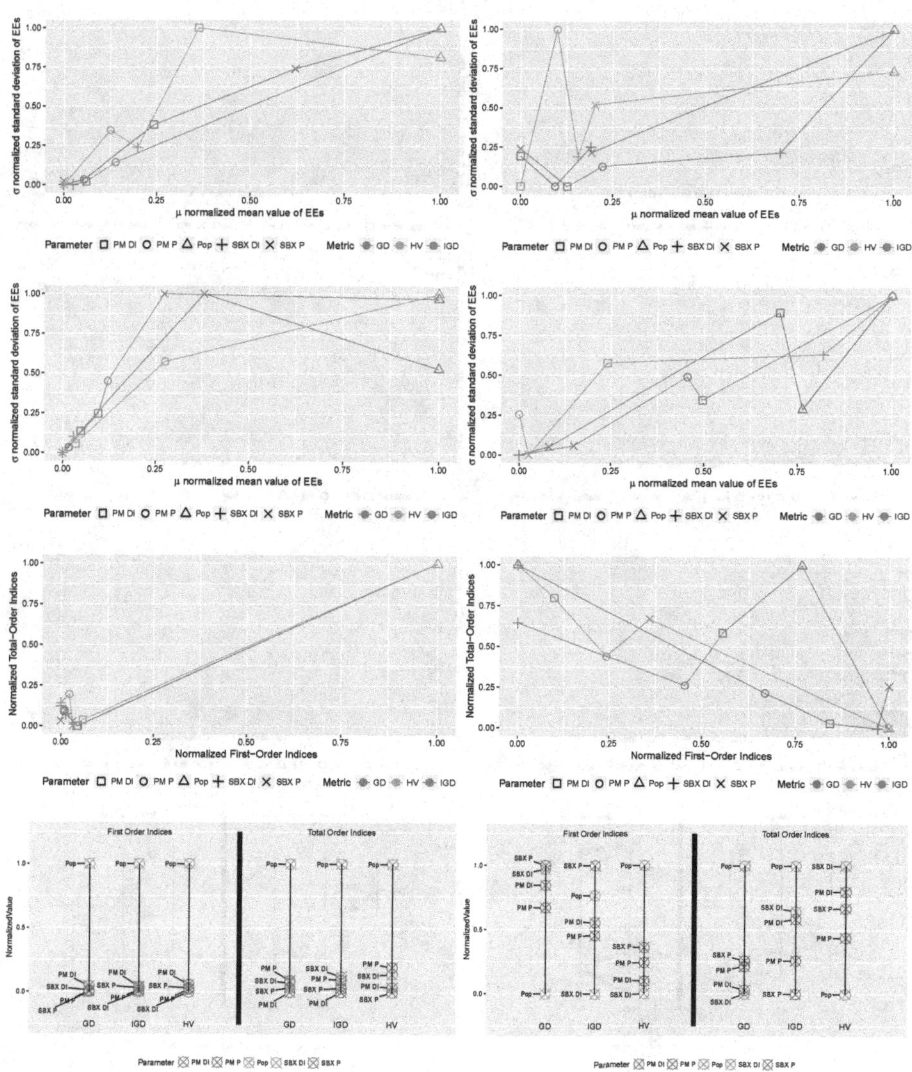

Fig. 8. These plots show results of the implemented Sensitivity Analysis techniques with data obtained from the solution of DTLZ1 with 8 Variables and 3 Objectives. First row: Morris, NSGA-III (left) and MOEA/D (right). Second row: Morris LHS, NSGA-III (left) and MOEA/D (right). Third row: VBSA, NSGA-III (left) and MOEA/D (right). Fourth row: VBSA, NSGA-III (left) and MOEA/D (right).

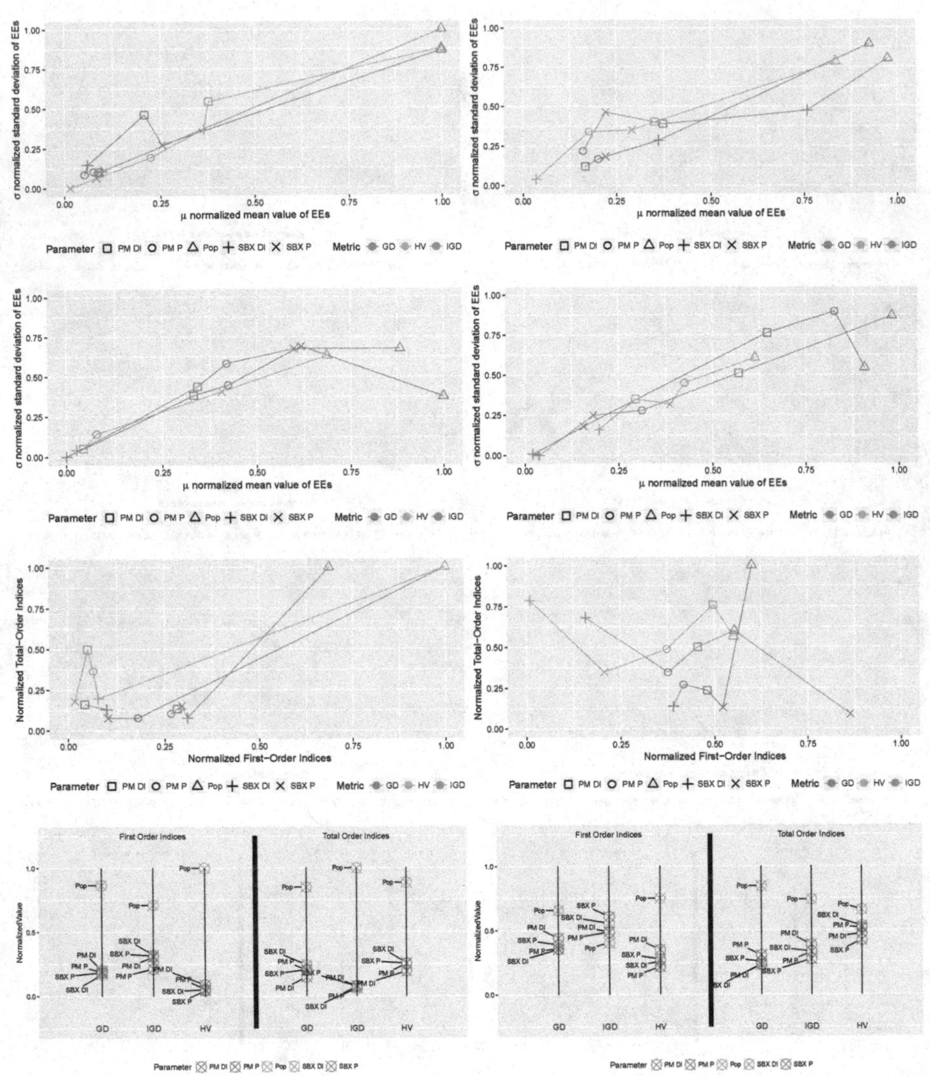

Fig. 9. These plots show Sensitivity Analysis techniques results for different Evolutionary Algorithms, whereas the sampling strategy is the same for each row. We show the mean results for all considered multi-objective optimization problems. First row: Morris, NSGA-III (left) and MOEA/D (right). Second row: Morris LHS, NSGA-III (left) and MOEA/D (right). Third row: VBSA, NSGA-III (left) and MOEA/D (right). Fourth row: VBSA, NSGA-III (left) and MOEA/D (right).

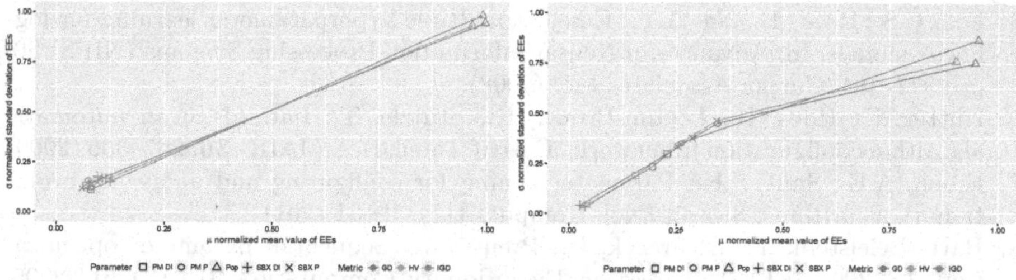

Fig. 10. These plots show Sensitivity Analysis techniques results for different sampling strategies. We show the mean results for both EAs (namely NSGA-III and MOEA/D) and for all considered multi-objective optimization problems. Morris (left) and Morris LHS (right).

4 Conclusions

In this study we demonstrated how to perform Sensitivity Analysis techniques on Evolutionary Algorithms, namely on NSGA-III and MOEA/D. Results indicated that usually two clusters of parameters appear: one of them takes the position of being the parameter with both higher μ, σ, First and Total Order Indices, with the second precise conformation varying a lot; the distance of those two clusters seems to be shorter in the case of MOEA/D. Only in the EET, though, some of the comparative plots between the two algorithms do show some similarity.

The obtained results do not give a unique ranking for each hyperparameter, but they indeed show that an appropriate choice of the population size has a great deal of effect on the performance of the algorithm.

First we must deal with the setting of the primary parameters (those that belong to the first class, i.e. the most sensitive parameters) and afterwards to the setting of the parameters of the remaining classes. For the algorithms dealt with in this study, the primary class includes the size of the population, the most sensitive parameter for the overall performances; immediately afterwards we must consider the remaining parameters that are certainly important but less sensitive than the population size.

References

1. Bergstra, J., Bengio, Y.: Random search for hyper-parameter optimization. J. Mach. Learn. Res. **13**, 281–305 (2012)
2. Wang, Z., Hutter, F., Zoghi, M., Matheson, D., de Freitas, N.: Bayesian optimization in a billion dimensions via random embeddings. J. Artif. Intell. Res. **55**, 361–387 (2016)
3. Snoek, J., Larochelle, H., Adams, R.P.: Practical bayesian optimization of machine learning algorithms. In: Advances in Neural Information Processing Systems, vol. 4 (2012)
4. Chapelle, O., Vapnik, V., Bousquet, O., Mukherjee, S.: Choosing multiple parameters for support vector machines. Mach. Learn. **46**, 131–159 (2001)

5. Foo, C.S., Do, C.B., Ng, A.Y.: Efficient multiple hyperparameter learning for log-linear models. In: Advances in Neural Information Processing Systems (NIPS) 20, pp. 377–384. Curran Associates Inc. (2008)
6. Hutter, F., Hoos, H., Leyton-Brown, K., Stützle, T.: ParamILS: an automatic algorithm configuration framework. J. Artif. Intell. Res. (JAIR) **36**, 267–306 (2009)
7. Eiben, A.E., Smit, S.K.: Parameter tuning for configuring and analyzing evolutionary algorithms. Swarm Evol. Comput. **1**(1), 19–31 (2011)
8. Bartz-Beielstein, T., Lasarczyk, C., Preuss, M.: Sequential parameter optimization. In: 2005 IEEE Congress on Evolutionary Computation, IEEE CEC 2005. Proceedings, vol. 1, pp. 773–780 (2005)
9. Wu, F., Weimer, W., Harman, M., Jia, Y., Krinke, J.: Deep parameter optimisation. In: GECCO 2015: Proceedings of the 2015 Annual Conference on Genetic and Evolutionary Computation, pp. 1375–1382. ACM, Madrid (2015)
10. Conca, P., Stracquadanio, G., Nicosia, G.: Automatic tuning of algorithms through sensitivity minimization. In: Pardalos, P., Pavone, M., Farinella, G.M., Cutello, V. (eds.) MOD 2015. LNCS, vol. 9432, pp. 14–25. Springer, Cham (2015). https://doi.org/10.1007/978-3-319-27926-8_2
11. Al-Salami, N.M.A.: Evolutionary algorithm definition. Am. J. Eng. Appl. Sci. **2**(6), 789–795 (2009)
12. Vikhar, P.A.: Evolutionary algorithms: a critical review and its future prospects. In: 2016 International Conference on Global Trends in Signal Processing, Information Computing and Communication (ICGTSPICC), pp. 261–265. Jalgaon (2016)
13. Deb, K., Jain, H.: An evolutionary many-objective optimization algorithm using reference-point-based nondominated sorting approach part i: solving problems with box constraints. IEEE Trans. Evol. Comput. **18**(4), 577–601 (2014)
14. Zhang, Q., Li, H.: MOEA/D: a multiobjective evolutionary algorithm based on decomposition. IEEE Trans. Evol. Comput. J. **11**(6), 712–731 (2007)
15. Yuen, T.J., Ramli, R.: Comparison of computational efficency of MOEA/D and NSGA-II for passive vehicle suspension optimization. In: 24th European Conference on Modelling and Simulation, Kuala Lumpur, Malaysia (2010)
16. Okabe, T., Jin, Y., Sendhoff, B.: A critical survey of performance indices for multi-objective optimisation. In: The 2003 Congress on Evolutionary Computation, vol. 4, pp. 2262–2269. IEEE Press (2003)
17. Zitzler, E., Thiele, L., Laumanns, M., Fonseca, C.M., Grunert da Fonseca, V.: Performance assessment of multiobjective optimizers: an analysis and review. IEEE Trans. Evol. Comput. **7**(2), 117–132 (2003)
18. Van Veldhuizen, D. A., Lamont, G. B.: Evolutionary computation and convergence to a pareto front. In: Late Breaking Papers on the Genetic Programmming 1998 Conference, pp. 221–228 (1998)
19. Van Veldhuizen, D.A.: Multiobjective Evolutionary Algorithms: Classifications, Analyses, and New Innovations. Faculty of the Graduate School of Engineering of the Air Force Institute of Technology, Air University (1999)
20. Zitzler, E., Thiele, L.: Multiobjective optimization using evolutionary algorithms—a comparative case study. In: Eiben, A.E., Bäck, T., Schoenauer, M., Schwefel, H.-P. (eds.) PPSN 1998. LNCS, vol. 1498, pp. 292–301. Springer, Heidelberg (1998). https://doi.org/10.1007/BFb0056872
21. Saltelli, A.: Sensitivity analysis for importance assessment. Risk Anal. **22**(3), 1–12 (2002)
22. Saltelli, A., et al.: Global Sensitivity Analysis: The Primer. Wiley, Chichester (2008)

23. Carapezza, G., et al.: Efficient behavior of photosynthetic organelles via pareto optimality, identifiability and sensitivity analysis. ACS Synthetic Biol. J. **2**(5), 274–288 (2013)
24. Costanza, J., Carapezza, G., Angione, C., Liò, P., Nicosia, G.: Multi-objective optimisation, sensitivity and robustness analysis in FBA modelling. In: Gilbert, D., Heiner, M. (eds.) CMSB 2012. LNCS, pp. 127–147. Springer, Heidelberg (2012). https://doi.org/10.1007/978-3-642-33636-2_9
25. Morris, M.D.: Factorial sampling plans for preliminary computational experiments. Technometrics **33**(2), 161–174 (1991)
26. Campolongo, F., Cariboni, J., Saltelli, A.: An effective screening design for sensitivity analysis of large models. Environ. Modell. Software **22**(10), 1509–1518 (2007)
27. Sobol, I.M.: Sensitivity estimates for nonlinear mathematical models. Math. Modell. Comput. Exp. **1**(4), 407–414 (1993)
28. Deb, K., Thiele, L., Laumanns, M., Zitzler, E.: Scalable multi-objective optimization test problems. In: The 2002 Congress on Evolutionary Computation, pp. 825–830. IEEE Press (2002)
29. Huband, S., Hingston, P., Barone, L., While, L.: A review of multiobjective test problems and a scalable test problem toolkit. IEEE Trans. Evol. Comput. **10**(5), 477–506 (2006)
30. Tian, Y., Ran Cheng, R., Zhang, X., Jin, Y.: PlatEMO: A MATLAB platform for evolutionary multi-objective optimization. IEEE Comput. Intell. Mag. **12**(4), 73–87 (2017)
31. Pianosi, F., Sarrazin, F., Wagener, T.: A matlab toolbox for global sensitivity analysis. Environ. Modell. Software **70**, 80–85 (2015)
32. Sun, Y., Xue, B., Zhang, M., Yen, G.G.: A new two-stage evolutionary algorithm for many-objective optimization. IEEE Trans. Evol. Comput. (2018)
33. Nicosia, G., Cutello, V.: The clonal selection principle for in silico and in vitro computing. In: de Castro, L.N., Von Zuben, F.J. (eds.) Recent Developments in Biologically Inspired Computing (2004)
34. Narzisi, G., Nicosia, G., Stracquadanio, G.: Robust bio-active peptide prediction using multi-objective optimization. In: The I International Conference on Advances in Bioinformatics and Applications - BIOINFO 2010, 7–13 March, 2010, Cancun, Mexico, pp. 44–50. IEEE Press (2010)
35. Alden, K., Read, M., Timmis, J., Andrews, P.S., Veiga-Fernandes, H., Coles, M.: Spartan: a comprehensive tool for understanding uncertainty in simulations of biological systems. PLOS Comput. Biol. **9**(2), e1002916 (2013)

A Machine Learning Framework for Volume Prediction

Umutcan Önal[1(✉)] and Zafeirakis Zafeirakopoulos[2]

[1] Gebze Technical University, Gebze, Turkey
umutcanonal@gmail.com
[2] Institute of Information Technologies, Gebze Technical University, Gebze, Turkey

Abstract. Computing the exact volume of a polytope is a #P-hard problem, which makes the computation for high dimensional polytopes computationally expensive. Due to this cost of computation, randomized approximation algorithms is an acceptable solution in practical applications. On the other hand, machine learning techniques, such as neural networks, saw a lot of success in recent years. We propose machine learning approaches to volume prediction and volume comparison. We employ various network architectures such as feed-forward networks, autoencoders and end-to-end networks. We develop different types of models with these architectures that emphasize different parts of the problem, such as representation of polytopes, volume comparison between polytopes and volume prediction. Our results have varying rate of success depending on model and experimentation parameters. This work intends to start the discussion about applying machine learning techniques to computationally hard geometric problems.

Keywords: Machine learning · Autoencoders · Neural networks · Polytope · Volume

1 Introduction

Recent success of machine learning and deep learning brought a lot of new applications to various fields, where machine learning was not used until recently. Computational geometry and algebra is one of the fields where machine learning is not applied sufficiently.

In this work, we employ random forests and neural networks for a geometric problem, namely volume computation. Among neural networks, autoencoders play a special role in the proposed framework. We focus on two problems related to polytope volume computation, namely volume comparison and volume prediction. The first problem is a binary classification problem in machine learning terminology. We consider pairs of polytopes and label them according to whether the first or the second polytope have larger volume. The second problem is a typical regression problem, where the model tries to predict a continuous value (namely the polytope's volume) for given input.

I. Kotsireas et al. (Eds.): SEA[2] 2019, LNCS 11544, pp. 408–423, 2019.
https://doi.org/10.1007/978-3-030-34029-2_27

Random Forest. Random forests are ensembles of multiple tree predictors which are trained with randomly sampled independent features from the original feature set. Every predictor generates an output based on their own feature set and the output of the forest is decided by majority voting from predictor outputs. If the tree is for a regression problem instead of a classification problem, the output will be the mean of the tree outputs. It is a flexible algorithm that has a good potential to generalize the data. However, it also has some caveats such as depth of the trees. One of the major problems with decision trees is that they tend to overfit if the depth is not limited to some degree. Random forests can overcome this problem to some extend, however it is still one of the important parameters to decide during training.

Neural Networks. Neural Networks or Artificial Neural Networks is a machine learning technique that uses a network of neuron-like units to learn a model. After Hinton et al. [12] introduced the back-propagation method, it became possible to construct multiple layered networks, which are called Deep Neural Networks . A basic form of these networks is called multilayer perceptron or Feed-forward Neural Network. Layers are densely connected in this type of networks which means every perceptron is connected to all of the perceptrons in the previous and the next layer in the network. A big success came for neural networks when Krizhevsky et al. [10] improved considerably the accuracy on ImageNet database with Convolutional Neural Network (CNN) in 2012 . This type of network skips some of the connections in order to simulate the convolution operation extracting features from images. The success of this model attracted a lot of attention and new convolutional networks such as GoogleNet [15], VGGNet [13], ResNet [6] followed. The same year, Hinton et al. [7] presented another state-of-the-art Deep Neural Network application for speech recognition.

Recurrent neural networks (RNN) [8], another important type of neural networks models, rely on using a feedback mechanism in order to remember previous states. This mechanism allows the network to sequentially process data and learn spatial or time based information. Even though it is used mostly for sequential problems, such as speech recognition or music composition, it can be used for other types of tasks to process data sequentially.

At first glance an RNN does not look like a Deep Neural Network. But when unfolded its depth becomes visible. This is due to the feedback mechanism that gives the previous output as an input as well.

Recurrent Neural Networks are powerful tools in theory, however, they are not working as well in practice. Back-Propagation Through Time algorithms apply the idea of back-propagation of errors in feed forward networks. Errors are back-propagated in time instead of between the layers of the RNN. This propagated error tends to blow up or vanish if the sequence is too long, i.e., the network is too deep. The Long-Short Term Memory (LSTM) architecture is proposed as remedy for this problem [8]. The LSTM architecture helps to prevent losing information due to long time lag, thus allows such networks to work with long sequences.

Autoencoders. Autoencoders are a type of neural networks which is used to extract features from data in a unsupervised fashion. A particularly powerful aspect of neural networks is feature extraction without any human supervision. This unsupervised feature extraction enables inference capacity of neural networks given enough data, although it obscures the exact decision making process. This idea is presented by Dai and Le [3] for a sentiment analysis model.

A network can be considered as an autoencoder when the input and the output of the network is the same. The network only learns the patterns and relations in the data itself and the weights or the intermediate output of this network can be used as a set of extracted features.

An encoder-decoder model takes an input of variable length and encodes it in the encoder layer into a fixed-size vector. Later, this fixed-size vector is decoded to output which has variable length again. Sutskever et al. [14] and Cho et al. [2] proposed this model (with slightly different implementations) for problems which require to convert a sequence to another sequence, e.g., machine translation or speech recognition.

Sequential problems, usually do not have fixed-length input, while most neural networks require fixed-length input. Since the representation of a polytope is a sequence, encoder-decoder based models are applicable for the problem we study. If the input and output of an encoder-decoder model is the same, it is called autoencoder and is one of the methods for feature extraction.

1.1 Polytopes

Polytopes have two different and equivalent representations. They can be represented as the intersection of a finite set of half-spaces (H-polytope) or as the convex hull of a finite set of vertices (V-polytope). A detailed description of the related theory is out of the scope of this paper, for an excellent presentation see [17].

In this work we consider V-polytopes, namely polytopes given by their vertices. Given the vertices, one can represent the polytope as a list of lists. The length of this list is the number of vertices and the length of each element is the (ambient) dimension of the polytope. The fact that both dimension and number of vertices varies, is one of the main obstacles in using machine learning for polytope related problems. As discussed above, a common practice is the use of autoencoders to extract a feature set that represents the data well enough (or even better) in order to improve the success of the machine learning model. We will present different approaches on how to encode polytopes (and their vertices).

One of the fundamental problems in computational geometry is to compute the volume of a polytope. Volume computation has many applications ranging from financial models [1] to systems biology [9]. Dyer et al. [4] proved that exact volume computation is #P-hard. This lead to the development of randomized and approximation algorithms in order to compute the volume of high dimensional polytopes. The current state-of-the-art is presented in Emiris et al. [5].

We choose volume computation as the problem of interest in this work because:

– an approximation is the best we expect in polynomial time,
– it has natural prediction (regression) and comparison (classification) versions
– there are reliable means to produce data.

1.2 Our Goal and Structure of the Paper

The goal of this work is is to propose a framework allowing the use of machine learning for geometric and algebraic problems. Among the abundance of geometric and algebraic problems, that can be considered for such applications, we choose to focus on polytopes and the computation of their volume. The reason is that this problem already encounters some of the fundamental obstacles in applying machine learning to geometric and algebraic problems, namely that the dimensions of the input are not fixed.

The rest of this paper is structure as follows: In Sect. 2 we present the proposed models and in Sect. 3 we describe the datasets we used. Section 4 contains an analysis of the experiments and the experimental comparison of the proposed models. Finally, in Sect. 5 we conclude the paper.

2 Description of the Models

In this work we consider V-polytopes, namely polytopes given by their vertices. Given the vertices, one can represent the polytope as a list of lists. But a list of lists is ordered, while the vertices of the polytope do not have a natural order. Even though it is possible to give an arbitrary order, since this will not be natural, it may cause failure in our models. For the rest of this work, we consider the polytope given by the list of its vertices. We will present different approaches on how to encode polytopes (and their vertices).

2.1 Encoding a Polytope

An essential difficulty of the encoding of a polytope is the variable length of input both in terms of how many vertices a polytope has and what is the dimension of the polytope.

We will indicate whether a model assumes the dimension is fixed or variable, by the first letter of the model name, i.e., `F` or `V` respectively.

Fixed Dimension. As we saw in Sect. 1, traditionally the dimensions of the input tensor are fixed, while with RNNs we can have one varying dimension if we think that dimension as timesteps. In order to exploit existing models, we have to restrict the generality of the problem.

If we fix an upper bound on the dimension of the polytope, then we can pad with zeroes the vertices of any polytope of dimension smaller than the bound. This allows us to use standard RNN/LSTM solutions and build end-to-end models with ease.

Variable Dimension - Autoencoder for Vertices. One way to achive encoding without fixing the dimension is by using an autoencoder for vertices. This autoencoder works in a way similar to word embeddings in NLP. It is used to encode vertices into a fixed size representation, in order to fix the vertex dimension. This way we don't lose dimension information. However the lack of context, i.e., that a set of vertices belongs to a polytope, may hinder the performance. After encoding, we use the encoded vertices to construct polytopes again. We use this approach as a preporcessing step, before training an end-to-end model for fixed dimensional polytopes.

Polytopes with Fixed Dimensions and Number of Vertices. Inspired by our fixed dimensional model, we also train a model with both fixed dimension and number of vertices. We achieve this by padding data with zeros up to some maximum number of vertices or dimension. One of the important aspects of training with this data is that it removes the need of sequential processing with RNNs. We only used feed forward networks with 2 inputs for different polytopes. This approach is much simpler in many ways compared to the previously mentioned models, but it lacks generality and can only be used for polytopes within its predefined limits.

Flatten. Another way to use RNN/LSTM to encode a polytope is to flatten its list of vertices. One of the major problems with this approach is the loss of dimension information. We use this approach in modular models.

2.2 The Problems

In this work we consider two problems related to volumes of polytopes. The first is the regression problem of approximating the volume of a given V-polytope. The second is the classification problem of deciding if polytope A has larger volume than polytope B, given two V-polytopes A and B.

Volume Prediction. Volume prediction is the main problem we are trying to solve in this work. This is a regression problem. In the first set of models we predict the volume of a given polytope, while in the second set of models we predict simultaneously the volumes of a pair of polytopes.

Volume Comparison. In order to define a classification problem related to the regression problem of volume prediction, we study the volume comparison problem. The goal is to observe the performance of the models in comparing volumes. Given two polytopes as input, the model compares the volumes and labels the pair of polytopes as 1 if the volume of the first polytope is larger and 0 otherwise. This is a binary classification problem.

Coupling Prediction with Comparison. We also want observe the effect of comparison and prediction on each other, by combining the two problems in one model. This was the original motivation for studying the classification problem. Generally, the classification problem is easier, and the question is whether predicting and comparing simultaneously, improves the performance of one or both tasks. Having a model performing multiple tasks may be a useful tool.

2.3 Modular Vs End-to-End

Concerning the training of the models, in general, there are two approaches we employ. The first is to train an autoencoder for polytopes and then separately train a network for prediction/comparison. We will mark such solutions by `APC_`. The second approach is to perform end-to-end learning and we will use `E2E_` to mark models following this approach.

An important effect of using modular encoder and regressor models is that it may work with different machine learning algorithms like Random Forest as well. As we will see in Sect. 4, those algorithms may produce better results. However, it is expected to have overall better results with `E2E_` neural network models.

2.4 The Models

In what follows we present different solutions to these problems and for clarity we will name the models in the following way:

- the first part declares if the dimension is fixed or variable by `F` or `V` respectively,
- the second part is `F` if input is flatened, `L` if a list of lists, and `P` if completely padded,
- the third part declares the training scheme by `E2E` or `APC` and
- the fourth part is `C` for volume comparison, `P` volume prediction or `PC` for a coupled model.

For every model, we describe the coupled version `*_*_***_PC` and in the corresponding figures we give also the outline of the comparison only (`*_*_***_C`) and prediction only (`*_*_***_P`) architectures.

In Fig. 1 we see model `F_L_E2E_PC`. The input of the model is a fixed dimensional polytope. The dimensions are fixed by padding the input vectors to the highest dimension in the data. However, it still employs an encoder to generate a fixed representation of a polytope because the number of vertices still variable.

The `F_P_E2E_PC` model (see Fig. 2) requires a fixed number of vertices as well as fixed dimensions. Thus, it only uses dense layers and requires no RNNs to process the input. After the input polytopes are encoded in a fixed size representation, the representation vectors of the two input polytopes are concatenated and fed to the next layer (Fig. 3).

The `V_L_E2E_PC` model (see Fig. 4) differentiates from previous models by keeping both dimension and number of vertices variable. We use two encoders,

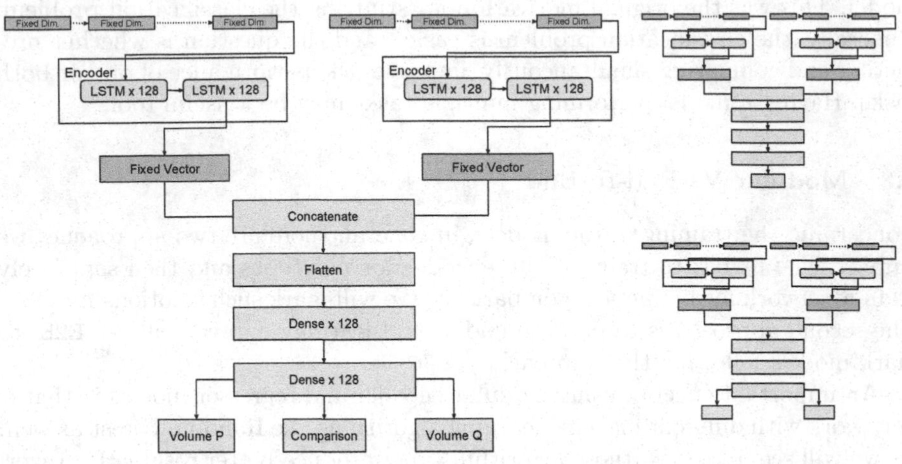

Fig. 1. Fixed dimensional end-to-end models

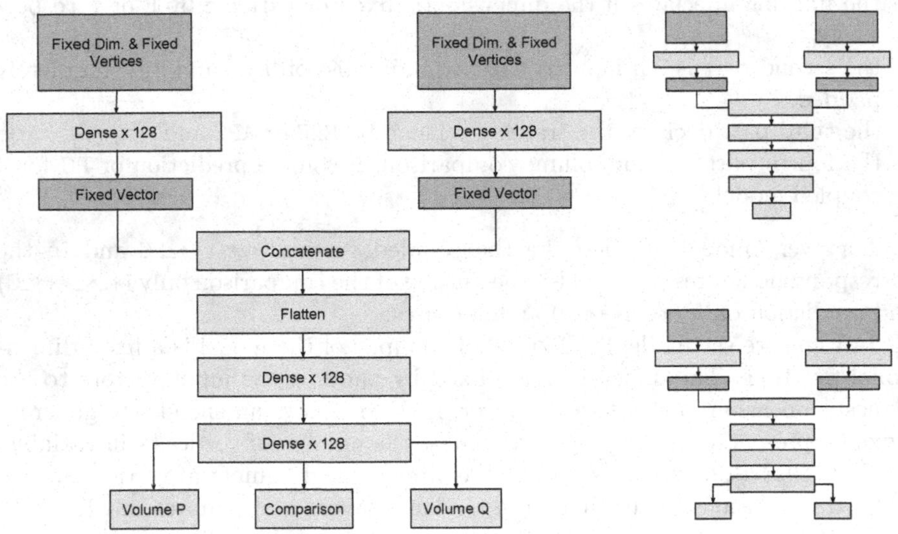

Fig. 2. Padded model

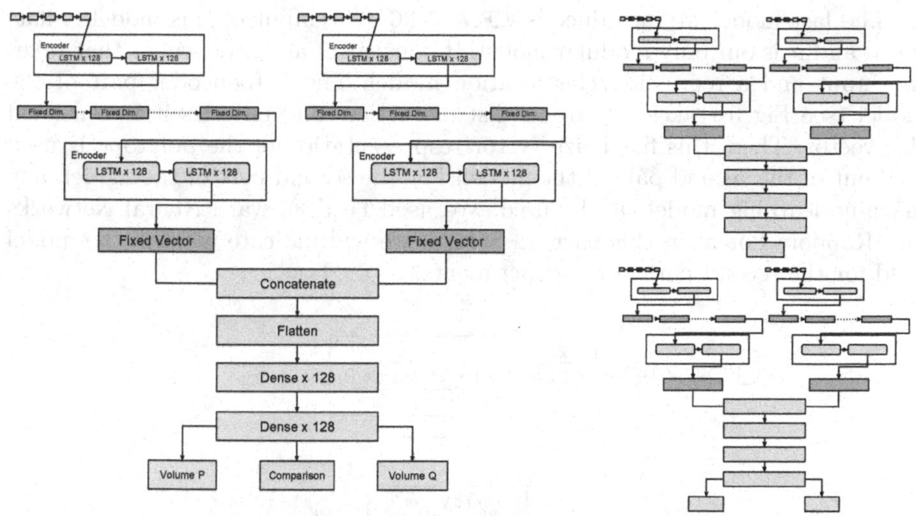

Fig. 3. End-to-end model for both volume predictions and comparison.

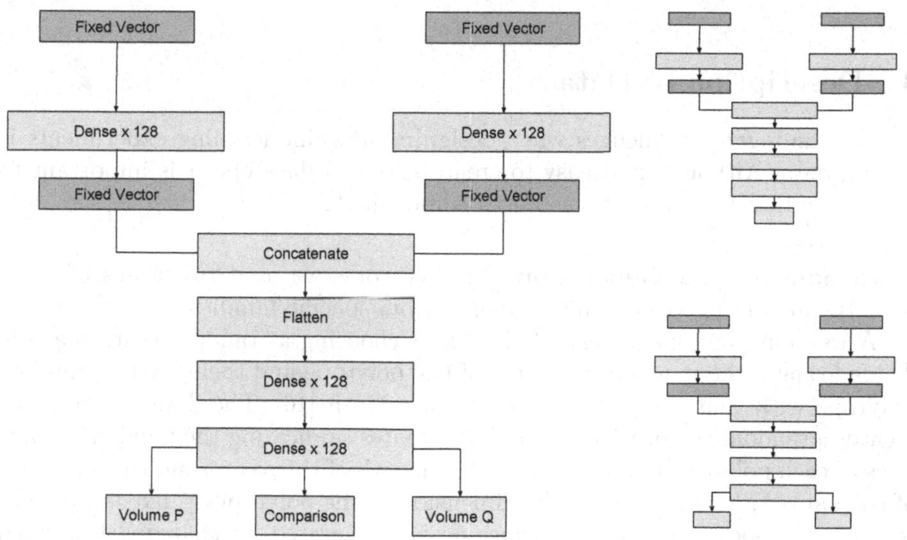

Fig. 4. Feed-forward Neural Network for modular solution.

one for vertices and one for polytopes, in order to achieve this. Apart from having one more encoder for input, this model is similar to F_L_E2E_PC.

The last model we introduce is V_F_APC_PC. The input of this model is flattened and it is our only modular model. It consists of an autoencoder that takes the input and a regression/classification model. The autoencoder part of the model (see Fig. 5) takes a flattened list of vertices and encodes it into a fixed size vector. Then, this fixed size vector respresentation of the polytope is used as input of the second part of the model. For the second part of the model, any machine learning model can be used. We used Feed-Forward Neural Networks and Random Forest in this part. In Sect. 4, we will indicate what is the model used for the second part in the experiments.

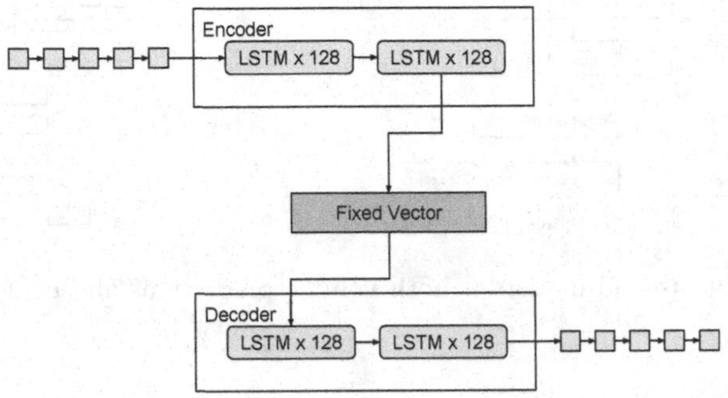

Fig. 5. The autoencoder model using LSTM

3 Description of Data

One of the biggest challenges when designing machine learning experiments is to find data. Although it is easy to create artificial datasets, it is important to make sure that the choice of instances is meaningful.

Data Sources and Generation. In this work, we used two types of data, namely, random polytopes and polytopes from specific families.

A random polytope is generated by first choosing at random (within a predefined range) the number of vertices of the polytope and then create a random polytope with that many vertices using SageMath [16]. The dataset generator creates a random vector with integer coordinates (indicating the numbers of vertices of each polytope in the dataset). The length of this vector indicates the size of the dataset. In this process, the dimension of the polytopes is fixed. Different sets of polytopes of different dimensions are generated by changing this fixed value.

Concerning polytopes from specific families, we use cubes, hypercubes, cross-polytopes and Fano polytopes. We use SageMath to generate cubes, hypercubes, cross-polytopes, while Fano polytopes are taken from [11].

Datasets. We use Fano polytopes paired with each other based on dimension in our experiments. This is a requirement of the comparison problem we described before. This also increases the the number of data instances even though we start with a limited number of polytopes from this data set. We used 200 polytopes and this produced more than 19.000 pairs in the end.

Our second data generator uses predefined families of polytopes in SageMath. We chose several base polytopes, such as unit cube, and we create new polytopes by applying random transformation to these base polytopes. Random transformations include scaling up to 10 times and replacing the center from the origin to another point and applying slight rotations.

A summary of the data can be seen in Table 2. We have three sets of polytopes coming from randomly generated polytopes and a set of polytopes only consists of Fano polytopes. We name these sets as A, B, C an D. Table 2 shows which data set includes which family of polytopes.

In Table 1 we give a statistical analysis of volumes in the datasets we used.

Table 1. Volume statistics

Polytope set	Dimensions	Instances	Min	Max	Mean	Variance
Set A	3,4,5	15000	8.0	1889568.0	113619.3272	62416457013.52322
Set B	3,4,5	7500	8.0	1889568.0	151488.8789	120202038886.83537
Set C	3	2500	8.0	5832.0	1845.8047	3814934.1211
Set D	3,4,5,6	200	6.0	123.0166	28.35704	510.43171

We split all the data sets into three sets before training phases. 30% of the data goes into validation and 10% goes to test set as rest of the data goes into training set. We make this split once and use the same splits in all of the experiments later. We use training and validation during training but we use the test set only for the evaluation of the models.

Table 2. Dataset summary

Polytope family	Dimension	Set A	Set B	Set C	Set D
Cube	3	yes	yes	yes	no
Hypercube	4	yes	yes	no	no
Hypercube	5	yes	yes	no	no
Cross-polytope	4	yes	no	no	no
Cross-polytope	5	yes	no	no	no
Random	3	yes	no	no	no
Fano	3,4,5,6	no	no	no	yes

3.1 Normalization

Normalization of data is very important for machine learning in general. Feature values are usually compressed into the $[-1, 1]$ range in order to remove range differences. However, actual values have importance when dealing with geometric and algebraic problems. This lead us to use different normalizations on volume by multiplying it by a scalar value. We have variations of out data sets with different normalization multipliers applied before training (see Table 3).

4 Experimental Results

In this section we present the training setup and results of the experiments.

Metrics and Parameters. Volume prediction is a regression problem and there are several different metrics to measure the performance of a regression model. A standard metric is the mean squared error (MSE). Although we use MSE during training as a loss function, it is hard to compare the performance of different models based on MSE. For example, applying different normalizations results in vastly different values for MSE. The same is true for mean absolute error. Since it is important that the metric is not affected by such changes, we use R^2 (R-squared) as regression measure. In R^2, the best score is 1, while 0 means the model produces the same output regardless of the input. It is possible to get negative values, meaning that the model performs worse than giving always the same output.

For the classification problem, the label distribution is totally balanced, since for every pair of polytopes (A, B), we also include (B, A) in the dataset. The balanced distribution of both labels makes accuracy a good and simple enough metric to measure performance. During training, we use Binary Cross Entropy as loss function.

Another important evaluating criterion to consider is overfitting. Overfitting occurs when a model memorizes too much from the training data and cannot perform well with new inputs. Datasets are split into training, validation and test sets in order to evaluate the fit of the model to the data. We do not see overfit in the experiments as the results from training and testing are similar.

Concerning the rest of the training parameters (unless stated otherwise), we use Adam as optimizer with a learning rate of 0.001. All of the models are trained for 20 epochs with the exception of autoencoders when a model has one. Autoencoders are trained for 30 epochs, with the same optimizer and learning rate.

Modular Framework. The first part of experiments uses randomly generated polytopes and the `V_F_APC_P` model to predict volume. Note that this part of experiments is significantly different from the following part, since it does not perform volume comparison and the datasets used have a wide range of volumes between 0 and 10^6. We use stochastic gradient descent (SGD) as optimizer with learning rate of 0.321 for autoencoders and Adam with learning rate of 0.001 for feed-forward networks. We train the autoencoder for 10 epochs and the feed-forward networks for 30 epochs. The results can be found in Table 3.

The main observation from this set of experiments is that it is hard to predict volume when the dataset contains polytopes with a wide range of volumes. Even when we use Dataset C, which only consists of cubes, scores are still lower than 0.5 due to the wide range of the volume values. However, results are better compared to results from other datasets as the range of the volume is relatively limited. We also tried to train the models without volume normalization, but the models failed to learn.

Using the insights gained from the first set of experiments, we design a new experiment. In addition to volume prediction, we consider volume comparison as a related problem and we also consider the two problems simultaneously (see Sect. 2). In terms of data, in dataset D we restrict to a specific family of polytopes, with a smaller range of volume values (see Sect. 3).

The modular framework performs better with this new experimentation setup. The coupled model `V_F_APC_PC` has better comparison accuracy compared to `V_F_APC_C` due to the effect of volume prediction. Both `V_F_APC_PC` and `V_F_APC_P` give bad results for volume prediction. When normalization is removed from input data, although volume prediction improves, there is no improvement for volume comparison.

End-to-End. We also experiment with end-to-end network architectures in addition to our modular model with the second experimentation setup.

Fixed dimensional models obtain decent performance in accuracy, but volume prediction is not successful with normalization. An interesting observation is that volume prediction is worse in `F_L_E2E_PC` compared to `F_L_E2E_P`. However, both of them are still below zero which means none of them succeed in volume prediction. Another interesting situation we observe is that `F_L_E2E_PC` has slightly better accuracy for comparison than `F_L_E2E_C`. This becomes even more obvious when we change the normalization multiplier to 0.08 and 1 and leave other parameters such as optimizer and learning rate unchanged. For unnormalized data, `F_L_E2E_PC` reaches 0.9 accuracy and 0.97 volume score. This allows us to assume that the more the model learns from volume prediction the more accurate it will get on comparison. The accuracy and volume scores make it a feasible model for such problems when limiting the dimension is an option.

`F_P_E2E_PC` and its variations are the models with the best performance regardless of normalization. Again, the coupled model `F_P_E2E_PC` has slightly better accuracy compared to the comparison only model `F_P_E2E_C`. However, this time `F_L_E2E_PC` has a volume score above 0, which means that it learns about volume as well (even though not enough). The previous observation, that learning from volume makes comparison better, is supported by the high score of `F_P_E2E_P`. `F_P_E2E_PC` produces close results to `F_P_E2E_P` with different normalization multipliers. The overall performance of the model also improves when there is no normalization on volume. However, an interesting point to consider is that F_P_E2E_P produce good results even with normalization. This model may be a good solution when the data is within the limits.

The training of V_L_E2E_PC includes an additional encoder, but this time the autoencoder works in a way similar to word embeddings in natural language processing (NLP). We used an autoencoder trained independently to encode our vertices in order to fix their dimension. It is trained for 30 epochs with Adam as optimizer and MSE as loss function. The rest of the model is trained similarly to F_L_E2E_PC as dimensions are fixed now. The reason of such training method instead of a complete end-to-end is that we could not find a way to train such model with LSTM.

V_L_E2E_PC and V_L_E2E_C offer acceptable performance for the comparison problem. However the results for volume are not good for either V_L_E2E_P or V_L_E2E_PC. The results do not seem to change much unless normalization is removed. When normalization is removed V_L_E2E_PC has a decrease in its comparison accuracy and increase in volume score. However, these are not especially good results.

4.1 Comparison of the Models

We propose four different approaches to solve volume prediction and volume comparison for polytopes.

We can divide our models into two groups based on their performances. The F_L_E2E_PC and F_P_E2E_PC models produce good results with over 0.90 accuracy and over 0.97 prediction score. On the other hand, V_L_E2E_PC and V_F_APC_PC models produce relatively worse results with around 0.80 accuracy and 0.7 prediction score. One key difference between these two groups is their input data format. Better results come when we fix the data size instead of trying to keep it variable. It is also possible that autoencoders do not learn a good feature set for these problems.

A general observation we can make is that almost all models produce better comparison accuracy and worse R^2 score when both of the problems are combined in a model (coupled models). The only exception is V_L_E2E_PC. We conclude that the two problems can affect each other, improving comparison while learning to predict volumes.

Another interesting result is the effect of volume normalization on all of the models. In contrast to our initial failure without normalization, our second set of experiments without normalization on volume produces better results. Training with lower normalization multiplier produces better results overall and the best results came when normalization was removed by setting the multiplier to 1. All the models can have significant increase in their R^2 scores due to this.

Table 3. Results from experiments with different model, normalization, encoding data, prediction data combinations. We will name the models in the following way: the first part declares if the dimension is fixed or variable by F or V respectively,the second part is F if input is flatened, L if a list of lists, and P if completely padded, the third part declares the training scheme by E2E or APC and the fourth part is C for volume comparison, P volume prediction or PC for a coupled model.

Model	Normalization	Enc. Data	Pred. Data	Accuracy	Volume Q	Volume P
V_F_APC_P+NN 2 Layer	0.000001	A	A	N/A	0.10	N/A
V_F_APC_P+NN 2 Layer	0.000001	A	B	N/A	0.40	N/A
V_F_APC_P+NN 2 Layer	0.000001	A	C	N/A	−2.91	N/A
V_F_APC_P+NN 2 Layer	0.000001	B	B	N/A	0.15	N/A
V_F_APC_P+NN 2 Layer	0.000001	B	C	N/A	−0.01	N/A
V_F_APC_P+NN 2 Layer	0.000001	C	C	N/A	0.35	N/A
V_F_APC_P+NN 4 Layer	0.000001	A	A	N/A	0.11	N/A
V_F_APC_P+NN 4 Layer	0.000001	A	B	N/A	0.42	N/A
V_F_APC_P+NN 4 Layer	0.000001	A	C	N/A	0.22	N/A
V_F_APC_P+NN 4 Layer	0.000001	B	B	N/A	0.30	N/A
V_F_APC_P+NN 4 Layer	0.000001	B	C	N/A	−0.11	N/A
V_F_APC_P+NN 4 Layer	0.000001	C	C	N/A	0.30	N/A
V_F_APC_P+RF(100,5)	0.000001	A	A	N/A	0.05	N/A
V_F_APC_P+RF(100,5)	0.000001	A	B	N/A	0.15	N/A
V_F_APC_P+RF(100,5)	0.000001	A	C	N/A	0.39	N/A
V_F_APC_P+RF(100,5)	0.000001	B	B	N/A	0.21	N/A
V_F_APC_P+RF(100,5)	0.000001	B	C	N/A	0.38	N/A
V_F_APC_P+RF(100,5)	0.000001	C	C	N/A	0.36	N/A
V_F_APC_P+RF(100,15)	0.000001	A	A	N/A	0.16	N/A
V_F_APC_P+RF(100,15)	0.000001	A	B	N/A	0.40	N/A
V_F_APC_P+RF(100,15)	0.000001	A	C	N/A	0.49	N/A
V_F_APC_P+RF(100,15)	0.000001	B	B	N/A	0.45	N/A
V_F_APC_P+RF(100,15)	0.000001	B	C	N/A	0.42	N/A
V_F_APC_P+RF(100,15)	0.000001	C	C	N/A	0.46	N/A
F_L_E2E_PC	0.0001	N/A	D	0.8624	−42.48	−43.58
F_L_E2E_C	0.0001	N/A	D	0.8401	N/A	N/A
F_L_E2E_P	0.0001	N/A	D	N/A	−0.1268	−0.0089
F_P_E2E_PC	0.0001	N/A	D	0.9266	0.1986	0.1419
F_P_E2E_C	0.0001	N/A	D	0.9106	N/A	N/A
F_P_E2E_P	0.0001	N/A	D	N/A	0.9785	0.9925
V_L_E2E_PC	0.0001	D	D	0.8510	−4.5031	−8.0108
V_L_E2E_C	0.0001	D	D	8559	N/A	N/A
V_L_E2E_P	0.0001	D	D	N/A	−0.5204	−0.0005
V_F_APC_PC	0.0001	D	D	0.8152	0.1260	−0.5238
V_F_APC_C	0.0001	D	D	0.7849	N/A	N/A
V_F_APC_P	0.0001	D	D	N/A	−0.0301	−0.4150
F_L_E2E_PC	0.008	N/A	D	0.7957	0.0163	0.0115
F_L_E2E_PC	1	N/A	D	0.9084	0.9774	0.9975
F_P_E2E_PC	0.008	N/A	D	0.9111	0.9269	0.9547
F_P_E2E_PC	1	N/A	D	0.9631	0.9980	0.9982
V_L_E2E_PC	0.008	N/A	D	0.8223	−0.1644	−0.2701
V_L_E2E_PC	1	N/A	D	0.7226	0.7013	0.7135
V_F_APC_PC	1	D	D	0.8277	0.5857	0.7754

5 Conclusion

In this paper we propose a framework allowing the use of machine learning for geometric and algebraic problems. The framework has similarities to models developed for other sequential problems employing RNNs, and in particular autoencoders.

We focus on two volume related problems, namely the prediction of polytope volume and the comparison of volumes of a pair of polytopes. The choice of problems is convenient since it provides naturally a regression problem and a related classification problem.

One of the important differences between the problems and solutions in the literature and the ones presented in this work is the effect of data normalization. In general, data normalization is essential for the success of most machine learning models. On the contrary, for the volume prediction problem and the proposed models, data normalization is catastrophic. If the range of volumes in the training data is too wide and the dataset not large enough, then the model fails to train due to lack of normalization. Nevertheless, as shown in our experiments, if the range is not too wide, unnormalized training data provide the best results.

To the best of our knowledge, this is the first attempt to employ machine learning in this type of problem. Our goal was to show that it is possible to train a model with input of two variable dimensions in a meaningful way. Although we are far from the ideal solution, the results presented here are very encouraging. We detect some important parameters that differentiate the problem from classical machine learning problems and we show how to overcome the issues raised by these differences.

Naturally, we will also continue testing our better performing models with different families of polytopes and more diverse datasets. Such experiments will provide a better understanding on how to solve geometric and algebraic problems with machine learning.

Acknowledgements. This work was supported by the project 117E501 under the program 3001 of the Scientific and Technological Research Council of Turkey.

References

1. Calès, L., Chalkis, A., Emiris, I.Z., Fisikopoulos, V.: Practical volume computation of structured convex bodies, and an application to modeling portfolio dependencies and financial crises. In: Speckmann, B., Tóth, C.D. (eds.) 34th International Symposium on Computational Geometry, SoCG 2018, 11–14 June 2018, Budapest, Hungary. LIPIcs, vol. 99, pp. 19:1–19:15. Schloss Dagstuhl - Leibniz-Zentrum fuer Informatik (2018). https://doi.org/10.4230/LIPIcs.SoCG.2018.19
2. Cho, K., et al.: Learning phrase representations using RNN encoder-decoder for statistical machine translation. In: Proceedings of the 2014 Conference on Empirical Methods in Natural Language Processing (EMNLP), pp. 1724–1734. Association for Computational Linguistics (2014). https://doi.org/10.3115/v1/D14-1179, http://aclweb.org/anthology/D14-1179

3. Dai, A.M., Le, Q.V.: Semi-supervised sequence learning. In: Cortes, C., Lawrence, N.D., Lee, D.D., Sugiyama, M., Garnett, R. (eds.) Advances in Neural Information Processing Systems, vol. 28, pp. 3079–3087. Curran Associates, Inc. (2015). http://papers.nips.cc/paper/5949-semi-supervised-sequence-learning.pdf
4. Dyer, M.E., Frieze, A.M.: On the complexity of computing the volume of a polyhedron. SIAM J. Comput. **17**(5), 967–974 (1988). https://doi.org/10.1137/0217060
5. Emiris, I.Z., Fisikopoulos, V.: Efficient random-walk methods for approximating polytope volume. In: Proceedings of the Thirtieth Annual Symposium on Computational Geometry, SOCG 2014, pp. 318:318–318:327. ACM, New York (2014). https://doi.org/10.1145/2582112.2582133, http://doi.acm.org/10.1145/2582112.2582133
6. He, K., Zhang, X., Ren, S., Sun, J.: Deep residual learning for image recognition, pp. 770–778, June 2016. https://doi.org/10.1109/CVPR.2016.90
7. Hinton, G., et al.: Deep neural networks for acoustic modeling in speech recognition: the shared views of four research groups. IEEE Signal Process. Mag. **29**(6), 82–97 (2012)
8. Hochreiter, S., Schmidhuber, J.: Long short-term memory. Neural Comput. **9**(8), 1735–1780 (1997). https://doi.org/10.1162/neco.1997.9.8.1735
9. Jaekel, U.: A monte carlo method for high-dimensional volume estimation and application to polytopes. In: Sato, M., Matsuoka, S., Sloot, P.M.A., van Albada, G.D., Dongarra, J.J. (eds.) Proceedings of the International Conference on Computational Science, ICCS 2011. Procedia Computer Science. Nanyang Technological University, Singapore, 1–3 June 2011, vol. 4, pp. 1403–1411. Elsevier (2011). https://doi.org/10.1016/j.procs.2011.04.151
10. Krizhevsky, A., Sutskever, I., Hinton, G.E.: ImageNet classification with deep convolutional neural networks. In: Pereira, F., Burges, C.J.C., Bottou, L., Weinberger, K.Q. (eds.) Advances in Neural Information Processing Systems, vol. 25, pp. 1097–1105. Curran Associates, Inc. (2012). http://papers.nips.cc/paper/4824-imagenet-classification-with-deep-convolutional-neural-networks.pdf
11. Paffenholzr, A.: Smooth reflexive lattice polytopes. https://polymake.org/polytopes/paffenholz/www/fano.html
12. Rumelhart, D.E., Hinton, G.E., Williams, R.J.: Learning representations by back-propagating errors. Nature **323**, 533 (1986). https://doi.org/10.1038/323533a0
13. Simonyan, K., Zisserman, A.: Very deep convolutional networks for large-scale image recognition. CoRR abs/1409.1556 (2014)
14. Sutskever, I., Vinyals, O., Le, Q.V.: Sequence to sequence learning with neural networks. In: Proceeding of the NIPS. Montreal, CA (2014). http://arxiv.org/abs/1409.3215
15. Szegedy, C., et al.: Going deeper with convolutions. In: Computer Vision and Pattern Recognition (CVPR) (2015). http://arxiv.org/abs/1409.4842
16. The Sage Developers: SageMath, the Sage Mathematics Software System (Version 8.3.0) (2019). https://www.sagemath.org
17. Ziegler, G.M.: Lectures on polytopes. Springer, New York (1995). https://doi.org/10.1007/978-1-4613-8431-1

Faster Biclique Mining in Near-Bipartite Graphs

Blair D. Sullivan, Andrew van der Poel(✉), and Trey Woodlief

North Carolina State University, Raleigh, NC 27607, USA
{blair_sullivan,ajvande4,adwoodli}@ncsu.edu

Abstract. Identifying dense bipartite subgraphs is a common graph data mining task. Many applications focus on the enumeration of all maximal bicliques (MBs), though sometimes the stricter variant of maximal induced bicliques (MIBs) is of interest. Recent work of Kloster et al. introduced a MIB-enumeration approach designed for "near-bipartite" graphs, where the runtime is parameterized by the size k of an odd cycle transversal (OCT), a vertex set whose deletion results in a bipartite graph. Their algorithm was shown to outperform the previously best known algorithm even when k was logarithmic in $|V|$. In this paper, we introduce two new algorithms optimized for near-bipartite graphs - one which enumerates MIBs in time $O(M_I|V||E|k)$, and another based on the approach of Alexe et al. which enumerates MBs in time $O(M_B|V||E|k)$, where M_I and M_B denote the number of MIBs and MBs in the graph, respectively. We implement all of our algorithms in open-source C++ code and experimentally verify that the OCT-based approaches are faster in practice than the previously existing algorithms on graphs with a wide variety of sizes, densities, and OCT decompositions.

Keywords: Bicliques · Odd cycle transversal · Bipartite · Enumeration algorithms · Parameterized complexity

1 Introduction

Bicliques (complete bipartite graphs) naturally arise in many data mining applications, including detecting cyber communities [18], data compression [1], epidemiology [23], artificial intelligence [30], and gene co-expression analysis [15,16]. In many settings, the bicliques of interest are *maximal* (not contained in any larger biclique) and/or *induced* (each side of the bipartition is independent in the host graph), and there is a large body of literature giving algorithms for enumerating all such subgraphs [3,5,6,20,22,23,26,32]. Many of these approaches make strong structural assumptions on the host graph; the case when the host graph is bipartite has been particularly well-studied, and the `iMBEA` algorithm

This work was supported by the Gordon & Betty Moore Foundation's Data-Driven Discovery Initiative under Grant GBMF4560 to Blair D. Sullivan and the NC State College of Engineering REU program.

I. Kotsireas et al. (Eds.): SEA[2] 2019, LNCS 11544, pp. 424–453, 2019.
https://doi.org/10.1007/978-3-030-34029-2_28

of Zhang et al. has been empirically established to be state-of-the-art [32]. Until recently, the only known non-trivial algorithm for enumerating maximal induced bicliques (MIBs) in general graphs was that of Dias et al. which did so in lexicographic order [5]. In [17], Kloster et al. presented a new algorithm for enumerating MIBs in general graphs, OCT-MIB, which extended ideas from iMBEA to work on non-bipartite graphs by using an *odd cycle transversal* (OCT set): a set of nodes O such that $G[V \setminus O]$ is bipartite. This yielded an algorithm with runtime $O(M_I nmn_O^2 3^{n_O/3})$ where $n_O = |O|$, M_I is the number of MIBs in $G = (V, E)$, and n and m denote $|V|$ and $|E|$, respectively. The $3^{n_O/3}$ term arises from OCT-MIB's dependence on the number of maximal independent sets (MISs) in O. In this paper, we give new algorithms for enumerating both MIBs and maximal, not necessarily induced bicliques (MBs) in general graphs. We first present OCT-MIB-II which again leverages odd cycle transversals to enumerate MIBs in time $O(M_I nmn_O)$. In contrast to OCT-MIB, the worst-case runtime of OCT-MIB-II is not dependent on the number of MISs in O, making it better than OCT-MIB when $n_O \in \omega(1)$. We also give a second algorithm for MIB-enumeration, Enum-MIB, which has runtime $O(M_I nm)$. Enum-MIB is essentially a modified version of the algorithm of Dias et al. [5], which achieves a faster runtime by dropping the lexicographic output requirement.

In the setting considering non-induced bicliques, the state-of-the-art approach is MICA of Alexe et al. [3]. MICA employs a consensus mechanism to iteratively find maximal bicliques by combining them together, resulting in an $O(M_B n^3)$ algorithm, where M_B is the number of MBs. We introduce a new algorithm OCT-MICA which leverages odd cycle transversals and runs in $O(M_B(n^2 n_O + mn))$ time.

Since all graphs have OCT sets (although they can be size $O(n)$, as in cliques), OCT-MIB, OCT-MIB-II, and OCT-MICA can all be run in the general case; their correctness does not require minimality or optimality of the OCT set. Further, we implement OCT-MIB-II, Enum-MIB and OCT-MICA in open source C++ code, and evaluate their performance on a suite of synthetic graphs with known OCT decompositions. Our experiments show that OCT-MICA and OCT-MIB-II are the dominant algorithms for their respective problems in many settings. Their efficiencies allow us to run on larger graphs than in [17].

We begin with preliminaries and a brief discussion of related work in Sect. 2, then describe each of our three new algorithms and provide proofs of their correctness and runtimes in Sect. 3. We highlight several implementation details in Sect. 4, before presenting our experimental evaluation in Sect. 5.

2 Preliminaries

2.1 Related Work

The complexity of finding bicliques is well-studied, beginning with the results of Garey and Johnson [7] which establish that in bipartite graphs, finding the largest balanced biclique is NP-hard but the largest biclique can be found in polynomial time. Particularly relevant to the mining setting, Kuznetsov showed that enumerating MBs in a bipartite graph is #P-complete [19]. Finding the biclique with the largest number of edges was shown to be NP-complete in general graphs [31], but the case of bipartite graphs remained open for many years. Several variants (including the weighted version) were proven NP-complete in [4] and in 2000, Peeters finally resolved the problem, proving the edge maximization variant is NP-complete in bipartite graphs [25].

For the problem of enumerating MIBs, the best known algorithm in general graphs is due to Dias et al. [5]; in the non-induced setting, approaches include a consensus algorithm `MICA` [3], an efficient algorithm for small arboricity [6], and a general framework for enumerating maximal cliques and bicliques [8], with `MICA` the most efficient among them, running in $O(M_B n^3)$. We note that, as described, the method in [5] may fail to enumerate all MIBs; a modified, correct version was given in [17].

There has also been significant work on enumerating MIBs in bipartite graphs. We note that since all bicliques in a bipartite graph are necessarily induced, non-induced solvers for general graphs (such as `MICA`) can be applied, and have been quite competitive. The best known algorithm however, is due to Zhang et al. [32] and directly exploits the bipartite structure. Other approaches in bipartite graphs include frequent closed itemset mining [20] and transformations to the maximal clique problem [22]; faster algorithms are known when a lower bound on the size of bicliques to be enumerated is assumed [23,26].

Kloster et al. [17] extended techniques for bipartite graphs to the general setting using odd cycle transversals, a form of "near-bipartiteness" which arises naturally in many applications [10,24,27]. This work resulted in `OCT-MIB`, an algorithm for enumerating MIBs in a general graph, parameterized by the size of a given OCT set. Although finding a minimum size OCT set is NP-hard, the problem of deciding if an OCT set with size k exists is fixed parameter tractable (FPT) with algorithms in [21] and [14] running in times $O(3^k kmn)$ and $O(4^k n)$, respectively. We note non-optimal OCT sets only affect the runtime (not correctness) of our algorithms, allowing us to use heuristic solutions. Recent implementations [9] of a heuristic ensemble alongside algorithms from [2,12] alleviate concerns about finding an OCT decomposition creating a barrier to usability.

2.2 Notation and Terminology

Let $G = (V, E)$ be a graph; we set $n = |V|$ and $m = |E|$. We define $N(v)$ to be the neighborhood of $v \in V$ and write $\overline{N}(v)$ for v's non-neighbors. An independent set $T \subseteq V(G)$ is a *maximal independent set* (MIS) if T is not contained in any other independent set of G. Unless otherwise noted, we assume without loss of generality that G is connected.

A biclique $A \times B$ in a graph $G = (V, E)$ consists of non-empty disjoint sets $A, B \subset V$ such that every vertex of A is neighbors with every vertex of B. We say a biclique $A \times B$ is *induced* if both A and B are independent sets in G. A *maximal biclique* (MB) in G is a biclique not properly contained in any other;

a *maximal induced biclique* (MIB) is analogous among induced bicliques. We use M_B and M_I to denote the number of MBs and MIBs in G, respectively. If O is an OCT set in G, we denote the corresponding OCT decomposition of G by $G[L, R, O]$, where the induced subgraph $G[L \cup R]$ is bipartite. We write n_L, n_R, and n_O for $|L|, |R|$, and $|O|$, respectively.

3 Algorithms

In this section we provide three novel algorithms, two of which of solve Maximal Induced Biclique Enumeration (`Enum-MIB` and `OCT-MIB-II`) and the other of which solves Maximal Biclique Enumeration (`OCT-MICA`). Both `Enum-MIB` and `OCT-MIB-II` follow the same general framework, which we now describe.

3.1 MIB Algorithm Framework

The MIB-enumeration algorithms both use two subroutines, `MakeIndMaximal` and `AddTo`. `MakeIndMaximal` takes in (C, S), where C is an induced biclique and $S \subseteq V$, and either returns a MIB C^+ where $C \subseteq C^+$, $C^+ \subseteq C \cup S$, $C \neq \emptyset$, or returns $\emptyset$. If it returns $\emptyset$ and $C \neq \emptyset$ then there is another MIB D which contains C and $v \in (V \setminus S) \setminus C$. `AddTo` takes in (C, v) where $C = C_1 \times C_2$ is an induced biclique and $v \in V \setminus (C_1 \cup C_2)$, and returns the induced biclique where v is added to C_1, $N(v)$ is removed from C_1, and $\overline{N}(v)$ is removed from C_2 if $C_2 \cap N(v) \neq \emptyset$; otherwise, $\emptyset$ is returned. Both `MakeIndMaximal` and `AddTo` operate in $O(m)$ time. We defer algorithmic details and proofs of the complexity and correctness for these routines to the Appendix.

The MIB-enumeration framework (shown in Algorithm 1) begins by finding a seed set of MIBs $\mathcal{C}_S$. At a high level, it operates by attempting to add vertices from the designated set I_S to previously found MIBs to make them maximal. We utilize a dictionary $\mathcal{D}$ to track which MIBs have already been found and a queue $\mathcal{Q}$ to store bicliques which have not yet been explored. We now prove two technical lemmas used to show the correctness of this framework.

Lemma 1. *Let $X \times Y$ be a MIB in graph G which contains a non-empty subset of $R \times S$, another MIB in G. Running `AddTo` with parameters $X \times Y$ and $v \in R \setminus (X \cup Y)$ returns a biclique which contains $R \cap X$, $S \cap Y$, and v if $Y \cap N(v) \neq \emptyset$.*

Proof. By construction, v must be independent from R and completely connected to S. Thus, none of $R \cap X$ will be removed from X and all of $S \cap Y$ will remain in Y, as required. Therefore, as long as $Y \cap N(v) \neq \emptyset$, the desired biclique is returned.

Lemma 2. *In Algorithm 1, if there exists a MIB $A' \times B'$ in $\mathcal{D}$ such that $A \setminus I_S \subseteq A'$, $B \setminus I_S \subseteq B'$ and $(A \cup B) \cap (A' \cup B') \neq \emptyset$, for each MIB $A \times B$ in G, then all MIBs in G are included in $\mathcal{D}$.*

Algorithm 1. MIB-enumeration algorithm framework

```
1: Input: G = (V, E), I_S
2: C_S = FindSeedSet(G)                          ▷ set of initial MIBs
3: Add each C ∈ C_S to D and Q
4: while Q is not empty do
5:     X × Y ← pop(Q)
6:     for j ∈ I_S \ (X ∪ Y) do
7:         C_1 = AddTo(X × Y, j)
8:         C'_1 = MakeIndMaximal(C_1, I_S)
9:         if C'_1 is not in D then
10:            Add C'_1 to D and Q
11:        C_2 = AddTo(Y × X, j)
12:        C'_2 = MakeIndMaximal(C_2, I_S)
13:        if C'_2 is not in D then
14:            Add C'_2 to D and Q
15: return D
```

Proof. Assume not. Let $A \times B$ be a MIB in G which is not in $\mathcal{D}$ with $|(A \cup B) \setminus I_S|$ maximum. Let $A' \times B'$ be the MIB in $\mathcal{D}$ such that $A \setminus I_S \subseteq A'$, $B \setminus I_S \subseteq B'$ and $(A \cup B) \cap (A' \cup B') \neq \emptyset$ and let $v \in ((A \cup B) \setminus (A' \cup B')) \subseteq I_S$. Without loss of generality assume $B \cap B' \neq \emptyset$ and $v \in A$.

Consider the iteration of Algorithm 1 when $X \times Y = A' \times B'$ and $j = v$ (lines 5–6). By Lemma 1, one of the calls to AddTo returns an induced biclique C which contains $A \setminus I_S$, $B \setminus I_S$, and v. Both sides of C are non-empty (since $B \cap B' \neq \emptyset$ and $v \in A$). If $C = A \times B$ we obtain a contradiction, as MakeIndMaximal (C, I_S) would return C, resulting in its addition to $\mathcal{D}$. Otherwise, either MakeIndMaximal returns $\emptyset$ or a biclique $C' = A' \times B'$ which is added to $\mathcal{D}$. Since both sides of C are nonempty, if MakeIndMaximal returns $\emptyset$, there exists a MIB in G containing C and $x \in (V \setminus I_S) \setminus C$. Let $A' \times B'$ be such a MIB; since it has more vertices in $V \setminus I_S$ than C, it must be in $\mathcal{D}$, and we set $C' = A' \times B'$. In either case, $C \subseteq (A' \cup B')$, $|(A \cup B) \setminus (A' \cup B')| < |(A \cup B) \setminus (X \cup Y)|$. We can repeat this argument for the new $A' \times B'$, noting that $(A \cup B) \cap (A' \cup B')$ will include vertices on both sides. Thus, the argument still holds without any assumption on the non-empty side of the intersection and $|(A \cup B) \setminus (A' \cup B')|$ will strictly decrease; when it reaches 0, $A' \times B' = A \times B$, a contradiction.

Note that as MakeIndMaximal only returns MIBs, this framework will only include MIBs in $\mathcal{D}$. Together with Lemma 2, this yields the following corollary.

Corollary 1. *If for every MIB $A \times B \in G$ there is a MIB $A' \times B' \in \mathcal{C}_S$ such that $A \setminus I_S \subseteq A'$, $B \setminus I_S \subseteq B'$ and $(A \cup B) \cap (A' \cup B') \neq \emptyset$, then upon completion of Algorithm 1, $\mathcal{D}$ will contain exactly the MIBs in G.*

Recall that AddTo and MakeIndMaximal each run in $O(m)$ time. Combining this with the fact that each MIB in G is popped at most once from $\mathcal{Q}$ we have:

Corollary 2. *The time complexity of this framework is $O(M_I mn + INIT)$, where $INIT$ is the time needed by FindSeedSet to compute $\mathcal{C}_S$.*

3.2 Enum-MIB

We now present Enum-MIB, which follows the MIB-enumeration framework. To form $\mathcal{C}_S$, for each vertex $v \in V$ we run MakeIndMaximal $(\{v\} \times \{x\}, V)$ where $x \in N(v)$ and add it to $\mathcal{C}_S$. We also let $I_S = V$. To show the correctness of this approach, we note that $V \setminus V = \emptyset$ and any MIB contains the empty set. Thus all that remains to show is that for each MIB there is a MIB in $\mathcal{C}_S$ with which it has a non-empty intersection. As every $v \in V$ is in some MIB in $\mathcal{C}_S$, this condition is met. Thus, via Corollary 1, Enum-MIB will find all MIBs. There may be $O(n)$ duplicates in $\mathcal{C}_S$ which can be removed in $O(n)$ time per duplicate. As MakeIndMaximal runs in $O(m)$ time, by Corollary 2, the time complexity of Enum-MIB is $O(M_I mn)$. We note that Enum-MIB is essentially a simplified version of the LexMIB algorithm from [17] which does not guarantee lexicographic order on output.

3.3 OCT-MIB-II

Next we describe OCT-MIB-II, an algorithm for enumerating all MIBs in a graph with a given OCT decomposition $G[L, R, O]$. OCT-MIB-II also makes use of the MIB-enumeration framework described in Sect. 3.1. In the calls to MakeIndMaximal we let $I_S = O$. To form $\mathcal{C}_S$, we begin by running iMBEA [32] to find the set $\mathcal{C}_B$ of MIBs in $G[L \cup R]$. For each $C_B \in \mathcal{C}_B$ we run MakeIndMaximal on (C_B, O). This creates a set X_B of MIBs in G.

Then for each node $o \in O$, we find the set of MISs in $N(o)$. This can be done in $O(mn)$ time per MIS using the algorithm of Tsukiyama et al. [28]. For each MIS I_o found, run MakeIndMaximal on the induced biclique $\{o\} \times I_o$. Let the multiset of all MIBs produced by this process be denoted X_Q. Note that a MIB may be in X_Q up to $O(n_O)$ times (once per $o \in O$, stemming from an MIS in $N(o)$), but we can remove duplicates from X_Q in $O(n)$ per MIB, forming X'_M. We then let $\mathcal{C}_S = X_B \cup X'_M$. Thus, FindSeedSet runs in $O(mnn_O)$ per unique MIB found, and by Corollary 2, the total time complexity of OCT-MIB-II is $O(M_I mnn_O)$.

To show the correctness of OCT-MIB-II, we must show that for every MIB in G, we include a MIB in $\mathcal{C}_S$ which includes all of its non-OCT nodes and a node in the MIB if the MIB is completely contained in O. If an entire MIB C is contained in O, then any MIB containing $\{o\} \times I_o$ for $o \in C$ suffices. If a MIB has non-OCT nodes on both sides, then there must be a MIB in X_B which contains these non-OCT nodes because there is a MIB in $G[L \cup R]$ containing them. If a MIB has all of its non-OCT nodes on one side, then there is an OCT node o which is neighbors with all of the non-OCT nodes, which thus must be contained in an MIS in $N(o)$. Thus, by Corollary 1, we find all of the MIBs in G.

3.4 OCT-MICA

OCT-MICA is an algorithm for enumerating the maximal bicliques (MBs) in a general graph with a given OCT decomposition $G[L, R, O]$. We adapt the approach of MICA [3], which relies on a seed set of bicliques which "cover" the graph.

Algorithm 2. OCT-MICA

```
1: procedure ENUMERATE(G = (L, R, O))
2:     M_B' = BIPARTITESOLVE(L, R)          ▷ Implementation of iMBEA, O(m'nM'_B)
3:     for B ∈ M_B' do                      ▷ O(M'_B)
4:         B = MAKEMAXIMAL(B)               ▷ Extend in place, O(m)
5:     C_O = {}
6:     for v in O do                        ▷ Initialize Bicliques from stars, O(n_O)
7:         B = MAKEMAXIMAL(v × N(v))        ▷ O(m)
8:         C_O.ADD(B)
9:     C = M_B' ∪ C_O.
10:    SORT(C)                              ▷ O(M'_B log(M'_B))
11:    found = true
12:    while found do
13:        found = false
14:        for B_1 in C_O do                ▷ O(n_O)
15:            for B_2 in C do              ▷ O(M_B)
16:                for B_3 in CONSENSUS(B_1, B_2) do
17:                    B_4 = MAKEMAXIMAL(B_3)          ▷ O(m)
18:                    if B_4 not in C then            ▷ O(n log(M_B))
19:                        found = true
20:                        C.INSERTINSORTEDORDER(B_4)
21:    return C
```

Specifically, we restrict `MICA`'s coverage requirement for the seed set to only the OCT set and leverage `iMBEA` [32] to enumerate the MBs entirely within $G[L \cup R]$. This reduces the runtime from $O(n^3 M_B)$ to $O(n^2 n_O M_B)$.

`OCT-MICA` begins by running `iMBEA` (line 2 in Algorithm 2) to get $\mathcal{M}_{\mathcal{B}}'$, the MBs in $G[L \cup R]$, in time $O(nm'M'_B)$, where m' is the number of edges in $G[L \cup R]$ and $M'_B = |\mathcal{M}_{\mathcal{B}}'|$. Using `MakeMaximal`, we convert elements of $\mathcal{M}_{\mathcal{B}}'$ to be maximal with respect to G (lines 3–4). `MakeMaximal` runs in $O(m)$ time and its algorithmic details are deferred to the Appendix. `OCT-MICA` then initializes its seed set of size $O(n_O)$ consisting of bicliques from the stars of the OCT set (lines 6–8), and adds these to the working set C of all identified MBs (line 9). Similar to `MICA`, the remainder of the algorithm builds new bicliques by combining (via `Consensus`, see Appendix) pairs of elements from the seed set C_O and previously identified MBs C (lines 11–20), until no new bicliques are generated. This runs in time $O(n^2 n_O M_B)$.

Lemma 3. *`OCT-MICA` returns exactly $\mathcal{M}_{\mathcal{B}}$, the set of maximal bicliques in G.*

Proof. Running `iMBEA` and `MakeMaximal` ensures all maximal bicliques from $G[L \cup R]$ were found and added to C. Thus, we restrict our attention to maximal bicliques with at least one node from O, and proceed similarly to the proof of Theorem 3 in [3]. We say that a biclique $B_1 = X_1 \times Y_1$ *absorbs* a biclique $B_2 = X_2 \times Y_2$ if $X_2 \subseteq X_1$ and $Y_2 \subseteq Y_1$ or $Y_2 \subseteq X_1$ and $X_2 \subseteq Y_1$.

We show that every biclique $B^* = X^* \times Y^*$ in G is absorbed by some biclique in C by induction on k, the number of OCT vertices in B^*. In the base case ($k = 0$), B^* is contained in $G[L \cup R]$ and is absorbed by a biclique in $\mathcal{M}_{\mathcal{B}}' \subseteq C$. We now consider $k \geq 1$; without loss of generality, assume X^* contains some OCT vertex v. Then $B' = \{v\} \times Y^*$ is absorbed by some biclique $B_1 = X_1 \times Y_1, v \in X_1, Y^* \subseteq Y_1$, where $B_1 \in C_0$ is formed from the star centered on v. Further, $B'' = (X^* \setminus \{v\}) \times Y^*$ has fewer vertices from OCT than B^*, so by induction it is absorbed by some biclique $B_2 = X_2 \times Y_2, (X \setminus \{v\}) \subseteq X_2, Y^* \subseteq Y_2$, where $B_2 \in C$. Now B^* is a consensus of B' and B'', and will be absorbed by the corresponding consensus of B_1 and B_2, guaranteeing absorption by a biclique in C.

Lemma 4. *The runtime of `OCT-MICA` after `iMBEA` is* $O(n^2 n_O M_B)$.

Proof. We begin by noting that $M_B \leq 2^n$, so $\log(M_B)$ is $O(n)$.

Finding the bicliques in $\mathcal{M}_{\mathcal{B}}'$ requires time $O(m'nM'_B)$ for `iMBEA` (line 2); making them maximal (lines 3–4) is $O(mM'_B)$. The bicliques generated by the OCT stars (lines 6–8) can be found in $O(mn_O)$. Sorting the initial set C (line 10) incurs an additional $O(M'_B \log(M'_B))$. Since $\log(M_B)$ is $O(n)$, the total runtime for our initialization (lines 2–10) is $O(mnM'_B + mn_O)$.

The consensus-building stage of `OCT-MICA` contains nested loops over C_0 (line 14) and C (line 15), which execute at most $O(n_O)$ and $O(M_B)$ times, respectively. The `Consensus` operation (line 16) executes in $O(n)$, and produces a constant number of candidate bicliques to check. Each execution of the inner loop incurs a cost of $O(m)$ for `MakeMaximal` (line 17) and $O(n \log(M_B))$ to insert the new MB in sorted order (lines 18–20). We note that the runtime of `Consensus` is dominated by the cost of the loop. Thus, the total runtime of consensus-building is $O(n_O M_B n \log(M_B))$, or $O(n^2 n_O M_B)$.

This analysis leads to an overall runtime of $O(m'nM'_B + n^2 n_O M_B)$, as desired. We note that for $n_O \in \Theta(n)$, `OCT-MICA`'s runtime degenerates to the $O(n^3 M_B)$ of `MICA`. Additionally, the stronger results for incremental polynomial time described for `MICA` in [3] still apply; the proofs are similar and are omitted for space. For bipartite graphs ($n_O = 0$), `OCT-MICA` is effectively `iMBEA`, which was empirically shown to be more efficient than `MICA` on bipartite graphs [32].

4 Implementation

In this section we describe several relevant implementation details and design decisions.

4.1 Algorithm Framework

We always (re-label and) store vertices as $\{0, 1, \ldots n\}$ and maintain internal dictionaries as needed to recover original labels – e.g. when taking subgraphs. This allows us to leverage native data types and structures; vertices are stored as `size_t`.

For efficiency in subroutines, we utilize two representations of G. One representation is as adjacency lists, stored as sorted vectors (to improve union and intersection relative to dictionaries or unsorted vectors). This representation is essential in the performance of `Consensus` in `MICA`/ `OCT-MICA` and `MakeIndMaximal` and `AddTo` in `OCT-MIB` or `OCT-MIB-II`. We also store the graph as a dictionary of dictionaries which is more amenable to taking subgraphs (as when finding MISs in `OCT-MIB`, `OCT-MIB-II`). Deleting a node requires time $O(N(v))$ as compared to $O(N(v)\Delta(G))$, where $\Delta(G)$ is the maximum degree, in the adjacency list representation.

4.2 MICA

The public implementation of `MICA` used in [32] is available at [13]. However, this implementation is only suitable for bipartite graphs as it makes certain efficiency improvements in storage, etc. which assume bipartite input. As such, we implemented `MICA` from scratch in the same framework as `OCT-MIB` and `OCT-MICA`, etc., using the data structures discussed above. This is incompatible with the technique described in [3] for storing only one side of each biclique (since in the non-induced case, maximality completely determines the other side). We note this could improve efficiency of both `MICA` and `OCT-MICA` in a future version of our software, and should not significantly affect their relative performance as analyzed in this work.

5 Experiments

5.1 Data and Experimental Setup

We implemented `OCT-MIB-II`, `Enum-MIB`, `MICA`, and `OCT-MICA` in C++, and used the implementation of `OCT-MIB` from [17]. All code is open source under a BSD 3-clause license and publicly available as part of MI-Bicliques at [11].

Data. For convenience, throughout this section, we assume $n_L \geq n_R$ and let $n_B = n_L + n_R$. Our synthetic data was generated using a modified version of the random graph generator of Zhang et al. [32] that augments random bipartite graphs to have OCT sets of known size. The generator allows a user to specify the sizes of L, R, and O (n_L, n_R, and n_O), the expected edge densities between L and R, O and $L \cup R$, and within O, and the coefficient of variation (cv; the standard deviation divided by the mean) of the expected number of neighbors in L over R and in $L \cup R$ over O. The generator is seeded for replicability. We use the naïve OCT decomposition $[L, R, O]$ returned by the generator for our

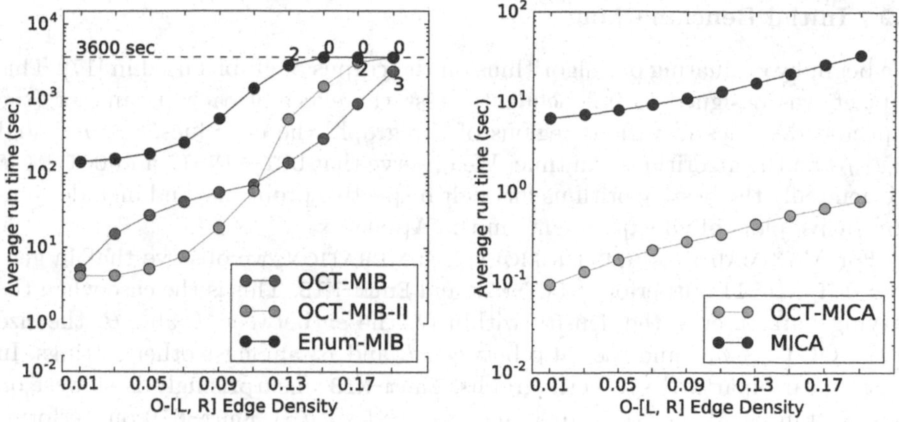

Fig. 1. Runtimes of the MIB-enumerating (left) and MB-enumerating (right) algorithms on graphs where $n_B = 1000$, $n_L/n_R = 10$, and $n_O = 10$. The expected edge density between O and $L \cup R$ was varied; all other densities were 0.05.

algorithm evaluation, but the techniques mentioned in Sect. 2 could also be used to find alternative OCT sets. Unless otherwise specified, the following default parameters are used: expected edge density $\bar{d} = 5\%$, $cv = 0.5$, $n_B = 1000$ and $n_L/n_R = 1/10$; additionally, the edge density between O and $L \cup R$ is the same as that between L and R.

To add the edges between L and R, the edge density and cv values are used to assign vertex degrees to R, and then neighbors are selected from L uniformly at random; this was implemented in the generator of [32]. Edges are added between O and $L \cup R$ via the same process, only with the corresponding edge density and cv values. Finally, we add edges within O with an Erdős-Rényi process based on expected density (no cv value is used here).

In most experiments we limit n_O to be $O(3 \log_3 n_B)$, and use a timeout of one hour (3600 s). Unless otherwise noted we run each parameter setting with five seeds and plot the average over these instances, using the time-out value as the runtime for instances that don't finish. If not all instances used for a plot point finished, we annotate it with the number of instances that did not time out.

We began by running our algorithms on the same corpus of graphs as in [17] (see Sect. 5.2). As the new algorithms finished considerably faster than those in [17], we were able to scale up both n_B and n_O to create new sets of experiments, discussed in Sect. 5.3. We also ran our algorithms on computational biology graphs from [29], which have been shown to be near-bipartite; these results are in Sect. 5.4.

Hardware. All experiments were run on identical hardware; each server had four Intel Xeon E5-2623 v3 CPUs (3.00 GHz) and 64 GB DDR4 memory. The servers ran Fedora 27 with Linux kernel 4.16.7-200.fc27.x86_64. The C/C++ codes were compiled using gcc/g++ 7.3.1 with optimization flag - O3.

5.2 Initial Benchmarking

We begin by evaluating our algorithms on the corpus of graphs used in [17]. This dataset was designed to independently test the effect of each parameter (the expected densities in various regions of the graph, the *cv* values, n_O, n_B, and n_L/n_R) on the algorithms' runtime. We observe that `OCT-MIB-II` and `OCT-MICA` are generally the best algorithms for their respective problems, and include comprehensive plots of all experiments in the Appendix.

For Maximal Induced Biclique Enumeration, we observe that in general, `OCT-MIB-II` outperforms `OCT-MIB` and `Enum-MIB`. This is the case when the varying parameter is the density within O, the *cv* between L and R, the size of the OCT set n_O, and the ratio between L and R, amongst other settings. In these "near-bipartite" synthetic graphs, `Enum-MIB` unsurprisingly is slowest on most instances. When $n_B = 1000$ and $n_O = 3\log_3(n_B)$, `Enum-MIB` outperforms `OCT-MIB` when the density within O increases above 0.05. This is likely due to the adverse effect of the number of MISs in the OCT set on `OCT-MIB`. The most interesting observation occurs when varying the edge density between O and $L \cup R$ (left panel of Fig. 1). In the $n_O = 10$ case, `OCT-MIB-II` is the fastest algorithm until the density exceeds 0.11, when `OCT-MIB` becomes faster. We believe this is likely due to `OCT-MIB` efficiently pruning away attempted expansions which are guaranteed to fail, while the number of MISs in O does not increase. This behavior is also seen in the case where $n_O = 3\log_3 n_B$, though the magnitude of the difference is not as extreme.

In the non-induced setting of Maximal Biclique Enumeration, `OCT-MICA` consistently outperforms `MICA` on this corpus, typically by at least an order of magnitude. The more interesting takeaway is that both MB-enumerating algorithms run considerably faster than their MIB-enumerating counterparts (e.g. right panel of Fig. 1), mostly because the number of MIBs is often one to two orders of magnitude larger than the number of MBs in these instances.

5.3 Larger Graphs

Given the much faster runtimes achieved in Sect. 5.2 we created a new corpus of larger synthetic graphs. For Maximal Induced Biclique Enumeration, we scaled up n_B to 10,000 and varied n_O in two settings, increasing the timeout to 7200 s.

When the expected density was 0.03 and $n_L/n_R = 100$, `OCT-MIB-II` outperformed `OCT-MIB` for all values of n_O by at least an order of magnitude and finished on all instances, whereas `OCT-MIB` timed out on all instances with $n_O \geq 13$ (left panel of Fig. 2). However, when the expected density was 0.01 and $n_L/n_R = 9$, `OCT-MIB` was faster (right panel of Fig. 2). We speculated that this was due to the sparsity of O, allowing for a speed-up due to the efficient pruning of `OCT-MIB` similar to what was seen in Sect. 5.2. To test this theory, we increased expected edge density within O to 0.05 while leaving the other parameters the same (right panel of Fig. 2), and observed that once $n_O \geq 9$, `OCT-MIB-II` outperforms `OCT-MIB`, confirming our hypothesis.

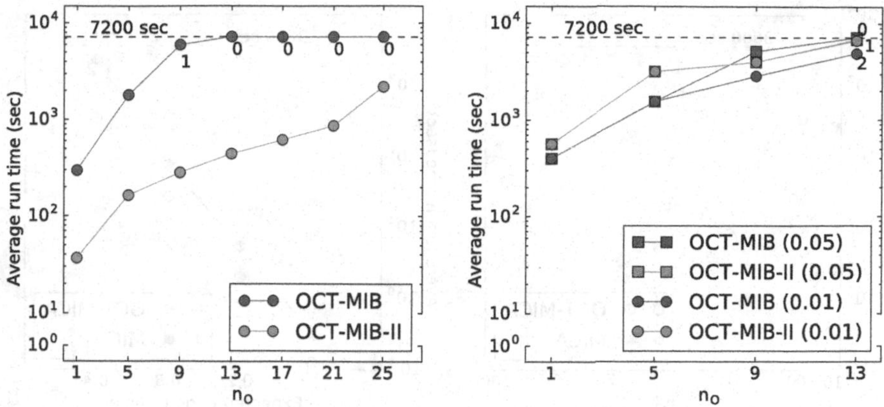

Fig. 2. Runtimes of the OCT-based MIB-enumeration algorithms on graphs where $n_B = 10000$ and n_O varies. In the left panel, $n_L = 9901, n_R = 99$ ($n_L/n_R \approx 100$) and the expected edge density is 0.03. In the right panel, $n_L = 9091, n_R = 909$ ($n_L/n_R \approx 10$) and the expected edge density (excluding within O) is 0.01; the marker-type denotes the expected edge density within O (see legend). For these larger instances we used 3 seeds and a 7200 s timeout.

For Maximal Biclique Enumeration, we also designed a new experiment where $n_B = 10000$ and n_O was scaled up to 1000 (left panel of Fig. 3). `OCT-MICA` finished on all instances, whereas `MICA` finished on none when n_O was 1000. We also tested how large we could scale the expected density between L and R (right panel of Fig. 3). When $n_B = 100$, `OCT-MICA` finished on all instances with density at most 0.4, while `MICA` finished on two of five when density is 0.4. Neither algorithm finished in less than the timeout of an hour when the density was 0.5 or greater, exhausting the hardware's memory in many cases. Thus `OCT-MICA` is able to scale to graphs with considerably larger OCT sets and higher density than both `MICA` and the MIB-enumerating algorithms.

We additionally created graphs with $n_O > 3\log_3 n_B$, which was not done in [17], and ran the algorithms for both MIBs and MBs (Fig. 4). These graphs had n_B values up to 4000 and for each value of n_B, we used three values of n_O; $10, 3\log_3 n_B$, and $\sqrt{n_B}$. The results were most interesting for the MIB-enumerating algorithms (Fig. 4 top). `OCT-MIB` performed the worst of the three algorithms when $n_O = \sqrt{n_B}$, but outperformed `Enum-MIB` in the other settings. This verifies the analysis from [17] on the range in which `OCT-MIB` is most effective. In general, `OCT-MIB-II` once again was the fastest algorithm and did best when n_O was smaller. The impact of n_O on `OCT-MIB-II` and `Enum-MIB` appeared comparable. In the MB-enumeration case, `OCT-MICA` consistently outperforms `MICA`, and there is a distinguishable difference in the runtime based on the value of n_O (Fig. 4 bottom). The value of n_O has far less effect on `MICA`, which does not finish on any graphs with $n_B = 4000$.

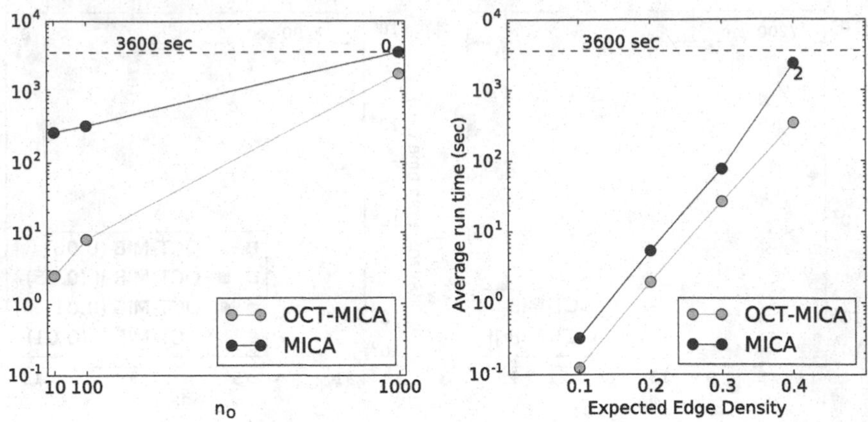

Fig. 3. Runtimes of the MB-enumerating algorithms on graphs with larger n_B and expected edge density. In the left panel, $n_L = 9091, n_R = 909$ ($n_L/n_R \approx 10$), the expected edge density is 0.05, and n_O varied. In the right panel, $n_L = 91, n_R = 9$ ($n_L/n_R \approx 10$), $n_O = 50$, and the expected edge density varied.

5.4 Computational Biology Data

Finally, we tested performance on real-world data using the graphs from [29], which come from computational biology. These graphs have previously been exhibited to have small OCT sets [12], and we used the implementation from [9] of Hüffner's iterative compression algorithm [12] to find the OCT decompositions. Computing the OCT decomposition for each graph ran in less than ten seconds, and often in less than one second. As can be seen in Table 1, `OCT-MIB-II` performs the best of the MIB-enumerating algorithms and `OCT-MICA` is faster than `MICA`. Full results are in the Appendix.

Table 1. A sampling of the runtimes of the biclique-enumeration algorithms on the Wernicke-Hüffner computational biology data [29].

G	n_B	m	n_O	$\|M_I\|$	OCT-MIB-II	OCT-MIB	Enum-MIB	$\|M_B\|$	OCT-MICA	MICA
aa-24	258	1108	21	3890	2.108	9.140	14.167	1334	0.237	2.477
aa-30	39	71	4	56	0.002	0.007	0.006	36	0.002	0.007
aa-41	296	1620	40	11705	16.519	82.439	50.205	20375	9.059	47.789
aa-50	113	468	18	1272	0.322	0.778	1.098	1074	0.132	0.612
j-20	241	640	1	274	0.013	0.065	0.484	228	0.009	0.188
j-24	142	387	4	150	0.013	0.027	0.089	104	0.007	0.025

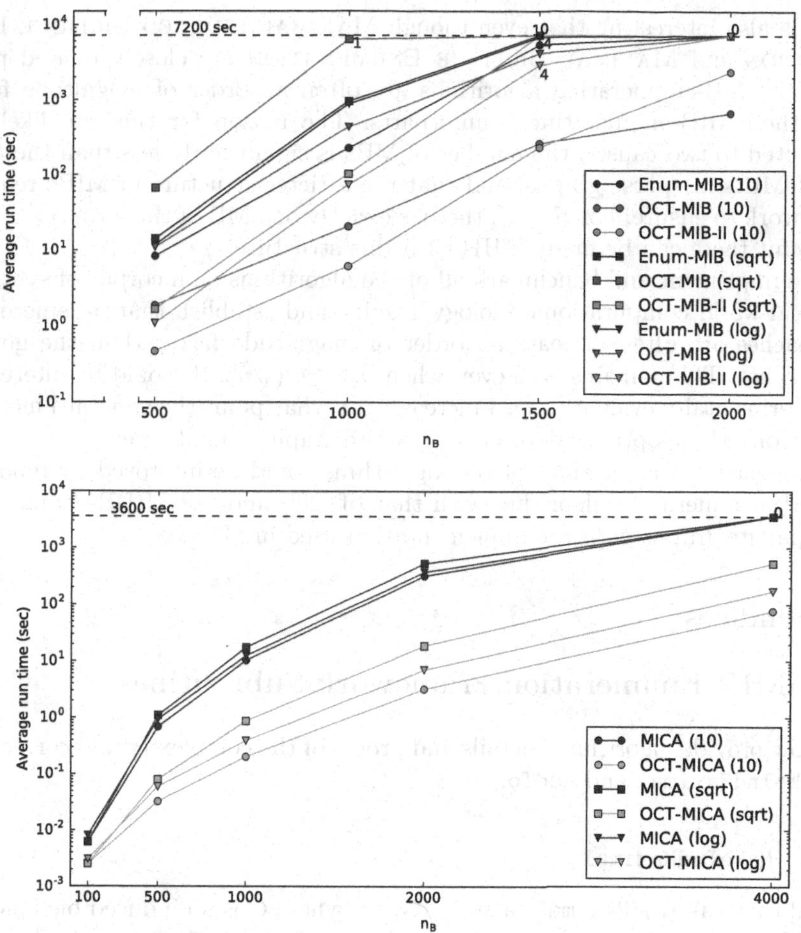

Fig. 4. Runtimes of the MIB-enumerating (top) and MB-enumerating (bottom) algorithms on graphs where $n_L/n_R = 9$ and all expected edge densities are 0.05. n_B is varied (x-axis) and the marker-type denotes the value of $n_O \in \{10, \sqrt{n_B}, 3\log_3(n_B)\}$ (see legend). The time-out value is set to 7200 s for the MIB-enumerating algorithms and 3600 s for the MB-enumerating algorithms.

6 Conclusion

We present a suite of new algorithms for enumerating maximal (induced) bicliques in general graphs, two of which are parameterized by the size of an odd cycle transversal. It is particularly noteworthy that the parameterized algorithms empirically outperform the general approaches even when their asymptotic worst-case complexities are worse. This highlights a weakness of standard complexity analysis, as many aspects of an algorithm get "swept under the rug".

It is also interesting that even though Maximal Induced Biclique Enumeration and Maximal Biclique Enumeration are closely related problems, the MB-enumerating algorithms are often an order of magnitude faster than their MIB-enumerating counterparts. The reason for this can likely be attributed to two causes: the number of MBs is significantly less than the number of MIBs in sparse graphs, and that the stricter structure of MIBs requires more work to ensure. For $S \subseteq V$, there is exactly one MB of the form $S \times T \subseteq V$ in G, but there can be many MIBs with this structure.

We implement and benchmark all of the algorithms on a corpus of synthetic and real-world computational biology graphs, and establish that parameterized approaches are often at least an order of magnitude faster than the general approaches. This remains true even when $n_O \in O(\sqrt{n})$. It would be interesting to experimentally evaluate as n_O increases, at what point the standard methods outperform those optimized for near-bipartite graphs. Finally, we note as in [17], the current implementations of the algorithms could be improved by replacing the MIS-enumeration algorithm with that of [28], and the M(I)B-enumeration on bipartite graphs with the implementation used in [32].

Appendices

A MIB-Enumeration Framework Subroutines

We now provide algorithmic details and proofs of the complexity and correctness of `MakeIndMaximal` and `AddTo`.

A.1 `MakeIndMaximal`

Recall that `MakeIndMaximal` takes in (C, S), where C is an induced biclique and $S \subseteq V$, and either returns a MIB C^+ where $C \subseteq C^+$, $C^+ \subseteq C \cup S$, $C \neq \emptyset$, or returns $\emptyset$. If it returns $\emptyset$ and $C \neq \emptyset$ then there is another MIB D which contains C and $v \in (V \setminus S) \setminus C$. We give pseudo-code of `MakeIndMaximal` in Algorithm 3.

Lemma 5. *`MakeIndMaximal` returns a MIB C^+ where $C \subseteq C^+$, $C^+ \subseteq C \cup S$, $C \neq \emptyset$, or returns $\emptyset$.*

Proof. Referring to the pseudo-code in Algorithm 3, it is clear that $C \subseteq C^+$, as no vertices are ever removed from the input biclique C. Furthermore, the only vertices added to C^+ are from S, so $C^+ \subseteq C \cup S$ and C^+ is the only biclique returned by `MakeIndMaximal`. Note that neither side of C is empty and the only vertices added are independent from the side of the biclique which they are added to, so if we do not return $\emptyset$ the object returned is an induced biclique. If no node from outside of S can be added to C^+, then we will not return $\emptyset$ and thus C^+ is maximal.

Lemma 6. *If `MakeIndMaximal` returns $\emptyset$ and $C \neq \emptyset$ then there is another MIB D in G which contains C and $v \in (V \setminus S) \setminus C$.*

Algorithm 3. `MakeIndMaximal`

1: Input: $G = (V, E)$, $C = C_1 \times C_2$, S
2: Let $C_S = S \setminus (C_1 \cup C_2)$
3: **if** $C == \emptyset$ **then**
4:     **return** $\emptyset$
5: **for** $v \in C_S$ **do**
6:     **if** $|N(v) \cap C_1| == |C_1|$ & $|N(v) \cap C_2| == 0$ **then**
7:         $C_2 = C_2 \cup \{v\}$
8:         $C_S \setminus \{v\}$
9: **for** $v \in C_S$ **do**
10:     **if** $|N(v) \cap C_2| == |C_2|$ & $|N(v) \cap C_1| == 0$ **then**
11:         $C_1 = C_1 \cup \{v\}$
12: $V_S = V \setminus (S \cup C_1 \cup C_2)$
13: **for** $v \in V_S$ **do**
14:     **if** $|N(v) \cap C_1| == |C_1|$ & $|N(v) \cap C_2| == 0$ **then**
15:         **return** $\emptyset$
16: **for** $v \in V_S$ **do**
17:     **if** $|N(v) \cap C_2| == |C_2|$ & $|N(v) \cap C_1| == 0$ **then**
18:         **return** $\emptyset$
19: **return** $C^+ = C_1 \times C_2$

Proof. Note that $C \subseteq C^* = C_1 \times C_2$ at line 12. As `MakeIndMaximal` returns $\emptyset$ there must be a vertex $v \in V_S = V \setminus (S \cup C^*)$ which can be added to C^*. Let D be a MIB containing C^* and v, thus D suffices to prove the lemma.

Lemma 7. *`MakeIndMaximal` runs in $O(m)$ time.*

Proof. Note that because G is connected, $n \in O(m)$. Setting C_S and V_S can be done in $O(n)$ time. In each for loop, we can scan all of the edges incident to each v in the iterated-over set and keep count of how many nodes from C_i have been seen (checking for inclusion can be done in $O(1)$ time with an $O(n)$ initialization step). Thus, each edge is scanned at most once per for loop.

A.2 `AddTo`

Recall that `AddTo` takes in (C, v) where $C = C_1 \times C_2$ is an induced biclique and $v \in V \setminus (C_1 \cup C_2)$, and returns the induced biclique where v is added to C_1, $N(v)$ is removed from C_1, and $\overline{N}(v)$ is removed from C_2 if $C_2 \setminus \overline{N}(v) \neq \emptyset$ and $\emptyset$ otherwise. We give pseudo-code of `AddTo` in Algorithm 4.

Lemma 8. *`AddTo` returns the induced biclique where v is added to C_1, $N(v)$ is removed from C_1, and $\overline{N}(v)$ is removed from C_2 if $C_2 \setminus \overline{N}(v) \neq \emptyset$, and $\emptyset$ otherwise.*

Proof. Referring to the pseudo-code in Algorithm 4, it is clear that v is added to C_1 and $N(v)$ is removed from C_1. Additionally v's non-neighbors are effectively removed from C_2 by intersecting it with $N(v)$. If $C_2' = \emptyset$ then $C_2 \setminus \overline{N}(v) = \emptyset$

Algorithm 4. AddTo

```
Input: G = (V, E), C = C1 × C2, v ∈ V \ (C1 ∪ C2)
C1' = (C1 ∪ {v}) \ N(v)
C2' = C2 ∩ N(v)
if C2' == ∅ then
    return ∅
return C1' × C2'
```

and $\emptyset$ is returned. Otherwise $C_1' \neq \emptyset$ since it includes v and thus $C_1' \times C_2'$ is a biclique. $C_1' \times C_2'$ must be an induced biclique as $C_2' \subseteq C_2$, $C_1' \setminus \{v\} \subseteq C_1$, and $C_1 \times C_2$ is an induced biclique and $(N(v) \cap C_1') = \emptyset$ by definition.

Lemma 9. *AddTo runs in $O(m)$ time.*

Proof. Note that because G is connected, $n \in O(m)$. AddTo can be completed by scanning all of v's $O(m)$ incident edges in tandem with an $O(n)$ preprocessing step to allow for constant-time look-ups when checking for inclusion in a set.

B MB-Enumeration Framework Subroutines

We give a detailed description of the MakeMaximal and Consensus subroutines used in OCT-MICA, along with arguments of their correctness and complexity.

B.1 MakeMaximal

Extending a biclique to be maximal is different in the non-induced case from the induced case, since MBs are completely characterized by one side of the biclique.

Algorithm 5. MakeMaximal

```
Input: G = (V, E), B = X × Y
X* = ∩_{i∈Y} N(i)
Y* = ∩_{i∈X*} N(i)
return B* = X* × Y*
```

Lemma 10. *MakeMaximal runs in $O(m)$ time.*

Proof. In order to form X^*, we can scan the edges incident to each $v \in Y$ and keep count of how many nodes from X^* have been seen (checking for inclusion can be done in $O(1)$ time with an $O(n)$ initialization step). The same can be done for Y^*, where instead we scan the edges incident to each $v \in X^*$. Thus, each edge is scanned at most twice in MakeMaximal.

B.2 Consensus

The MICA section of OCT-MICA relies heavily on the Consensus operation introduced in [3] for finding new candidate bicliques. For each pair of bicliques, there are four candidate bicliques which form the *consensus* of the pair. Note that any of the four candidates may be empty and if so discarded. Consensus runs in $O(n)$ time using standard techniques for set union and intersection.

Algorithm 6. Consensus

Input: $G = (V, E)$, $B_\alpha = X_\alpha \times Y_\alpha$, $B_\beta = X_\beta \times Y_\beta$
$B_1 = (X_\alpha \cup X_\beta) \times (Y_\alpha \cap Y_\beta)$
$B_2 = (X_\alpha \cap X_\beta) \times (Y_\alpha \cup Y_\beta)$
$B_3 = (Y_\alpha \cup X_\beta) \times (X_\alpha \cap Y_\beta)$
$B_4 = (X_\alpha \cap Y_\beta) \times (Y_\alpha \cup X_\beta)$
$S = \{\}$
for $B_i = X_i \times Y_i \in \{B_1, B_2, B_3, B_4\}$ **do**
    if $|X_i| > 0$ & $|Y_i| > 0$ **then**
        S.ADD(B_i)
return S

C Additional Enumeration Experiments

Here we include figures corresponding to additional experimental results of our initial benchmarking and on the computation biology data from [29] described in Sects. 5.2 and 5.4 respectively (Figs. 5, 6, 7, 8, 9, 10, 11, 12, 13, 14, 15, 16, 17, 18, 19, 20, 21, 22 and Tables 2, 3).

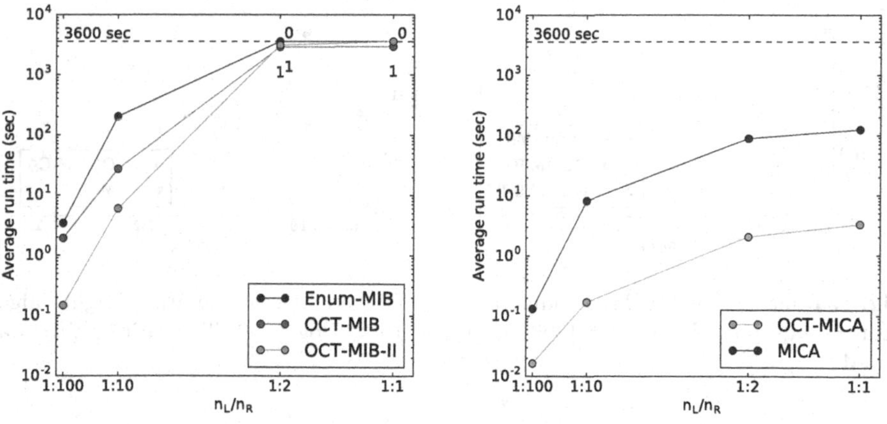

Fig. 5. Runtimes of the MIB-enumerating (left) and MB-enumerating (right) algorithms on graphs where $n_B = 1000$ and $n_O = 10$. The ratio n_L/n_R was varied.

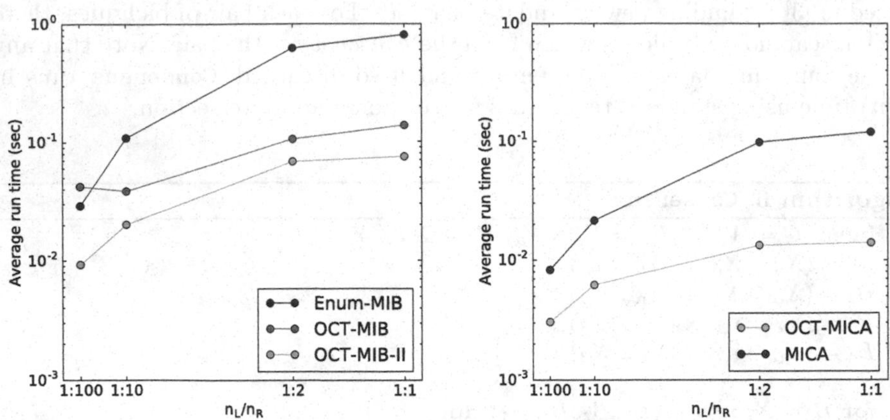

Fig. 6. Runtimes of the MIB-enumerating (left) and MB-enumerating (right) algorithms on graphs where $n_B = 200$ and $n_O = 10$. The ratio n_L/n_R was varied.

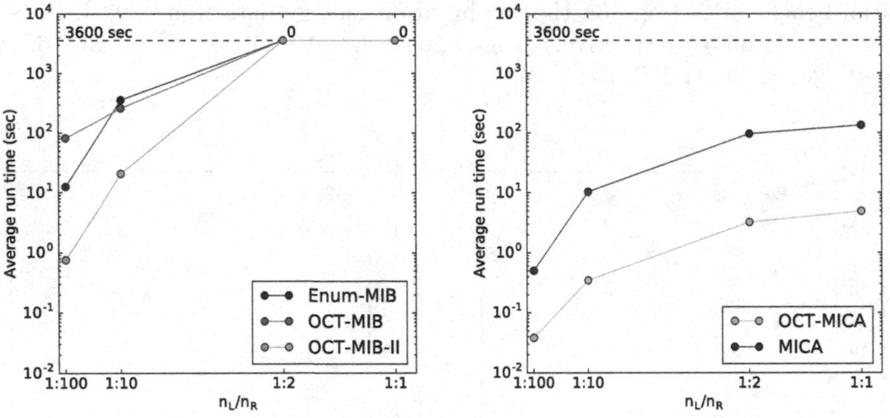

Fig. 7. Runtimes of the MIB-enumerating (left) and MB-enumerating (right) algorithms on graphs where $n_B = 1000$ and $n_O = 19 \approx 3\log_3(n_B)$. The ratio n_L/n_R was varied.

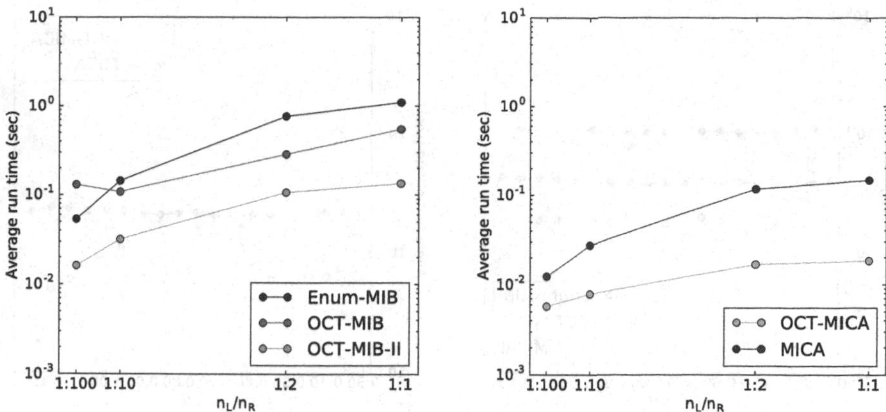

Fig. 8. Runtimes of the MIB-enumerating (left) and MB-enumerating (right) algorithms on graphs where $n_B = 200$ and $n_O = 14 \approx 3\log_3(n_B)$. The ratio n_L/n_R was varied.

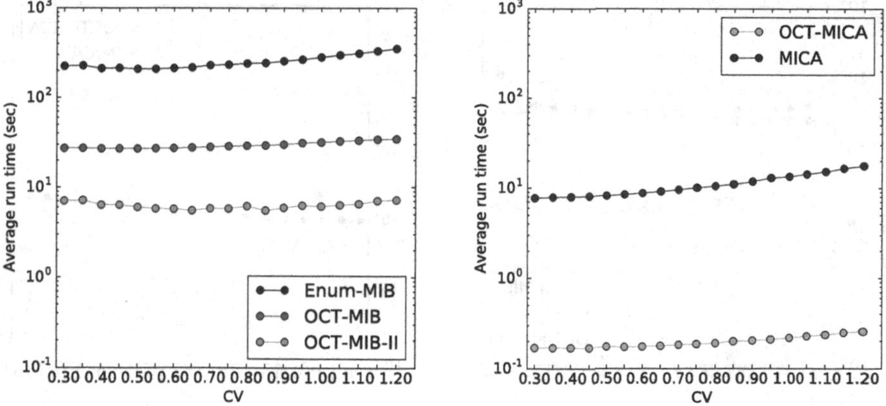

Fig. 9. Runtimes of the MIB-enumerating (left) and MB-enumerating (right) algorithms on graphs where $n_B = 1000$ and $n_O = 10$. The coefficient of variation between L and R was varied.

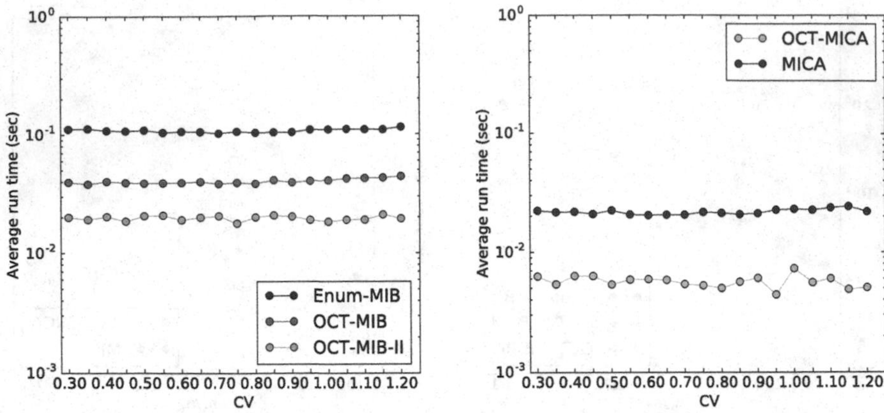

Fig. 10. Runtimes of the MIB-enumerating (left) and MB-enumerating (right) algorithms on graphs where $n_B = 200$ and $n_O = 10$. The coefficient of variation between L and R was varied.

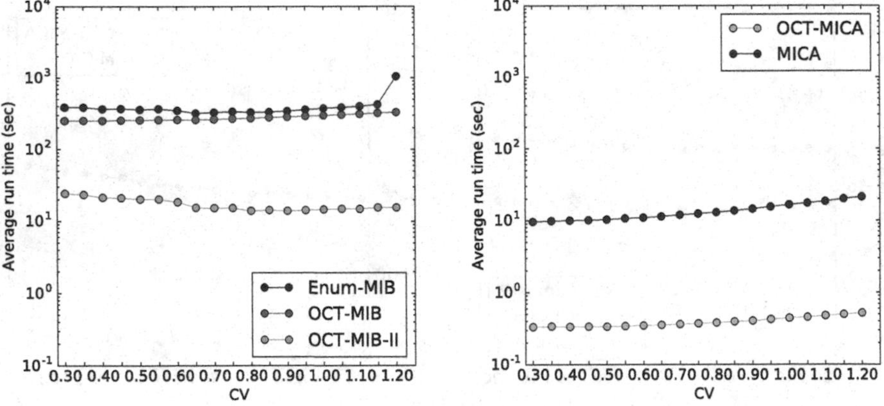

Fig. 11. Runtimes of the MIB-enumerating (left) and MB-enumerating (right) algorithms on graphs where $n_B = 1000$ and $n_O = 19 \approx 3\log_3(n_B)$. The coefficient of variation between L and R was varied.

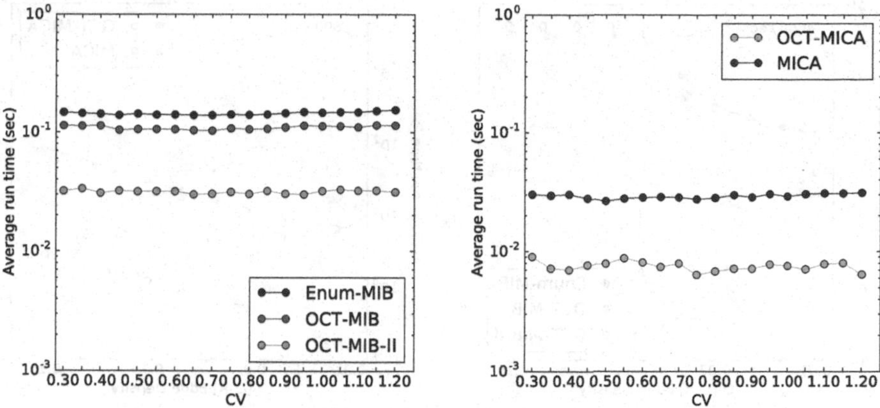

Fig. 12. Runtimes of the MIB-enumerating (left) and MB-enumerating (right) algorithms on graphs where $n_B = 200$ and $n_O = 14 \approx 3\log_3(n_B)$. The coefficient of variation between L and R was varied.

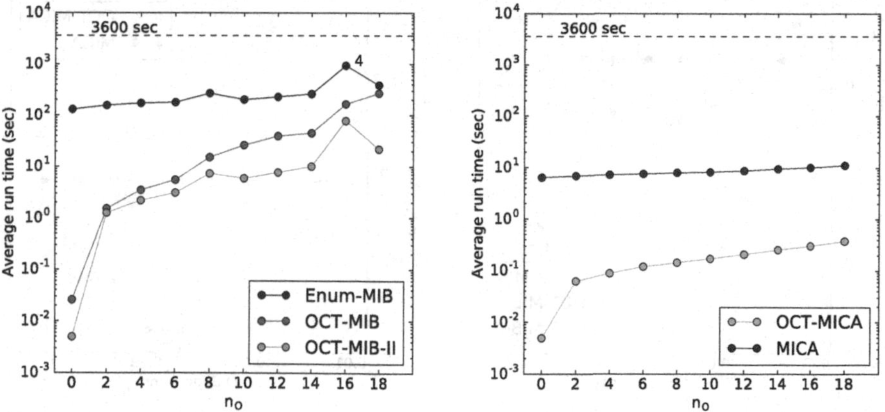

Fig. 13. Runtimes of the MIB-enumerating (left) and MB-enumerating (right) algorithms on graphs where $n_B = 1000$ and n_O was varied.

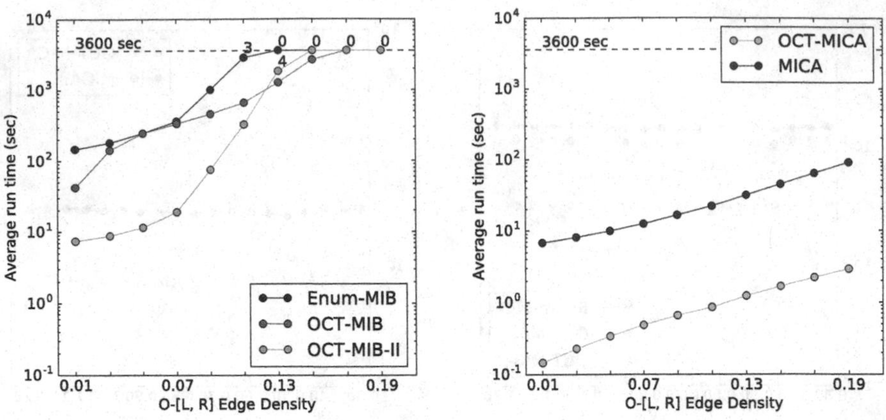

Fig. 14. Runtimes of the MIB-enumerating (left) and MB-enumerating (right) algorithms on graphs where $n_B = 1000$ and $n_O = 19 \approx 3\log_3(n_B)$. The expected edge density between O and $\{L, R\}$ was varied.

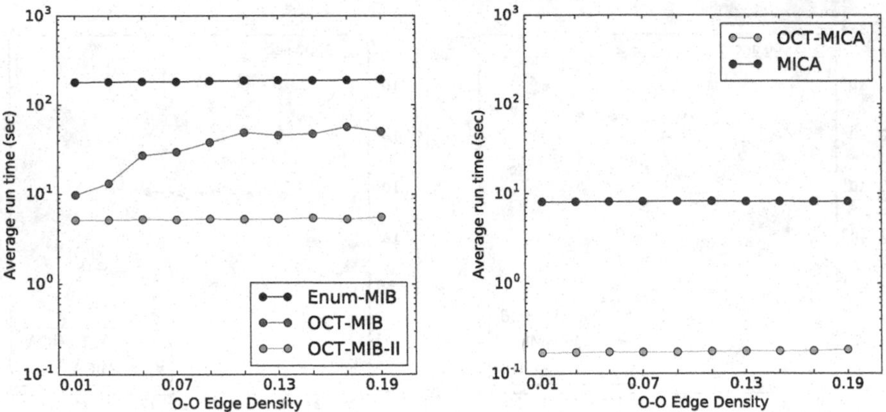

Fig. 15. Runtimes of the MIB-enumerating (left) and MB-enumerating (right) algorithms on graphs where $n_B = 1000$ and $n_O = 10$. The expected edge density within O was varied.

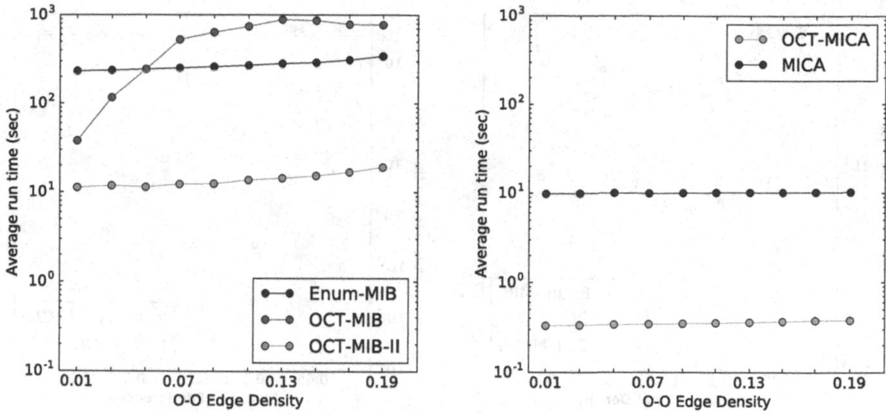

Fig. 16. Runtimes of the MIB-enumerating (left) and MB-enumerating (right) algorithms on graphs where $n_B = 1000$ and $n_O = 19 \approx 3\log_3(n_B)$. The expected edge density within O was varied.

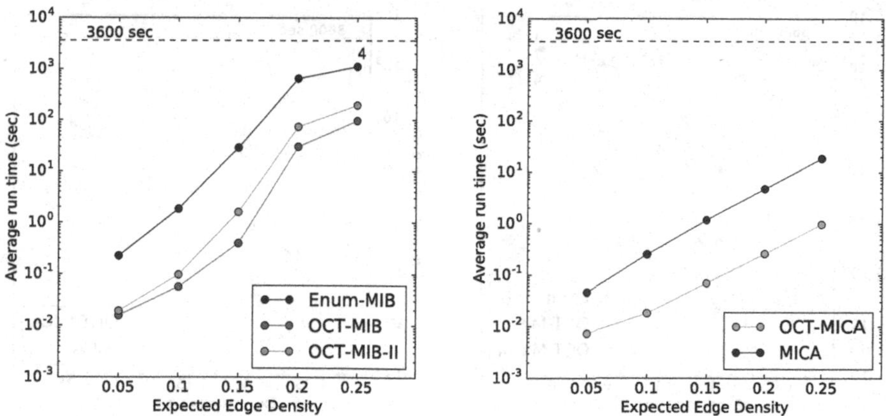

Fig. 17. Runtimes of the MIB-enumerating (left) and MB-enumerating (right) algorithms on graphs where $n_B = 150$, $n_L = n_R$ and $n_O = 5$. The expected edge density in the graph was varied except for the expected edge density within O which was fixed to 0.05.

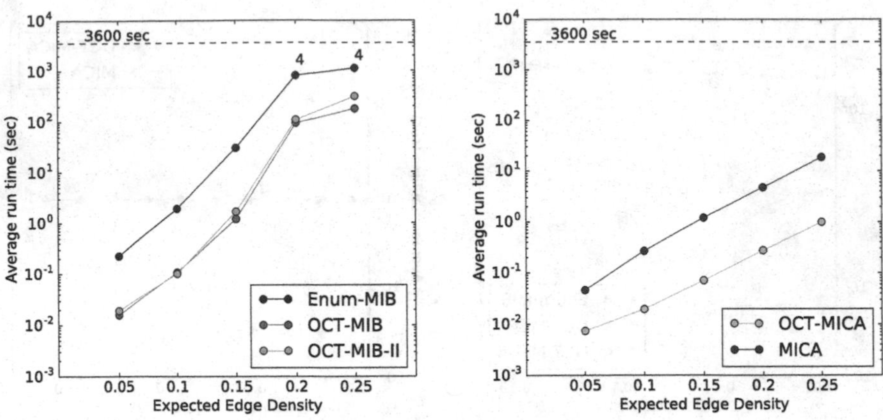

Fig. 18. Runtimes of the MIB-enumerating (left) and MB-enumerating (right) algorithms on graphs where $n_B = 150$, $n_L = n_R$ and $n_O = 5$. The expected edge density in the graph was varied, including the expected edge density within O.

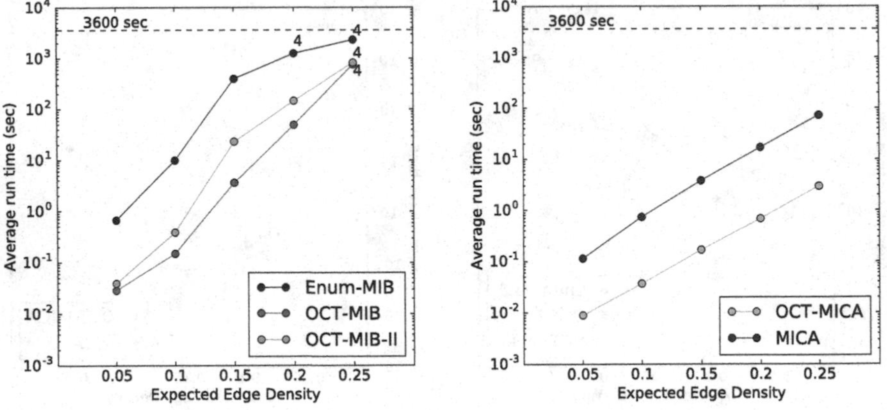

Fig. 19. Runtimes of the MIB-enumerating (left) and MB-enumerating (right) algorithms on graphs where $n_B = 200$, $n_L = n_R$ and $n_O = 5$. The expected edge density in the graph was varied except for the expected edge density within O which was fixed to 0.05.

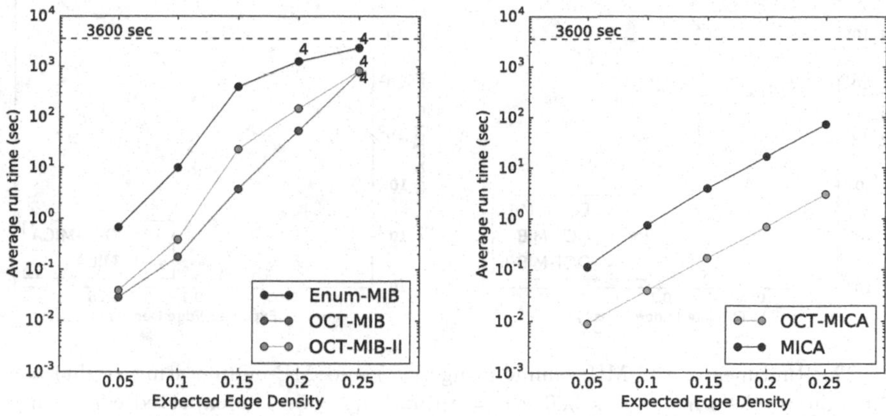

Fig. 20. Runtimes of the MIB-enumerating (left) and MB-enumerating (right) algorithms on graphs where $n_B = 200$, $n_L = n_R$ and $n_O = 5$. The expected edge density in the graph was varied, including the expected edge density within O.

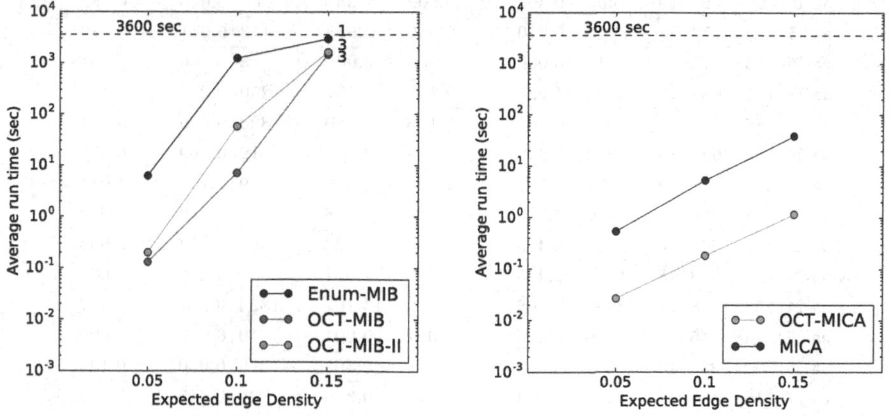

Fig. 21. Runtimes of the MIB-enumerating (left) and MB-enumerating (right) algorithms on graphs where $n_B = 300$, $n_L = n_R$ and $n_O = 5$. The expected edge density in the graph was varied except for the expected edge density within O which was fixed to 0.05.

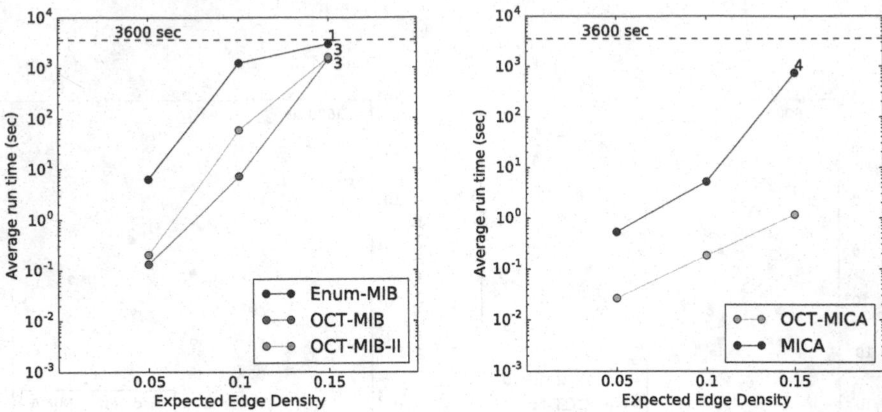

Fig. 22. Runtimes of the MIB-enumerating (left) and MB-enumerating (right) algorithms on graphs where $n_B = 300$, $n_L = n_R$ and $n_O = 5$. The expected edge density in the graph was varied, including the expected edge density within O.

Table 2. The runtimes (rounded to nearest thousandth-of-a-second) of the biclique-enumeration algorithms on the Afro-American subset of the Wernicke-Hüffner computational biology data [29].

G	n_B	m	n_O	$\|M_I\|$	OCT-MIB-II	OCT-MIB	Enum-MIB	$\|M_B\|$	OCT-MICA	MICA
aa-10	69	191	6	178	0.008	0.023	0.057	98	0.007	0.031
aa-11	102	307	11	424	0.055	0.115	0.259	206	0.018	0.120
aa-13	129	383	12	523	0.083	0.239	0.470	269	0.028	0.166
aa-14	125	525	19	1460	0.366	0.902	1.254	605	0.090	0.485
aa-15	66	179	7	206	0.010	0.019	0.053	113	0.011	0.030
aa-16	13	15	0	15	0.000	0.000	0.000	8	0.000	0.000
aa-17	151	633	25	2252	1.023	2.132	3.457	1250	0.242	1.137
aa-18	87	381	14	660	0.100	0.173	0.389	823	0.090	0.351
aa-19	191	645	19	1262	0.449	1.569	2.385	519	0.069	0.450
aa-20	224	766	19	1607	0.705	2.431	3.809	949	0.154	1.061
aa-21	28	90	9	116	0.006	0.013	0.008	213	0.019	0.030
aa-22	167	641	16	1520	0.423	1.387	2.629	560	0.074	0.638
aa-23	139	508	18	1766	0.435	0.788	1.651	1530	0.210	1.000
aa-24	258	1108	21	3890	2.108	9.140	14.167	1334	0.237	2.477
aa-25	14	15	1	10	0.000	0.001	0.001	10	0.000	0.000
aa-26	92	284	13	583	0.084	0.186	0.309	370	0.030	0.128
aa-27	118	331	11	458	0.054	0.270	0.343	229	0.015	0.114
aa-28	167	854	27	2606	1.464	2.201	4.162	2814	0.755	3.250
aa-29	276	1058	21	3122	1.909	8.418	10.707	1924	0.382	3.344
aa-30	39	71	4	56	0.002	0.007	0.006	36	0.002	0.007
aa-31	30	51	2	37	0.002	0.002	0.002	22	0.001	0.002
aa-32	143	750	30	4167	2.286	7.694	5.290	3154	0.684	2.635
aa-33	193	493	4	578	0.046	0.204	0.993	218	0.012	0.218
aa-34	133	451	13	705	0.132	0.316	0.756	275	0.031	0.226

Table 2. (*continued*)

G	n_B	m	n_O	$\|M_I\|$	OCT-MIB-II	OCT-MIB	Enum-MIB	$\|M_B\|$	OCT-MICA	MICA
aa-35	82	269	10	459	0.037	0.108	0.178	215	0.019	0.081
aa-36	111	316	7	248	0.015	0.076	0.155	143	0.011	0.078
aa-37	72	170	5	135	0.005	0.018	0.054	82	0.005	0.022
aa-38	171	862	26	4270	2.428	5.223	7.586	4964	1.136	5.179
aa-39	144	692	23	2153	0.872	1.574	3.034	1177	0.237	1.009
aa-40	136	620	22	2727	1.022	2.086	2.973	1911	0.301	1.324
aa-41	296	1620	40	11705	16.519	82.439	50.205	20375	9.059	47.789
aa-42	236	1110	30	6967	5.646	45.560	21.244	8952	2.428	13.479
aa-43	63	308	18	905	0.137	0.294	0.311	875	0.116	0.302
aa-44	59	163	10	211	0.014	0.024	0.051	158	0.008	0.037
aa-45	80	386	20	1768	0.336	0.775	0.859	1716	0.244	0.796
aa-46	161	529	13	719	0.157	0.438	0.922	374	0.036	0.257
aa-47	62	229	14	572	0.057	0.082	0.138	451	0.051	0.127
aa-48	89	343	17	896	0.144	0.338	0.497	519	0.060	0.230
aa-49	26	62	5	50	0.004	0.002	0.003	74	0.006	0.013
aa-50	113	468	18	1272	0.322	0.778	1.098	1074	0.132	0.612
aa-51	78	274	11	429	0.035	0.082	0.174	250	0.020	0.078
aa-52	65	231	14	690	0.073	0.135	0.200	431	0.040	0.122
aa-53	88	232	12	340	0.036	0.186	0.162	199	0.011	0.052
aa-54	89	233	12	286	0.027	0.063	0.113	177	0.015	0.039

Table 3. The runtimes (rounded to nearest thousandth-of-a-second) of the biclique-enumeration algorithms on the Japanese subset of the Wernicke-Hüffner computational biology data [29].

G	n_B	m	n_O	$\|M_I\|$	OCT-MIB-II	OCT-MIB	Enum-MIB	$\|M_B\|$	OCT-MICA	MICA
j-10	55	117	3	52	0.002	0.009	0.010	39	0.001	0.010
j-11	51	212	5	63	0.003	0.014	0.011	36	0.003	0.012
j-13	78	210	6	224	0.015	0.028	0.074	90	0.009	0.032
j-14	60	107	4	44	0.004	0.007	0.003	38	0.003	0.003
j-15	44	55	1	13	0.001	0.000	0.004	10	0.001	0.000
j-16	9	10	0	10	0.000	0.000	0.000	3	0.000	0.000
j-17	79	322	10	317	0.025	0.051	0.127	126	0.014	0.056
j-18	71	296	9	154	0.011	0.038	0.053	91	0.012	0.028
j-19	84	172	3	105	0.002	0.010	0.019	46	0.002	0.013
j-20	241	640	1	274	0.013	0.065	0.484	228	0.009	0.188
j-21	33	102	9	107	0.006	0.012	0.008	197	0.017	0.024
j-22	75	391	9	221	0.020	0.051	0.080	113	0.009	0.048
j-23	76	369	19	682	0.095	0.404	0.217	459	0.057	0.132
j-24	142	387	4	150	0.013	0.027	0.089	104	0.007	0.025
j-25	14	14	0	14	0.000	0.000	0.000	3	0.000	0.000
j-26	63	156	6	156	0.007	0.019	0.035	67	0.003	0.013
j-28	90	567	13	492	0.073	0.130	0.244	416	0.044	0.193

References

1. Agarwal, P., Alon, N., Aronov, B., Suri, S.: Can visibility graphs be represented compactly? Discret. Comput. Geom. **12**, 347–365 (1994)
2. Akiba, T., Iwata, Y.: Branch-and-reduce exponential/FPT algorithms in practice: a case study of vertex cover. Theoret. Comput. Sci. **609**, 211–225 (2016)
3. Alexe, G., Alexe, S., Crama, Y., Foldes, S., Hammer, P., Simeone, B.: Consensus algorithms for the generation of all maximal bicliques. Discret. Appl. Math. **145**, 11–21 (2004)
4. Dawande, M., Keskinocak, P., Swaminathan, J., Tayur, S.: On bipartite and multipartite clique problems. J. Algorithms **41**, 388–403 (2001)
5. Dias, V., De Figueiredo, C., Szwarcfiter, J.: Generating bicliques of a graph in lexicographic order. Theoret. Comput. Sci. **337**, 240–248 (2005)
6. Eppstein, D.: Arboricity and bipartite subgraph listing algorithms. Inf. Process. Lett. **51**, 207–211 (1994)
7. Garey, M., Johnson, D.: Computers and Intractability: A Guide to NP-Completeness. Freeman, San Fransisco (1979)
8. Gély, A., Nourine, L., Sadi, B.: Enumeration aspects of maximal cliques and bicliques. Discret. Appl. Math. **157**(7), 1447–1459 (2009)
9. Goodrich, T., Horton, E., Sullivan, B.: Practical graph bipartization with applications in near-term quantum computing,. arXiv preprint arXiv:1805.01041, 2018
10. Gülpinar, N., Gutin, G., Mitra, G., Zverovitch, A.: Extracting pure network submatrices in linear programs using signed graphs. Discret. Appl. Math. **137**, 359–372 (2004)
11. Horton, E., Kloster, K., Sullivan, B.D., van der Poel, A., Woodlief, T.: MI-bicliques: Version 2.0, August 2019. https://doi.org/10.5281/zenodo.3381532
12. Hüffner, F.: Algorithm engineering for optimal graph bipartization. In: Nikoletseas, S.E. (ed.) WEA 2005. LNCS, vol. 3503, pp. 240–252. Springer, Heidelberg (2005). https://doi.org/10.1007/11427186_22
13. Chang, W.: Maximal biclique enumeration, December 2004. http://genome.cs.iastate.edu/supertree/download/biclique/README.html
14. Iwata, Y., Oka, K., Yoshida, Y.: Linear-time FPT algorithms via network flow. In: SODA, pp. 1749–1761 (2014)
15. Kaytoue-Uberall, M., Duplessis, S., Napoli, A.: Using formal concept analysis for the extraction of groups of co-expressed genes. In: Le Thi, H.A., Bouvry, P., Pham Dinh, T. (eds.) MCO 2008. CCIS, vol. 14, pp. 439–449. Springer, Heidelberg (2008). https://doi.org/10.1007/978-3-540-87477-5_47
16. Kaytoue, M., Kuznetsov, S., Napoli, A., Duplessis, S.: Mining gene expression data with pattern structures in formal concept analysis. Inf. Sci. **181**, 1989–2011 (2011)
17. Kloster, K., Sullivan, B., van der Poel, A.: Mining maximal induced bicliques using odd cycle transversals. In: Proceedings of the 2019 SIAM International Conference on Data Mining (2019, to appear)
18. Kumar, R., Raghavan, P., Rajagopalan, S., Tomkins, A.: Trawling the web for emerging cyber-communities. Comput. Netw. **31**, 1481–1493 (1999)
19. Kuznetsov, S.: On computing the size of a lattice and related decision problems. Order **18**, 313–321 (2001)
20. Li, J., Liu, G., Li, H., Wong, L.: Maximal biclique subgraphs and closed pattern pairs of the adjacency matrix: a one-to-one correspondence and mining algorithms. IEEE Trans. Knowl. Data Eng. **19**, 1625–1637 (2007)

21. Lokshtanov, D., Saurabh, S., Sikdar, S.: Simpler parameterized algorithm for OCT. In: Fiala, J., Kratochvíl, J., Miller, M. (eds.) IWOCA 2009. LNCS, vol. 5874, pp. 380–384. Springer, Heidelberg (2009). https://doi.org/10.1007/978-3-642-10217-2_37
22. Makino, K., Uno, T.: New algorithms for enumerating all maximal cliques. In: Hagerup, T., Katajainen, J. (eds.) SWAT 2004. LNCS, vol. 3111, pp. 260–272. Springer, Heidelberg (2004). https://doi.org/10.1007/978-3-540-27810-8_23
23. Mushlin, R., Kershenbaum, A., Gallagher, S., Rebbeck, T.: A graph-theoretical approach for pattern discovery in epidemiological research. IBM Syst. J. **46**, 135–149 (2007)
24. Panconesi, A., Sozio, M.: Fast hare: a fast heuristic for single individual SNP haplotype reconstruction. In: Jonassen, I., Kim, J. (eds.) WABI 2004. LNCS, vol. 3240, pp. 266–277. Springer, Heidelberg (2004). https://doi.org/10.1007/978-3-540-30219-3_23
25. Peeters, R.: The maximum edge biclique problem is NP-complete. Discret. Appl. Math. **131**, 651–654 (2003)
26. Sanderson, M., Driskell, A., Ree, R., Eulenstein, O., Langley, S.: Obtaining maximal concatenated phylogenetic data sets from large sequence databases. Mol. Biol. Evol. **20**, 1036–1042 (2003)
27. Schrook, J., McCaskey, A., Hamilton, K., Humble, T., Imam, N.: Recall performance for content-addressable memory using adiabatic quantum optimization. Entropy **19**, 500 (2017)
28. Tsukiyama, S., Ide, M., Ariyoshi, H., Shirakawa, I.: A new algorithm for generating all the maximal independent sets. SIAM J. Comput. **6**, 505–517 (1977)
29. Wernicke, S.: On the algorithmic tractability of single nucleotide polymorphism (SNP) analysis and related problems (2014)
30. Wille, R.: Restructuring lattice theory: an approach based on hierarchies of concepts. In: Rival, I. (ed.) Ordered Sets. NATO Advanced Study Institutes Series (Series C– Mathematical and Physical Sciences), vol. 83, pp. 445–470. Springer, Dordrecht (1982). https://doi.org/10.1007/978-94-009-7798-3_15
31. Yannakakis, M.: Node-and edge-deletion NP-complete problems. In: STOC, pp. 253–264 (1978)
32. Zhang, Y., Phillips, C.A., Rogers, G.L., Baker, E.J., Chesler, E.J., Langston, M.A.: On finding bicliques in bipartite graphs: a novel algorithm and its application to the integration of diverse biological data types. BMC Bioinform. **15**, 110 (2014)

k-Maximum Subarrays for Small k: Divide-and-Conquer Made Simpler

Ovidiu Daescu and Hemant Malik(✉)

University of Texas at Dallas, Richardson, TX 75080, USA
{daescu,malik}@utdallas.edu

Abstract. Given an array A of n real numbers, the maximum subarray problem is to find a contiguous subarray which has the largest sum. The k-maximum subarrays problem is to find k such subarrays with the largest sums. For the 1−maximum subarray the well known divide-and-conquer algorithm, presented in most textbooks, although suboptimal, is easy to implement and can be made optimal with a simple change that speeds up the combine phase. On the other hand, the only known divide-and-conquer algorithm for $k > 1$, that is efficient for small values of k, is difficult to implement, due to the intricacies of the combine phase. In this paper, we show how to simplify the combine phase considerably while preserving the overall running time.

In the process of designing the combine phase of the algorithm we provide a simple, sublinear, $O(\sqrt{k}\log^3 k)$ time algorithm, for finding the k largest sums of $X + Y$, where X and Y are sorted arrays of size n and $k \leq n^2$. The k largest sums are implicitly represented and can be enumerated with an additional $O(k)$ time.

Our solution relies on simple operations such as merging sorted arrays, binary search and selecting the kth smallest number in an array. We have implemented our algorithm and report excellent performance as compared to previous results.

Keywords: k-Maximum subarrays · Divide and conquer · X + Y · Sublinear

1 Introduction

The well-known problem of finding the maximum sum (contiguous) subarray of a given array of real numbers has been used in various applications and received much attention over time. Some of the applications are in data mining [1,13], pattern recognition [14], and image processing and communication [6].

Given an array A of n real numbers and an integer k, such that $1 \leq k \leq n(n+1)/2$, the k-maximum subarrays problem is to find k contiguous subarrays with the largest sums (not necessarily in sorted order of the sums). If $k=1$, Kadane's algorithm [5] solves the maximum subarray problem in O(n) time using an iterative method. On the other hand, the well-known divide and conquer

I. Kotsireas et al. (Eds.): SEA² 2019, LNCS 11544, pp. 454–472, 2019.
https://doi.org/10.1007/978-3-030-34029-2_29

algorithm [10], found in virtually all algorithms textbooks, has a suboptimal $O(n \log n)$ running time. An $O(n)$ time divide-and-conquer algorithm is briefly presented in [4].

For $k > 1$, Bengtsson and Chen [3] presented an algorithm that takes time $O(\min\{k+n\log^2 n, n\sqrt{k}\})$, where the second term, $O(n\sqrt{k})$, comes from a divide-and-conquer solution. That divide-and-conquer algorithm is difficult to implement, due to the intricacies of the combine phase.

In this paper we propose a competitive, $O(n\sqrt{k})$ time divide and conquer solution to find the k-maximum subarrays, which is optimal for $k = O(1)$ and $k = \theta(n^2)$. Our algorithm is much simpler than the one in [3] due to a more direct way of performing the combine phase. Specifically, the combine phase we propose is itself a simple recursive procedure. To this end, we also address the following subproblem: *Given two sorted arrays of real numbers, X and Y, each of size n, let S be the set $S = \{(x, y)|x \in X$ and $y \in Y\}$, with the value of each pair in S defined as $Val(x, y) = x + y$. Find the k pairs from S with largest values.* This problem is closely related to the famous pairwise sum $(X + Y)$ problem [11,12], that asks to sort all pairwise sums. Our main contribution is a sublinear, $O(\sqrt{k}\log^3 k)$ time algorithm, for finding the k largest sums of $X+Y$. The k largest sums are implicitly represented and can be enumerated with an additional $O(k)$ time. A key feature of our solution is its simplicity, compared to previous algorithms [11,12], that find and report the k largest sums in $O(k+\sqrt{k})$ time. Our algorithm uses only operations such as merging sorted arrays, binary search, and selecting the kth largest number of an array.

We have implemented our algorithms in JAVA and performed extensive experiments on macOS High Sierra with 3.1 GHz Intel i5 processor and 8 GB of RAM, reporting excellent performance. For example, on random arrays of size 10^6, with $k = 10^6$, we can find the k maximum subarrays in about 52 s.

The rest of the paper is organized as follows. In Sect. 2 we discuss previous results. In Sect. 3 we describe the divide and conquer algorithm for k maximum subarray and continue on to present a $O(n\sqrt{k})$ time algorithm, in Sect. 4. We also describes a $O(\sqrt{k}\log^3 k)$ time solution for finding the k largest sums of $X + Y$. We discuss the implementation details, experimental results and the comparison with previous results in Sect. 5.

2 Previous Work

Bengtsson, and Chen [3] provided a complex, $O(min\{k + n\log^2 n, n\sqrt{k}\})$ time algorithm to solve the k-maximum subarray problem. Their main algorithm, for general k, has five phases. First, the problem is reduced to finding the top k maximum values over all the "good" elements in some matrix of size $n \times n$. In the second phase, repeated constraint searches are performed, which decrease the number of candidate elements to $O(\min\{kn, n^2\})$. In the third phase, a range reduction procedure is performed to reduce the number of candidates further to $\theta(k)$. In the fourth phase, a worst-case linear-time selection algorithm is used on the remaining candidates, resulting in an element x, that is the kth largest sum. The final phase involves finding the "good" elements with values not less than x. The $O(n\sqrt{k})$ part of the running time comes from a divide-and-conquer solution, and is useful for small values of k. The combine phase of the divide and conquer algorithm uses the $O(\sqrt{k})$ time algorithm from [12] to find the kth largest element in a sorted matrix, which is fairly difficult to understand and tedious to implement. A trivial lower bound for this problem is $O(n+k)$.

In the same year (2006), Bae and Takaoka [2] provided an $O((n+k)\log k)$ solution that reports the k maximum subarrays in sorted order.

Still in 2006, Cheng, Chen, Tien, and Chao [9] provided an algorithm with $O(n+k\log(\min\{n,k\}))$ running time. The authors adapted an iterative strategy where the table of possible subarray sums is built partially after every iteration, and the algorithm terminates in $O(\log n)$ iterations, which yields a time complexity of $O(n+k\log(\min\{n,k\}))$.

Finally, in 2007, Bengtsson and Chen [4] provided a solution that takes time $O(n+k\log n)$ in the worst case to compute and rank all k subsequences. They also proved that their algorithm is optimal for $k=O(n)$, by providing a matching lower bound. Their approach is different from the previous ones. In particular, although only briefly described, their solution provides an O(n) time algorithm for the maximum subarray ($k=1$) problem. They give a tree-based algorithm that uses a full binary tree, augmented with information about prefix sums, suffix sums, sums, and ranking among subsequences concerning their sums. There are two phases of this algorithm. In the first phase, initial information (prefix sum, suffix sum, sum, largest elements) is computed and stored in the tree. The tree is constructed in a bottom-up fashion. The algorithm is based on the well-known observation that the maximum sum can be obtained from the left branch or the right branch, or from a subsequence spanning over the left and right branches (subarrays). The second phase is the query phase which uses a binary heap to compute the k-maximum subarrays. A special property of this algorithm is that if l largest sums are already computed then the $(l+1)^{th}$ largest sum can be found in $O(\log n)$ time.

Frederickson and Johnson [12] provided an efficient algorithm to find the kth maximum element of a matrix with sorted rows and columns. When the sorted matrix has k rows and k columns, their algorithm finds the kth largest element in O($\sqrt{k}$) time. It can then be used to find and report the k largest values in the matrix in an additional $O(k)$ time. This corresponds to finding and reporting the k largest values of $X+Y$. The algorithm is not simple and is tedious to implement.

Very recently, in 2018, Kaplan et al. [15] provided a simple, comparison-based, output sensitive algorithm to report the kth smallest element from a collection of sorted lists, and from $X+Y$, where both X and Y are two unordered sets. They show that we only need $O(m+\sum_{i=1}^{m}\log(k_i+1))$ comparisions for selecting the kth smallest item or finding the set of k smallest items from a collection of m sorted lists, where the ith list has k_i items that belong to the overall set of k smallest items. They use the "soft heap" data structure introduced by Chazelle [8] in 2000. A soft heap is a simple variant of the priority queue. The data structure supports all operations of the standard heap and every operation can be done in constant time except for insert, which takes $O(\log 1/\epsilon)$ time, where ϵ is an error rate ($0 \leq \epsilon \leq 1/2$) ensuring that, at any time, at most ϵn elements have their keys raised.

3 k-Maximum Subarrays by Divide and Conquer

Given an array A of n numbers and an integer k, where $1 \leq k \leq n(n+1)/2$, the k-maximum subarrays problem is to find the k contiguous subarrays with the largest sums. A detailed description of a simple, linear time divide and conquer algorithm to find the maximum subarray (k = 1) is provided in Appendix 1. In this section, we provide an $O(n\sqrt{k}\log(1/\epsilon))$ time divide and conquer algorithm to address the k-maximum subarrays problem. The divide part of the algorithm is similar as of [3] where the array is recursively divided into two subarrays with equal number of elements until a base case (of size $\sqrt{k}$) is reached. The main difference between our algorithm and [3] is that recursive calls return information about k largest subarrays from corresponding subproblems, including k largest sums from the left and the right, and we are finding k largest subarray values in the combine step. A detailed description of the generic algorithm is given in [3]. Our primary goal is to simplify the combine phase.

In this paper, notations like max_left, max_right, max_cross, and max_sub refer to arrays of size k, holding the corresponding k largest sum values. Except for max_cross, these arrays are sorted in non-increasing order.

Consider the left and right subarrays, A_l and A_r, of some internal node v in the recursion tree. The k largest sums at v are among the k largest sums from A_l, the k largest sums from A_r, and the k largest sums of contiguous subarrays that cross between A_l and A_r (we call these last sums *crossing sums*). The difficult part is to efficiently compute the k crossing sums and the various k largest sums that need to be passed up to the parent node.

In this section, we provide a solution for the combine step that is simple yet efficient, easy to implement. The function **MERGE** used in the following algorithms is similar to the one in the *merge-sort* sorting algorithm, except that we stop after finding the largest k values, and takes O(k) time. By a slight abuse, we allow the MERGE function to work with a constant number of arrays in the input, rather than just two arrays. If there are more than two arrays passed to the MERGE function we perform a pairwise merge to find the k largest numbers. Similarly, function **SELECT**, whenever mentioned, is the standard linear time selection function [7], that finds the kth (and thus k) largest number(s) of a given set of ($O(k)$ in our case) numbers.

The function **SUM (a, A)** used below takes in the input an array A of size k and an integer a and adds a to each entry of A. It is used to add the value of the sum of elements of the left (or right) child of v to the k largest sum prefix (suffix) values of the right (resp., left) child of v.

The function **MAX_SUM_CROSS (A, B)** takes as input two arrays, A and B, each of size k, sorted in non-increasing order and outputs the k-maximum sums of the pairwise addition of A and B. We can use priority queues to find k maximum sums which takes $O(k \log k)$ time as mentioned in Appendix 2. However, soft heap [8] recently used by Kaplan [15] can find the kth largest sum, and the k maximum sums in $O(k \log(1/\epsilon))$ time.

The **Max-k** algorithm below computes the k-maximum sums (subarrays) of the given array A. The values *low* and *high* correspond to the start index and end index of the subarray $A[low \dots high]$. The following algorithm is also used in [3] with a different combine phase, MAX_SUM_CROSS.

Algorithm 1. Max-k (A, low, high)

if $(low + \sqrt{k} \geq high)$ then find $max\_left, max\_right, sum, max\_sub$ by brute force and return $(max\_left, max\_right, sum, max\_sub)$
mid $= \lfloor \frac{low+high}{2} \rfloor$
$(max\_left1, max\_right1, sum1, max\_sub1) =$ **Max-k** (A, low, mid)
$(max\_left2, max\_right2, sum2, max\_sub2) =$ **Max-k** $(A, mid+1, high)$
$max\_left =$ **MERGE**$(max\_left1,$ **SUM** $(sum1, max\_left2))$;
$max\_right =$ **MERGE**$(max\_right2,$ **SUM** $(sum2, max\_right1))$
$sum = sum1 + sum2$
$max\_cross =$ **MAX_SUM_CROSS**(max_right1, max_left2)
$max\_sub =$ **MERGE**$(max\_cross, max\_sub1, max\_sub2)$
return $(max\_left, max\_right, sum, max\_sub)$

Running Time of Max-k: It can be easily seen that the running time of algorithm **Max-k** is described by the recurrence:
$T(n,k) = 2T(n/2,k) + O(k\log(1/\epsilon))$, with $T(\sqrt{k},k) = O(k)$.
Using substitution method, we have

$$T(n,k) = 2^i T(n/2^i, k) + O(\sum_{j=0}^{i-1} 2^j k \log 1/\epsilon)$$

Letting $(n/2^i)^2 = k$ results in $2^i = n/\sqrt{k}$.

$$T(n,k) = (n/\sqrt{k})T(\sqrt{k},k) + O(n\sqrt{k}\log 1/\epsilon)$$

Since $T(\sqrt{k},k) = O(k)$ the overall time complexity is $O(n\sqrt{k}\log 1/\epsilon)$. The algorithm is an $O(\log 1/\epsilon)$ factor slower than the one in [3], while being simple to describe and implement.

In the next section we provide a simple, $O(k)$ time prune-and-search algorithm for the *MAX_SUM_CROSS* procedure, which improves the overall time complexity of algorithm **Max-k** to $O(n\sqrt{k})$ and also finds the kth maximum sum in $O(\sqrt{k}\log^3 k)$.

4 An Improved Algorithm for k-Maximum Subarrays

In this section, we improve the results in the previous section by providing an $O(k)$ time divide-and-conquer solution for the combine phase. To this end, we first find the kth largest element x of the pairwise sum $A+B$ [11,12], and then scan A and B for elements in $A+B$ greater than or equal to x. If this output would be sorted, then it leads to an $O(k \log k)$ running time. However, as explained later in this section, there is no need to sort these elements, that correspond to the values of the crossing sums.

Frederickson and Johnson [12] provided an algorithm that can find the kth maximum element of a matrix consisting of k rows and k columns, each sorted in nonincreasing order, in $O(\sqrt{k})$ time. Given a sorted matrix M, the algorithm:

1. Extracts a set S of submatrices of different shapes which guarantee to contain all elements greater than or equal to the kth largest element of M. These matrices also contain elements which are less than the kth largest element.
2. Given a set of sorted submatrices, the algorithm forms a new matrix with the help of dummy matrices (matrices where all entries are $-\infty$). The new matrix is also a sorted matrix. The submatrices are referred to as *cells*, and for each cell C, $\min(C)$ and $\max(C)$ represent the smallest and largest elements in this cell. Initially, there is a single cell which is the matrix formed from dummy matrices and the set S.
3. After each iteration, a cell is divided into four subcells. From all the subcells formed in the previous steps, the algorithm computes some values that allow to discard a few cells guaranteed not to contain the kth largest element.

The algorithm is not easy to follow especially in the second step where sorted submatrices are combined with dummy matrices to form another sorted matrix. In step 1, the authors are creating submatrices from a given matrix. This is done efficiently by storing the start and end indices of each submatrix. For the second step, new dummy matrices are combined with the submatrices from previous steps to create a sorted matrix. Storing indices for all these sub-matrices and finding the sorted matrix is not practical. In contrast, for our algorithm, mentioned in the rest of the section, the only operations required are binary search and sorting which are easy to implement. We do not create new sub-matrices and our algorithm does not involve large amounts of matrix manipulations.

Intuitively, for our problem, the rows and columns of the matrix are generated by the sums in $A+B$, where A and B are sorted arrays of size k each. In row i, $A[i]$ is summed over the entries in array B. Similarly, in column j, $B[j]$ is summed over the entries in the array A. The matrix does not have to be explicitly stored as the matrix entries can be generated as needed from the values in A and B. Thus, using the algorithm in [12] one can compute the kth maximum element x of $A+B$ in $O(\sqrt{k})$ time. Retrieving the elements of $A+B$ that are greater than (or equal) to x takes an additional $O(k)$ time. This makes the algorithm **MAX_SUM_CROSS** in previous section run in $O(k)$ time. Since the k largest crossing sum values are no longer sorted, we replace the **MERGE** call in line 9 of algorithm **Max-k** with a **SELECT** call. The algorithm, as presented above, has been described in [3].

The only place where the k largest crossing sum values are used as an internal node u of the recursion tree is in the calculation of the k largest sum values at u, given the k largest sum values from the left and the right children of u. Let v be the parent node of u. Node u needs to pass up to v the k largest sum values of subarrays that start at the leftmost entry, and the k largest sum values of subarrays that start at the rightmost entry (max_left and max_right arrays at u) and these subarrays are either distinct from the crossing subarrays at u or computed independently of those subarrays by function **MAX_SUM**. See Fig. 3 in Appendix 1.

The following lemma is implicitly used in [3].

Lemma 1. *The k largest crossing sum values do not need to be sorted for algorithm **Max-k** to correctly report the k largest sum values of A.*

Then, the running time of the **Max-k** algorithm is now described by the recurrence

$$T(n,k) = 2T(n/2,k) + O(k)$$

with $T(\sqrt{k},k) = O(k)$. As described in [3], using the substitution method, we have

$$T(n,k) = 2^i T(n/2^i,k) + O(\sum_{j=0}^{i-1} k2^j)$$

Letting $(n/2^i)^2 = k$ results in $2^i = n/\sqrt{k}$.

$$T(n,k) = (n/\sqrt{k})T(\sqrt{k},k) + O(n\sqrt{k})$$

Thus, $T(n) = O(n\sqrt{k})$.

As mentioned earlier, the algorithm for finding the kth largest entry in $A+B$ presented in [12], is complex and tedious to implement. In what follows, we provide a simple algorithm to find the kth largest element in $A+B$, which takes $O(\sqrt{k}\log^3 k)$ time and is easy to implement. Moreover, unlike the algorithm in [12], our algorithm is a simple prune-and-search procedure. Also, unlike in [12], our algorithm implicitly finds the k largest elements of A + B in the process. The total time needed to report all k largest elements in $A+B$ is then $O(k+\sqrt{k}\log^3 k)$ which is still $O(k)$, and thus the final time complexity of algorithm **Max-k** remains $O(n\sqrt{k})$.

Let A and B be arrays of size n and let k be an integer such that $k \leq n^2$. We now show how to find the kth-largest element of A + B and an implicit representation of the k largest elements of A + B in sublinear, $O(\sqrt{k}\log^3 k)$ time.

Consider a matrix M with n rows and n columns, such that the element $M[i,j]$ in matrix is the sum $A[i] + B[j]$. Without loss of generality, assume that $k \leq n$. We call M a *sorted matrix*. Note that for the k-maximum subarray

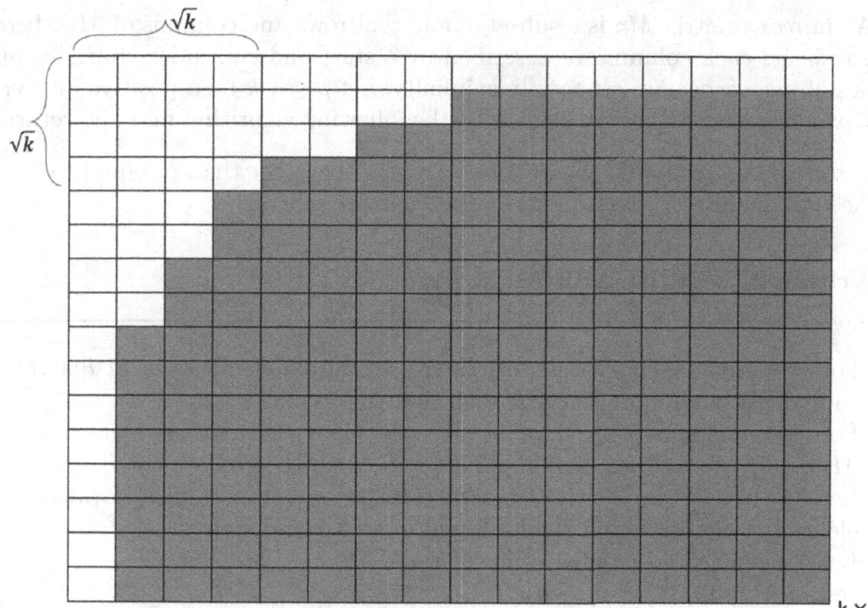

Fig. 1. Sorted matrix M of size $k \times k$ where $k = 16$. k^{th} largest element will not lie in shaded region; therefore the shaded region of M is irrelevant.

problem, matrix M is of size $k \times k$. The notation $M[i][0 : j]$ denotes the entries in row i of the matrix M, columns 0 to j. The notation $M[0 : i][j]$ denotes the entries in column j of matrix M, rows 0 to i.

Matrix M is only considered for better understanding of the algorithms presented in this section, but there is no need to store it explicitly. Instead, it's entries are computed only as needed. The only information required is the start and end index of each row $\in [1, \sqrt{k}]$ and each column $\in [1, \sqrt{k}]$, that define the "active" entries of M at a given step. Let s_i^r, e_i^r be the start and end index of row i and s_j^c, e_j^c be the start and end index of column j. For each row $i \in [1, \sqrt{k}]$, initialize value of s_i^r with 1 and e_i^r with $\lceil k/i \rceil$). Similarly, for each column $j \in [1, \sqrt{k}]$, initialize value of s_j^c with 1 and e_j^c with $\lceil k/j \rceil$. It is easy to observe that all rows and columns in matrix M are sorted and for each row $i \in [1, \sqrt{k}]$ and,

1. The k largest values will not lie in $M[i][e_i^r + 1, \min\{n, k\}]$.
2. For each column $j \in [1, \sqrt{k}]$, the k largest values will not lie in $M[e_j^c + 1, \min\{n, k\}][j]$.
3. The k largest values will not lie in submatrix $M[\sqrt{k}, \sqrt{k}][\min\{k, n\}, \min\{k, n\},]$.

This irrelevant region is shaded in Fig. 1. We can globally store the start and end indexes of these rows and columns and update them as necessary which takes $O(\sqrt{k})$ time.

A *staircase* matrix M_S is a subset of adjacent rows and columns of M, where each row and each column are described by a start and end index. Initially, all entries of max_cross are set to minus infinity. Let $p \leq \sqrt{k}$ be a positive integer.

Given a matrix M and an element x, the following algorithm find and return:

1. a staircase matrix of M where all elements are greater than or equal to x
2. the total number T of elements in the staircase matrix.

Algorithm 2. STAIRCASE(M, x, p)

1. For each row $i \in [1, p]$ of M, use binary search on elements in $M[i][s_i^r, e_i^r]$ to find the maximum index α_i such that $M[i][s_i^r : \alpha_i] \geq x$.
2. For each column $j \in [1, p]$ of M, use binary search on elements in $M[s_j^c, e_j^c][j]$ to find the maximum index β_j such that $M[s_j^c : \beta_j][j] \geq x$.
3. Let M_S be the (implicitly defined, staircase) submatrix of M formed by elements larger or equal than x found in step 2 and step 3.
4. Let T be the total number of elements in M_S
5. return M_S, T

In algorithm **STAIRCASE**, binary search on each row i requires O(log($e_i^r - s_i^r$)) time while binary search on each column j requires O(log($e_j^c - s_j^c$)) time which is bounded by $O(\log k)$. There are p rows and p columns, and $p \leq \sqrt{k}$. The staircase matrix M_S is defined implicitly, by start-end pairs for rows and columns. Therefore the total time of algorithm **STAIRCASE** is $O(p \log k)$.

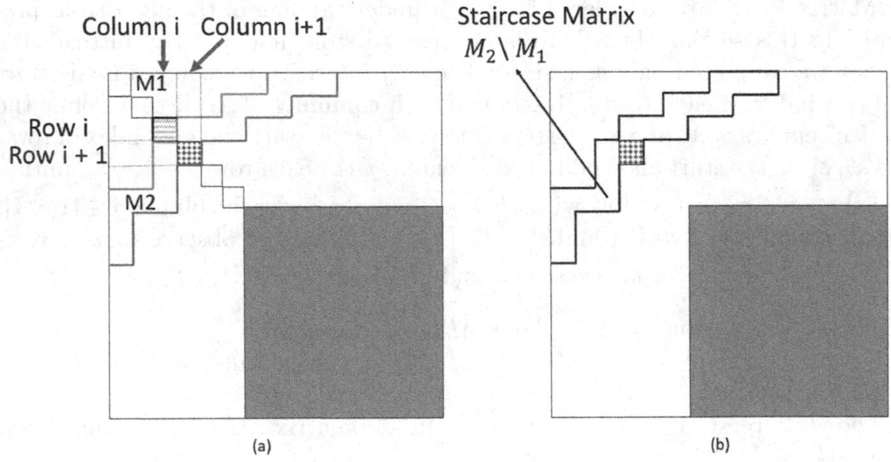

Fig. 2. (a) Two staircase matrices M_1 and M_2. Matrix M_1 has total number of elements less than or equal to k while matrix M_2 has more than k elements.(b) Illustrating $M_2 \setminus M_1$

Notice that we need to pay attention to not double count the entries in $M[0:p][0:p]$, which can be quickly done in $O(p)$ time.

The idea behind the following algorithm is to find two consecutive diagonal index d and $d+1$ of matrix M such that the staircase matrix computed by $STAIRCASE(M, M[d,d], d)$ contain only elements which are greater than or equal to the kth largest element of matrix M, while elements which are not part of the staircase matrix $STAIRCASE(M, M[d+1, d+1], d+1)$ are guaranteed to be less than the kth largest element of M. It is easy to notice that the kth largest element lies in the subtraction of the matrices $STAIRCASE(M, M[d+1, d+1], d+1)$ and $STAIRCASE(M, M[d,d], d)$.

The following algorithm takes as input a sorted matrix M and computes a staircase submatrix of M containing the k largest entries in M. As we will see, it does that in $O(p \log^3 k)$ time, using an implicit representation of submatrices of M. The k largest entries can then be reported in an additional $O(k)$ time.

Algorithm 3. MAX_SUM_CROSS-1(M)

1. Use binary search on $1, 2, \ldots, \sqrt{k}$ to find index d such that the total number of elements returned by STAIRCASE($M, M[d,d], d$) is at most k and the total number of elements returned by STAIRCASE($M, M[d+1, d+1], d+1$) is greater than k (This binary search on the diagonal of matrix M is illustrated in Figure 2-a).
   (a) Let $M_1, T_1 =$ STAIRCASE($M, M[d,d], d$).
   (b) Let $M_2, T_2 =$ STAIRCASE($M, M[d+1, d+1], d+1$)
2. if $T_1 = k$
3. return M_1
4. totalElementsLeft $= k - T_1$ $//elements$ to be $found$ in $M_2 \setminus M_1$
5. $M_S = M_2 \setminus M_1$
6. $M_{final} =$ FIND_INDEX(M_S, totalElementsLeft, $d+1$)
7. return M_{final}

In algorithm **MAX_SUM_CROSS-1**, $M_2 \setminus M_1$ corresponds to the staircase matrix formed by deleting elements of matrix M_1 from M_2. We don't need to create matrix M_S, instead update the value of s_i^r, e_i^r for each row $i \in [1, d+1]$ and s_j^c, e_j^c for each column $j \in [1, d+1]$. Whenever we subtract two matrices, we update the start and end index of each row and column in $[1, d+1]$ which takes $O(d)$ time where $d \leq \sqrt{k}$. Step 1 requires $O(\sqrt{k}\log^2 k)$ and finds the tuples (M_1, T_1) and (M_2, T_2). In step 5, we store $O(d)$ indexed pairs into M_S, which takes $O(d)$ time. Let Γ be the running time for algorithm **FIND_INDEX** (step 6). The total time taken by algorithm **MAX_SUM_CROSS-1** is then $O(\max\{\sqrt{k}\log^2 k, \Gamma\})$.

Let M_S be the staircase matrix which corresponds to $M_2 \setminus M_1$. M_S is implicitly defined and stored. The median value of each row $i \in [1, d+1]$ can be found in constant time, at entry $(s_i^r + e_i^r)/2$. Similarly, we can find the median value of each column of M_S.

Algorithm **FIND_INDEX** below takes as input M_S, an integer which stores the rank of the element we need to find in M_S, and an index p useful for computing staircase matrices, and returns the kth largest elements of matrix M in an implicit representation.

Algorithm 4. FIND_INDEX(M_S, totalElementsLeft, p)

1. Find the median value in each row 1 to p and in each column 1 to p of M_S and place them into an array X. Let size of array X be m where $m = O(p)$.
2. Sort X in non-increasing order. For element x_i at i-th position in array X, let α_i be the total number of elements of M_S greater than or equal to x_i which can be found via function $STAIRCASE(M_S, x_i, p)$ and let β_i be the total number of elements of M_S strictly greater than x_i which can be found via function $STAIRCASE(M_S, x_i + 1, p)$.
3. Use binary search on X together with the STAIRCASE function to find the maximum index i and the minimum index j in array X such that $totalElementsLeft - \alpha_i \geq 0$ and $totalElementsLeft - \alpha_j < 0$. Find corresponding β_i and β_j. Notice that $j = i + 1$. When searching, the last argument passed to the STAIRCASE function is p.
4. if $\exists\, i, j$ in step 3, //k^{th} largest element lie in range $[x_i, x_j)$
   (a) if $totalElementsLeft - \alpha_i = 0$ //k^{th} largest element is x_i
      i. Let $M_{final} = STAIRCASE(M, x_i, p)$
      ii. Return staircase matrix M_{final}
   (b) else if $totalElementsLeft - \beta_j = 0$ //k-maximum values consist of all elements greater than x_j
      i. Let $M_{final} = STAIRCASE(M, x_j + 1, p)$
      ii. Return staircase matrix M_{final}
   (c) else if $totalElementsLeft - \beta_j < 0$ //k^{th} largest element lies in range (x_i, x_j)
      i. Let $M' = STAIRCASE(M_S, x_i, p)$ and $M'' = STAIRCASE(M_S, x_j + 1, p)$
      ii. $M_{new} = M' \setminus M''$
      iii. totalElementsLeft = totalElementsLeft - α_i
      iv. FIND_INDEX(M_{new}, totalElementsLeft, p)
   (d) else if $totalElementsLeft - \beta_j > 0$ //k largest values contain all elements greater than x_j
      i. totalElementsLeft = totalElementsLeft - β_j
      ii. Let $M_{final} = STAIRCASE(M, x_j + 1, p)$
      iii. Let M_{new} be new matrix formed by adding $totalElementsLeft$ number of elements equal to x_j to M_{final}
      iv. return M_{new}
5. else if $\exists\, j$ and $\nexists\, i$ //k^{th} largest element is less than x_0
   (a) if $totalElementsLeft - \beta_j = 0$

i. Let $M_{final} = STAIRCASE(M, x_j + 1, p)$
ii. Return staircase matrix M_{final}
(b) else if $totalElementsLeft - \beta_j < 0$
i. Let $M' = STAIRCASE(M_S, x_j + 1, p)$ //Update end indexes for each row and column in $[1, p]$
ii. FIND_INDEX(M', totalElementsLeft, p)
(c) else if $totalElementsLeft - \beta_j > 0$ //k largest values contain all elements greater than x_j
i. totalElementsLeft = totalElementsLeft - β_j
ii. Let $M_{final} = STAIRCASE(M, x_j + 1, p)$
iii. Let M_{new} be new matrix formed by adding $totalElementsLeft$ number of element equal to x_j to M_{final}
iv. return M_{new}
6. else if $\exists\ i$ and $\nexists\ j$ //k^{th} largest element is greater than x_m
(a) if $totalElementsLeft - \alpha_i = 0$
i. $M_{final} = STAIRCASE(M, x_i, p)$
ii. Return staircase matrix M_{final}
(b) else
i. Let $M' = STAIRCASE(M_S, x_i, p)$
ii. totalElementsLeft = totalElementsLeft - α_i
iii. FIND_INDEX($M_S \setminus M'$, totalElementsLeft, p)

In algorithm **FIND_INDEX**, we update the value of start index of each row and column with range in $[1, p]$ to 1 before computing the $STAIRCASE$ matrix M_{final}. We now analyze the running time of algorithm **FIND_INDEX**. Step 1 requires O(p) time, Step 2 requires O($p \log p$) time, and Step 3 requires O($p \log^2 k$) time. Computing the staircase matrix and the total elements in steps 4, 5, and 6 requires O($p \log k$) time. In steps 4, 5, and 6, half of elements are removed before function **FIND_INDEX** is called recursively. With $p = O(\sqrt{k})$, the time complexity is then described by

$$T(k, p) = T(k/2, p) + p \log^2 k,$$

with $T(1, p) = O(1)$

Using substitution method, we have

$$T(k, p) = T(k/2^i, p) + O(\sum_{j=0}^{i-1} p \log^2(k/2^j))$$

Letting $k/2^i = 1$ results in $i = \log k$ and

$$T(k, p) = T(1, p) + O(\sum_{j=0}^{\log k - 1} p \log^2(k/2^j))$$

which solves for

$$T(k,p) = O(p \log^3 k)$$

With $\Gamma = O(p \log^3 k)$, algorithm **MAX_SUM_CROSS-1** thus takes $O(p \log^3 k)$ time where $p = O(\sqrt{k})$. We summarize our result below.

Theorem 1. *Algorithm **MAX_SUM_CROSS-1** finds the k largest elements in $A+B$ (and thus the k largest crossing sums) in $O(k \log^3 k)$ time. The k sums are implicitly represented and can be report with an additional $O(k)$ time.*

Recall that there is no need to sort the elements in the *max_cross* since the arrays contributing in the combine step are *max_left* and *max_right*. Therefore, instead of using *MERGE*, we can use the *SELECT* algorithm in line 9 of algorithm **Max-k** to find *k max_sub* in $O(k)$ time.

Plugging in the new procedure for finding crossing sums, **MAX_SUM_CROSS-1**, the running time of the divide and conquer algorithm **Max-k** is now described by the recurrence

$$T(n,k) = 2T(n/2,k) + O(k),$$

with $T(\sqrt{k},k) = O(k)$ which solves for $T(n) = O(n\sqrt{k})$.

Theorem 2. *Algorithm **MAX-k** finds the k largest subarrays of an array A of size n in $O(n\sqrt{k})$ time, where $1 \le k \le n(n+1)/2$.*

5 Implementation and Experiments

We have implemented our algorithms and performed multiple experiments, reporting excellent results. For comparison, we also implemented the algorithm presented in [12] which is used in the combine phase in [3]. The implementation is in JAVA on macOS High Sierra with 3.1 GHz Intel i5 processor and 8 GB of RAM, while the data sets have been randomly generated.

For experimentation, we generated two random sorted arrays of size n which consist of integer values and defined an integer k. For comparison both n and k are power of 4 as assumed in [12]. Table 1 shows the comparison results between the two algorithms and it is clear that our algorithm outperformed [12]. The time complexity of our algorithm depends upon the value of $p \le \sqrt{k}$. At each step of our algorithm, we eliminate many elements which are not candidates for the kth largest element. For example, while computing a staircase matrix, we know each row and column (elements in submatrix $M[1,m;1,r]$ are greater than or equal to $M[m,r]$).

For better accessing our algorithm, we performed further experiments by varying the size of the input array (10^1 to 10^6). Results of the MAX SUM CROSS- 1 procedure are shown in Table 2. The input arrays are generated randomly for each iteration. It is easy to notice that even for the large value of k, our algorithm takes only a few milliseconds. Tables 3, 4 and 5 shows the time taken to compute the k-maximum subarray. As it can be seen, our algorithm is very efficient: for arrays with size 10^6 and k = 10^6, we can obtain the k maximum subarrays in about 52 s.

Table 1. Comparison between our algorithm and [12] ($k = n$).

Size of input array	Average time for 10^2 test cases (in milliseconds)	
	Algorithm in [12]	Our algorithm
4^4	0.18	0.03
4^5	0.91	0.12
4^6	1.2	0.14
4^7	2.6	0.17

Table 2. Average time taken to find the k-maximum values of $A + B$ ($k = n$).

Size of input array	Average time (in milliseconds)		
	Number of tests = 10^2	Number of tests = 10^3	Number of tests = 10^4
10	0.03	0.01	0.02
10^2	0.07	0.03	0.04
10^3	0.18	0.08	0.09
10^4	0.37	0.23	0.27
10^5	1.35	1.44	1.39
10^6	16.21	15.78	15.85

Table 3. Average time taken to find the k-maximum subarrays when k = n.

Size of input array	Average time (in milliseconds)		
	Number of tests = 10^2	Number of tests = 10^3	Number of tests = 10^4
10	0.08	0.03	0.01
10^2	0.44	0.363	0.10
10^3	4.19	3.63	1.60
10^4	91.28	98.56	93.43
10^5	1729.78	1890.03	1780.13
10^6	52101.62	-	-

Table 4. Average time taken to find the k-maximum subarrays for small k.

Size of input array	Average time (in milliseconds)				
	k = 5	k = 15	k = 25	k = 35	k = 45
10	0.07	0.04	0.02	0.03	0.02
10^2	0.31	0.33	0.24	0.31	0.28
10^3	0.81	2.03	1.45	1.21	0.88
10^4	4.03	4.78	6.2	5.39	5.02
10^5	34.63	45.11	46.33	51.58	46.94

Table 5. Average time taken to find the k-maximum subarrays.

Size of input array	Average time (in milliseconds)					
	k = 105	k = 205	k = 405	k = 605	k = 805	k = 1005
10^2	0.54	0.4	0.69	0.24	0.31	0.93
10^3	2.52	1.63	2.06	1.61	1.39	2.68
10^4	10.29	8.4	9.96	14.24	13.65	15.16
10^5	71.06	76.49	114.97	123.29	160.73	192.22

6 Conclusion

In this paper, we studied the k-maximum subarray problem and proposed a simple divide-and-conquer algorithm for small values of k. Our algorithm matches the best known divide-and-conquer algorithm, while considerably simplifying the combine step. As part of our solution, we provided a simple prune-and-search procedure for finding the largest k values of $X + Y$, where X and Y are sorted arrays of size n each. These values are computed and stored implicitly in $O(\sqrt{k} \log^3 k)$ time, and can be reported in additional $O(k)$ time. Our solutions benefit from simplicity and fast execution time, even for large values of n and k. We implemented our algorithms and reported excellent results.

Appendix-1: Linear Time Divide-and-Conquer Maximum Subarray

In this section we give a detailed description of a simple, linear time divide and conquer algorithm to find the maximum subarray ($k = 1$), by placing the algorithm in [4] in a standard divide-and-conquer framework.

Given an array A of n real numbers, the maximum subarray problem is to find a contiguous subarray whose sum of elements is maximum over all possible subarrays, including A itself. The divide and conquer algorithm divides A into two subarrays of equal size, makes two recursive calls, and then proceeds with the combine step while keeping track of the maximum subarray sum found in the process.

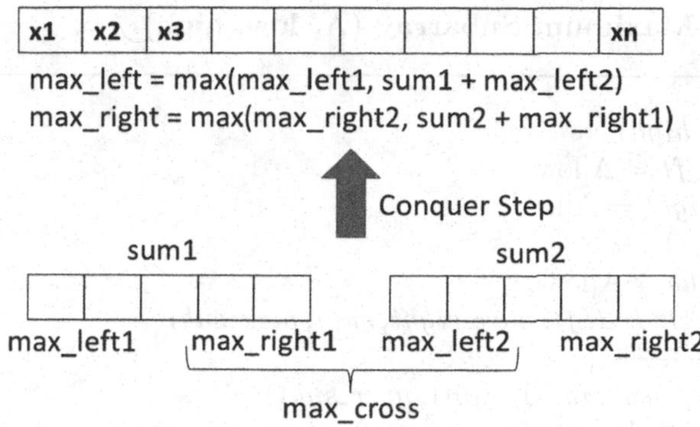

Fig. 3. Illustration of combine phase

In the combine phase, at an internal node, we have two subarrays, A_1 (from the left child) and A_2 (from the right child). We define the following variables which are used to find the maximum subarray (see also Fig. 3):

$max\_left \leftarrow -\inf$	maximum subarray starting from leftmost index
$max\_right \leftarrow -\inf$	maximum subarray starting from rightmost index
$sum \leftarrow 0$	sum of all elements in array
$max\_cross \leftarrow -\inf$	maximum crossing subarray
$max\_sub \leftarrow -\inf$	maximum subarray

The idea is to make the combine phase run in $O(1)$ time instead of the $O(n)$ time, as described in [10]. For that, the values (and corresponding array indexes) of max_left, max_right, and sum must also be passed up from the recursive calls. The sum value at a given node can be found by adding up the sums from the children. The value max_left is either the max_left from the left child or the sum value from the left plus the max_left value from the right child. Similarly, the value max_right is either the max_right from the right child or the sum value from the right plus the max_right value from the left child. The following divide and conquer algorithm, $Maximum\_Subarray$, takes in the input an array A of size n and two integers, low and $high$, which correspond to the start index and end index of subarray $A[low \ldots high]$, and finds and returns the maximum subarray of $A[low, high]$.

Algorithm 5. Maximum_Subarray (A, low, high)

```
1. if (low == high)
2.     max_left = A[low];
3.     max_right = A[low];
4.     sum = A[low];
5.     max_sub = A[low];
6.     return (max_left, max_right, sum, max_sub)
7. mid = ⌊(low+high)/2⌋
8. (max_left1, max_right1, sum1, max_sub1)=
   Maximum_Subarray(A, low, mid)
9. (max_left2, max_right2, sum2, max_sub2)=
   Maximum_Subarray(A, mid + 1, high)
10. max_left = max(max_left1, sum1 + max_left2);
11. max_right = max(max_right2, sum2 + max_right1)
12. sum = sum1 + sum2
13. max_cross = max_right1 + max_left2
14. max_sub = max(max_cross, max_sub1, max_sub2)
15. return (max_left, max_right, sum, max_sub)
```

In above algorithm, steps 1-7 take O(1) time. Steps 8-9 correspond to the recursive calls. Steps 10-15 take O(1) time. Therefore, the time taken by Algorithm 1 is:

$$T(n) = 2T(n/2) + O(1) = O(n)$$

Appendix-2: An $O(k \log K)$ Algorithm for X + Y

Given two input arrays, A and B, each of size k, sorted in non-increasing order, and outputs the k-maximum sums of the pairwise addition of A and B. For our purpose, A would contain the k largest sums of A_l for subarrays starting at the rightmost entry of A_l, while B would contain the k largest sums of A_r for subarrays starting at the leftmost entry of A_r. We use a priority queue Q implemented as a binary heap to store pairwise sums, as they are generated. An AVL tree T is also used, to avoid placing duplicate pairs (i, j) in Q.

Algorithm 6. MAX_SUM_CROSS (A, B)

```
1. k = sizeof(A)
2. Q ← null                                        //Max Priority Queue
3. M[k] ← null;                                    //Output Array
4. T ← null;                                       //AVL Tree
5. m ← 0;
6. add (0, 0) to Q with priority A[0] + B[0]
7. store (0, 0) in T
8. while k > 0:
9.      (i, j) = pop Q
10.     M[m] = A[i] + B[j]
11.     m = m + 1; k = k - 1
12.     if (i < k and (i + 1, j) ∉ T)
13.          store (i+1, j) in T
14.          add (i + 1, j) to Q with priority (A[i + 1] + B[j])
15.     if (j < k and (i, j + 1) ∉ T)
16.          store (i, j + 1) in T
17.          add (i, j + 1) to Q with priority (A[i] + B[j + 1])
18. return M
```

Time Complexity of Algorithm MAX_SUM_CROSS: Lines 12, 15 take $O(\log k)$ time for searching T, lines 13, 16 take $O(\log k)$ time to store indices in T, lines 9, 14, 17 take $O(\log k)$ to add or remove an element in the priority queue, and the while loop in line 8 runs k times. Therefore, the time complexity for algorithm MAX_SUM_CROSS is $O(k \log k)$.

References

1. Agrawal, R., Imieliński, T., Swami, A.: Mining association rules between sets of items in large databases. In: Acm SIGMOD Record, vol. 22, pp. 207–216. ACM (1993)
2. Bae, S.E., Takaoka, T.: Improved algorithms for the k-maximum subarray problem. Comput. J. **49**(3), 358–374 (2006)
3. Bengtsson, F., Chen, J.: Efficient algorithms for k maximum sums. Algorithmica **46**(1), 27–41 (2006)
4. Bengtsson, F., Chen, J.: Ranking k maximum sums. Theor. Comput. Sci. **377**(1–3), 229–237 (2007)
5. Bentley, J.: Algorithm design techniques. Commun. ACM **27**(9), 865–871 (1984)
6. Bentley, J.: Programming pearls: algorithm design techniques. Commun. ACM **27**(9), 865–873 (1984)
7. Blum, M., Floyd, R.W., Pratt, V.R., Rivest, R.L., Tarjan, R.E.: Time bounds for selection. J. Comput. Syst. Sci. **7**(4), 448–461 (1973)
8. Chazelle, B.: The soft heap: an approximate priority queue with optimal error rate. J. ACM (JACM) **47**(6), 1012–1027 (2000)

9. Cheng, C.H., Chen, K.Y., Tien, W.C., Chao, K.M.: Improved algorithms for the k maximum-sums problems. Theor. Comput. Sci. **362**(1–3), 162–170 (2006)
10. Cormen, T.H.: Introduction to Algorithms. MIT press, Cambridge (2009)
11. Frederickson, G.N., Johnson, D.B.: The complexity of selection and ranking in x+ y and matrices with sorted columns. J. Comput. Syst. Sci. **24**(2), 197–208 (1982)
12. Frederickson, G.N., Johnson, D.B.: Generalized selection and ranking: sorted matrices. SIAM J. Comput. **13**(1), 14–30 (1984)
13. Fukuda, T., Morimoto, Y., Morishita, S., Tokuyama, T.: Data mining using two-dimensional optimized association rules: scheme, algorithms, and visualization. ACM SIGMOD Record **25**(2), 13–23 (1996)
14. Grenander, U.: Pattern analysis: lectures in pattern theory 2. Appl. Math. Sci. **24** (1978)
15. Kaplan, H., Kozma, L., Zamir, O., Zwick, U.: Selection from heaps, row-sorted matrices and $x + y$ using soft heaps. arXiv preprint arXiv:1802.07041 (2018)

A Faster Convex-Hull Algorithm via Bucketing

Ask Neve Gamby[1] and Jyrki Katajainen[2,3](✉)

[1] National Space Institute, Technical University of Denmark, Centrifugevej, 2800 Kongens Lyngby, Denmark
aknvg@space.dtu.dk
[2] Department of Computer Science, University of Copenhagen, Universitetsparken 5, 2100 Copenhagen East, Denmark
[3] Jyrki Katajainen and Company, 3390 Hundested, Denmark
jyrki@di.ku.dk
http://hjemmesider.diku.dk/~jyrki/

Abstract. In the convex-hull problem, in two-dimensional space, the task is to find, for a given sequence S of n points, the smallest convex polygon for which each point of S is either in its interior or on its boundary. In this paper, we propose a variant of the classical bucketing algorithm that (1) solves the convex-hull problem for any multiset of points, (2) uses $O(\sqrt{n})$ words of extra space, (3) runs in $O(n)$ expected time on points drawn independently and uniformly from a rectangle, and (4) requires $O(n \lg n)$ time in the worst case. Also, we perform experiments to compare BUCKETING to other alternatives that are known to work in linear expected time. In our tests, in the integer-coordinate setting, BUCKETING was a clear winner compared to the considered competitors (PLANE-SWEEP, DIVIDE & CONQUER, QUICKHULL, and THROW-AWAY).

Keywords: Computational geometry · Convex hull · Algorithm · Bucketing · Linear expected time · Experimental evaluation

1 Introduction

Bucketing is a practical method to improve the efficiency of algorithms. As an example, consider the sorting problem under the assumption that the elements being sorted are integers. In BUCKETSORT [1], the elements are sorted as follows: (1) Find the minimum and maximum of the elements. (2) Divide the closed interval between the two extrema into equal-sized subintervals (*buckets*). (3) Distribute the points into these buckets. (4) Sort the elements in each bucket. (5) Concatenate the sorted buckets to form the final output. A data structure is needed to keep track of the elements inside the buckets, so this is not an in-place sorting method. The key is to use a worst-case optimal sorting algorithm when processing the buckets. This way the worst-case performance remains unchanged since the bucketing overhead is linear.

In our exploratory experiments, we could confirm that for integer sorting BUCKETSORT, which used the C++ standard-library `std::sort` to sort the buckets,

I. Kotsireas et al. (Eds.): SEA^2 2019, LNCS 11544, pp. 473–489, 2019.
https://doi.org/10.1007/978-3-030-34029-2_30

was often faster than `std::sort` itself. In our implementation we followed the guidelines given by Nevalainen and Raita [27]. Their advice can be summed up as follows:

1. Do not make the distribution table too large! According to the theory, the number of buckets should be proportional to the number of elements. Often this is too large and may lead to bad cache behaviour. According to our experience, the memory footprint of the distribution table should not be much larger than the size of the second-level cache in the underlying computer.
2. Do not use any extra space for the buckets when distributing the elements into the buckets! Permute the elements inside the input sequence instead.

In geometric applications, in two-dimensional space, bucketing can be used in a similar manner: (1) Find the smallest rectangle covering the input points. (2) Divide this rectangle into equal-sized rectangles (*cells*). (3) Solve the problem by moving from cell to cell while performing some local computations. The details depend on the application in question.

In the 1980s, bucketing was a popular technique to speed up geometric algorithms (for two surveys, see [5,13]). In the 1990s, the technique was declared dead, because of the change in computer architectures: caching effects started to dominate the computational costs. Today, bucketing can again be used, provided that the distribution table is kept small. In this work, we only study bucketing algorithms that use a small distribution table and that avoid explicit linking by permuting the elements inside the sequence.

More specifically, we consider the problem of finding the convex hull for a sequence S of n points in the plane. Each point p is specified by its Cartesian coordinates (p_x, p_y). The *convex hull* $\mathcal{H}(S)$ of S is the boundary of the smallest convex set enclosing all the points of S. The goal is to find the smallest description of the convex hull, i.e. the output is the boundary of a convex polygon, the vertices of which—so-called *extreme points*—are from S.

Throughout the paper, we use n to denote the size of the input, h the size of the output, and $\lg x$ as a shorthand for $\log_2(\max\{2, x\})$. Our specific goal is to develop a convex-hull algorithm that (1) requires $O(n \lg n)$ time in the worst case, (2) runs in $O(n)$ time in the average case, (3) uses as little extra space as possible, and (4) is efficient in practice. In the integer domain, good average-case performance may be due to various reasons [7]:

1. The points are chosen uniformly and independently at random from a bounded domain.
2. The distribution of the points is *sparse-hulled* [11, Exercise 33-5], meaning that in a sample of n points the expected number of extreme points is $O(n^{1-\varepsilon})$ for some constant $\varepsilon > 0$.

In the book by Devroye [14, Section 4.4], a bucketing algorithm for finding convex hulls was described and analysed. It could improve the efficiency of any worst-case-efficient algorithm such that it runs in linear expected time without sacrificing the worst-case behaviour. According to Devroye, this algorithm is due to Shamos. This algorithm needs a distribution table of size $\Theta(n)$ which may lead to bad cache behaviour. We describe a variant of this algorithm that reduces the consumption of extra memory to $O(\sqrt{n})$, which was low enough to not contribute significantly to cache misses in our computing environment.

Many convex-hull algorithms are known to run in linear expected time. Therefore, we also compared the practical performance of the new bucketing algorithm to that of other algorithms. The competitors considered were PLANE-SWEEP [4] (using BUCKETSORT as proposed in [3]; the average-case analysis of BUCKETSORT can be found, e.g., in [14, Chapter 1]); DIVIDE & CONQUER [29] (for the analysis, see [7]); QUICKHULL [10,16,20] (for the analysis, see [28]); and THROW-AWAY [2,4,12] (for the analysis, see [12]).

The contributions of this paper can be summarized as follows:

- We describe a space-efficient bucketing algorithm that solves the convex-hull problem for any multiset of points in the plane (Sect. 2).
- We perform micro-benchmarks to show which operations are expensive and which are not (Sect. 3). The results of these micro-benchmarks give an indication of how different algorithms should be implemented.
- We provide a few enhancements to most algorithms to speed up their straightforward implementations (Sect. 4). It turns out that, with careful programming, most known algorithms can be made fast.
- We perform experiments, in the integer-coordinate setting, to find out what is the state of the art when computing the convex hulls in the plane (Sect. 5).
- We report the lessons learned while doing this study (Sect. 6). Many of the guidelines are common-sense rules that can be found from the texts discussing experimentation (see, e.g. [8, Chapter 8]).

2 Bucketing

In one-dimensional bucketing, when the values come from the interval $[\texttt{min}, \texttt{max}]$ and the distribution table has $\texttt{m}$ entries, the bucket index i, $0 \le i \le \texttt{m} - 1$, of value v is computed using the formula

$$i = \left\lfloor \frac{(v - \texttt{min}) * (\texttt{m} - 1)}{\texttt{max} - \texttt{min}} \right\rfloor .$$

In two-dimensional bucketing, such a formula is needed for both x- and y-coordinates. Normally, geometric primitives only use addition, subtraction, and multiplication, but here we also need whole-number division.

Old Version. The BUCKETING algorithm described in [14, Section 4.4] for solving the convex-hull problem works as follows: (1) Determine a bounding rectangle of the n input points. (2) Divide this rectangle into rectangular cells using a grid of size $\lceil\sqrt{n}\rceil \times \lceil\sqrt{n}\rceil$ and distribute the points into these cells. (3) Mark all outer-layer cells that may contain extreme points. (4) Collect all points in the marked cells as the extreme-point candidates. (5) Finally, use any known algorithm to compute the convex hull of the candidates.

In a naive implementation, the data at each cell are stored in a linked list and a two-dimensional array is used to store the headers to these lists. In the book [14, Section 4.4], the computation of the outer-layer cells was described as follows. (1) Find the leftmost non-empty column of cells and mark all the occupied cells in this column. Recall the row index j of the northernmost occupied cell. (2) After processing column i, mark one or more cells in column $i+1$ as follows: (a) Mark the cell at row number j. (b) Mark all cells between this cell and the northernmost occupied cell on that column provided that its row number is at least $j+1$. (c) Update j if we moved upwards. This way we get a staircase of at most $2\lceil\sqrt{n}\rceil$ marked cells. As to the correctness, all extreme points in the north-west quadrant must be in a marked cell. The other three quadrants are processed in a similar manner. Eventually at most $8\lceil\sqrt{n}\rceil$ cells are marked.

Devroye proved [14] that this algorithm runs in linear expected time when the points are independently and uniformly distributed in a rectangle. The worst-case running time depends on the algorithm used in the last step; with the PLANE-SWEEP algorithm this is $O(n \lg n)$ [4].

New Version. We call the columns *slabs*. There are $\lceil\sqrt{n}\rceil$ slabs, each storing two y-values; we call them `min` and `max`, but their meaning depends on in which step of the algorithm we are. We only use bucketing for the x-coordinates when determining in which slab a point is; we do not materialize the cells. In the $\pm y$ directions, we maintain a staircase of y-values instead of a staircase of cells.

In detail, the algorithm works as follows:

(1) Find the minimum and maximum x-coordinate values of the points. These values are needed in the formula determining the slab of a point.
(2) If all points are on a vertical line, solve this special case by finding the points with the extreme y-coordinates, move these two points (or one point) to the beginning of the input sequence, and report them as the answer.
(3) Allocate space for the slab structure and initialize it so that in each slab the `max` value is $-\infty$ and the `min` value $+\infty$.
(4) Determine the extreme y-coordinates in each slab by scanning the points, calculating their slab index, and updating the stored `min` and `max` values within the slabs whenever necessary (random access needed).
(5) Determine the indices of the slabs where the topmost and bottommost points lie. This information can be extracted by examining the slab structure. After this the quadrants are uniquely determined. For example, when processing the west-north quadrant, the slabs are visited from the leftmost slab to the one that contains the topmost point and the `max` values are reset.
(6) In the west-north quadrant, form a staircase of y-values that specifies in each slab where the extreme points can lie. Initially, the `roof`, keeping track of the highest y-value seen so far, is set to the `max` value of the first slab and the `max` value at that slab is reset to $-\infty$. When visiting the following slabs, both the `max` value and the `roof` are updated such that the `roof` becomes the maximum of itself and `max` of the current slab, while `max` takes the value of the `roof` at the previous slab.

Table 1. Hardware and software in use.

Processor. Intel® Core™ i7-6600U CPU @ 2.6 GHz (turbo-boost up to 3.6 GHz) × 4
Word size. 64 bits
First-level data cache. 8-way set-associative, 64 sets, 4 × 32 KB
First-level instruction cache. 8-way set-associative, 64 sets, 4 × 32 KB
Second-level cache. 4-way set-associative, 1 024 sets, 256 KB
Third-level cache. 16-way set-associative, 4 096 sets, 4.096 MB
Main memory. 8.038 GB
Operating system. Ubuntu 18.04.1 LTS
Kernel. Linux 4.15.0-43-generic
Compiler. g++ 8.2.0
Compiler options. `-O3 -std=c++2a -Wall -Wextra -fconcepts -DNDEBUG`

(7) The other quadrants are treated in the same way to form the staircases there.
(8) Partition the input by moving all the points outside the region determined by the staircases (above or below) to the beginning of the input sequence, and the points that were inside and could be eliminated to the end of the sequence. For each point, its slab index must be computed and the slab structure must be consulted (which involves random access).
(9) Release the space allocated for the slab structure.
(10) Apply any space-efficient convex-hull algorithm for the remaining points and report the convex hull first in the sequence.

Since the region that is outside the staircases computed by the new algorithm is smaller than the region covered by the marked cells in the original algorithm, the runtime analysis derived for the old version also applies for the new version. The critical region covers at most $8\lceil\sqrt{n}\,\rceil$ cells (that are never materialized) and the expected value for the maximum number of points in a cell is $O(\lg n/\lg\lg n)$. Hence, the work done in the last step is asymptotically insignificant. The amount of space required by the slab structure is $O(\sqrt{n})$. All the other computations can be carried out using a workspace of constant size (*in place*) or logarithmic size (*in situ*); for a space-efficient variant of PLANE-SWEEP, see [17].

3 Micro-benchmarking

When tuning our implementations, we based our design decisions on micro-benchmarks. These benchmarks should be understood as sanity checks. We encourage the reader to redo some of the tests to see whether the same conditions are valid in his or her computer system.

Test Environment. All the experiments were run on a Linux computer. The programs were written in C++ and the code was compiled using the g++ compiler. The hardware and software specifications of the test computer are summarized in Table 1. In the micro-benchmarks the same data set was used:

Square Data Set. The coordinates of the points were integers drawn randomly from the range $[\![-2^{31} \,..\, 2^{31})\!)$ (i.e. random **ints**).

The points were stored in a C array. We report the test results for five values of n: 2^{10}, 2^{15}, 2^{20}, 2^{25}, and 2^{30}. All the reported numbers are scaled: For every performance indicator, if X is the observed measurement result, we report X/n. That is, a constant means linear performance. To avoid the problems with inadequate clock precision, a test for $n = 2^i$ was repeated $\lceil 2^{27}/2^i \rceil$ times; each repetition was done with a new input array.

Orientation Tests. As an example of a geometric computation, where correctness is important, let us consider Graham's scan [19] as it appears in the PLANE-SWEEP algorithm [4]: We are given n points *sorted* according to their x-coordinates. The task is to perform a scan over the sequence by repeatedly considering triples (p, q, r) of points and eliminate q if there is not a right turn at q when moving from p to r via q. After this computation the points on the upper hull are gathered together at the beginning of the input. The scan is carried out in place. Typically, about $2n$ orientation tests are done in such a scan.

For three points $p = (p_x, p_y)$, $q = (q_x, q_y)$, and $r = (r_x, r_y)$, the orientation test boils down to the question of determining the sign of a 3×3 determinant:

$$\begin{bmatrix} p_x & p_y & 1 \\ q_x & q_y & 1 \\ r_x & r_y & 1 \end{bmatrix}$$

If the sign of this determinant is positive, then r is on left of the oriented line determined by p and q. If the sign of the determinant is negative, then r is on right of the oriented line. And if the sign is zero, then r is on that line.

Formulated in another way, one can calculate $lhs = (q_x - p_x) * (r_y - p_y)$ and $rhs = (r_x - p_x) * (q_y - p_y)$, and then determine whether $lhs > rhs$, $lhs < rhs$, or $lhs == rhs$. When doing these calculations, we considered three alternatives:

Multiple-precision arithmetic. It is clear that (1) a construction of a signed value from an unsigned one may increase the length of the representation by one bit, (2) a subtraction may increase the needed precision by one bit, and (3) a multiplication may double the needed precision. Hence, if the coordinates of the points are `w` bits wide, it is sufficient to use $(2\mathtt{w}+4)$-bit integers to get the correct value of *lhs* and *rhs*. For this purpose, we used the multiple-precision integers available at the CPH STL [24].

Double-precision arithmetic. By converting the coordinates to double-length integers and by handling the possible overflows in an ad-hoc manner, the calculations can be done with 2`w`-bit numbers and some **if** statements.

Floating-point filter. Many computers have additional hardware to accelerate computations on floating-point numbers. Therefore, it might be advantageous to perform the calculations on floating-point numbers and, only if the result is not known to be correct due to accumulation of rounding errors, use one of the above-mentioned exact methods to redo the calculations. In the present case, we employed Shewchuk's filter coded in [31].

In this micro-benchmark the coordinates were 32-bit integers of type **int**. Hence, in the multiple-precision solution the numbers were 68 bits wide. In the double-precision solution we could rely on 64-bit built-in integers of types **long long** and **unsigned long long**. For randomly generated data, the floating-point filter worked with 100 % accuracy. Table 2 shows how Graham's scan performed for different right-turn predicates. Generally, this linear-time computation is by no means critical compared to the cost of sorting. Since the floating-point filter gave the best results, we used it in all subsequent experiments.

Table 2. Performance of Graham's scan for different right-turn predicates for the x-sorted square data set [ns per point]. The width `w` of the coordinates was 32 bits. In the multiple-precision solution, intermediate calculations were done with numbers of type `cphstl::`$\mathbb{Z}$`<2 * w + 4>`.

n	Multiple precision	Double precision	Floating-point filter
2^{10}	30.8	17.7	15.7
2^{15}	30.8	17.8	15.3
2^{20}	30.7	17.7	15.3
2^{25}	30.9	17.9	15.4
2^{30}	55.8	39.8	48.8

Distribution. To explore different options when implementing distributive methods, consider the task of producing a histogram of bucket sizes along the x-axis when we are given n points and a distribution table of size `m`. This involves a scan over the input and, for each point. the calculation of its bucket index and an increment of the counter at that bucket. In the micro-benchmark we varied the table size and the data type used in arithmetic operations. For each of the considered types, the x-coordinates were first cast to this type, the calculations were done with them, and the result was cast back to an index type.

The results of these experiments are reported in Table 3 for several different table sizes and data types. For multiple-precision integer arithmetic, the package from the CPH STL was used [24]. These results confirm two things:

1. The distribution table should not be large; otherwise, caching effects will become visible.
2. It is preferable to do the bucket calculations using floating-point numbers.

Scanning. Next, let us consider what is the cost of sequential scanning. Three tasks reappear in several algorithms: (1) Find the minimum and maximum of the points according to their lexicographic order. For n points, this task can be accomplished with about $(3/2)n$ point comparisons using the standard-library function `std::minmax_element`. (2) Use two points p and r to partition a sequence of n points into two parts so that the points above and on the line determined

Table 3. Performance of the histogram creation along the x-axis for the square data set [ns per point]. In all runs, n was fixed to 2^{20}. The coordinates were of type **int**; their width w was 32 bits. With integers of width 2w + 3, all arithmetic overflows could be avoided. Integers of width 2w were also safe since $\lg \mathtt{m} \leq \mathtt{w} - 3$.

m	double	cphstl::$\mathbb{Z}$<2 * w + 3>	cphstl::$\mathbb{Z}$<2 * w>
2^{10}	3.90	37.91	10.96
2^{11}	3.85	39.04	11.05
2^{12}	3.90	38.99	10.87
2^{13}	3.92	38.97	10.76
2^{14}	4.19	39.14	11.07
2^{15}	4.07	38.90	10.96
2^{16}	4.74	39.10	11.81
2^{17}	5.03	39.27	11.98
2^{18}	5.21	38.97	12.09
2^{19}	7.77	48.81	18.86
2^{20}	13.31	75.94	35.82

by p and r come on the left and those below the line on the right. For this task, the standard-library function `std::partition` can be employed. The orientation predicate is needed to determine on which side of the line a point lies. (3) Copy a sequence of size n to an array. The function `std::copy` is designed for this task.

Table 4 shows the performance of these functions in our test environment: `std::minmax_element` is very fast, whereas `std::partition` is slower since it involves point moves and orientation tests. Especially, for the largest instance, when the size of the input is close to the maximum capacity of main memory, the slowdown is noticeable. The performance is linear up to a certain point, but after that point the memory operations became more expensive. In the test computer, the saturation point was reached for $n = 2^{28}$; for $n = 2^{27}$, the cost per point was still about the same as that for $n = 2^{25}$. For the largest instance, copying failed since there was not space for two point arrays of size 2^{30} in main memory.

Table 4. Performance of scanning for the square data set [ns per point].

n	std::minmax_element	std::partition	std::copy
2^{10}	0.22	11.2	0.73
2^{15}	0.20	10.3	0.80
2^{20}	0.20	10.4	1.41
2^{25}	0.20	11.6	3.13
2^{30}	0.20	34.5	out of memory

Sorting. Most industry-strength sorting algorithms are hybrids. The C++ standard-library INTROSORT [26] is a typical example: It uses MEDIAN-OF-THREE QUICKSORT [32] (see also [23]) for rough sorting and it finishes its job by a final INSERTIONSORT scan. If the recursion stack used by QUICKSORT becomes too deep, the whole input will be processed by HEAPSORT [33]. Now, small inputs are processed fast due to INSERTIONSORT, the worst case is $O(n \lg n)$ for an input of size n due to HEAPSORT, and the performance is good due to QUICKSORT.

We wanted to test whether a combination, where the input elements are distributed into buckets and the buckets sorted by INTROSORT, can improve the performance even further. In our implementation we followed closely the guidelines given in [27]. We name the resulting algorithm ONE-PHASE BUCKETSORT. This algorithm has two drawbacks: (1) Its interface is not the same as that of the library sort. Namely, it has one additional functor as a parameter that is used to map every element to a numerical value. In our application this is not a problem since the coordinates of the points are integers. (2) For an input of size n, the algorithm—as implemented in [27]—requires an extra array for n elements and a header array for $\mathtt{m} = \min\{n/5, \mathtt{max\_m}\}$ integers. The bucket headers are used as counters to keep track of the size of the buckets and as cursors when placing the elements into the buckets. We selected the constant `max_m` such that the whole header array could be stored in the second-level cache.

To make the distributive approach competitive with respect to space usage, we also implemented TWO-PHASE BUCKETSORT that distributes the input elements into $O(\sqrt{n}\,)$ buckets and sorts each of them using ONE-PHASE BUCKETSORT. This version permutes the elements inside the input sequence before sorting the buckets. For the two-phase version the linear running time is valid as long as $\sqrt{n}$ is not significantly larger than `max_m`.

In the benchmark we sorted an array of n points according to their x-coordinates using the above-mentioned sorting algorithms. As seen from Table 5, both versions of BUCKETSORT worked reasonably well compared to INTROSORT until the size of the input reached that of main memory. In this situation, most of the other memory intensive programs had troubles as well—either they failed due to excessive memory usage or became slow.

Table 5. Performance of sorting for the square data set [ns per point].

n	INTROSORT	ONE-PHASE BUCKETSORT	TWO-PHASE BUCKETSORT
2^{10}	36.5	14.2	33.4
2^{15}	53.5	19.9	29.4
2^{20}	70.7	33.1	43.7
2^{25}	89.1	58.9	63.0
2^{30}	173	out of memory	324

4 Competitors

Any algorithm solving the convex-hull problem should read the whole input and report the extreme points in the output in sorted angular order. Thus, if $scan(n)$ denotes the cost of scanning a sequence of size n sequentially and $sort(h)$ the cost of sorting a sequence of size h, $\Omega(scan(n) + sort(h))$ is a lower bound for the running time of any convex-hull algorithm.

Many algorithms have been devised for the convex-hull problem, but none of them is known to match the above-mentioned lower bound on the word RAM [21]—when the coordinates of the points are integers that fit in one word each. The best deterministic algorithms are known to run in $O(n \lg \lg n)$ worst-case time (for example, the PLANE-SWEEP algorithm [4] combined with fast integer sorting [22]) or in $O(n \lg h)$ worst-case time (the MARRIAGE-BEFORE-CONQUEST algorithm [25]). On the other hand, when the input points are drawn according to some random distribution, which is a prerequisite for the analysis, there are algorithms that can solve the convex-hull problem in linear expected time. This is not in conflict with the $\Omega(scan(n) + sort(h))$ bound, since integers drawn independently at random from a uniform distribution in a bounded interval can be sorted in linear expected time by BUCKETSORT.

The most noteworthy alternatives for a practical implementation are:

Plane sweep. The PLANE-SWEEP algorithm [4] is a variation of ROTATIONAL-SWEEP [19] where the points are sorted according to their x-coordinates. The problem is solved by computing the upper-hull and lower-hull chains separately. To start with, two extreme points are found—one on the left and another on the right. Then the line segment determined by these two is used to partition the input into upper-hull candidates and lower-hull candidates. Finally, the upper-hull candidates are scanned from left to right and the lower-hull candidates from the right to left as in Graham's algorithm [19]. To work in linear expected time, BUCKETSORT (see, for example, [1]) could be used when sorting the candidate collections. If INTROSORT was used to sort the buckets, the worst-case running time would be $O(n \lg n)$.

Divide and conquer. As is standard in the DIVIDE & CONQUER scheme [29], if the number of given points is less than some constant, the problem is solved directly using some straightforward method. Otherwise, the problem is divided into two subproblems of about equal size, these subproblems are solved recursively, and the resulting convex hulls of size h_1 and h_2 are merged in $O(h_1 + h_2)$ worst-case time. An efficient merging procedure guarantees that the worst-case running time of the algorithm is $O(n \lg n)$ and, for sparse-hulled distributions, the average-case running time is $O(n)$ [7]. For the theoretical analysis to hold, two properties are important: (1) the division step must be accomplished in $O(1)$ time and (2) the points in the subproblems must obey the same distribution as the original points. This can be achieved by storing the points in an array and shuffling the input randomly at the beginning of the computation.

Quickhull. This algorithm, which mimics QUICKERSORT [30], has been reinvented several times (see, e.g. [10,16,20]). It also starts by finding two extreme points p and r, one on the left and another on the right, and computes the upper-hull chain from p to r and the lower-hull chain from r to p separately. For concreteness, consider the computation of the upper chain from p to r. In the general step, when the problem is still large enough, the following is done: (1) Find the extreme point q with the largest distance from the line segment $\overline{pr}$. (2) Eliminate the points inside the triangle pqr from further consideration. (3) Compute the chain from p to q recursively by considering the points above the line segment $\overline{pq}$ and (4) the chain from q to r by considering the points above the line segment $\overline{qr}$. (5) Concatenate the chains produced by the recursive calls and return that chain.

Eddy [16] proved that in the worst case QUICKHULL runs in $O(nh)$ time. However, Overmars and van Leeuwen [28] proved that in the average case the algorithm runs in $O(n)$ expected time. Furthermore, if the coordinates of the points are integers drawn from a bounded universe of size U, the worst-case running time of QUICKHULL is $O(n \lg U)$ [17].

Throw-away elimination. When computing the convex hull for a sequence of points, one has to find the extreme points at all directions. A rough approximation of the convex hull is obtained by considering only a few predetermined directions. As discussed in several papers (see, e.g. [2,4,12]), by eliminating the points falling inside such an approximative hull, the problem size can often be reduced considerably. After this preprocessing, any of the above-mentioned algorithms could be used to process the remaining points.

Akl and Toussaint [2] used four equispaced directions—those determined by the coordinate axes; and Devroye and Toussaint [12] used eight equispaced directions. The first of these papers demonstrated the usefulness of this idea experimentally and the second paper verified theoretically that for certain random distributions the number of points left will be small. Unfortunately, the result depends heavily on the shape of the domain, from where the points are drawn at random. A rectangle is fine, but a circle is not.

Our implementations of these algorithms are available from the website of the CPH STL [http://www.cphstl.dk] in the form of a `pdf` file and a `tar` archive [18]. To ensure the reproducibility of the experimental results, the package also contains the driver programs used in the experiments.

5 Experiments

As the micro-benchmarks indicated, the performance of the memory system became an issue when the problem size reached the capacity of main memory. Therefore, it became a matter of honour for us to ensure that our programs can also handle large problem instances—an "out-of-memory" signal was not acceptable if the problem instance fitted in internal memory.

Briefly stated, the improvements made to the algorithms were as follows:

Plane sweep. As our starting point, we used the in-situ version described in [17]. To achieve the linear expected running time, TWO-PHASE BUCKETSORT was used when sorting the points according to their x-coordinates.

Divide and conquer. As in PLANE-SWEEP, the upper and lower hulls were computed separately. When merging two sorted chains of points, we used an in-place merging algorithm (`std::inplace_merge`). The library routine was adaptive: If there was free memory available, it was used; if not, an in-place merging routine was employed. It must be pointed out that the emergency routine did not run in linear worst-case time [15]—although it relied on sequential access. Therefore, when the memory limit was hit, the worst-case running time of the DIVIDE & CONQUER algorithm was $O(n(\lg n)^2)$.

Quickhull. We implemented this algorithm recursively by letting the runtime system handle the recursion stack. Otherwise, we relied on many of the same in-place routines as those used in the other algorithms. The extra space required by the recursion stack is $O(\lg U)$ words when the coordinates come from a bounded universe of size U [17].

Bucketing. The bucketing was done by maintaining information on the `min` and `max` y-values in each of the $\lceil \sqrt{n} \rceil$ slabs. After determining the staircases, the points outside them were moved to the beginning of the sequence and in-situ PLANE-SWEEP was used to finish the job. The elimination overhead was three scans: one min-max scan to find the extreme points in the $\pm x$ directions, one scan to determine the outermost points at each slab in the $\pm y$ directions, and one partitioning scan to eliminate the points that were inside the region determined by the staircases. The slab structure used $O(\sqrt{n})$ words of space and, except the above-mentioned scans, its processing cost was $O(\sqrt{n})$. Bucket indices were calculated using floating-point numbers.

Throw-away elimination. We found the extreme points in eight predetermined directions, eliminated all the points inside the convex polygon determined by them, and processed the remaining points with in-situ PLANE-SWEEP. Thus, the elimination overhead was two scans: one max-finding scan to find the extrema and another partitioning scan to do the elimination.

We wanted to test the performance of these heuristics in their home ground when the input points were drawn according to some random distribution. We run the experiments on the following data sets—the first one was already used in the micro-benchmarks:

Square data set. The coordinates of the points were integers drawn randomly from the range $[\![-2^{31} \,..\, 2^{31})\!)$. The expected size of the output is $O(\lg n)$ [6].

Disc data set. As above, the coordinates of the points were random **ints**, but only the points inside the circle centred at the origin with the radius $2^{31} - 1$ were accepted to the input. Here the expected size of the output is $O(n^{1/3})$.

In earlier studies, the number of orientation tests and that of coordinate comparisons have been targets for optimization. Since the algorithms reorder the points in place, the number of coordinate moves is also a relevant performance indicator. In our first experiments, we measured the performance of the algorithms with respect to these indicators. The purpose was to confirm that the algorithms execute a linear number of basic operations. Due to hardware effects, it might be difficult to see the linearity from the CPU-time measurements.

Table 6. The number of orientation tests executed [per n] for the square data set.

n	PLANE-SWEEP	DIVIDE & CONQUER	QUICKHULL	BUCKETING	THROW-AWAY
2^{10}	2.31	2.76	4.95	0.16	2.04
2^{15}	2.34	2.79	4.86	0.02	2.01
2^{20}	2.34	2.80	4.82	0.00	2.00
2^{25}	2.39	2.85	4.70	0.00	2.00
2^{30}	2.09	2.56	5.37	0.00	2.00

Table 7. The number of coordinate comparisons done [per n] for the square data set.

n	PLANE-SWEEP	DIVIDE & CONQUER	QUICKHULL	BUCKETING	THROW-AWAY
2^{10}	11.0	8.81	2.26	6.31	15.0
2^{15}	9.45	8.84	2.25	5.65	15.0
2^{20}	9.01	8.88	2.25	5.53	15.0
2^{25}	8.91	8.80	2.25	5.51	15.1
2^{30}	8.88	8.88	2.25	5.50	15.2

Table 8. The number of coordinate moves performed [per n] for the square data set.

n	PLANE-SWEEP	DIVIDE & CONQUER	QUICKHULL	BUCKETING	THROW-AWAY
2^{10}	26.5	32.4	5.14	2.49	1.21
2^{15}	29.1	32.7	4.59	0.42	0.20
2^{20}	29.0	32.6	4.51	0.08	0.04
2^{25}	29.0	32.6	4.72	0.01	0.00
2^{30}	29.1	32.7	4.91	0.00	0.00

The operation counts are shown in Table 6 (orientation tests), Table 7 (coordinate comparisons), and Table 8 (coordinate moves). The results are unambiguous: (1) There are no significant fluctuations in the results. For all competitors, the observed performance is linear—as the theory predicts. For the elimination strategies, the number of coordinate moves is sublinear; for BUCKETING this is even the case for orientation tests. (2) For BUCKETING some of the observed values are frighteningly small compared to others albeit casts to floating-point numbers and operations on them were not measured.

Both BUCKETING and THROW-AWAY eliminate some points before applying in-situ PLANE-SWEEP. To understand better how efficient these elimination strategies are, we measured the average fraction of points left after elimination. The figures are reported in Table 9. For the square data set, both methods show extremely good elimination efficiency. For the disc data set, BUCKETING will be better since the expected number of points in the outer layer is at most $O(\sqrt{n}\,(\lg n/\lg\lg n))$ [14, Section 4.4], whereas for THROW-AWAY the expected

Table 9. The average fraction of points left after elimination [%] for the two data sets.

n	BUCKETING (square)	BUCKETING (disc)	THROW-AWAY (square)	THROW-AWAY (disc)
2^{10}	8.07	12.2	4.29	12.2
2^{15}	1.18	2.07	0.71	10.2
2^{20}	0.19	0.36	0.12	9.99
2^{25}	0.03	0.06	0.01	9.97
2^{30}	0.00	0.01	0.00	9.97

number of points left is $\Omega(n)$ [17, Fact 3]. Although the elimination efficiency can vary, the performance of these algorithms can never be much worse than that of PLANE-SWEEP since the elimination overhead is linear.

Table 10. The running time of the competitors for the square data set [ns per point].

n	PLANE-SWEEP	DIVIDE & CONQUER	QUICKHULL	BUCKETING	THROW-AWAY
2^{10}	60.5	73.2	49.5	14.3	20.4
2^{15}	56.9	70.0	46.0	6.98	16.4
2^{20}	62.6	69.7	45.7	5.88	16.4
2^{25}	85.9	69.8	48.4	5.50	15.2
2^{30}	189.9	114.2	124.5	75.4	45.5

Table 11. The running time of the competitors for the disc data set [ns per point].

n	PLANE-SWEEP	DIVIDE & CONQUER	QUICKHULL	BUCKETING	THROW-AWAY
2^{10}	57.7	71.0	60.4	16.8	29.2
2^{15}	53.7	68.4	57.8	7.64	27.4
2^{20}	56.9	67.0	57.6	5.86	28.5
2^{25}	80.2	66.7	57.8	5.53	30.1
2^{30}	183.9	116.9	128.3	75.9	77.2

In our final experiments, we measured the CPU time used by the competitors for the considered data sets. The results are shown in Table 10 (square) and Table 11 (disc). We can say that (1) PLANE-SWEEP had mediocre performance; (2) DIVIDE & CONQUER worked at about the same efficiency; (3) THROW-AWAY improved the performance with its preprocessing; (4) QUICKHULL was not effective because it performed many expensive orientation tests; and (5) BUCKETING showed outstanding performance compared to its competitors thanks to its effective preprocessing. However, THROW-AWAY that uses less space and has better memory-access locality behaved well when almost all memory was in use.

6 Reflections

When we started this work, we were afraid of that this would become a study on sorting in new clothes. But the essence turned out to be how to avoid sorting.

When writing the programs and performing the experiments, we made many mistakes. We have collected the following checklist of the most important issues to prevent ourselves and others from repeating these mistakes.

Compiler options. Switch all compiler warnings on when developing the programs and use full optimization when running the experiments.

Library facilities. Use the available library resources; do not reinvent them.

Techniques. Keep bucketing in your toolbox. While the other convex-hull algorithms are busily partitioning the input into upper-hull and lower-hull candidates, BUCKETING has already solved the problem. Partitioning is also needed in BUCKETING, but it is fast since the partitioning criterion is simpler and only a small fraction of the points will be moved.

Robustness. Implement geometric primitives in a robust manner. The simplest way to achieve this is to do intermediate calculations with multiple-precision numbers. We found it surprising that multiple-precision arithmetic is not provided by the C++ standard library.

Floating-point acceleration. Use floating-point numbers wisely. These can speed up things; we used them in bucket calculations and orientation tests.

Space efficiency. Do not waste space. As one of the micro-benchmarks showed, an innocent copying can generate an "out-of-memory" signal when the problem size reaches the capacity of main memory. It is known that the PLANE-SWEEP algorithm can be implemented in place (see [9,17]). On the other hand, a program can be useful even though its memory usage is not optimal; it is quite acceptable to use $O(\sqrt{n})$ words of extra space.

Correctness. Use an automated test framework. Our first checkers could verify the correctness of the output in $O(n^2)$ worst-case time. After refactoring, the checkers were improved to carry out this task in $O(n \lg n)$ worst-case time (for more details, see [17,18]). But the checkers do some copying which means that they cannot be used for the largest problem instances.

Quality assurance. Try several alternatives for the same task to be sure about the quality of the chosen alternative.

References

1. Akl, S.G., Meijer, H.: On the average-case complexity of "bucketing" algorithms. J. Algorithms **3**(1), 9–13 (1982). https://doi.org/10.1016/0196-6774(82)90003-7
2. Akl, S.G., Toussaint, G.T.: A fast convex hull algorithm. Inf. Process. Lett. **7**(5), 219–222 (1978). https://doi.org/10.1016/0020-0190(78)90003-0
3. Allison, D.C.S., Noga, M.T.: Some performance tests of convex hull algorithms. BIT **24**(1), 2–13 (1984). https://doi.org/10.1007/BF01934510
4. Andrew, A.M.: Another efficient algorithm for convex hulls in two dimensions. Inf. Process. Lett. **9**(5), 216–219 (1979). https://doi.org/10.1016/0020-0190(79)90072-3

5. Asano, T., Edahiro, M., Imai, H., Iri, M., Murota, K.: Practical use of bucketing techniques in computational geometry. In: Toussaint, G.T. (ed.) Computational Geometry. North-Holland (1985)
6. Bentley, J.L., Kung, H.T., Schkolnick, M., Thompson, C.D.: On the average number of maxima in a set of vectors and applications. J. ACM **25**(4), 536–543 (1978). https://doi.org/10.1145/322092.322095
7. Bentley, J.L., Shamos, M.I.: Divide and conquer for linear expected time. Inf. Process. Lett. **7**(2), 87–91 (1978). https://doi.org/10.1016/0020-0190(78)90051-0
8. Berberich, E., Hagen, M., Hiller, B., Moser, H.: Experiments. In: Müller-Hannemann, M., Schirra, S. (eds.) Algorithm Engineering: Bridging the Gap between Algorithm Theory and Practice. Springer-Verlag (2010)
9. Brönnimann, H., Iacono, J., Katajainen, J., Morin, P., Morrison, J., Toussaint, G.: Space-efficient planar convex hull algorithms. Theoret. Comput. Sci. **321**(1), 25–40 (2004). https://doi.org/10.1016/j.tcs.2003.05.004
10. Bykat, A.: Convex hull of a finite set of points in two dimensions. Inf. Process. Lett. **7**(6), 296–298 (1978). https://doi.org/10.1016/0020-0190(78)90021-2
11. Cormen, T.H., Leiserson, C.E., Rivest, R.L., Stein, C.: Introduction to Algorithms. The MIT Press, 3rd edn. (2009)
12. Devroye, L., Toussaint, G.T.: A note on linear expected time algorithms for finding convex hulls. Computing **26**(4), 361–366 (1981). https://doi.org/10.1007/BF02237955
13. Devroye, L.: Expected time analysis of algorithms in computational geometry. In: Toussaint, G.T. (ed.) Computational Geometry. North-Holland (1985)
14. Devroye, L.: Lecture Notes on Bucket Algorithms. Birkhäuser Boston, Inc. (1986)
15. Dvořák, S., Ďurian, B.: Stable linear time sublinear space merging. Comput. J. 30(4), 372–375 (1987). https://doi.org/10.1093/comjnl/30.4.372
16. Eddy, W.F.: A new convex hull algorithm for planar sets. ACM Trans. Math. Software **3**(4), 398–403 (1977). https://doi.org/10.1145/355759.355766
17. Gamby, A.N., Katajainen, J.: Convex-hull algorithms: Implementation, testing, and experimentation. Algorithms 11(12) (2018). https://doi.org/10.3390/a11120195
18. Gamby, A.N., Katajainen, J.: Convex-hull algorithms in C++. CPH STL report 2018–1, Dept. Comput. Sci., Univ. Copenhagen (2018–2019), http://www.diku.dk/~jyrki/Myris/GK2018S.html
19. Graham, R.L.: An efficient algorithm for determining the convex hull of a finite planar set. Inf. Process. Lett. **1**(4), 132–133 (1972). https://doi.org/10.1016/0020-0190(72)90045-2
20. Green, P.J., Silverman, B.W.: Constructing the convex hull of a set of points in the plane. Comput. J. **22**(3), 262–266 (1979). https://doi.org/10.1093/comjnl/22.3.262
21. Hagerup, T.: Sorting and searching on the word RAM. In: Morvan, M., Meinel, C., Krob, D. (eds.) STACS 1998. LNCS, vol. 1373, pp. 366–398. Springer, Heidelberg (1998). https://doi.org/10.1007/BFb0028575
22. Han, Y.: Deterministic sorting in $O(n \log \log n)$ time and linear space. J. Algorithms **50**(1), 96–105 (2004). https://doi.org/10.1016/j.jalgor.2003.09.001
23. Hoare, C.A.R.: Quicksort. Comput. J. **5**(1), 10–16 (1962). https://doi.org/10.1093/comjnl/5.1.10
24. Katajainen, J.: Class templates `cphstl::`$\mathbb{N}$ and `cphstl::`$\mathbb{Z}$ for fixed-precision arithmetic. Work in progress (2017–2019)

25. Kirkpatrick, D.G., Seidel, R.: The ultimate planar convex hull algorithm? SIAM J. Comput. **15**(1), 287–299 (1986). https://doi.org/10.1137/0215021
26. Musser, D.R.: Introspective sorting and selection algorithms. Software Pract. Exper. **27**(8), 983–993 (1997)
27. Nevalainen, O., Raita, T.: An internal hybrid sort algorithm revisited. Comput. J. **35**(2), 177–183 (1992). https://doi.org/10.1093/comjnl/35.2.177
28. Overmars, M.H., van Leeuwen, J.: Further comments on Bykat's convex hull algorithm. Inf. Process. Lett. **10**(4–5), 209–212 (1980). https://doi.org/10.1016/0020-0190(80)90142-8
29. Preparata, F.P., Hong, S.J.: Convex hulls of finite sets of points in two and three dimensions. Commun. ACM **20**(2), 87–93 (1977). https://doi.org/10.1145/359423.359430
30. Scowen, R.S.: Algorithm 271: Quickersort. Commun. ACM **8**(11), 669–670 (1965). https://doi.org/10.1145/365660.365678
31. Shewchuk, J.R.: Adaptive precision floating-point arithmetic and fast robust predicates for computational geometry (1996), http://www.cs.cmu.edu/~quake/robust.html
32. Singleton, R.C.: Algorithm 347: An efficient algorithm for sorting with minimal storage [M1]. Commun. ACM **12**(3), 185–187 (1969). https://doi.org/10.1145/362875.362901
33. Williams, J.W.J.: Algorithm 232: Heapsort. Commun. ACM **7**(6), 347–348 (1964). https://doi.org/10.1145/512274.512284

Fixed Set Search Applied to the Minimum Weighted Vertex Cover Problem

Raka Jovanovic[1] and Stefan Voß[2,3](✉)

[1] Qatar Environment and Energy Research Institute (QEERI), Hamad bin Khalifa University, PO Box 5825, Doha, Qatar
rjovanovic@hbku.edu.qa

[2] Institute of Information Systems, University of Hamburg, Von-Melle-Park 5, 20146 Hamburg, Germany
stefan.voss@uni-hamburg.de

[3] Escuela de Ingenieria Industrial, Pontificia Universidad Católica de Valparaíso, Valparaíso, Chile

Abstract. Fixed set search (FSS) is a novel metaheuristic adding a learning mechanism to enhanced greedy approaches. In this paper we use FSS for solving the Minimum Weighted Vertex Cover Problem (MWVCP). First we define a Greedy Randomized Adaptive Search Procedure (GRASP) by randomizing the standard greedy constructive algorithm and combine it with a local search. The used local search is based on a simple downhill procedure. It checks if substituting a single or a pair of elements from a solution with ones that need to be added to keep the solution a vertex cover decreases the value of the objective function. The performance of the GRASP algorithm is improved by extending it towards FSS. Computational experiments performed on standard test instances from literature show that the proposed FSS algorithm for the MWVCP is highly competitive with state-of-the-art methods. Further, it is shown that the FSS manages to significantly improve the GRASP algorithm it is based on.

Keywords: Metaheuristics · Minimum Weighted Vertex Cover Problem · GRASP · Fixed set search

1 Introduction

The Minimum Vertex Cover Problem (MVCP) is one of the standard combinatorial optimization problems that has been extensively researched. The decision version of the MVCP is one of Karp's 21 NP-complete problems [9]. It is defined for a graph $G(V, E)$ having a set of vertices V and a set of edges E. A vertex set $C \subset V$ is called a vertex cover if for every edge $\{u, v\} \in E$ at least one of the vertices u or v is an element of C. The objective of the MVCP is to find a vertex cover C that has minimum cardinality. In this paper, we focus on solving the Minimum Weighted Vertex Cover Problem (MWVCP) which is a variation of the MVCP in which for each node $u \in V$ there is a corresponding weight w_u.

I. Kotsireas et al. (Eds.): SEA[2] 2019, LNCS 11544, pp. 490–504, 2019.
https://doi.org/10.1007/978-3-030-34029-2_31

The objective in the MWVCP is to find the vertex cover having the minimum total weight. Formally, the objective is to find a set $C \subset V$ which minimizes:

$$\sum_{u \in V} w_u x_u \tag{1}$$

In (1), variables of type x_u are equal to 1 if $u \in C$ and zero otherwise. The variables x_u need to satisfy the following constraints:

$$x_u + x_v \geq 1 \qquad (\{u, v\} \in E) \tag{2}$$

The MWVCP well represents a large number of real-world problems related to wireless communication, circuit design, and network flow [11,15] which resulted in an extensive amount of research dedicated to finding optimal and near optimal solutions. It should be noted that the vast majority of research solves the MWVCP with positive coefficients. It has been shown that it can be solved as fast as the unweighted vertex cover in $O(1.2738^p + pN_V)$, with exponential memory use [3,4] (here N_V is the size of the vertex set and p the size of the prospective cover, if it exists). Due to the NP-Hardness of the MWVCP a wide range of methods have been developed for finding near optimal solutions ranging from greedy algorithms to different types of metaheuristics.

In [12], an ant colony optimization (ACO) is presented. The performance of the ACO method has been further improved using a pheromone correction strategy as presented in [7]. The problem has also been solved using genetic algorithms combined with a greedy heuristic [13], a population-based iterated greedy (PBIG) algorithm [1] and a reactive tabu search hybridized with simulated annealing [14]. One of the most successful approaches is the multi-start iterated tabu search (MS-ITS) algorithm [16]. The most successful methods incorporate some types of local searches [10]. In this paper a dynamic scoring strategy is incorporated to improve the local search performance, which produces a computationally very effective method being capable to solve problem instances having hundreds of thousands of nodes and edges. Another method designed to solve problem instances on massive graphs can be found in [2], in which first an initial vertex cover is generated that is later improved using an advanced local search.

Due to the fact that the use of local searches has proven very efficient in case of the MWVCP, in this paper the potential effectiveness of the Greedy Randomized Adaptive Search Procedure (GRASP) [5] is explored. To be more precise, our objective is to see the effectiveness of combining a simple to implement greedy algorithm and local search. Two local searches are presented based on a downhill procedure using swap operations which remove one or two vertices from the solutions and add necessary vertices. The basic idea of the swap operations is very similar to the ones used in [10,14]. The performance of the proposed GRASP algorithm is further improved by extending it to the novel Fixed Set Search metaheuristic [8], which has previously been successfully applied to the Traveling Salesman Problem (TSP). The FSS uses a simple approach to add a learning mechanism to GRASP based on elements frequently appearing in high quality solutions. The performed computational experiments show that the FSS is highly competitive with the state-of-the-art methods in the quality of found solutions. Further, we show that the FSS produces a significant improvement when compared to the GRASP algorithm on which it is based.

The paper is organized as follows. In Sect. 2, we give a brief description of the randomized greedy algorithm for the MWVCP. In the following section details of the local searches are presented. In Sect. 4, an outline of the GRASP metaheuristic is given. In the next section, we give details of the FSS and how it is applied to the MWVCP. In Sect. 6, we discuss the performed computational experiments. The paper is finalized with concluding remarks.

2 Greedy Algorithm

In this section, the standard greedy constructive algorithm for the MWVCP is presented. The basic idea of the method is to start from a partial solution $S = \emptyset$ and at each step expand it with the vertex that has the most desirable properties based on a heuristic function h. To be more precise, it is preferable to expand the solution with a vertex u that covers the largest number of non-covered edges and has the minimal weight w_u. Formally, the heuristic function for a partial solution S and a node n has the following form.

$$Cov(n, S) = \{\{n, v\} \mid (\{n, v\} \in E) \wedge (v \notin S)\} \tag{3}$$

$$h(n, S) = \frac{|Cov(n, S)|}{w_n} \tag{4}$$

In (3), $Cov(n, S)$ is the set of edges in E that contain node n but are not already covered by S. An edge is covered if at least on of its vertices is in S. The heuristic function, defined in (4), is proportional to the number of elements of $Cov(S, n)$, and reversely proportional to the weight w_n of the vertex n.

Since, our goal is to use the presented greedy algorithm as a part of the GRASP metaheuristic it is necessary to include randomization. In the proposed algorithm we use the standard approach of a restricted candidate list (RCL) as follows. Let us define R as the set of N elements from $v \in V \setminus S$ that have the largest value of $h(v, S)$. Now, we can expand the partial solution with a random element of set R. The pseudocode for the proposed randomized greedy constructive algorithm (RGC) can be seen in Algorithm 1. In it, the partial solution S is initially set to an empty set. At each iteration S is expanded with a random element from the RCL. This is repeated until all the edges E are covered. The proposed RGC has computational complexity of $|S||V|$, where S is the generated solution.

3 Local Searches

In this section, two local searches based on a correction procedure are presented. The basic idea of the proposed local searches is based on the concept of swapping elements of a solution S with elements of $V \setminus S$ that produce a vertex cover but decrease the objective function. This approach has proven to be very successful on the closely related dominating set problem.

Algorithm 1. Pseudocode for the RGC for the MWVCP

```
S = ∅
while Not all edges covered do
    Generate RCL based on h and S
    Select random element n ∈ RCL
    S = S ∪ n
end while
```

3.1 Element Swap

Assume that we aim to improve a solution S. Since S is a vertex cover of G, for each edge $\{u, v\}$ at least one of u or v is an element of S. Let us define $Un(v, S)$, for a solution S and vertex $v \in S$, as the set of vertices that correspond to edges that are uniquely covered by vertex vS as

$$Un(v, S) = \{u \mid u \notin S \wedge \{u, v\} \in E\} \tag{5}$$

It is evident that if we swap a vertex v with all the elements $Un(v, S)$ a new vertex cover will be created. For simplicity of notation let us define the swap operation for a vertex v as

$$Swap(v, S) = (S \cup Un(v, S)) \setminus \{v\} \tag{6}$$

Now, a swap operation for a vertex v can be evaluated as

$$EvSwap(v, S) = w_v - \sum_{i \in Un(v,S)} w_i \tag{7}$$

In Eq. (7), $EvSwap(v)$ gives the change in the solution weight when $v \in S$ is swapped. More precisely, it is equal to the weight w_v of vertex v that is removed from the solution minus the total sum of weights of vertices that are added to the solution. Now, we can define $Imp(S)$ as the set of all vertices of S for which a swap operation produces an improvement

$$Imp(S) = \{v \mid v \in S \wedge EvSwap(v, S) > 0\} \tag{8}$$

3.2 Pair Swap

The basic idea of the swap operation can be extended to pairs of vertices. In case of a local search based on swap pairs it generally is the case that the computational cost will increase $|S|$ times, where S is the solution being improved. Although this cannot be changed asymptotically, it can be greatly decreased in practical applications. It is important to note that in case of the MWVCP swap operations of this type are more effective than for other problems since the elements that are used for substitution are uniquely defined. In designing the local search based on swap pairs, we focus on two objectives. Firstly, to have a very small overlap with a local search based on element swaps and secondly to increase computational efficacy.

In our application, we assume that in a pair swap operation involving $\{u, v\}$ both elements will be removed and none of them will be re-added. In case this constraint is not used the same effect can be achieved using an element swap. In case such a constraint exists, if $\{u, v\} \in E$ such a pair can never be swapped since the edge $\{u, v\}$ will not be covered. Additional positive effects of a pair swap $\{u, v\}$ can only occur if u and v have overlapping neighborhoods, or in other words in case there is a node w that is adjacent to both u and v. Using this idea, let us formally define the set of improving swap pairs for a solution S. Based on the previous discussion the set of all vertex pairs that should be tested for a graph G can be defined as follows:

$$C_p = \{\{u, v\} \mid (u, v \in V) \wedge (N(v) \cap N(u) \neq \emptyset)\} \setminus E \tag{9}$$

In (9), the notation $N(v)$ is used for the open neighborhood of v, i.e., all nodes adjacent to v not including itself. Using this set of candidate swap pairs for graph G, we can define the set of improving swap pairs in a similar way as a set of improving elements using the following set of equations.

$$Un(u, v, S) = Un(u, S) \cup Un(v, S) \tag{10}$$

$$Swap(u, v, S) = (S \cup Un(u, v, S)) \setminus \{u, v\} \tag{11}$$

$$EvSwap(u, v, S) = w_u + w_v - \sum_{i \in Un(u,v,S)} w_i \tag{12}$$

$$\begin{aligned} ImpPair(S) = \{\{u, v\} \mid &(\{u, v\} \in C_p \cap S^2) \\ &\wedge (EvSwap(u, v, S) > 0)\} \end{aligned} \tag{13}$$

In (10), $Un(u, v, S)$ corresponds to the set of nodes that correspond to the set of edges that are uniquely covered by one of the vertices u or v. Note, that this set excludes vertices u and v. In (11), the effect of the swapping elements u and v from a solution S is given. To be more precise, the vertices u and v are removed from the solution S, and all the nodes corresponding to uniquely covered edges are added. $EvSwap(u, v, S)$, given in (12), is equal to the change on the weight of the solution if the vertex pair $\{u, v\}$ is swapped. Finally, in the next equation $ImpPair(S)$ is the set of all swap pairs that improve the quality of the solution from the set of the restricted list of candidate pairs. The restricted set of candidate pairs is equal to the intersection of the unordered product of set S with itself S^2 and the set of all candidate pairs for graph G.

3.3 Local Search

In this subsection, we present the local search based on the presented improvement using element swaps and pair swaps. It should be noted that these two types of improvement explore different neighborhoods of a solution S. Because of this, as in the case of the variable neighborhood search [6], it is advantageous

Algorithm 2. Pseudocode for the local search based on swap operations

```
repeat
    while Imp(S) ≠ ∅ do
        Select random v ∈ Imp(S)
        S = Swap(v, S)
    end while
    if ImpPair(S) ≠ ∅ then
        Select random {v, u} ∈ ImpPair(S)
        S = Swap(u, v, s)
    end if
until (Imp(S) = ∅) ∧ (ImpPair(S) = ∅)
```

to use both of them interchangeably. The pseudocode for the local search based on swap operations can be seen in Algorithm 2.

In Algorithm 2, a solution S is interchangeably improved based on swap elements and swap pairs. Firstly, all the possible improvements are performed using swap elements since this operation is computationally less expensive. This is done by repeatedly performing swap element improvements until no further improvement of this type is possible. Next, we test if an improvement can be achieved using swap pairs. If this is true, the improvement is performed. As there is a possibility, that after applying a swap pair improvement new element swaps can produce improvement, the main loop is repeated until no such improvement exists. It should be noted that for both types of improvements there are several different ways to select the swap that will be performed; in the proposed implementation we simply select a random one.

4 GRASP

To enhance the performance of the proposed greedy algorithm and local search, we extend them to the GRASP metaheuristic as illustrated in Algorithm 3. In the main loop of Algorithm 3, a new solution S to the MWVCP is generated using the *RGC* algorithm. The local search is applied to the solution S and tested if it is the new best solution. This procedure is repeated until some stopping criterion is satisfied, usually a time limit or a maximal allowed number of solutions has been generated.

Algorithm 3. Pseudocode for the GRASP

```
while Not Stop Criteria Satisfied do
    Generate Solutions S using randomized greedy algorithm
    Apply local search to S
    Check if S is the new best
end while
```

5 Fixed Set Search

The fixed set search (FSS) is a novel metaheuristic that adds a learning mechanism to the GRASP. Literally it uses elite solutions, consistent solution elements or alike to direct the search. It has previously been successfully applied to the TSP [8]. The FSS has several important positive traits. Firstly, there is a wide range of problems on which it can possibly be applied (this paper tries to put evidence on it) since the only requirement is that the solution of the problem is represented in a form of a set. The learning mechanism is simple to implement and many existing GRASP algorithms can easily be extended to this form. In this section the general concepts used in the FSS are presented as well as details of its application to the MWVCP. A more detailed explanation of the concepts used in the FSS can be found in [8].

The main inspiration for the FSS is the fact that generally many high quality solutions for a combinatorial optimization problem contain some common elements. The idea is to use such elements to steer the search of the solution space. To be more precise, we wish to force such elements in a newly generated solution and dedicate computational effort to finding optimal or near optimal solutions in the corresponding subset of the solution space. The selected set of common elements will be called the *fixed set*. In the FSS, we are trying to find the additional elements to complete the partial solution, corresponding to the fixed set, or in other words to "fill in the gaps." In practice, we are intensifying the search around such fixed sets. This can be achieved through the following steps. Firstly, a method for generating fixed sets needs to be implemented. Next, the randomized greedy algorithm used in the corresponding GRASP needs to be adapted in a way to be able to use a preselected set of elements. Lastly, the learning mechanism which gains experience from previously generated solutions needs to be specified.

5.1 Fixed Set

Let us first define a method that will make it possible to generate random fixed sets. As previously stated the FSS can be applied to a problem for which a solution can be represented in a form of a set S having elements in some set W, or in other words $S \subset W$. In case of the MWVCP this concerns a solution $S \subset V$. In the following the notation $\mathcal{P}$ will be used for the set of all the generated solutions (population). Next, let us define $\mathcal{P}_n \subset \mathcal{P}$ as the set of n solutions having the best value of the objective function inside $\mathcal{P}$.

One of the requirements of the FSS is that the method used to generate a fixed set F has the ability to control its size $|F|$. Further, such fixed sets need to be able to produce high quality feasible solutions. This can be achieved using a base solution $B \in \mathcal{P}_m$. If the fixed set satisfies $F \subset B$, it can be used to generate the base solution. In practice this means it can generate a feasible solution at least of the same quality as B, and F can contain arbitrary elements of B. It is preferable for F to contain elements that frequently occur in some group of high quality solutions. To achieve this, let us define $\mathcal{S}_{kn}$ as the set of k randomly selected solutions out of the n best ones P_n.

Using these building blocks it is possible to define a function $Fix(B, \mathcal{S}_{kn}, Size)$ for generating a fixed set $F \subset B$ that consists of $Size$ elements of the base solution $B = \{v_1, ...v_l\}$ that most frequently occur in $\mathcal{S}_{kn} = \{S_1, .., S_k\}$. Let use define the function $C(v_x, S)$, for an element $v_x \in V$ and a solution $S \subset V$, which is equal to 1 if $v_x \in S$ and 0 otherwise. We can define a function that counts the number of occurrences of element v_x in the elements of the set $\mathcal{S}_{kn}$ using the function $C(v_x, S)$ as follows.

$$O(v_x, \mathcal{S}_{kn}) = \sum_{S \in \mathcal{S}_{kn}} C(v_x, S) \tag{14}$$

Now, we can define $Fix(B, \mathcal{S}_{kn}, Size)$ as the set of $Size$ elements $v_x \in B$ that have the largest value of $O(v_x, \mathcal{S}_{kn})$.

5.2 Learning Mechanism

The learning mechanism in the FSS is implemented through the use of fixed sets. To achieve this it is necessary to adapt the RGC algorithm used in the corresponding GRASP to a setting where some elements are preselected (the newly generated solution must contain them). Let us use the notation $RGF(F)$ for the solution generated using such an algorithm with a preselected (fixed) set of elements F. In case of the MWVCP, the RGC algorithm is trivially adapted to a $RGF(F)$ by setting the initial partial solution S to F instead of an empty set.

In the FSS, as in the case of the GRASP, solutions are repeatedly generated and a local search is applied to each of them. The first step is generating an initial population of solutions $\mathcal{P}$ by performing N iterations of the corresponding GRASP algorithm. The initial population is used to generate a random fixed set F having some size $Size$, using the method from the previous section. The fixed set F is used to generate a new solution $S = RGF(F)$ and the local search is applied to it. The population of solutions is expanded using the newly generated locally optimal solutions. This procedure is repeated until no new best solutions are found for a long period by some criteria, or in other words until stagnation has occurred. In case of stagnation the size of the fixed set is increased. In case the maximal allowed size of the fixed set is reached, the size of the fixed set is reset to the minimal allowed value. This procedure is repeated until some stopping criterion is reached. An important part of the algorithm is defining the array of allowed fixed set sizes, which is related to the part of the solution that is fixed. In our implementation this array is defined as follows:

$$Sizes[i] = (1 - \frac{1}{2^i}) \tag{15}$$

The size of the used fixed sets is proportional to the used base solution B. More precisely, at the i-th level it is equal to $|B| \cdot Size[i]$.

The pseudocode for FSS can be seen in Algorithm 4. In it, the first step is initializing the sizes of fixed sets using (15). The current size of the fixed

Algorithm 4. Pseudocode for the Fixed Set Search

```
Initialize Sizes
Size = Sizes.Next
Generate initial population P using GRASP(N)
while (Not termination condition) do
    Set S_kn to random k elements of P_n
    Set B to a random solution in P_m
    F = Fix(B, S_kn, Size|B|)
    S = RGF(F)
    Apply local search to S
    P = P ∪ {S}
    if Stagnant Best Solution then
        Size = Sizes.Next
    end if
end while
```

set *Size* is set to the smallest value. The next part of the initialization stage is generating the initial population of N solutions by performing N iterations of the basic GRASP algorithm. Each iteration of the main loop consists of the following steps. A random set of solutions $\mathcal{S}_{kn}$ is generated by selecting k elements from $\mathcal{P}_n$ and a random base solution B is selected from the set $\mathcal{P}_m$. Next, the function $Fix(B, \mathcal{S}_{kn}, Size|B|)$ is used to generate a fixed set F. A new solution $S = RGF(F)$ is generated using the randomized greedy algorithm with preselected elements and the local search is applied to it. Next, we check if S is the new best solution and add it to the set of generated solutions $\mathcal{P}$. In case stagnation has occurred, the value of *Size* is set to the next value in *Sizes*. Let us note, that the next size is the next larger element of array *Sizes*. In case *Size* is already the largest size, we select the smallest element in *Sizes*. This procedure is repeated until some termination criterion is satisfied.

In our implementation of the proposed algorithm for the MWVCP, the criterion for stagnation was that no new best solution has been found in the last, say, *Stag* iterations. As previously stated the adaptation of the randomized constructive greedy algorithm to the $RGF(F)$ consists of simply setting the initial partial solution to the fixed set instead of an empty set. The set of candidate swap pairs C_p is calculated in the initialization stage. At this time the set of all neighboring vertices for valid candidate pairs $\{u, v\}$ are also calculated with the intention of speeding the calculation of $ImpPair(S)$.

6 Results

In this section we give details of the performed computational experiments. Their objective is to evaluate the performance of the proposed GRASP and FSS in combination with the element- (GRASP-E and FSS-E) and pair- (GRASP-P and FSS-P) based local searches. Note that the pair-based local search is only used in combination with the element-based one. This has been done in comparison with the ACO algorithm from [12] and its improvement version (ACO-SEE) [7]. Further, a comparison is with the population-based iterated greedy (PBIG) algorithm [1], the multi-start iterated tabu search (MS-ITS) algorithm [16] and the Diversion Local Search based on Weighted Configuration Checking (DLSWC) [10] which are the best performing methods. Note that the reactive tabu search hybrid produces about the same quality of results than DLSWC, but in [14] results are only presented for a small subset of instances.

The comparison is done on the set of test instances introduced in [12], that have been also used to evaluate the other mentioned methods. The test instances are divided into three groups: small, medium and large. In case of the small and medium test instances, random graphs having 10–300 nodes and 10–5000 edges are used for evaluation. For each pair (N_V, N_E) with N_V and N_E being the number of vertices and edges, respectively, there are ten different graph instances. The test instances are divided into Type 1 where there is no correlation between the weight of a vertex and number on incident edges, and Type 2 where some weak correlation exists; details can be found in [12]. In case of large test instances the graphs have between 500 and 1000 vertices and between 500 and 20 000 edges, and there is only one instance for each pair (N_V, N_E).

The used parameters for FSS are the following, $k = 10$ random solutions are selected from the best $n = 100$ ones for the set of solutions $\mathcal{S}_{kn}$. The base solution is selected from the $m = 100$ best solutions. The size of the initial population is 100. The stagnation criterion is that no new best solution is found in the last $Stag = 100$ iterations for the current fixed set size. The used size of the RCL in the randomized greedy algorithm is 10. The stopping criterion for all the proposed methods is that 5000 solutions are generated or a time limit of 10 minutes has been reached. The FSS and GRASP have been implemented in C# using Microsoft Visual Studio 2017. The calculations have been done on a machine with Intel(R) Core(TM) i7-2630 QM CPU 2.00 Ghz, 4 GB of DDR3-1333 RAM, running on Microsoft Windows 7 Home Premium 64-bit.

In Tables 1 and 2 the results for the medium-size problem instances are given for graphs of Type 1 and Type 2, respectively. For each pair (N_V, N_E), the average weight of all the vertex covers of this type are evaluated. With the intention of having a clearer presentation, the average value of the objective function is only given for DLSWC, while for the other methods only the difference to this value is presented. The values for the methods used for comparison are taken from the corresponding papers. Note that we did not include the results for small problem instances since all the methods except the two ACO methods manage to find all the optimal solutions. From the results in these tables it can be seen that the two ACO algorithms have a substantially worse performance than the other methods. Further, the FSS-P had the overall best performance of all the methods except DLSWC, having on average only 0.8 and 0.1 higher value of the objective function. It should be noted that although FSS overall has a worse performance than DLSWC in case of two pairs (N_V, N_E) it managed to find higher quality average solutions. The experiments performed on large test

Table 1. Comparison of the methods for medium-size problem instances of Type 1.

$N_V \times N_E$	ACO		Element		Pair		PBIG	MS-ITS	DLSWC
	Basic	SEE	GRASP	FSS	GRASP	FSS			
50 × 50	2.1	0.9	0.0	0.0	0.0	0.0	0.0	0.0	1280.0
50 × 100	5.8	5.4	0.0	0.0	0.0	0.0	0.0	0.0	1735.3
50 × 250	15.1	8.3	0.0	0.0	0.0	0.0	0.0	0.0	2272.3
50 × 500	17.1	7.4	0.0	0.0	0.0	0.0	0.0	0.0	2661.9
50 × 750	8.0	6.3	0.0	0.0	0.0	0.0	0.0	0.0	2951.0
50 × 1000	17.5	6.1	0.0	0.0	0.0	0.0	0.0	0.0	3193.7
100 × 100	18.7	9.8	0.0	0.0	0.0	0.0	3.4	0.0	2534.2
100 × 250	24.8	13.3	0.0	0.0	0.0	0.0	1.1	0.0	3601.6
100 × 500	91.5	35.8	0.0	0.0	0.0	0.0	0.0	0.0	4600.6
100 × 750	30.9	37.3	0.0	0.0	0.0	0.0	0.0	0.0	5045.5
100 × 1000	25.9	14.5	0.0	0.0	0.0	0.0	1.2	0.0	5508.2
100 × 2000	43.8	16.4	0.0	0.0	0.0	0.0	0.0	0.0	6051.9
150 × 150	18.0	9.9	1.4	0.4	0.9	0.0	0.4	0.1	3666.9
150 × 250	49.8	35.0	1.8	0.0	0.0	0.0	0.4	0.0	4719.9
150 × 500	58.6	63.3	6.7	0.0	0.0	0.0	0.3	0.0	6165.4
150 × 750	58.3	39.9	7.6	6.8	0.0	0.0	7.3	10.6	6956.4
150 × 1000	82.1	23.9	8.0	1.6	1.6	1.6	9.1	0.0	7359.7
150 × 2000	81.8	47.8	0.0	0.2	0.0	0.0	12.6	0.0	8549.4
150 × 3000	50.4	40.4	0.0	0.0	0.0	0.0	0.0	0.0	8899.8
200 × 250	37.1	20.8	3.6	0.0	0.0	0.0	0.3	0.0	5551.6
200 × 500	67.3	41.8	2.6	1.1	0.0	0.0	0.5	3.2	7191.9
200 × 750	79.9	30.4	6.6	2.5	1.2	1.2	4.6	0.0	8269.9
200 × 1000	116.7	62.9	23.6	4.5	5.3	1.8	5.1	4.5	9145.5
200 × 2000	86.5	61.1	10.8	3.9	0.3	0.4	1.0	0.0	10830.0
200 × 3000	93.3	84.6	0.2	0.0	0.0	0.0	4.4	3.8	11595.8
250 × 250	49.1	20.5	4.6	0.0	0.0	0.0	0.0	0.0	6148.7
250 × 500	102.6	59.7	20.9	7.0	6.7	3.1	4.5	2.6	8436.2
250 × 750	123.5	69.6	24.8	6.4	1.9	-0.3	6.9	0.0	9745.9
250 × 1000	114.9	39.3	14.7	1.1	1.7	0.0	2.0	0.4	10751.7
250 × 2000	166.2	75.5	25.8	3.1	3.2	2.2	6.1	4.4	12751.5
250 × 3000	159.2	107.3	23.2	6.4	0.0	0.0	0.2	0.0	13723.3
250 × 5000	132.1	66.2	8.0	0.0	0.0	0.0	7.0	0.0	14669.7
300 × 300	46.9	30.8	7.0	0.2	0.0	0.0	0.2	0.0	7295.8
300 × 500	114.3	88.8	55.1	11.4	8.3	7.7	0.0	7.7	9403.1
300 × 750	137.6	127.2	60.9	14.0	13.3	2.2	8.8	2.7	11029.3
300 × 1000	143.2	65.2	40.2	8.9	4.6	4.4	10.4	9.2	12098.5
300 × 2000	162.7	102.4	49.7	15.4	6.7	2.7	17.7	5.5	14732.2
300 × 3000	213.3	69.7	42.3	1.7	1.0	1.0	7.4	0.6	15840.8
300 × 5000	202.5	136.9	31.6	19.3	2.3	1.9	7.7	0.0	17342.9
Average	78.18	45.70	12.35	2.97	1.51	0.77	3.35	1.42	
Found Best	0	0	14	19	24	27	12	25	

Table 2. Comparison of the methods for medium-size problem instances of Type 2.

$N_V \times N_E$	ACO		Element		Pair		PBIG	MS-ITS	DLSWC
	Basic	SEE	GRASP	FSS	GRASP	FSS			
50 × 50	0.2	0.2	0.0	0.0	0.0	0.0	0.0	0.0	83.7
50 × 100	5.0	3.2	0.0	0.0	0.0	0.0	0.0	0.0	271.2
50 × 250	33.4	16.9	0.0	0.0	0.0	0.0	0.0	0.0	1853.4
50 × 500	90.8	51.6	0.0	0.0	0.0	0.0	0.0	0.0	7825.1
50 × 750	55.1	8.6	0.0	0.0	0.0	0.0	0.0	0.0	20079.0
100 × 50	0.2	0.0	0.0	0.0	0.0	0.0	0.0	0.0	67.2
100 × 100	2.5	1.2	0.0	0.0	0.0	0.0	0.0	0.0	166.6
100 × 250	15.2	8.8	0.0	0.0	0.0	0.0	0.0	0.0	886.5
100 × 500	33.1	13.4	0.0	0.0	0.0	0.0	0.0	0.0	3693.6
100 × 750	74.3	62.1	0.0	0.0	0.0	0.0	0.0	0.0	8680.2
150 × 50	0.0	0.1	0.0	0.0	0.0	0.0	0.0	0.0	65.8
150 × 100	0.7	0.1	0.0	0.0	0.0	0.0	0.0	0.0	144.0
150 × 250	9.9	9.0	0.2	0.0	0.0	0.0	0.2	0.0	615.8
150 × 500	43.5	27.1	1.1	0.0	0.0	0.0	0.0	0.0	2331.5
150 × 750	100.7	8.5	0.9	0.0	0.0	0.0	0.2	0.0	5698.5
200 × 50	0.0	0.0	0.0	0.0	0.0	0.0	0.0	0.0	59.6
200 × 100	0.2	0.1	0.0	0.0	0.0	0.0	0.0	0.0	134.5
200 × 250	5.6	4.8	0.3	0.0	0.0	0.0	0.0	1.4	483.1
200 × 500	39.7	14.8	0.2	0.0	0.0	0.1	0.4	0.0	1803.9
200 × 750	69.3	33.5	0.0	0.2	0.2	0.2	0.1	0.0	4043.5
250 × 250	4.2	2.2	0.0	0.0	0.0	0.0	0.0	0.0	419.0
250 × 500	23.2	20.1	1.9	0.4	0.5	0.0	1.5	0.5	1434.2
250 × 750	59.8	33.3	3.1	0.0	0.0	0.0	4.9	0.3	3256.1
250 × 1000	71.8	53.6	7.7	−0.1	−0.3	−0.3	3.0	1.8	5986.4
250 × 2000	512.6	295.6	42.1	6.2	0.0	0.0	22.0	9.9	25636.5
250 × 5000	1648.2	1231.7	120.2	28.9	0.1	0.1	0.1	0.1	170269.0
300 × 250	4.5	3.3	0.4	0.0	0.1	0.0	0.1	0.2	399.4
300 × 500	22.7	20.9	2.1	0.0	0.2	0.0	0.0	0.8	1216.4
300 × 750	38.9	34.8	4.5	0.6	0.4	0.0	0.1	1.3	2639.3
300 × 1000	100.5	72.9	17.2	1.8	6.6	1.8	1.3	1.2	4795.0
300 × 2000	413.9	226.4	60.6	6.6	0.0	2.5	10.3	5.1	20881.3
300 × 5000	2023.1	1072.2	109.9	11.5	4.8	4.8	44.9	6.4	141220.4
Average	171.96	104.09	11.64	1.75	0.39	0.29	2.78	0.91	
Found Best	2	2	16	23	24	26	18	20	

Table 3. Comparison of best found solutions over 10 independent runs for large problem instances by different methods.

$N_V \times N_E$	ACO	Element		Pair		PBIG	MS-ITS	DLSWC
	SEE	GRASP	FSS	GRASP	FSS			
500 × 500	59.0	43.0	5.0	5.0	0.0	0.0	7.0	12616.0
500 × 1000	51.0	36.0	2.0	1.0	0.0	0.0	15.0	16465.0
500 × 2000	137.0	204.0	10.0	0.0	0.0	0.0	0.0	20863.0
500 × 5000	53.0	387.0	0.0	21.0	0.0	77.0	0.0	27241.0
500 × 10000	0.0	165.0	0.0	0.0	0.0	0.0	0.0	29573.0
800 × 500	24.0	44.0	0.0	0.0	0.0	0.0	21.0	15025.0
800 × 1000	45.0	99.0	0.0	15.0	0.0	0.0	13.0	22747.0
800 × 2000	379.0	472.0	144.0	102.0	16.0	54.0	8.0	31301.0
800 × 5000	277.0	518.0	159.0	163.0	62.0	112.0	0.0	38553.0
800 × 10000	148.0	290.0	45.0	41.0	6.0	45.0	0.0	44351.0
1000 × 1000	133.0	288.0	36.0	34.0	9.0	23.0	12.0	24723.0
1000 × 5000	243.0	460.0	62.0	79.0	61.0	52.0	27.0	45203.0
1000 × 10000	497.0	742.0	61.0	92.0	0.0	0.0	0.0	51378.0
1000 × 15000	400.0	670.0	133.0	169.0	61.0	20.0	20.0	57994.0
1000 × 20000	359.0	523.0	128.0	84.0	27.0	139.0	24.0	59651.0
Average	187.00	329.40	52.33	53.73	16.13	34.80	9.80	
Found Best	1	0	4	3	8	7	6	

Table 4. Comparison of average quality of found solutions over 10 independent runs for large problem instances by different methods.

$N_V \times N_E$	ACO	Element		Pair		PBIG	MS-ITS	DLSWC
	SEE	GRASP	FSS	GRASP	FSS			
500 × 500	71.7	68.0	5.0	5.0	2.0	4.0	19.0	12616.0
500 × 1000	109.9	56.8	5.6	2.0	0.0	5.1	18.1	16465.0
500 × 2000	226.8	226.4	12.2	−3.2	−3.2	4.6	0.7	20866.2
500 × 5000	344.5	411.0	91.2	64.8	0.0	187.2	0.0	27241.0
500 × 10000	223.4	199.4	86.8	3.0	0.0	93.8	0.0	29573.0
800 × 500	44.9	53.6	0.0	0.0	0.0	0.0	29.1	15025.0
800 × 1000	105.1	119.4	0.0	19.0	0.0	16.0	13.0	22747.0
800 × 2000	481.9	502.6	227.4	157.2	17.6	117.6	40.7	31305.0
800 × 5000	337.6	561.7	149.3	171.9	72.5	149.6	−12.0	38569.1
800 × 10000	337.8	350.9	42.1	51.1	15.1	43.9	6.0	44353.9
1000 × 1000	202.4	315.6	45.8	34.0	29.0	40.1	43.1	24723.0
1000 × 5000	349.8	469.3	73.9	92.3	25.3	56.5	18.0	45238.9
1000 × 10000	724.6	755.2	88.4	128.0	−2.4	160.5	42.6	51380.4
1000 × 15000	659.8	723.8	132.0	223.4	77.0	150.2	73.9	57995.0
1000 × 20000	612.9	597.3	148.7	149.5	80.3	192.6	64.6	59655.3
Average	322.21	360.73	73.89	73.20	20.88	81.45	23.79	

instances can be seen in Tables 3 and 4 where the values of the best found and average weight over 10 runs of each algorithm are given, respectively. In case of problems instances of this size a similar behavior can be seen as for medium-size instances.

From the computational results it is evident that the methods FSS-P and GRASP-P in which the local search includes pair swaps manages to find significantly better solutions than the element-based ones. The use of pair swaps in the local search produces a more significant improvement than the addition of the learning mechanism used in the FSS. It should be noted that GRASP-P has a better performance than FSS-E for medium-size instances, while FSS-E manages to have a slightly better performance for large instances with FSS-P being consistently better in both cases. The improvement that is achieved by FSS is more significant in case of the weaker local search based on element swaps. However, it is most important to note that the improvement achieved by FSS compared to the corresponding GRASP is very consistent, and it only has worse average quality of found solutions for 1 or 3 of the (N_V, N_E) pairs, when the used local search was based on elements or pairs, respectively.

The convergence speed of the FSS-P is competitive to other methods, in case of medium-size instances it needs an average time of 0.76 and 0.43 s to find the best solution for Type 1 and Type 2 graphs, respectively. This is a very similar result to MS-ITS which needed between 0.51 and 0.45 s to solve instances of Type 1 and Type 2, and better than PBIG which needs 2.49 and 4.23 s. DLSWC has a substantially better performances; on average it needs only 0.03 and 0.4 s. The FSS-P scales well, and for large graphs needs an average of 5.05 s to find the best solution for an instance which is similar to 5.20 of DLWSC, but it should be noted that the quality of solutions is of lower quality. The scaling of PBIG and MS-ITS is significantly worse and the methods on average need 126.94 and 74.80 s to solve large problem instances, respectively. It is interesting to point out that although the asymptotic computational cost of the local search based on pairs is greater than the one based on elements, the time for finding the best solutions for GRASP-P and FSS-P is generally 2–5 times lower than for GRASP-E and FSS-E. The FSS, on average, needs around half the time of GRASP with the same type of local search to find the best solution. The pair swap local search proves to be very efficient; for graphs having up to 100 nodes GRASP-P generally needs less than 20 iterations to find the best known solutions.

7 Conclusion

In this paper we have presented an efficient easy to implement method for finding near optimal solutions for the MWVCP. This has been done by developing two local searches based on a correction procedures which switches one or two vertices from a solution with new ones which produces a new vertex cover having a lower weight. These local searches have been used as a part of a GRASP algorithm. The performance of the developed GRASP has been improved by extending it to the novel Fixed Set Search metaheuristic. The conducted computational experiments have shown that the proposed FSS is highly competitive with the state-of-the-art methods. The results also indicate that the learning mechanism in the FSS manages to significantly enhance its performance when compared to the GRASP on which it is based. Importantly, the positive effect is most significant on large-scale problem instances on which the effectiveness of GRASP algorithms is generally decreased. For future research we aim to extend the application of the FSS to other types of problems.

References

1. Bouamama, S., Blum, C., Boukerram, A.: A population-based iterated greedy algorithm for the minimum weight vertex cover problem. Appl. Soft Comput. **12**(6), 1632–1639 (2012)
2. Cai, S., Li, Y., Hou, W., Wang, H.: Towards faster local search for minimum weight vertex cover on massive graphs. Inf. Sci. **471**, 64–79 (2019)
3. Chen, J., Kanj, I.A., Xia, G.: Improved upper bounds for vertex cover. Theor. Comput. Sci. **411**(40–42), 3736–3756 (2010)
4. Cygan, M., Kowalik, Ł., Wykurz, M.: Exponential-time approximation of weighted set cover. Inf. Process. Lett. **109**(16), 957–961 (2009)
5. Feo, T.A., Resende, M.G.: Greedy randomized adaptive search procedures. J. Glob. Optim. **6**(2), 109–133 (1995). https://doi.org/10.1007/BF01096763
6. Hansen, P., Mladenović, N.: Variable neighborhood search: principles and applications. Eur. J. Oper. Res. **130**(3), 449–467 (2001)
7. Jovanovic, R., Tuba, M.: An ant colony optimization algorithm with improved pheromone correction strategy for the minimum weight vertex cover problem. Appl. Soft Comput. **11**(8), 5360–5366 (2011)
8. Jovanovic, R., Tuba, M., Voß, S.: Fixed set search applied to the traveling salesman problem. In: Blesa Aguilera, M.J., Blum, C., Gambini Santos, H., Pinacho-Davidson, P., Godoy del Campo, J. (eds.) HM 2019. LNCS, vol. 11299, pp. 63–77. Springer, Cham (2019). https://doi.org/10.1007/978-3-030-05983-5_5
9. Karp, R.M.: Reducibility among combinatorial problems. In: Miller, R.E., Thatcher, J.W., Bohlinger, J.D. (eds.) Complexity of computer computations, pp. 85–103. Springer, Boston (1972). https://doi.org/10.1007/978-1-4684-2001-2_9
10. Li, R., Hu, S., Zhang, H., Yin, M.: An efficient local search framework for the minimum weighted vertex cover problem. Inf. Sci. **372**, 428–445 (2016)
11. Pullan, W.: Optimisation of unweighted/weighted maximum independent sets and minimum vertex covers. Discrete Optim. **6**(2), 214–219 (2009)
12. Shyu, S.J., Yin, P.Y., Lin, B.M.: An ant colony optimization algorithm for the minimum weight vertex cover problem. Ann. Oper. Res. **131**(1–4), 283–304 (2004). https://doi.org/10.1023/B:ANOR.0000039523.95673.33
13. Singh, A., Gupta, A.K.: A hybrid heuristic for the minimum weight vertex cover problem. Asia-Pac. J. Oper. Res. **23**(02), 273–285 (2006)
14. Voß, S., Fink, A.: A hybridized tabu search approach for the minimum weight vertex cover problem. J. Heuristics **18**(6), 869–876 (2012)
15. Wang, L., Du, W., Zhang, Z., Zhang, X.: A PTAS for minimum weighted connected vertex cover P3 problem in 3-dimensional wireless sensor networks. J. Comb. Optim. **33**(1), 106–122 (2017)
16. Zhou, T., Lü, Z., Wang, Y., Ding, J., Peng, B.: Multi-start iterated tabu search for the minimum weight vertex cover problem. J. Comb. Optim. **32**(2), 368–384 (2016)

Automated Deep Learning for Threat Detection in Luggage from X-Ray Images

Alessio Petrozziello and Ivan Jordanov(✉)

University of Portsmouth, Portsmouth, UK
{Alessio.petrozziello,ivan.jordanov}@port.ac.uk

Abstract. Luggage screening is a very important part of the airport security risk assessment and clearance process. Automating the threat objects detection from x-ray scans of passengers' luggage can speed-up and increase the efficiency of the whole security procedure. In this paper we investigate and compare several algorithms for detection of firearm parts in x-ray images of travellers' baggage. In particular, we focus on identifying steel barrel bores as threat objects, being the main part of the weapon needed for deflagration. For this purpose, we use a dataset of 22k double view x-ray scans, containing a mixture of benign and threat objects. In the pre-processing stage we apply standard filtering techniques to remove noisy and ambiguous images (i.e., smoothing, black and white thresholding, edge detection, etc.) and subsequently employ deep learning techniques (Convolutional Neural Networks and Stacked Autoencoders) for the classification task. For comparison purposes we also train and simulate shallow Neural Networks and Random Forests algorithms for the objects detection. Furthermore, we validate our findings on a second dataset of double view x-ray scans of courier parcels. We report and critically discuss the results of the comparison on both datasets, showing the advantages of our approach.

Keywords: Baggage screening · Deep learning · Convolutional neural networks · Image filtering · Object detection algorithms · X-ray images

1 Introduction

Identifying and detecting dangerous objects and threats in baggage carried on board of aircrafts plays important role in ensuring and guaranteeing passengers' security and safety. The security checks relay mostly on X-ray imaging and human inspection, which is a time consuming, tedious process performed by human experts assessing whether threats are hidden or occluded by other objects in a closely packed bags. Furthermore, a variety of challenges makes this process tedious, among those: very few bags actually contain threat items; the bags can include a wide range of items, shapes and substances (e.g., metals, organic, etc.); the decision needs to be made in few seconds (especially in rush hours); and the objects can be rotated, thus presenting a difficult to recognize view. Due to the complex nature of the task, the literature suggests that human expert detection performance is only about 80-90% accurate [1]. Automating the screening process through incorporating intelligent techniques for image processing and object detection can increase the efficiency, reduce the time, and improve the overall accuracy of dangerous objects recognition.

I. Kotsireas et al. (Eds.): SEA^2 2019, LNCS 11544, pp. 505–512, 2019.
https://doi.org/10.1007/978-3-030-34029-2_32

Research on threat detection in luggage security can be grouped based on three imaging modalities: single-view x-ray scans [2], multi-view x-ray scans [3, 4], and computed tomography (CT) [5]. Classification performance usually shows improvements with the number of utilised views, with detection performance ranging from 89% true positive rate (TPR) with 18% false positive rate (FPR) for single view imaging [2] to 97.2% TPR and 1.5% FPR in full CT imagery [5].

The general consensus in the baggage research community is that the classification of x-ray images is more challenging than the visible spectrum data, and that direct application of methods used frequently on natural images (such as SIFT, RIFT, HoG, etc.) does not always perform well when applied to x-ray scans [6]. However, identification performance can be improved by exploiting the characteristics of x-ray images by: augmenting multiple views; using a coloured material image or employing simple (gradient) density histogram descriptors [7–9]. Also, the authors of [10] discuss some of the potential difficulties when learning features using deep learning techniques, on varying size images with out-of-plane rotations.

This work aims to develop a framework to automatically detect firearms from x-ray scans using deep learning techniques. The classification task focusses on the detection of steel barrel bores to determine the likelihood of firearms being present within an x-ray image, using a variety of classification approaches. Two datasets of dual view x-ray scans are used to assess the performance of the classifiers: the first dataset contains images of hand-held travel luggage, while the second dataset comprises scans of courier parcels. We handle the varying image size problem by combining the two views in one unique sample, while we do not explicitly tackle the out-of-plane rotation problem, instead, we rely on data augmentation techniques and on a dataset containing the threat objects recorded in different poses.

We investigate two deep learning techniques, namely Convolutional Neural Networks (CNN) and Stacked Autoencoders, and two widely used classification models (Feedforward Neural Networks and Random Forests) and the results from their implementation are critically compared and discussed.

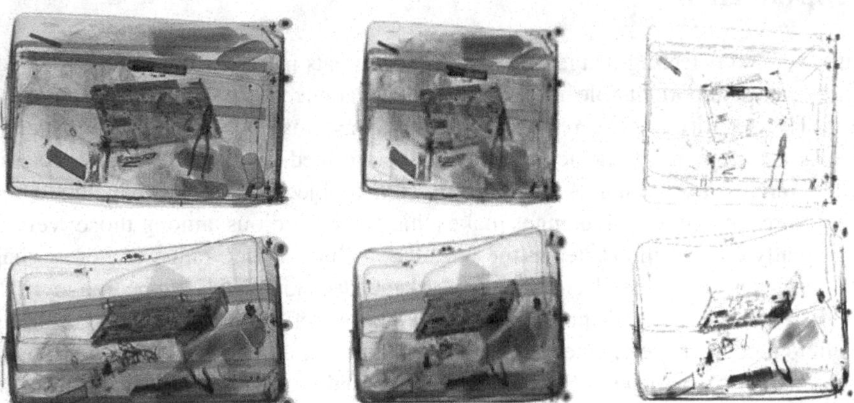

Fig. 1. A sample image containing a steel barrel bores (top left cylinder in the top row) from the baggage dataset. The left image (in both rows) is the raw dual view x-ray scan, in the middle, the grey scale smoothed one, and on the right, the b/w thresholded one.

The remainder of the paper is organized as follows. Section 2 describes the datasets used in the empirical experimentation and illustrates the proposed framework; Sect. 3 reports details on the carried experiments and results; while conclusion and future work are given in Sect. 4.

2 Threat Identification Framework

The proposed framework for automated weapon detection consists of three modules: pre-processing, data augmentation and threat detection. The pre-processing stage comprises four steps: green layer extraction, greyscale smoothing, black and white (b/w) thresholding and data augmentation.

The original dataset consists of over 22000 images of which approximately 6000 contain a threat item (i.e., a whole firearm or a component). The threat images are produced by a dual view x-ray machine: one view from above, and one from the side. Each image contains metadata about the image class (i.e., benign or threat), and firearm component (i.e., barrel only, full weapon, set of weapons, etc.). From the provided image library, a sample of 3546 threat images were selected containing a firearm barrel (amongst the other items), and 1872 benign images only containing allowed objects. The aim of the classification is to discriminate only the threat items - as common objects are displayed in both 'benign' and 'threat' samples (e.g., Figs. 1 and 2). During the pre-processing phase, each image is treated separately and the two views are combined before the classification stage.

The raw x-ray scans are imported in the framework as a 3-channel images (RGB) and scaled to 128 × 128 pixels in order to have images of same size for the machine learning procedure, and to meet memory constraints during training.

From the scaled image, the green colour channel is extracted as the one found to have the greatest contrast in dense material.

The resulting greyscale image is intended to reflect more accurately the raw x-ray data (i.e., measure of absorption). This step is performed to enable subsequent filtering and better identification of a threshold for dense material and eventually to facilitate the recognition of the barrel.

Fig. 2. A sample image containing a steel barrel bores (top right cylinder in the top row) from the parcel dataset. The left image (both rows) is the raw dual view x-ray scan, in the middle, the grey scale smoothed one, and on the right, b/w thresholded one. The parcel dataset usually contains a higher amount of steel objects and the barrels are better concealed.

A smoothing algorithm is applied on the greyscale image in order to reduce the low-level noise within it, while preserving distinct object edges. A number of smoothing algorithms were tested and a simple 3 × 3 kernel Gaussian blur was found to generate the best results. Then, on the smoothed image we apply a thresholding technique to isolate any dense material (e.g., steel). The chosen threshold is approximated within the algorithm to the equivalent of 2 mm of steel, which ensures that metal objects, such as firearm barrels and other components are kept. This step removes much of the benign background information within the image, such as organic materials and plastics. The resulting image is normalised to produce a picture where the densest material is black and the image areas with intensity below the threshold are white. At this point, the instances for which the produced image lacks any significant dense material, can be directly classified as benign. From cursory examination of the operational benign test set, this is a significant proportion of the samples for the *baggage* dataset, while only filtering out a small portion of images on the *parcels* one (mainly because in the courier parcels there is a higher variety of big and small metallic objects compared to the hand-held travel luggage). When applying deep learning techniques on images, it is often useful to increase the robustness of the classification by adding realistic noise and variation to the training data (i.e., augmentation), especially in the case of high imbalance between the classes [11]. There are several ways in which this can be achieved: object volume scaling: scaling the object volume V by a factor v; object flips/shifts: objects can be flipped/shifted in the x or y direction to increase appearance variation. This way, for every image in the training set, multiple instances are generated, combining different augmentation procedures and these are subsequently used by the models during the learning phase. Lastly, the two views of each sample are vertically stacked to compose one final image (Figs. 1 and 2).

The four machine learning methods incorporated and critically compared in this work include two from the deep learning area, namely Convolutional Neural Networks (CNN) and Stacked Autoencoders; and two shallow techniques: Neural Networks and Random Forests.

The CNN are considered state-of-the-art neural network architectures for image recognition, having the best results in different applications, e.g.: from a variety of problems related to image recognition and object detection [12], to control of unmanned helicopters [13], x-ray cargo inspection [7], and many others. A CNN is composed of an input layer (i.e., the pixels matrix), an output layer (i.e., the class label) and multiple hidden layers. Each hidden layer usually includes convolution, activation, and pooling functions, and the last few layers are fully connected, usually with a softmax output function. A convolutional layer learns a representation of the input applying a 2D sliding filters on the image and capturing the information of contingent patches of pixels. The pooling is then is used to reduce the input size, aggregating (e.g., usually using a max function) the information learned by the filters (e.g., a 3 × 3 pixels patch is passed in the learned filter and the 3 × 3 output is then pooled taking the maximum among the nine values). After a number of hidden layers (performing convolution, activation, and pooling), the final output is flattened into an array and passed to a classic fully connected layer to classify the image.

Stacked Autoencoders, also called auto-associative neural networks, are machine learning technique used to learn features at different level of abstraction in an unsupervised fashion. The autoencoder is composed of two parts: an encoder, which maps the input to a reduced space; and a decoder which task is to reconstruct the initial input from the lower dimensional representation. The new learned representation of the raw features can be used as input to another autoencoder (hence the name stacked). Once each layer is independently trained to learn a hierarchical representation of the input space, the whole network is fine-tuned (by performing backpropagation) in a supervised fashion to discriminate among different classes. In this work we use sparse autoencoders, that rely on heavy regularization to learn a sparse representation of the input.

3 Experimentation and Results

After the pre-processing and filtering off the images not containing enough dense material, we ended with 1848 and 1764 samples for classification of the *baggage* and *parcel* datasets respectively. The *baggage* dataset comprises 672 images from the benign class and 1176 containing threats; while the *parcel* dataset 576 and 1188 samples for the benign and threat classes respectively. Each dataset was split in 70% for training and 30% as independent test set. Due to their different operational environments, the baggage and parcel scans were trained and tested separately.

In this experiment we used a three layer stacked autoencoder with 200, 100, 50 neurons respectively, followed by a *softmax* output function to predict the classes probability. For the CNN we emploed a topology with three convolutional layers (with 128, 64 and 32 neurons) followed by a fully connected neural network and a *softmax* output function.

The RF was trained with 200 trees while the shallow NN had a topology of ***n-n-2***, where ***n*** was the input size. Since both RF and shallow NN cannot be directly trained on raw pixels, a further step of feature extraction was performed. In particular, we used histograms of *oriented Basic Image Features* (*oBIFs*) as a texture descriptor (as suggested in [6]), which has been applied successfully in many machine vision tasks. The *Basic Image Features* is a scheme for classification of each pixel of an image into one of seven categories, depending on local symmetries. These categories are: flat (no strong symmetry), slopes (e.g., gradients), blobs (dark and bright), lines (dark and bright), and saddle-like. *Oriented BIFs* are an extension of the *BIFs,* that include the quantized orientation of rotationally asymmetric features [14], which encode a compact representation of images. The *oBIF* feature vector is then fed as input into the RF and the shallow NN classifiers.

To evaluate the classification performance we employ three metrics: area under the ROC curve (AUC), the false positive rate at 90% true positive rate (FPR@90%TPR), and the F1-score. The AUC is a popular metric for classification tasks and the FPR@90%TPR is one cut-off point from the AUC, which describes the amount of false positives we can expect when correctly identifying 90% of all threats. The cut-off at 90% is suggested by [6] for the classification of x-ray images in a similar context. The F1-score is also a widely used metric for classification of imbalanced datasets that takes into account the precision (the number of correctly identified threats divided by the number of all threats identified by the classifier) and the recall (the number of correctly identified threats divided by the number of all threat samples).

Table 1. Baggage dataset results for the AUC, FPR@90%TPR and F1-Score metrics. The results are reported for the four classification techniques and three pre-processing step: raw data, grey scale smoothing and b/w thresholding.

Metric	Technique	Raw	Smoothing	B/w thresholding
AUC	CNN	**93**	**95**	**96**
	Autoencoder	75	78	90
	oBIFs + NN	85	87	94
	oBIFs + RF	66	72	80
FPR @ 90% TPR	CNN	**9**	**7**	**6**
	Autoencoder	70	60	26
	oBIFs + NN	50	31	14
	oBIFs + RF	86	66	53
F1-Score	CNN	**91**	**93**	**93**
	Autoencoder	60	65	81
	oBIFs + NN	64	67	79
	oBIFs + RF	36	41	56

As it can be seen from Table 1, the CNN outperformed the other methods with AUC ranging between 93% and 96%, depending on the pre-processing stage. The second best method was the shallow NN with AUC values between 85% and 94%, while the worst performance was achieved by the RF with 66%–80% AUC. Similar results were achieved when considering the FPR@90%TPR and F1-score metrics. The CNN reached the best FPR (6%) when trained on the b/w thresholded images, while still having only 9% FPR when using raw data. On the other hand, while achieving 14% FPR with the last stage of pre-processing, the NN performance dropped drastically when employing the raw and the smoothed data, with 50% and 31% FPR respectively. The same can be observed when using the F1-score: the CNN achieving up to 93%, followed by the Stacked Autoencoders and the shallow NN with 81% and 79% respectively. Once again, it is worth noticing that the CNN was the only technique able to score high classification accuracy across all used pre-processing approaches, while the other methods needed more time spent on the features engineering and extracting steps.

Table 2. Parcel dataset results for the AUC, FPR@90%TPR and F1-Score metrics. The results are reported for the four classification techniques and three pre-processing step: raw data, grey scale smoothing and b/w thresholding

Metric	Technique	Raw	Smoothing	B/w Thresholding
AUC	CNN	**80**	**79**	**84**
	Autoencoder	65	66	75
	oBIFs + NN	65	69	84
	oBIFs + RF	63	63	79

(*continued*)

Table 2. (*continued*)

Metric	Technique	Raw	Smoothing	B/w Thresholding
FPR @ 90% TPR	CNN	**46**	**46**	**37**
	Autoencoder	66	69	70
	oBIFs + NN	71	75	40
	oBIFs + RF	91	88	56
F1-Score	CNN	**86**	**83**	**87**
	Autoencoder	40	43	55
	oBIFs + NN	36	32	63
	oBIFs + RF	34	42	58

Table 2 shows the performance metrics on the *parcel* dataset, illustrating generally lower performance across all techniques. This can be explained by the larger variety of metal items contained in the courier parcels, when compared to the objects contained in a hand-held airport luggage. Again, the CNN outperformed the other considered methods, with an AUC ranging from 79% to 84%, followed by the NN with 65% to 84%, RF with 63% to 79%, and the Stacked Autoencoders with 65% to 75%. The AUC achieved on the *parcel* dataset by the shallow NN, RF and Stacked Autoencoders are much closer than those achieved on the *baggage* one, where the best performing method outstands more.

Yet again, the CNN achieved the lowest FPR (37%), followed by the shallow NN with 40% FPR, the RF with 56% FPR and the Stacked Autoencoders with 70% FPR. Lastly, the F1-score metric produced the largest difference in values across the methods, with the CNN achieving up to 87% F1-score, followed by shallow NN with 63%, RF with 58% and Stacked Autoencoders with 55%. Also, in this case the CNN was the only technique able to classify threats with high accuracy, just using the raw images, where all other techniques performed very poorly (e.g., the AUC on raw data for the CNN was 15 percentage points better than the NN, while holding similar performance on the b/w thresholded one; 20 percentage points better in FPR@90% TPR when compared to the second best (Autoencoder); and even 46 percentage points better than the Autoencoder for the F1-score).

4 Conclusion

In this work we investigated a deep learning framework for automated identification of steel barrel bores in datasets of X-ray images in operational settings such as airport security clearance process and courier parcel inspections. In particular we compare two deep learning methods (Convolutional Neural Networks and Stacked Autoencoders), and two widely used classification techniques (shallow Neural Networks and Random Forest) on two datasets of X-ray images (*baggage* and *parcel* datasets). We evaluated the methods performance using three commonly accepted metrics for classification tasks: area under the ROC curve (AUC), the false positive rate at 90% true positive rate (FPR@90%TPR), and the F1-score. The obtained results showed that the CNN is not only able to consistently outperform all other compared techniques over the three metrics and on both datasets, but it is also able to achieve good prediction accuracy when using the raw data (whether the other techniques need multiple steps of data pre-processing and feature extraction to improve their performance). Furthermore, the CNN also achieved higher accuracy than the reported in literature results from human screening [1] (although, the employed datasets have not been screened by human experts, so an accurate direct comparison cannot be reported). Future work will explore application of different architectures for the CNN and Stacked Autoencoders, based on simulations on larger datasets to further investigate the result of this initial experimentation.

References

1. Michel, S., Koller, S.M., de Ruiter, J.C., Moerland, R., Hogervorst, M., Schwaninger, A.: Computer-based training increases efficiency in X-ray image interpretation by aviation security screeners. In: 41st Annual IEEE International Carnahan Conference on Security Technology (2007)
2. Riffo, V., Mery, D.: Active X-ray testing of complex objects. Insight-Non-Destructive Test. Condition Monit. **54**(1), 28–35 (2012)
3. Mery, D., et al.: The database of X-ray images for nondestructive testing. J. Nondestr. Eval. **34**(4), 1–12 (2015)
4. Mery, D., Riffo, V., Zuccar, I., Pieringer, C.: Automated X-ray object recognition using an efficient search algorithm in multiple views. In: IEEE Conference on Computer Vision and Pattern Recognition Workshops (2013)
5. Flitton, G., Mouton, A., Breckon, T.: Object classification in 3D baggage security computed tomography imagery using visual codebooks. Pattern Recogn. **48**(8), 2489–2499 (2015)
6. Jaccard, N., Rogers, T.W., Morton, E.J., Griffin, L.D.: Tackling the X-ray cargo inspection challenge using machine learning. In: Anomaly Detection and Imaging with X-Rays (ADIX) (2016)
7. Rogers, T.W., Jaccard, N., Griffin, L.D.: A deep learning framework for the automated inspection of complex dual-energy x-ray cargo imagery. In: Anomaly Detection and Imaging with X-Rays (ADIX) II (2017)
8. Li, G., Yu, Y.: Contrast-oriented deep neural networks for salient object detection. IEEE Trans. Neural Networks Learn. Syst. **29**(1), 6038–6051 (2018)
9. Shen, Y., Ji, R., Wang, C., Li, X., Li, X.: Weakly supervised object detection via object-specific pixel gradient. IEEE Trans. Neural Networks Learn. Syst. **29**(1), 5960–5970 (2018)
10. Bastan, M., Byeon, W., Breuel, T.M.: Object recognition in multi-view dual energy x-ray images. In: BMVC (2013)
11. Zhang, C., Tan, K.C., Li, H., Hong, G.S.: A cost-sensitive deep belief network for imbalanced classification. IEEE Trans. Neural Networks Learn. Syst. **30**(1), 109–122 (2019)
12. Zhao, Z.-Q., Zheng, P., Xu, S.-T., Wu, X.: Object detection with deep learning: a review. IEEE Trans. Neural Networks Learn. Syst. 1–21 (2019)
13. Kang, Y., Chen, S., Wang, X., Cao, Y.: Deep convolutional identifier for dynamic modeling and adaptive control of unmanned helicopter. IEEE Trans. Neural Networks Learn. Syst. **30** (2), 524–538 (2019)
14. Newell, A.J., Griffin, L.D.: Natural image character recognition using oriented basic image features. In: International Conference on Digital Image Computing Techniques and Applications (2011)

Algorithmic Aspects on the Construction of Separating Codes

Marcel Fernandez[1] and John Livieratos[2(✉)]

[1] Department of Network Engineering, Universitat Politecnica de Catalunya, Barcelona, Spain
[2] Department of Mathematics, National and Kapodistrian University of Athens, Athens, Greece
jlivier89@math.uoa.gr

Abstract. In this paper, we discuss algorithmic aspects of separating codes, that is, codes where any two subsets (of a specified size) of their code words have at least one position with distinct elements. More precisely we focus on the (non trivial) case of binary 2-separating codes. Firstly, we use the Lovász Local Lemma to obtain a lower bound on the existence of such codes that matches the previously best known lower bound. Then, we use the algorithmic version of the Lovász Local Lemma to construct such codes and discuss its implications regarding computational complexity. Finally, we obtain explicit separating codes, with computational complexity polynomial in the length of the code and with rate larger than the well-known Simplex code.

Keywords: Separating codes · Lovász Local Lemma · Moser-Tardos constructive proof

1 Introduction

Separating codes [16] have a long tradition of study in the areas of coding theory and combinatorics. This type of codes have been proven useful in applications in the areas of technical diagnosis, construction of hash functions, automata synthesis and traitor tracing.

We can represent a code as a matrix of symbols over a finite alphabet. The *length* of a code is the number of its columns, while its *size* that of its rows. The *rate* of a code is the fraction of (the logarithm of) its size over its length. A code is called c-separating if for any two sets of at most c disjoint rows each, there is a column where the symbols in the first set are different from the symbols in the second. This turns out to be a really strong requirement. Consequently, although there is a vast research focused on obtaining lower and upper bounds to the rates of such codes, these bounds are weak. Also, explicit constructions of such codes are very scarce.

The work of Marcel Fernandez has been supported by TEC2015-68734-R (MINECO/FEDER) "ANFORA" and 2017 SGR 782.

I. Kotsireas et al. (Eds.): SEA² 2019, LNCS 11544, pp. 513–526, 2019.
https://doi.org/10.1007/978-3-030-34029-2_33

For instance, for the already non trivial case of binary 2-separating codes, the best lower bound for the rate obtained so far is 0.064 [2,16,19], whereas the best upper bound is 0.2835 given by Korner and Simonyi in [12]. One immediately sees that these bounds are not by any means tight. The lower bound is obtained by the elegant technique of random coding with expurgation. The drawback of the random coding strategy is that it gives no clue whatsoever about how to obtain an explicit code matching the bound.

An alternate route one can take in trying to show the existence of combinatorial objects, is the Lovász Local Lemma (LLL). This powerful result was first stated by Erdos and Lovász in [4]. In its simple and *symmetric* form, the LLL gives a necessary condition on the chance of avoiding a certain number of undesirable events, given a constant upper bound on the probability of each event to occur and on the number of dependencies between them.

Our interest in the LLL lies in its algorithmic proofs. The first such proof was given by Moser [14] and applied solely to the satisfiability problem. Later, Moser and Tardos [15] gave a proof for the stronger *asymmetric* version of the lemma in the general case, based on the *entropic method* (see [21]). More recently, Giotis et al. [7,9] applied a direct probabilistic approach to prove various forms of the LLL (for an analytic exposition of the various forms of the LLL, see Szegedy's review in [20]). Both of the above approaches (as most algorithmic approaches to the LLL), use what is known as the *variable framework*, where all the undesirable events are assumed to depend on a number of *independent random variables*. For algorithmic approaches not in the variable framework, see for example the work of Harvey and Vondrak [10] and Achlioptas and Illiopoulos [1].

The LLL has been frequently used to prove the existence of combinatorial objects with desired properties. To name some examples, one can see the work of Gebauer et al. [5,6] in bounding the number of literals in each clause of a formula, in order for it to be satisfiable, and the work of Giotis et al. [8] in bounding the number of colors needed to properly color the edges of a graph so that no bichromatic cycle exists. Closer to the subject of the present work, we find the work of Sarkar and Colbourn [17] on covering arrays and that of Deng et al. [3] on perfect and separating hash families.

Our Contribution: In this paper we focus on 2-separating binary codes. The generalization to larger alphabets and larger separating sets is straightforward and will be done elsewhere.

In Sect. 3, we prove a lower bound of such codes using the Lovász Local Lemma in the style of Deng et al. in [3]. In Sect. 4, we move on and provide explicit constructions of such codes, using the work of Giotis et al. [7] and Kirousis and Livieratos [11]. Although the straightforward application of those results leads to constructions of exponential complexity, we show how we can explicitly obtain codes better than the current known constructions, by appropriately changing the conditions required in the LLL.

2 Definitions and Previous Results

In this section we provide some definitions and results, about both separating codes and the Lovász Local Lemma, that will be used throughout the paper.

2.1 Separating Codes

We start with some basic definitions. Let $\mathcal{Q}$ denote a finite alphabet of size q. Then $\mathcal{Q}^n$ denotes the set of vectors of length n, of elements from $\mathcal{Q}$. We call such vectors *words*. An $(n, M)_q$ *code* $\mathcal{C} \subset \mathcal{Q}^n$ is a subset of size M. A word in $\mathcal{C}$ will be called a *code word*. The *Hamming distance* between two code words is the number of positions where they differ. The *minimum distance* of $\mathcal{C}$, denoted by d, is defined as the smallest distance between two different code words.

In algebraic coding theory, $\mathcal{Q}$ is usually $\mathbb{F}_q$, the finite field with q elements. In this case, a code $\mathcal{C}$ is a *linear* if it forms a subspace of $\mathbb{F}_q^n$. An (n, k, d)-code is a code with length n, dimension k and minimum distance d.

Let $U = \{\mathbf{u}^1, \dots, \mathbf{u}^c\} \subset \mathcal{C}$ be a subset of size $|U| = c$, where $\mathbf{u}^i = (u_1^i, \dots, u_n^i)$, $i = 1, \dots, c$. Then $U_j = \{u_j^1, \dots . u_j^c\}$ is the set of the alphabet elements in the j-th coordinate of the words in U. Consider now the following definitions.

Definition 1 (Sagalovich [16]). *A code $\mathcal{C}$ is a c-separating code if for any two disjoint sets of code words U and V with $|U| \leq c$, $|V| \leq c$ and $U \cap V = \emptyset$, there exists at least one coordinate j such that U_j and V_j are disjoint, i.e. $U_j \cap V_j = \emptyset$. We say that the coordinate j separates U and V.*

Throughout this paper, we will be interested in the case where $c = 2$.

Definition 2. *Let $\mathcal{C}$ be an $(n, M)_q$ code over $\mathcal{Q}$. The rate R of $\mathcal{C}$ is defined as*

$$R = \frac{\log_q M}{n}. \tag{1}$$

Let $R(n, c)_q$ be the optimal rate of a c-separating $(n, M)_q$ code. We are interested in the asymptotic rate:

$$\underline{R}_q(c) = \liminf_{n \to \infty} R_q(n, c). \tag{2}$$

For the binary case there exist codes of positive asymptotic rate, as shown in [2]. For completeness we provide the proof here:

Proposition 1 (Barg et al. [2]). *There exist binary c-separating codes of length n and size $\frac{1}{2}(1 - 2^{-(2c-1)})^{-(n/(2c-1))}$, i.e.*

$$\underline{R}(n, 2)_2 \geq -\frac{\log_2(1 - 2^{-(2c-1)})}{2c - 1} - \frac{1}{n}. \tag{3}$$

Proof. Let $\mathcal{C}$ be a random binary $(n, 2M)$ code. We consider pairs of code words of $\mathcal{C}$. The probability $\Pr[U \not\leftrightarrow V]$ that a pair of pairs of code words U and V are not separated is:

$$\Pr[U \not\leftrightarrow V] = \left(1 - 2^{-(2c-1)}\right)^n.$$

Then the expected number $E(N_s)$ of pairs of pairs in $\mathcal{C}$ that are not separated is:

$$E(N_s) \leq \binom{2M}{c}\binom{M-c}{c}\Pr[U \not\leftrightarrow V].$$

By taking:

$$2M = \left(\frac{c!c!}{-n} \cdot \frac{1}{\Pr[U \not\leftrightarrow V]}\right)^{1/(2c-1)}$$

we have:

$$E(N_s) < \frac{(2M)^{2c}}{c!c!}\Pr[U \not\leftrightarrow V] = M.$$

Now, from the random binary $(n, 2M)$ code $\mathcal{C}$, we remove at most M code words in order to destroy the non separated pairs. Finally, since we have that $(c!c!/2)^{(1/(1-2c)} \geq 1$, the result on the size of the code follows.

Focusing on the rate of the binary 2-separating codes, we have the following corollary:

Corollary 1 (Sagalovich [16]). *There exist binary 2-separating codes of rate:*

$$\underline{R}_2(2) \geq 1 - \log_2(7/8) = 0.0642$$

The Simplex Code. A well known example of a 2-separating code is the *Simplex code*:

Definition 3 (MacWilliams and Sloane [13]). *The binary* simplex code $\mathcal{S}_k$ *is a* $(2^k - 1, k, 2^{k-1})$*-code which is the dual of the* $(2^k - 1, 2^k - 1 - k, 3)$*-Hamming code.*

$\mathcal{S}_k$ *consists of* $\mathbf{0}$ *and* $2^k - 1$ *code words of weight* 2^{k-1}. *It is called a simplex code, because every pair of code words is at the same distance apart.*

For this family of codes, we have the following lemma, which we give here without proof.

Lemma 1. *The binary simplex code is 2-separating.*

2.2 Algorithmic Lovász Local Lemma

Let $E_1, \ldots, E_m$ be events defined on a common probability space Ω, which are considered undesirable. We assume the events are ordered according to their indices. Consider a (simple) graph G with vertex set $[m] := \{1, \ldots, m\}$ and where two vertices $i, j \in [m]$ are connected by an (undirected) edge if E_i and E_j are dependent. In the literature, such graphs are called *dependency graphs*.

Let Γ_j be the *neighborhood* of vertex j in G and assume that no vertex belongs to its neighborhood ($j \notin \Gamma_j$, $j = 1, \ldots, m$). We will sometimes say that an event E_i such that $i \in \Gamma_j$ is an event in the neighborhood of E_j. Let also $s \geq 1$ be the maximum degree of G (thus $|\Gamma_j| \leq s$ for $j = 1, \ldots, m$) and suppose that there is a number $p \in (0,1)$ such that $\Pr[E_j] \leq p$, $j = 1, \ldots, m$.

In its simple, *symmetric* version, the Lovász Local Lemma provides a sufficient condition, depending on p and s, for avoiding all the events $E_1, \ldots, E_m$.

Theorem 1 (Symmetric Lovász Local Lemma). *Suppose $E_1, \ldots, E_m$ are events, whose dependency graph has degree s and such that there exists a $p \in (0,1)$ such that $Pr[E_j] \leq p$, $j = 1, \ldots, m$. If*

$$ep(s+1) \leq 1, \tag{4}$$

then

$$\Pr\left[\bigcap_{i=1}^{m} \overline{E}_i\right] > 0.$$

We are interested in constructive approaches to the LLL. We will thus use what is known as the *variable framework*, which first appeared in a work by Moser and Tardos [15]. Let X_i, $i \in [t]$ be *mutually independent* random variables, defined on the probability space Ω and taking values in a finite set $\mathcal{Q}$. An assignment of values to the random variables is a t-ary vector $\alpha = (a_1, \ldots, a_t)$, with $a_i \in \mathcal{Q}$, $i = 1, \ldots, t$. In what follows, we assume that $\Omega = \mathcal{Q}^t$.

Recall the events $E_1, \ldots, E_m$ defined on Ω. We assume each event depends only on a subset of the random variables, which we call its *scope*. The scope of E_j is denoted by $\mathrm{sc}(E_j)$.

The LLL provides a sufficient condition for the existence of a point in the probability space Ω, that is, an assignment of values to the random variables, such that none of the events occurs. We are interested in finding this assignment efficiently. Consider Algorithm 1 below:

Algorithm 1. M-ALGORITHM.

```
Sample the variables X_i, i = 1,...,t and let α be the resulting assignment.
while there exists an event that occurs under the current assignment, let E_j be
  the least indexed such event and do
    RESAMPLE(E_j)
end while
Output current assignment α.
```

<u>RESAMPLE(E_j)</u>

```
Resample the variables in sc(E_j).
while some event in Γ_j ∪ {j} occurs under the current assignment, let E_k be the
  least indexed such event and do
    RESAMPLE(E_k)
end while
```

Using M-Algorithm, Giotis et al. [7] proved the following Theorem:

Theorem 2 (Algorithmic LLL). *Assuming p and s are constants such that $\left(1+\frac{1}{s}\right)^{s} ps < 1$ (and therefore if $ep(s+1) \leq 1$), then there exists an integer N_0, which depends linearly on m, and a constant $t \in (0,1)$ (depending on p and s) such that if $N/\log N \geq N_0$, then the probability that* M-Algorithm *lasts for at least N rounds is $< c^N$ (is* inverse exponential in N*).*

We can deduce two things from the above result. First, since by the **while**-loop of line 2, if and when M-Algorithm terminates we have an assignment such that no undesirable event occurs, Theorem 2 implies the existence of such an assignment. Furthermore, M-Algorithm finds such an assignment in time polynomial in N.

3 A Lower Bound on the Rate of 2-Separating Binary Codes

Our aim is to use the Lovász Local Lemma to obtain a lower bound on the rate of 2-separating binary codes. The bound we obtain is of the same order of magnitude that the bound in Proposition 1. However, as we will see in the following section, the use of the LLL will allow us to make the construction explicit.

Let X_{ij} $1 \leq i \leq M$, $1 \leq j \leq n$, be nM independent random variables, following the Bernoulli distribution, where:

$$\Pr(X_{ij} = 0) = \Pr(X_{ij} = 1) = \frac{1}{2}.$$

Let also $\Omega = \{0,1\}^{nM}$ be the set of all (nM)-ary binary vectors. It will be convenient to think about Ω as the set of $M \times n$ matrices with binary entries. This will allow for the matrix obtained by assigning values to the random variables X_{ij} $1 \leq i \leq M$, $1 \leq j \leq n$, to be seen as an $(n.M)$ binary code. Let us denote such a code by $\mathcal{C}$. Recall that M is the size of the code, i.e. the number of code words it contains and n is the length of these code words.

Let $\mathbf{u} = \{u^1, u^2\}$ be a set of two distinct code words $u^i = (u_1^i, \dots, u_n^i)$ of $\mathcal{C}$, $i = 1, 2$, and let:

$$\mathcal{P}_{\mathcal{C}} := \{\{\mathbf{u}, \mathbf{v}\} \mid \mathbf{u} \cap \mathbf{v} = \emptyset\},$$

be the set of disjoint pairs of distinct code words of $\mathcal{C}$. For each $\{\mathbf{u}, \mathbf{v}\} \in \mathcal{P}$, we define the event $E_{\mathbf{u},\mathbf{v}}$ to occur when $\mathbf{u}, \mathbf{v}$ are *not* separated. There are $m = \binom{M}{2}\binom{M-2}{2}$ such events, which we assume to be ordered arbitrarily.

We will need two lemmas. We begin by computing the probability of each event to occur.

Lemma 2. *The probability of any event $E_{\mathbf{u},\mathbf{v}}$ is:*

$$\Pr[E_{\mathbf{u},\mathbf{v}}] = \left(\frac{7}{8}\right)^{n}. \tag{5}$$

Proof. Consider the j-th indices of u^1, u^2, v^1 and v^2. The probability of the event $E^j_{\mathbf{u},\mathbf{v}}$, which is the event that $\mathbf{u}$ and $\mathbf{v}$ are not separated in the j-th coordinate, is equal to the probability that $u^1_j \neq u^2_j$ plus the probability that $u^1_j = u^2_j \neq v^i_j$, for at least one $i \in \{1, 2\}$:

$$\Pr[E^j_{\mathbf{u},\mathbf{v}}] = \frac{1}{2} + \frac{1}{2} \cdot \frac{3}{4} = \frac{7}{8}.$$

Now, since each coordinate of a code word takes values independently, we have the required result. □

It can be easily seen that two events are dependent if they have at least one common code word. Thus:

Lemma 3. *The number of events depending on $E_{\mathbf{u},\mathbf{v}}$ is at most:*

$$s = 5M^3 - 1. \tag{6}$$

Proof. By subtracting the number of events that share no common code word with $E_{\mathbf{u},\mathbf{v}}$, we get that the number of dependent events of $E_{\mathbf{u},\mathbf{v}}$ is equal to:

$$\binom{M}{2}\binom{M-2}{2} - \binom{M-4}{2}\binom{M-6}{2} - 1.$$

Now, by elementary operations, this number is bounded by $s = 5M^3 - 1$. □

Armed with the previous lemmas and Theorem 1, we can state the following theorem:

Theorem 3. *For every $n > 0$ there exists a binary 2-separating code of size:*

$$M \leq \frac{1}{\sqrt[3]{5e}} \left(\frac{8}{7}\right)^{n/3}. \tag{7}$$

Proof. Indeed, (4) requires that:

$$ep(s + 1) \leq 1.$$

Substituting p and s we get, by Lemmas 2 and 3, and by solving for M, that:

$$e\left(\frac{7}{8}\right)^n (5M^3) \leq 1 \Leftrightarrow$$

$$M \leq \frac{1}{\sqrt[3]{5e}} \left(\frac{8}{7}\right)^{n/3}.$$

Thus, by Theorem 1, we get the required result. □

Theorem 3 implies the following corollary:

Corollary 2. *There exist binary 2-separating codes of rate $R \approx 0.064$.*

Proof. By Definition 2 and since the code is binary, we have that:

$$\begin{aligned} R =& \frac{\log_2 M}{n} \\ =& \frac{\log_2\left(\frac{1}{\sqrt[3]{5e}}\left(\frac{8}{7}\right)^{n/3}\right)}{n} \\ =& \frac{\log_2\left(\frac{1}{\sqrt[3]{5e}}\right)}{n} + \frac{n\log_2(\frac{8}{7})}{3n} \\ =& \frac{\log_2\left(\frac{1}{\sqrt[3]{5e}}\right)}{n} + \frac{\log_2(\frac{8}{7})}{3}, \end{aligned}$$

which, for $n \to \infty$, gives $R \approx 0.064$. □

Remark 1. Note that there is a strong symmetry in our problem. First, by Lemma 2, all the events have exactly the same probability of occurring. Furthermore, the number s of Lemma 3 is again the same for every event, although in reality, it could be refined to a polynomial with a lower coefficient to M^3, containing also terms of degree 1 and 2. As this would not drastically change our results, we opted for the simpler bound of $5M^3 - 1$.

Due to this symmetry we described, we know that the stronger asymmetric version of the lemma (see for example [15]) cannot provide any improvement to our result. There our however other versions of the LLL that could, in principle, be applied here in hopes of producing better results.

4 Explicit Constructions

Now that we have established a lower bound for 2-separating binary codes, we turn our attention to obtaining explicit constructions of such codes. We first see that to obtain a code of positive rate, the computational complexity of our algorithm turns out to necessarily be exponential in the code length (see Remark 2). We then show that we can tune the algorithm to be polynomial in the code length, at the cost of having non positive code rate. Nevertheless, the code we construct has a better rate than that of a Simplex code of equivalent length.

4.1 Direct Application of the Algorithmic LLL for Constructing 2-Separating Codes

Let $t = nM$ and $m = \binom{M}{2}\binom{M-2}{2}$. By the discussion above and by renaming the random variables X_ij and events $E_{\mathbf{u},\mathbf{v}}$, we have t random variables $X_1, \ldots, X_t$ (in some arbitrary ordering) and m events $E_1, \ldots, E_m$ (again ordered arbitrarily), with $p = (\frac{7}{8})^n$ and $s = 5M^3$. Thus, we can directly apply the M-ALGORITHM and the analysis of Giotis et al. [7] to algorithmically obtain the results of Sect. 3. Here, we briefly highlight some parts of this analysis, based on the proof of [11] (the discussion there concerning lopsidependency can be omitted).

Theorem 4. *For every $n > 0$, there is a randomized algorithm, such that the probability of it lasting for at least N rounds is inverse exponential in N, and that outputs a 2-separating code of size:*

$$M \leq \frac{1}{\sqrt[3]{5e}} \left(\frac{8}{7}\right)^{n/3}.$$

Proof. Consider the M-ALGORITHM (Algorithm 1). First observe that if the algorithm terminates, then by line 2, it produces an assignment of values to the random variables such that no undesirable event occurs. This translates, as we have already seen, to a 2-separating code $\mathcal{C}$.

We say that a *root* call of RESAMPLE is any call made from line 3 of the main algorithm, while a *recursive* call is one made from line 3 of another RESAMPLE call. A round is the duration of any RESAMPLE call.

It can be shown that any event that did not occur at the beginning of a RESAMPLE(E_j), i.e., any disjoint pair of distinct code words that were separated, continues to be separated if and when that call terminates: any event that is made to happen at any point during RESAMPLE(E_j), will be subsequently checked and resampled by some RESAMPLE sub-routine called from within RESAMPLE(E_j). Also, by line 2, it is straightforward to see that if and when RESAMPLE(E_j) terminates, E_j does not occur, although it did at the beginning. Thus, there is some progress made from the algorithm in every round, that is not lost in subsequent ones (for the full proof the reader is referred to [11]).

Given an execution of M-ALGORITHM, we construct a labeled rooted forest $\mathcal{F}$ (i.e. forest comprised of rooted trees), in the following way:

(i) For each RESAMPLE(E_j) call, we construct a node labeled by E_j.
(ii) If RESAMPLE(E_r) is called from line 3 of RESAMPLE(E_j), then the corresponding node labeled by E_r is a child of that labeled by E_j.

It is not difficult to see that the roots of $\mathcal{F}$ correspond to root calls of RESAMPLE, while the rest of the nodes to recursive calls. Furthermore, by the above discussion, we have that the labels of the roots are pair-wise district. The same holds for the labels of siblings. Finally, if a node labeled by E_r is a child of one labeled by E_j, then $r \in \Gamma_j$.

We call the forest created in the above way, the *witness forest* of the algorithm's execution. Given an execution that lasts for N steps, its witness forest has N nodes. We order the nodes of the forest in the following way: (i) trees and siblings are ordered according to the indices of their labels (ii) the nodes of a tree are ordered in pre-order, respecting the ordering of siblings. Thus, from each witness forest $\mathcal{F}$ with N nodes, we can obtain its *label-sequence* $(E_{j_1}, \ldots, E_{j_N})$.

Now, letting P_N be the probability that M-ALGORITHM lasts for at least N rounds, we have that:

$$\mathrm{P}_N = \Pr[\text{some witness forest } \mathcal{F} \text{ with } N \text{ nodes is constructed}]. \tag{8}$$

Consider now the following *validation* algorithm, that takes as input the label sequence of a witness forest:

Algorithm 2. VALALG.

```
Input: Label sequence (E_{j_1}, ..., E_{j_N}) of F
Sample the variables X_i, i = 1, ..., t.
for i = 1, ..., N do
    if E_{j_i} occurs under the current assignment then
        Resample the variables in sc(E_{j_i})
    else
        Return failure and exit
    end if
end for
Return success
```

Observe that the success or failure of VALALG has nothing to do with whether or not all the events have been avoided and thus on whether we have a separable code.

By observing that the event that an execution of M-ALGORITHM produced the witness forest $\mathcal{F}$, implies the event that VALALG succeeded on input $\mathcal{F}$ (VALALG can make the same random choices as M-ALGORITHM), we can bound the probability that M-ALGORITHM lasts for at least N rounds by:

$$\mathrm{P}_N \leq \sum_{|\mathcal{F}|=n} \Pr[\text{VALALG succeeds on input } \mathcal{F}], \tag{9}$$

where $|\mathcal{F}|$ denotes the number of nodes of $\mathcal{F}$. For the rhs of (9), let $V_i(\mathcal{F})$ be the event that VALALG does not fail on round i, on input $\mathcal{F}$. It can be shown that, given a forest $\mathcal{F}$ whose label sequence is $(E_{j_1}, \ldots, E_{j_N})$, it holds that:

$$\begin{aligned}\Pr[\text{VALALG succeeds on input } \mathcal{F}] &= \prod_{i=1}^{N} \Pr[V_i(\mathcal{F}) \mid \bigcap_{r=1}^{i-1} V_r(\mathcal{F})] \\ &= \prod_{i=1}^{N} \Pr[E_{j_i}] \\ &= p^N.\end{aligned}$$

Thus, to bound the rhs of (9), we need to count the number of forests with N internal nodes. It can be shown that to do that, we can instead count the number f_N of *rooted planar forests* with N internal nodes, comprised of m *full* $(s+1)$-ary rooted planar trees (for the necessary details, the reader is again referred to [11]).

Denote the number of full $(s+1)$-ary rooted planar trees with N internal nodes by t_N. It holds that $t_N = \frac{1}{sN+1}\binom{(s+1)N}{N}$ (see [18, Theorem 5.13]), which, by Stirling's approximation gives that there is some constant A, depending only on s, such that:

$$t_n < A\left(\left(1+\frac{1}{s}\right)^s (s+1)\right)^N. \tag{10}$$

Finally, by (10), we get:

$$f_N = \sum_{\substack{N_1+\cdots+N_m=N\\N_1,\ldots,N_m\geq 0}} t_{N_1}\cdots t_{N_m} < (AN)^m \left(\left(1+\frac{1}{s}\right)^s (s+1)\right)^N. \tag{11}$$

Thus, taking Eqs. (9) and (11), we have that:

$$\mathrm{P}_N < (AN)^m \left(\left(1+\frac{1}{s}\right)^s (s+1)p\right)^N, \tag{12}$$

which concludes the proof. □

Remark 2. Consider line 2 of M-ALGORITHM. For the algorithm to find the least indexed event, it must go over all the approximately $2M^4$ elements of $\mathcal{P}$ and check if they are separated. Accordingly, in line 2 of a RESAMPLE(E_j) call of M-ALGORITHM, the algorithm must check all the approximately $5M^3$ events in the neighborhood of E_j. Given the bound we proved for M, it is easy to see that in both cases, the number of events that need to be checked is exponentially large in n. In the next subsection, we will deal, in a way, with this problem.

4.2 Constructions of Polynomial Complexity

In Remark 2 above, we have exposed the drawback of applying in a straightforward way the algorithmic version of the LLL. Since we are aiming for a code with asymptotic positive rate, this means that the number of code words has to be exponential in the code length, which implies that the algorithmic complexity is exponential in the code length too. If we insist in building positive rate 2-separating codes and use the algorithmic version of the LLL for it, this exponential dependence seems to be unavoidable.

For explicit binary 2-separating codes, one can refer to [2] and the references therein. Known constructions are somewhat particular and rare. For instance, there exists a 2-separating binary [35,6] code, there exists a 2-separating binary [126,14] code, and of course as stated in Lemma 1 the Simplex code is also 2-separated.

We now take a step into constructing 2-separating binary codes with rate better than the Simplex code for any code length and in polynomial time to their length. The following lemma is a weaker result than Theorem 3 and follows from the fact that in Theorem 1 one only needs an upper bound of the probability of the bad events.

Lemma 4. *For every $n > 0$ and any $\alpha > 0$, there exists a binary 2-separating code of size:*

$$M \leq \frac{1}{\sqrt[3]{5e}} n^{\alpha/3}. \tag{13}$$

Proof. The proof follows the lines of the proof of Theorem 3 by taking $p = \frac{1}{n^\alpha}$.

Referring to Sect. 2, the rate of a Simplex code of length n is

$$R_{Simplex}(n) = \frac{\log_2(n+1)}{n}. \tag{14}$$

With little algebraic manipulation one can see that for any $n = 2^k - 1$ with $k > 0$, and any $\alpha > 0$ such that:

$$n^\alpha > 5e(n+1)^3,$$

the codes in Lemma 4 have better rate than the Simplex code of the same length and with a polynomial number of code words.

Recall the observations made in Remark 2. In M-Algorithm we have to:

- go over the approximately $2M^4$ elements of $\mathcal{P}$ in line 2 and
- check all the approximately $5M^3$ events in the neighborhood of E_j in line 2 of a Resample(E_j) call.

Observe that, with the value of $M \leq \frac{1}{\sqrt[3]{5e}} n^{\alpha/3}$ given by Lemma 4, this can be done in polynomial time in the code length. Furthermore, by taking this value for M, Theorem 4 applies verbatim. Thus, we have proven the following:

Theorem 5. *For every $n > 0$, there is a randomized algorithm such that the probability of it lasting for at least N rounds is inverse exponential in N and that outputs a 2-separating code of length n and size:*

$$M \leq \frac{1}{\sqrt[3]{5e}} n^{\alpha/3},$$

with rate larger than the Simplex code of the same length. The computational complexity is $O(n^{4\alpha/3})$ for any α such that $n^\alpha > 5e(n+1)^3$.

5 Conclusions

Let us summarize our results. We have first shown that the Lovász Local Lemma can be used to establish bounds on separating codes, and that for this matter, it is as effective as the random coding with expurgation technique.

Moreover, it has also been shown that, using the algorithmic version of the LLL, the construction of a separating code with positive rate has exponential complexity in the length of the code. Looking at the results in Sect. 3 it seems difficult to move away from such a complexity.

Finally, by relaxing the restriction of obtaining positive rate codes, we show that 2-separating codes with rates better than Simplex codes of the same length can be built. This time, with complexity polynomial in the code length.

References

1. Achlioptas, D., Iliopoulos, F.: Random walks that find perfect objects and the Lovász local lemma. J. ACM (JACM) **63**(3), 22 (2016)
2. Barg, A., Blakley, G.R., Kabatiansky, G.A.: Digital fingerprinting codes: problem statements, constructions, identification of traitors. IEEE Trans. Inf. Theory **49**(4), 852–865 (2003)
3. Deng, D., Stinson, D.R., Wei, R.: The Lovász local lemma and its applications to some combinatorial arrays. Des. Codes Crypt. **32**(1–3), 121–134 (2004)
4. Erdős, P., Lovász, L.: Problems and results on 3-chromatic hypergraphs and some related questions. Infin. Finite Sets **10**, 609–627 (1975)
5. Gebauer, H., Moser, R.A., Scheder, D., Welzl, E.: The Lovász local lemma and satisfiability. In: Albers, S., Alt, H., Näher, S. (eds.) Efficient Algorithms. LNCS, vol. 5760, pp. 30–54. Springer, Heidelberg (2009). https://doi.org/10.1007/978-3-642-03456-5_3
6. Gebauer, H., Szabó, T., Tardos, G.: The local lemma is tight for SAT. In: Proceedings 22nd Annual ACM-SIAM Symposium on Discrete Algorithms (SODA), pp. 664–674. SIAM (2011)
7. Giotis, I., Kirousis, L., Psaromiligkos, K.I., Thilikos, D.M.: On the algorithmic Lovász local lemma and acyclic edge coloring. In: Proceedings of the Twelfth Workshop on Analytic Algorithmics and Combinatorics. Society for Industrial and Applied Mathematics (2015). http://epubs.siam.org/doi/pdf/10.1137/1.9781611973761.2
8. Giotis, I., Kirousis, L., Psaromiligkos, K.I., Thilikos, D.M.: Acyclic edge coloring through the Lovász local lemma. Theoret. Comput. Sci. **665**, 40–50 (2017)
9. Giotis, I., Kirousis, L., Livieratos, J., Psaromiligkos, K.I., Thilikos, D.M.: Alternative proofs of the asymmetric Lovász local lemma and Shearer's lemma. In: Proceedings of the 11th International Conference on Random and Exhaustive Generation of Combinatorial Structures, GASCom (2018). http://ceur-ws.org/Vol-2113/paper15.pdf
10. Harvey, N.J., Vondrák, J.: An algorithmic proof of the Lovász local lemma via resampling oracles. In: Proceedings 56th Annual Symposium on Foundations of Computer Science (FOCS), pp. 1327–1346. IEEE (2015)
11. Kirousis, L., Livieratos, J.: A simple algorithmic proof of the symmetric lopsided Lovász local lemma. In: Battiti, R., Brunato, M., Kotsireas, I., Pardalos, P.M. (eds.) LION 12 2018. LNCS, vol. 11353, pp. 49–63. Springer, Cham (2019). https://doi.org/10.1007/978-3-030-05348-2_5
12. Körner, J., Simonyi, G.: Separating partition systems and locally different sequences. SIAM J. Discrete Math. **1**(3), 355–359 (1988)
13. MacWilliams, F.J., Sloane, N.J.A.: The Theory of Error-Correcting Codes, vol. 16. Elsevier, Amsterdam (1977)
14. Moser, R.A.: A constructive proof of the Lovász local lemma. In: Proceedings 41st Annual ACM Symposium on Theory of Computing (STOC), pp. 343–350. ACM (2009)
15. Moser, R.A., Tardos, G.: A constructive proof of the general Lovász local lemma. J. ACM (JACM) **57**(2), 11 (2010)
16. Sagalovich, Y.L.: Separating systems. Problems Inform. Transmission **30**(2), 105–123 (1994)
17. Sarkar, K., Colbourn, C.J.: Upper bounds on the size of covering arrays. SIAM J. Discrete Math. **31**(2), 1277–1293 (2017)

18. Sedgewick, R., Flajolet, P.: An Introduction to the Analysis of Algorithms. Addison-Wesley, Boston (2013)
19. Staddon, J.N., Stinson, D.R., Wei, R.: Combinatorial properties of frameproof and traceability codes. IEEE Trans. Inf. Theory **47**(3), 1042–1049 (2001)
20. Szegedy, M.: The Lovász local lemma – a survey. In: Bulatov, A.A., Shur, A.M. (eds.) CSR 2013. LNCS, vol. 7913, pp. 1–11. Springer, Heidelberg (2013). https://doi.org/10.1007/978-3-642-38536-0_1
21. Tao, T.: Moser's entropy compression argument (2009). https://terrytao.wordpress.com/2009/08/05/mosers-entropy-compression-argument/

Lagrangian Relaxation in Iterated Local Search for the Workforce Scheduling and Routing Problem

Hanyu Gu(✉), Yefei Zhang, and Yakov Zinder

School of Mathematical and Physical Sciences, University of Technology, Sydney, Australia
hanyu.gu@uts.edu.au

Abstract. The efficiency of local search algorithms for vehicle routing problems often increases if certain constraints can be violated during the search. The penalty for such violation is included as part of the objective function. Each constraint, which can be violated, has the associated parameters that specify the corresponding penalty. The values of these parameters and the method of their modification are usually a result of computational experiments with no guarantee that the obtained values and methods are equally suitable for other instances. In order to make the optimisation procedure more robust, the paper suggests to view the penalties as Lagrange multipliers and modify them as they are modified in Lagrangian relaxation. It is shown that such modification of the Xie-Potts-Bektaş Algorithm for the Workforce Scheduling and Routing Problem permits to achieve without extra tuning the performance comparable with that of the original Xie-Potts-Bektaş Algorithm.

Keywords: Iterated local search · Lagrangian relaxation · Workforce Scheduling and Routing

1 Introduction

It is known that the efficiency of local search algorithms often can be improved if the violation of some constraints is allowed [1,2,6,8]. Such violation attracts a certain penalty, which is part of the augmented objective function. The values of the parameters, specifying the penalty for the violation of constraints, are changed in the course of optimisation. The choice of these values and the algorithms of their modification usually are a result of computational experiments. This approach often involves tedious computational experimentation with no guarantee that these values and the methods of their modification will be equally suitable for other instances. In order to make the optimisation procedure more robust, this paper suggests to view the parameters, specifying the penalties, as Lagrange multipliers and to modify them using the methods of the modification of Lagrange multipliers in Lagrangian relaxation [4]. This replaces the

I. Kotsireas et al. (Eds.): SEA² 2019, LNCS 11544, pp. 527–540, 2019.
https://doi.org/10.1007/978-3-030-34029-2_34

largely subjective tuning by an algorithm which is based of the mathematical programming formulation, which remains the same for all instances of the problem on hand and therefore reflects its the specific features. Furthermore, since the optimisation procedure utilises predefined algorithms of the modification of Lagrange this approach significantly reduces the burden of tuning, although the Lagrangian relaxation itself requires certain tuning and the corresponding computational experimentation.

The merits of this approach are demonstrated below by a Lagrangian relaxation modification of the highly efficient algorithm, presented in [8] for the Workforce Scheduling and Routing Problem (WSRP). In this modification, the parameters, specifying the penalty for the violation of the constraints, are computed using the subgradient method. According to Workforce Scheduling and Routing Problem, a group of technicians should be assigned to a set of tasks at different locations. Each task has the associated time window and skill requirement. Each technician has certain skills and can be assigned to a task only if these skills satisfy the task's requirement. Since the tasks have time windows, which restrict the time when the service can be provided, and since the tasks are at different locations with a given travel time between locations, the problem requires not only to allocate the tasks to the technicians but also to determine the order in which each technician should attend the allocated tasks. In addition, there exists a restriction on the shift duration for each technician. More detailed description of this problem will be given in Sect. 2. Some versions of the Workforce Scheduling and Routing Problem and alternative solution algorithms can be found in [3] and [5]. The optimisation procedure, presented in [8], is an implementation of the iterated local search where the restrictions imposed by the time windows and the maximal permissible shift duration can be violated. At each iteration of the local search, this violation attracts a certain penalty which is part of the augmented objective function. The penalty for the violation of the time windows is the total violation of this restriction multiplied by some coefficient (weight). Similarly, the penalty for the violation of the shift duration is the total violation of this restriction multiplied by some coefficient (weight). These coefficients are modified during the optimisation depending on the level of violation.

In what follows, the algorithm, presented in [8], will be referred to as the Xie-Potts-Bektaş Algorithm. The Lagrangian relaxation based modification of this algorithm relaxes the same constraints as in the Xie-Potts-Bektaş Algorithm, that is, the constraints imposed by the time windows and the constraints imposed by the restriction on the duration of shifts. The resultant relaxation is solved by the iterated local search procedure in [8], but the penalty weights are modified according to the subgradient method commonly used in Lagrangian relaxation.

The remaining part of the paper is structured as follows. A mathematical programming formulation of WSRP is given in Sect. 2. Section 3 describes the Xie-Potts-Bektaş Algorithm. Section 4 presents a Lagrangian relaxation based modification of the Xie-Potts-Bektaş Algorithm. Section 5 presents the computation experiments. Section 6 concludes the paper.

2 Problem Description

Following [8], consider a complete graph $G = \{V, A\}$, where $V = \{0, 1, ..., n, n+1\}$ is the set of vertices and $A = \{(i,j) : i,j \in V, i \neq j\}$ is the set of arcs. Vertex 0 represents the depot and vertex $n+1$ is a copy of this depot. The set $C = V \setminus \{0, n+1\}$ represents the set of customers. Each arc $(i,j) \in A$ has the associated travel cost $c_{i,j}$ and travel time $t_{i,j}$. The service of each customer $i \in C$ should commence within the time window $[e_i, l_i]$, specified for this customer, and its duration will be denoted by d_i.

The customers are to be served by technicians. Let K be the set of these technicians. A technician can not depart from the depot (vertex 0) earlier than e_0 and can not return to the depot (vertex $n+1$) later than l_{n+1}. For each technician, the length of the time interval between the departure from the depot and the return to the deport can not exceed a given time D.

For each technician $k \in K$ and each customer $i \in C$, a given binary parameter q_i^k determines whether or not technician k can be assigned to customer i. If $q_i^k = 1$, then technician k can be assigned to customer i, whereas if $q_i^k = 0$, then this assignment is not allowed. If no technician is assigned to a customer i, then the corresponding service is outsourced at the cost o_i.

As has been shown in [8], the problem can be formulated as a mixed integer program as follows. Let

$$x_{i,j}^k = \begin{cases} 1 & \text{if customers } i \text{ and } j \text{ are consecutive customers visited by technician } k, \\ 0 & \text{otherwise,} \end{cases}$$

$$y_i = \begin{cases} 1 & \text{if service of customer } i \text{ is outsourced,} \\ 0 & \text{otherwise.} \end{cases}$$

For each customer $i \in C$ and each technician $k \in K$, let b_i^k be the time when the service commences, if technician k is assigned to customer i, and be any number in $[e_i, l_i]$ otherwise. For each technician $k \in K$, let b_0^k be the time when technician k leaves the depot (vertex 0), and let b_{n+1}^k be the time when technician k returns to the depot (vertex $n+1$).

$$\text{minimise } f = \sum_{k\in K} \sum_{(i,j)\in A} c_{i,j} x_{i,j}^k + \sum_{i\in C} o_i y_i \tag{1}$$

Subject to:

$$\sum_{k\in K} \sum_{j\in V} x_{i,j}^k + y_i = 1, \quad \forall i \in C \tag{2}$$

$$\sum_{j\in V} x_{i,j}^k \leq q_i^k, \quad \forall k \in K, \forall i \in C \tag{3}$$

$$\sum_{j\in V} x_{0,j}^k = 1, \quad \forall k \in K \tag{4}$$

$$\sum_{i\in V} x_{i,n+1}^k = 1, \quad \forall k \in K \tag{5}$$

$$\sum_{i \in V} x_{i,h}^k - \sum_{j \in V} x_{h,j}^k = 0, \quad \forall k \in K, \forall h \in C \tag{6}$$

$$b_i^k + (d_i + t_{i,j}) x_{i,j}^k \leq b_j^k + l_i (1 - x_{i,j}^k), \quad \forall k \in K, \forall (i,j) \in A \tag{7}$$

$$e_i \leq b_i^k \leq l_i, \quad \forall k \in K, \forall i \in V \tag{8}$$

$$b_{n+1}^k - b_0^k \leq D, \quad \forall k \in K \tag{9}$$

$$x_{i,j}^k \in \{0,1\}, \quad \forall k \in K, \forall (i,j) \in A \tag{10}$$

$$y_i \in \{0,1\}, \quad \forall i \in C \tag{11}$$

$$b_i^k \geq 0, \quad \forall k \in K, \forall i \in V \tag{12}$$

Constraints (2) guarantee that each customer either is assigned to a technician or is outsourced. By virtue of Constraints (3), each technician is assigned to a customer only if this is permissible. Constraints (4) and (5) ensure that each technician departs from the depot and returns to the depot. Constraints (6) stipulate that after visiting a customer a technician must travel to another customer or to the depot. Constraints (7) ensure that if customer i and customer j are two consecutive customers for some technician, then the difference between arrival times is not less than the duration of service, required by customer i, plus the travel time between thesc two customers. Constraints (8) enforce time windows. According to the Constraints (9), any technician does not work longer than D.

3 Xie-Potts-Bektaş Algorithm

Xie-Potts-Bektaş Algorithm starts with a randomly generated initial feasible solution, which is to be improved by iterated local search. This process is repeated several times in the hope of finding the global optimal solution. The implementation details of this approach are described in Algorithm 1. At step 8 a local search procedure is utilised to greedily improve the current solution, while perturbation is exploited at step 13 to escape a locally optimal solution.

The local search procedure alternates between the Inter-route search and Intra-route search. During the Inter-route search, a Swap-and-Relocate operation is applied repeatedly until a local optimum is found. The Swap-and-Relocate operation exchanges sub-paths from two different route, and each sub-path can contain at most 2 customers. During the Intra-route search, three commonly used operations (Op1, Opt2 and 2-Opt) are applied repeatedly until a local optimum is found. Opt1 removes a single customer from its current position and inserts it back into a different position within the same route; Opt2 selects two consecutive customers and inserts them back into a different position within the same route; 2-Opt selects a sub-path of any length, reverse its order, and then insert it back into the same position. Different search strategies can be applied with all these operations, but it is recommended in [8] that intra-route search should be applied as a post-optimisation when a local optimum has been found by the inter-route search.

Algorithm 1. Xie-Potts-Bektaş Algorithm Procedure [8]

1: $f(s^*) \leftarrow +\infty$ $\{s^*$ records the best solution found so far$\}$
2: $i \leftarrow 0$
3: **while** $i < 5$ **do**
4:     construct solution s_i
5:     $\bar{s} \leftarrow s_i$, $\hat{s} \leftarrow s_i$ $\{\hat{s}$ records the local best solution found at each iteration of ILS$\}$
6:     $It_{Non-Imp} = 0$
7:     **while** $It_{Non-Imp} \leq Max_{NII}$ **do**
8:         $s' \leftarrow$Local Search$(\bar{s})$
9:         **if** (s' is feasible)AND($f(s') < f(\hat{s})$) **then**
10:             $\hat{s} \leftarrow s'$
11:             $It_{NonImp} = 0$
12:         **end if**
13:         $\bar{s} \leftarrow$Perturb$(\hat{s})$
14:         $It_{Non-Imp} + +$
15:     **end while**
16:     **if** $f(\hat{s}) < f(s^*)$ **then**
17:         $s^* \leftarrow \hat{s}$
18:     **end if**
19: **end while**
20: return s^*

The local search procedure at step 9 of Algorithm 1 allows the violation of time window constraints and working duration constraints. However, the objective function is modified as

$$f'(x) = f(x) + \alpha * TW(x) + \beta * WD(x) \tag{13}$$

where $f(x)$ is the original objective function; $TW(x)$ is the total time window violation by solution x, and $WD(x)$ is the total working duration violation; α and β are the penalty weights for violating the corresponding constraints. At the beginning of local search, α and β are initialised to 1, and are updated at each iteration until local search gets trapped at a local optimum. The formula for updating the weights at iteration i is as follows:

$$\alpha_{i+1} = \begin{cases} \alpha_i * (1+\delta), TW(x) > 0 \\ \frac{\alpha_i}{(1+\delta)}, \text{otherwise} \end{cases} \qquad \beta_{i+1} = \begin{cases} \beta_i * (1+\delta), WD(x) > 0 \\ \frac{\beta_i}{(1+\delta)}, \text{otherwise} \end{cases}$$

where δ is a parameter which controls the strength of updating penalty.

The details of the local search procedure are described in Algorithm 2.

Algorithm 2. Xie-Potts-Bektaş Local Search Procedure

```
α_o ← 1, β_0 ← 1
ŝ ← input solution
local optimal = false
while local optimal == false do
  s' ←neighbourhood search(ŝ)
  if f'(s') < f'(ŝ) then
    ŝ ← s'
    compute TW(s'), WD(s') and update α, β
  else
    local optimal =true
  end if
end while
```

4 Lagrangian Relaxation Based Modification of the Xie-Potts-Bektaş Algorithm

To present the Lagrangian relaxation based modification of the Xie-Potts-Bektaş Algorithm, we first reformulate the original mathematical formulation as follows:

$$\text{minimise } f = \sum_{k\in K} \sum_{(i,j)\in A} c_{i,j} x_{i,j}^k + \sum_{i\in C} o_i y_i \tag{14}$$

Subject to:

$$\sum_{k\in K} \sum_{j\in V} x_{i,j}^k + y_i = 1, \quad \forall i \in C \tag{15}$$

$$\sum_{j\in V} x_{i,j}^k \leq q_i^k, \quad \forall k \in K, \forall i \in C \tag{16}$$

$$\sum_{j\in V} x_{0,j}^k = 1, \quad \forall k \in K \tag{17}$$

$$\sum_{i\in V} x_{i,n+1}^k = 1, \quad \forall k \in K \tag{18}$$

$$\sum_{i\in V} x_{i,h}^k - \sum_{j\in V} x_{h,j}^k = 0, \quad \forall k \in K, \forall h \in C \tag{19}$$

$$b_i^k + (d_i + t_{i,j}) x_{i,j}^k \leq b_j^k + l_i(1 - x_{i,j}^k), \quad \forall k \in K, \forall (i,j) \in A \tag{20}$$

$$e_i \leq b_i^k \leq l_i + \rho_i, \quad \forall k \in K, \forall i \in V \tag{21}$$

$$b_{n+1}^k - b_0^k \leq D + d^k, \quad \forall k \in K \tag{22}$$

$$\sum_{i\in V} \rho^i = 0, \quad \forall i \in V \tag{23}$$

$$\sum_{k\in K} d^k = 0, \forall k \in K \tag{24}$$

$$x_{i,j}^k \in \{0,1\}, \qquad \forall k \in K, \forall (i,j) \in A \tag{25}$$

$$y_i \in \{0,1\}, \qquad \forall i \in C \tag{26}$$

$$b_i^k \geq 0, \quad \forall k \in K, \forall i \in V \tag{27}$$

$$\rho_i \geq 0, \qquad \forall i \in V \tag{28}$$

$$d^k \geq 0, \qquad \forall k \in K \tag{29}$$

where ρ_i is the time window violation for late arrival at customer $i \in C$; d^k is the working duration violation for technician k. Constraints (21) are modified from Constraints (7) to allow the violation of time window constraints; Constraints (22) are modified from Constraints (8) to allow the violation of working duration constraints as well. Constraints (23) and (24) guarantee that total violations are zero.

Constraints (23) and (24) can be dualised, and the objective function of the corresponding Lagrangian relaxation problem is

$$f' = \sum_{k\in K} \sum_{(i,j)\in A} c_{i,j} x_{i,j}^k + \sum_{i\in C} o_i y_i + \alpha \sum_{i\in V} \rho^i + \beta \sum_{k\in K} d^k \tag{30}$$

where α and β are the Lagrangian multipliers for the time window and work duration constraints respectively.

While the time window and working duration constraints are relaxed, the resulting Lagrangian relaxation problem remains NP-hard. And it is impossible to decompose it into easier sub-problem like ordinary Lagrangian relaxation approach. In our Lagrangian relaxation based modification of the Xie-Potts-Bektaş Algorithm, we use the same iterated local search procedure described in Xie-Potts-Bektaş Algorithm. However, the local search inside iterated local search is modified so that the penalty weights are updated according to the approximate subgradient algorithm. The modified local search is described in Algorithm 3.

At the beginning of the procedure, the approximate subgradient guided local search will initialise Lagrangian multipliers to 0. Instead of applying local search when multipliers are 0, a greedy heuristic Phase1 is run to insert all unallocated customers. Phase1 should not dramatically change the input solution in order that the effect of perturbation at step 14 of Algorithm 1 is not lost. At each iteration, if the current solution is infeasible, the approximate subgradient and step size will be computed to update Lagrangian multipliers. $\lambda_k = \frac{\eta*(f(x^*) - f(x))}{(\sum_{i\in V}\rho^i)^2 + (\sum_{k\in K} d^k)^2}$ is the formula to calculate the step size that suggested by [4]. η is a parameter which controls the step length, $f(x^*)$ is an upper bound which record the objective value for current best feasible solution. $f(x)$ is the lower bound obtained from solving the Lagrangian problem at current iteration. Since local search is an approximate approach, there is no guarantee

Algorithm 3. Approximate Subgradient Guided Local Search

$s \leftarrow$Input solution
feasible solution= false
$\alpha_0 \leftarrow 0$, $\beta_0 \leftarrow 0$
$\hat{s} \leftarrow$Phase1(s)
if $\hat{s}$ is not feasible **then**
  compute $\sum_{i \in V} \rho^i$, $\sum_{k \in K} d^k$ for $\hat{s}$ and step size λ_0
  $\alpha_0 = \lambda_0 * \sum_{i \in V} \rho^i$; $\beta_0 = \lambda_0 * \sum_{k \in K} d^k$
end if
$j \leftarrow 1$
while (j <100)AND(feasible solution==false) **do**
  $s' \leftarrow$ Search Strategy ($\hat{s}$)
  compute $\sum_{i \in V} \rho^i$, $\sum_{k \in K} d^k$ for s' and step size λ_j
  $\alpha_{j+1} = \alpha_j + \lambda_j * \sum_{i \in V} \rho^i$; $\beta_{j+1} = \beta_j + \lambda_j * \sum_{k \in K} d^k$
  $\hat{s} \leftarrow s'$
  if s' is feasible **then**
    feasible solution = true
  end if
  $j++$
end while
return $\hat{s}$

that the Lagrangian relaxation problem can be exactly solved. It is even possible that the obtained $f(x)$ is larger than the upper bound. Therefore we replace $(f(x^*) - f(x))$ with an estimation of this difference. In what follows, the formula to calculate step size becomes, $\lambda_k = \frac{\eta * 0.1 * f(x^*)}{(\sum_{i \in V} \rho^i)^2 + (\sum_{k \in K} d^k)^2}$. The approximate subgradient guided local search uses the same neighbourhood structures and search strategy recommended by [8]. Our goal is to study the behaviour of these two different ways of adjusting penalty.

5 Computational Experiment

Both, Xie-Potts-Bektaş Algorithm and Lagrangian relaxation based modification of the Xie-Potts-Bektaş Algorithm, were implemented in C++ by the second author. The computational experiments were run on a computer with Intel Xeon CPU E5-2697 v3 2.60 GHz and 8 GB Memory (RAM).

The instances, used in the computational experiment, originate from [5]. They are an adaptation of Solomon's benchmark instances [7] for Vehicle Routing Problem with Time Window (VRPTW) and ROADEF 2007 challenge [3] for Technician and Task Scheduling Problem (TTSP). The base of the computational experiments was data R101, R103, R201, R203, C101, C103, C201, C203, RC101, RC103, RC201, RC203 in [7]. The number of customers and the number of technicians used in the computer experiments are given in the tables below that report the results of these experiments.

For each customer i and each technician k, the parameters q_i^k was specified using the skill requirement matrices in [5]. The rows of such matrix correspond to the different skills that may be required by a customer or a technician can possess. The columns correspond to different levels of skills. So, a customer may require a certain level of a certain skill. A technician k can be assigned to a customer i (in this case q_i^k is given value 1) only if this technician possesses the required skills each at the required or higher level.

The cost of service in the case when this service is outsourced was also calculated using the information on what skills and at what levels the corresponding customer requires.

5.1 Setting of the Algorithms

In the course of computational experimentation, both, the Xie-Potts-Bektaş Algorithm and the Lagrangian relaxation based modification of the Xie-Potts-Bektaş Algorithm, used the same settings that are recommended in [8]. More specifically:

- The maximal permissible number of iterations that failed to improve the value of the objective function (the parameter MAX-NII) was chosen as follows

$$\text{MAX-NII} = |C| + 10 * |K|,$$

  where $|C|$ is the number of customers that are to be served and $|K|$ is the number of technicians who should provide this service.
- The maximum number of swaps of the randomly chosen segments of the routes in a perturbation was five.
- The minimal number of swaps of the randomly chosen segments of the routes in a perturbation was set initially to one and was increased by one (if this number becomes five it ceases increasing) each time when twenty consecutive applications of the local search subroutine fail to produce a feasible solution with a better value of the objective function.

5.2 Evaluation of Different Step Length η

Table 1 presents the results of our Lagrangian relaxation based approach for the instances with 25 customers. Different step length η is used ranging from 0.2 to 3. The algorithm is run five times for each instance and the averages of the best found solutions are taken for comparison. We report the average objective value under the column titled "Avg". Column $|C^*|$ has the average number of allocated customers. Column $|K^*|$ contains the average number of used technicians. The average of the objective value over all the instances ranges from 2100.18 to 2120.06, which suggests that our approach is not sensitive to the parameter η. In practice this means that our algorithm is easy to tune the parameter. For the 25-customer instances, the best average objective value is achieved when $\eta = 2.0$. Therefore, we will use this setting for the following studies.

5.3 Comparison of the Performance

In this section we compare the performance of our Lagrangian relaxation based approach with the Xie-Potts-Bektaş Algorithm. Columns below heading "ILS" are the Xie-Potts-Bektaş Algorithm's results reported in [8]. We also implemented Xie-Potts-Bektaş Algorithm by ourselves. Columns below heading "Self-imp ILS" are the results produced from our self-implemented of the Xie-Potts-Bektaş Algorithm. Columns below heading "Mod-ILS" are the results produce from Lagrangian relaxation based modification of the Xie-Potts-Bektaş Algorithm.

Table 1. Evaluation of different η setting

			$\eta = 0.2$			$\eta = 0.6$			$\eta = 1.0$		
Instances	$\|C\|$	$\|K\|$	Avg	$\|C^*\|$	$\|K^*\|$	Avg	$\|C^*\|$	$\|K^*\|$	Avg	$\|C^*\|$	$\|K^*\|$
C101 5x4	50	6	830.00	49	6	830.00	49	6	830.00	49	6
C201 5x4	50	4	859.54	49	4	859.54	49	4	859.54	49	4
R101 5x4	50	6	4515.37	31	6	4507.87	31	6	4507.87	31	6
R201 5x4	50	4	1107.51	49	4	1107.51	49	4	1107.51	49	4
C101 6x6	50	6	1154.84	47	5	1154.84	47	5	1154.84	47	5
C201 6x6	50	4	1203.93	47	3	1203.93	47	3	1203.93	47	3
R101 6x6	50	6	5194.89	28	6	5190.32	28	6	5190.32	28	6
R201 6x6	50	4	1647.7	47	3	1647.7	47	3	1647.7	47	3
C101 7x4	50	6	1449.55	46	6	1356.54	47	6	1453.39	46.2	6
C201 7x4	50	4	1312.21	47	3	1312.21	47	3	1312.21	47	3
R101 7x4	50	6	4611.95	31	6	4505.3	31.8	6	4501.81	31.8	6
R201 7x4	50	4	1553.23	47	4	1553.23	47	4	1553.23	47	4
Average			2120.06	43.17	4.67	2102.42	43.32	4.67	2110.19	43.25	4.67
			$\eta = 1.4$			$\eta = 2.0$			$\eta = 3.0$		
Instances	$\|C\|$	$\|K\|$	Avg	$\|C^*\|$	$\|K^*\|$	Avg	$\|C^*\|$	$\|K^*\|$	Avg	$\|C^*\|$	$\|K^*\|$
C101 5x4	50	6	830.00	49	6	830.00	49	6	830.00	49	6
C201 5x4	50	4	859.54	49	4	859.54	49	4	859.54	49	4
R101 5x4	50	6	4507.87	31	6	4510.6	31	6	4507.87	31	6
R201 5x4	50	4	1107.51	49	4	1107.51	49	4	1107.51	49	4
C101 6x6	50	6	1154.84	47	5	1154.84	47	5	1154.84	47	5
C201 6x6	50	4	1203.93	47	3	1203.93	47	3	1203.93	47	3
R101 6x6	50	6	5197.67	28	6	5190.32	28	6	5195.89	28	6
R201 6x6	50	4	1647.7	47	3	1647.7	47	3	1647.7	47	3
C101 7x4	50	6	1413.04	46.6	6	1356.54	47	6	1356.54	47	6
C201 7x4	50	4	1312.21	47	3	1312.21	47	3	1312.21	47	3
R101 7x4	50	6	4483.18	32	6	4475.75	32	6	4498.31	31.8	6
R201 7x4	50	4	1553.23	47	4	1553.23	47	4	1553.23	47	4
Average			2105.89	43.3	4.67	2100.18	43.33	4.67	2102.30	43.32	4.67

The results for the 25-customer instances are reported in Table 2. The result from self-implemented Xie-Potts-Bektaş Algorithm is worse than the results listed in [8]. However, using the same implementation skill and same computing environment. It is clear that the overall solution quality produced by our Lagrangian relaxation based modification of the Xie-Potts-Bektaş Algorithm is very competitive. More important, our Lagrangian relaxation based approach can actually find the optimal solution for instance "R201 6x6" on each of the five runs. The solution is proved to be optimal from CPLEX. The solution obtains from [8] require 2 technicians, where our approach only cost 1 technician.

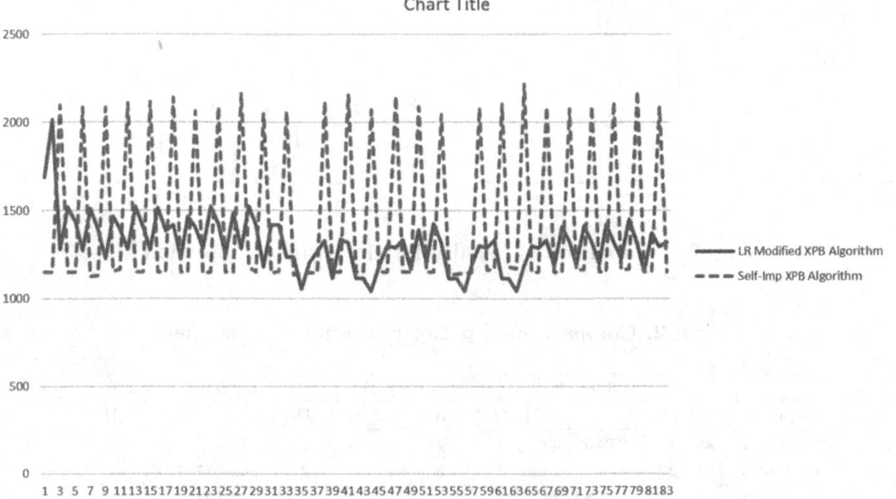

Fig. 1. Comparison of the objective value

Table 3 presents the results for 50-customer instances. Our Lagrangian relaxation based approach can find the optimal solution for instance "R201 5x4" on each of the five runs. And compare the overall solution quality with results reported in [8], the Lagrangian relaxation based approach produce competitive results.

Overall, it is worth noting that, although the same neighbourhood structures and search strategy are used in our implementation of the Xie-Potts-Bektaş Algorithm, it performs much worse than the other two algorithms. It could possibly indicate that the parameters in the Xie-Potts-Bektaş Algorithm are sensitive to the computing environment.

Figures 1 and 2 present an investigation of the behaviour for self-implemented Xie-Potts-Bektaş Algorithm and Lagrangian relaxation based approach. These figures plot the value for the first 100 iterations of local search for Instance "R201 5x4".

Figure 1 presents how the objective value behaves during the iterations of the self-implemented Xie-Potts-Bektaş Algorithm and Lagrangian relaxation based algorithm. Figure 2 presents how the penalty for time window violation behaves during the iterations of the self-implemented Xie-Potts-Bektaş Algorithm and Lagrangian relaxation based algorithm.

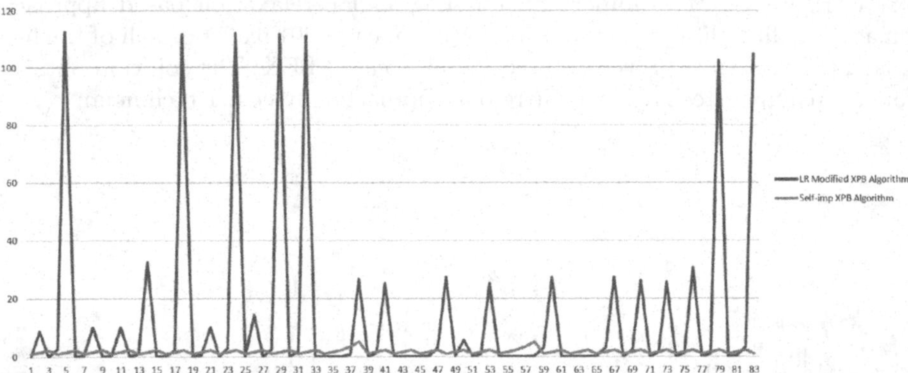

Fig. 2. Comparison of penalty for time window violation

Table 2. Comparison of performance for 25 customers

			ILS [8]			Self-imp ILS			Mod-ILS ($\eta = 2.0$)		
Instances	$\|C\|$	$\|K\|$	Avg	$\|C^*\|$	$\|K^*\|$	Avg	$\|C^*\|$	$\|K^*\|$	Avg	$\|C^*\|$	$\|K^*\|$
C101 5x4	25	4	272.96	25	4	271.7	25	4	271.7	25	4
C201 5x4	25	2	863.08	22	2	863.08	22	2	863.077	22	2
C203 5x4	25	2	835.83	22	1	843.33	22	1.4	835.828	22	1
R101 5x4	25	4	2195.04	16	4	2301.1	15.6	4	2195.04	16	4
R201 5x4	25	2	1091.07	22	2	1091.07	22	2	1091.07	22	2
RC101 5x4	25	4	862.21	23	4	869.56	23	4	868.386	23	4
RC201 5x4	25	3	465.31	25	3	465.25	25	3	465.254	25	3
C101 6x6	25	4	927.35	22	3	927.35	22	3	927.349	22	3
C201 6x6	25	2	1217.1	21	1	1217.1	21	1	1217.1	21	1
C203 6x6	25	2	930.6	22	1	930.60	22	1	930.598	22	1
R101 6x6	25	4	2868.19	13	4	3102.81	12	4	2864.96	13	4
R201 6x6	25	2	1422.57	21	2	1530.04	20.6	2	**1377.42**	**21**	**1**
RC101 6x6	25	4	1361.80	21	4	1837.02	18.6	4	1361.8	21	4
RC201 6x6	25	3	1228.89	22	2	1228.89	22	2	1228.89	22	2
C101 7x4	25	4	789.08	23	4	789.08	23	4	815.6608	22.8	4
C103 7x4	25	4	671.06	23	3	673.13	23	3	673.99	23	3
C201 7x4	25	2	738.35	23	2	738.35	23	2	738.347	23	2
C203 7x4	25	2	684.98	23	2	684.98	23	2	684.977	23	2
R101 7x4	25	4	2447.74	15	4	2459.48	15	4	2447.74	15	4
R201 7x4	25	2	959.51	23	2	964.52	23	2	964.52	23	2
R203 7x4	25	2	849.47	23	2	849.47	23	2	849.465	23	2
RC101 7x4	25	4	1669.63	19	4	1900.694	18	4	1669.63	19	4
RC201 7x4	25	3	967.6	23	3	968.16	23	3	967.60	23	3
Average			1144.32	21.39	2.74	1195.95	21.17	2.76	1143.93	21.38	2.70

Table 3. Comparison of performance for 50 customers

			ILS [8]			Self-imp ILS			Mod-ILS ($\eta = 2.0$)		
Instances	$\|C\|$	$\|K\|$	Avg	$\|C^*\|$	$\|K^*\|$	Avg	$\|C^*\|$	$\|K^*\|$	Avg	$\|C^*\|$	$\|K^*\|$
C101 5x4	50	6	830.00	49	6	886.01	49	6	830.00	49	6
C201 5x4	50	4	859.54	49	4	859.54	49	4	859.54	49	4
R101 5x4	50	6	4511.36	31	6	5757.17	25	6	4510.6	31	6
R201 5x4	50	4	1112.25	49	4	1142.70	49	4	1107.51	49	4
C101 6x6	50	6	1154.84	47	5	1398.802	46.4	6	1154.84	47	5
C201 6x6	50	4	1203.93	47	3	1203.93	47	3	1203.93	47	3
R101 6x6	50	6	5190.32	28	6	5567.07	26.2	6	5190.32	28	6
R201 6x6	50	4	1649.95	47	3	1683.25	47	3	1647.7	47	3
C101 7x4	50	6	1367.75	47	6	1899.54	44	6	1356.54	47	6
C201 7x4	50	4	1312.21	47	3	1312.21	47	3	1312.21	47	3
R101 7x4	50	6	4469.31	32	6	5089.86	28.8	6	4475.75	32	6
R201 7x4	50	4	1553.23	47	4	1594.364	47	4	1553.23	47	4
Average			2101.22	43.33	4.67	2366.95	42.12	5	2100.18	43.33	4.67

According to the Xie-Potts-Bektaş Algorithm, the penalty is adjusted by a constant factor. Each time the local search is trapped at local optimum, the algorithm initialises the penalty to 1. In our approach, the penalty is adjusted according to the information received from solving the corresponding Lagrangian relaxation problem. The algorithm sets the penalty to 0 in the case of local optimum.

In general, our approach has much larger values for the penalty, but initialise the penalties less frequently. The implication of this behaviour requires further investigation.

6 Conclusion

This paper present an approach aimed at making more robust the local search algorithms that permit, during the search, violation of some constraints by introducing penalties for such violation. The key idea is to treat the parameters, specifying the penalty for the violation of constraints, as Lagrange multipliers and to modify these parameters using one of the techniques of Lagrangian relaxation.

The merits of the suggested approach are demonstrated by the results of computational experiments with the highly efficient algorithm, presented in [8] for the Workforce Scheduling and Routing Problem, and its Lagrangian relaxation based modification. This modification utilises the subgradient method which is commonly used in Lagrangian relaxation. The future research will include other optimisation problems and algorithms and other Lagrangian relaxation techniques.

References

1. Cordeau, J.F., Gendreau, M., Laporte, G.: A tabu search heuristic for periodic and multi-depot vehicle routing problems. Netw.: Int. J. **30**(2), 105–119 (1997)

2. Cordeau, J.F., Laporte, G., Mercier, A.: A unified tabu search heuristic for vehicle routing problems with time windows. J. Oper. Res. Soc. **52**(8), 928–936 (2001)
3. Cordeau, J.F., Laporte, G., Pasin, F., Ropke, S.: Scheduling technicians and tasks in a telecommunications company. J. Sched. **13**(4), 393–409 (2010)
4. Fisher, M.L.: The Lagrangian relaxation method for solving integer programming problems. Manag. Sci. **27**(1), 1–18 (1981)
5. Kovacs, A.A., Parragh, S.N., Doerner, K.F., Hartl, R.F.: Adaptive large neighborhood search for service technician routing and scheduling problems. J. Sched. **15**(5), 579–600 (2012)
6. Nagata, Y., Bräysy, O., Dullaert, W.: A penalty-based edge assembly memetic algorithm for the vehicle routing problem with time windows. Comput. Oper. Res. **37**(4), 724–737 (2010)
7. Solomon, M.M.: Algorithms for the vehicle routing and scheduling problems with time window constraints. Oper. Res. **35**(2), 254–265 (1987)
8. Xie, F., Potts, C.N., Bektaş, T.: Iterated local search for workforce scheduling and routing problems. J. Heuristics **23**(6), 471–500 (2017)

Approximation Algorithms and an Integer Program for Multi-level Graph Spanners

Reyan Ahmed(✉), Keaton Hamm, Mohammad Javad Latifi Jebelli, Stephen Kobourov, Faryad Darabi Sahneh, and Richard Spence

University of Arizona, Tucson, USA
abureyanahmed@email.arizona.edu

Abstract. Given a weighted graph $G(V, E)$ and $t \geq 1$, a subgraph H is a t–*spanner* of G if the lengths of shortest paths in G are preserved in H up to a multiplicative factor of t. The *subsetwise spanner* problem aims to preserve distances in G for only a subset of the vertices. We generalize the minimum-cost subsetwise spanner problem to one where vertices appear on multiple levels, which we call the *multi-level graph spanner* (MLGS) problem, and describe two simple heuristics. Applications of this problem include road/network building and multi-level graph visualization, especially where vertices may require different grades of service.

We formulate a 0–1 integer linear program (ILP) of size $O(|E||V|^2)$ for the more general minimum *pairwise spanner problem*, which resolves an open question by Sigurd and Zachariasen on whether this problem admits a useful polynomial-size ILP. We extend this ILP formulation to the MLGS problem, and evaluate the heuristic and ILP performance on random graphs of up to 100 vertices and 500 edges.

Keywords: Graph spanners · Integer programming · Multi-level graph representation

1 Introduction

Given an undirected edge-weighted graph $G(V, E)$ and a real number $t \geq 1$, a subgraph $H(V, E')$ is a (multiplicative) t–*spanner* of G if the lengths of shortest paths in G are preserved in H up to a multiplicative factor of t; that is, $d_H(u, v) \leq t \cdot d_G(u, v)$ for all $(u, v) \in V \times V$, where $d_G(u, v)$ is the length of the shortest path from u to v in G. We refer to t as the *stretch factor* of H. Peleg et al. [12] show that determining if there exists a t–spanner of G with m or fewer edges is NP–complete. Further, it is NP–hard to approximate the (unweighted) t–spanner problem to within a factor of $O(\log |V|)$, even when restricted to bipartite graphs [15].

In the *pairwise spanner* problem [11], distances only need to be preserved for a subset $\mathcal{P} \subseteq V \times V$ of pairs of vertices. Thus, the classical t–spanner problem is

This work was supported in part by NSF grants CCF-1740858, CCF-1712119, and DMS-1839274.

I. Kotsireas et al. (Eds.): SEA^2 2019, LNCS 11544, pp. 541–562, 2019.
https://doi.org/10.1007/978-3-030-34029-2_35

a special case of the pairwise spanner problem where $\mathcal{P} = V \times V$. The *subsetwise spanner* problem is a special case of the pairwise spanner problem where $\mathcal{P} = S \times S$ for some $S \subset V$; that is, distances need only be preserved between vertices in S [11]. The case $t = 1$ is known as the *pairwise distance preserver* or *sourcewise distance preserver* problem, respectively [10]. The subsetwise spanner problem where t is arbitrarily large is known as the *Steiner tree* problem on graphs.

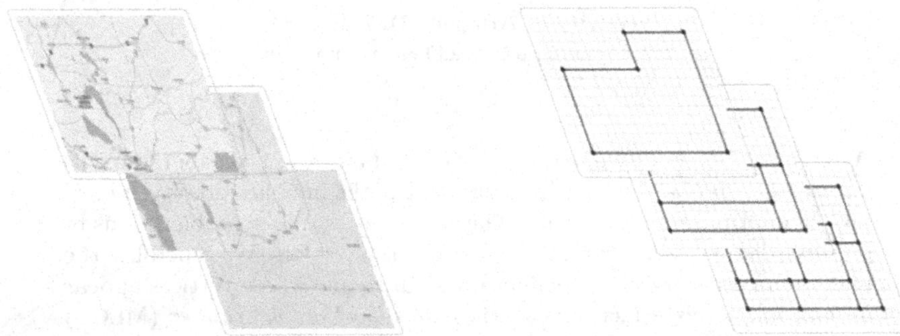

Fig. 1. An interactive road map serves as a good analogy for the MLGS problem, where the top level graph G_ℓ represents the network of major highways, and zooming in to $G_{\ell-1}$ shows a denser network of smaller roads.

1.1 Multi-level Graph Spanners

In many network design problems, vertices or edges come with a natural notion of priority, grade of service, or level; see Fig. 1. For example, consider the case of rebuilding a transportation infrastructure network after a natural disaster. Following such an event, the rebuilding process may wish to prioritize connections between important buildings such as hospitals or distribution centers, making these higher level terminals, while ensuring that no person must travel an excessive distance to reach their destination. Such problems have been referred to by names such as hierarchical network design, grade of service problems, multi-level, multi-tier, and have applications in network routing and visualization.

Similar to other graph problems which generalize to multiple levels or grades of service [8], we extend the subsetwise spanner problem to the *multi-level graph spanner (MLGS)* problem:

Definition 1. *[Multi-level graph spanner (MLGS) problem] Given a graph $G(V, E)$ with positive edge weights $c : E \to \mathbb{R}_+$, a nested sequence of terminals, $T_\ell \subseteq T_{\ell-1} \subseteq \ldots \subseteq T_1 \subseteq V$, and a real number $t \geq 1$, compute a minimum-cost sequence of spanners $G_\ell \subseteq G_{\ell-1} \subseteq \ldots \subseteq G_1$, where G_i is a subsetwise $(T_i \times T_i)$-spanner for G with stretch factor t for $i = 1, \ldots, \ell$. The cost of a solution is defined as the sum of the edge weights on each graph G_i, i.e., $\sum_{i=1}^{\ell} \sum_{e \in E(G_i)} c_e$.*

We refer to T_i and G_i as the terminals and the graph on level i. A more general version of the MLGS problem can involve different stretch factors on each level or a more general definition of cost, but for now we use the same stretch factor t for each level.

An equivalent formulation of the MLGS problem which we use interchangeably involves *grades of service*: given $G = (V, E)$ with edge weights, and required grades of service $R : V \to \{0, 1, \dots, \ell\}$, compute a single subgraph $H \subseteq G$ with varying grades of service on the edges, with the property that for all $u, v \in V$, if u and v each have required a grade of service greater than or equal to i, then there exists a path in H from u to v using edges with a grade of service greater than or equal to i, and whose length is at most $t \cdot d_G(u, v)$. Thus, $T_\ell = \{v \in V \mid R(v) = \ell\}$, $T_{\ell-1} = \{v \in V \mid R(v) \geq \ell - 1\}$, and so on. If y_e denotes the grade of edge e (or the number of levels e appears in), then the cost of a solution is equivalently $\sum_{e \in H} c_e y_e$, that is, edges with a higher grade of service incur a greater cost. This interpretation makes it clear that more important vertices (e.g., hubs) are connected with higher quality edges; see example instance and solution in Fig. 2.

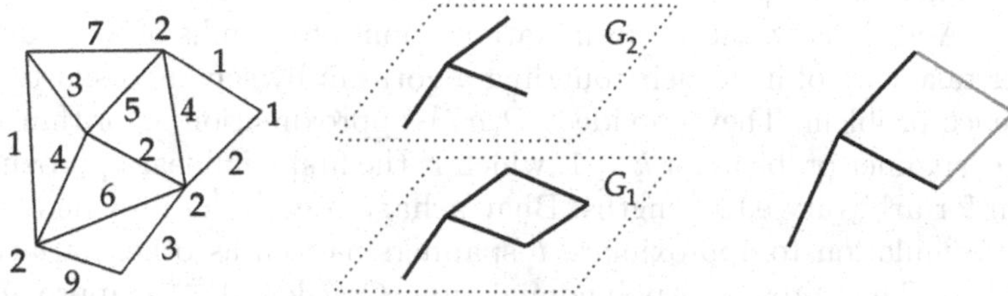

Fig. 2. *Left:* Input graph G with edge weights, $\ell = 2$, $|T_2| = 4$, $|T_1| = 3$, and $t = 3$. Required grades of service $R(v)$ are shown in red. *Center:* A valid MLGS $G_2 \subseteq G_1 \subseteq G$ is shown. *Right:* The equivalent solution, where dark edges e have $y_e = 2$ and light edges have $y_e = 1$. The cost of this solution is $2 \times (4 + 2 + 5) + 1 \times (1 + 2) = 25$. (Color figure online)

If t is arbitrarily large, the MLGS problem reduces to the *multi-level Steiner tree* (MLST) problem [1]. However it is worth noting that the problem of computing or approximating spanners is significantly harder than that of computing Steiner trees, and that a Steiner tree of G may be an arbitrarily poor spanner; a cycle on $|V|$ vertices with one edge removed is a possible Steiner tree of G, but is only a $(|V| - 1)$-spanner of G. The techniques used here have similarities to those used in the MLST problem, but more sophisticated methods are needed as well, including the use of approximate distance preservers and a new ILP formulation for the pairwise spanner problem.

1.2 Related Work

Spanners and variants thereof have been studied for at least three decades, so we focus on results relating to pairwise or subsetwise spanners. Althöfer et al. [2] provide a simple greedy algorithm that constructs a multiplicative r–spanner given a graph G and a real number $r > 0$. The greedy algorithm sorts edges in E by nondecreasing weight, then for each $e = \{u, v\} \in E$, computes the shortest path $P(u, v)$ from u to v in the current spanner, and adds the edge to the spanner if the weight of $P(u, v)$ is greater than $r \cdot c_e$. The resulting subgraph H is a r–spanner for G. The main result of [2] is that, given a weighted graph G and $t \geq 1$, there is a greedy $(2t+1)$–spanner H containing at most $n\lceil n^{1/t}\rceil$ edges, and whose weight is at most $w(MST(G))(1 + \frac{n}{2t})$ where $w(MST(G))$ denotes the weight of a minimum spanning tree of G.

Sigurd and Zachariasen [17] present an ILP formulation for the minimum-weight pairwise spanner problem (see Sect. 3), and show that the greedy algorithm [2] performs well on sparse graphs of up to 64 vertices. Álvarez-Miranda and Sinnl [3] present a mixed ILP formulation for the tree t^*–spanner problem, which asks for a spanning tree of a graph G with the smallest stretch factor t^*.

Dinitz et al. [13] provide a flow-based linear programming (LP) relaxation to approximate the directed spanner problem. Their LP formulation is similar to that in [17]; however, they provide an approximation algorithm which relaxes their ILP, whereas the previous formulation was used to compute spanners to optimality. Additionally, the LP formulation applies to graphs of unit edge cost; they later take care of it in their rounding algorithm by solving a shortest path arborescence problem. They provide a $\tilde{O}(n^{\frac{2}{3}})$–approximation algorithm for the directed k–spanner problem for $k \geq 1$, which is the first sublinear approximation algorithm for arbitrary edge lengths. Bhattacharyya et al. [6] provide a slightly different formulation to approximate t–spanners as well as other variations of this problem. They provide a polynomial time $O((n \log n)^{1-\frac{1}{k}})$–approximation algorithm for the directed k–spanner problem. Berman et al. [5] provide an alternative randomized LP rounding schemes that lead to better approximation ratios. They improved the approximation ratio to $O(\sqrt{n} \log n)$ where the approximation ratio of the algorithm provided by Dinitz et al. [13] was $O(n^{\frac{2}{3}})$. They have also improved the approximation ratio for the important special case of directed 3–spanners with unit edge lengths.

There are several results on multi-level or grade-of-service Steiner trees, e.g., [1,4,8,9,16], while multi-level spanner problems have not been studied yet.

2 Approximation Algorithms for MLGS

Here, we assume an oracle subroutine that computes an optimal $(S \times S)$–spanner, given a graph G, subset $S \subseteq V$, and t. The intent is to determine if approximating MLGS is significantly harder than the subsetwise spanner problem. We formulate simple bottom-up and top-down approaches for the MLGS problem.

2.1 Oracle Bottom-Up Approach

The approach is as follows: compute a minimum subsetwise $(T_1 \times T_1)$–spanner of G with stretch factor t. This immediately induces a feasible solution to the MLGS problem, as one can simply copy each edge from the spanner to every level (or, in terms of grades of service, assign grade ℓ to each spanner edge). We then prune edges that are not needed on higher levels. It is easy to show that the solution returned has cost no worse than ℓ times the cost of the optimal solution. Let OPT denote the cost of the optimal MLGS $G_\ell^* \subseteq G_{\ell-1}^* \subseteq \ldots \subseteq G_1^*$ for a graph G. Let MIN_i denote the cost of a minimum subsetwise $(T_i \times T_i)$–spanner for level i with stretch t, and let BOT denote the cost computed by the bottom-up approach. If no pruning is done, then $\text{BOT} = \ell\text{MIN}_1$.

Theorem 1. *The oracle bottom-up algorithm described above yields a solution that satisfies* $BOT \leq \ell \cdot OPT$.

Proof. We know $\text{MIN}_1 \leq \text{OPT}$, since the lowest-level graph G_1^* is a $(T_1 \times T_1)$–spanner whose cost is at least MIN_1. Further, we have $\text{BOT} = \ell\text{MIN}_1$ if no pruning is done. Then $\text{MIN}_1 \leq \text{OPT} \leq \text{BOT} = \ell \cdot \text{MIN}_1$, so $\text{BOT} \leq \ell \cdot \text{OPT}$. □

The ratio of ℓ is asymptotically tight; an example can be constructed by letting G be a cycle containing t vertices and all edges of cost 1. Let two adjacent vertices in G appear in T_ℓ, while all vertices appear in T_1, as shown in Fig. 3. As $t \to \infty$, the ratio $\frac{\text{BOT}}{\text{OPT}}$ approaches ℓ. Note that in this example, no edges can be pruned without violating the t–spanner requirement.

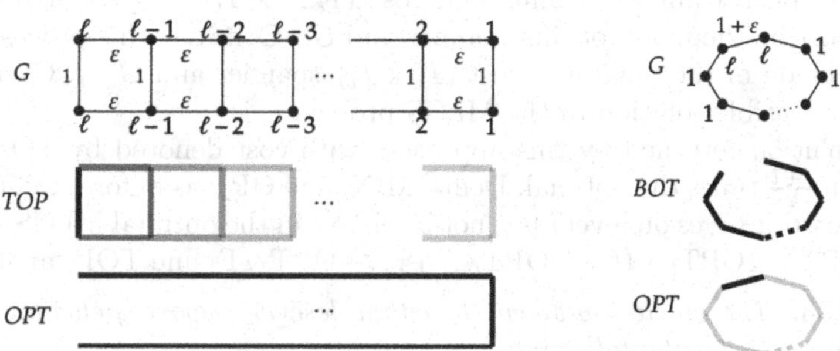

Fig. 3. *Left:* Tightness example of the top-down approach. Consider the lattice graph G with pairs of vertices of grade ℓ ($|T_\ell| = 2$), $\ell - 1$, and so on. The edge connecting the two vertices of grade i has weight 1, and all other edges have weight ε, where $0 < \varepsilon \ll 1$. Set $t = 2$. The top-down solution (middle) has cost $\text{TOP} \approx \ell + (\ell-1) + \ldots + 1 = \frac{\ell(\ell+1)}{2}$, while the optimal solution (bottom) has cost $\text{OPT} \approx \ell$. *Right:* Tightness example of the bottom-up approach. Consider a cycle G containing two adjacent vertices of grade ℓ, and the remaining vertices of grade 1. The edge connecting the two vertices of grade ℓ is $1 + \varepsilon$, while the remaining edges have weight 1. Setting $t = |E|$ yields $\text{BOT} = \ell|E|$ while $\text{OPT} = (1 + \varepsilon)\ell + 1(|E| - 1) \approx |E| + \ell$.

We give a simple heuristic that attempts to "prune" unneeded edges without violating the t–spanner requirement. Note that any pruning strategy may not prune any edges, as a worst case example (Fig. 3) cannot be pruned. Let G_1 be the $(T_1 \times T_1)$–spanner computed by the bottom-up approach. To compute a $(T_2 \times T_2)$–spanner G_2 using the edges from G_1, we can compute a distance preserver of G_1 over terminals in T_2. One simple strategy is to use shortest paths as explained below.

Even more efficient pruning is possible through the *distant preserver* literature [7,10]. A well-known result of distant preservers is due to the following theorem:

Theorem 2. ([10]). *Given $G = (V, E)$ with $|V| = n$, and $P \subset \binom{V}{2}$, there exists a subgraph G' with $O(n + \sqrt{n}|P|)$ edges such that for all $(u, v) \in P$ we have $d_{G'}(u, v) = d_G(u, v)$.*

The above theorem hints at a *sparse* construction of G_2 simply by letting $P = T_2 \times T_2$. Given G_1, let G_i be a distance preserver of G_{i-1} over the terminals T_i, for all $i = 2, \ldots, \ell$. An example is to let G_2 be the union of all shortest paths (in G_1) over vertices $v, w \in G_2$. The result is clearly a feasible solution to the MLGS problem, as the shortest paths are preserved exactly from G_1, so each G_i is a $(T_i \times T_i)$–spanner of G with stretch factor t.

2.2 Oracle Top-Down Approach

A simple top-down heuristic that computes a solution is as follows: let G_ℓ be the minimum-cost $(T_\ell \times T_\ell)$–spanner over terminals T_ℓ with stretch factor t, and cost MIN_ℓ. Then compute a minimum cost $(T_{\ell-1} \times T_{\ell-1})$–spanner over $T_{\ell-1}$, and let $G_{\ell-1}$ be the union of this spanner and G_ℓ. Continue this process, where G_i is the union of the minimum cost $(T_i \times T_i)$–spanner and G_{i+1}. Clearly, this produces a feasible solution to the MLGS problem.

The solution returned by this approach, with cost denoted by TOP, is not worse than $\frac{\ell+1}{2}$ times the optimal. Define MIN_i and OPT as before. Define OPT_i to be the cost of edges on level i but not level $i+1$ in the optimal MLGS solution, so that $\text{OPT} = \ell\text{OPT}_\ell + (\ell-1)\text{OPT}_{\ell-1} + \ldots + \text{OPT}_1$. Define TOP_i analogously.

Theorem 3. *The oracle top-down algorithm described above yields an approximation that satisfies the following:*

(i) $TOP_\ell \leq OPT_\ell$,
(ii) $TOP_i \leq OPT_i + OPT_{i+1} + \ldots + OPT_\ell, \quad i = 1, \ldots, \ell-1,$
(iii) $TOP \leq \frac{\ell+1}{2} OPT$.

Proof. Inequality (i) is true by definition, as we compute an optimal $(T_\ell \times T_\ell)$–spanner whose cost is TOP_ℓ, while OPT_ℓ is the cost of some $(T_\ell \times T_\ell)$-spanner. For (ii), note that $\text{TOP}_i \leq \text{MIN}_i$, with equality when the minimum-cost $(T_i \times T_i)$–spanner and G_{i+1} are disjoint. The spanner of cost $\text{OPT}_i + \text{OPT}_{i+1} + \ldots + \text{OPT}_\ell$ is a feasible $(T_i \times T_i)$–spanner, so $\text{MIN}_i \leq \text{OPT}_i + \ldots + \text{OPT}_\ell$, which shows (ii).

To show (iii), note that (i) and (ii) imply

$$\begin{aligned}
\text{TOP} &= \ell\text{TOP}_\ell + (\ell-1)\text{TOP}_{\ell-1} + \ldots + \text{TOP}_1 \\
&\leq \ell\text{OPT}_\ell + (\ell-1)(\text{OPT}_{\ell-1} + \text{OPT}_\ell) + \ldots + (\text{OPT}_1 + \text{OPT}_2 + \ldots + \text{OPT}_\ell) \\
&= \frac{\ell(\ell+1)}{2}\text{OPT}_\ell + \frac{(\ell-1)\ell}{2}\text{OPT}_{\ell-1} + \ldots + \frac{1 \cdot 2}{2}\text{OPT}_1 \\
&\leq \frac{\ell+1}{2}\text{OPT},
\end{aligned}$$

as by definition $\text{OPT} = \ell\text{OPT}_\ell + (\ell-1)\text{OPT}_{\ell-1} + \ldots + \text{OPT}_1$. □

The ratio $\frac{\ell+1}{2}$ is tight as illustrated in Fig. 3, left.

2.3 Combining Top-Down and Bottom-Up

Again, assume we have access to an oracle that computes a minimum weight $(S\times S)$–spanner of an input graph G with given stretch factor t. A simple combined method, similar to [1], is to run the top-down and bottom-up approaches for the MLGS problem, and take the solution with minimum cost. This has a slightly better approximation ratio than either of the two approaches.

Theorem 4. *The solution whose cost is* $\min(TOP, BOT)$ *is not worse than* $\dfrac{\ell+2}{3}$ *times the cost OPT of the optimal MLGS.*

The proof is given in Appendix A.

2.4 Heuristic Subsetwise Spanners

So far, we have assumed that we have access to an optimal subsetwise spanner given by an oracle. Here we propose a heuristic algorithm to compute subsetwise spanner. The key idea is to apply the greedy spanner to an auxiliary complete graph with terminals as its vertices and the shortest distance between terminals as edge weights. Then, we apply the distance preserver discussed in Theorem 2 to construct a subsetwise spanner.

Theorem 5. *Given graph* $G(V,E)$, *stretch factor* $t\geq 1$, *and subset* $T\subset V$, *there exists a* $(T\times T)$*–spanner for* G *with stretch factor* t *and* $O(n+\sqrt{n}|T|^{1+\frac{2}{t+1}})$ *edges.*

Proof. The spanner may be constructed as follows:

1. Construct the terminal complete graph $\bar{G}$ whose vertices are $\bar{V} := T$, such that the weight of each edge $\{u,v\}$ in $\bar{G}$ is the length of the shortest path connecting them in G, i.e., $w(u,v) = d_G(u,v)$.
2. Construct a greedy t–spanner $\bar{H}(\bar{V},\bar{E}')$ of $\bar{G}$. According to [2], this graph has $|T|^{1+\frac{2}{t+1}}$ edges. Let $P=\bar{E}'$.
3. Apply Theorem 2 to obtain a subgraph H of G such that for all $(u,v)\in P$ we have $d_H(u,v)=d_G(u,v)$. Therefore, for arbitrary $u,v\in T$ we get $d_H(u,v)\leq t\,d_G(u,v)$.
4. Finally, let shortest-path(u,v) be the collection of edges in the shortest path from u to v in H, and

$$E = \bigcup_{(u,v)\in P} \{e\in E \mid e\in \text{shortest-path}(u,v).\}$$

According to Theorem 2, the number of edges in the constructed spanner $H(V, E)$ is $O(n + \sqrt{n}|P|) = O\left(n + \sqrt{n}|T|^{1+\frac{2}{t+1}}\right)$. □

Hence, we may replace the oracle in the top-down and bottom-up approaches (Sects. 2.1 and 2.2) with the above heuristic; we call the resulting algorithms heuristic top-down and heuristic bottom-up. We analyze the performance of all algorithms on several types of graphs.

Incorporating the heuristic subsetwise spanner in our top-down and bottom up heuristics has two implications. First, the size of the final MLGS is dominated by the size of the spanner at the bottom level, i.e., $O(n + \sqrt{n}|P|) = O\left(n + \sqrt{n}|T_1|^{1+\frac{2}{t+1}}\right)$. Second, since the greedy spanner algorithm used in the above subsetwise spanner can produce spanners that are $O(n)$ more costly than the optimal solution, the same applies to the subsetwise spanner. Our experimental results, however, indicate that the heuristic approaches are very close to the optimal solutions obtained via our ILP.

3 Integer Linear Programming (ILP) Formulations

We describe the original ILP formulation for the pairwise spanner problem [17]. Let $K = \{(u_i, v_i)\} \subset V \times V$ be the set of vertex pairs; recall that the t–spanner problem is a special case where $K = V \times V$. Here we will use unordered pairs of distinct vertices, so in the t–spanner problem we have $|K| = \binom{|V|}{2}$ instead of $|V|^2$. This ILP formulation uses *paths* as decision variables. Given $(u, v) \in K$, denote by P_{uv} the set of paths from u to v of cost no more than $t \cdot d_G(u, v)$, and denote by P the union of all such paths, i.e., $P = \bigcup_{(u,v)\in K} P_{uv}$. Given a path $p \in P$ and edge $e \in E$, let $\delta_p^e = 1$ if e is on path p, and 0 otherwise. Let $x_e = 1$ if e is an edge in the pairwise spanner H, and 0 otherwise. Given $p \in P$, let $y_p = 1$ if path p is in the spanner, and zero otherwise. An ILP formulation for the pairwise spanner problem is given below.

$$\text{Minimize} \sum_{e\in E} c_e x_e \text{ subject to} \tag{1}$$

$$\sum_{p\in P_{uv}} y_p \delta_p^e \leq x_e \qquad \forall e \in E; \forall (u, v) \in K \tag{2}$$

$$\sum_{p\in P_{uv}} y_p \geq 1 \qquad \forall (u, v) \in K \tag{3}$$

$$x_e \in \{0, 1\} \qquad \forall e \in E \tag{4}$$

$$y_p \in \{0, 1\} \qquad \forall p \in P \tag{5}$$

Constraint (3) ensures that for each pair $(u, v) \in K$, at least one t–spanner path is selected, and constraint (2) enforces that on the selected u-v path, every edge along the path appears in the spanner. The main drawback of this ILP is that the number of path variables is exponential in the size of the graph. The authors use delayed column generation by starting with a subset $P' \subset P$

of paths, with the starting condition that for each $(u, v) \in K$, at least one t–spanner path in P_{uv} is in P'. The authors leave as an open question whether this problem admits a useful polynomial-size ILP.

We introduce a 0–1 ILP formulation for the pairwise t–spanner problem based on multicommodity flow, which uses $O(|E||K|)$ variables and constraints, where $|K| = O(|V|^2)$. Define t, c_e, $d_G(u, v)$, K, and x_e as before. Note that $d_G(u, v)$ can be computed in advance, using any all-pairs shortest path (APSP) method.

Direct the graph by replacing each edge $e = \{u, v\}$ with two edges (u, v) and (v, u) of weight c_e. Let E' be the set of all directed edges, i.e., $|E'| = 2|E|$. Given $(i, j) \in E'$, and an unordered pair of vertices $(u, v) \in K$, let $x^{uv}_{(i,j)} = 1$ if edge (i, j) is included in the selected u-v path in the spanner H, and 0 otherwise. This definition of path variables is similar to that by Álvarez-Miranda and Sinnl [3] for the tree t^*–spanner problem. We select a total order of all vertices so that the path constraints (8)–(9) are well-defined. This induces $2|E||K|$ binary variables, or $2|E|\binom{|V|}{2} = 2|E||V|(|V|-1)$ variables in the standard t–spanner problem. Note that if u and v are connected by multiple paths in H of length $\leq t \cdot d_G(u, v)$, we need only set $x^{uv}_{(i,j)} = 1$ for edges along some path. Given $v \in V$, let $In(v)$ and $Out(v)$ denote the set of incoming and outgoing edges for v in E'. In (7)–(11) we assume $u < v$ in the total order, so spanner paths are from u to v. An ILP formulation for the pairwise spanner problem is as follows.

$$\text{Minimize} \sum_{e \in E} c_e x_e \text{ subject to} \tag{6}$$

$$\sum_{(i,j) \in E'} x^{uv}_{(i,j)} c_e \leq t \cdot d_G(u, v) \qquad \forall (u, v) \in K; e = \{i, j\} \tag{7}$$

$$\sum_{(i,j) \in Out(i)} x^{uv}_{(i,j)} - \sum_{(j,i) \in In(i)} x^{uv}_{(j,i)} = \begin{cases} 1 & i = u \\ -1 & i = v \\ 0 & \text{else} \end{cases} \qquad \forall (u, v) \in K; \forall i \in V \tag{8}$$

$$\sum_{(i,j) \in Out(i)} x^{uv}_{(i,j)} \leq 1 \qquad \forall (u, v) \in K; \forall i \in V \tag{9}$$

$$x^{uv}_{(i,j)} + x^{uv}_{(j,i)} \leq x_e \qquad \forall (u, v) \in K; \forall e = \{i, j\} \in E \tag{10}$$

$$x_e, x^{uv}_{(i,j)} \in \{0, 1\} \tag{11}$$

Constraint (7) requires that for all $(u, v) \in K$, the sum of the weights of the selected edges corresponding to the pair (u, v) is not more than $t \cdot d_G(u, v)$. Constraints (8)–(9) require that the selected edges corresponding to $(u, v) \in K$ form a simple path beginning at u and ending at v. Constraint (10) enforces that, if edge (i, j) or (j, i) is selected on some u-v path, then its corresponding undirected edge e is selected in the spanner; further, (i, j) and (j, i) cannot both be selected for some pair (u, v). Finally, (11) enforces that all variables are binary.

The number of variables is $|E| + 2|E||K|$ and the number of constraints is $O(|E||K|)$, where $|K| = O(|V|^2)$. Note that the variables $x^{uv}_{(i,j)}$ can be relaxed to be continuous in $[0, 1]$.

3.1 ILP Formulation for the MLGS Problem

Recall that the MLGS problem generalizes the subsetwise spanner problem, which is a special case of the pairwise spanner problem for $K = S \times S$. Again, we use unordered pairs, i.e., $|K| = \binom{|S|}{2}$.

We generalize the ILP formulation in (6)–(11) to the MLGS problem as follows. Recall that we can encode the levels in terms of required grades of service $R : V \to \{0, 1, \ldots, \ell\}$. Instead of 0–1 indicators x_e, let y_e denote the grade of edge e in the multi-level spanner; that is, $y_e = i$ if e appears on level i but not level $i+1$, and $y_e = 0$ if e is absent. The only difference is that for the MLGS problem, we assign grades of service to all u-v paths by assigning grades to edges along each u-v path. That is, for all $u, v \in T_1$ with $u < v$, the *selected* path from u to v has grade $\min(R(u), R(v))$, which we denote by m_{uv}. Note that we only need to require the existence of a path for terminals $u, v \in T_1$, where $u < v$. An ILP formulation for the MLGS problem is as follows.

$$\text{Minimize} \sum_{e \in E} c_e y_e \text{ subject to} \tag{12}$$

$$\sum_{(i,j) \in E'} x^{uv}_{(i,j)} c_e \leq t \cdot d_G(u, v) \qquad \forall u, v \in T_1; e = \{i, j\} \tag{13}$$

$$\sum_{(i,j) \in Out(i)} x^{uv}_{(i,j)} - \sum_{(j,i) \in In(i)} x^{uv}_{(j,i)} = \begin{cases} 1 & i = u \\ -1 & i = v \\ 0 & \text{else} \end{cases} \qquad \forall u, v \in T_1; \forall i \in V \tag{14}$$

$$\sum_{(i,j) \in Out(i)} x^{uv}_{(i,j)} \leq 1 \qquad \forall u, v \in T_1; \forall i \in V \tag{15}$$

$$y_e \geq m_{uv} x^{uv}_{(i,j)} \qquad \forall u, v \in T_1; \forall e = \{i, j\} \tag{16}$$

$$y_e \geq m_{uv} x^{uv}_{(j,i)} \qquad \forall u, v \in T_1; \forall e = \{i, j\} \tag{17}$$

$$x^{uv}_{(i,j)} \in \{0, 1\} \tag{18}$$

Constraints (16)–(17) enforce that for each pair $u, v \in V$ such that $u < v$, the edges along the selected u-v path (not necessarily every u-v path) have a grade of service greater than or equal to the minimum grade of service needed to connect u and v, that is, m_{uv}. If multiple pairs (u_1, v_1), (u_2, v_2), $\ldots$, (u_k, v_k) use the same edge $e = \{i, j\}$ (possibly in opposite directions), then the grade of edge e should be $y_e = \max(m_{u_1 v_1}, m_{u_2 v_2}, \ldots, m_{u_k v_k})$. It is implied by (16)–(17) that $0 \leq y_e \leq \ell$ in an optimal solution.

Theorem 6. *An optimal solution to the ILP given in* (6)–(11) *yields an optimal pairwise spanner of* G *over a set* $K \subset V \times V$.

Theorem 7. *An optimal solution to the ILP given in* (12)–(18) *yields an optimal solution to the MLGS problem.*

We give the proofs in Appendices B and C.

3.2 Size Reduction Techniques

We can reduce the size of the ILP using the following shortest path tests, which works well in practice and also applies to the MLGS problem. Note that we are concerned with the total cost of a solution, not the number of edges.

If $d_G(i,j) < c(i,j)$, for some edge $\{i,j\} \in E$, then we can remove $\{i,j\}$ from the graph, as no min-weight spanner of G uses edge $\{i,j\}$. If H^* is a min-cost pairwise spanner that uses edge $\{i,j\}$, then we can replace $\{i,j\}$ with a shorter i-j path p_{ij} without violating the t–spanner requirement. In particular, if some u-v path uses both edge $\{i,j\}$ as well as some edge(s) along p_{ij}, then this path can be rerouted to use only edges in p_{ij} with smaller cost.

We reduce the number of variables needed in the single-level ILP formulation ((6)–(11)) with the following test: given $u,v \in K$ with $u < v$ and some directed edge $(i,j) \in E'$, if $d_G(u,i) + c(i,j) + d_G(j,v) > t \cdot d_G(u,v)$, then (i,j) cannot possibly be included in the selected u-v path, so set $x^{uv}_{(i,j)} = 0$. If (i,j) or (j,i) cannot be selected on any u-v path, we can safely remove $\{i,j\}$ from E.

Conversely, given some directed edge $(i,j) \in E'$, let G' be the directed graph obtained by removing (i,j) from E' (so that G' has $2|E| - 1$ edges). For each $u,v \in K$ with $u < v$, if $d_{G'}(u,v) > t \cdot d_G(u,v)$, then edge (i,j) must be in any u-v spanner path, so set $x^{uv}_{(i,j)} = 1$. For its corresponding undirected edge e, $x_e = 1$.

4 Experimental Results

4.1 Setup

We use the Erdős–Rényi [14] and Watts–Strogatz [18] models to generate random graphs. Given a number of vertices, n, and probability p, the model ER(n,p) assigns an edge to any given pair of vertices with probability p. An instance of ER(n,p) with $p = (1+\varepsilon)\frac{\ln n}{n}$ is connected with high probability for $\varepsilon > 0$ [14]). For our experiments we use $n \in \{20, 40, 60, 80, 100\}$, and $\varepsilon = 1$.

In the Watts-Strogatz model, WS(n,K,β), initially we create a ring lattice of constant degree K, and then rewire each edge with probability $0 \leq \beta \leq 1$ while avoiding self-loops and duplicate edges. The Watts-Strogatz model generates small-world graphs with high clustering coefficients [18]. For our experiments we use $n \in \{20, 40, 60, 80, 100\}$, $K = 6$, and $\beta = 0.2$.

An instance of the MLGS problem is characterized by four parameters: graph generator, number of vertices $|V|$, number of levels ℓ, and stretch factor t. As there is randomness involved, we generated 3 instances for every choice of parameters (e.g., ER, $|V| = 80$, $\ell = 3$, $t = 2$).

We generated MLGS instances with 1, 2, or 3 levels ($\ell \in \{1,2,3\}$), where terminals are selected on each level by randomly sampling $\lfloor |V| \cdot (\ell - i + 1)/(\ell + 1) \rfloor$ vertices on level i so that the size of the terminal sets decreases linearly. As the terminal sets are nested, T_i can be selected by sampling from T_{i-1} (or from V if $i = 1$). We used four different stretch factors in our experiments, $t \in \{1.2, 1.4, 2, 4\}$. Edge weights are randomly selected from $\{1, 2, 3, \ldots, 10\}$.

Algorithms and Outputs. We implemented the bottom-up (BU) and top-down (TD) approaches from Sect. 2 in Python 3.5, as well as the combined approach that selects the better of the two (Sect. 2.3). To evaluate the approximation algorithms and the heuristics, we implemented the ILPs described in Sect. 3 using CPLEX 12.6.2. We used the same high-performance computer for all experiments (Lenovo NeXtScale nx360 M5 system with 400 nodes).

For each instance of the MLGS problem, we compute the costs of the MLGS returned using the bottom-up (BU), the top-down (TD), and the combined (min(BU, TD)) approaches, as well as the minimum cost MLGS using the ILP in Sect. 3.1. The three heuristics involve a (single-level) subroutine; we used both the heuristic described in Sect. 2.4, as well as the flow formulation described in Sect. 3 which computes subsetwise spanners to optimality. We compare the algorithms with and without the oracle to assess whether computing (single-level) spanners to optimality significantly improves the overall quality of the solution.

We show the performance ratio for each heuristic in the y-axis (defined as the heuristic cost divided by OPT), and how the ratio depends on the input parameters (number of vertices $|V|$, number of levels ℓ, and stretch factors t). Finally, we discuss the running time of the ILP. All box plots show the minimum, interquartile range and maximum, aggregated over all instances using the parameter being compared.

4.2 Results

We first discuss the results for Erdős–Rényi graphs. Figures 4, 5, 6 and 7 show the results of the oracle top-down, bottom-up, and combined approaches. We show the impact of different parameters (number of vertices $|V|$, number of levels ℓ, and stretch factors t) using line plots for three heuristics separately in Figs. 4, 5 and 6. Figure 7 shows the performance of the three heuristics together in box plots. In Fig. 4 we can see that the bottom-up heuristic performs slightly worse for increasing $|V|$, while the top-down heuristic performs slightly better. In Fig. 5 we see that the heuristics perform worse when ℓ increases, consistent with the ratios discussed in Sect. 2. In Fig. 6 we show the performance of the heuristics with respect to the stretch factor t. In general, the performance becomes worse as t increases.

The most time consuming part of the experiment is the execution time of the ILP for solving MLGS instances optimally. The running time of the heuristics is significantly smaller compared to that of the ILP. Hence, we first show the running times of the exact solution of the MLGS instances in Fig. 8. We show the running time with respect to the number of vertices $|V|$, number of levels ℓ, and stretch factors t. For all parameters, the running time tends to increase as the size of the parameter increases. In particular, the running time with stretch factor 4 (Fig. 8, right) was much worse, as there are many more t-spanner paths to consider, and the size reduction techniques in Sect. 3.2 are less effective at reducing instance size. We show the running times of for computing oracle bottom-up, top-down and combined solutions in Fig. 9.

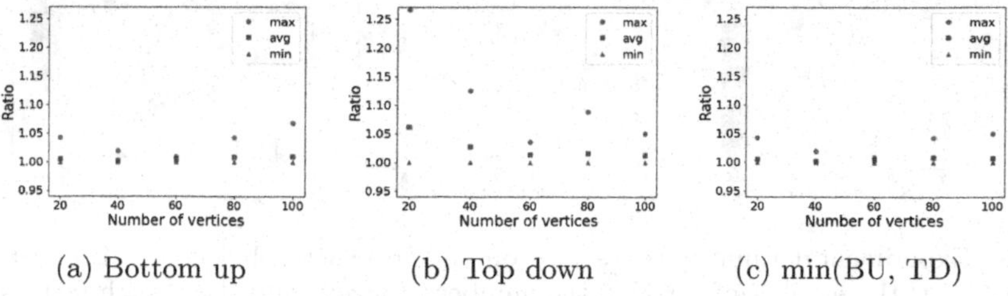

(a) Bottom up (b) Top down (c) min(BU, TD)

Fig. 4. Performance with oracle on Erdős–Rényi graphs w.r.t. $|V|$. Ratio is defined as the cost of the returned MLGS divided by OPT.

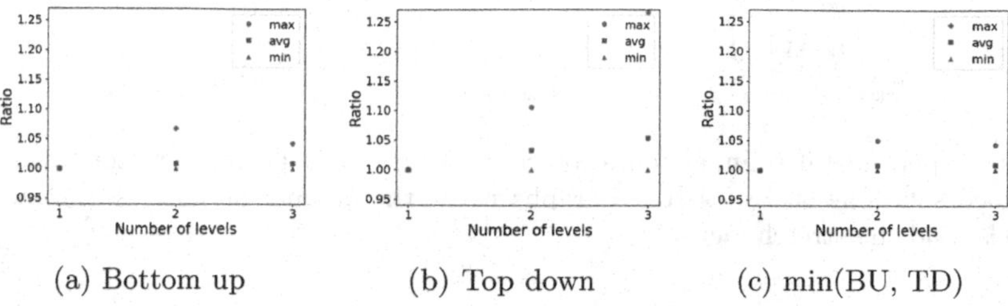

(a) Bottom up (b) Top down (c) min(BU, TD)

Fig. 5. Performance with oracle on Erdős–Rényi graphs w.r.t. the number of levels

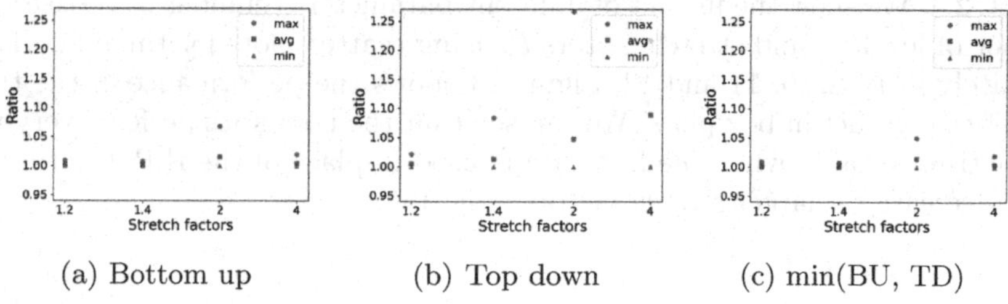

(a) Bottom up (b) Top down (c) min(BU, TD)

Fig. 6. Performance with oracle on Erdős–Rényi graphs w.r.t. stretch factor

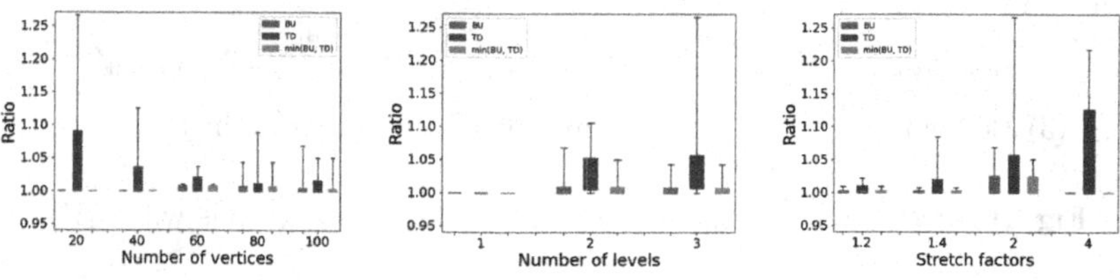

Fig. 7. Performance with oracle on Erdős–Rényi graphs w.r.t. the number of vertices, the number of levels, and the stretch factors

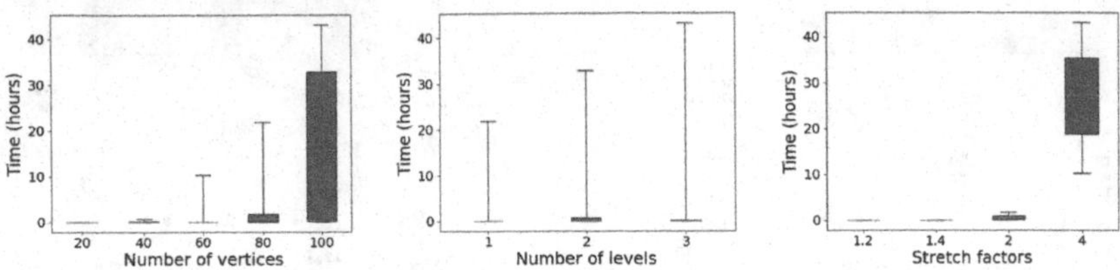

Fig. 8. Experimental running times for computing exact solutions on Erdős–Rényi graphs w.r.t. the number of vertices, the number of levels, and the stretch factors

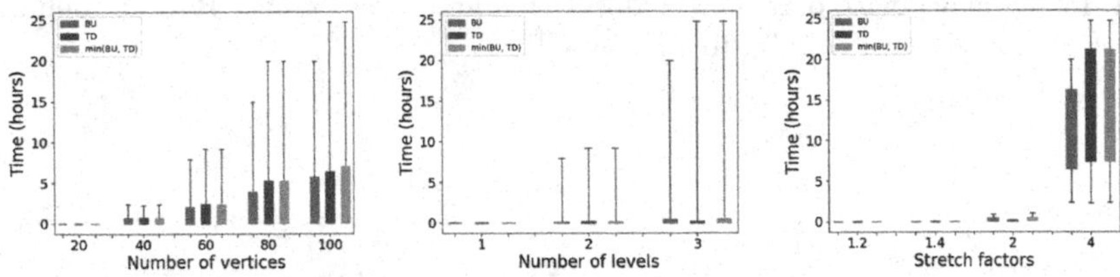

Fig. 9. Experimental running times for computing oracle bottom-up, top-down and combined solutions on Erdős–Rényi graphs w.r.t. the number of vertices, the number of levels, and the stretch factors

The ILP is too computationally expensive for larger input sizes and this is where the heurstic can be particularly useful. We now consider a similar experiment using the heuristic to compute subsetwise spanners, as described in Sect. 2.4. We show the impact of different parameters (number of vertices $|V|$, number of levels ℓ, and stretch factors t) using scatter plots for three heuristics separately in Figs. 10, 11 and 12. Figure 13 shows the performance of the three heuristics together in box plots. We can see that the heuristics perform very well in practice. Notably when the heuristic is used in place of the ILP, the running times decrease for larger stretch factors (Fig. 14).

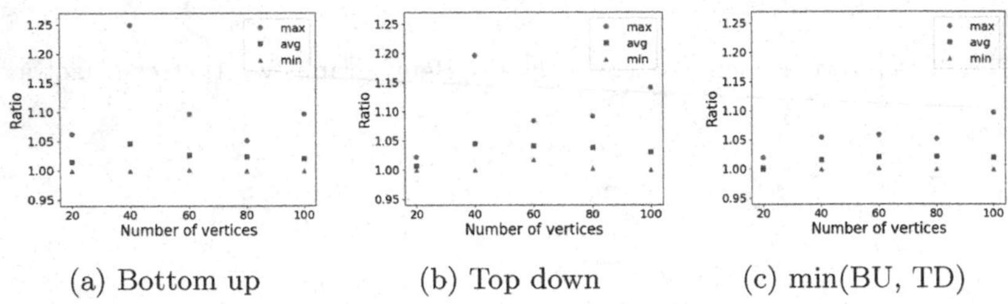

(a) Bottom up (b) Top down (c) min(BU, TD)

Fig. 10. Performance without oracle on Erdős–Rényi graphs w.r.t. $|V|$

We also analyzed graphs generated from the Watts–Strogatz model and the results are shown in Appendix D.

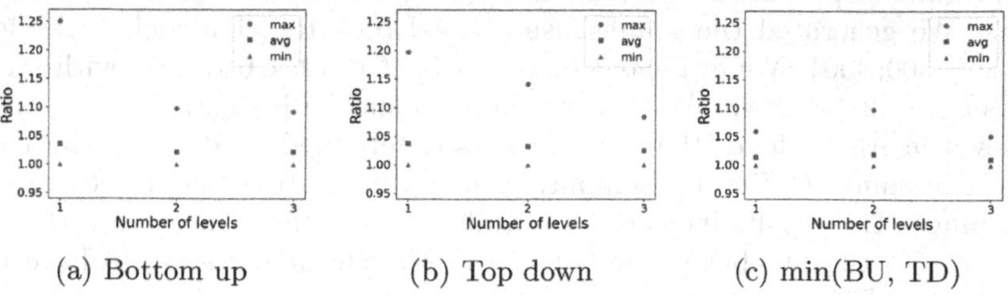

(a) Bottom up (b) Top down (c) min(BU, TD)

Fig. 11. Performance without oracle on Erdős–Rényi graphs w.r.t. the number of levels

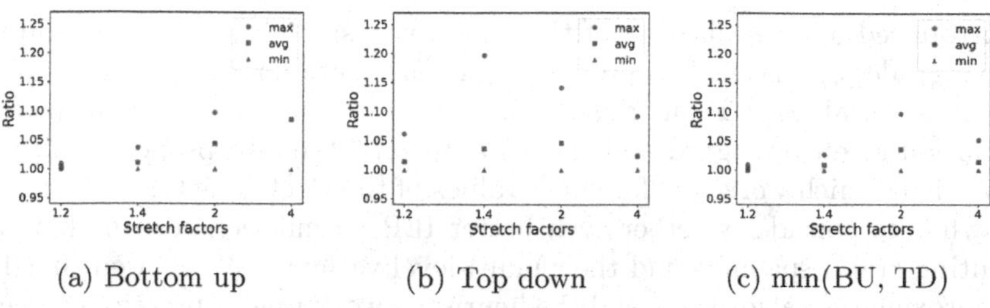

(a) Bottom up (b) Top down (c) min(BU, TD)

Fig. 12. Performance without oracle on Erdős–Rényi graphs w.r.t. the stretch factors

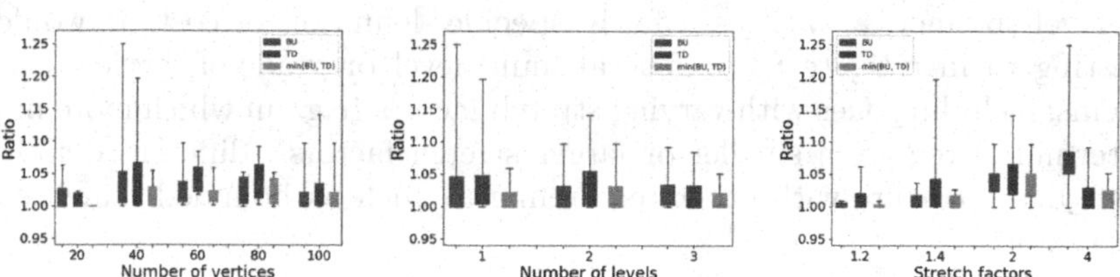

Fig. 13. Performance without oracle on Erdős–Rényi graphs w.r.t. the number of vertices, the number of levels, and the stretch factors

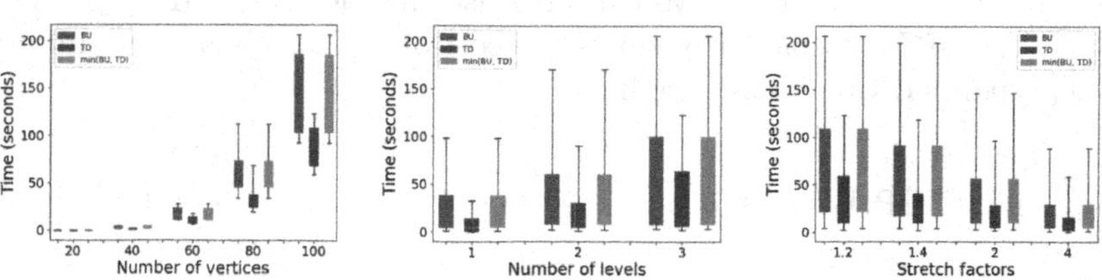

Fig. 14. Experimental running times for computing heuristic bottom-up, top-down and combined solutions on Erdős–Rényi graphs w.r.t. the number of vertices, the number of levels, and the stretch factors

Our final experiments test the heuristic performance on a set of larger graphs. We generated the graphs using the Erdős–Rényi model, with $|V| \in \{100, 200, 300, 400\}$. We evaluated more levels ($\ell \in \{2, 4, 6, 8, 10\}$) with stretch factors $t \in \{1.2, 1.4, 2, 4\}$. We show the performance of heuristic bottom-up and top-down in Appendix E. Here, the ratio is determined by dividing the BU or TD cost by $\min(BU, TD)$ (as computing the optimal MLGS would be too time-consuming). The results indicate that while running times increase with larger input graphs, the number of levels and the stretch factors seem to have little impact on performance.

5 Discussion and Conclusion

We introduced a generalization of the subsetwise spanner problem to multiple levels or grades of service. Our proposed ILP formulation requires only a polynomial size of variables and constraints, which is an improvement over the previous formulation given by Sigurd and Zachariasen [17]. We also proposed improved formulations which work well for small values of the stretch factor t. It would be worthwhile to consider whether even better ILP formulations can be found for computing graph spanners and their multi-level variants. We showed that both the approximation algorithms and the heuristics work well in practice on several different types of graphs, with different number of levels and different stretch factors.

We only considered a stretch factor t that is the same for all levels in the multi-level spanner, as well as a fairly specific definition of cost. It would be interesting to investigate more general multi-level or grade-of-service spanner problems, including ones with varying stretch factors (e.g., in which more important terminals require a smaller or larger stretch factors), different definitions of cost, and spanners with other requirements, such as bounded diameters or degrees.

A Proof of Theorem 4

Proof. We use the simple algebraic fact that $\min\{x, y\} \leq \alpha x + (1 - \alpha)y$ for all $x, y \in \mathbb{R}$ and $\alpha \in [0, 1]$. Here, we can also use the fact that $\text{MIN}_1 \leq \text{OPT}_1 + \text{OPT}_2 + \ldots + \text{OPT}_\ell$, as the RHS equals the cost of G_1^*, which is some subsetwise $(T_1 \times T_1)$-spanner. Combining, we have

$$\begin{aligned}
\min(\text{TOP}, \text{BOT}) &\leq \alpha \sum_{i=1}^{\ell} \frac{i(i+1)}{2} \text{OPT}_i + (1-\alpha)\ell \sum_{i=1}^{\ell} \text{OPT}_i \\
&= \sum_{i=1}^{\ell} \left[\left(\frac{i(i+1)}{2} - \ell \right) \alpha + \ell \right] \rho \text{OPT}_i
\end{aligned}$$

Since we are comparing $\min\{\text{TOP}, \text{BOT}\}$ to $r \cdot \text{OPT}$ for some approximation ratio $r > 1$, we can compare coefficients and find the smallest $r \geq 1$ such that the system of inequalities

$$\left(\frac{\ell(\ell+1)}{2} - \ell\right)\alpha + \ell\rho \leq \ell r$$
$$\left(\frac{(\ell-1)\ell}{2} - \ell\right)\alpha + \ell\rho \leq (\ell-1) r$$
$$\vdots$$
$$\left(\frac{2\cdot 1}{2} - \ell\right)\alpha + \ell\rho \leq r$$

has a solution $\alpha \in [0,1]$. Adding the first inequality to $\ell/2$ times the last inequality yields $\frac{\ell^2+2\ell}{2} \leq \frac{3\ell r}{2}$, or $r \geq \frac{\ell+2}{3}$. Also, it can be shown algebraically that $(r,\alpha) = (\frac{\ell+2}{3}, \frac{2}{3})$ simultaneously satisfies the above inequalities. This implies that $\min\{\text{TOP}, \text{BOT}\} \leq \frac{\ell+2}{3}\rho \cdot \text{OPT}$. □

B Proof of Theorem 6

Proof. Let H^* denote an optimal pairwise spanner of G with stretch factor t, and let OPT denote the cost of H^*. Let OPT_{ILP} denote the minimum cost of the objective in the ILP (6). First, given a minimum cost t–spanner $H^*(V, E^*)$, a solution to the ILP can be constructed as follows: for each edge $e \in E^*$, set $x_e = 1$. Then for each unordered pair $(u,v) \in K$ with $u < v$, compute a shortest path p_{uv} from u to v in H^*, and set $x^{uv}_{(i,j)} = 1$ for each edge along this path, and $x^{uv}_{(i,j)} = 0$ if (i,j) is not on p_{uv}.

As each shortest path p_{uv} necessarily has cost $\leq t \cdot d_G(u,v)$, constraint (7) is satisfied. Constraints (8)–(9) are satisfied as p_{uv} is a simple u-v path. Constraint (10) also holds, as p_{uv} should not traverse the same edge twice in opposite directions. In particular, every edge in H^* appears on some shortest path; otherwise, removing such an edge yields a pairwise spanner of lower cost. Hence $\text{OPT}_{ILP} \leq \text{OPT}$.

Conversely, an optimal solution to the ILP induces a feasible t–spanner H. Consider an unordered pair $(u,v) \in K$ with $u < v$, and the set of decision variables satisfying $x^{uv}_{(i,j)} = 1$. By (8) and (9), these chosen edges form a simple path from u to v. The sum of the weights of these edges is at most $t \cdot d_G(u,v)$ by (7). Then by constraint (10), the chosen edges corresponding to (u,v) appear in the spanner, which is induced by the set of edges e with $x_e = 1$. Hence $\text{OPT} \leq \text{OPT}_{ILP}$.

Combining the above observations, we see that $\text{OPT} = \text{OPT}_{ILP}$. □

C Proof of Theorem 7

Proof. Given an optimal solution to the ILP with cost OPT_{ILP}, construct an MLGS by letting $G_i = (V, E_i)$ where $E_i = \{e \in E \mid y_e \geq i\}$. This clearly gives a nested sequence of subgraphs. Let u and v be terminals in T_i (not necessarily of required grade $R(\cdot) = i$), with $u < v$, and consider the set of all variables of the form $x_{(i,j)}^{uv}$ equal to 1. By (13)–(15), these selected edges form a path from u to v of length at most $t \cdot d_G(u, v)$, while constraints (16)–(17) imply that these selected edges have grade at least $m_{uv} \geq i$, so the selected path is contained in E_i. Hence G_i is a subsetwise $(T_i \times T_i)$–spanner for G with stretch factor t, and the optimal ILP solution gives a feasible MLGS.

Given an optimal MLGS with cost OPT, we can construct a feasible ILP solution with the same cost in a way similar to the proof of Theorem 6. For each $u, v \in T_1$ with $u < v$, set $m_{uv} = \min(R(u), R(v))$. Compute a shortest path in $G_{m_{uv}}$ from u to v, and set $x_{(i,j)}^{uv} = 1$ for all edges along this path. Then for each $e \in E$, consider all pairs $(u_1, v_1), \ldots, (u_k, v_k)$ that use either (i, j) or (j, i), and set $y_e = \max(m_{u_1v_1}, m_{u_2v_2}, \ldots, m_{u_kv_k})$. In particular, y_e is not larger than the grade of e in the MLGS, otherwise this would imply e is on some u-v path at grade greater than its grade of service in the actual solution. □

D Experimental Results on Graphs Generated Using Watts-Strogatz

The results for graphs generated from the Watts–Strogatz model are shown in Figs. 15, 16, 17, 18, 19, 20, 21, 22 and 23, which are organized in the same way as for Erdős–Rényi.

E Experimental Results on Large Graphs Using Erdős-Rényi

Figure 24 shows a rough measure of performance for the bottom-up and top-down heuristics on large graphs using the Erdős-Rényi model, where the ratio is defined as the BU or TD cost divided by min(BU, TD). Figure 25 shows the aggregated running times per instance, which significantly worsen as $|V|$ is large.

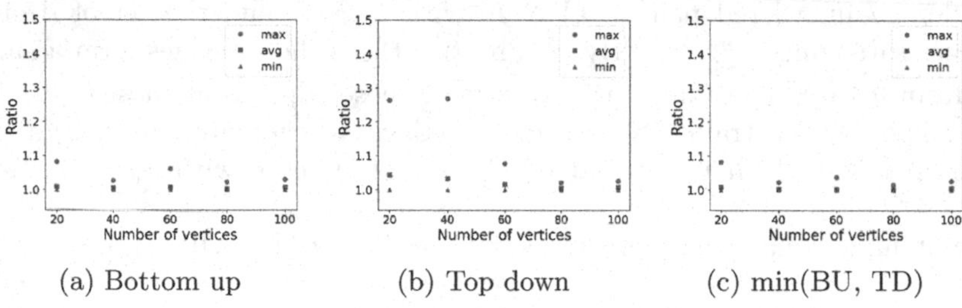

(a) Bottom up (b) Top down (c) min(BU, TD)

Fig. 15. Performance with oracle on Watts–Strogatz graphs w.r.t. the number of vertices

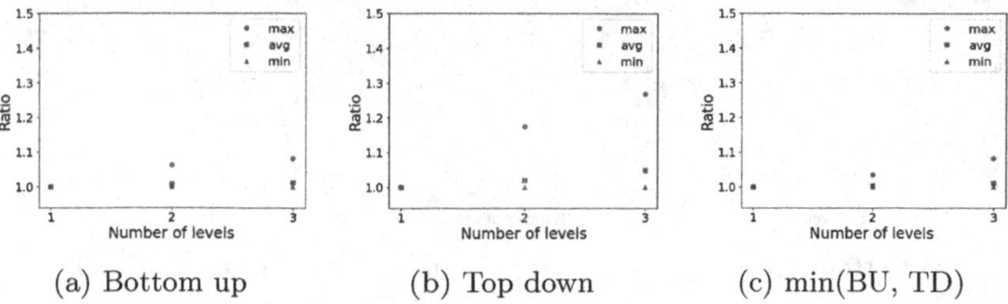

(a) Bottom up (b) Top down (c) min(BU, TD)

Fig. 16. Performance with oracle on Watts–Strogatz graphs w.r.t. the number of levels

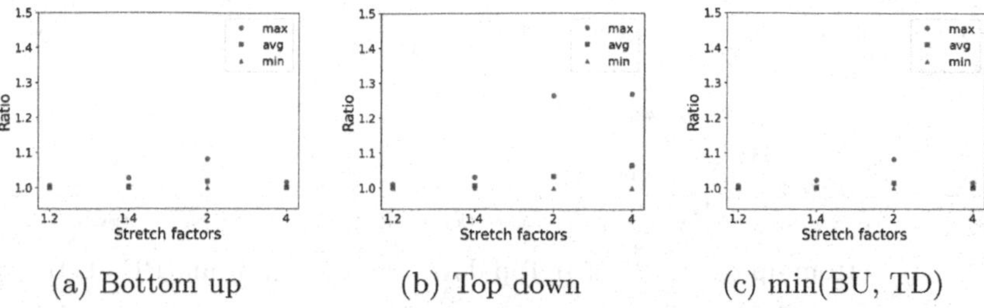

(a) Bottom up (b) Top down (c) min(BU, TD)

Fig. 17. Performance with oracle on Watts–Strogatz graphs w.r.t. the stretch factors

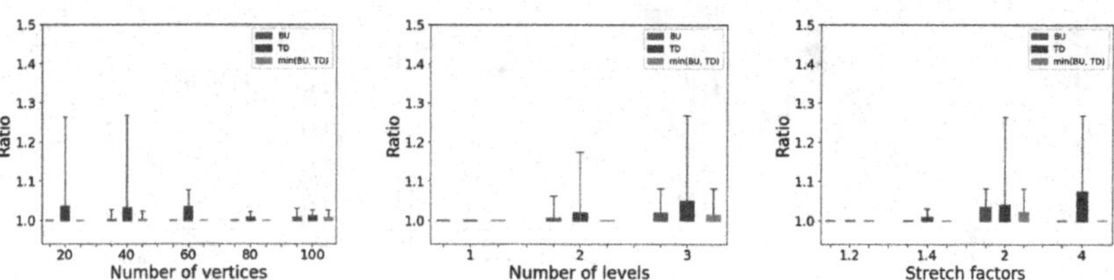

Fig. 18. Performance with oracle on Watts–Strogatz graphs w.r.t. the number of vertices, the number of levels, and the stretch factors

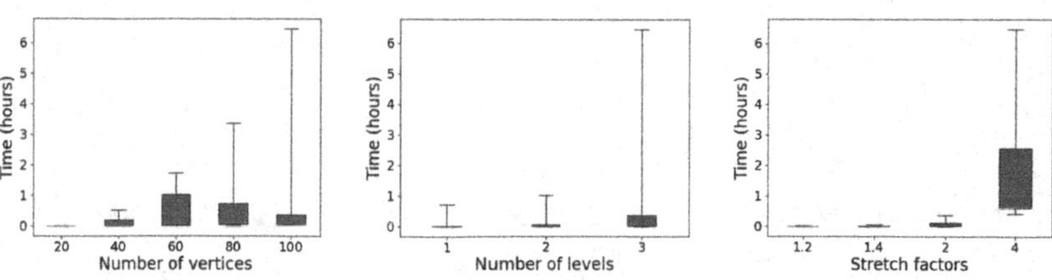

Fig. 19. Experimental running times for computing exact solutions on Watts–Strogatz graphs w.r.t. the number of vertices, the number of levels, and the stretch factors

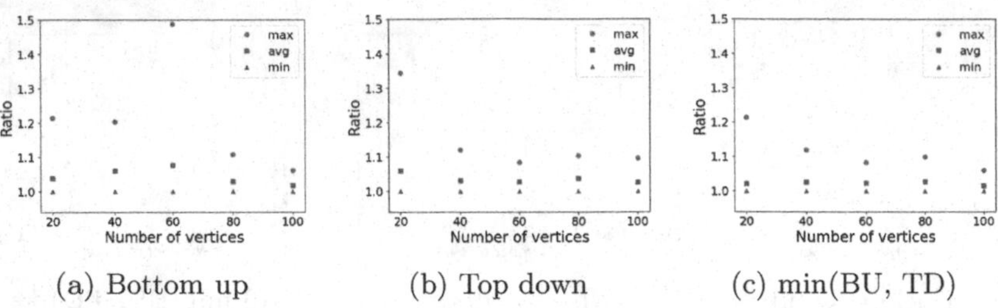

(a) Bottom up (b) Top down (c) min(BU, TD)

Fig. 20. Performance without oracle on Watts–Strogatz graphs w.r.t. the number of vertices

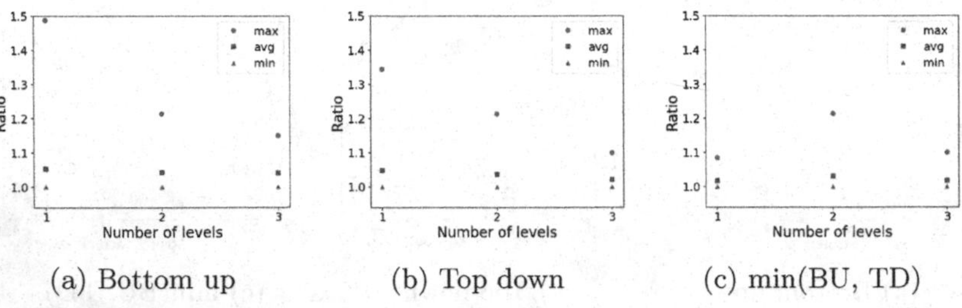

(a) Bottom up (b) Top down (c) min(BU, TD)

Fig. 21. Performance without oracle on Watts–Strogatz graphs w.r.t. the number of levels

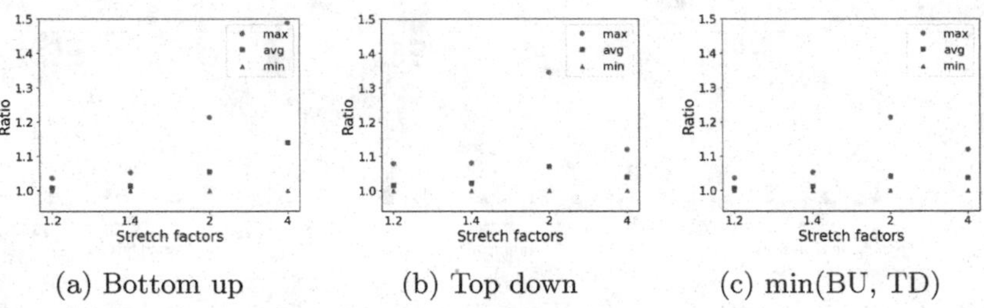

(a) Bottom up (b) Top down (c) min(BU, TD)

Fig. 22. Performance without oracle on Watts–Strogatz graphs w.r.t. the stretch factors

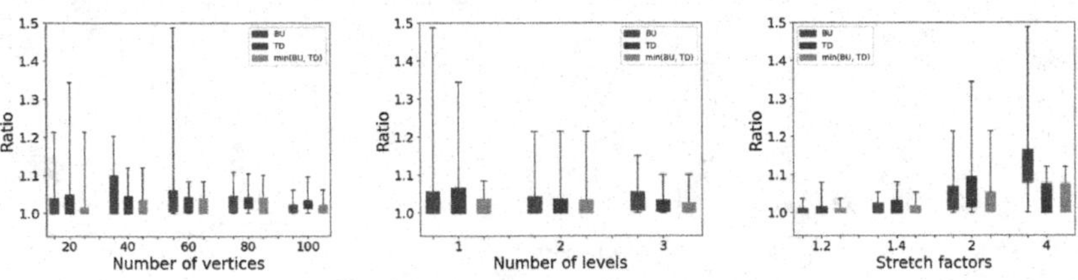

Fig. 23. Performance without oracle on Watts–Strogatz graphs w.r.t. the number of vertices, the number of levels, and the stretch factors

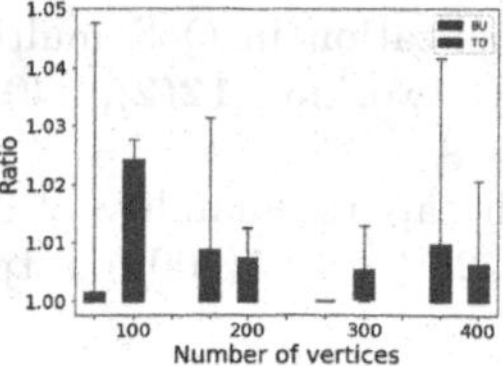

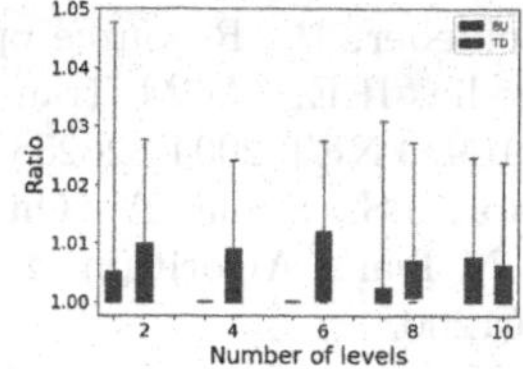

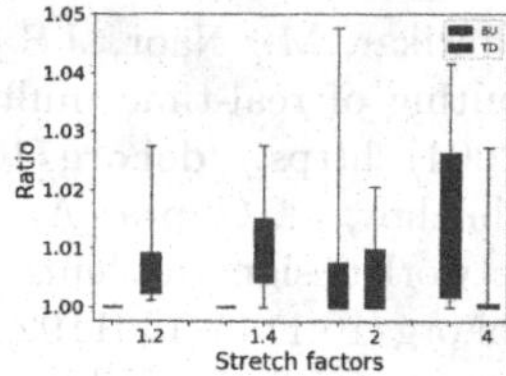

Fig. 24. Performance of heuristic bottom-up and top-down on large Erdős–Rényi graphs w.r.t. the number of vertices, the number of levels, and the stretch factors. The ratio is determined by dividing the objective value of the combined (min(BU, TD)) heuristic.

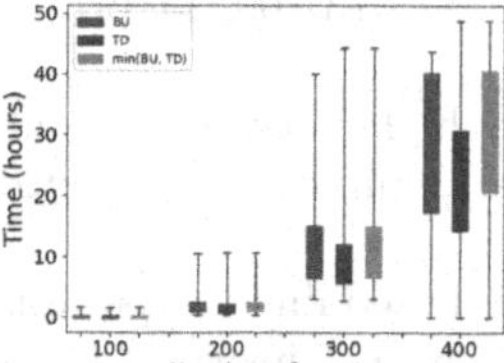

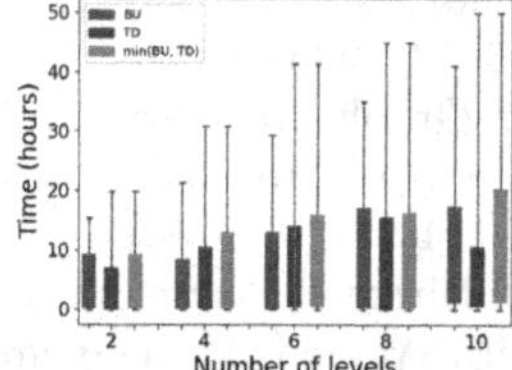

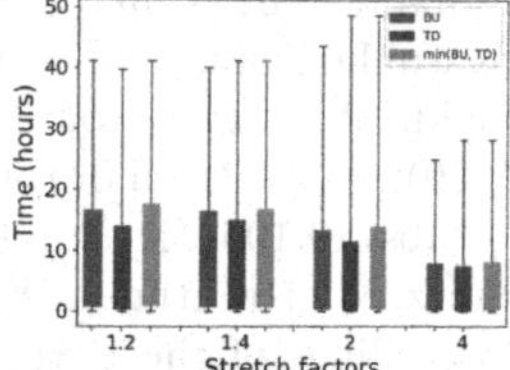

Fig. 25. Experimental running times for computing heuristic bottom-up, top-down and combined solutions on large Erdős–Rényi graphs w.r.t. the number of vertices, the number of levels, and the stretch factors.

References

1. Ahmed, A.R., et al.: Multi-level Steiner trees. In: 17th International Symposium on Experimental Algorithms, (SEA), pp. 15:1–15:14 (2018). https://doi.org/10.4230/LIPIcs.SEA.2018.15
2. Althöfer, I., Das, G., Dobkin, D., Joseph, D.: Generating sparse spanners for weighted graphs. In: Gilbert, J.R., Karlsson, R. (eds.) SWAT 90, pp. 26–37. Springer, Heidelberg (1990)
3. Álvarez-Miranda, E., Sinnl, M.: Mixed-integer programming approaches for the tree t*-spanner problem. Optimization Letters (2018). https://doi.org/10.1007/s11590-018-1340-0
4. Balakrishnan, A., Magnanti, T.L., Mirchandani, P.: Modeling and heuristic worst-case performance analysis of the two-level network design problem. Manag. Sci. **40**(7), 846–867 (1994). https://doi.org/10.1287/mnsc.40.7.846
5. Berman, P., Bhattacharyya, A., Makarychev, K., Raskhodnikova, S., Yaroslavtsev, G.: Approximation algorithms for spanner problems and directed steiner forest. Inf. Comput. **222**, 93–107 (2013). 38th International Colloquium on Automata, Languages and Programming (ICALP 2011), https://doi.org/10.1016/j.ic.2012.10.007, http://www.sciencedirect.com/science/article/pii/S0890540112001484
6. Bhattacharyya, A., Grigorescu, E., Jung, K., Raskhodnikova, S., Woodruff, D.P.: Transitive-closure spanners. SIAM J. Comput. **41**(6), 1380–1425 (2012). https://doi.org/10.1137/110826655
7. Bodwin, G.: Linear size distance preservers. In: Proceedings of the Twenty-Eighth Annual ACM-SIAM Symposium on Discrete Algorithms, pp. 600–615. Society for Industrial and Applied Mathematics (2017)

8. Charikar, M., Naor, J.S., Schieber, B.: Resource optimization in QoS multicast routing of real-time multimedia. IEEE/ACM Trans. Networking **12**(2), 340–348 (2004). https://doi.org/10.1109/TNET.2004.826288
9. Chuzhoy, J., Gupta, A., Naor, J.S., Sinha, A.: On the approximability of some network design problems. ACM Trans. Algorithms **4**(2), 23:1–23:17 (2008). https://doi.org/10.1145/1361192.1361200
10. Coppersmith, D., Elkin, M.: Sparse sourcewise and pairwise distance preservers. SIAM J. Discrete Math. **20**(2), 463–501 (2006)
11. Cygan, M., Grandoni, F., Kavitha, T.: On pairwise spanners. In: Portier, N., Wilke, T. (eds.) 30th International Symposium on Theoretical Aspects of Computer Science (STACS 2013). Leibniz International Proceedings in Informatics (LIPIcs), vol. 20, pp. 209–220. Schloss Dagstuhl-Leibniz-Zentrum fuer Informatik, Dagstuhl, Germany (2013). https://doi.org/10.4230/LIPIcs.STACS.2013.209, http://drops.dagstuhl.de/opus/volltexte/2013/3935
12. David, P., Alejandro, S.A.: Graph spanners. J. Graph Theory **13**(1), 99–116 (1989). https://doi.org/10.1002/jgt.3190130114. https://onlinelibrary.wiley.com/doi/abs/10.1002/jgt.3190130114
13. Dinitz, M., Krauthgamer, R.: Directed spanners via flow-based linear programs. In: Proceedings of the Forty-third Annual ACM Symposium on Theory of Computing. STOC 2011, pp. 323–332. ACM, New York (2011). https://doi.org/10.1145/1993636.1993680, http://doi.acm.org/10.1145/1993636.1993680
14. Erdős, P., Rényi, A.: On random graphs, i. Publicationes Mathematicae (Debrecen) **6**, 290–297 (1959)
15. Kortsarz, G.: On the hardness of approximating spanners. Algorithmica **30**(3), 432–450 (2001). https://doi.org/10.1007/s00453-001-0021-y. https://doi.org/10.1007/s00453-001-0021-y
16. Mirchandani, P.: The multi-tier tree problem. INFORMS J. Comput. **8**(3), 202–218 (1996)
17. Sigurd, M., Zachariasen, M.: Construction of minimum-weight spanners. In: Albers, S., Radzik, T. (eds.) ESA 2004. LNCS, vol. 3221, pp. 797–808. Springer, Heidelberg (2004). https://doi.org/10.1007/978-3-540-30140-0_70
18. Watts, D.J., Strogatz, S.H.: Collective dynamics of 'small-world' networks. Nature **393**(6684), 440 (1998)

Author Index

Ahmed, Reyan 541
Ameranis, Konstantinos 343

Baldo, Fabiano 202
Bonnet, Édouard 167
Borradaile, Glencora 98

Çalık, Çağdaş 332
Cleve, Jonas 317

Daescu, Ovidiu 454
Denzumi, Shuhei 265
Dufossé, Fanny 248
Dworkin, Morris 332
Dykas, Nathan 332

Ekim, Tınaz 21
Emiris, Ioannis Z. 1

Fălămaş, Diana-Elena 167
Fernandez, Marcel 513
Fotakis, Dimitris 343
Funke, Stefan 158

Gamby, Ask Neve 473
Gelashvili, Koba 114
Geppert, Hanna 359
Ghosh, Anirban 142
Godinho, Noé 69
Grdzelidze, Nikoloz 114
Greco, Alessio 392
Gu, Hanyu 527

Hamm, Keaton 541
Hicks, Brian 142
Hu, Qifu 82

Jordanov, Ivan 505
Jovanovic, Raka 490

Kampel, Ludwig 300
Katajainen, Jyrki 376, 473
Katsamaki, Christina 1
Kawahara, Jun 125
Kaya, Kamer 248
Kobourov, Stephen 541
Krishnaa, Prem 51
Kumar, Neeraj 35

Latifi Jebelli, Mohammad Javad 541
Le, Hung 98
Li, Angsheng 82
Liu, Jun 82
Livieratos, John 513

M. Perera, Sirani 184
Malik, Hemant 454
Malík, Josef 283
Matsuda, Kotaro 265
Mendel, Thomas 158
Mulzer, Wolfgang 317

Nakamura, Kengo 265
Nasre, Meghana 51
Nicosia, Giuseppe 392
Nishino, Masaaki 265

Ogle, Austin 184
Önal, Umutcan 408

Pan, Yicheng 82
Panagiotas, Ioannis 248
Paquete, Luís 69
Parpinelli, Rafael S. 202
Peralta, Rene 332
Petrozziello, Alessio 505
Phan, Duc-Minh 237

Riccio, Salvatore Danilo 392

Sahneh, Faryad Darabi 541
Saitoh, Toshiki 125
Schmitt, João P. 202
Şeker, Oylum 21
Shalom, Mordechai 21
Shevchenko, Ronald 142

Silverio, Daniel 184
Simos, Dimitris E. 300
Somani, Vedant 51
Spence, Richard 541
Suchý, Ondřej 283
Sullivan, Blair D. 424
Suzuki, Hirofumi 125

Tamaki, Hisao 219
Timmis, Jon 392
Tutberidze, Mikheil 114

Uçar, Bora 248
Utture, Akshay 51

Valla, Tomáš 283
van der Poel, Andrew 424
Vathis, Nikolaos 343
Viennot, Laurent 237
Voß, Stefan 490

Wagner, Michael 300
Watrigant, Rémi 167
Wilhelm, Martin 359
Woodlief, Trey 424

Yasuda, Norihito 265
Yoshinaka, Ryo 125

Zafeirakopoulos, Zafeirakis 408
Zhang, Yefei 527
Zheng, Baigong 98
Zinder, Yakov 527

Printed in the United States
By Bookmasters